CAMBRIDGE LIBRARY COLLECTION

Books of enduring scholarly value

Mathematical Sciences

From its pre-historic roots in simple counting to the algorithms powering modern desktop computers, from the genius of Archimedes to the genius of Einstein, advances in mathematical understanding and numerical techniques have been directly responsible for creating the modern world as we know it. This series will provide a library of the most influential publications and writers on mathematics in its broadest sense. As such, it will show not only the deep roots from which modern science and technology have grown, but also the astonishing breadth of application of mathematical techniques in the humanities and social sciences, and in everyday life.

The Collected Mathematical Papers

Arthur Cayley (1821-1895) was a key figure in the creation of modern algebra. He studied mathematics at Cambridge and published three papers while still an undergraduate. He then qualified as a lawyer and published about 250 mathematical papers during his fourteen years at the Bar. In 1863 he took a significant salary cut to become the first Sadleirian Professor of Pure Mathematics at Cambridge, where he continued to publish at a phenomenal rate on nearly every aspect of the subject, his most important work being in matrices, geometry and abstract groups. In 1882 he spent five months at Johns Hopkins University, and in 1883 became president of the British Association for the Advancement of Science. Publication of his Collected Papers - 967 papers in 13 volumes plus an index volume - began in 1889 and was completed after his death under the editorship of his successor in the Sadleirian Chair. This volume contains 74 papers, mostly published between 1874 and 1877.

Cambridge University Press has long been a pioneer in the reissuing of out-of-print titles from its own backlist, producing digital reprints of books that are still sought after by scholars and students but could not be reprinted economically using traditional technology. The Cambridge Library Collection extends this activity to a wider range of books which are still of importance to researchers and professionals, either for the source material they contain, or as landmarks in the history of their academic discipline.

Drawing from the world-renowned collections in the Cambridge University Library, and guided by the advice of experts in each subject area, Cambridge University Press is using state-of-the-art scanning machines in its own Printing House to capture the content of each book selected for inclusion. The files are processed to give a consistently clear, crisp image, and the books finished to the high quality standard for which the Press is recognised around the world. The latest print-on-demand technology ensures that the books will remain available indefinitely, and that orders for single or multiple copies can quickly be supplied.

The Cambridge Library Collection will bring back to life books of enduring scholarly value across a wide range of disciplines in the humanities and social sciences and in science and technology.

The Collected Mathematical Papers

VOLUME 9

ARTHUR CAYLEY

CAMBRIDGE UNIVERSITY PRESS

Cambridge New York Melbourne Madrid Cape Town Singapore São Paolo Delhi

Published in the United States of America by Cambridge University Press, New York

www.cambridge.org
Information on this title: www.cambridge.org/9781108005012

This edition first published 1896
This digitally printed version 2009

ISBN 978-1-108-00501-2

MATHEMATICAL PAPERS.

London: C. J. CLAY AND SONS,
CAMBRIDGE UNIVERSITY PRESS WAREHOUSE,
AVE MARIA LANE.
Glasgow: 263, ARGYLE STREET.

Leipzig: F. A. BROCKHAUS.
New York: MACMILLAN AND CO.

THE COLLECTED

MATHEMATICAL PAPERS

OF

ARTHUR CAYLEY, Sc.D., F.R.S.,

LATE SADLERIAN PROFESSOR OF PURE MATHEMATICS IN THE UNIVERSITY OF CAMBRIDGE.

VOL. IX.

CAMBRIDGE:
AT THE UNIVERSITY PRESS.
1896

CAMBRIDGE:
PRINTED BY J. AND C. F. CLAY,
AT THE UNIVERSITY PRESS.

ADVERTISEMENT.

THE present volume contains 74 papers, numbered 556 to 629, published for the most part in the years 1874 to 1877.

The Table for the nine volumes is

Vol.		Numbers			
Vol.	I.	Numbers	1	to	100,
,,	II.	,,	101	,,	158,
,,	III.	,,	159	,,	222,
,,	IV.	,,	223	,,	299,
,,	V.	,,	300	,,	383,
,,	VI.	,,	384	,,	416,
,,	VII.	,,	417	,,	485,
,,	VIII.	,,	486	,,	555,
,,	IX.	,,	556	,,	629.

A. R. FORSYTH.

17 *December* 1895.

CONTENTS.

CLASSIFICATION.

ANALYSIS:

556.

ON STEINER'S SURFACE.

[From the *Proceedings of the London Mathematical Society*, vol. v. (1873—1874), pp. 14—25. Read December 11, 1873.]

I HAVE constructed a model and drawings of the symmetrical form of Steiner's Surface, viz. that wherein the four singular tangent planes form a regular tetrahedron, and consequently the three nodal lines (being the lines joining the mid-points of opposite edges) a system of rectangular axes at the centre of the tetrahedron. Before going into the analytical theory, I describe as follows the general form of the surface: take the tetrahedron, and inscribe in each face a circle (there will be, of course, two circles touching at the mid-point of each edge of the tetrahedron; each circle will contain, on its circumference at angular distances of 120°, three mid-points, and the lines joining these with the centre of the tetrahedron, produced beyond the centre, meet the opposite edges, and are in fact the before-mentioned lines joining the mid-points of opposite edges). Now truncate the tetrahedron by planes parallel to the faces so as to reduce the altitudes each to three-fourths of the original value, and from the centre of each new face round off symmetrically up to the adjacent three circles; and within each circle scoop down to the centre of the tetrahedron, the bounding surface of the excavation passing through the three right lines, and the sections (by planes parallel to the face) being in the neighbourhood of the face nearly circular, but as they approach the centre, assuming a trigonoidal form, and being close to the centre an indefinitely small equilateral triangle. We have thus the surface, consisting of four lobes united only by the lines through the mid-points of opposite edges, these lines being consequently nodal lines; the mid-points being pinch-points of the surface, and the faces singular planes, each touching the surface along the inscribed circle. The joining lines, produced indefinitely both ways, belong as nodal lines to the surface; but they are, outside the tetrahedron, mere acnodal lines not traversed by any real sheet of the surface.

We may imagine the tetrahedron placed in two different positions, (1) resting with one of its faces on the horizontal plane, (2) with two opposite edges horizontal, or say with the horizontal plane passing through the centre of the tetrahedron and being parallel to two opposite edges; or, what is the same thing, the nodal lines form a system of rectangular axes, one of them, say that of z, being vertical. And I proceed to consider, in the two cases respectively, the horizontal sections of the surface.

In the first case, the coordinates x, y, z, w may be taken to be the perpendicular distances of a point from the faces of the tetrahedron, w being the distance from the base. We have*, if the altitude be h,

$$x + y + z + w = h;$$

an equation which may be used to homogenize any equation not originally homogeneous; thus, for the plane $w = \lambda$, of altitude λ, we have

$$w = \frac{\lambda}{h}(x + y + z + w),$$

or, what is the same thing,

$$w = \frac{\lambda}{h - \lambda}(x + y + z).$$

The equation of the surface is

$$\sqrt{x} + \sqrt{y} + \sqrt{z} + \sqrt{w} = 0,$$

and if we herein consider w as having the last-mentioned value, the equation will belong to the section by the plane $w = \lambda$. I remark that the section of the tetrahedron, by this plane, is an equilateral triangle, the side of which is to an edge of the tetrahedron as $h - \lambda : h$. For a point in the plane of the triangle, if X, Y, Z are the perpendiculars on the sides, then

$$X + Y + Z = P,$$

(if for a moment P is the perpendicular from a vertex on the opposite side of the triangle, viz. we have $P = \frac{h - \lambda}{h} p$, if p be the perpendicular for a face of the tetrahedron). And it is clear that x, y, z are proportional to X, Y, Z; we consequently have, for the equation of the section,

$$\sqrt{X} + \sqrt{Y} + \sqrt{Z} + \sqrt{\frac{\lambda}{h - \lambda}(X + Y + Z)} = 0,$$

* I take the opportunity of remarking that in a regular tetrahedron, if s be the length of an edge, p the perpendicular from a summit on an edge (or altitude of a face), h the perpendicular from a summit on a face (or altitude of the tetrahedron), and q the distance between the mid-points of opposite edges, then

$$s = \frac{\sqrt{3}}{\sqrt{2}} h, \quad p = \frac{3}{2\sqrt{2}} h, \quad q = \frac{\sqrt{3}}{2} h.$$

The tetrahedron can, by means of planes through the mid-points of the edges at right angles thereto, be divided into four hexahedral figures (8 summits, 6 faces, 12 edges, each face a quadrilateral); viz. in each such figure there are, meeting in a summit of the tetrahedron, three edges, each $= \frac{1}{2}s$; meeting in the centre three edges, each $= \frac{3}{4}h$; and six other edges, each $= \frac{1}{3}p$.

where the coordinates X, Y, Z are the perpendicular distances from the sides of the triangle which is the section of the tetrahedron. To simplify, I write

$$\frac{\lambda}{h-\lambda}=2q+1,$$

that is,

$$q=\frac{2\lambda-h}{2h-2\lambda};$$

the equation then is

$$\sqrt{X}+\sqrt{Y}+\sqrt{Z}+\sqrt{(2q+1)(X+Y+Z)}=0;$$

or, proceeding to rationalize, we have first

$$q(X+Y+Z)=\sqrt{YZ}+\sqrt{ZX}+\sqrt{XY},$$

and thence

$$q^2(X+Y+Z)^2-(YZ+ZX+XY)=2\sqrt{XYZ}(\sqrt{X}+\sqrt{Y}+\sqrt{Z});$$

and finally

$$\{q^2(X+Y+Z)^2-YZ-ZX-XY\}^2=4(2q+1)XYZ(X+Y+Z).$$

This is a quartic curve, having for double tangents the four lines $X=0$, $Y=0$, $Z=0$, $X+Y+Z=0$, the last of these being the line infinity touching the curve in two imaginary points, since obviously the whole real curve lies within the triangle. This is as it should be: the double tangents are the intersections of the plane $w=\lambda$ by the singular planes of the surface.

To find the points of contact, writing for instance $Z=0$, the equation becomes

$$q^2(X+Y)^2-XY=0,$$

that is,

$$X^2+\left(2-\frac{1}{q^2}\right)XY+Y^2=0;$$

whence

$$X=\left(-1+\frac{1}{2q^2}\pm\sqrt{\frac{1}{4q^4}-\frac{1}{q^2}}\right)Z,$$

giving the two points of contact equi-distant from the centre; these are imaginary if $q>\frac{1}{2}$, but otherwise real, which agrees with what follows. (See the Table afterwards referred to.)

The nodal lines of the surface are $(x-y=0,\ z-w=0)$, $(y-z=0,\ x-w=0)$, $(z-x=0,\ y-w=0)$. Considering the first of these, we have for its intersection with the plane $w=\lambda$,

$$X=Y,\quad Z=\frac{\lambda}{h-\lambda}(X+Y+Z),\ =(2q+1)(X+Y+Z),$$

and the last equation gives

$$Z=(2q+1)(2X+Z),$$

that is,

$$0=(2q+1)X+qZ,$$

so that for the point in question we have $X : Y : Z = -q : -q : 2q+1$; and taking the perpendicular from the vertex on a side as unity, the values $-q$, $-q$, $2q+1$ will be absolute magnitudes. We thus see that the curve must have the three nodes $(2q+1, -q, -q)$, $(-q, 2q+1, -q)$, $(-q, -q, 2q+1)$; and it is easy to verify that this is so.

The curve will pass through the centre $X = Y = Z$, if

$$(9q^2 - 3)^2 = 12(2q+1),$$

that is, if

$$3(3q^2 - 1)^2 - 4(2q+1) = 0,$$

or if

$$(3q+1)^2 (q-1) = 0.$$

If $q = 1$, that is, $\lambda = 3(h - \lambda)$, or $\lambda = \frac{3}{4}h$, the equation is

$$(X^2 + Y^2 + Z^2 + YZ + ZX + XY)^2 - 12XYZ(X+Y+Z) = 0,$$

where the curve is, in fact, a pair of imaginary conics meeting in the four real points $(3, -1, -1)$, $(-1, 3, -1)$, $(-1, -1, 3)$, $(\frac{1}{3}, \frac{1}{3}, \frac{1}{3})$. To verify this, observe that, writing

$$\begin{aligned} A &= (Y - Z)(2X + Y + Z), \\ B &= (Z - X)(X + 2Y + Z), \\ C &= (X - Y)(X + Y + 2Z), \end{aligned}$$

and therefore

$$A + B + C = 0,$$

the function in (X, Y, Z) is $= \frac{1}{2}(A^2 + B^2 + C^2)$, and thus the equation may be written in the equivalent forms

$$B^2 + BC + C^2 = 0, \quad C^2 + CA + A^2 = 0, \quad A^2 + AB + B^2 = 0,$$

each of which shows that the curve breaks up into two imaginary conics. The foregoing value $q = 1$, or $\lambda = \frac{3}{4}h$, belongs to the summit or highest real point of the surface.

In the case $3q + 1 = 0$, that is,

$$q = -\tfrac{1}{3} = \frac{2\lambda - h}{2h - 2\lambda}, \quad \text{or } \lambda = \tfrac{1}{4}h,$$

the equation is

$$\{(X + Y + Z)^2 - 9(YZ + ZX + XY)\}^2 = 108XYZ(X + Y + Z),$$

which is, in fact, the equation of a curve having the centre, or point $X = Y = Z$, for a triple point.

To verify this, write

$$\begin{aligned} X &= \beta - \gamma + u, \\ Y &= \gamma - \alpha + u, \\ Z &= \alpha - \beta + u; \end{aligned}$$

also

$$2\Delta = (\beta-\gamma)^2 + (\gamma-\alpha)^2 + (\alpha-\beta)^2,$$
$$\Omega = (\beta-\gamma)(\gamma-\alpha)(\alpha-\beta).$$

Then we have

$$X + Y + Z = 3u,$$
$$YZ + ZX + XY = 3u^2 - \Delta,$$
$$XYZ = u^3 - u\Delta + \Omega,$$

and the equation is

$$\{9u^2 - 9(3u^2 - \Delta)\}^2 - 324u(u^3 - u\Delta + \Omega) = 0,$$

that is,

$$(-2u^2 + \Delta)^2 - 4u(u^3 - u\Delta + \Omega) = 0,$$

or finally

$$\Delta^2 - 4u\Omega = 0,$$

where the lowest terms in $\beta-\gamma$, $\gamma-\alpha$, $\alpha-\beta$ are of the order 3, and the theorem is thus proved. The case in question, $q=-\frac{1}{3}$ or $\lambda=\frac{1}{4}h$, is where the plane passes through the centre of the tetrahedron.

When $q = \frac{1}{2} = \dfrac{2\lambda - h}{2h - 2\lambda}$, or $\lambda = \frac{2}{3}h$, the equation is

$$(X^2 + Y^2 + Z^2 - 2YZ - 2ZX - 2XY)^2 = 128XYZ(X + Y + Z).$$

Here each of the lines $X=0$, $Y=0$, $Z=0$ is an osculating tangent having with the curve a 4-pointic intersection.

When $q = 0 = \dfrac{2\lambda - h}{2h - 2\lambda}$, or $\lambda = \frac{1}{2}h$, the equation is

$$(YZ + ZX + XY)^2 - 4XYZ(X + Y + Z) = 0,$$

that is,

$$Y^2Z^2 + Z^2X^2 + X^2Y^2 - 2XYZ(X + Y + Z) = 0;$$

viz. each angle of the triangle is here a cusp.

When $q = -\frac{1}{2}$, or $\lambda = 0$, the curve is

$$\{X^2 + Y^2 + Z^2 - 2(YZ + ZX + XY)\}^2 = 0,$$

viz. the plane is here the base of the tetrahedron, and the section is the inscribed circle taken twice.

For tracing the curves, it is convenient to find the intersections with the lines $Y-Z=0$, $Z-X=0$, $X-Y=0$ drawn from the centre of the triangle to the vertices; each of these lines passes through a node, and therefore besides meets the curve in two points. Writing, for instance, $Y=X$, the equation becomes

$$\{q^2(2X + Z)^2 - 2XZ - X^2\}^2 - 4(2q + 1)X^2Z(2X + Z) = 0;$$

viz. this is

$$\{qZ + (2q + 1)X\}^2\{q^2Z^2 + (4q^2 - 2q - 4)XZ + (4q^2 - 4q + 1)X^2\} = 0,$$

where the first factor gives the node. Equating to zero the second factor, we have

$$\left\{qZ + \left(2q - 1 - \frac{2}{q}\right) X\right\}^2 = X^2 \left\{\left(2q - 1 - \frac{2}{q}\right)^2 - 4q^2 + 4q - 1\right\}$$

$$= X^2 \frac{4}{q^2}(1 - q)(1 + 2q);$$

or, finally,

$$qZ = \left\{-2q + 1 + \frac{2}{q} \pm \frac{2}{q}\sqrt{(1 - q)(1 + 2q)}\right\} X,$$

giving two real values for all values of q from $q = 1$ to $q = -\frac{1}{2}$. (See the Table afterwards referred to.)

We may recapitulate as follows:

$q > 1$, or $\lambda > \frac{3}{4}h$; the curve is imaginary, but with three real acnodes, answering to the acnodal parts of the nodal lines:

$q = 1$, or $\lambda = \frac{3}{4}h$; the summit appears as a fourth acnode:

$q < 1 > \frac{1}{2}$, or $\lambda < \frac{3}{4}h > \frac{2}{3}h$; the curve consists of three acnodes and a trigonoid lying within the triangle and having the sides of the triangle for bitangents of imaginary contact:

$q = \frac{1}{2}$, or $\lambda = \frac{2}{3}h$; the curve consists of three acnodes and a trigonoid having the sides of the triangle for osculating tangents:

$q < \frac{1}{2} > 0$, or $\lambda < \frac{2}{3}h > \frac{1}{2}h$; the curve consists of three conjugate points and an indented trigonoid having the sides of the triangle for bitangents of real contact:

$q = 0$, or $\lambda = \frac{1}{2}h$; curve has the summits of the triangle for cusps:

$q < 0 > -\frac{1}{2}$, or $\lambda < \frac{1}{2}h > \frac{1}{4}h$; curve has three crunodes, or say it is a cis-centric trifolium:

$q = -\frac{1}{3}$, or $\lambda = \frac{1}{4}h$; curve has a triple point, or say it is a centric trifolium:

$q < -\frac{1}{3} > -\frac{1}{2}$, or $\lambda < \frac{1}{4}h > 0$; curve has three crunodes, or say it is a trans-centric trifolium:

$q = -\frac{1}{2}$, or $\lambda = 0$; curve is a two-fold circle:

$q < -\frac{1}{2}$, or $\lambda < 0$; the curve becomes again imaginary, consisting of three acnodes answering to the acnodal parts of the nodal lines.

For the better delineation of the series of curves, I calculated the following Table, wherein the first column gives a series of values of $\lambda : h$; the second the corresponding values of q, $= \dfrac{2\lambda - h}{2h - 2\lambda}$; the third the positions of the point of contact, say with the side $Z = 0$, the value of $X : Y$ being calculated from the foregoing formula,

$$X \div Y = -1 + \frac{1}{2q^2} \pm \sqrt{\frac{1}{4q^4} - \frac{1}{q^2}};$$

and the fourth the apsidal distances, say for the radius vector $X=Y$, the value of $Z : X$ being calculated from the foregoing formula

$$Z \div X = -2 + \frac{1}{q} + \frac{2}{q^2} \pm \frac{2}{q}\sqrt{\left(\frac{1}{q}-1\right)\left(2+\frac{1}{q}\right)}.$$

The Table is:

	$\lambda : h$	q	Contact, $Z=0$, $X : Y=$			Apses, $X=Y$; $X : Z=$		
*	·75	1·00	imposs.			1·		
	·70	·6666	imposs.			·032	or	7·968
*	·666	·5	1·			0·		16·
	·65	·4285	·320	or	3·124	·006		22·44
	·6	·25	·0721		13·9279	·059		67·941
	·55	·1111	·011		78·988	·15		337·85
*	·5	0·	0	or	∞	·25		∞
	·45	− ·0909	·005		118·99	·39		457·61
	·4	− ·1666	·029		33·971	·496		127·504
	·35	− ·2308	·060		16·72	·648		61·79
*	·3	− ·2857	·099		10·151	·812		37·187
*	·25	− ·3333	·141		6·854	1·		25
	·2	− ·375	·207		4·904	1·218		17·89
	·15	− ·4118	·276		3·622	1·48		13·25
	·10	− ·4444	·372		2·690	1·813		9·937
	·05	− ·4737	·515	or	1·941	3·15	or	6·46
*	·0	− ·5	1·			4·		

where the asterisks show the critical values of $\lambda : h$.

It is worth while to transform the equation to new coordinates X', Y', Z' such that $X'=0$, $Y'=0$, $Z'=0$ represent the sides of the triangle formed by the three nodes. Writing for shortness $X+Y+Z=P$, $YZ+ZX+XY=Q$, $XYZ=R$, the equation is

$$(q^2P-Q)^2 = 4(2q+1)PR.$$

The expressions of X, Y, Z in terms of the new coordinates are of the form $X'+\theta P'$, $Y'+\theta P'$, $Z'+\theta P'$, where $P'=X'+Y'+Z'$; writing also $Q'=Y'Z'+Z'X'+X'Y'$, $R'=X'Y'Z'$, then the values of P, Q, R are

$$(1+3\theta)P', \quad Q'+(2\theta+3\theta^2)P'^2, \quad R'+\theta P'Q'+(\theta^2+\theta^3)P',$$

and the transformed equation is

$$[\{q^2(1+3\theta)^2-2\theta-3\theta^2\}P'^2-Q']^2=4(2q+1)(1+3\theta)P\{(\theta^2+\theta^3)P'^3+\theta P'Q'+R'\},$$

which is satisfied by $Q'=0$, $R'=0$, if only

$$\{q^2(1+3\theta)^2-2\theta-3\theta^2\}^2=4(2q+1)(1+3\theta)(\theta^2+\theta^3),$$

or, if for a moment $q(1+3\theta)=\Omega$, the equation is

$$(\Omega^2-2\theta-3\theta^2)^2=4(\theta^2+\theta^3)(2\Omega+1+3\theta),$$

that is,

$$\Omega^4+\Omega^2(-6\theta^2-4\theta)+\Omega(-8\theta^3-8\theta^2)-3\theta^4-4\theta^3=0,$$

that is,

$$(\Omega+\theta)^2(\Omega^2-2\theta\Omega-3\theta^2-4\theta)=0.$$

If the new axes pass through the nodes, then $\Omega+\theta=0$; that is, $q(1+3\theta)+\theta=0$, which equation gives the value of θ for which the new axes have the position in question; substituting in the first instance for q the value $\frac{-\theta}{3\theta+1}$, the equation becomes

$$\{2\theta(1+\theta)P'^2+Q'\}^2=4(1+\theta)P'\{\theta^2(1+\theta)P'^3+\theta P'Q'+R'\},$$

that is,

$$Q'^2=4(1+\theta)P'R';$$

or, finally, substituting for θ its value in terms of q, the required equation is,

$$Q'^2=4\frac{2q+1}{3q+1}P'R',$$

that is,

$$(Y'Z'+Z'X'+X'Y')^2=4\frac{2q+1}{3q+1}X'Y'Z'(X'+Y'+Z').$$

In particular, for $q=0$ the equation is

$$(Y'Z'+Z'X'+X'Y')^2-4X'Y'Z'(X'+Y'+Z')=0,$$

which is right, since, in the case in question (the tricuspidal curve), we have

$$X,\ Y,\ Z=X',\ Y',\ Z'.$$

I remark, in passing, that, taking the equation to be

$$(Y'Z'+Z'X'+X'Y')^2=mX'Y'Z'(X'+Y'+Z'),$$

we may write herein

$$Z'=\tfrac{1}{3}-\ x,$$

$$X'=\tfrac{1}{3}+\tfrac{1}{2}x-\frac{\sqrt{3}}{2}y,$$

$$Y'=\tfrac{1}{3}+\tfrac{1}{2}x+\frac{\sqrt{3}}{2}y;$$

where

$$x = \frac{2\sqrt{m(m-3)}}{9}\cos\theta - \frac{2(m-3)}{9}\cos 2\theta,$$

$$y = \frac{2\sqrt{m(m-3)}}{9}\sin\theta + \frac{2(m-3)}{9}\sin 2\theta,$$

which are the formulæ for the description of the trinodal quartic as a unicursal curve.

I consider now the second position; viz. the horizontal plane now passes through the centre of the tetrahedron, and is parallel to two opposite edges. The equations of the nodal lines are here $(y=0,\ z=0)$, $(z=0,\ x=0)$, $(x=0,\ y=0)$; and if for convenience we assume the distance of the mid-points of opposite edges to be $=2$, or the half of this $=1$, then the equations of the faces are

$$\begin{aligned} X &= \quad x+y+z-1=0, \\ Y &= -x-y+z-1=0, \\ Z &= \quad x-y-z-1=0, \\ W &= -x+y-z-1=0, \end{aligned}$$

and the equation of the surface is

$$\sqrt{X}+\sqrt{Y}+\sqrt{Z}+\sqrt{W}=0.$$

Proceeding to rationalise, this is

$$X+Y+2\sqrt{XY}=Z+W+2\sqrt{ZW},$$

viz.

$$2z+\sqrt{XY}=\sqrt{ZW};$$

we thence have

$$4z^2+4z\sqrt{XY}+XY=ZW;$$

or, since

$$ZW-XY=4z+4xy,$$

this is

$$z+xy-z^2=z\sqrt{XY};$$

whence

$$(z+xy-z^2)^2=z^2\{(z-1)^2-(x+y)^2\};$$

or reducing,

$$2xyz+y^2z^2+z^2x^2+x^2y^2=0,$$

a form which puts in evidence the nodal lines. Considering z as constant, we have the equation of the section; this is a quartic having the node $(x=0,\ y=0)$, and two other nodes at infinity on the two axes respectively; moreover, the curve has for bitangents the intersections of its plane with the faces of the tetrahedron; or what is the same thing, attributing to z its constant value, the equations of the bitangents are

$$\begin{aligned} x+y+z-1&=0, \\ -x-y+z-1&=0, \\ x-y-z-1&=0, \\ -x+y-z-1&=0. \end{aligned}$$

These lines form a rectangle which is the section of the tetrahedron; observe that this is inscribed in the square the corners of which are $x=\pm 1$, $y=\pm 1$; viz. $z=+1$ (highest section), this is the dexter diagonal (considered as an indefinitely thin rectangle), and as z diminishes, the longer side decreases and the shorter increases until for $z=0$ (central section) the rectangle becomes a square; after which, for z negative it again becomes a rectangle in the conjugate direction, and finally, for $z=-1$ (lowest section) it becomes the sinister diagonal (considered as an indefinitely thin rectangle). But on account of the symmetry it is sufficient to consider the upper sections for which z is positive. The sides $\pm(x+y)+z-1=0$ parallel to the dexter diagonal of the square may for convenience be termed the dexter sides, and the others the sinister sides. In what follows I write c to denote the constant value of z.

We require to know whether the bitangents have real or imaginary contacts; and for this purpose to find the coordinates of the points of contact.

Take first a dexter bitangent $x+y+c-1=0$; the coordinates of any point hereof are

$$x=\tfrac{1}{2}(1-c+\theta),\quad y=\tfrac{1}{2}(1-c-\theta),$$

where θ is arbitrary; and substituting in the equation of the curve, we should have for θ a twofold quadric equation, giving the values of θ for the two points of contact respectively. We have

$$x^2+y^2=\tfrac{1}{2}\{(1-c)^2+\theta^2\},\quad xy=\tfrac{1}{4}\{(1-c)^2-\theta^2\},$$

and thence

$$8c^2\{(1-c)^2+\theta^2\}+8c\{(1-c)^2-\theta^2\}+\{(1-c)^2-\theta^2\}^2=0,$$

viz. this equation is

$$\{\theta^2-(1-c)(1+3c)\}^2=0,$$

a twofold quadric equation, as it should be; and the values of θ being $=\pm\sqrt{(1-c)(1+3c)}$, we see that these, and therefore the contacts, are real from $c=1$ to $c=-\tfrac{1}{3}$.

In exactly the same way for a sinister bitangent $\pm(x-y)-c-1=0$, we have

$$x=\tfrac{1}{2}(1+c+\phi),\quad -y=\tfrac{1}{2}(1+c-\phi),\ \text{and}\ \phi=\pm\sqrt{(1-3c)(1+c)},$$

viz. the values of ϕ, and therefore the contacts, are real from $c=\tfrac{1}{3}$ to $c=-1$.

That is,

	Contacts of Dexter Bitangents.	Contacts of Sinister Bitangents.
$c=\ \ 1$ to $\tfrac{1}{3}$	real,	imaginary,
$c=\ \ \tfrac{1}{3}$ to $-\tfrac{1}{3}$	real,	real,
$c=-\tfrac{1}{3}$ to -1	imaginary,	real;

or say $c=1$ to $\tfrac{1}{3}$, the contacts are real, imaginary; but $c=\tfrac{1}{3}$ to 0, they are real, real. In the transition case, $c=\tfrac{1}{3}$, the sinister bitangents become osculating (4-pointic) tangents touching at points on the dexter diagonal. This can be at once verified.

Observe that when $c = 1$, we have

$$(x+y)^2 + x^2y^2 = 0,$$

so that the only real point is $x = 0$, $y = 0$; viz. this is a tacnode, having the real tangent $x + y = 0$. For $c = 0$ (central section) the equation becomes $x^2y^2 = 0$; viz. the curve is here the two nodal lines each twice.

It is now easy to trace the changes of form.

$c = 1$; curve is a tacnode, as just mentioned, tangent the dexter diagonal.

$c < 1 > \frac{1}{3}$; curve is a figure of 8 inside the rectangle, having real contacts with the dexter sides, but imaginary contacts with the sinister sides.

$c = \frac{1}{3}$; curve is a figure of 8 having real contacts with the dexter sides, and osculating (4-pointic) contacts with the sinister sides.

$c < \frac{1}{3} > 0$; curve is an indented figure of 8 having real contacts as well with the sinister as the dexter sides.

$c = 0$; curve is squeezed up into a finite cross, being the crunodal parts of the nodal lines; and joined on to these we have the acnodal parts, so that the whole curve consists of the lines $x = 0$, $y = 0$ each as a twofold line.

For tracing the curve, it is convenient to turn the axes through an angle of $45°$; viz. writing $\frac{y+x}{\sqrt{2}}$, $\frac{y-x}{\sqrt{2}}$ in place of x, y respectively, the equation becomes

$$c\,(y^2 - x^2) + c^2(y^2 + x^2) + \tfrac{1}{4}(y^2 - x^2)^2 = 0;$$

$$x = 0 \text{ gives } y^2 = 0 \text{ or } y^2 = -4c\,(1+c),\text{*}$$

$$y = 0 \text{ gives } x^2 = 0 \text{ or } x^2 = \quad 4c\,(1-c).$$

Moreover, we have

$$4\,(c - c^2)(y^2 - x^2) + 8c^2y^2 + (y^2 - x^2)^2 = 0,$$

viz.

$$(x^2 - y^2 + 2c^2 - 2c)^2 = 4c^2\,\{(c-1)^2 - 2y^2\},$$

and similarly

$$(y^2 - x^2 + 2c^2 + 2c)^2 = 4c^2\,\{(c+1)^2 - 2x^2\},$$

putting in evidence the bitangents, now represented by the equations $c - 1 = \pm y\sqrt{2}$ and $c + 1 = \pm x\sqrt{2}$ respectively. And for the first of these, or $c - 1 = \pm y\sqrt{2}$, we have for the points of contact $x^2 = \frac{1}{2}(1-c)(1+3c)$; and for the second of them, or $c + 1 = \pm x\sqrt{2}$, the points of contact are $y^2 = \frac{1}{2}(1+c)(1-3c)$.

I consider the circumscribed cone having its vertex at a point $(0, 0, \gamma)$ on the nodal line $(x = 0, y = 0)$. Writing in the equation of the surface $x = \lambda(z - \gamma)$, $y = \mu(z - \gamma)$, the equation, throwing out the factor $(z - \gamma)^2$, becomes

$$2\lambda\mu z + (\lambda^2 + \mu^2)\,z^2 + \lambda^2\mu^2(z - \gamma)^2 = 0,$$

that is,

$$\begin{aligned}&(\lambda^2\mu^2 + \lambda^2 + \mu^2)\,z^2\\ &+ 2\,(-\gamma\lambda\mu + 1)\,z\lambda\mu\\ &+ \quad \gamma^2 \;.\; \lambda^2\mu^2 = 0;\end{aligned}$$

* y always imaginary when c is positive.

and equating to zero the discriminant in regard to z, we have

$$\gamma^2(\lambda^2\mu^2+\lambda^2+\mu^2)-(-\gamma\lambda\mu+1)^2=0,$$

that is,

$$\gamma^2(\lambda^2+\mu^2)+2\gamma\lambda\mu-1=0;$$

and substituting herein the values $\lambda=\dfrac{x}{z-\gamma}$ and $\mu=\dfrac{y}{z-\gamma}$, we have the equation of the cone, viz. this is

$$\gamma^2(x^2+y^2)+2\gamma xy-(z-\gamma)^2=0,$$

or, what is the same thing,

$$\gamma^2(x^2+y^2-1)+2\gamma(xy+z)-z^2=0;$$

viz. this is a quadric cone having for its principal planes $z-\gamma=0$, $x+y=0$, $x-y=0$, these last being the planes through the nodal line and the two edges of the tetrahedron. In the particular case $\gamma=\infty$, the cone becomes the circular cylinder $x^2+y^2-1=0$.

The cone intersects the plane $z=0$ in the conic

$$\gamma^2(x^2+y^2-1)+2\gamma xy=0,$$

which is a conic passing through the corners of the square ($x=0$, $y=\pm 1$), ($x=\pm 1$, $y=0$). For $\gamma>1$, that is, for an exterior point, the conic is an ellipse having for the squares of the reciprocals of the semi-axes $1+\dfrac{1}{\gamma}$, $1-\dfrac{1}{\gamma}$ (this at once appears by writing in the equation $\dfrac{x+y}{\sqrt{2}}$, $\dfrac{x-y}{\sqrt{2}}$ in place of x, y respectively). In particular, for $\gamma=\infty$, the curve becomes the circle $x^2+y^2-1=0$. We have thus the apparent contour of the surface as seen from the point $z=\gamma$ on the nodal line, projected on the plane $z=0$ of the other two nodal lines.

To find the curve of contact of the cone and surface, or say the surface-contour from the same point, write for a moment

$$V=\gamma(x^2+y^2-1)+2\gamma(xy+z)-z^2,$$
$$U=(xy+z)^2+z^2(x^2+y^2-1);$$

then, substituting for x^2+y^2-1 its value in terms of V from the first equation, we find

$$U=\left(xy+z-\frac{z^2}{\gamma}\right)^2+\frac{z^2}{\gamma^2}V,$$

and the equations $U=0$, $V=0$ give therefore $xy+z-\dfrac{z^2}{\gamma}=0$, or say $\gamma(xy+z)-z^2=0$. The cone and surface therefore touch along the quadriquadric curve

$$\gamma^2(x^2+y^2-1)+2\gamma(xy+z)-z^2=0,$$
$$\gamma(xy+z)-z^2=0,$$

equations which may be replaced by

$$\gamma\,(x^2+y^2-1)+xy+z=0,$$
$$\gamma^2(x^2+y^2-1)+z^2\quad=0.$$

In the case $\gamma=\infty$, the equations are $x^2+y^2-1=0$, $xy+z=0$, viz. the curve is the intersection of the hyperbolic paraboloid $xy+z=0$ by the cylinder $x^2+y^2-1=0$.

557.

ON CERTAIN CONSTRUCTIONS FOR BICIRCULAR QUARTICS.

[From the *Proceedings of the London Mathematical Society*, vol. v. (1873—1874), pp. 29—31. Read March 12, 1874.]

I CALL to mind that if F, G are any two points and F', G' their antipoints; then the circle on the diameter FG and that on the diameter $F'G'$ are concentric orthotomics, viz. they have the same centre, and the sum of the squared radii is $=0$. Moreover, if the circles B, B' are concentric orthotomics, and the circle A is orthotomic to B, then it is a bisector of B', viz. it cuts B' at the extremities of a diameter of B'; and B' is then said to be a bifid circle in regard to A.

Given two real circles, these have an axial orthotomic, the circle, centre on the line of centres at its intersection with the radical axis, which cuts at right angles the given circles; viz. this axial orthotomic is real if the circles have no real intersection; but if the intersections are real, then the axial orthotomic is a pure imaginary, and instead thereof we may consider its concentric orthotomic, viz. this is the axial bifid of the two circles, or circle having its centre on the line of centres at the intersection thereof with the radical axis or common chord of the two circles, and having this common chord for its diameter.

If one of the circles is a pure imaginary, then we have still an axial orthotomic; viz. the pure imaginary circle is replaced by the concentric orthotomic; and the axial orthotomic is a bisector of the substituted circle; and so if each of the circles is a pure imaginary, then we have still an axial orthotomic, viz. each circle is replaced by the concentric orthotomic, and the axial orthotomic is a bisector of the substituted circles. And in either case the axial orthotomic of the original circles (one or each of them pure imaginary) *is real;* viz. this is given either as the axial bisector of one real circle and orthotomic of another real circle; or as the axial bisector of two circles, from which the reality thereof easily appears. Or we may verify it thus: Suppose that the two circles are $(x-\alpha)^2+y^2=\beta^2$, $(x-\alpha')^2+y^2=\beta'^2$, and their axial orthotomic $(x-m)^2+y^2=k^2$, then we have $(m-\alpha)^2=\beta^2+k^2$, $(m-\alpha')^2=\beta'^2+k^2$; subtracting, it appears that m is real; and then if either β^2 or β'^2 is negative, the equation containing this quantity shows that k^2 is positive; viz. the circle $(x-m)^2+y^2=k^2$ is real.

The above remarks have an obvious application to the theory of bicircular quartics; viz. a bicircular quartic is the envelope of a variable circle, having its centre on a conic, and orthotomic to a circle: it may be that this circle is a pure imaginary. We then replace it by the concentric orthotomic, and say that the curve is the envelope of a variable circle having its centre on a conic and bisecting a circle. We have thus a real form for cases which originally present themselves under an imaginary form.

The Bicircular Quartic with given vertices.

First, if the vertices are real; let the vertices taken in order be F, G, H, K.

First construction: On FG as diameter describe a circle, and on HK as diameter a circle; on the line terminated by the two centres (as transverse or conjugate axis) describe a conic Θ_1, and describe the axial orthotomic circle Σ_1 of the two circles (viz. the centre of Σ_1 is on the axis of symmetry at its intersection with the radical axis of the two circles); then the curve is the envelope of a variable circle having its centre on Θ_1 and orthotomic to Σ_1.

Second construction: On FH as diameter describe a circle, and on GK as diameter a circle. On the line terminated by the two centres (as transverse or conjugate axis) describe a conic Θ_2, and describe the axial bifid circle Σ_2' of the two circles (viz. the centre of Σ_2' is on the axis of symmetry at its intersection with the radical axis or common chord of the two circles, and its diameter is this common chord); then the curve is the envelope of a variable circle having its centre on Θ_2 and bisecting Σ_2'.

Third construction: On FK as diameter describe a circle, and on GH as diameter a circle; and then, as in the first construction, a conic Θ_3 and a circle Σ_3; the curve is the envelope of a variable circle having its centre on Θ_3 and orthotomic to Σ_3.

Observe that in the three constructions the conics have always the same centre; and if the three conics are taken with the same foci, then the three constructions give one and the same bicircular quartic. The first and third constructions form a pair, and there is no reason for selecting one of them in preference to the other; but the second construction is unique; it is on this account natural to make use of it in discussing the series of curves with the given vertices.

In the particular case where the points F, G and H, K are situate symmetrically on opposite sides of a centre O ($OF=OK$, $OG=OH$), then in the third construction the centres each coincide with O, or the axis of the conic vanishes; hence the construction fails: the first and second constructions hold good, and in each of them the conic and circle are concentric. The curve is in this case quadrantal: having, besides the original axis of symmetry, another axis of symmetry through O, at right angles thereto.

Secondly, if the vertices are two real, two imaginary, say f, $g = \alpha \pm \beta\iota$; h, k, we modify the first or third construction; viz. if F', G' are the antipoints of F, G; then on $F'G'$ as diameter describe a circle, and on HK as diameter a circle. On the line terminated by the two centres (as transverse or conjugate axis) describe a conic Θ_1, and describe the axial bisector-orthotomic circle Σ_1 of the two circles; viz. this is the circle (centre on the axis of symmetry) which bisects the circle $F'G'$, and cuts at right angles the circle HK; then the curve is the envelope of the variable circle having its centre on Θ_1 and orthotomic to Σ_1.

Thirdly, if the vertices are all imaginary, say f, $g = \alpha \pm \beta\iota$; h, $k = \gamma \pm \delta\iota$, we modify the first or third construction. Take F', G' the antipoints of F, G, and H', K' the antipoints of H, K; then on $F'G'$ as diameter describe a circle, and on $H'K'$ as diameter a circle; on the line terminated by the two centres (as transverse or conjugate axis) describe a conic Θ, and describe the axial bisector-circle Σ of the two circles (viz. this is a circle, centre on the axis of symmetry, bisecting each of the circles): the curve is the envelope of a variable circle, centre on the conic Θ and cutting at right angles the circle Σ.

558.

A GEOMETRICAL INTERPRETATION OF THE EQUATIONS OBTAINED BY EQUATING TO ZERO THE RESULTANT AND THE DISCRIMINANTS OF TWO BINARY QUANTICS.

[From the *Proceedings of the London Mathematical Society*, vol. v. (1873—1874), pp. 31—33. Read March 12, 1874.]

CONSIDER the equations

$$U = (a, b, \ldots \between t, 1)^{\lambda} = 0,$$
$$U' = (a', b', \ldots \between t, 1)^{\lambda'} = 0;$$

and equating to zero the discriminants of the two functions respectively, and also the resultant of the two functions, let the equations thus obtained be

$$\Delta = (a, b, \ldots)^{2\lambda-2} = 0,$$
$$\Delta' = (a', b', \ldots)^{2\lambda'-2} = 0,$$
$$R = (a, b, \ldots)^{\lambda'} (a, b, \ldots)^{\lambda} = 0.$$

I take $(a, b, \ldots)$, $(a', b', \ldots)$ to be linear functions of the coordinates (x, y, z); and t to be an indeterminate parameter. Hence $U=0$ represents a line the envelope whereof is the curve $\Delta=0$, or, what is the same thing, the equation $U=0$ represents any tangent of the curve $\Delta=0$; this is a unicursal curve of the order $2\lambda-2$ and class λ, with $3(\lambda-2)$ cusps and $\frac{1}{2}(\lambda-2)(\lambda-3)$ nodes. Similarly $U'=0$ represents a line the envelope of which is the curve $\Delta'=0$: this is a unicursal curve of the order $2\lambda'-2$ and class λ', with $3(\lambda'-2)$ cusps and $\frac{1}{2}(\lambda'-2)(\lambda'-3)$ nodes; the equation $U'=0$ represents any tangent of this curve.

The equations $U=0$, $U'=0$ considered as existing simultaneously with the same value of t, establish a (1, 1) correspondence between the tangents (or if we please, between the points) of the two curves. The locus of the intersection of the corresponding tangents is the curve $R=0$, a unicursal curve of the order $\lambda+\lambda'$, with $\frac{1}{2}(\lambda+\lambda'-1)(\lambda+\lambda'-2)$ nodes and no cusps; consequently of the class $2(\lambda+\lambda'-1)$.

It is to be shown that the curve $R=0$ touches the curve $\Delta=0$ in $\lambda'+2\lambda-2$ points, and similarly the curve $\Delta'=0$ in $2\lambda'+\lambda-2$ points.

In fact, consider any tangent T' of the curve Δ'; let this meet the curve Δ in a point A, and let Q be the tangent at A to the curve Δ; suppose, moreover, that T is the tangent of Δ corresponding to the tangent T' of Δ'. Then if Q and T coincide, the corresponding tangent of T' will be Q, and the curve R will pass through A. It is easy to see that in this case the curves R, Δ will touch at A. Again, if P be a tangent from A to the curve Δ, then, if P and T coincide, the corresponding tangent of T' will be P, and the curve R will pass through A; but in this case the point A will be a mere intersection, not a point of contact, of the two curves.

The tangents T, Q each correspond to T', and they consequently correspond to each other. For a given position of T we have a single position of T', and therefore $2\lambda-2$ positions of A, or, what is the same thing, of Q; that is, for a given position of T we have $2\lambda-2$ positions of Q. Again, to a given position of Q corresponds a single position of A, therefore λ' positions of T', therefore also λ' positions of T; that is, for a given position of Q we have λ' positions of T. The correspondence between T, Q is thus a $(\lambda', 2\lambda-2)$ correspondence, and the number of united tangents is therefore $\lambda'+2\lambda-2$, or the curves R, Δ touch in $\lambda'+2\lambda-2$ points.

The tangents T, P each correspond to T', and they therefore correspond to each other. For a given position of T we have a single position of T', and therefore $2\lambda-2$ positions of A, and thence $(2\lambda-2)(\lambda-2)$ positions of P; that is, for a given position of T we have $(2\lambda-2)(\lambda-2)$ positions of P. Again, to a given position of P correspond $2\lambda-4$ positions of A, therefore $(2\lambda-4)\lambda'$ positions of T' or of T; that is, for a given position of P we have $(2\lambda-4)\lambda'$ positions of T. The correspondence between T, P is thus a $[2\lambda'(\lambda-2), 2(\lambda-1)(\lambda-2)]$ correspondence, and the number of united tangents is $2(\lambda+\lambda'-1)(\lambda-2)$; or the curves R, Δ meet in $2(\lambda+\lambda'-1)(\lambda-2)$ points.

Reckoning the contacts twice, the total number of intersections of R, Δ is

$$2\lambda'+4\lambda-4+2(\lambda+\lambda'-1)(\lambda-2), \ =(\lambda+\lambda')(2\lambda-2),$$

as it should be.

In the particular case $\lambda=\lambda'=2$, the curves Δ, Δ' are conics, and the curve R is a quartic curve touching each of the conics 4 times; this is at once verified, since the equations here are

$$ac-b^2=0, \quad a'c'-b'^2=0, \quad 4(ac-b^2)(a'c'-b'^2)-(ac'+a'c-2bb')^2=0.$$

559.

[NOTE ON INVERSION.]

[From the *Proceedings of the London Mathematical Society*, vol. v. (1873—1874), p. 112.]

THE inverse of the anchor ring (in the foregoing paper* called the cyclide) is in fact the general binodal cyclide or binodal bicircular quartic; viz. assuming it to be a cyclide (bicircular quartic), to see that it is binodal, it need only be observed that the anchor ring is binodal (has two real or imaginary conic points, viz. these are the intersections of the circles in the several axial planes); and to see that it is the general binodal cyclide, we have only to count the constants; viz. the general cyclide or surface

$$(x^2+y^2+z^2)^2+(x^2+y^2+z^2)(\alpha x+\beta y+\gamma z)+(a, b, c, d, f, g, h, l, m, n)(x, y, z, 1)^2=0$$

contains 13 constants, and therefore the binodal cyclide $13-2, =11$ constants. But the anchor ring, irrespective of position, contains 2 constants; centre of inversion, taken in given axial plane, has 2 constants; radius of inversion, 1 constant; in all $2+2+1, =5$ constants; or taking the inverse surface in an arbitrary position, the number of constants is $5+6, =11$.

* By Mr H. M. Taylor: Inversion, with special reference to the Inversion of an Anchor Ring or Torus, (*Lond. Math. Soc. Proc.*, same volume, pp. 105—112).

560.

[ADDITION TO LORD RAYLEIGH'S PAPER "ON THE NUMERICAL CALCULATION OF THE ROOTS OF FLUCTUATING FUNCTIONS."]

[From the *Proceedings of the London Mathematical Society*, vol. v. (1873—1874), pp. 123, 124. November 22.]

PROF. CAYLEY, to whom Lord Rayleigh's paper was referred, pointed out that a similar result may be attained by a method given in a paper by Encke, "Allgemeine Auflösung der numerischen Gleichungen," *Crelle*, t. XXII. (1841), pp. 193—248, as follows:

Taking the equation

$$0 = 1 - ax + bx^2 - cx^3 + dx^4 - ex^5 + fx^6 - gx^7 + hx^8 - \dots;$$

if the equation whose roots are the squares of these is

$$0 = 1 - a_1x + b_1x^2 - c_1x^3 + \dots,$$

then

$$\begin{aligned} a_1 &= a^2 - 2b, \\ b_1 &= b^2 - 2ac + 2d, \\ c_1^2 &= c^2 - 2bd + 2ae - 2f, \\ d_1^2 &= d^2 - 2ce + 2bf - 2ag + 2h, \text{ \&c.}; \end{aligned}$$

and we may in the same way derive a_2, b_2, c_2, &c. from a_1, b_1, c_1, &c., and so on.

As regards the function

$$J_n(z) = \frac{z^n}{2^n . \Gamma(n+1)} \left\{1 - \frac{z^2}{2 . 2n+2} + \frac{z^4}{2 . 4 . 2n+2 . 2n+4} - \dots\right\},$$

we have as follows:

$$a^{-1} = 2^2 . n+1,$$
$$b^{-1} = 2^5 . n+1 . n+2,$$
$$c^{-1} = 2^7 . 3 . n+1 \ldots n+3,$$
$$d^{-1} = 2^{11} . 3 . n+1 \ldots n+4,$$
$$e^{-1} = 2^{13} . 3 . 5 . n+1 \ldots n+5,$$
$$f^{-1} = 2^{16} . 3^2 . 5 . n+1 \ldots n+6,$$
$$g^{-1} = 2^{18} . 3^2 . 5 . 7 . n+1 \ldots n+7,$$
$$h^{-1} = 2^{23} . 3^2 . 5 . 7 . n+1 \ldots n+8,$$
$$a_1^{-1} = 2^4 . (n+1)^2 . n+2,$$
$$b_1^{-1} = 2^9 . (n+1 . n+2)^2 . n+3 . n+4,$$
$$c_1^{-1} = 2^{13} . 3 . (n+1 \ldots n+3)^2 . n+4 \ldots n+6,$$
$$d_1^{-1} = 2^{19} . 3 . (n+1 \ldots n+4)^2 . n+5 \ldots n+8,$$

$$a_2 = \frac{5n+11}{2^8 . (n+1)^4 (n+2)^2 n+3 . n+4},$$

$$b_2 = \frac{25n^2+231n+542}{2^{17} . (n+1 . n+2)^4 (n+3 . n+4)^2 n+5 \ldots n+8},$$

$$a_3 = \frac{429n^5+7640n^4+53752n^3+185430n^2+311387n+202738}{2^{16}(n+1)^8 (n+2)^4 (n+3 . n+4)^2 n+5 . n+6 . n+7 . n+8}.$$

If $n=0$,

$$\Sigma p^{-16} = a_3 = \frac{101369}{2^{27} . 3^3 . 5 . 7} = p_1^{-16}, \text{ suppose};$$

whence

$$p_1 = 2{\cdot}404825.$$

[The quantities $p_1, p_2, \ldots$ are the roots of the function $J_n(x)$ in increasing order of magnitude, so that, as these roots are all real, it follows that for $J_0(x)$,

$$a = \Sigma p_1^{-2}, \quad a_1 = \Sigma p_1^{-4}, \quad a_2 = \Sigma p_1^{-8}, \quad a_3 = \Sigma p_1^{-16}, \ldots]$$

561.

ON THE GEOMETRICAL REPRESENTATION OF CAUCHY'S THEOREMS OF ROOT-LIMITATION.

[From the *Transactions of the Cambridge Philosophical Society*, vol. XII. Part II. (1877), pp. 395—413. Read February 16, 1874.]

THERE is contained in Cauchy's Memoir "Calcul des Indices des Fonctions," *Journ. de l'École Polytech.* t. XV. (1837) a general theorem, which, though including a well-known theorem in regard to the imaginary roots of a numerical equation, seems itself to have been almost lost sight of. In the general theorem (say Cauchy's two-curve theorem) we have in a plane two curves $P=0$, $Q=0$, and the real intersections of these two curves, or say the "roots," are divided into two sets according as the Jacobian

$$d_xP\,.\,d_yQ-d_xQ\,.\,d_yP$$

is positive or negative, say these are the Jacobian-positive and the Jacobian-negative roots: and the question is to determine for the roots within a given contour or circuit, the difference of the numbers of the roots belonging to the two sets respectively.

In the particular theorem (say Cauchy's rhizic theorem) P and Q are the real part and the coefficient of i in the imaginary part of a function of $x+iy$ with, in general, imaginary coefficients (or, what is the same thing, we have $P+iQ=f(x+iy)+i\phi(x+iy)$, where f, ϕ are real functions of $x+iy$): the roots of necessity are of the same set: and the question is to determine the number of roots within a given circuit.

In each case the required number is theoretically given by the same rule, viz. considering the fraction $\frac{P}{Q}$, it is the excess of the number of times that the fraction changes from + to − over the number of times that it changes from − to +, as the point (x, y) travels round the circuit, attending only to the changes which take place on a passage through a point for which P is $=0$.

In the case where the circuit is a polygon, and most easily when it is a rectangle the sides of which are parallel to the two axes respectively, the excess in question can be actually determined by means of an application of Sturm's theorem successively to each side of the polygon, or rectangle.

In the present memoir I reproduce the whole theory, presenting it under a completely geometrical form, viz. I establish between the two sets of roots the distinction of *right-* and *left-handed:* and (availing myself of a notion due to Prof. Sylvester*) I give a geometrical form to the theoretic rule, making it depend on the "intercalation" of the intersections of the two curves with the circuit: I also complete the Sturmian process in regard to the sides of the rectangle: the memoir contains further researches in regard to the curves in the case of the particular theorem, or say as to the rhizic curves $P=0$, $Q=0$.

The General Theory. Articles Nos. 1 to 19.

1. Consider in a plane two curves $P=0$, $Q=0$ (P and Q each a rational and integral function of x, y), which to fix the ideas I call the *red* curve and the *blue* curve respectively†: the curve $P=0$ divides the plane into two sets of regions, say a positive set for each of which P is positive, and a negative set for each of which P is negative: it is of course immaterial which set is positive and which negative, since writing $-P$ for P the two sets would be interchanged: but taking P to be given, the two sets are distinguished as above. And we may imagine the negative regions to be coloured red, the positive ones being left uncoloured, or say they are white. Similarly the curve $Q=0$ divides the plane into two sets of regions, the negative regions being coloured blue, and the positive ones being left uncoloured, or say they are white. Taking account of the twofold division, and considering the coincidence of red and blue as producing black, there will be four sets of regions, which for convenience may be spoken of as *sable, gules, argent, azure:* viz. in the figures we have

P	Q	
$-$	$-$	sable, shown by cross lines,
$-$	$+$	gules, „ „ vertical lines,
$+$	$+$	argent, left white,
$+$	$-$	azure, shown by horizontal lines,

sable and argent ($-\ -$ and $+\ +$) being thus positive colours, and gules and azure ($-\ +$ and $+\ -$) negative colours. See figures [pp. 32, 38] towards end of Memoir.

* See his memoir, "A theory of the Syzygetic relations, &c." *Phil. Trans.*, 1853. The Sturmian process is by Sturm and Cauchy applied to two independent functions ϕx, fx of a variable x; but the notion of an intercalation as applied to the order of succession of the roots of the equations $\phi(x)=0$, $f(x)=0$ is due to Sylvester, and it was he who showed that what the Sturmian process determined was in fact the intercalation of these roots: but, not being concerned with circuits, he was not led to consider the intercalation of a circuit.

† It is assumed throughout that the two curves have no points (or at least no real points) of multiple intersection; i.e. they nowhere touch each other, and neither curve passes through a multiple point of the other curve.

2. Consider any point of intersection of the two curves. There will be about this point four regions, sable and argent being opposite to each other, as also gules and azure; whence selecting an order

sable, gules, argent, azure;

if to have the colours in this order we have to go about the point, or root, right-handedly, the root is right-handed: but if left-handedly, then the root is left-handed: or, what is more convenient, going always right-handedly, then, if the order of the colours is

sable, gules, argent, azure,

the root is right-handed: but if the order is

sable, azure argent, gules,

the root is left-handed.

3. The distinction of right- and left-handed corresponds to the sign of the Jacobian

$$\frac{d(P, Q)}{d(x, y)} (= d_xP \,.\, d_yQ - d_xQ \,.\, d_yP);$$

we may (reversing if necessary the original sign of one of the functions) assume that for a right-handed root the Jacobian is positive, for a left-handed one, negative.

4. I consider a trajectory which may be either an unclosed curve not cutting itself, or else a circuit, viz. this is a closed curve not cutting itself. A circuit is considered as described right-handedly: an unclosed trajectory is considered as described according to a currency always determinate *pro hâc vice:* viz. one extremity is selected as the beginning and the other as the end of the trajectory: but the currency may if necessary or convenient be reversed: thus if an unclosed trajectory forms part of a circuit the currency is thereby determined: but the same unclosed trajectory may form part of two opposite circuits, and as such may have to be taken with opposite currencies. It is assumed that a trajectory does not pass through any intersection of the P and Q curves.

5. A trajectory has its P- and Q-sequence, viz. considering in order its intersections with the two curves, we write down a P for each intersection with the red curve and a Q for each intersection with the blue curve, thus obtaining an intermingled series of P's and Q's, which is the sequence in question. In the case of a circuit, the sequence is considered as a circuit, viz. the first and last terms are considered as contiguous, and it is immaterial at what point the sequence commences. The sequence will of course vanish if the trajectory does not meet either of the curves.

6. A P- and Q-sequence gives rise to an "intercalation," viz. if in the sequence there occur together any even number of the same letter, these are omitted (whence also any odd number of the same letter is reduced to the letter taken once): and if by reason of an omission there again occur an even number of the same letter, these

are omitted: and so on. The intercalation contains therefore only the letters P and Q alternately: viz. in the case of an unclosed trajectory the intercalation may contain an even number of letters, beginning with the one and ending with the other letter, and so containing the same number of each letter—or it may contain an odd number of letters, beginning and ending with the same letter, and so containing one more of this than of the other letter; say the intercalation is PQ or QP, or else PQP or QPQ. The intercalation may vanish altogether; thus if the sequence were $QPPQ$, this would be the case.

7. In the case of a circuit the intercalation cannot begin and end with the same letter, for these, as contiguous letters, would be omitted; and since any letter thereof may be regarded as the commencement it is PQ or QP indifferently. A little consideration will show that the whole number of letters must be evenly even, or, what is the same thing, the number of each letter must be even. Thus imagine the circuit beginning in sable, and let the intercalation begin with PQ; viz. P we pass from sable to azure, and Q we pass from azure to argent: in order to get back into sable we must either return the same way (Q argent to azure, P azure to sable), but then the sequence is $PQQP$, and the intercalation vanishes: here the number of letters is 0, an evenly even number: or else we must complete the cycle of colours P argent to gules, Q gules to sable: and the sequence and therefore also the intercalation then is $PQPQ$, where the number of letters is 4, an evenly even number.

8. In the case of any trajectory whatever, the half number of letters in the intercalation is termed the "index," viz. this is either an integer or an integer $+\frac{1}{2}$. But in the case of a circuit the index is an even integer, and the half-index is therefore an integer. The index may of course be $=0$.

9. But we require a further distinction: instead of a P- and Q-sequence we have to consider a $\pm P$- and Q-sequence. To explain this, observe that a passage over the red curve may be from a negative to a positive colour (azure to sable or gules to argent), this is $+P$, or from a positive to a negative colour (sable to azure or argent to gules), this is $-P$. And so the passage over the blue curve may be from a negative to a positive colour (gules to sable or azure to argent), this is $+Q$, or else from a positive to a negative colour (sable to gules or argent to azure), this is $-Q$. The sequence will contain the P and Q intermingled in any manner, but the signs will always be $+\,-$ *alternately;* for $+(P$ or $Q)$, denoting the passage into a positive colour, must always be immediately succeeded by $-(P$ or $Q)$, denoting the passage into a negative colour. Whence, knowing the sequence independently of the signs, we have only to prefix to the first letter the sign $+$ or $-$ as the case may be, and the sequence is then completely determined.

10. Passing to a $\pm$ intercalation, observe that in omitting any even number of P's or Q's, the omitted signs are always $+\,-\,+\,-$ &c. or else $-\,+\,-\,+$ &c., viz. the omitted signs begin with one sign and end with the opposite sign. Hence the signs being in the first instance alternate, they will after any omission remain alternate: and the letters being also alternate, the intercalation can contain only $+P$ and $-Q$

or else $-P$ and $+Q$. Hence in the case of a circuit the intercalation is either $(+P-Q)$, say this is a *positive* circuit, or else $(-P+Q)$, say this is a *negative* circuit. There is of course the *neutral* circuit $(PQ)_0$ for which the intercalation vanishes.

11. Consider a circuit not containing within it any root; as a simple example let the circuit lie wholly in one colour, or wholly in two adjacent colours, say sable and gules: in the former case the sequence, and therefore also the intercalation, vanishes: in the latter case the sequence is $+Q-Q$, and therefore the intercalation vanishes: viz. in either case the intercalation is $(PQ)_0$.

12. Consider next a circuit containing within it one right-handed root; for instance let the circuit lie wholly in the four regions adjacent to this root, cutting the two curves each twice; the sequence and therefore also the intercalation is $+P-Q+P-Q$; viz. this is a positive circuit $(+P-Q)_1$, where the subscript number is the half-index, or half of the number of P's or of Q's. Similarly if a circuit contains within it one left-handed root, for instance if the circuit lies wholly in the four regions adjacent to this root, cutting the two curves each twice, the sequence and therefore also the intercalation is $-P+Q-P+Q$, viz. this is a negative circuit $(-P+Q)_1$: and the consideration of a few more particular cases leads easily to the general and fundamental theorem:

13. *A circuit is positive* $(+P-Q)_\delta$ *or negative* $(-P+Q)_\delta$ *according as it contains within it more right-handed or more left-handed roots; and in either case the half-index* δ *is equal to the excess of the number of one over that of the other set of roots. If the circuit is neutral* $(PQ)_0$, *then there are within it as many left-handed as right-handed roots.*

14. The proof depends on a composition of circuits, but for this some preliminary considerations are necessary.

Imagine two unclosed trajectories forming a circuit, and write down in order the intercalation of each. The whole number of letters must be even: viz. the numbers for the two intercalations respectively must be both even or both odd. I say that if the terminal letter of the first intercalation and the initial letter of the second intercalation are different, then also the initial letter of the first intercalation and the terminal letter of the second intercalation will be different: if the same, then the same. In fact, the intercalations may be each PQ or each QP, or one PQ and the other QP: or each PQP, or each QPQ, or one PQP and the other QPQ. Supposing the letters in question are different, then the intercalations may be termed similar; but if the same, then the intercalations may be termed contrary.

15. In the first case, that is when the intercalations are similar, the two together form the intercalation of the circuit; the sum of their numbers of letters (that is, twice the sum of their indices) will be evenly even, and the half of this, or the sum of the indices, will be the index of the circuit; each intercalation will be $(+P-Q)$ or else each will be $(-P+Q)$; and the circuit will be $(+P-Q)$ or $(-P+Q)$ accordingly.

In the second case, that is, when the intercalations are contrary, they counteract each other in forming the intercalation of the circuit: it is the *difference* of the numbers of letters, or twice the difference of the indices, which is evenly even, and the half of this, or the difference of the indices, which is the index of the circuit: one intercalation is $(+P-Q)$, and the other is $(-P+Q)$: and the circuit will agree with that which has the larger index.

In particular if the circuit consist of a single unclosed trajectory, taken forwards and backwards; then the trajectory taken one way is $(+P-Q)$, taken the other way it is $(-P+Q)$; the number of terms is of course equal, and the circuit is $(PQ)_0$.

16. Consider now two circuits $ABCA$ and $ACDA$, having a common portion CA, or, more accurately, the common portions AC and CA: write down in order the intercalations of

$$ABC, \quad CA, \quad AC, \quad CDA:$$

the two mean terms destroy each other, and we can hence deduce the intercalation of the entire circuit $ABCDA$.

Suppose *first*, that ABC and CDA are similar; then if CA is similar to ABC it is also similar to CDA, that is, AC is contrary to CDA: and so if CA is contrary to ABC, then AC is similar to CDA.

To fix the ideas suppose CA similar to ABC, but AC contrary to CDA, then $ABCA$ is similar to CA; but $ACDA$ will be similar or contrary to AC, that is, contrary or similar to CA, that is, to $ABCA$, according as index of $AC >$ or $<$ index of CDA.

Suppose Ind. $AC <$ Ind. CDA, then $ACDA$ is similar to $ABCA$.

Now
$$\text{Ind. } ABCDA = \text{Ind. } ABC + \text{Ind. } CDA,$$
$$\text{Ind. } ABCA = \text{Ind. } ABC + \text{Ind. } AC,$$
$$\text{Ind. } ACDA = \text{Ind. } CDA - \text{Ind. } AC,$$
and thence
$$\text{Ind. } ABCDA = \text{Ind. } ABCA + \text{Ind. } ACDA,$$
the whole circuit being in this case similar to each of the component ones.

But if Ind. $AC >$ Ind. CDA, then $ACDA$ is contrary to $ABCA$.

And
$$\text{Ind. } ABCDA = \text{Ind. } ABC + \text{Ind. } CDA,$$
$$\text{Ind. } ABCA = \text{Ind. } ABC + \text{Ind. } CA,$$
$$\text{Ind. } ACDA = -\text{Ind. } CDA + \text{Ind. } AC,$$
and thence
$$\text{Ind. } ABCDA = \text{Ind. } ABCA - \text{Ind. } ACDA;$$
and the investigation is like hereto if CA is contrary to ABC but AC similar to CDA.

17. *Secondly,* if ABC and CDA are contrary, then if CA is similar to ABC it is contrary to CDA, that is, AC is similar to CDA; and so if CA is contrary to ABC it is similar to CDA, that is, AC is contrary to CDA.

Suppose CA similar to ABC, and AC similar to CDA; then $ABCA$ is also similar to ABC, and $ACDA$ similar to CDA; viz. ABC, CA and $ABCA$ are similar to each other, and contrary to AC, CDA, $ACDA$ which are also similar to each other.

Also
$$\begin{aligned}\text{Ind. } ABCDA &= \text{Ind. } ABC \sim \text{Ind. } CDA,\\ \text{Ind. } ABCA &= \text{Ind. } ABC + \text{Ind. } CA,\\ \text{Ind. } ACDA &= \text{Ind. } CDA + \text{Ind. } AC,\end{aligned}$$
and thence
$$\text{Ind. } ABCDA = \text{Ind. } ABCA \sim \text{Ind. } ACDA,$$

and the investigation is like hereto if CA is contrary to ABC and AC contrary to CDA.

18. It thus appears that in every case
$$\begin{aligned}\text{Ind. } ABCDA &= \text{Ind. } ABCA + \text{Ind. } ACDA,\\ \text{or} &= \text{Ind. } ABCA \sim \text{Ind. } ACDA,\end{aligned}$$
according as the component circuits are similar or contrary, and in the latter case the entire circuit is similar to that which has the largest index.

Moreover, any circuit whatever can be broken up into two smaller circuits, and these again continually into smaller circuits until we arrive at the before-mentioned elementary circuits, and the theorem as to the number of roots within a circuit is true as regards these elementary circuits; wherefore the theorem is true as regards any circuit whatever.

19. In the case where a trajectory is a finite right line, y is a given linear function of x, or the coordinates x, y can if we please be expressed as linear functions of a parameter u, so that as the describing point passes along the line, u varies between given limits, say from $u=0$ to $u=1$. The functions P, Q thus become given rational and integral functions of a single variable u (or it may be x or y), and the question of the P- and Q-sequence and intercalation relates merely to the order of succession of the roots of the equations $P=0$, $Q=0$, where P and Q denote functions of a single variable as above. To fix the ideas, let the trajectory be a line parallel to the axis of x; and in this case taking x as the parameter, and supposing that y_0 is the given value of y, P and Q are the functions of x obtained by writing y_0 for y in the original expressions of these functions. Of course the theory will be precisely the same for a line parallel to the axis of y: and by combining two lines parallel to each axis we have the case of a rectangular circuit. We require, for each side of the rectangle considered according to its proper currency, the intercalation PQ, QP, PQP or QPQ as the case may be, and also the sign + or − of the initial letter of the first intercalation; for then writing down the intercalations in order, with the signs for the several letters, + and − alternately (the first sign being + or − as the case may be), we have or deduce the intercalation of the circuit, and thus obtain the value of the difference of the numbers of the included right- and left-handed roots. We thus see how the whole theory depends on the case where the trajectory is a right line.

Intercalation-theory for a right line. Articles Nos. 20 to 31.

20. Considering then the case where the trajectory is a line parallel to the axis of x, P and Q will denote given rational functions of x; the curves $P=0$, $Q=0$ being of course each of them a set of right lines parallel to the axis of y: the regions will be bands each of them included between two such lines; and colouring them as explained in the general case, the colours will be as before, sable, gules, argent, azure, each region having in the neighbourhood of the trajectory (what we are alone concerned with) the same colour that it had in the original case where P and Q were functions of (x, y). We may regard the trajectory as described according to the currency $x=-\infty$ to $x=+\infty$: we have in regard to the trajectory a P- and Q-sequence and intercalation, a $\pm P$- and Q-sequence, &c., as in the original case. The intercalation may be as before PQ, QP, PQP or QPQ, and in each of these cases it may be positive, that is, $(+P-Q)$, or else negative, that is, $(-P+Q)$.

21. The question of sign may in the present case be disposed of without difficulty. For the initial point of the trajectory, we know the signs of P, Q, that is, the colour of the region: suppose for example that we have $P=-$, $Q=+$, or that the region is gules: then if the intercalation begin with P, this means that we either first pass a red line, or before doing so we pass an even number of blue lines: but in the last case the colours are sable, gules, sable, gules, ... always ending in gules; and the passage over the red line is gules to argent, viz. this is $+P$; and so in general the initial P or Q of the intercalation has the sign opposite to that of the P or Q belonging to the commencement of the trajectory.

22. For the solution of the problem we connect with P, Q a set of functions R, S, T, &c.: the intercalation is in fact given by means of the gain or loss of changes of sign in these functions on substituting therein the initial and final values of the variable x. It is convenient to consider the functions as arranged in a column

$$\begin{matrix} P \\ Q \\ R \\ S \\ \vdots \end{matrix}$$

say this is the column $PQRS...$, and to connect therewith a signaletic bicolumn: viz. the left-hand column is here the series of signs of these functions for the initial value of x, and the right-hand column is the series of signs for the terminal value of x: the bicolumn thus consisting of as many rows each of two signs, as there are functions. But such a bicolumn may be considered apart from any series of functions, as a set of rows each of two signs taken at pleasure.

We say that the "gain" of a bicolumn is

$$= -\text{(No. of changes of sign in left-hand column)} + \text{(No. in right-hand ditto)},$$

the gain being of course positive or negative; and a negative gain being regarded as a loss. Also if a positive gain be converted into an equal negative gain or *vice versâ*, we may speak of the gain as *reversed*.

23. A bicolumn may be divided in any manner into parts, taking always the last row of any part as being also the first row of the next succeeding part. This being so, the gain of the whole bicolumn is equal to the sum of the gains of its parts.

In a bicolumn of two rows, if we reverse either row (that is, write therein $-$ for $+$ and $+$ for $-$), we reverse the gain: and hence dividing a bicolumn into bicolumns each of two rows, viz. first and second rows, second and third rows, and so on, it at once appears that if we reverse alternate rows (viz. either the first, third, fifth, &c., rows, or the second, fourth, sixth, &c., rows) we reverse the gain. It of course follows that reversing all the rows, we leave the gain unaltered.

24. If to any bicolumn we prefix at the top thereof the second row reversed, we either leave the gain unaltered or we alter it by ± 1. In fact, as regards either column, if this originally begin with a change, the process introduces no change therein; but if it begins with a continuation, then the process introduces a change. Hence if the columns begin each with a change or each with a continuation, the gain is unaltered: but if one begins with a change, and the other with a continuation, then the gain is altered by ± 1; viz. the left-hand column beginning with a continuation, the gain is altered by -1: and the right-hand column beginning with a continuation, the gain is altered by $+1$.

The column $PQRST...$ is taken to satisfy the following conditions: two consecutive terms never vanish together (that is, for the same value of the variable): if for a given value of the variable, any term vanishes, the preceding and succeeding terms have then opposite signs; the last term, say V, is of constant sign.

25. Considering P, Q as given functions without a common measure, such a column of functions is obtained by the well-known process of seeking for the greatest common measure, reversing at each step the sign of the remainder: viz. we thus derive a set of functions R, S, T,... where

$$
\begin{aligned}
P &= \lambda Q - R, \\
Q &= \mu R - S, \\
R &= \nu S - T, \\
S &= \rho T - U, \\
&\vdots
\end{aligned}
$$

the degrees of the successive functions R, S, T, ..., being successively less and less, so that the last of them, say V, is an absolute constant: or we may stop the process as soon as we arrive at a function V, the sign of which remains unaltered for all values between the initial and final values of the variable. It may be observed that the process may be regarded as applicable in the case where the degree of Q exceeds that of P: viz. we then have $\lambda = 0$, $R = -P$, and the column begins $(P, Q, -P, S, ...)$, the subsequent terms being, except as to sign, the same as if P, Q had been interchanged.

Reversing the sign of P or Q, we reverse in the bicolumn a set of alternate rows, and thus reverse the gain: and reversing both signs we reverse all the rows, and leave the gain unaltered—of course the intercalation (considered irrespectively of sign) is in each case unaltered. It is convenient to take the signs in such manner that for the initial value of x, the signs of P, Q shall be each positive: or, what is the same thing, taking P, Q with their proper signs, we may in the bicolumn, by reversing if necessary each or either set of alternate rows, make the left-hand column to begin with the signs $+\ +$.

26. The complete rule now is—for a given trajectory form the bicolumn for $PQRS\ldots$, and if necessary, by reversing each or either set of alternate rows, make the left-hand column to begin with $+\ +$: then if there is a gain the intercalation begins with P, if a loss with Q, the gain or loss showing the number of P's. To find the number of Q's prefix at the top of the bicolumn the second row reversed—then the gain or loss (equal to or differing by unity from the original value) shows the number of Q's. It may happen that for P the gain is $=0$; then for Q the gain is 0 or ± 1, and the intercalation vanishes or is Q.

27. I give some simple examples.

	0	2	4
$P = x - 1$	−	+	+
$Q = x - 3$	−	−	+
$R = \ -1$	−	−	−

0 P 2 Q 4

	0	2	4
$P = x - 3$	−	−	+
$Q = x - 1$	−	+	+
$R = +1$	+	+	+

0 Q 2 P 4

In the left-hand example taking the intervals to be successively $0-2$, $0-4$, $2-4$, the bicolumns modified as above are

0 − 2		0 − 4		2 − 4	
−	−	−	+	−	+
+	−	+	−	+	+
+	+	+	−	+	−
+	+	+	+	−	−

viz.

Interval $0-2$; for P gain $=1$, P first; for Q gain $=0$; Intercalation is P;

" $0-4$ " " $=1$, " " " $=1$; " " PQ;

" $2-4$ " " $=0$ ...... " loss $=1$; " " Q.

And similarly in the right-hand example we have

0 − 2		0 − 4		2 − 4	
−	+	−	+	−	−
+	+	+	−	+	−
+	−	+	−	+	+
−	−	−	−	−	−

Interval $0-2$	for P gain $=\ \ 0$,	......	for Q gain $=-1$;	Intercalation is	Q;
" $0-4$	" " $=-1$,	Q first,	" " $=-1$;	" "	QP;
" $2-4$	" " $=+1$,	P first,	" " $=\ \ 0$;	" "	P.

28. Or to take a slightly more complicated example,

	1	3	$5\pm\epsilon$	7	9
$P=x^2-\ \ 8x+12$	+	−	−	+	+
$Q=x^2-12x+32$	+	+	−	−	+
$R=-\ \ \ \ \ x+\ 5$	+	+	∓	−	−
$S=\ \ \ \ \ \ \ \ \ \ \ \ +\ 1$	+	+	+	+	+

		P		Q		P		Q	
0	1	2	3	4	5	6	7	8	9

And hence for the several intervals,

	$1-3$	$1-5$	$1-7$	$1-9$	$3-5$	$3-7$	$3-9$	$5-7$	$5-9$	$7-9$
	− −	− +	− +	− −	− +	− +	− −	− −	− +	− +
	+ −	+ −	+ +	+ +	+ +	+ −	+ −	+ −	+ −	+ +
	+ +	+ −	+ −	+ +	+ −	+ −	+ +	+ +	+ −	+ −
	+ +	+ ∓	+ −	+ −	− ±	− +	− +	± +	± +	− −
	+ +	+ +	+ +	+ +	+ +	+ +	+ +	− −	− −	− −
Showing	P	PQ	PQP	$PQPQ$	Q	QP	QPQ	P	PQ	Q

For instance:—

Interval $1-9$ for P gain $=2$, P first, for Q gain $=2$: Intercalation is $PQPQ$.

It may be added that P being + for $x=1$, the ± intercalation is $+PQPQ$.

29. As an example of circuits take the following: curves are $P=0$, $Q=0$, where

$$P=x^2+y^2-4,$$
$$Q=y\ -x\ -1;$$

viz. $P=0$ (see figure) is a circle, centre the origin, radius $=2$: the inside hereof ($P=-$) being coloured red: and $Q=0$ is a right line cutting the axes of x, y at the points $(-1,\ 0)$ and $(0,\ 1)$ respectively, or say running N.E. and S.W., the lower region ($Q=-$) being coloured blue: the square is an arbitrary circuit ($x=\pm 3$, $y=\pm 3$) surrounding the circle, and the regions within the square are coloured by what precedes sable, gules, azure, argent, as shown in the figure: the line and circle intersect in two points M, N. Going right-handedly round these respectively, for M the order is sable, gules, argent, azure, viz. M is a right-handed root; while for N the order is sable, azure, argent, gules, viz. N is a left-handed root: the two points are accordingly in the figure denoted by $+M$ and $-N$ respectively.

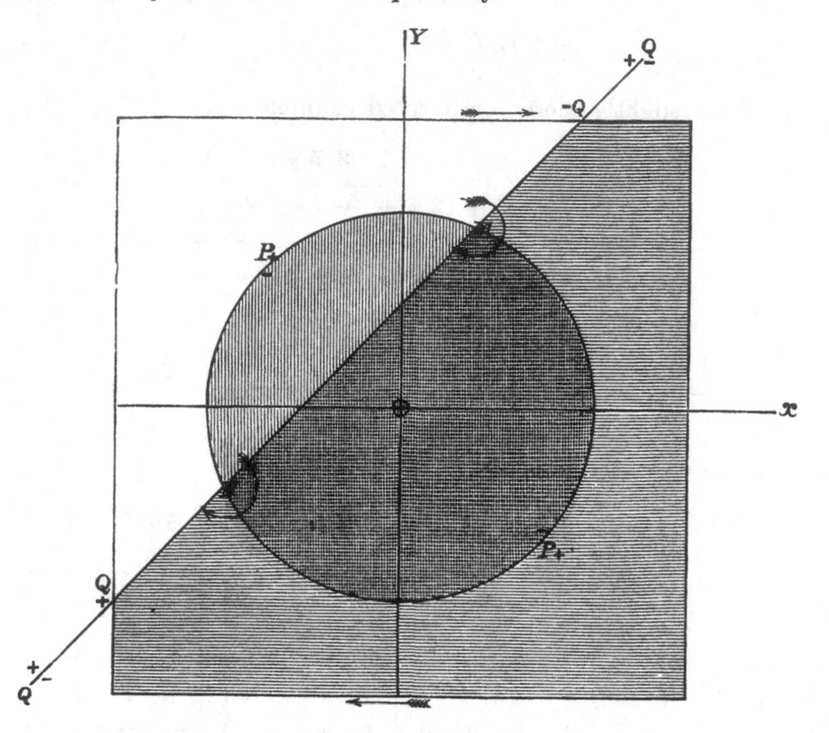

30. Now considering successively the four smaller squares of the figure, say these are the squares N.E., S.E., S.W., N.W.: and going right-handedly round each of these:

In the square N.E., the sequence and therefore also the intercalation is $+P-Q+P-Q$, viz. this is an intercalation $(+P-Q)$, showing an excess 1 of right-handed roots, and of course consisting with the single right-handed root M.

In the square S.E., the sequence is $-P+P$, viz. this is an intercalation $(PQ)_0$, showing an equality of right- and left-handed roots, and consisting with no root.

In the square S.W., the sequence and therefore also the intercalation is $-P+Q-P+Q$: viz. this is an intercalation $(-P+Q)_1$, showing an excess 1 of left-handed roots, and consisting with the single left-handed root N.

And in the square N.W., the sequence is $-Q+P-P+Q$, viz. this is an intercalation $(PQ)_0$, showing an equality of right- and left-handed roots, and consisting with no root.

Again take the whole large square: the sequence is $-Q+Q$: viz. the intercalation is $(PQ)_0$, showing an equality of right- and left-handed roots, and consisting with there being one of each.

So taking the squares N.E. and N.W. conjointly, the sequence and therefore also the intercalation is $-Q+P-Q+P$, viz. this is an intercalation $(+P-Q)_1$, as for the single square N.E.

31. As regards the analytical determination it will be sufficient to consider a single square, say N.E.: going round right-handedly, the trajectories will be

(1) $x=0$, $y=0$ to 3;

(2) $y=3$, $x=0$ to 3;

(3) $x=3$, $y=3$ to 0; or if $y'=-y$, then $y'=-3$ to 0;

(4) $y=0$, $x=3$ to 0; or if $x'=-x$, then $x'=-3$ to 0.

And we thus have

(1)

		0	3
$P=$	y^2-4	$-$	$+$
$Q=$	$y-1$	$-$	$+$
$R=$	-1	$+$	$-$

, that is,

0	3
$-$	$+$
$+$	$-$
$+$	$-$
$-$	$-$

, for P gain $=-1$, Q begins, $1P$;
" Q " $=-1$, " $1Q$.
Intercalation is QP, or since at origin $P=-$, $R=-$, or region is sable, it is $-Q+P$.

(2)

		0	3
$P=$	x^2+5	$+$	$+$
$Q=$	$-x+2$	$+$	$-$
$R=$	-1	$-$	$-$

, that is,

0	3
$-$	$+$
$+$	$+$
$+$	$-$
$-$	$-$

, for P gain $=0$,
" Q " $=-1$.
Intercalation is $-Q$.

(3)

		-3	0
$P=$	y'^2+5	$+$	$+$
$Q=$	$-y'+4$	$-$	$-$
$R=$	-1	$-$	$-$

, that is,

-3	0
$-$	$-$
$+$	$+$
$+$	$+$
$-$	$-$

, for P gain $=0$,
" Q " $=0$.
Intercalation vanishes.

(4)

		-3	0
$P=$	x'^2-4	$+$	$-$
$Q=$	$x'-1$	$-$	$-$
$R=$	$+1$	$+$	$+$

, that is,

-3	0
$-$	$-$
$+$	$-$
$+$	$+$
$+$	$+$

, for P gain $=+1$, P first,
" Q " $=0$.
Intercalation is $+P$.

Hence for the four sides, combining the intercalations, we have $-Q+P-Q+P$, and since there are no terms to be omitted, this is the intercalation of the N.E. square: which is right.

The Rhizic Theory. Articles Nos. 32 to 38.

32. Consider now $F(z) = (*)(z, 1)^n$ a rational and integral function of z, of the order n with in general imaginary (complex) coefficients, or, what is the same thing, let $F(z) = f(z) + i\phi(z)$, where the functions f, ϕ are real*. Writing herein $z = x + iy$, let P, Q be the real part and the coefficient of the imaginary part in the function $F(x+iy)$: or, what is the same thing, assume

$$P + iQ = f(x+iy) + i\phi(x+iy),$$

then it is clear that to any root $\alpha + i\beta$ (real or imaginary) of the equation $F(z) = 0$, there corresponds a real intersection, or root, $x = \alpha$, $y = \beta$, of the curves $P = 0$, $Q = 0$. The functions, P, Q, as thus serving for the determination of the roots of the equation $F(z) = 0$, are termed "rhizic functions," and similarly the curves $P = 0$, $Q = 0$ are "rhizic curves." The assumed equation shows at once that we have

$$d_y(P + iQ) = i\, d_x(P + iQ),$$

or, what is the same thing,

$$d_y P = -d_x Q,\ d_x P = d_y Q.$$

And we hence see that

$$\frac{d(P, Q)}{d(x, y)},\ = (d_x P)^2 + (d_y P)^2, \text{ or } (d_x Q)^2 + (d_y Q)^2,$$

is positive: viz. that the roots $P = 0$, $Q = 0$ are all of them right-handed (the essential thing is that they are same-handed; for by reversing the signs of P and Q they might be made left-handed: but it is convenient to take them as right-handed): hence the theorem—which in the general case, where P and Q are arbitrary functions, serves to determine the difference of the numbers of the right- and left-handed roots—in the particular case, where P and Q are rhizic functions, serves to determine the number of intersections of the curves $P = 0$, $Q = 0$: or, what is the same thing, the number of the (real or imaginary) roots of the equation $F(z) = 0$: viz. we thus determine the number of roots within a given circuit.

33. The rhizic curves $P = 0$, $Q = 0$ have various properties. 1°. Each curve has n real points at infinity, or, what is the same thing, n real asymptotes: and the P and Q points at infinity succeed each other, a P-point and then a Q-point, and so on, alternately.

In fact, from the equation

$$P + iQ = (a' + ia'')(x + iy)^n + \ldots + (k' + k''i),$$

writing herein $a' + ia'' = a(\cos\alpha + i\sin\alpha)$, and $x + iy = \rho(\cos\theta + i\sin\theta)$, we have

$$P + iQ = a\rho^n[\cos(n\theta + \alpha) + i\sin(n\theta + \alpha)] + \ldots + k' + k''i.$$

* It is assumed that the equation $F(z) = 0$ has no equal roots: this being so, the curves $P = 0$, $Q = 0$, will have no point of multiple intersection; which accords with the assumption made in the general case of two arbitrary curves.

It thus appears that for the curve $P=0$, the points at infinity are given by the equation $\cos(n\theta+\alpha)=0$; while for the curve $Q=0$, they are given by the equation $\sin(n\theta+\alpha)=0$: which proves the theorem.

Representing infinity as a closed curve or circuit, each point at infinity must be represented by two opposite points on the circuit; so that writing down P for each P-point and Q for each Q-point, we have $2n$ P's and $2n$ Q's succeeding each other, a P-point and then a Q-point, and so on, alternately.

It may be assumed that taking the circuit right-handedly, the P's are + and the Q's −, (this depends only on the colouring, but it corresponds with the foregoing assumption that the roots $P=0$, $Q=0$ are right-handed): the theorem just obtained then really is that for the circuit infinity, the intercalation is $(+P-Q)_n$: and we have herein a proof of the theorem that a numerical equation of the order n with real or imaginary coefficients has precisely n real or imaginary roots. But the force of this will more distinctly appear presently.

34. 2°. Neither of the curves $P=0$, $Q=0$ can include as part of itself a closed curve or circuit.

The foregoing relations between the differential coefficients give

$$d_x{}^2P+d_y{}^2P=0,\quad d_x{}^2Q+d_y{}^2Q=0,$$

which equations for the two curves respectively lead to the theorem in question. For as regards the curve $P=0$, take z a coordinate perpendicular to the plane of xy, and consider the surface $z=P$: if the curve $P=0$ included as part of itself a closed curve, then corresponding to some point (x, y) within the curve we should have z a proper maximum or minimum, viz. there would be a summit or an imit; at the point in question we should have $d_xP=0$, $d_xQ=0$; and also (as the condition of a summit or imit) $d_x{}^2P\,.\,d_y{}^2P-(d_xd_yP)^2=+$, implying that $d_x{}^2P$ and $d_y{}^2P$ have at this point the same sign: but this is inconsistent with the foregoing relation $d_x{}^2P+d_y{}^2P=0$.

35. 3°. The curves $P=0$, $Q=0$ have not in general any double (or higher multiple) points. A point which is a double (or higher multiple) point on one of these curves is not of necessity a point on the other curve: but being a point on the other curve it is on that curve a point of the same multiplicity. For changing if necessary the coordinates, the point in question may be taken to be at the origin: forming the equation

$$P+iQ=(a'+a''i)(x+iy)^n+\ldots+(k'+k''i)(x+iy)^2+(l'+l''i)(x+iy)+m'+m''i=0,$$

the point $x=0$, $y=0$ will not be a double point on the curve $P=0$, unless we have $m'=0$, $l'=0$, $l''=0$; these conditions being satisfied, it will not be a point on the curve $Q=0$ unless also $m''=0$; but this being so, it will be a double point on the curve $Q=0$: and the like for points of higher multiplicity. But a point which is a multiple point on each curve, represents four or more coincident intersections of the curves $P=0$, $Q=0$, that is, four or more equal roots of the equation $F(z)=0$; so that assuming that the equation has no equal roots, the case does not arise: and we in fact exclude it from consideration.

To fix the ideas assume that the curves $P=0$, $Q=0$ are each of them without double points. As already seen, neither of them includes as part of itself a closed curve. Hence in the figure the curve $P=0$ must consist of n branches each drawn from a point P in the circuit (viz. the circuit infinity) to another point P in the circuit; and in such manner that no two branches intersect each other: this implies that the two points P of the same branch must include between them an even number (which may of course be $=0$) of points P. And the like as regards the curve $Q=0$.

36. 4°. No branch of the P-curve can meet a branch of the Q-curve more than once. In fact, drawing the two branches to meet twice, the colouring would at once show that of the two intersections or roots, one must be right, the other left-handed: whence, the roots being all right-handed, the branches do not meet twice. And in exactly the same way it appears that no P-branch can meet two Q-branches, or any Q-branch meet two P-branches. And under these restrictions it requires only a consideration of a few successive cases to show that the n P-branches, and the n Q-branches can only be drawn on the condition that each P-branch shall intersect once and only once a single Q-branch; which of course implies that each Q-branch intersects once and once only a single P-branch: and further, that there shall be precisely n intersections: viz. the n P-branches and the n Q-branches must satisfy the conditions just stated. And the theorem of the n roots is thus obtained as a consequence of the impossibility (except under the same conditions) of drawing the n P-branches and the n Q-branches, so as to give rise to right-handed roots only. But the case of double or higher multiple points would need to be specially considered.

37. It is interesting for a given value of n to consider $\phi(n)$, the number of different ways in which the P-branches and the Q-branches can be drawn. We have $2n$ points P and $2n$ points Q, in all $4n$ points: starting from any point P, these may be numbered in order $1, 2, 3, \ldots, 4n$, the points P bearing odd numbers and the points Q even numbers. We may consider the P-branch which joins 1 with some P-point β, and (intersecting this) the Q-branch which joins some two Q-points α and γ: the numbers $1\alpha\beta\gamma$ are then in order of increasing magnitude: and excluding these four points there remain the points corresponding to numbers between 1 and α, between α and β, between β and γ, and between γ and 1. Now since the P-branch 1β meets the Q-branch $\alpha\gamma$, no branch from a point between 1 and α can meet either of these curves; hence these points form a system by themselves, capable of being connected together by P-branches and Q-branches: the number of them must therefore be a multiple of 4: and the like as to the points between α and β, between β and γ, and between γ and 1. Taking the number of the points in the four systems to be $4x$, $4y$, $4z$, and $4w$ respectively, we have $x+y+z+w=n-1$, and the first-mentioned four points bear the numbers

$$\begin{aligned} &1, \\ \alpha &= 4x+2, \\ \beta &= 4x+4y+3, \\ \gamma &= 4x+4y+4z+4. \end{aligned}$$

For the four systems the number of ways of drawing the P- and Q-branches are ϕx, ϕy, ϕz, ϕw respectively: that is, x, y, z, w being any partition whatever of $n-1$ (order attended to), and $\phi(0)$ being $=1$, we have

$$\phi(n) = \Sigma\phi(x)\,\phi(y)\,\phi(z)\,\phi(w),$$

which is the condition for the determination of ϕn.

Taking then θ for the value of the generating function

$$1 + t\phi(1) + t^2\phi(2) \ldots + t^n\phi(n) + \ldots,$$

it hereby appears that we have

$$\theta = 1 + t\theta^4;$$

or writing this for a moment $\theta = u + t\theta^4$, and expanding by Lagrange's theorem, but putting finally $u=1$, we have the value of θ, that is of the generating function,

$$= 1 + [4]^0\frac{t}{1} + [8]^1\frac{t^2}{1.2} + [12]^2\frac{t^3}{1.2.3} \ldots\ldots + [4n]^{n-1}\frac{t^n}{1.2\ldots n} + \ldots$$

$$= 1 + t + 4t^2 + 22t^3 + 140t^4 + \ldots,$$

that is,

$$\phi(1)=1,\quad \phi(2)=4,\quad \phi(3)=22,\quad \phi(4)=140,\ldots$$

and generally

$$\phi(n) = \frac{[4n]^{n-1}}{[n]^n},\ = \frac{4n.4n-1\ldots 3n+2}{2.3\ldots n}.$$

The results are easily verified for the successive particular cases; thus $n=1$, the points are 1, 2, 3, 4, and the P- and Q-branches respectively are 13, 24: $\phi(1)=1$. Again $n=2$, the points are 1, 2, 3, 4, 5, 6, 7, 8: we may join 13, 24 or 13, 28 or 17, 28 or 17, 68, leaving in each case four contiguous numbers which may be joined in a single manner: that is, $\phi(2)=4$. Or, what is the same thing, the partitions of 1 are 0001, 0010, 0100, 1000, whence $\phi(2) = 4\{\phi(0)\}^3\phi(1) = 4$. Again $n=3$, the partitions of 2 are 0002, &c. (4 of this form) and 1100 (6 of this form): that is, $\phi(3) = 4\{\phi(0)\}^3\phi(2) + 6\{\phi(0)\}^2\{\phi(1)\}^2$, $= 4.4 + 6.1 = 22$, and so on.

38. Starting from the $4n$ points P and Q, and joining them in any manner subject to the foregoing conditions, we have a diagram representing two rhizic curves; and colouring the regions we verify that the n roots are all of them right-handed. We have for instance the annexed figure ($n=3$).

Having drawn such a figure we may, by a continuous variation of the several lines, in a variety of ways introduce a double point in the P-curve, or in the Q-curve: and by a continued repetition of the process introduce double points in each or either curve: thus for instance we may from the last figure derive a new figure in which the P-curve has a node at N. It will be observed that here it is no longer the case that each P-branch intersects one and only one Q-branch: the P-branch $1-9$ does not meet any Q-branch, but the P-branch $7-11$ meets two Q-branches. But looking at the figure in a different manner, and considering the P-branches through N as being either $11-N-1$ and $7-N-9$, or $1-N-7$ and $9-N-11$, then in either case each P-branch intersects one and only one Q-branch: and in this way, in a

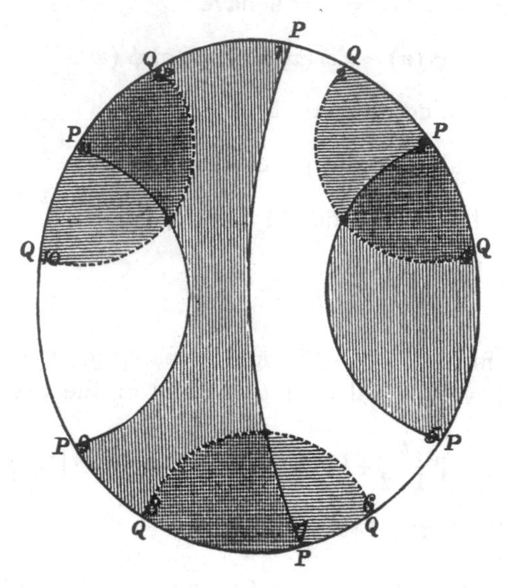

diagram in which the two curves have each or either of them double points, but neither curve passes through a double point of the other curve, the theorem may be regarded as remaining true—we in fact consider the diagram as the limit of a diagram wherein the curves have no double points. It will be recollected that, the equation

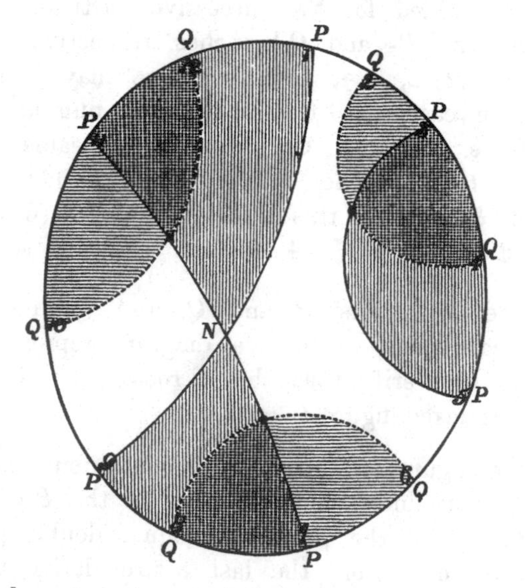

$F(z)$ being without equal roots, we cannot have either curve passing through a multiple point of the other curve. And we thus see that the various figures drawn as above without double points are, so to speak, the types of all the different forms of a system of rhizic curves $P=0$, $Q=0$.

In connexion with the present paper I give the following list of Memoirs:—

Cauchy. Calcul des Indices des fonctions. *Jour. de l'École Polyt.* t. XV. (1837), pp. 176—229. First part seems to have been written in 1833: second part is dated 20th June, 1837. Refers to a memoir presented to the Academy of Turin the 17th Nov. 1831, wherein the principles of the "Calcul des Indices des fonctions" are deduced from the theory of definite integrals: I have not seen this.

Sturm and Liouville. Démonstration d'un théorème de M. Cauchy relative aux racines imaginaires des équations. *Liouv.* t. I. (1836), pp. 278—289.

Sturm. Autres démonstrations du même théorème. *Liouv.* t. I. (1836), pp. 290—308.

These two papers contain proofs of the particular theorem relating to the roots of an equation $F(z)=0$, but do not refer to the general theorem relating to the intersection of the two curves $P=0$, $Q=0$: the special theorem of the existence of the n roots of the equation $F(z)=0$ is considered.

Sylvester. A theory of the syzygetic relations of two rational integral functions, comprising an application to the theory of Sturm's functions and that of the greatest algebraical common measure. *Phil. Trans.* t. CXLIII. (1853), pp. 407—548.

De Morgan. A proof of the existence of a root in every algebraic equation, with an examination and extension of Cauchy's theorem on imaginary roots, and remarks on the proofs of the existence of roots given by Argand and Mourey. *Camb. Phil. Trans.* t. X. (1858), pp. 261—270.

Contains the important remark that the two curves $P=0$, $Q=0$ are such that two branches, one of each curve, cannot inclose a space; also that the two curves always [i.e. at a simple intersection] intersect orthogonally.

Airy, G. B. Suggestion of a proof of the theorem that every algebraic equation has a root. *Camb. Phil. Trans.* t. X. (1859), pp. 283—289.

Cayley, A. Sketch of a proof of the theorem that every algebraic equation has a root. *Phil. Mag.* t. XVIII. (1859), [248], pp. 436—439.

Walton, W. On a theorem in maxima and minima. *Quart. Math. Jour.* t. X. (1870), pp. 253—262. **Cayley, A.** Addition thereto, [562], pp. 262, 263. (Relates to the curves $P=0$, $Q=0$.)

Walton, W. Note on rhizic curves. *Quart. Math. Jour.* t. XI. (1871), pp. 91—98. First use of the term "rhizic curves:" relates chiefly to the configuration of each curve at a multiple point, and of the two at a common multiple point.

Walton, W. On the spoke-asymptotes of rhizic curves. *Quart. Math. Jour.* t. XI. (1871), pp. 200—202.

Walton, W. On a property of the curvature of rhizic curves at multiple points. *Quart. Math. Jour.* t. XI. (1871), pp. 274—281.

Björling. Sur la séparation des racines d'équations algébriques. *Upsala, Nova Acta Soc. Sci.* (1870), pp. 1—35. (Contains delineations of some rhizic curves.)

562.

[ADDITION TO MR. WALTON'S PAPER "ON A THEOREM IN MAXIMA AND MINIMA."]

[From the *Quarterly Journal of Pure and Applied Mathematics*, vol. x. (1870), pp. 262, 263.]

IN what follows I write x, y, z in place of Mr Walton's u, v, w: (so that if $i = \sqrt{(-1)}$, as usual, we have

$$f(x+iy) = P + iQ):$$

and I attend exclusively to the case where the second differential coefficients of P, Q do not vanish.

There are not on the surface $z = P$ any proper maxima or minima; but only level points, such as at the top of a pass: say there are not any summits or imits, but only cruxes; and moreover at any crux, the two crucial (or level) directions intersect at right angles. Every node of the curve $Q = 0$ is subjacent to a crux of the surface $z = P$; and moreover the two directions of the curve $Q = 0$ at the node are at right angles to each other; hence, considering the intersection of the surface $z = P$ by the cylinder $Q = 0$, the path $Q = 0$ on the surface has a node at the crux; or say there are at the crux two directions of the path; these cross at right angles, and are consequently separated the one from the other by the crucial directions; that is to say, there is one path ascending, and another path descending, each way from the crux. And the complete statement is; that the elevation of the path is then only a maximum or minimum when the path passes through a crux; and that at any crux there are two paths, one ascending, the other descending, each way from the crux.

The analytical demonstration is exceeding simple; we have

$$\left(\frac{dP}{dy} + i\frac{dQ}{dy}\right) = i\left(\frac{dP}{dx} + i\frac{dQ}{dx}\right);$$

that is,

$$\frac{dP}{dy} = -\frac{dQ}{dx}, \quad \frac{dQ}{dy} = \frac{dP}{dx},$$

and passing thence to the second differential coefficients, we may write

$$\frac{dP}{dx} = \frac{dQ}{dy} = L, \quad \frac{dP}{dy} = -\frac{dQ}{dx} = M,$$

$$\frac{d^2P}{dx\,dy} = -\frac{d^2Q}{dx^2} = \frac{d^2Q}{dy^2} = a,$$

$$\frac{d^2Q}{dx\,dy} = \frac{d^2P}{dx^2} = -\frac{d^2P}{dy^2} = b,$$

so that we have

$$\delta P = L\delta x + M\delta y, \qquad \delta Q = -M\delta x + L\delta y,$$

$$\delta^2 P = (b,\ a,\ -b \between \delta x,\ \delta y)^2, \quad \delta^2 Q = (-a,\ b,\ a \between \delta x,\ \delta y)^2.$$

Hence, for the maximum or minimum elevation of the path, we have $0 = \delta P$, where $\delta Q = 0$; that is, $0 = \frac{L^2 + M^2}{L}\delta x$, and therefore $L^2 + M^2 = 0$; that is, $L = 0$, $M = 0$; and at any such point $\delta z = 0$, that is, there is a crux of the surface $z = P$; and $\delta Q = 0$, that is, there is a node of the curve $Q = 0$. Moreover the crucial directions for the surface $z = P$ are given by the equation $(b,\ a,\ -b \between \delta x,\ \delta y)^2 = 0$, or these are at right angles to each other; and the nodal directions for the curve $Q = 0$ are given by $(-a,\ b,\ a \between \delta x,\ \delta y)^2 = 0$; or these are likewise at right angles to each other.

563.

NOTE ON THE TRANSFORMATION OF TWO SIMULTANEOUS EQUATIONS.

[From the *Quarterly Journal of Pure and Applied Mathematics*, vol. XI. (1871), pp. 266, 267.]

WRITING in Mr Walton's equations (1) and (2)

$$\frac{a}{d},\ \frac{b}{d},\ \frac{c}{d},\ \frac{\alpha}{\delta},\ \frac{\beta}{\delta},\ \frac{\gamma}{\delta}$$

instead of a, b, c, α, β, γ respectively; and putting for shortness

$$\begin{aligned} A &= b\gamma - c\beta, & F &= a\delta - d\alpha,\\ B &= c\alpha - a\gamma, & G &= b\delta - d\beta,\\ C &= a\beta - b\alpha, & H &= c\delta - d\gamma, \end{aligned}$$

the equations become

$$\frac{a(b-c)}{F} + \frac{b(c-a)}{G} + \frac{c(a-b)}{H} = 0,$$

$$\frac{\alpha(\beta-\gamma)}{F} + \frac{\beta(\gamma-\alpha)}{G} + \frac{\gamma(\alpha-\beta)}{H} = 0.$$

Multiplying by FGH and effecting some obvious transformations, the equations become

$$\left.\begin{aligned} aAF + bBG + cCH &= 0\\ \alpha AF + \beta BG + \gamma CH &= 0 \end{aligned}\right\} \quad\ldots\ldots\ldots\ldots\ldots\ldots\ldots\ldots(18);$$

whence also

$$AF^2 + BG^2 + CH^2 = 0 \quad\ldots\ldots\ldots\ldots\ldots\ldots\ldots\ldots(19).$$

Now regarding $(\alpha, \beta, \gamma, \delta)$ as the coordinates of a point in space, the equations (18) and (19) represent each of them a cone having for vertex the point $\alpha : \beta : \gamma : \delta = a : b : c : d$, viz. (18) is a quadric cone, (19) a cubic cone; they intersect therefore in six lines; and it may be shown that these are

the line	$\alpha : \beta : \gamma = a : b : c$	(twice)	2
"	$\beta : \gamma : \delta = b : c : d$		1
"	$\gamma : \alpha : \delta = c : a : d$		1
"	$\alpha : \beta : \delta = a : b : d$		1
"	$\beta-\gamma : \gamma-\alpha : \alpha-\beta : \delta = b-c : c-a : a-b : d$		1
			6,

agreeing with Mr Walton's result.

564.

ON A THEOREM IN ELIMINATION.

[From the *Quarterly Journal of Pure and Applied Mathematics*, vol. XII. (1873), pp. 5, 6.]

I FIND among my papers the following example of a theorem in elimination communicated to me by Prof. Sylvester. Writing

$$\begin{aligned}
\phi &= ax^3 + 3bx^2y + 3cxy^2 + dy^3,\\
\phi_1 &= \quad bx^2 + 2cxy + dy^2,\\
\phi_2 &= \qquad cx + dy\,,\\
\phi_3 &= \qquad\quad d\;;\\
f &= bx^3 + 3cx^2y + 3dxy^2 + ey^3,\\
f_1 &= \quad cx^2 + 2dxy + ey^2,\\
f_2 &= \qquad dx + ey\,,\\
f_3 &= \qquad\quad e\;,
\end{aligned}$$

then we have

$$\Delta_a \,.\, R(f,\ \phi) = \Delta f \,.\, R(\phi_1,\ f_1)^2\, R(\phi_1,\ f_2)^2,$$

viz. $R(f,\ \phi)$ is the resultant of the functions $(f,\ \phi)$, and similarly $R(\phi_1,\ f_1)$, $R(\phi_1,\ f_2)$. Moreover, Δf is the discriminant of f; and $\Delta_a R(f,\ \phi)$ is the discriminant of $R(f,\ \phi)$ in regard to a. The equation thus is

$$\begin{aligned}
&\Delta_a\,[(ae - 4bd + 3c^2)^3 - 27\,(ace - ad^2 - b^2e - c^3 + 2bcd)^2]\\
&\qquad\qquad = (b^2e^2 + 4bd^3 + 4c^3e - 3c^2d^2 - 6bcde)^3\,(d^3 - 2cde + be^2)^2\,;
\end{aligned}$$

or, what is the same thing, reversing the order of the letters $(a,\ b,\ c,\ d,\ e)$, it is

$$\begin{aligned}
&\Delta_e\,[(ae - 4bd + 3c^2)^3 - 27\,(ace - ad^2 - b^2e - c^3 + 2bcd)]\\
&\qquad\qquad = (a^2d^2 + 4ac^3 + 4b^3d - 3b^2c^2 - 6abcd)^3\,(b^3 - 2abc + a^2d)^2,
\end{aligned}$$

viz. arranging in powers of e, the function is

$$\begin{aligned}
& e^3 \,.\quad a^3 \\
+\,& 3e^2 \,.\, -a^2(4bd-3c^2) - 9(ac-b^2)^2 \\
+\,& 3e \,.\quad a(4bd-3c^2)^2 + 18(ac-b^2)(ad^2-2bcd+c^3) \\
+\,& 1 \,.\, - (4bd-3c^2)^3 - 27(ad^2-2bcd+c^3)^2,
\end{aligned}$$

which last coefficient is

$$= -d^2(27a^2d^2 + 54ac^3 + 64b^3d - 36b^2c^2 - 108abcd),$$

and the discriminant of this cubic function of e is

$$= (a^2d^2 + 4ac^3 + 4b^3d - 3b^2c^2 - 6abcd)^3(b^3 - 2abc + a^2d)^2.$$

The occurrence of the factor

$$a^2d^2 + 4ac^3 + 4b^3d - 3b^2c^2 - 6abcd$$

is accounted for as the resultant in regard to e of the invariants I, J; we, in fact, have

$$\begin{aligned}
(ac-b^2)I - aJ &= (ac-b^2)(-4bd+3c^2) - a(-ad^2 - c^3 + 2bcd) \\
&= a^2d^2 + 4ac^3 + 4b^3d - 3b^2c^2 - 6abcd,
\end{aligned}$$

and the identity itself may be proved without any particular difficulty.

565.

NOTE ON THE CARTESIAN.

[From the *Quarterly Journal of Pure and Applied Mathematics*, vol. XII. (1873), pp. 16—19.]

THE following are doubtless known theorems, but the form of statement, and the demonstration of one of them, may be interesting.

A point P on a Cartesian has three "opposite" points on the curve, viz. if the axial foci are A, B, C, then the opposite points are P_a, P_b, P_c where

P_a is intersection of line PA with circle PBC,
P_b „ „ PB „ PCA,
P_c „ „ PC „ PAB.

And, moreover, supposing in the three circles respectively, the diameters at right angles to PA, PB, PC are $\alpha\alpha'$, $\beta\beta'$, $\gamma\gamma'$ respectively, then the points α, α', β, β', γ, γ' lie by threes in two lines passing through P, viz. one of these, say $P\alpha\beta\gamma$, is the tangent, and the other $P\alpha'\beta'\gamma'$ the normal, at P; and then the tangents and normals at the opposite points are $P_a\alpha$ and $P_a\alpha'$, $P_b\beta$ and $P_b\beta'$, $P_c\gamma$, and $P_c\gamma'$ respectively.

There exists a second Cartesian with the same axial foci A, B, C, and passing through the points P, P_a, P_b, P_c (which are obviously opposite points in regard thereto); the tangent at P is $P\alpha'\beta'\gamma'$ and the normal is $P\alpha\beta\gamma$; and the tangent and the normal at the other points are $P_a\alpha'$ and $P_a\alpha$, $P_b\beta'$ and $P_b\beta$, $P_c\gamma'$ and $P_c\gamma$ respectively: viz. the two curves cut at right angles at each of the four points.

Starting with the foci A, B, C and the point P, the points P_a, P_b, P_c are constructed as above, without the employment of the Cartesian; there are through P with the foci A, B, C two and only two Cartesians; and if it is shown that these pass through one of the opposite points, say P_b, they must, it is clear, pass through the other two points P_a, P_c. I propose to find the two Cartesians in question. To fix the ideas, let the points C, B, A be situate in order as shown in the figure, their

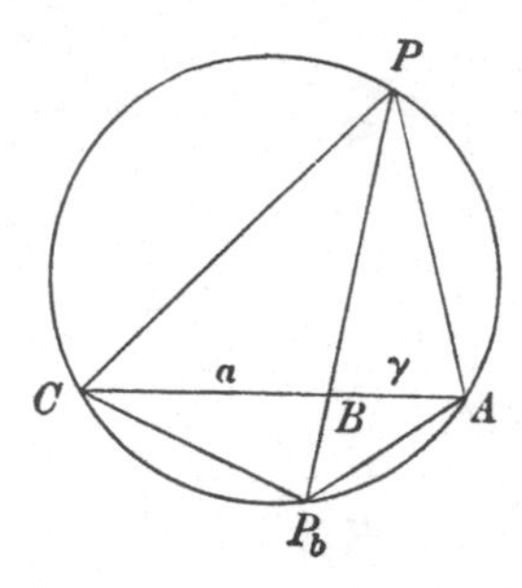

distances from a fixed point O being a, b, c, so that writing α, β, $\gamma = b-c$, $c-a$, $a-b$ respectively, we have $\alpha+\beta+\gamma=0$, and α, γ will represent the positive distances CB and BA respectively, and $-\beta$ the positive distance AC. Suppose, moreover, that the distances PA, PB, PC regarded as positive are R, S, T respectively; and that the distances P_bA, P_bB, P_bC regarded as positive are R', S', T' respectively.

Suppose that for a current point Q the distances QA, QB, QC regarded as indifferently positive, or negative, are r, s, t respectively; then the equation of a bicircular quartic having the points A, B, C for axial foci is

$$lr + ms + nt = 0,$$

where l, m, n are constants; and this will be a Cartesian if only

$$\frac{l^2}{\alpha} + \frac{m^2}{\beta} + \frac{n^2}{\gamma} = 0.$$

We have the same curve whatever be the signs of l, m, n, and hence making the curve pass through P, we may, without loss of generality, write

$$lR + mS + nT = 0,$$

R, S, T denoting the positive distances PA, PB, PC as above. We have thus for the ratios $l : m : n$, two equations, one simple, the other quadric; and there are thus two systems of values, that is, two Cartesians with the foci A, B, C, and passing through P.

I proceed to show that for one of these we have $-lR' + mS' + nT' = 0$, and for the other $lR' + mS' - nT' = 0$, or, what is the same thing, that the values of $l : m : n$ are

$$l : m : n = -(ST' + S'T) : TR' + T'R : RS' - R'S,$$

and

$$l : m : n = \quad (ST' - S'T) : -(TR' + T'R) : RS' + R'S;$$

viz. that the equations of the two Cartesians are

$$\begin{vmatrix} r, & s, & t \\ R, & S, & T \\ -R', & S', & T' \end{vmatrix} = 0, \text{ and } \begin{vmatrix} r, & s, & t \\ R, & S, & T \\ R', & S', & -T' \end{vmatrix} = 0,$$

respectively; this being so each of the Cartesians will, it is clear, pass through the point P_b, and therefore also through P_a and P_c.

The geometrical relations of the figure give

$$\begin{aligned} \alpha R^2 + \beta S^2 + \gamma T^2 &= -\alpha\beta\gamma, \\ \alpha R'^2 + \beta S'^2 + \gamma T'^2 &= -\alpha\beta\gamma, \\ RT' + R'T &= -\beta(S+S'), \\ \gamma\alpha &= SS', \\ \gamma TT' &= \alpha RR', \end{aligned}$$

to which might be joined

$$\begin{aligned} R'^2S + \gamma^2(S+S') + R^2S' &= SS'(S+S'), \\ T'^2S + \alpha^2(S+S') + T^2S' &= SS'(S+S'), \\ SR'T' &= S'RT, \\ SP'R' &= S'PR, \end{aligned}$$

but these are not required for the present purpose.

As regards the first Cartesian, we have to verify that

$$\frac{(ST'+S'T)^2}{\alpha} + \frac{(TR'+T'R)^2}{\beta} + \frac{(RS'-R'S)^2}{\gamma} = 0.$$

The left-hand side is

$$\frac{S^2T'^2 + S'^2T^2 + 2\gamma\alpha TT'}{\alpha} + \frac{\beta^2(S^2+S'^2+2\gamma\alpha)}{\beta} + \frac{S^2R'^2 + S'^2R^2 - 2\gamma\alpha RR'}{\gamma},$$

viz. this is

$$= S^2\left(\frac{T'^2}{\alpha} + \beta + \frac{R'^2}{\gamma}\right) + S'^2\left(\frac{T^2}{\alpha} + \beta + \frac{R^2}{\gamma}\right) + 2\alpha\beta\gamma + 2(\gamma TT' - \alpha RR'),$$

which is

$$= S^2\left(\frac{-\beta S'^2}{\gamma\alpha}\right) + S'^2\left(\frac{-\beta S^2}{\gamma\alpha}\right) + 2\alpha\beta\gamma + 2(\gamma TT' - \alpha RR'),$$

and since the first and second terms are together $= -2\frac{\beta}{\gamma\alpha}S^2S'^2$, that is, $= -2\alpha\beta\gamma$, the whole is as it should be $= 0$.

In precisely the same manner we have

$$\frac{(ST'-S'T)^2}{\alpha} + \frac{(TR'+T'R)^2}{\beta} + \frac{(RS'+R'S)^2}{\gamma} = 0,$$

which is the condition for the second Cartesian: and the theorem in question is thus proved.

566.

ON THE TRANSFORMATION OF THE EQUATION OF A SURFACE TO A SET OF CHIEF AXES.

[From the *Quarterly Journal of Pure and Applied Mathematics*, vol. XII. (1873), pp. 34—38.]

WE have at any point P of a surface a set of chief axes (PX, PY, PZ), viz. these are, say the axis of Z in the direction of the normal, and those of X, Y in the directions of the tangents to the two curves of curvature respectively. It may be required to transform the equation of the surface to the axes in question; to show how to effect this, take (x, y, z) for the original (rectangular) coordinates of the point P, $x+\delta x$, $y+\delta y$, $z+\delta z$ for the like coordinates of any other point on the surface, so that $(\delta x, \delta y, \delta z)$ are the coordinates of the point referred to the origin P; the equation of the surface, writing down only the terms of the first and second orders in the coordinates δx, δy, δz, is

$$A\delta x + B\delta y + C\delta z + \tfrac{1}{2}(a, b, c, f, g, h)(\delta x, \delta y, \delta z)^2 + \&c. = 0,$$

where (A, B, C) are the first derived functions and (a, b, c, f, g, h) the second derived functions of U for the values (x, y, z) which belong to the given point P, if $U=0$ is the equation of the surface in terms of the original coordinates (x, y, z); we have X, Y, Z linear functions of $(\delta x, \delta y, \delta z)$; say

	δx	δy	δz
X	α_1	β_1	γ_1
Y	α_2	β_2	γ_2
Z	α	β	γ

that is, $X = \alpha_1\delta x + \beta_1\delta y + \gamma_1\delta z$, &c. and $\delta x = \alpha_1 X + \alpha_2 Y + \alpha Z$, &c. where the coefficients satisfy the ordinary relations in the case of transformation between two sets of rectangular axes; and the transformed equation is therefore

$$A(\alpha_1 X + \alpha_2 Y + \alpha Z) + B(\beta_1 X + \beta_2 Y + \beta Z) + C(\gamma_1 X + \gamma_2 Y + \gamma Z)$$
$$+ (a, b, c, f, g, h)(\alpha_1 X + \alpha_2 Y + \alpha Z,\ \beta_1 X + \beta_2 Y + \beta Z,\ \gamma_1 X + \gamma_2 Y + \gamma Z)^2 = 0,$$

or, as this may be written,

$$\begin{aligned}X(A\alpha_1+B\beta_1+C\gamma_1)+Y(A\alpha_2+B\beta_2+C\gamma_2)+Z(A\alpha+B\beta+C\gamma)&\\ +\tfrac{1}{2}X^2\ (a,\ldots)(\alpha_1,\ \beta_1,\ \gamma_1)^2&\\ +\tfrac{1}{2}Y^2\ (a,\ldots)(\alpha_2,\ \beta_2,\ \gamma_2)^2&\\ +\ XY(a,\ldots)(\alpha_1,\ \beta_1,\ \gamma_1)(\alpha_2,\ \beta_2,\ \gamma_2)&\\ +\ XZ\ (a,\ldots)(\alpha_1,\ \beta_1,\ \gamma_1)(\alpha,\ \beta,\ \gamma)&\\ +\ YZ\ (a,\ldots)(\alpha_2,\ \beta_2,\ \gamma_2)(\alpha,\ \beta,\ \gamma)&\\ +\tfrac{1}{2}Z^2\ (a,\ldots)(\alpha,\ \beta,\ \gamma)^2\qquad +\&\text{c.}=0,&\end{aligned}$$

where the &c. refers to terms of the form $(X,\ Y,\ Z)^3$ and higher powers.

But in order that the new axes may be chief axes, we must have

$$\begin{aligned}A\alpha_1+B\beta_1+C\gamma_1&=0,\\ A\alpha_2+B\beta_2+C\gamma_2&=0,\\ (a,\ldots)(\alpha_1,\ \beta_1,\ \gamma_1)(\alpha_2,\ \beta_2,\ \gamma_2)&=0,\end{aligned}$$

so that putting for shortness

$$A\alpha+B\beta+C\gamma=\nabla,$$

the equation becomes

$$\begin{aligned}\nabla Z+\tfrac{1}{2}X^2\ (a,\ldots)(\alpha_1,\ \beta_1,\ \gamma_1)^2+\tfrac{1}{2}Y^2(a,\ldots)(\alpha_2,\ \beta_2,\ \gamma_2)^2&\\ +\ XZ(a,\ldots)(\alpha_1,\ \beta_1,\ \gamma_1)(\alpha,\ \beta,\ \gamma)&\\ +\ YZ\ (a,\ldots)(\alpha_2,\ \beta_2,\ \gamma_2)(\alpha,\ \beta,\ \gamma)&\\ +\tfrac{1}{2}Z^2\ (a,\ldots)(\alpha,\ \beta,\ \gamma)^2\qquad +\&\text{c.}=0.&\end{aligned}$$

We have

$$A\ :\ B\ :\ C=\beta_1\gamma_2-\beta_2\gamma_1\ :\ \gamma_1\alpha_2-\gamma_2\alpha_1\ :\ \alpha_1\beta_2-\alpha_2\beta_1,$$

that is,

$$=\quad \alpha\quad :\quad \beta\quad :\quad \gamma\quad ,$$

and thence

$$\alpha,\ \beta,\ \gamma=\frac{A}{\nabla},\ \frac{B}{\nabla},\ \frac{C}{\nabla};\ \ \nabla=\surd(A^2+B^2+C^2).$$

I write

$$\frac{1}{\rho_1}=(a,\ldots)(\alpha_1,\ \beta_1,\ \gamma_1)^2,$$

and also for a moment

$$\begin{aligned}P&=\left(a-\frac{1}{\rho_1},\quad h\quad ,\quad g\quad\right)(\alpha_1,\ \beta_1,\ \gamma_1),\\ Q&=\left(\quad h\quad ,\quad b-\frac{1}{\rho_1},\quad f\quad\right)(\alpha_1,\ \beta_1,\ \gamma_1),\\ R&=\left(\quad g\quad ,\quad f\quad ,\quad c-\frac{1}{\rho_1}\right)(\alpha_1,\ \beta_1,\ \gamma_1).\end{aligned}$$

We find

$$P\alpha_1 + Q\beta_1 + R\gamma_1 = (a, \ldots)(\alpha_1, \beta_1, \gamma_1)^2 - \frac{1}{\rho_1}, = 0,$$

$$P\alpha_2 + Q\beta_2 + R\gamma_2 = (a, \ldots)(\alpha_1, \beta_1, \gamma_1)(\alpha_2, \beta_2, \gamma_2) - \frac{1}{\rho_1}(\alpha_1\alpha_2 + \beta_1\beta_2 + \gamma_1\gamma_2), = 0,$$

and thence

$$\begin{aligned} P : Q : R &= \beta_1\gamma_2 - \beta_2\gamma_1 : \gamma_1\alpha_2 - \gamma_2\alpha_1 : \alpha_1\beta_2 - \alpha_2\beta_1 \\ &= \quad \alpha \quad : \quad \beta \quad : \quad \gamma \quad , \end{aligned}$$

or say

$$P, Q, R = \theta_1 A, \theta_1 B, \theta_1 C;$$

we have thus the equations

$$\left(a - \frac{1}{\rho_1}, \quad h \quad , \quad g \quad\right)(\alpha_1, \beta_1, \gamma_1) = \theta_1 A,$$

$$\left(\quad h \quad , b - \frac{1}{\rho_1}, \quad f \quad\right)(\alpha_1, \beta_1, \gamma_1) = \theta_1 B,$$

$$\left(\quad g \quad , \quad f \quad , c - \frac{1}{\rho_1}\right)(\alpha_1, \beta_1, \gamma_1) = \theta_1 C,$$

and joining hereto

$$(A, B, C)(\alpha_1, \beta_1, \gamma_1) = 0,$$

we eliminate $\alpha_1, \beta_1, \gamma_1$ and obtain the equation

$$\begin{vmatrix} a - \frac{1}{\rho_1}, & h \ , & g \ , & A \\ h \ , & b - \frac{1}{\rho_1}, & f \ , & B \\ g \ , & f \ , & c - \frac{1}{\rho_1}, & C \\ A \ , & B \ , & C \ , & 0 \end{vmatrix} = 0,$$

and in like manner writing

$$\frac{1}{\rho_2} = (a, \ldots)(\alpha_2, \beta_2, \gamma_2)^2,$$

we have the same equation for ρ_2; wherefore ρ_1, ρ_2 are the roots of the quadric equation

$$\begin{vmatrix} a - \frac{1}{\rho}, & h \ , & g \ , & A \\ h \ , & b - \frac{1}{\rho}, & f \ , & B \\ g \ , & f \ , & c - \frac{1}{\rho}, & C \\ A \ , & B \ , & C \ , & 0 \end{vmatrix} = 0.$$

Moreover, ρ_1, ρ_2 being thus determined, we have, α_1, β_1, γ_1, θ_1 proportional to the determinants formed with the matrix

$$\left|\begin{array}{cccc} a-\frac{1}{\rho_1}, & h\ , & g\ , & A \\ h\ , & b-\frac{1}{\rho_1}, & f\ , & B \\ g\ , & f\ , & c-\frac{1}{\rho_1}, & C \end{array}\right|,$$

say, α_1, β_1, γ_1, $\theta_1 = k\mathfrak{A}_1$, $k\mathfrak{B}_1$, $k\mathfrak{C}_1$, $k\Omega$, where $\mathfrak{A}_1$, $\mathfrak{B}_1$, $\mathfrak{C}_1$, Ω_1 are the determinants in question; and then $1 = k^2(\mathfrak{A}_1^2 + \mathfrak{B}_1^2 + \mathfrak{C}_1^2)$, or we have

$$\theta_1 = \frac{\Omega_1}{\sqrt{(\mathfrak{A}_1^2 + \mathfrak{B}_1^2 + \mathfrak{C}_1^2)}}.$$

But we find at once

$$(a, \ldots)(\alpha_1, \beta_1, \gamma_1)(\alpha, \beta, \gamma) = \theta_1(A\alpha + B\beta + C\gamma) = \theta_1\nabla,$$

that is,

$$(a, \ldots)(\alpha_1, \beta_1, \gamma_1)(\alpha, \beta, \gamma) = \frac{\nabla\Omega_1}{\sqrt{(\mathfrak{A}_1^2 + \mathfrak{B}_1^2 + \mathfrak{C}_1^2)}},$$

and in the same manner

$$(a, \ldots)(\alpha_2, \beta_2, \gamma_2)(\alpha, \beta, \gamma) = \frac{\nabla\Omega_2}{\sqrt{(\mathfrak{A}_2^2 + \mathfrak{B}_2^2 + \mathfrak{C}_2^2)}}.$$

Hence the transformed equation is

$$\begin{aligned} &\nabla Z + \tfrac{1}{2}\frac{X^2}{\rho_1} + \tfrac{1}{2}\frac{Y^2}{\rho_2} \\ &+ XZ\frac{\nabla\Omega_1}{\sqrt{(\mathfrak{A}_1^2 + \mathfrak{B}_1^2 + \mathfrak{C}_1^2)}} + YZ\frac{\nabla\Omega_2}{\sqrt{(\mathfrak{A}_2^2 + \mathfrak{B}_2^2 + \mathfrak{C}_2^2)}} \\ &+ \tfrac{1}{2}Z^2\frac{(a, \ldots)(A, B, C)^2}{\nabla^2} + \&c. = 0, \end{aligned}$$

where it will be recollected that $\nabla = \sqrt{(A^2 + B^2 + C^2)}$. The &c. refers as before to the terms $(X, Y, Z)^3$ and higher powers, which are obtained from the corresponding terms in δx, δy, δz, by substituting for these their values $\delta x = \alpha_1 X + \alpha_2 Y + \alpha Z$, &c., where the coefficients have the values above obtained for them. It will be observed, that the radii of curvature are $\nabla\rho_1$, $\nabla\rho_2$, and that the process includes an investigation of the values of these radii of curvature similar to the ordinary one; the novelty is in the terms in XZ, YZ, and Z^2. But regarding X, Y as small quantities of the first order, Z is of the second order, and the terms in XZ, YZ are of the third order, and that in Z^2 of the fourth order.

567.

ON AN IDENTICAL EQUATION CONNECTED WITH THE THEORY OF INVARIANTS.

[From the *Quarterly Journal of Pure and Applied Mathematics*, vol. XII. (1873), pp. 115—118.]

WRITE

$$a = g - h,$$
$$b = h - f,$$
$$c = f - g,$$

equations implying a fourth equation forming with them the system

$$\begin{array}{l} \cdot \quad -h+g-a=0, \\ h \quad \cdot \quad -f-b=0, \\ -g+f \quad \cdot \quad -c=0, \\ a+b+c \quad \cdot \quad =0, \end{array}$$

and also

$$af + bg + ch = 0.$$

Then, putting for shortness

$$P = (bg - ch)(ch - af)(af - bg),$$
$$Q = a^2g^2h^2 + b^2h^2f^2 + c^2f^2g^2 + a^2b^2c^2,$$
$$R = a^2f^2(a^2 + f^2) + b^2g^2(b^2 + g^2) + c^2h^2(c^2 + h^2),$$

we have

$$2P + Q - R = 0,$$

viz. substituting for a, b, c their values $g-h$, $h-f$, $f-g$, this is an identical equation.

The direct verification is however somewhat tedious, and the equation may be proved more easily as follows:

In the terms a^2+f^2, b^2+g^2, c^2+h^2 of R, substituting for a, b, c their values, we find

$$R = (f^2+g^2+h^2)(a^2f^2+b^2g^2+c^2h^2) - 2fgh(a^2f+b^2g+c^2h),$$

which may be written

$$R = -2(f^2+g^2+h^2)(bcgh+cahf+abfg) - 2fgh(a^2f+b^2g+c^2h).$$

We have then

$$2P = -2bcgh(bg-ch) - 2cahf(ch-af) - 2abfg(af-bg),$$

and thence

$$\begin{aligned} 2P-R = {} & 2bcgh(f^2+g^2+h^2-bg+ch) \\ & + 2cahf(f^2+g^2+h^2-ch+af) \\ & + 2abfg(f^2+g^2+h^2-af+bg) \\ & + 2fgh(a^2f+b^2g+c^2h), \end{aligned}$$

which is at once converted into

$$\begin{aligned} 2P-R = {} & 2bcgh\,\{a^2+f(f+g+h)\} \\ & + 2cahf\,\{b^2+g(f+g+h)\} \\ & + 2abfg\,\{c^2+h(f+g+h)\} \\ & + 2fgh(a^2f+b^2g+c^2h); \end{aligned}$$

or, what is the same thing,

$$2P-R = 2fgh\,\{(bc+ca+ab)(f+g+h)+a^2f+b^2g+c^2h\} + 2abc(agh+bhf+cfg),$$

where, since

$$agh+bhf+cfg = -abc,$$

the last term is

$$= -2a^2b^2c^2.$$

But from the equation last written down we deduce at once

$$Q = 2a^2b^2c^2 - 2fgh(bcf+cag+abh),$$

and we thence have

$$2P+Q-R = 2fgh\,\{(bc+ca+ab)(f+g+h)+(a^2f+b^2g+c^2h)-bcf-cag-abh\},$$

which is

$$= 2fgh(a+b+c)(af+bg+ch),$$

and consequently $=0$, the theorem in question.

Instead of a, b, c, f, g, h, I write $aW \div YZ$, $bW \div ZX$, $cW \div XY$, $f \div X$, $g \div Y$, $h \div Z$: we have therefore

$$\begin{array}{l} \quad . \quad -hY + gZ - aW = 0, \\ hX \quad . \quad -fZ - bW = 0, \\ -gX + fY \quad . \quad -cW = 0, \\ aX + bY + cZ \quad . \quad = 0, \end{array}$$

and as before

$$af + bg + ch = 0.$$

Moreover, omitting a common factor, the new values of P, Q, R are

$$P = XYZW(bg - ch)(ch - af)(af - bg),$$

$$Q = a^2g^2h^2X^4 + b^2h^2f^2Y^4 + c^2f^2g^2Z^4 + a^2b^2c^2W^4,$$

$$R = a^2f^2(a^2X^2W^2 + f^2Y^2Z^2) + b^2g^2(b^2Y^2W^2 + g^2Z^2X^2) + c^2h^2(c^2Z^2W^2 + h^2X^2Y^2),$$

and the identical equation is, as before,

$$2P + Q - R = 0.$$

Consider the operative symbols

$$\begin{array}{l} d_{x_1}, \ d_{x_2}, \ d_{x_3}, \ d_{x_4}, \\ d_{y_1}, \ d_{y_2}, \ d_{y_3}, \ d_{y_4}, \end{array}$$

and write $a = d_{x_1}d_{y_2} - d_{y_1}d_{x_2} = 12$, &c., that is

$$\begin{array}{ll} a = 23, & f = 14, \\ b = 31, & g = 24, \\ c = 12, & h = 34, \end{array}$$

and also $X = xd_{x_1} + yd_{y_1}$, &c. say

$$X = \nabla_1, \quad Y = \nabla_2, \quad Z = \nabla_3, \quad W = \nabla_4.$$

These values of $a, b, c, f, g, h, X, Y, Z, W$ satisfy the above written equations of connexion, and therefore the identical equation $2P + Q - R = 0$. Hence taking U to denote the quartic function $U = (a, b, c, d, e)(x, y)^4$, and therefore $U_1 = (a, \ldots)(x_1, y_1)^4$, &c., we have

$$(2P + Q - R)\, U_1U_2U_3U_4 = 0,$$

where, after the differentiations, $(x_1, y_1), \ldots, (x_4, y_4)$ are to be each of them replaced by (x, y).

Observe that P is the sum of three positive and three negative terms, but that after the omission of the suffixes each term taken with its proper sign becomes equal to the same quantity, and the value of P is $= 6$ times any one term thereof. Thus omitting for the moment the factor $\nabla_1\nabla_2\nabla_3\nabla_4$, two of the terms are $-(af)^2\, bg + af(bg)^2$, that is,

$$-(14 \,.\, 23)^2(24 \,.\, 31) + (14 \,.\, 23)(24 \,.\, 31)^2,$$

and, if in the first term we interchange 3 and 4, it becomes $-(13.24)^2(23.41)$, that is, $+(14.23)(24.31)^2$, viz. it becomes equal to the second term. As regards Q the terms are all positive and become equal to each other; and the like as regards R: hence we have

$$\{12\nabla_1\nabla_2\nabla_3\nabla_4(14.23)(24.31)^2+4\nabla_1^4(23)^2(34)^2(42)^2-6\nabla_1^2\nabla_4^2(43)^4(14)^2\}\, U_1U_2U_3U_4=0,$$

which, omitting a numerical factor $6.2.12^2.2.24^2.4, =3^5.2^{15}$, is in fact the well-known equation

$$\Omega+JU-IH=0,$$

where

$$U=(a,\ b,\ c,\ d,\ e)(x,\ y)^4,$$

$$\Omega=\text{disct.}\,(ax+by,\ bx+cy,\ cx+dy,\ dx+ey)(\xi,\ \eta)^3$$

$$=(ax+by)^2(dx+ey)^2+\&c.,$$

$$I=ae-4bd+3c^2,$$

$$J=ace-ad^2-b^2c-c^3+2bcd,$$

viz. attending only to the coefficient of x^4, this equation is

$$a^2d^2+4ac^3+4b^3d-3b^2c^2-6abcd+a(ace-ad^2-b^2e-c^3+2bcd)+(ac-b^2)(ae-4bd+3c^2)=0.$$

568.

NOTE ON THE INTEGRALS $\int_0^\infty \cos x^2\,dx$ AND $\int_0^\infty \sin x^2\,dx$.

[From the *Quarterly Journal of Pure and Applied Mathematics*, vol. XII. (1873), pp. 118—126.]

MR WALTON has raised, in relation to these integrals, a question which it is very interesting to discuss. Taking for greater convenience the limits to be $-\infty$, $+\infty$, and writing

$$2u = \int_{-\infty}^{\infty} \cos x^2 dx,\ 2v = \int_{-\infty}^{\infty} \sin x^2 dx,$$

then we have

$$4(u^2 - v^2) = \int_{-\infty}^{\infty}\int_{-\infty}^{\infty} \cos(x^2 + y^2)\,dx\,dy,$$

$$8uv = \int_{-\infty}^{\infty}\int_{-\infty}^{\infty} \sin(x^2 + y^2)\,dx\,dy,$$

and writing herein $x = r\cos\theta$, $y = r\sin\theta$, and therefore $dxdy = rdr\,d\theta$, it would thence appear that we have

$$4(u^2 - v^2) = \int_0^\infty \int_0^{2\pi} \cos r^2 .\, rdr\,d\theta = 2\pi \int_0^\infty \cos r^2 .\, rdr,$$

$$8uv = \int_0^\infty \int_0^{2\pi} \sin r^2 .\, rdr\,d\theta = 2\pi \int_0^\infty \sin r^2 .\, rdr,$$

or, finally

$$4(u^2 - v^2) = \pi \sin\infty,$$

$$8uv = \pi(1 - \cos\infty);$$

that is, either the integrals have their received values $\left\{\text{each } = \frac{\sqrt{(\pi)}}{2\sqrt{(2)}}\right\}$, and then $\sin\infty = 0$, $\cos\infty = 0$; or else the integrals, instead of having their received values, are indeterminate.

The error is in the assumption as to the limits of r, θ; viz. in the original expressions for $4(u^2 - v^2)$, $8uv$, we integrate over the area of an indefinitely large square (or rectangle); and the assumption is that we are at liberty, instead of this, to integrate over the area of an indefinitely large circle.

Consider in general in the plane of xy, a closed curve, surrounding the origin, depending on a parameter k, and such that each radius vector continually increases and becomes indefinitely large as k increases and becomes indefinitely large: the curve in question may be referred to as the bounding curve; and the area inside or outside this curve as the inside or outside area. And consider further an integral $\iint z\,dx\,dy$, where z is a given function of x, y, and the integration extends over the inside area. The function z may be such that, for a given form of the bounding curve, the integral, as k becomes indefinitely large, continually approaches to a determinate limiting value (this of course implies that z is indefinitely small for points at an indefinitely large distance from the origin); and we may then say that the integral taken over the infinite inside area has this determinate value; but it is by no means true that the value is independent of the form of the bounding curve; or even that, being determinate for one form of this curve, it is determinate for another form of the curve.

I remark, however, that if z is always of the same sign (say always positive) then the value, assumed to be determinate for a certain form of the bounding curve, is independent of the form of this curve and remains therefore unaltered when we pass to a different form of bounding curve. To fix the ideas, let the first form of bounding curve be a square ($x = \pm k$, $y = \pm k$), and the second form a circle ($x^2 + y^2 = k^2$). Imagine a square inside a circle which is itself inside another square; then z being always positive, the integral taken over the area of the circle is less than the integral over the area of the larger square, greater than the integral over the area of the smaller square. Let the sides of the two squares continually increase, then for each square the integral has ultimately its limiting value; that is, for the area included between the two squares the value is ultimately $= 0$, and consequently for the circle the integral has ultimately the same value that it has for the square. When z is not always of the same sign the proof is inapplicable; and although, for certain forms of z, it may happen that the value of the integral is independent of the form of the bounding curve, this is not in general the case.

We have thus a justification of the well known process for obtaining the value of the integral $\int_0^\infty e^{-x^2}\,dx$, viz. calling this u, or writing

$$2u = \int_{-\infty}^{\infty} e^{-x^2}\,dx,$$

then

$$4u^2 = \int_{-\infty}^{\infty}\int_{-\infty}^{\infty} e^{-(x^2+y^2)}\,dx\,dy = \int_0^{\infty}\int_0^{2\pi} e^{-r^2}\,r\,dr\,d\theta$$

$$= 2\pi \cdot \tfrac{1}{2}, \quad \text{or} \quad u = \tfrac{1}{2}\sqrt{(\pi)},$$

but in consequence of the alternately positive and negative values of $\cos x^2$ and $\sin x^2$, we cannot infer that the like process is applicable to the integrals of these functions.

To show that it is in fact inapplicable, it will be sufficient to prove that the integrals in question have determinate values; for this being so, the double integrals $\iint \cos(x^2+y^2)\,dx\,dy$ and $\iint \sin(x^2+y^2)\,dx\,dy$, taken over an infinite square (or, if we please, over a rectangle the sides of which are both infinite, the ratio having any value whatever), will have determinate values; whereas, by what precedes, the values taken over an infinite circle are indeterminate. The thing may be seen in a very general sort of way thus: consider the surface $z = \sin(x^2+y^2)$, and let the plane of xy be divided into zones by the concentric circles, radii $\sqrt{(\pi)}$, $\sqrt{(2\pi)}$, $\sqrt{(3\pi)}$, &c...., then in the several zones z is alternately positive and negative, the maximum (positive or negative) value being ± 1; and though the breadths of the successive zones decrease, the areas and values of the integral remain constant for the successive zones; the integral over the circle radius $\sqrt{(n\pi)}$ is thus given as a neutral series having no determinate sum. But if the plane xy is divided in like manner into squares by the lines $x = \pm\sqrt{(n\pi)}$, $y = \pm\sqrt{(n\pi)}$, then in each of the bands included between successive squares, z has a succession of positive and negative values; the breadths continually diminish, and although the areas remain constant, yet, on account of the succession of the positive and negative values of z, there is a continual diminution in the values of the integral for the successive bands respectively, and the value of the integral for the whole square is given as a series which may very well be, and which I assume is in fact, convergent. Observe that I have not above employed this mode of integration (but by considering the single integral have in effect divided the square into indefinitely thin slices, and considered each slice separately); it would be interesting to carry out the analytical division of the square into bands, and show that we actually obtain a convergent series; but I do not pursue this inquiry.

Consider the integral

$$v = \int_0^\infty \sin x^2\,dx,$$

and taking for a moment the superior limit to be $(n+1)\pi$, then the quantity under the integral sign is positive from $x^2=0$ to $x^2=\pi$, negative from $x^2=\pi$ to $x^2=2\pi$, and so on; we may therefore write

$$\int_0^{(n+1)\pi} \sin x^2\,dx = A_0 - A_1 + A_2 \ldots + (-)^n A_n,$$

where

$$A_r, = (-)^r \int_{r\pi}^{(r+1)\pi} \sin x^2\,dx,$$

is positive. Writing herein $x^2 = r\pi + u$, we have

$$A_r = \tfrac{1}{2}\int_0^\pi \frac{\sin u\,du}{\sqrt{(r\pi+u)}},$$

which, for r large, may be taken to be

$$=\tfrac{1}{2}\int_0^\pi \frac{\sin u du}{\sqrt{(r\pi)}},\ =\frac{1}{\sqrt{(r\pi)}},$$

viz. r being large, we have A_r differing from the above value $\frac{1}{\sqrt{(r\pi)}}$ by a quantity of the order $\frac{1}{r^{\frac{3}{2}}}$.

It is obviously immaterial whether we integrate from $x^2=0$ to $(n+1)\pi$ or to $(n+1)\pi+\epsilon$, where ϵ has any value less than π; for by so doing, we alter the value of the integral by a quantity less than A_{n+1}, and which consequently vanishes when n is indefinitely large. And similarly, it is immaterial whether we stop at an odd or an even value of n.

We have therefore

$$v=\int_0^{(n+1)\pi}\sin x^2\,dx=A_0-A_1+A_2\ldots+(-)^n A_n,$$

or, taking n to be odd, this is

$$=A_0-A_1+A_2\ldots-A_n,$$

or, say it is

$$=(A_0-A_1)+(A_2-A_3)\ldots+(A_{n-1}-A_n),$$

viz. n here denotes an indefinitely large odd integer.

If instead of $A_0-A_1+A_2-A_3+\&c.$, the signs had been all positive, then the term A being ultimately as $\frac{1}{\sqrt{(r)}}$, the series would have been divergent, and would have had no definite sum: but with the actual alternate signs, the series is convergent, and the sum has a determinate value. To show this more distinctly, observe that we have

$$A_{r-1}-A_r=(-)^{r-1}\,.\,\tfrac{1}{2}\int_{-\pi}^{\pi}\frac{\sin(r\pi+u)\,du}{\sqrt{(r\pi+u)}},\ =-\tfrac{1}{2}\int_{-\pi}^{\pi}\frac{\sin u dy}{\sqrt{(r\pi+u)}},$$

or, taking the integral from $-\pi$ to 0 and from 0 to π, and in the first integral writing $-u$ in place of u, then

$$A_{r-1}-A_r=\tfrac{1}{2}\int_0^\pi \sin u du\left\{\frac{1}{\sqrt{(r\pi-u)}}-\frac{1}{\sqrt{(r\pi+u)}}\right\},$$

where, r being large, expanding the term in $\{\ \}$ in ascending powers of u, then $A_{r-1}-A_r$ is of the order $\frac{1}{r^{\frac{3}{2}}}$: and the series $(A_0-A_1)+(A_2-A_3)\ldots+(A_{n-1}-A_n)$ is therefore convergent, and the sum as n is increased approaches a definite limit. Hence the integral v has a definite value: and similarly, the integral u has a definite value.

The values of u, v being shown to be determinate, I see no ground for doubting that these are the values of the more general integrals

$$\int_0^\infty e^{-ax^2}\cos x^2\,dx,\quad \int_0^\infty e^{-ax^2}\sin x^2\,dx,$$

(a real and positive) when a is supposed to continually diminish and ultimately become $=0$. We have, in fact, (a as above)

$$\int_0^\infty e^{(-a+bi)y}\,y^{n-1}\,dy = \frac{\Gamma(n)\,e^{in\theta}}{(a^2+b^2)^{\frac{1}{2}n}},\,*$$

where $\theta = \tan^{-1}\dfrac{b}{a}$, an angle included between the limits $-\frac{1}{2}\pi$, $+\frac{1}{2}\pi$. Writing herein $n=\frac{1}{2}$, $b=1$, $y=x^2$, then

$$\int_0^\infty e^{(-a+i)x^2}\,dx = \frac{\sqrt{(\pi)}\,e^{\frac{1}{2}i\theta}}{2\,(a^2+1)^{\frac{1}{4}}},$$

where $\theta=\tan^{-1}\dfrac{1}{a}$, an angle included between the limits $-\frac{1}{2}\pi$, $+\frac{1}{2}\pi$; or, putting herein $a=0$, we have $\theta=\frac{1}{2}\pi$, and therefore

$$\int_0^\infty e^{ix^2}\,dx = \tfrac{1}{2}\sqrt{(\pi)}\,e^{\frac{1}{4}i\pi};$$

that is, equating the real and imaginary parts,

$$u = v = \frac{\sqrt{\pi}}{2\sqrt{2}},$$

which are the received values of the integrals

$$u=\int_0^\infty \cos x^2\,dx,\quad v=\int_0^\infty \sin x^2\,dx.$$

An important instance of the general theory presents itself in the theory of elliptic functions, viz. the integral

$$\iint \frac{dx\,dy}{(\Omega x+\Upsilon y)^2},$$

the ratio $\Omega:\Upsilon$ being imaginary, will, if the bounding curve be symmetrical in regard to the two axes respectively, have a determinate value *dependent on the form of the bounding curve;* if for instance this is a rectangle $x=\pm ak$, $y=\pm bk$, then the value of the integral will depend on the ratio $a:b$ of the infinite sides; and so if the bounding curve be an infinite ellipse, the value of the integral will depend on the ratio and position of the axes. See as to this my papers "On the inverse elliptic

* For brevity I take the integral under this form, but the real and imaginary parts might have been considered separately; and there would have been some advantage in following that course. The like remark applies to a subsequent investigation.

functions," *Camb. Math. Jour.*, t. IV. (1845), pp. 257—277, [24]; and "Mémoire sur les fonctions doublement périodiques," *Liouv.* t. X. (1845), pp. 385—420, [25].

A like theory applies to series, viz. as remarked by Cauchy, although the series $A_0 + A_1 + A_2 + \ldots$ and $B_0 + B_1 + B_2 + \&c.$ are respectively convergent, then arranging the product in the form

$$\begin{aligned} &A_0B_0 + A_0B_1 + A_0B_2 + \ldots \\ &+ A_1B_0 + A_1B_1 + A_1B_2 + \ldots \\ &+ A_2B_0 + A_2B_1 + A_2B_2 + \ldots \\ &+ \ldots, \end{aligned}$$

say the general term is $C_{m,n}$, then if we sum this double series according to an assumed relation between the suffixes m, n (if, for instance, we include all those terms for which $m^2 + n^2 \overline{\gtrless} k^2$, making k to increase continually) it by no means follows that we approach a limit which is equal to the product of the sums of the original two series, or even that we approach a determinate limit.

Mr Walton, agreeing with the rest of the foregoing Note, wrote that he was unable to satisfy himself that the value of $\int_0^\infty e^{ix^2}dx$ is correctly deduced from that of $\int_0^\infty e^{(-a+bi)y}\,y^{n-1}\,dy$. Writing $n = \frac{1}{2}$, the question in fact is whether the formula

$$\int_0^\infty e^{(-a+bi)y}\,y^{-\frac{1}{2}}\,dy = \frac{\sqrt{(\pi)}\,e^{\frac{1}{2}i\theta}}{(a^2+b^2)^{\frac{1}{4}}}\left(\theta = \tan^{-1}\frac{b}{a},\ \text{angle between } \tfrac{1}{2}\pi,\ -\tfrac{1}{2}\pi\right),$$

which is true when a is an indefinitely small positive quantity, is true when $a = 0$; that is, taking b positive, whether we have

$$\int_0^\infty e^{iby}\,y^{-\frac{1}{2}}\,dy = \frac{\sqrt{(\pi)}\,e^{\frac{1}{4}i\pi}}{\sqrt{(b)}}.$$

Write in general

$$u = \int_0^\infty e^{(-a+bi)y}\,y^{-\frac{1}{2}}\,dy,$$

then, differentiating with respect to b, we have

$$\frac{du}{db} = \int_0^\infty iy^{\frac{1}{2}}\,e^{(-a+bi)y}\,dy,$$

or, integrating by parts,

$$\frac{du}{db} = \frac{i}{-a+bi}\,y^{\frac{1}{2}}\,e^{(-a+bi)y} - \frac{i}{2(-a+bi)}\int_0^\infty y^{-\frac{1}{2}}\,e^{(-a+bi)y}\,dy,$$

where the first term is to be taken between the limits ∞, 0; viz. this is

$$\frac{du}{db}=\left[\frac{i}{-a+bi}\,y^{\frac{1}{2}}e^{(-a+bi)y}\right]_0^\infty-\frac{i}{2(-a+bi)}\,u.$$

When a is not $=0$, the first term vanishes at each limit, and we have

$$\frac{du}{db}=\frac{-i}{2(-a+bi)}\,u.$$

The doubt was in effect whether this last equation holds good for the limiting value $a=0$. When a is $=0$, then in the original equation for $\frac{du}{db}$ the first term is indeterminate, and if the equation were true, it would follow that $\frac{du}{db}$ was indeterminate; the original equation for $\frac{du}{db}$ *is not true*, but we *truly* have

$$\frac{du}{db}=-\frac{1}{2b}\,u,$$

the same result as would be obtained from the general equation, rejecting the first term and writing $a=0$.

To explain this observe that for $a=0$, we have

$$u=\int_0^\infty y^{-\frac{1}{2}}\,e^{iby}\,dy,$$

which for a moment I write

$$u=\int_0^k y^{-\frac{1}{2}}\,e^{iby}\,dy,$$

where, as before, b is taken to be positive. Writing herein $by=x$, we have

$$u=\frac{1}{\sqrt{(b)}}\int_0^{bk} x^{-\frac{1}{2}}\,e^{ix}\,dx,$$

and assuming only that the integral $\int_0^M x^{-\frac{1}{2}}e^{ix}\,dx$ has a *determinate* limit as M becomes indefinitely large*, then supposing that k is indefinitely large, the integral in the last-mentioned expression for u has the value in question

$$\left(=\int_0^\infty x^{-\frac{1}{2}}\,e^{ix}\,dx\right),$$

* This is in fact the theorem $\int_0^\infty e^{ix^2}\,dx=$ a determinate value $\{=\frac{1}{2}\sqrt{(\pi)}\,e^{\frac{1}{4}i\pi}\}$, proved in the former part of the present Note.

which is independent of b, say this is

$$u = \frac{C}{\sqrt{(b)}},$$

and thence differentiating in regard to b, we find

$$\frac{du}{db} = -\frac{1}{2b}u,$$

the theorem in question.

But the value of $\frac{du}{db}$ *cannot be obtained by differentiating under the integral sign*, for this would give

$$\frac{du}{db} = \int_0^\infty iy^{\frac{1}{2}}\,e^{iby}\,dy,$$

and this integral is certainly indeterminate.

569.

ON THE CYCLIDE*.

[From the *Quarterly Journal of Pure and Applied Mathematics*, vol. XII. (1873), pp. 148—165.]

THE Cyclide, according to the original definition, is the envelope of a variable sphere which touches three given spheres, or, more accurately, the envelope of a variable sphere belonging to one of the four series of spheres which touch three given spheres. In fact, the spheres which touch three given spheres form four series, the spheres of each series having their centres on a conic; viz. if we consider the plane through the centres of the given spheres, and in this plane the eight circles which touch the sections of the given spheres, the centres of these circles form four pairs of points, or joining the points of the same pair, we have four chords which are the transverse axes of the four conics in question.

It thus appears, that one condition imposed on the variable sphere is, that its centre shall be in a plane; and a second condition, that the centre shall be on a conic in this plane; so that the original definition may be replaced first by the following one, viz.:

The cyclide is the envelope of a variable sphere having its centre on a given plane, and touching two given spheres.

Starting herefrom, it follows that the locus of the centre will be a conic in the given plane: the transverse axis of the conic being the projection on the given plane of the line joining the centres of the given spheres; and it, moreover, follows, that if in the perpendicular plane through the transverse axis we construct a conic having for vertices the foci, and for foci the vertices, of the locus-conic, then the conic so constructed will pass through the centres of the given spheres.

* I use the term in its original sense, and not in the extended sense given to it by Darboux, and employed by Casey in his recent memoir "On Cyclides and Spheroquartics," *Phil. Trans.* 1871, pp. 582—721. With these authors the Cyclide here spoken of is a Dupin's or tetranodal Cyclide.

Two conics related in the manner just mentioned are the flat-surfaces of a system of confocal quadric surfaces; they may for convenience be termed anti-conics (fig. 1); one of them is always an ellipse and the other a hyperbola; and the property of them is that, taking any two fixed points on the two branches, or on the same branch of the

Fig. 1.

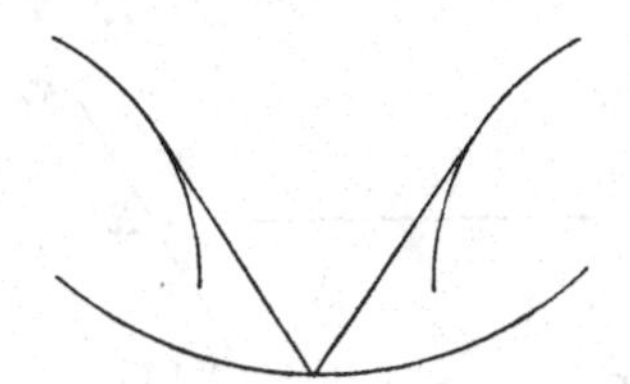

hyperbola, and considering their distances from a variable point of the ellipse: in the first case the sum, in the second case the difference, of these two distances is constant. And similarly taking any two fixed points on the ellipse, and considering their distances from a variable point of the hyperbola, then the difference, first distance *less* second distance is a constant, $+\alpha$ for one branch, $-\alpha$ for the other branch of the hyperbola.

And we thus arrive at a third, and simplified definition of the cyclide, viz. considering any two anti-conics, the cyclide is the envelope of a variable sphere having its centre on the first anti-conic, and touching a given sphere whose centre is on the second anti-conic.

And it is to be added, that the same cyclide will be the envelope of a variable sphere having its centre on the second anti-conic and touching a given sphere whose centre is on the first anti-conic, such given sphere being in fact any particular sphere of the first series of variable spheres. And, moreover, the section of the surface by the plane of either of the anti-conics is a pair of circles, the surface being thus (as will further appear) of the fourth order.

In the series of variable spheres the intersection of any two consecutive spheres is a circle, the centre of which is in the plane of the locus-anti-conic, and its plane perpendicular to that of the locus-anti-conic, this variable circle having for its diameter in the plane of the locus-anti-conic a line terminated by the two fixed circles in that plane. The cyclide is thus in two different ways the locus of a variable circle; and investigating this mode of generation, we arrive at a fourth definition as follows:—

Consider in a plane any two circles, and through either of the centres of symmetry draw a secant cutting the two circles; in the perpendicular plane through the secant, draw circles having for their diameters the chords formed by the two pairs of anti-parallel points on the secant (viz. each pair consists of two points, one on each circle, such that the tangents at the two points are not parallel to each other): the locus of the two variable circles is the cyclide.

Before going further it will be convenient to establish the definition of "skew anti-points": viz. if we have the points K_1, K_2 (fig. 2), mid-point R, and L_1, L_2, mid-point S, such that K_1K_2, RS and L_1L_2 are respectively at right angles to each other, and

$\overline{K_1R}^2+\overline{RS}^2+\overline{SL_1}^2=0$, &c.; or, what is the same thing, the distances $L_1K_1=L_1K_2=L_2K_1=L_2K_2$ are each $=0$, so that the points K_1, K_2 and L_1, L_2 are skew anti-points. Observe that the lines of the figure and the points R, S are taken to be real; but the distances $RK_1=RK_2$ and $SL_1=SL_2$ cannot be both real: it is assumed that one is real and

Fig. 2.

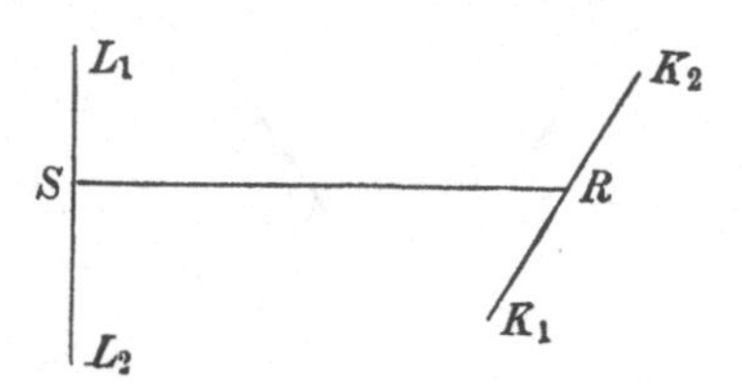

the other a pure imaginary, or else that they are both of them pure imaginaries. To fix the ideas we may in the figure consider the plane through K_1K_2, RS as horizontal, and that through RS, L_1L_2 as vertical.

Reverting now to the cyclide, suppose that we have (in the same plane) the two circles C, C' intersecting in K_1, K_2, and having S for a centre of symmetry, and let R be the mid-point of K_1, K_2.

The construction is:—through S draw a secant meeting the two circles in A, B and A', B' respectively, where A, A' and B, B' are parallel points, (therefore A, B' and A', B anti-parallel points), then the cyclide is the locus of the circles in the perpendicular plane on the diameters AB' and $A'B$ respectively.

The two circles have their radical axis passing through S, and not only so, but the points of intersection L_1, L_2 of the two circles are situate at a distance $SL_1=SL_2$, which is independent of the position of the secant: the points L_1, L_2 and K_1, K_2 being in fact a system of skew anti-points. And, moreover, the two circles have a centre of symmetry at the point where the plane of the two circles meets the line K_1K_2.

Consider in particular the two circles D, D' which are situate in the perpendicular plane through SR; these have the radical axis L_1L_2, and a centre of symmetry R; and if with these circles D, D' as given circles, and with R as the centre of symmetry, we obtain in a plane through K_1K_2 two circles having K_1K_2 for their radical axis, and having for a centre of symmetry the intersection of their plane with L_1L_2, the locus of these circles is the same cyclide as before; and, in particular, if their plane passes through RS, then the two circles are the before-mentioned circles C, C', having S for a centre of symmetry.

It will be noticed that, starting with the same two circles C, C' or D, D', we obtain two different cyclides according as we use in the construction one or other of the two centres of symmetry.

The cyclide is a quartic surface having the circle at infinity for a nodal line: viz. it is an anallagmatic or bicircular quartic surface; and it has besides the points

K_1, K_2, L_1, L_2, that is, a system of skew anti-points, for nodal points; these determine the cyclide save as to a single parameter. In fact, starting with the four points L_1, L_2, K_1, K_2, which give S, and therefore the plane of the circles C, C'; the circle C is then any one of the circles through K_1, K_2; and then drawing from S the two tangents to C, there is one other circle C' passing through K_1, K_2 and touching these tangents; C' is thus uniquely determined, and the construction is effected as above. Hence, with a given system of skew anti-points we have a single series of cyclides, say a series of conodal cyclides.

If in general we consider a quartic surface having a nodal conic and four nodes A, B, C, D, then it is to be observed that, taking the nodes in a proper order, we have a skew quadrilateral $ABCD$, the sides whereof AB, BC, CD, DA, lie wholly on the surface. In fact, considering the section by the plane ABC, this will be a quartic curve having the nodes A, B, C and two other nodes, the intersections of the plane with the nodal conic; the section is thus made up of a pair of lines and a conic; it follows that two of the sides of the triangle ABC, say the sides AB, BC, each meet the nodal conic, and that the section in question is made up of the lines AB, BC, and of a conic through the points A, C and the intersections of AB, BC with the nodal conic. Considering next the section by the plane through ACD, here (since AC is not a line on the surface) the lines CD, DA each meet the nodal conic, and the section is made up of the lines CD, DA and of a conic passing through the points A, C and the intersections of the lines CD, DA with the nodal conic. Thus the lines AB, BC, CD, DA each meet the nodal conic, and lie wholly on the surface; the lines AC, BD do not meet the conic or lie wholly on the surface.

A quartic surface depends upon 34 constants; it is easy to see that, if the surface has a given nodal conic, this implies 21 conditions, or say the postulation of a given nodal conic is $=21$, whence also the postulation of a nodal conic (not a given conic) is $=13$. Suppose that the surface has the given nodes A, B, C, D; the postulation hereof is $=16$; the nodal conic is then a conic meeting each of the lines AB, BC, CD, DA, viz. if the plane of the conic is assumed at pleasure, then the conic passes through 4 given points, and thus it still contains 1 arbitrary parameter; that is, in order that the nodal conic may be a given conic (satisfying the prescribed conditions) the postulation is $=4$. The whole postulation is thus $16+13+4, =33$, or the quartic surface which satisfies the condition in question (viz. which has for nodes the given points A, B, C, D, and for nodal conic a given conic meeting each of the lines AB, BC, CD, DA) contains still 1 arbitrary parameter: which agrees with the foregoing result in regard to the existence of a series of conodal cyclides.

It is to be added that, if a quartic surface has for a nodal line the circle at infinity and has four nodes, then the nodes form a system of skew anti-points and the surface is a cyclide. In fact, taking the nodes to be A, B, C, D, then each of the lines AB, BC, CD, DA meets the circle at infinity; but if the line AB meets the circle at infinity, then the distance AB is $=0$, and similarly the distances BC, CD, DA are each $=0$; that is, the nodes (A, C) and (B, D) are a system of skew anti-points.

Reverting to the cyclide, and taking (as before) the nodes to be K_1, K_2 and L_1, L_2, the line RS which joins the mid-points of K_1K_2 and L_1L_2 may be termed the axis of the cyclide, and the points where it meets the cyclide, or, what is the same thing, the circles C, C' or D, D', the vertices of the cyclide, say these are the points F, G, H, K. Supposing that the distances of these from a point on the axis are f, g, h, k, the origin may be taken so that $f+g+h+k=0$; the origin is in this case the "centre" of the cyclide. It is to be remarked, that given the vertices there are three series of cyclides; viz. we may in an arbitrary plane through the axis take for C, C' the circles standing on the diameters FG and HK respectively; and then, according as we take one or the other centre of symmetry, we have in the plane at right angles hereto for D, D' the circles on the diameters FH and GK, or else the circles on the diameters FK and GH respectively; there are thus three cases according as the two pairs of circles are the circles on the diameters

$$\begin{array}{lcl} FH,\ KG & \text{and} & FK,\ GH, \\ FK,\ GH & ,, & FG,\ HK, \\ FG,\ HK & ,, & FH,\ KG. \end{array}$$

The equation of the cyclide expressed in terms of the parameters f, g, h, k assumes a peculiarly simple form; in fact, taking the origin at the centre, so that $f+g+h+k=0$, the axis of x coinciding with the axis of the cyclide, and those of y, z parallel to the lines K_1K_2 and L_1L_2, or L_1L_2 and K_1K_2 respectively: writing also

$$fg+hk=G,$$
$$fh+kg=H,$$
$$fk+gh=K,$$

then the equation of one of the cyclides is

$$(y^2+z^2)^2+2x^2(y^2+z^2)+Gy^2+Hz^2+(x-f)(x-g)(x-h)(x-k)=0,$$

which we may at once partially verify by observing that for $z=0$ this equation becomes

$$[y^2+(x-f)(x-g)][y^2+(x-h)(x-k)]=0,$$

and for $y=0$ it becomes

$$[z^2+(x-f)(x-h)][z^2+(x-k)(x-g)]=0,$$

viz. the equations of the circles C, C' are

$$y^2+(x-f)(x-g)=0,\quad y^2+(x-h)(x-k)=0,$$

and those of D, D'

$$z^2+(x-f)(x-h)=0,\quad z^2+(x-k)(x-g)=0.$$

Starting from these equations of the four circles, the points K_1, K_2 are given by

$$Y^2=-(P-f)(P-g)=-(P-h)(P-k),$$

and the points L_1, L_2 by

$$Z^2 = -(Q-f)(Q-h) = -(Q-k)(Q-g).$$

Now writing for a moment

$$\begin{aligned}\beta &= f+g = -h-k,\\ \gamma &= f+h = -k-g,\\ \delta &= f+k = -g-h,\end{aligned}$$

we have $P = -\tfrac{1}{2}\dfrac{\gamma\delta}{\beta}$, $Q = -\tfrac{1}{2}\dfrac{\beta\delta}{\gamma}$, and thence $PQ = \tfrac{1}{4}\delta^2$. Moreover

$$\begin{aligned}&2Y^2 + 2Z^2 + 2(P-Q)^2\\ &= -(P-f)(P-g) - (P-h)(P-k)\\ &\quad - (Q-f)(Q-h) - (Q-k)(Q-g) + 2(P-Q)^2\\ &= -(fg+hk+fh+gk) - 4PQ\\ &= \delta^2 - 4PQ\\ &= 0,\end{aligned}$$

that is,

$$Y^2 + Z^2 + (P-Q)^2 = 0,$$

which equation expresses that the four points are a system of skew anti-points.

The point $x = Q$ should be a centre of symmetry of the circles C, C'; to verify that this is so, transforming to the point in question as origin, the equations are

$$y^2 + \{x + Q - \tfrac{1}{2}(f+g)\}^2 - \tfrac{1}{4}(f-g)^2 = 0,$$

$$y^2 + \{x + Q - \tfrac{1}{2}(h+k)\}^2 - \tfrac{1}{4}(k-h)^2 = 0,$$

that is,

$$y^2 + \left\{x - \tfrac{1}{2}\frac{\beta}{\gamma}(\delta+\gamma)\right\}^2 - \tfrac{1}{4}(f-g)^2 = 0,$$

$$y^2 + \left\{x - \tfrac{1}{2}\frac{\beta}{\gamma}(\delta-\gamma)\right\}^2 - \tfrac{1}{4}(k-h)^2 = 0.$$

But $\delta+\gamma = f-g$, $\delta-\gamma = k-h$, so that these equations are

$$y^2 + \left\{x - \tfrac{1}{2}\frac{\beta}{\gamma}(f-g)\right\}^2 = \tfrac{1}{4}(f-g)^2,$$

$$y^2 + \left\{x - \tfrac{1}{2}\frac{\beta}{\gamma}(k-h)\right\}^2 = \tfrac{1}{4}(k-h)^2,$$

which are of the form

$$\begin{aligned}y^2 + (x - \alpha)^2 &= c^2,\\ y^2 + (x - m\alpha)^2 &= m^2c^2,\end{aligned}$$

and consequently $x = Q$ is a centre of symmetry of the circles C, C'; and in like manner it would appear that $x = P$ is a centre of symmetry of the circles D, D'.

If in the last-mentioned equations of the circles C, C' we write $x = \Omega\cos\theta$, $y = \Omega\sin\theta$, and put for shortness

$$\rho = \alpha\cos\theta - \nabla,\quad \sigma = m(\alpha\cos\theta - \nabla),$$
$$\rho' = \alpha\cos\theta + \nabla,\quad \sigma' = m(\alpha\cos\theta + \nabla),$$

where $\nabla = \sqrt{(c^2 - \alpha^2\sin^2\theta)}$, then the values of Ω for the first circle are ρ, σ, and those for the second circle are ρ', σ'. Hence the equations of the generating circles are

$$z^2 + (r - \rho)(r - \sigma') = 0,$$
$$z^2 + (r + \rho')(r - \sigma) = 0,$$

where r is the abscissa in the plane of the circles, measured from the point $x = Q$. Attending say to the first of these equations, to find the equation of the cyclide, we must eliminate θ from the equations

$$z^2 + (r - \rho)(r - \sigma') = 0,\quad x = r\cos\theta,\quad y = r\sin\theta;$$

the first equation is

$$z^2 + r^2 + m(\alpha^2 - c^2) - r(\rho + \sigma') = 0,$$

and we have

$$\rho + \sigma' = (m+1)\alpha\cos\theta - (m-1)\sqrt{(c^2 - \alpha^2\sin^2\theta)},$$

and thence

$$(\rho + \sigma')r = (m+1)\alpha x - (m-1)\sqrt{\{c^2(x^2 + y^2) - \alpha^2 y^2\}},$$

so that we have

$$z^2 + x^2 + y^2 + m(\alpha^2 - c^2) - (m+1)\alpha x + (m-1)\sqrt{\{c^2(x^2 + y^2) - \alpha^2 y^2\}} = 0,$$

viz. this is the equation of the cyclide in terms of the parameters α, c, m, the origin being at the point $x = Q$, the centre of symmetry of the circles C, C'.

Reverting to the former origin at the centre of the cyclide, we must write $x - Q$ for x; the equation thus is

$$\{y^2 + z^2 + (x-Q)^2 - (m+1)\alpha(x-Q) + m(\alpha^2 - c^2)\}^2 - (m-1)^2[\{c^2(x-Q)^2 + (c^2 - \alpha^2)y^2\}] = 0,$$

where

$$Q = -\tfrac{1}{2}\frac{\beta\delta}{\gamma},\quad \alpha = \tfrac{1}{2}\frac{\beta}{\gamma}(f-g),\quad c^2 = \tfrac{1}{4}(f-g)^2,\quad m = \frac{k-h}{f-g},$$

whence also

$$m + 1 = \frac{2\delta}{f-g},\quad m - 1 = \frac{-2\gamma}{f-g},\quad \alpha^2 - c^2 = \tfrac{1}{4}(f-g)^2\frac{(f-k)(g-h)}{\gamma^2}.$$

After all reductions, the equation assumes the before-mentioned form

$$(y^2 + z^2)^2 + 2x^2(y^2 + z^2) + Gy^2 + Hz^2 + (x-f)(x-g)(x-h)(x-k) = 0.$$

The equation may be written

$$(x^2 + y^2 + z^2)^2 + (G + H + K)x^2 + Gy^2 + Hz^2 - \beta\gamma\delta x + fghk = 0,$$

and if we express everything in terms of β, γ, δ by the formulæ

$$\begin{aligned}
2f &= \beta+\gamma+\delta, & 2G &= \beta^2-\gamma^2-\delta^2,\\
2g &= \beta-\gamma-\delta, & 2H &= -\beta^2+\gamma^2-\delta^2,\\
2h &= -\beta+\gamma-\delta, & 2K &= -\beta^2-\gamma^2+\delta^2,\\
2k &= -\beta-\gamma+\delta, & 2(G+H+K) &= -\beta^2-\gamma^2-\delta^2;
\end{aligned}$$

then we have

$$(x^2+y^2+z^2)^2+\tfrac{1}{2}(-\beta^2-\gamma^2-\delta^2)\,x^2+\tfrac{1}{2}(\beta^2-\gamma^2-\delta^2)\,y^2+\tfrac{1}{2}(-\beta^2+\gamma^2-\delta^2)\,z^2$$
$$-\beta\gamma\delta x+\tfrac{1}{16}(\beta^4+\gamma^4+\delta^4-2\beta^2\gamma^2-2\beta^2\delta^2-2\gamma^2\delta^2)=0;$$

or, what is the same thing,

$$(x^2+y^2+z^2+\tfrac{1}{4}\beta^2+\tfrac{1}{4}\gamma^2-\tfrac{1}{4}\delta^2)^2-(\beta^2+\gamma^2)\,x^2-\gamma^2y^2-\beta^2z^2-\beta\gamma\delta x-\tfrac{1}{4}\beta^2\gamma^2=0.$$

An equivalent form of equation may be obtained very simply as follows: the surface

$$(x^2+y^2+z^2)^2+2Ax^2+2By^2+2Cz^2+2Kx+L=0$$

will be a cyclide if only the section by each of the planes $y=0$, $z=0$ breaks up into a pair of circles. Now for $y=0$ the equation is

$$(x^2+z^2)^2+2Ax^2+2Cz^2+2Kx+L=0,$$

that is,

$$z^4+2z^2(x^2+C)+x^4+2Ax^2+2Kx+L=0,$$

or

$$(z^2+x^2+C)^2=2(C-A)\,x^2-2Kx+C^2-L,$$

which will be a pair of circles if only

$$2(C-A)(C^2-L)=K^2;$$

and similarly writing $z=0$, we obtain

$$2(B-A)(B^2-L)=K^2.$$

These equations give

$$\begin{aligned}
L &= (B+C)^2-(BC+CA+AB),\\
K^2 &= -2(B-A)(C-A)(B+C),
\end{aligned}$$

so that L, K having these values the surface is a cyclide; there are two cyclides corresponding to the two different values of K, which agrees with a former result.

Reverting to the equation in terms of β, γ, δ this may be written

$$\beta^2-\gamma^2+\sqrt{\{(2\gamma x+\beta\delta)^2-4(\beta^2-\gamma^2)\,y^2\}}+\sqrt{\{(2\beta x+\gamma\delta)^2+4(\beta^2-\gamma^2)\,z^2\}}=0.$$

[Compare herewith Kummer's form

$$b^2=\sqrt{\{(ax-ek)^2+b^2y^2\}}+\sqrt{\{(ex-ak)^2-b^2z^2\}},\ \text{where } b^2=a^2-e^2.]$$

In fact, representing this for a moment by

$$\beta^2 - \gamma^2 + \sqrt{(\Theta)} + \sqrt{(\Phi)} = 0,$$

we have

$$(\beta^2 - \gamma^2)^2 + \Theta - \Phi = -2(\beta^2 - \gamma^2)\sqrt{(\Theta)},$$

or, substituting and dividing by $\beta^2 - \gamma^2$, we have

$$\beta^2 - \gamma^2 + \delta^2 - 4(x^2 + y^2 + z^2) + 2\sqrt{\{(2\gamma x + \beta\delta)^2 - 4(\beta^2 - \gamma^2)y^2\}} = 0,$$

or, similarly

$$\beta^2 - \gamma^2 - \delta^2 + 4(x^2 + y^2 + z^2) + 2\sqrt{\{(2\beta x + \gamma\delta)^2 + 4(\beta^2 - \gamma^2)z^2\}} = 0,$$

either of which leads at once to the rational form.

The irrational equation

$$\beta^2 - \gamma^2 + \sqrt{\{(2\gamma x + \beta\delta)^2 - 4(\beta^2 - \gamma^2)y^2\}} + \sqrt{\{(2\beta x + \gamma\delta)^2 + 4(\beta^2 - \gamma^2)z^2\}} = 0$$

is of the form

$$p + \sqrt{(qr)} + \sqrt{(st)} = 0,$$

which belongs to a quartic surface having the nodal conic $p = 0$, $qr - st = 0$ (in the present case the circle at infinity), and also the four nodes ($q = 0$, $r = 0$, $p^2 - st = 0$) and ($s = 0$, $t = 0$, $p^2 - qr = 0$), viz. these are

$$x = -\tfrac{1}{2}\frac{\beta\delta}{\gamma}, \quad y = 0, \quad z = \pm\tfrac{1}{2}\frac{1}{\gamma}\sqrt{\{(\beta^2 - \gamma^2)(\gamma^2 - \delta^2)\}},$$

and

$$x = -\tfrac{1}{2}\frac{\gamma\delta}{\beta}, \quad y = \pm\tfrac{1}{2}\frac{1}{\beta}\sqrt{\{(\gamma^2 - \beta^2)(\beta^2 - \delta^2)\}}, \quad z = 0,$$

and we hence again verify that the nodes form a system of skew anti-points, viz. the condition for this is

$$\delta^2\left(\frac{\beta}{\gamma} - \frac{\gamma}{\beta}\right)^2 + (\beta^2 - \gamma^2)\left(1 - \frac{\delta^2}{\gamma^2}\right) - (\beta^2 - \gamma^2)\left(1 - \frac{\delta^2}{\beta^2}\right) = 0,$$

that is,

$$\delta^2(\beta^2 - \gamma^2) + \beta^2(\gamma^2 - \delta^2) - \gamma^2(\beta^2 - \delta^2) = 0,$$

which is satisfied identically.

The cyclide has on the nodal conic or circle at infinity four pinch-points, viz these are the intersections of the circle at infinity with the planes $\beta^2y^2 + \gamma^2z^2 = 0$.

If $\beta = 0$, the equation becomes

$$\tfrac{1}{2}\gamma + \sqrt{(x^2 + y^2)} + \sqrt{(\tfrac{1}{4}\delta^2 - z^2)} = 0,$$

viz. the cyclide has in this case become a torus; there are here two nodes on the axis ($x = 0$, $y = 0$), and two other nodes on the circle at infinity, viz. these are the circular points at infinity of the sections perpendicular to the axes, and the pinch-points coincide in pairs with the last-mentioned two nodes; viz. each of the circular points at infinity = node + two pinch-points.

The Parabolic Cyclide.

One of the circles C, C' and one of the circles D, D' may become each of them a line; the cyclide is in this case a cubic surface. The easier way would be to treat the case independently, but it is interesting to deduce it from the general case. For this purpose, starting from the equation

$$(y^2+z^2)^2+2x^2(y^2+z^2)+Gy^2+Hz^2+(x-f)(x-g)(x-h)(x-k)=0,$$

where $f+g+h+k=0$, $G=fg+hk$, $H=fh+gk$, I write $x-\alpha$ for x, and assume $\alpha+f$, $\alpha+g$, $\alpha+h$, $\alpha+k$, equal to f', g', h', k' respectively; whence $4\alpha=f'+g'+h'+k'$; and the equation is

$$(y^2+z^2)^2+2(x-\alpha)^2(y^2+z^2)+(f'g'+h'k'-2\alpha^2)y^2+(f'h'+g'k'-2\alpha^2)z^2$$
$$+(x-f')(x-g')(x-h')(x-k')=0,$$

or, what is the same thing,

$$(y^2+z^2)^2+(2x^2-4\alpha x)(y^2+z^2)+(f'g'+h'k')y^2+(f'h'+g'k')z^2$$
$$+(x-f')(x-g')(x-h')(x-k')=0.$$

Now assuming $k'=\infty$, we have $4\alpha=k'=\infty$, or writing 4α instead of k', and attending only to the terms which contain α, we have

$$x(y^2+z^2)-h'y^2-g'z^2+(x-f')(x-g')(x-h')=0,$$

or, what is the same thing,

$$(x-f')(x-g')(x-h')+(x-h')y^2+(x-g')z^2=0,$$

where by altering the origin we may make $f'=0$.

It is somewhat more convenient to take the axis of z (instead of that of x) as the axis of the cyclide; making this change, and writing also 0, β, γ in place of the original constants, I take the equation to be

$$z(z-\beta)(z-\gamma)+(z-\gamma)y^2+(z-\beta)x^2=0,$$

viz. this is a cubic surface having upon it the right lines ($z=\gamma$, $x=0$), ($z=\beta$, $y=0$); the section by a plane through either of these lines is the line itself and a circle; and in particular the circle in the plane $x=0$ is $z(z-\beta)+y^2=0$, and that in the plane $y=0$ is $z(z-\gamma)+x^2=0$. And it is easy to see how the surface is generated: if, to fix the ideas, we take β positive, γ negative, the lines and circles are as shown in fig. 3; and if we draw through Cy a plane cutting the circle CO and the line Bx in P, Q respectively, then the section is a circle on the diameter PQ; and similarly for the sections by the planes through Bx. It is easy to see that the whole surface is included between the planes $z=\beta$, $z=\gamma$; considering the sections parallel to these planes (that is, to the plane of xy) $z=\beta$, the section is the two-fold line $y=0$; $z=$ any smaller positive value, it is a hyperbola having the axis of y for its transverse axis; $z=0$, it is the pair of real lines $\gamma y^2+\beta x^2=0$; z negative and less in absolute magnitude than $-\gamma$, it is a hyperbola having the axis of x for its transverse axis; and finally $z=\gamma$, it is the two-fold line $x=0$. It is easy to see the forms of the cubic curves which are the sections by any planes $x=$ const. or $y=$ const.

Fig. 3.

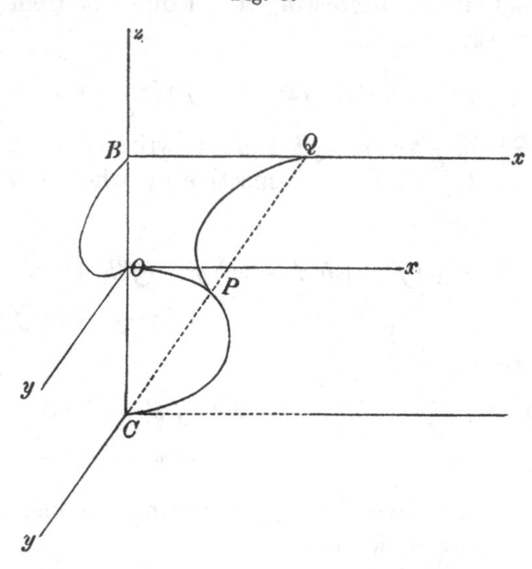

The before-mentioned circles are curves of curvature of the surface; to verify this *à posteriori*, write

$$U=z(z-\beta)(z-\gamma)+(z-\gamma)y^2+(z-\beta)x^2=0$$

for the equation of the surface; and put for shortness $P=3z^2-2z(\beta+\gamma)+\beta\gamma$, $P+x^2+y^2=L$, so that $d_zU=P+x^2+y^2$, $=L$. The differential equation for the curves of curvature is

$$\begin{vmatrix} 2x(z-\beta) & , & 2y(z-\gamma) & , & P+x^2+y^2 \\ xdz+(z-\beta)\,dx, & & ydz+(z-\gamma)\,dy, & & \tfrac{1}{2}P'dz+xdx+ydy \\ dx & , & dy & , & dz \end{vmatrix}=0,$$

or, say this is

$$\begin{aligned}\Omega = dx^2\,.\,2xy\,(z-\gamma)-dy^2\,.\,2xy\,(z-\beta)+dz^2\,.\,2xy\,(\gamma-\beta)\\ +\,dzdy\,.\,x\,[-\,2\,(z-\beta)\,(2z-\beta)+L]\\ +\,dxdz\,.\,y\,[\quad 2\,(z-\gamma)\,(2z-\gamma)-L]\\ +\,dxdy\,.\quad[(\gamma-\beta)\,P+(2z-\beta-\gamma)\,(y^2-x^2)]=0.\end{aligned}$$

But in virtue of the equation $U=0$, we have identically

$$\{2\,(z-\beta)\,x\,dx+2\,(z-\gamma)\,y\,dy+Ldz\}\times\left\{-\frac{z-\gamma}{z-\beta}y\,dx+\frac{z-\beta}{z-\gamma}x\,dy+\frac{xy\,(\gamma-\beta)}{(z-\beta)\,(z-\gamma)}dz\right\}$$

$$+\,\Omega$$

$$=(\gamma-\beta)\left\{2-\frac{L}{(z-\beta)\,(z-\gamma)}\right\}\times\{xydz^2-y(z-\gamma)\,dzdx-x\,(z-\beta)\,dzdy+(z-\beta)\,(z-\gamma)\,dxdy\}.$$

Hence in virtue of the equations $U=0$, $dU=0$ the equation $\Omega=0$ becomes

$$xy\,dz^2 - y(z-\gamma)\,dz\,dx - x(z-\beta)\,dz\,dy + (z-\beta)(z-\gamma)\,dx\,dy = 0,$$

that is,

$$\{x\,dz - (z-\gamma)\,dx\}\,\{y\,dz - (z-\beta)\,dy\} = 0,$$

whence either $x - C(z-\gamma) = 0$ or $y - C'(z-\beta) = 0$; viz. the section of the surface by a plane of either series (which section is a circle) is a curve of curvature of the surface.

The equation of the cyclide can be elegantly expressed in terms of the ellipsoidal coordinates (λ, μ, ν) of a point (x, y, z); viz. writing for shortness $\alpha = b^2 - c^2$, $\beta = c^2 - a^2$, $\gamma = a^2 - b^2$, the coordinates (λ, μ, ν) are such that

$$\begin{aligned} -\beta\gamma x^2 &= (a^2+\lambda)(a^2+\mu)(a^2+\nu), \\ -\gamma\alpha y^2 &= (b^2+\lambda)(b^2+\mu)(b^2+\nu), \\ -\alpha\beta z^2 &= (c^2+\lambda)(c^2+\mu)(c^2+\nu), \end{aligned}$$

(see Roberts, *Comptes Rendus*, t. LIII. (Dec., 1861), p. 1119), whence

$$x^2+y^2+z^2 = a^2+b^2+c^2+\lambda+\mu+\nu,$$

$$(b^2+c^2)\,x^2 + (c^2+a^2)\,y^2 + (a^2+b^2)\,z^2 = b^2c^2 + c^2a^2 + a^2b^2 - \mu\nu - \nu\lambda - \lambda\mu.$$

The equation of the cyclide then is

$$\sqrt{(a^2+\lambda)} + \sqrt{(a^2+\mu)} + \sqrt{(a^2+\nu)} = \sqrt{(\delta)}.$$

In fact, starting from this equation and rationalising, we have

$$\begin{aligned} (3a^2+\lambda+\mu+\nu-\delta)^2 &= 4\,[\sqrt{\{(a^2+\mu)(a^2+\nu)\}} + \sqrt{\{(a^2+\nu)(a^2+\lambda)\}} + \sqrt{\{(a^2+\lambda)(a^2+\mu)\}}]^2 \\ &= 4\,[3a^4 + 2a^2(\lambda+\mu+\nu) + \mu\nu + \nu\lambda + \lambda\mu + 2\sqrt{\{(a^2+\lambda)(a^2+\mu)(a^2+\nu)\}}\sqrt{(\delta)}], \end{aligned}$$

which, substituting for

$$\lambda+\mu+\nu,\ \ \mu\nu+\nu\lambda+\lambda\mu \ \text{ and } \ \sqrt{\{(a^2+\lambda)(a^2+\mu)(a^2+\nu)\}}$$

their values, is

$$(x^2+y^2+z^2+\gamma-\beta-\delta)^2 = 4\,\{(\gamma-\beta)\,x^2 - \beta y^2 + \gamma z^2 - \beta\gamma - 2x\sqrt{(-\beta\gamma\delta)}\},$$

or, writing $-\tfrac{1}{4}\gamma^2$, $\tfrac{1}{4}\beta^2$, $\tfrac{1}{4}\delta^2$ in place of β, γ, δ respectively, this is

$$(x^2+y^2+z^2+\tfrac{1}{4}\gamma^2+\tfrac{1}{4}\beta^2-\tfrac{1}{4}\delta^2)^2 = (\gamma^2+\beta^2)\,x^2 + \beta^2z^2 + \gamma^2y^2 + \tfrac{1}{4}\beta^2\gamma^2 + \beta\gamma\delta x,$$

which agrees with a foregoing form of the equation.

The generating spheres of the cyclide cut at right angles each of a series of spheres; viz. each of these spheres passes through one and the same circle in the plane of, and having double contact with, the conic which contains the centres of the generating spheres; the centres of the orthotomic spheres being consequently in a line meeting an axis, and at right angles to the plane of the conic in question. Or, what is the same thing, starting with a conic, and a sphere having double contact therewith, the cyclide is the envelope of a variable sphere having its centre on the conic and cutting at right angles the fixed sphere.*

* I am indebted for this mode of generation of a Cyclide to the researches of Mr Casey.

It may be remarked, that if we endeavour to generalize a former generation of the cyclide, and consider the envelope of a variable sphere having its centre on a conic, and *touching* a fixed sphere, this is in general a surface of an order exceeding 4; it becomes a surface of the fourth order, viz. a cyclide, *only* in the case where the fixed sphere has its centre on the anti-conic. But if we consider the envelope of a variable sphere having its centre on a conic and cutting at right angles a fixed sphere, this is always a quartic surface having the circle at infinity for a double line; the surface has moreover two nodes, viz. these are the anti-points of the circle which is the intersection of the sphere by the plane of the conic. If the sphere touches the conic, then there is at the point of contact a third node; and similarly, if it has double contact with the conic, then there is at each point of contact a node; viz. in this case the surface has four nodes, and it is in fact a cyclide.

There is no difficulty in the analytical proof: consider the envelope of a variable sphere having its centre on the conic $Z=0$, $-\frac{X^2}{\beta}+\frac{Y^2}{\alpha}=1$, and which cuts at right angles the sphere $(x-l)^2+(y-m)^2+(z-n)^2=k^2$.

Take the equation of the variable sphere to be

$$(x-X)^2+(y-Y)^2+z^2=c^2,$$

then the orthotomic condition is

$$(X-l)^2+(Y-m)^2+n^2=c^2+k^2,$$

or, substituting this value of c^2, the equation of the variable sphere is

$$(x-X)^2+(y-Y)^2+z^2=-k^2+(X-l)^2+(Y-m)^2+n^2,$$

all which spheres pass through the points

$$x=l, \quad y=m, \quad z=\pm\sqrt{(n^2-k^2)};$$

that is,

$$x^2+y^2+z^2+k^2-l^2-m^2-n^2-2(x-l)X-2(y-m)Y=0,$$

and considering Y, Y as variable parameters connected by the equation $-\frac{X^2}{\beta}+\frac{Y^2}{\alpha}=1$, the equation of the envelope is

$$(x^2+y^2+z^2+k^2-l^2-m^2-n^2)^2+4\beta(x-l)^2-4\alpha(y-m)^2=0,$$

viz. this is a bicircular quartic, having the two nodes $x=l$, $y=m$, $z=\pm\sqrt{(n^2-k^2)}$; these are the anti-points of the circle $(x-l)^2+(y-m)^2=k^2-n^2$, which is the intersection of the sphere $(x-l)^2+(y-m)^2+(z-n)^2=k^2$ by the plane of the conic.

The constants might be particularised so that the equation should represent a cyclide; but I treat the question in a somewhat different manner, by showing that the generating spheres of a cyclide cut at right angles each of a series of fixed spheres. Write $\alpha, \beta, \gamma=b^2-c^2, c^2-a^2, a^2-b^2$; then if

$$\frac{X^2}{-\beta}+\frac{Y^2}{\alpha}=1; \quad \frac{X_1^2}{\gamma}-\frac{Z_1^2}{\alpha}=1,$$

the points $(X, Y, 0)$ and $(X_1, 0, Z_1)$ will be situate on a pair of anti-conics.

Consider the fixed sphere

$$(x-X_1)^2+y^2+(z-Z_1)^2=c_1^2,$$

then if this is touched by the variable sphere

$$(x-X)^2+(y-Y)^2+z^2=c^2,$$

the last-mentioned sphere will be a generating sphere of the cyclide. The condition of contact is

$$(X-X_1)^2+Y^2+Z_1^2=(c+c_1)^2,$$

that is,

$$\begin{aligned}(c+c_1)^2 &= X^2-2XX_1+X_1^2+\alpha\left(1+\frac{X^2}{\beta}\right)+\alpha\left(\frac{X_1^2}{\gamma}-1\right)\\ &= -\frac{\gamma}{\beta}X^2-2XX_1-\frac{\beta}{\gamma}X_1^2\\ &= \Omega^2,\end{aligned}$$

if for a moment

$$\Omega=X\sqrt{\left(\frac{-\gamma}{\beta}\right)}+X_1\sqrt{\left(\frac{-\beta}{\gamma}\right)},$$

that is, $c=-c_1+\Omega$, and the equation of the variable sphere is

$$(x-X)^2+(y-Y)^2+z^2=(c_1-\Omega)^2,$$

where X, Y are variable parameters connected by

$$\frac{X^2}{-\beta}+\frac{Y^2}{\alpha}=1.$$

Suppose that the variable sphere is orthotomic to

$$(x-X_2)^2+y^2+(z-Z_2)^2=c_2^2,$$

the condition for this is

$$(X-X_2)^2+Y^2+Z_2^2=c^2+c_2^2,$$

or combining with the identical equation

$$(X-X_1)^2+Y^2+Z_1^2=(c+c_1)^2,$$

we have

$$\begin{aligned}-2X(X_2-X_1)+X_2^2-X_1^2+Z_2^2-Z_1^2 &= c_2^2-c_1^2-2c_1(-c_1+\Omega)\\ &= c_2^2+c_1^2-2c_1\Omega,\end{aligned}$$

or, substituting for Z_1, Ω their values, this is

$$-2X(X_2-X_1)+X_2^2-X_1^2+Z_2^2-\alpha\left(\frac{X_1^2}{\gamma}-1\right)=c_2^2+c_1^2-2c_1\left\{X\sqrt{\left(-\frac{\gamma}{\beta}\right)}+X_1\sqrt{\left(-\frac{\beta}{\gamma}\right)}\right\},$$

viz. this will be identically true if

$$X_2 = X_1 + c_1\sqrt{\left(-\frac{\gamma}{\beta}\right)},$$

$$X_2^2 + Z_2^2 - c_2^2 = -\frac{\beta}{\gamma}X_1^2 + 2c_1X_1\sqrt{\left(-\frac{\beta}{\gamma}\right)} - \alpha + c_1^2,$$

or, as this last equation may be written

$$Z_2^2 - c_2^2 = Z_1^2 - c_1^2\frac{\alpha}{\beta} + 2c_1X_1\left\{\sqrt{\left(-\frac{\beta}{\gamma}\right)} + \sqrt{\left(-\frac{\gamma}{\beta}\right)}\right\}.$$

The equation of the orthotomic sphere is thus found to be

$$\left\{x - X_1 - c_1\sqrt{\left(-\frac{\gamma}{\beta}\right)}\right\}^2 + y^2 + z^2 - 2zZ_2 + Z_1^2 - c_1^2\frac{\alpha}{\beta} + 2c_1X_1\left\{\sqrt{\left(-\frac{\beta}{\gamma}\right)} + \sqrt{\left(-\frac{\gamma}{\beta}\right)}\right\} = 0;$$

or, what is the same thing,

$$x^2 + y^2 + z^2 - 2zZ_2 - 2x\left\{X_1 + c_1\sqrt{\left(-\frac{\gamma}{\beta}\right)}\right\} - X_1^2\frac{\beta}{\gamma} - 2c_1X_1\sqrt{\left(-\frac{\beta}{\gamma}\right)} + c_1^2 - \alpha = 0,$$

or, as this may be written

$$\alpha\left(-\frac{x^2}{\beta} + \frac{y^2}{\alpha} - 1\right) + z^2 - 2zZ_2 - \frac{\gamma}{\beta}x^2 - 2xX_1 - \frac{\beta}{\gamma}X_1^2 - 2c_1x\sqrt{\left(-\frac{\beta}{\gamma}\right)} - 2c_1X_1\sqrt{\left(-\frac{\beta}{\gamma}\right)} + c_1^2 = 0,$$

viz. this is

$$\alpha\left(-\frac{x^2}{\beta} + \frac{y^2}{\alpha} - 1\right) + z^2 - 2zZ_2 + \left\{x\sqrt{\left(-\frac{\gamma}{\beta}\right)} + X_1\sqrt{\left(-\frac{\beta}{\gamma}\right)} - c_1\right\}^2 = 0,$$

where Z_2 is arbitrary. We have thus a series of orthotomic spheres; viz. taking any one of these, the envelope of a variable sphere having its centre on the conic $-\frac{x^2}{\beta} + \frac{y^2}{\alpha} - 1 = 0$, and cutting at right angles the orthotomic sphere, is a cyclide. The centre of the orthotomic sphere is a point at pleasure on the line

$$x = X_1 + c_1\sqrt{\left(-\frac{\gamma}{\beta}\right)},\ y = 0;$$

and the sphere passes through the circle $z = 0$,

$$\left\{x - X_1 - c_1\sqrt{\left(-\frac{\gamma}{\beta}\right)}\right\}^2 + y^2 + Z_1^2 - c_1^2\frac{\alpha}{\beta} + 2c_1X_1\left\{\sqrt{\left(-\frac{\beta}{\gamma}\right)} + \sqrt{\left(-\frac{\gamma}{\beta}\right)}\right\} = 0,$$

viz. this is a circle having double contact with the conic $-\frac{x^2}{\beta} + \frac{y^2}{\alpha} = 1$; or, what is the same thing, the orthotomic sphere is a sphere having its centre on the line in question, and having double contact with the conic $-\frac{x^2}{\beta} + \frac{y^2}{\alpha} = 1$.

570.

ON THE SUPERLINES OF A QUADRIC SURFACE IN FIVE-DIMENSIONAL SPACE.

[From the *Quarterly Journal of Pure and Applied Mathematics*, vol. XII. (1873), pp. 176—180.]

IN ordinary or three-dimensional space a quadric surface has upon it two singly infinite systems of lines, such that each line of the one system intersects each line of the other system, but that two lines of the same system do not intersect.

In five-dimensional space* a quadric surface has upon it two triply infinite systems of superlines, such that each superline of either system intersects each superline of the same system; a superline of the one system does not in general intersect a superline of the opposite system, but it may do so, and then it intersects it not in a mere point, but in a line.

The theory will be established by an independent analysis, but it is, in fact, a consequence of the correspondence which exists between the lines of ordinary space and the points of a quadric surface in five-dimensional space. Thus the correspondence is

In ordinary space.	In five-dimensional space.
Line.	Point on quadric surface.
Lines meeting a given line.	Points which lie in tangent plane at given point.
Pair of intersecting lines.	Two points such that each lies in the tangent plane at the other, or say, pair of harmonic points.
Lines meeting each of two given lines.	Points lying in the sub-plane common to the tangent planes at two given points.

* In explanation of the nomenclature, observe that in 5 dimensional geometry we have: space, surface, subsurface, supercurve, curve, and point-system, according as we have between the six coordinates 0, 1, 2, 3, 4, or 5 equations: and so when the equations are linear, we have: space, plane, subplane, superline, line, and point. Thus in the text a quadric surface is the locus determined by a single quadric equation between the coordinates; and the superline and line are the loci determined by three linear equations and four linear equations respectively.

But in ordinary space if the two given lines intersect, then the system of lines meeting these, breaks up into two systems, viz. that of the lines which pass through the point of intersection, and that of the lines which lie in the common plane of the two given lines. It follows that in the five-dimensional space the intersection of the quadric surface by the subplane common to the tangent planes at two harmonic points must break up into a pair of superlines, viz. that we have on the quadric two systems of superlines; a superline of the one kind answering in ordinary space to the lines which pass through a given point, and a superline of the other kind answering to the lines which lie in a given plane. (Observe that, as regards the five-dimensional geometry, this is no distinction of nature between the two kinds of superlines, they are simply correlative to each other, like the two systems of generating lines of a quadric in ordinary space.)

Moreover, considering two superlines of the first kind, then answering thereto in ordinary space we have the lines through one given point, and the lines through another given point; and these systems have a common line, that joining the two given points; whence the two superlines have a common point. And, similarly, two superlines of the second kind have a common point. But taking two superlines of opposite kinds, then in ordinary space we have the lines through a given point, and the lines in a given plane: and the two systems have not in general any common line; that is, the two superlines have no common point. If, however, the given point lies in the given plane, then there is not one common line, but a singly infinite series of common lines, viz. all the lines in the given plane and through the given point; and corresponding hereto we have as the intersection of the two superlines, not a mere point, but a line.

Passing now to the independent theory, I consider, for comparison, first the case of the lines on a quadric surface in ordinary space; the equation of the surface may be taken to be

$$u^2 + v^2 - x^2 - y^2 = 0,$$

(u, v, x, y ordinary quadriplanar coordinates) and the equations of a line on the surface are

$$u = \alpha x + \beta y,$$
$$v = \alpha' x + \beta' y,$$

where α, β, α', β' are coefficients of a rectangular transformation, viz. we have $\alpha^2 + \beta^2 = 1$, $\alpha'^2 + \beta'^2 = 1$, $\alpha\alpha' + \beta\beta' = 0$; and therefore $(\alpha\beta' - \alpha'\beta)^2 = 1$, consequently $\alpha\beta' - \alpha'\beta = \pm 1$; and the lines will be of one or the other kind, according as the sign is $+$ or $-$. It is rather more convenient to assume always $\alpha\beta' - \alpha'\beta = +1$, and write the equations

$$u = \alpha x + \beta y,$$
$$v = k(\alpha' x + \beta' y),$$

k denoting ± 1, and the lines being of the one kind or of the other kind, according as the sign is $+$ or $-$.

Thus considering any two lines, the equations may be written

$$u = \alpha x + \beta y, \quad u = - \quad (ax + by),$$
$$v = \alpha' x + \beta' y, \quad v = -k(a'x + b'y),$$

where the lines will be of the same kind or of different kinds, according as k is $= +1$ or $= -1$. Observe that k is introduced into one equation only; if it had been introduced into both, there would be no change of kind. If the lines intersect we have

$$(\alpha + \ a)x + (\beta + \ b)y = 0,$$
$$(\alpha' + ka')x + (\beta' + kb')y = 0,$$

viz. the condition of intersection is

$$\begin{vmatrix} \alpha + \ a, & \beta + \ b \\ \alpha' + ka', & \beta' + kb' \end{vmatrix} = 0,$$

that is,

$$\alpha\beta' - \alpha'\beta + k(ab' - \alpha'b) + a\beta' - a'\beta + k(ab' - a'b) = 0,$$

or, what is the same thing,

$$1 + \alpha\beta' - a'\beta + k(1 + ab' - \alpha'b) = 0.$$

But we have, say

$$\alpha = \ \cos\theta, \quad \beta = \sin\theta, \quad a = \ \cos\phi, \quad b = \sin\phi,$$
$$\alpha' = -\sin\theta, \quad \beta' = \cos\theta, \quad a' = -\sin\phi, \quad b' = \cos\phi,$$

and thence

$$a\beta' - a'\beta = \cos(\theta - \phi) = \alpha b' - \alpha' b,$$

and the equation is

$$(1 + k)\{1 + \cos(\theta - \phi)\} = 0,$$

viz. this is satisfied if $k = -1$, i.e. if the lines are of opposite kinds, but not if $k = +1$. And it is important to remark that there is no exception corresponding to the other factor, viz. if $k = +1$, and $1 + \cos(\theta - \phi) = 0$, for we then have $\theta - \phi = \pi$, $\cos\phi = -\cos\theta$, $\sin\phi = -\sin\theta$, and consequently the two sets of equations for u, v become identical; that is, for lines of the same kind a line meets itself only.

Passing to the five-dimensional space, the equation of the quadric surface may be taken to be

$$u^2 + v^2 + w^2 - x^2 - y^2 - z^2 = 0,$$

and for a superline on the surface we have

$$u = \alpha x \ + \beta y \ + \gamma z,$$
$$v = \alpha' x \ + \beta' y \ + \gamma' z,$$
$$w = \alpha'' x + \beta'' y + \gamma'' z,$$

where (α, β, γ), &c., are the coefficients of a rectangular transformation; the determinant formed with these coefficients is $= \pm 1$, and the superline is of the one kind or the other, according as the sign is + or −. It is more convenient to take the determinant to be always +, and to write the equations in the form

$$\begin{aligned} u &= k(\alpha x + \beta y + \gamma z),\\ v &= k(\alpha' x + \beta' y + \gamma' z),\\ w &= k(\alpha'' x + \beta'' y + \gamma'' z),\end{aligned}$$

where $k = \pm 1$, and the superline is of the one or the other kind, according as the sign is + or −.

Now considering two superlines, we may write

$$\begin{aligned} u &= \alpha x + \beta y + \gamma z, & u &= -k(ax + by + cz),\\ v &= \alpha' x + \beta' y + \gamma' z, & v &= -k(a'x + b'y + c'z),\\ w &= \alpha'' x + \beta'' y + \gamma'' z, & w &= -k(a''x + b''y + c''z).\end{aligned}$$

If the superlines intersect, then

$$\begin{aligned} (\alpha + ka)\,x + (\beta + kb)\,y + (\gamma + kc)\,z &= 0,\\ (\alpha' + ka')\,x + (\beta' + kb')\,y + (\gamma' + kc')\,z &= 0,\\ (\alpha'' + ka'')\,x + (\beta'' + kb'')\,y + (\gamma'' + kc'')\,z &= 0,\end{aligned}$$

viz. the determinant formed with these coefficients must be $=0$. The condition is at once reduced to

$$1 + k^3 + (k + k^2)(a\alpha + b\beta + c\gamma + a'\alpha' + b'\beta' + c'\gamma' + a''\alpha'' + b''\beta'' + c''\gamma'') = 0,$$

viz. it is satisfied when $k = -1$, that is, when the superlines are of the *same* kind; but not in general when $k = +1$.

If $k = +1$ the condition will be satisfied if

$$1 + a\alpha + b\beta + c\gamma + a'\alpha' + b'\beta' + c'\gamma' + a''\alpha'' + b''\beta'' + c''\gamma'' = 0,$$

and it is to be shown that then the three equations reduce themselves not to two equations, but to a single equation.

It is allowable to take the second set of equations to be simply $u = -kx$, $v = -ky$, $w = -kz$; for this comes to replacing the analytically rectangular system $ax + by + cz$, $a'x + b'y + c'z$, $a''x + b''y + c''z$ by x, y, z. Writing also $k = +1$, the theorem to be proved is that the equations

$$\begin{aligned} (\alpha + 1)\,x + \beta y + \gamma z &= 0,\\ \alpha' x + (\beta' + 1)\,y + \gamma' z &= 0,\\ \alpha'' x + \beta'' y + \gamma'' z &= 0,\end{aligned}$$

reduce themselves to a single equation, provided only $1 + \alpha + \beta' + \gamma'' = 0$; or, what is the same thing, we have to prove that the expressions $\beta'' - \gamma'$, $\gamma - \alpha''$, $\alpha' - \beta$ each vanish, provided only $1 + \alpha + \beta' + \gamma'' = 0$. This is a known theorem depending on the theory of the resultant axis, viz. the rotation round the resultant axis is then 180°, and we have $OX = OX'$, $OY = OY'$, $OZ = OZ'$, and thence we have evidently $YZ' = Y'Z$, $ZX' = Z'X$, $XY' = X'Y$.

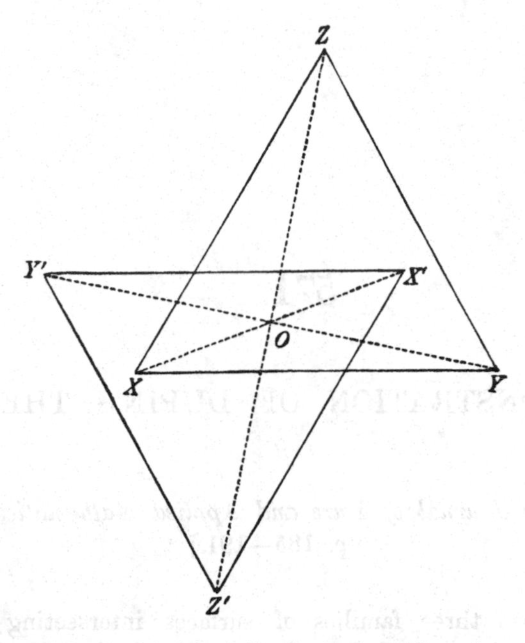

But to prove it analytically, writing P, Q, R for $\beta'' - \gamma'$, $\gamma - \alpha''$, $\alpha' - \beta$ respectively, and Ω for $1 + \alpha + \beta' + \gamma''$, observe that we have identically

$$\begin{aligned}
(\beta'' + \gamma\ )\,\Omega &= QR,\\
(\gamma\ + \alpha'\ )\,\Omega &= RP,\\
(\alpha'\ + \beta'')\,\Omega &= PQ,\\
(\beta'' + \gamma')\,P = (\gamma + \alpha'')\,Q &= (\alpha' + \beta)\,R,\\
(\alpha\ - 1)\,\Omega &= -\ \ \gamma Q\ \ +\ \ \beta R\ ,\\
\alpha'\Omega\ &= \ \ \gamma' Q\ + (1 + \beta')\,R,\\
\alpha''\Omega\ &= -(1 + \gamma'')\,Q + \ \beta'' R\ ,\\
\beta\Omega\ &= -(1 + \alpha\ )\,R + \ \gamma P\ ,\\
(\beta' - 1)\,\Omega &= -\ \ \alpha' R\ \ +\ \ \gamma' P\ ,\\
\beta''\Omega\ &= -\ \ \alpha'' R\ + (1 + \gamma'')\,P,\\
\gamma\Omega\ &= -\ \ \beta P\ + (1 + \alpha\ )\,Q,\\
\gamma'\Omega\ &= -(1 + \beta')\,P + \ \alpha' Q\ ,\\
(\gamma'' - 1)\,\Omega &= -\ \ \beta'' P\ \ +\ \ \alpha'' Q\ ,
\end{aligned}$$

whence Ω being $= 0$, we have also $P = 0$, $Q = 0$, $R = 0$. The final conclusion is that the two superlines of opposite kinds, when they intersect, intersect in a line.

571.

A DEMONSTRATION OF DUPIN'S THEOREM.

[From the *Quarterly Journal of Pure and Applied Mathematics*, vol. XII. (1873), pp. 185—191.]

THE theorem is that three families of surfaces intersecting everywhere at right angles intersect along their curves of curvature. The following demonstration puts in evidence the geometrical ground of the theorem.

I remark that it was suggested to me by the perusal of a most interesting paper by M. Lévy, "Mémoire sur les coordonnées curvilignes orthogonales et en particulier sur celles qui comprennent une famille quelconque de surfaces de second degré," (*Jour. de l'École Polyt.*, Cah. 43 (1870), pp. 157—200). It was known that a family of surfaces $\rho = f(x, y, z)$ where the function is arbitrary, does not in general form part of an orthogonal system, but that ρ considered as a function of (x, y, z) must satisfy a partial differential equation of the third order. M. Lévy obtains a theorem which, in fact, enables the determination of this partial differential equation; he does not himself obtain it, although he finds what the equation becomes on writing therein $\frac{d\rho}{dx} = 0$, $\frac{d\rho}{dy} = 0$; but I have, in a recent communication to the French Academy, found this equation.

Proceeding to the consideration of Dupin's theorem, on a surface of the first family take a point A and through it two elements of length on the surface, AB, AC, at right angles to each other; draw at A, B, C the normals meeting the consecutive surface in A', B', C' and join $A'B'$, $A'C'$. It is to be shown that the condition in order that $B'A'C'$ may be a right angle is the same as the condition for the intersection of the normals AA' and BB' (or of the normals AA' and CC'); for this being so, since by hypothesis $B'A'C'$ is a right angle, it follows that AA', BB' intersect;

that is, that AB is an element of one of the curves of curvature through the point A of the surface. And, similarly, that AA', CC' intersect; that is, that AC is an element of the other of the curves of curvature through the point A on the surface.

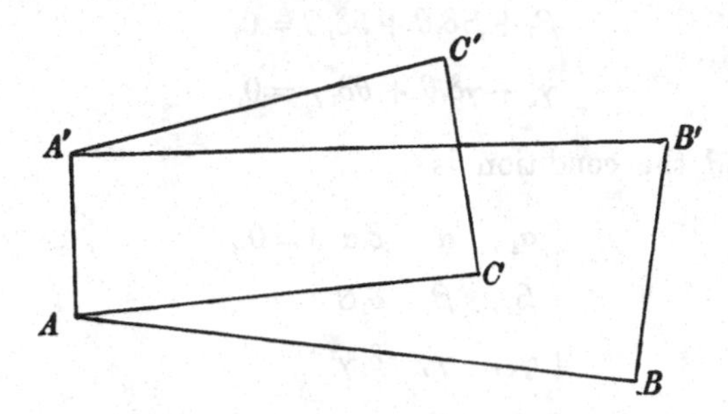

Take x, y, z for the coordinates of the point A; α, β, γ for the cosine inclinations of AA'; α_1, β_1, γ_1 for those of AB; and α_2, β_2, γ_2 for those of AC. Write also

$$\delta = \alpha\, d_x + \beta\, d_y + \gamma\, d_z,$$
$$\delta_1 = \alpha_1 d_x + \beta_1 d_y + \gamma_1 d_z,$$
$$\delta_2 = \alpha_2 d_x + \beta_2 d_y + \gamma_2 d_z;$$

then it will be shown that the condition for the intersection of the normals AA', BB' is

$$\alpha_2\delta_1\alpha + \beta_2\delta_1\beta + \gamma_2\delta_1\gamma = 0,$$

the condition for the intersection of the normals AA', CC' is

$$\alpha_1\delta_2\alpha + \beta_1\delta_2\beta + \gamma_1\delta_2\gamma = 0,$$

and that these are equivalent to each other, and to the condition for the angle $B'A'C'$ being a right angle.

Taking l, l_1, l_2 for the lengths AA', AB, AC, the coordinates of A', B, C measured from the point A are

$$(l\alpha,\ l\beta,\ l\gamma),\quad (l_1\alpha_1,\ l_1\beta_1,\ l_1\gamma_1),\quad (l_2\alpha_2,\ l_2\beta_2,\ l_2\gamma_2) \text{ respectively.}$$

The equations of the normal at A may be written

$$X = x + \theta\alpha,$$
$$Y = y + \theta\beta,$$
$$Z = z + \theta\gamma,$$

where X, Y, Z are current coordinates, and θ is a variable parameter. Hence for the normal at B, passing from the coordinates x, y, z to $x + l_1\alpha_1$, $y + l_1\beta_1$, $z + l_1\gamma_1$, the equations are

$$X = x + l_1\alpha_1 + l_1\delta_1(\theta\alpha),$$
$$Y = y + l_1\beta_1 + l_1\delta_1(\theta\beta),$$
$$Z = z + l_1\gamma_1 + l_1\delta_1(\theta\gamma),$$

and if the two normals intersect in the point (X, Y, Z), then

$$\alpha_1 + \alpha\delta_1\theta + \theta\delta_1\alpha = 0,$$
$$\beta_1 + \beta\delta_1\theta + \theta\delta_1\beta = 0,$$
$$\gamma_1 + \gamma\delta_1\theta + \theta\delta_1\gamma = 0,$$

viz. eliminating θ and $\delta_1\theta$ the condition is

$$\begin{vmatrix} \alpha_1, & \alpha, & \delta_1\alpha \\ \beta_1, & \beta, & \delta_1\beta \\ \gamma_1, & \gamma, & \delta_1\gamma \end{vmatrix} = 0;$$

or, since

$$\alpha_2, \beta_2, \gamma_2 = \beta\gamma_1 - \beta_1\gamma, \ \gamma\alpha_1 - \gamma_1\alpha, \ \alpha\beta_1 - \alpha_1\beta,$$

this is

$$\alpha_2\delta_1\alpha + \beta_2\delta_1\beta + \gamma_2\delta_1\gamma = 0.$$

Similarly the condition for the intersection of the normals AA', CC' is

$$\alpha_1\delta_2\alpha + \beta_1\delta_2\beta + \gamma_1\delta_2\gamma = 0.$$

We have

$$\alpha_2\delta_1\alpha + \beta_2\delta_1\beta + \gamma_2\delta_1\gamma = \alpha_1\delta_2\alpha + \beta_1\delta_2\beta + \gamma_1\delta_2\gamma;$$

in fact, this equation is

$$(\alpha_2\delta_1 - \alpha_1\delta_2)\,\alpha + (\beta_2\delta_1 - \beta_1\delta_2)\,\beta + (\gamma_2\delta_1 - \gamma_1\delta_2)\,\gamma = 0,$$

which I proceed to verify.

In the first term the symbol $\alpha_2\delta_1 - \alpha_1\delta_2$ is

$$\alpha_2(\alpha_1 d_x + \beta_1 d_y + \gamma_1 d_z) - \alpha_1(\alpha_2 d_x + \beta_2 d_y + \gamma_2 d_z),$$

viz. this is

$$(\alpha_2\beta_1 - \alpha_1\beta_2)\,d_y + (\gamma_1\alpha_2 - \gamma_2\alpha_1)\,d_z;$$

or, what is the same thing, it is

$$\beta d_z - \gamma d_y,$$

and the equation to be verified is

$$(\beta d_z - \gamma d_y)\,\alpha + (\gamma d_x - \alpha d_z)\,\beta + (\alpha d_y - \beta d_x)\,\gamma = 0,$$

viz. writing

$$\alpha, \beta, \gamma = \frac{X}{R}, \frac{Y}{R}, \frac{Z}{R},$$

where if $\rho = f(x, y, z)$ is the equation of the surface X, Y, Z are the derived functions $\frac{df}{dx}, \frac{df}{dy}, \frac{df}{dz}$, and $R = \sqrt{(X^2 + Y^2 + R^2)}$, the function on the left-hand consists of two parts; the first is

$$\frac{1}{R}\{(\beta d_z - \gamma d_y)\,X + (\gamma d_x - \alpha d_z)\,Y + (\alpha d_y - \beta d_x)\,Z\},$$

that is,

$$\frac{1}{R}\{\alpha(d_yZ-d_zY)+\beta(d_zX-d_xZ)+\gamma(d_xY-d_yX)\},$$

which vanishes; and the second is

$$-\frac{1}{R}\{\alpha(\beta d_z-\gamma d_y)+\beta(\gamma d_x-\alpha d_z)+\gamma(\alpha d_y-\beta d_x)\}\,R,$$

which also vanishes; that is, we have identically

$$\alpha_2\delta_1\alpha+\beta_2\delta_1\beta+\gamma_2\delta_1\gamma=\alpha_1\delta_2\alpha+\beta_1\delta_2\beta+\gamma_1\delta_2\gamma,$$

and the vanishing of the one function implies the vanishing of the other.

Proceeding now to the condition that the angle $B'A'C'$ shall be a right angle, the coordinates of B' are what those of A' become on substituting therein $x+l_1\alpha_1$, $y+l_1\beta_1$, $z+l_1\gamma_1$ in place of x, y, z; that is, these coordinates are

$$x+l\alpha+l_1\alpha_1+l_1\delta_1(l\alpha),\ \&\text{c.},$$

or, what is the same thing, measuring them from A' as origin, the coordinates of B' are

$$l_1(\alpha_1+l\delta_1\alpha+\alpha\,\delta_1 l),$$
$$l_1(\beta_1+l\delta_1\beta+\beta\delta_1 l),$$
$$l_1(\gamma_1+l\delta_1\gamma+\gamma\,\delta_1 l),$$

and similarly those of C' measured from the same origin A' are

$$l_2(\alpha_2+l\delta_2\alpha+\alpha\,\delta_2 l),$$
$$l_2(\beta_2+l\delta_2\beta+\beta\delta_2 l),$$
$$l_2(\gamma_2+l\delta_2\gamma+\gamma\,\delta_2 l).$$

Hence the condition for the right angle is

$$\begin{aligned}&(\alpha_1+l\delta_1\alpha+\alpha\delta_1 l)(\alpha_2+l\delta_2\alpha+\alpha\delta_2 l)\\&+(\beta_1+l\delta_1\beta+\beta\delta_1 l)(\beta_2+l\delta_2\beta+\beta\delta_2 l)\\&+(\gamma_1+l\delta_1\gamma+\gamma\delta_1 l)(\gamma_2+l\delta_2\gamma+\gamma\delta_2 l)=0.\end{aligned}$$

Here the terms independent of l, $\delta_1 l$, $\delta_2 l$ vanish; and writing down only the terms which are of the first order in these quantities, the condition is

$$\begin{aligned}&\alpha_1(l\delta_2\alpha+\alpha\delta_2 l)+\alpha_2(l\delta_1\alpha+\alpha\delta_1 l)\\&+\beta_1(l\delta_2\beta+\beta\delta_2 l)+\beta_2(l\delta_1\beta+\beta\delta_1 l)\\&+\gamma_1(l\delta_2\gamma+\gamma\delta_2 l)+\gamma_2(l\delta_1\gamma+\gamma\delta_1 l)=0,\end{aligned}$$

where the terms in $\delta_1 l$, $\delta_2 l$ vanish; the remaining terms divide by l, and throwing out this factor, the condition is

$$(\alpha_1\delta_2\alpha+\beta_1\delta_2\beta+\gamma_1\delta_2\gamma)+(\alpha_2\delta_1\alpha+\beta_2\delta_1\beta+\gamma_2\delta_1\gamma)=0,$$

viz. by what precedes, this may be written under either of the forms

$$\alpha_1\delta_2\alpha + \beta_1\delta_2\beta + \gamma_1\delta_2\gamma = 0,$$

$$\alpha_2\delta_1\alpha + \beta_2\delta_1\beta + \gamma_2\delta_1\gamma = 0,$$

and the theorem is thus proved.

It may be remarked that if we had simply the first surface, and two other surfaces, or say a second and a third surface, cutting the first surface and each other at right angles, that is, cutting each other in AA' the element of the normal at A, and cutting the first surface in the elements AB, AC at right angles to each other, then the tangent plane of the second surface will be the plane $A'AB$, not in general passing through B'; and the tangent plane of the third surface will be the plane $A'AC$, not in general passing through C'. The condition, that the elements $A'B'$ and $A'C'$ on the surface consecutive to the first surface are at right angles, makes CC' and BB' each intersect AA'; and we then have, the tangent plane of the second surface is the plane through the elements AA', BB', the tangent plane of the second surface is the plane through the elements AA', CC'.

As already remarked, a family of surfaces $\rho = f(x, y, z)$ where the function is arbitrary cannot form part of an orthogonal system. In fact, if the surfaces do belong to an orthogonal system, we have AA', BB' in the same plane, and consequently AB and $A'B'$ intersect; and, similarly, AC and $A'C'$ intersect; that is, if from a point A on a given surface of the family we pass along the normal to the point A' on the consecutive surface; and if the lines AB, AC are the tangents to the curves of curvature at A, and $A'B'$, $A'C'$ the tangents to the curves of curvature at A', then AB intersects $A'B'$, or, what is the same thing, AC intersects $A'C'$; and, conversely, when this condition is satisfied in general (that is, for every surface of the family and the surface consecutive thereto), then the family forms part of an orthogonal system; this is, in fact, the fundamental theorem of M. Lévy's memoir. The analytical form of the condition, viewed in this manner, is

$$\alpha_2\delta\alpha_1 + \beta_2\delta\beta_1 + \gamma_2\delta\gamma_1 = 0, \quad \text{or} \quad \alpha_1\delta\alpha_2 + \beta_1\delta\beta_2 + \gamma_1\delta\gamma_2 = 0;$$

or, as it is convenient to write it,

$$\alpha_2\delta\alpha_1 + \beta_2\delta\beta_1 + \gamma_2\delta\gamma_1 - (\alpha_1\delta\alpha_2 + \beta_1\delta\beta_2 + \gamma_1\delta\gamma_2) = 0;$$

and it was by means of it that I obtained the partial differential equation of the third order above referred to. The condition written in the form

$$X_2\delta X_1 + Y_2\delta Y_1 + Z_2\delta Z_1 = 0, \quad \text{or} \quad X_1\delta X_2 + Y_1\delta Y_2 + Z_1\delta Z_2 = 0,$$

presents itself in the proof of Dupin's Theorem by R. L. Ellis, (given in Gregory's *Examples*, Cambridge, 1841), but the geometrical signification of it is not explained.

Closely connected with Dupin's, we have the following theorem: if two surfaces intersect at right angles along a curve which is a curve of curvature of one of them, it is a curve of curvature of the other of them. I remark hereon as follows:

Let the intersection be a curve of curvature on the first surface; the successive normals intersect, giving rise to a developable, and the intersection of the two surfaces, say I, is an involute of the edge of regression of this developable, say of the curve C. The successive normals of the second surface are the lines at the different points of I at right angles to the planes of the developable, that is, to the osculating planes of C; or, what is the same thing, they are lines parallel to the binormals of C (the line at any point of a curve, at right angles to the osculating plane, is termed the "binormal"). But if the intersection I is a curve of curvature on the second surface, then the successive lines intersect; that is, starting from the curve C, the theorem in effect is that at each point of the involute drawing a line parallel to the binormal of the corresponding point of the curve, the successive lines intersect, giving rise to a developable. To prove this, let the arc s be measured from any fixed point of the curve, and the coordinates x, y, z be considered as functions of s; and let x', x'', x''' denote $\frac{dx}{ds}, \frac{d^2x}{ds^2}, \frac{d^3x}{ds^3}$, and the like as regards y and z. Measuring off on the tangent at the point (x, y, z) a length $l-s$, the locus of the extremity is the involute; that is, for the point (x, y, z) on the curve, the coordinates of the corresponding point on the involute are $x+(l-s)x'$, $y+(l-s)y'$, $z+(l-s)z'$. Moreover, the cosine inclinations of the binormal are as $y'z''-y''z'$, $z'x''-z''x'$, $x'y''-x''y'$. Hence taking X, Y, Z as current coordinates, the equations of the line parallel to the binormal may be written

$$X = x+(l-s)x' + \theta(y'z''-y''z'),$$
$$Y = y+(l-s)y' + \theta(z'x''-z''x'),$$
$$Z = z+(l-s)z' + \theta(x'y''-x''y'),$$

and the condition of intersection is therefore

$$\begin{vmatrix} x'', & y'z''-y''z', & (y'z''-y''z')' \\ y'', & z'x''-z''x', & (z'x''-z''x')' \\ z'', & x'y''-x''y', & (x'y''-x''y')' \end{vmatrix} = 0.$$

Form a minor out of the first and second columns, e.g.

$$y''(x'y''-x''y')-z''(z'x''-z''x'),$$

this is,

$$x'(x''^2+y''^2+z''^2)-x''(x'x''+y'y''+z'z''),$$

or the last term being $=0$, and the factor $x''^2+y''^2+z''^2$ being common, the minors are as $x' : y' : z'$. Moreover $(y'z''-y''z')' = y'z'''-y'''z'$, &c., hence the determinant is

$$x'(y'z'''-y'''z')+y'(z'x'''-z'''x')+z'(x'y'''-x'''y'),$$

viz. this is $=0$, or the theorem is proved.

572.

THEOREM IN REGARD TO THE HESSIAN OF A QUATERNARY FUNCTION.

[From the *Quarterly Journal of Pure and Applied Mathematics*, vol. XII. (1873), pp. 193—197.]

I WISH to put on record the following expression for the Hessian of $P^k + \lambda P'^{k'}$, where P, P' are quaternary functions of (x, y, z, w) of the degrees l, l' respectively, and λ is a constant; the demonstration is tedious enough, but presents no particular difficulty.

I write (A, B, C, D) for the first derived functions of P; and $(a, b, c, d, f, g, h, l, m, n)$ for the second derived functions; and similarly for P'. The Hessian of P is thus

$$\begin{vmatrix} a, & h, & g, & l \\ h, & b, & f, & m \\ g, & f, & c, & n \\ l, & m, & n, & d \end{vmatrix},$$

which is denoted by $(abcd)$; moreover, if in this determinant we substitute the accented letters for the letters of each line successively, the result is denoted by $(abcd')$; and so if we substitute the accented letters for the letters of each pair of lines successively, the result is denoted by $(abc'd')$. Observe that

$$abcd' = (a'\delta_a + b'\delta_b + \ldots)\, abcd \text{ and } abc'd' = \tfrac{1}{2}(a'\delta_a + b'\delta_b + \ldots)^2\, abcd.$$

The notation $(abcD'^2)$ is used to denote the determinant

$$-\begin{vmatrix} & A', & B', & C', & D' \\ A', & a, & h, & g, & l \\ B', & h, & b, & f, & m \\ C', & g, & f, & c, & n \\ D', & l, & m, & n, & d \end{vmatrix},$$

and from it we derive the expression $(abc'D'^2)$, viz.

$$abc'D'^2 = (a'\delta_a + b'\delta_b + \dots)\, abcD'^2.$$

The final result is expressed in terms of the several functions $abcd$, $abcd'$, $abc'd'$, $ab'c'd'$, $a'b'c'd'$, $abcD'^2$, $a'b'c'D^2$, $abc'D'^2$, $a'b'cD^2$, viz. we have

$$\begin{aligned}
\mathfrak{H}(P^k + \lambda P'^{k'}) = \;& k^4(k-1)\left(\frac{1}{k-1} + \frac{l}{l-1}\right) P^{4k-4} . abcd \\[6pt]
&+ \lambda \left\{ \begin{array}{l} k^3(k-1)\left(\dfrac{1}{k-1} + \dfrac{l}{l-1}\right) k' P^{3k-3} \left\{ \begin{array}{r} P'^{k'-1} . abcd' \\ + (k'-1) P'^{k'-2} . abcD'^2 \end{array} \right\} \\[6pt] - \dfrac{k^3(k-1)k'}{(l-1)^2} [l'(l'-1) + l^2(k-1)]\, P^{3k-4} P'^{k'} . abcd \end{array} \right\} \\[6pt]
&+ \lambda^2 \left\{ \begin{array}{l} \quad k^2k'^2 \qquad P^{2k-2} P'^{2k'-2}\, abc'd' \\ + k^2k'^2(k'-1)\, P^{2k-2} P'^{2k'-3}\, abc'D'^2 \\ + k^2k'^2(k-1)\, P^{2k-3} P'^{2k'-2}\, a'b'cD^2 \\ + k^2k'^2(k-1)(k'-1)\, P^{2k-3} P'^{2k'-3} \left\{ \begin{array}{l} \dfrac{l(l-1)}{(l'-1)^2} a'b'c'D^2 . P \\[4pt] + \dfrac{l'(l'-1)}{(l-1)^2} abcD'^2 . P' \\[4pt] - \dfrac{l^2}{(l'-1)^2} ab'c'd' . P^2 \\[4pt] + \dfrac{ll'}{(l-1)(l'-1)} abc'd' . PP' \\[4pt] - \dfrac{l'^2}{(l-1)^2} abcd' . P'^2 \end{array} \right\} \end{array} \right\} \\[6pt]
&+ \lambda^3 \left\{ \begin{array}{l} k'^3(k'-1)\left(\dfrac{1}{k'-1} + \dfrac{l'}{l'-1}\right) k P'^{3k'-3} \left\{ \begin{array}{r} P^{k-1} . a'b'c'd \\ + (k-1) P^{k-2} . a'b'c'D^2 \end{array} \right\} \\[6pt] - \dfrac{k'^3(k'-1)k}{(l'-1)^2} [l(l-1) + l'^2(k'-1)]\, P^k P'^{3k'-4} . a'b'c'd' \end{array} \right\} \\[6pt]
&+ \lambda^4 . k'^4(k'-1)\left(\frac{1}{k'-1} + \frac{l'}{l'-1}\right) P'^{4k'-4} . a'b'c'd'.
\end{aligned}$$

In verification, I remark that, $\lambda = 0$, the formula becomes

$$\mathfrak{H}(P^k) = k^4(k-1)\left(\frac{1}{k-1} + \frac{l}{l-1}\right) P^{4k-4} . abcd,$$

that is

$$= \frac{k^4(kl-1)}{l-1} P^{4k-4} . abcd.$$

Hence, writing $P' = P$, which implies $k' = k$ and $l' = l$, we ought to have

$$\mathfrak{H}\{(1+\lambda)P^k\} = (1+\lambda)^4 . \frac{k^4(kl-1)}{l-1} P^{4k-4} . abcd.$$

But writing in the formula $P' = P$, it is to be observed that $abcd' = 4abcd$, $abc'd' = 6abcd$, $ab'c'd' = 4abcd$, $a'b'c'd' = abcd$: moreover that $abcD'^2$ and $a'b'c'D^2$ are each $= abcD^2$, but that $abc'D'^2$ and $a'b'cD^2$ are each $= 3abcD^2$, and (as is easily shown to be the case)

$$abcD^2 = \frac{l}{l-1} P . abcd.$$

Thus the whole coefficient of λ becomes

$$\left\{\begin{array}{l} k^4(k-1)\left(\dfrac{1}{k-1} + \dfrac{l}{l-1}\right)\left\{4 + \dfrac{(k-1)l}{l-1}\right\} \\ - k^4 \dfrac{(k-1)}{(l-1)^2}\{l(l-1) + l^2(k-1)\} \end{array}\right\} P^{4k-4} . abcd,$$

where the numerical factor is

$$= k^4(k-1)^2\left(\frac{1}{k-1} + \frac{l}{l-1}\right)\left(\frac{4}{k-1} + \frac{l}{l-1} - \frac{l}{l-1}\right)$$

$$= 4k^4(k-1)\left(\frac{1}{k-1} + \frac{l}{l-1}\right);$$

or, finally, it is

$$= \frac{4k^4(kl-1)}{l-1}.$$

The coefficient of λ^2 is

$$= k^4\left\{6 - \frac{2l^2(k-1)^2}{(l-1)^2}\right\} P^{4k-4} . abcd$$

$$+ k^4\left\{6(k-1) + \frac{2(k-1)^2 l}{l-1}\right\} P^{4k-5} . abcD^2;$$

or, substituting for $abcD^2$ its value $= \dfrac{l}{l-1} P . abcd$, the expression is equal to $P^{4k-4}abcd$ into a numerical coefficient, which is

$$k^4\left\{6 - \frac{2l^2(k-1)^2}{(l-1)^2} + \left(\frac{6(k-1)l}{l-1} + \frac{2(k-1)^2 l^2}{(l-1)^2}\right)\right\},$$

viz. this is

$$6k^4\left\{1 + \frac{(k-1)l}{l-1}\right\}$$

$$= 6\frac{k^4(kl-1)}{l-1},$$

and the coefficients of λ^3, and λ^4 are equal to those of λ and λ^0 respectively. Hence the formula gives, as it should do,

$$\mathfrak{H}\{(1+\lambda)P^k\} = (1+\lambda)^4 \frac{k^4(kl-1)}{l-1} P^{4k-4} . abcd.$$

Attending only to the form of the result, and representing the numerical factors by A, B, &c., we may write

$$\begin{aligned}
\mathfrak{H}(P^k + \lambda P'^k) = \quad & A \quad P^{4k-4}\, abcd \\
+\lambda\,.\; & B \begin{Bmatrix} P^{3k-3}\, P'^{k'-1}\, abcd' \\ +(k'-1)\, P^{3k-3}\, P'^{k'-2}\, abcD'^2 \end{Bmatrix} \\
+ & C \quad P^{3k-4}\, P'^{k'}\, abcd \\
+\lambda^2.\; & D \quad P^{2k-2}\, P'^{2k'-2}\, abc'd' \\
+ & E \quad P^{2k-2}\, P'^{2k'-3}\, abc'D'^2 \\
+ & E' \quad P^{2k-3}\, P'^{2k'-2}\, a'b'cD^2 \\
+ & F \quad P^{2k-3}\, P'^{2k'-3}\, (\Lambda P + \Lambda' P') \\
+\lambda^3.\; & C' \quad P^k\, P'^{3k'-4}\, a'b'c'd' \\
+ & B' \begin{Bmatrix} P^{k-1}\, P'^{3k'-3}\, a'b'c'd \\ +(k-1)\, P^{k-2}\, P'^{3k'-3}\, a'b'c'D^2 \end{Bmatrix} \\
+\lambda^4.\; & A' \quad P'^{4k'-4}\, a'b'c'd',
\end{aligned}$$

where, for shortness, certain terms in λ^2 have been represented by $\Lambda P + \Lambda' P'$.

Suppose $k = k' = 2$; then attending only to the terms of the lowest order in P, P' conjointly, we have

$$\begin{aligned}
\mathfrak{H}(P^2 + \lambda P'^2) = \quad & \lambda B \,.\, P^3 \,.\, abcD'^2 \\
& + \lambda^2 \quad .\, PP'(\Lambda P + \Lambda' P') \\
& + \lambda^3 B' \,.\, P'^3 \,.\, a'b'c'D^2.
\end{aligned}$$

If the function operated upon with $\mathfrak{H}$ had been $UP^2 + U'P'^2$, the lowest terms in P, P' would have been of the like form; and it thus appears that for a surface of the form $UP^2 + U'P'^2 = 0$, the nodal curve $P = 0$, $P' = 0$ is a triple curve on the Hessian surface.

If $k = 2$, $k' = 3$, then attending only to the terms of the lowest order in P, P' conjointly, we have

$$\begin{aligned}
\mathfrak{H}(P^2 + \lambda P'^3) = \quad & A \,.\, P^4 \,.\, abcd \\
& + \lambda \,.\, 2B \,.\, P^3 P' \;.\, abcD'^2;
\end{aligned}$$

and the like result would be obtained if the function operated upon with $\mathfrak{H}$ had been $UP^2 + U'P'^3$. It thus appears that for a surface of the form $UP^2 + U'P'^3 = 0$, the cuspidal curve $P = 0$, $P' = 0$ is a 4-tuple curve on the Hessian surface, the form in the vicinity of this line, or direction of the tangent plane, being given by

$$P^3 (A \,.\, P \,.\, abcd + 2B\lambda \,.\, P' \,.\, abcD'^2) = 0,$$

viz. there is a triple sheet $P^3 = 0$, coinciding with the direction of the surface in the vicinity of the cuspidal line; and a single sheet

$$A \,.\, P \,.\, abcd + 2B\lambda \,.\, P' \,.\, abcD'^2 = 0.$$

At the points for which the osculating plane of the curve $P = 0$, $P' = 0$ coincides with the tangent plane of $P = 0$ (or, what is the same thing, with that of the surface), we have $abcD'^2 = 0$, and the triple and single sheets then coincide in direction.

573.

NOTE ON THE (2, 2) CORRESPONDENCE OF TWO VARIABLES.

[From the *Quarterly Journal of Pure and Applied Mathematics*, vol. XII. (1873), pp. 197, 198.]

IN connection with my paper "On the porism of the in-and-circumscribed polygon and the (2, 2) correspondence of points on a conic," *Quar. Math. Jour.*, t. XI. (1871), pp. 83—91, [489], I remark that if (θ, ϕ) have a symmetrical (2, 2) correspondence, and also (ϕ, χ) the same symmetrical (2, 2) correspondence, then (θ, χ) will have a (not in general the same) symmetrical (2, 2) correspondence. In fact, to a given value θ there correspond, say the values ϕ_1, ϕ_2 of ϕ; then to ϕ_1 correspond the values θ, χ_1 of χ (viz. one of the two values is $=\theta$), and to ϕ_2 the values θ, χ_2 of χ (viz. one of the values is here again $=\theta$); that is, to the given value θ there correspond the two values χ_1, χ_2 of χ; and similarly to any value of χ there correspond two values of θ; viz. to χ_1 the value θ and say θ_1; to χ_2 the value θ and say θ_2; that is, the correspondence of θ, χ is a (2, 2) correspondence and is symmetrical.

Analytically, if we have

$$(a, b, c, f, g, h \between \theta\phi, \theta+\phi, 1)^2 = 0,$$

and

$$(a, b, c, f, g, h \between \phi\chi, \phi+\chi, 1)^2 = 0,$$

then writing

$$(a, \ldots \between \phi u, \phi + u, 1)^2 = 0,$$

the roots hereof are $u=\theta$, $u=\chi$; i.e. we have

$$(a, \ldots \between \phi u, \phi + u, 1)^2 = (a, \ldots \between \phi, 1, 0)^2 (u-\theta)(u-\chi);$$

or, what is the same thing, we have

$$1 : -(\theta+\chi) : \theta\chi = (a, \ldots \between \phi, 1, 0)^2 : 2\,(a, \ldots \between \phi, 1, 0 \between 0, \phi, 1) : (a, \ldots \between 0, \phi, 1)^2$$

$$= a\phi^2 + 2h\phi + b \quad : \quad 2\,(h\phi^2 + \overline{b+g}\,\phi + f) \quad : \quad b\phi^2 + 2f\phi + c,$$

giving $\phi^2 : \phi : 1$ proportional to linear functions of $1,\ \theta+\chi,\ \theta\chi$, and therefore a quadric relation $(* \between \theta\chi,\ \theta+\chi,\ 1)^2 = 0$, with coefficients which are not in general (a, b, c, f, g, h).

Suppose, however, that the coefficients have these values, or that the correspondence is

$$(a, b, c, f, g, h \between \theta\chi,\ \theta+\chi,\ 1)^2 = 0,$$

we must have

$$(a, b, c, f, g, h \between a\phi^2 + 2h\phi + b,\ \ -2\,(h\phi^2 + \overline{b+g}\,\phi + f),\ \ b\phi^2 + 2f\phi + c)^2 = 0,$$

that is,

$$(ac + b^2 + 2bg - 4fh)\,(a, b, c, f, g, h \between \phi^2,\ -2\phi,\ 1)^2 = 0,$$

or, we have

$$ac + b^2 + 2bg - 4fh = 0,$$

as the condition in order that the symmetrical (2, 2) correspondence between θ and χ may be the same correspondence as that between θ and ϕ, or between ϕ and χ.

574.

ON WRONSKI'S THEOREM.

[From the *Quarterly Journal of Pure and Applied Mathematics*, vol. XII. (1873), pp. 221—228.]

THE theorem, considered by the author as an answer to the question "En quoi consistent les Mathématiques? N'y aurait-il pas moyen d'embrasser par un seul problème, tous les problèmes de ces sciences et de resoudre généralement ce problème universel?" is given without demonstration in his *Réfutation de la Théorie de Fonctions Analytiques de Lagrange*, Paris, 1812, p. 30, and reproduced (with, I think, a demonstration) in the *Philosophie de la Technie*, Paris, 1815; and it is also stated and demonstrated in the *Supplément à la Réforme de la Philosophie*, Paris, 1847, p. CIX *et seq.*; the theorem, but without a demonstration, is given in *Montferrier's Encyclopédie Mathématique* (Paris, no date), t. III. p. 398.

The theorem gives the development of a function Fx of the root of an equation

$$0 = fx + x_1 f_1 x + x_2 f_2 x + \&c.,$$

but it is not really more general than that for the particular case $0 = fx + x_1 f_1 x$; or say when the equation is $0 = \phi x + \lambda fx$.* Considering then this equation

$$\phi x + \lambda fx = 0,$$

let a be a root of the equation $\phi x = 0$; the theorem is

$$
\begin{aligned}
Fx = {} & F \\
& - \frac{\lambda}{1} \frac{1}{\phi'} \left| \, (\textstyle\int fF')' \, \right| \\
& + \frac{\lambda^2}{1 \,.\, 2} \frac{1}{\phi'^3} \begin{vmatrix} \phi', & (\int f^2 F')' \\ \phi'', & (\int f^2 F')'' \end{vmatrix} \begin{matrix} 1 \\ 1 \end{matrix}
\end{aligned}
$$

* For in the result, as given in the text, instead of λfx write $x_1 f_1 x + x_2 fx + \&c.$, then expanding the several powers of this quantity, each determinant is replaced by a sum of determinants of the same order, and we have the expansion of Fx in powers of x_1, x_2, ...

$$-\frac{\lambda^3}{1.2.3}\frac{1}{\phi'^6}\left|\begin{array}{lll}\phi', & (\phi^2)', & (\int f^3F')' \\ \phi'', & (\phi^2)'', & (\int f^3F')'' \\ \phi''', & (\phi^2)''', & (\int f^3F')'''\end{array}\right|\frac{1}{1.1.2}$$

$$+\ \&\text{c.},$$

where F, f, F', &c. denote Fa, fa, $F'a$, &c. and the accents denote differentiation in regard to a; the integral sign $\int$ is written instead of $\int_a$; this is introduced for symmetry only, and obviously disappears; in fact, we may equally well write

$$Fx = F$$

$$-\frac{\lambda}{1}\frac{1}{\phi'}fF'$$

$$+\frac{\lambda^2}{1.2}\frac{1}{\phi'^3}\left|\begin{array}{ll}\phi', & f^2F' \\ \phi'', & (f^2F')'\end{array}\right|\frac{1}{1}$$

$$-\frac{\lambda^3}{1.2.3}\frac{1}{\phi'^6}\left|\begin{array}{lll}\phi', & (\phi^2)', & f^3F' \\ \phi'', & (\phi^2)'', & (f^3F')' \\ \phi''', & (\phi^2)''', & (f^3F')''\end{array}\right|\frac{1}{1.1.2}$$

$$+\ \&\text{c.}$$

I stop for a moment to remark that Laplace's theorem is really equivalent to Lagrange's; viz. in the first mentioned theorem we have $x = \phi(a + \lambda fx)$, that is $\phi^{-1}x = a + \lambda f\phi \,.\, \phi^{-1}x$, and then $Fx = F\phi \,.\, \phi^{-1}x$, viz. by Lagrange's theorem

$$Fx = F\phi + \frac{\lambda}{1}F\phi' \,.\, f\phi + \frac{\lambda^2}{1.2}\{F\phi' \,.\, (f\phi)^2\}' + \&\text{c.},$$

where on the right hand $F\phi$ and $f\phi$ are each regarded as one symbol, the argument being always a and the accents denoting differentiation in regard to a, thus $F\phi'$ is

$$d_a \,.\, F\phi a = F'\phi a \,.\, \phi' a,\ \&\text{c.},$$

viz. this is Laplace's theorem.

Suppose in Wronski's theorem $\phi x = x - a$; that is, let the equation be

$$x - a + \lambda\phi x = 0,$$

then each determinant reduces itself to a single term: thus the determinant of the third order is

$$\left|\begin{array}{lll}(x-a)', & \{(x-a)^2\}', & f^3F' \\ (x-a)'', & \{(x-a)^2\}'', & (f^3F')' \\ (x-a)''', & \{(x-a)^2\}''', & (f^3F')''\end{array}\right|,$$

where in the first and second columns the accents denote differentiation in regard to x, which variable is afterwards put $= a$; the determinant is thus

$$= \left|\begin{array}{lll}1, & *, & * \\ 0, & 1.2, & * \\ 0, & 0, & (f^3F')''\end{array}\right|,$$

viz. it is

$$=1.1.2(f^3F')'',$$

and so in other cases; the formula is thus

$$Fx=F-\frac{\lambda}{1}fF'+\frac{\lambda^2}{1.2}(f^2F')'-\frac{\lambda^3}{1.2.3}(f^3F')''+\&c.,$$

agreeing with Lagrange's theorem.

Suppose in general $\phi x=(x-a)\psi x$, or let the equation be

$$(x-a)\psi x+\lambda fx=0,$$

that is,

$$x-a+\lambda\frac{fx}{\psi x}=0:$$

we have then by Lagrange's theorem

$$Fx=F-\frac{\lambda}{1}F'\frac{f}{\psi}+\frac{\lambda^2}{1.2}\left\{F'\left(\frac{f}{\psi}\right)^2\right\}'-\frac{\lambda^3}{1.2.3}\left\{F'\left(\frac{f}{\psi}\right)^3\right\}''+\&c.$$

Consider for example the term $\left\{F'\left(\frac{f}{\psi}\right)^3\right\}''$; this is

$$=\left\{F'x.\frac{(x-a)^3(fx)^3}{(\phi x)^3}\right\}'',$$

the accents denoting differentiation in regard to x, and x being ultimately put $=a$; or, what is the same thing, it is

$$=\left(\frac{d}{d\theta}\right)^2\left[F'(a+\theta)\frac{\theta^3\{f(a+\theta)\}^3}{\{\phi(a+\theta)\}^3}\right],$$

the accents now denoting differentiation in regard to θ, and this being ultimately put $=0$. This is

$$\left(\frac{d}{d\theta}\right)^2\left[F'(a+\theta)\frac{\{f(a+\theta)\}^3}{\left(\phi'a+\frac{\theta}{1.2}\phi''a+\ldots\right)^3}\right].$$

This may be written $\left(F'f^3\frac{1}{A^3}\right)''$, where

$$A=\phi'+\tfrac{1}{2}\theta\phi''+\tfrac{1}{6}\theta^2\phi''+\ldots,$$

it being understood that as regards $F'f^3$, which is expressed as a function of a only (θ having been therein put $=0$), the exterior accents denote differentiations in respect to a, whereas in regard to A, $=\phi'+\frac{1}{2}\theta\phi''+\&c.$, they denote differentiation in regard to θ, which is afterwards put $=0$. And the theorem thus is

$$Fx=F-\frac{\lambda}{1}\left(F'f.\frac{1}{A}\right)+\frac{\lambda^2}{1.2}\left(F'f^2.\frac{1}{A^2}\right)'-\frac{\lambda^3}{1.2.3}\left(F'f^3.\frac{1}{A^3}\right)''+\&c.$$

This must be equivalent to Wronski's theorem; it is in a very different, and, I think, a preferable form; but the results obtained from the comparison are very interesting, and I proceed to make this comparison.

Taking the foregoing coefficient $\left(F'f^3\frac{1}{A^3}\right)''$ this should be equal to Wronski's term

$$\frac{1}{1.1.2}\frac{1}{\phi'^6}\begin{vmatrix}\phi', & (\phi^2)', & f^3F' \\ \phi'', & (\phi^2)'', & (f^3F')' \\ \phi''', & (\phi^2)''', & (f^3F')''\end{vmatrix};$$

or, what is the same thing, the determinant should be

$$=1.1.2\phi'^6\left(\frac{1}{A^3}f^3F'\right)''$$

$$=1.1.2\phi'^6\left\{f^3F'\left(\frac{1}{A^3}\right)''+2(f^3F')'\left(\frac{1}{A^3}\right)'+(f^3F')''\frac{1}{A^3}\right\},$$

that is, the values of

$$1.1.2\phi'^6\frac{1}{A^3},\quad 1.1.2\phi'^6 2\left(\frac{1}{A^3}\right)',\quad 1.1.2\phi'^6\left(\frac{1}{A^3}\right)''$$

should be

$$=\phi'(\phi^2)''-\phi''(\phi^2)',\quad \phi'''(\phi^2)'-\phi'(\phi^2)''',\quad \phi''(\phi^2)'''-\phi'''(\phi^2)''$$

respectively. Or, what is the same thing, if

$$\frac{1}{\left(\phi'+\frac{\theta}{2}\phi''+\frac{\theta^2}{2.3}\phi'''+\dots\right)^3}=A_0+\frac{1}{1}A_1\theta+\frac{1}{1.2}A_2\theta^2+\dots,$$

then the last mentioned functions should be

$$1.1.2\phi'^6A_0,\quad 1.1.2\phi'^6 2A_1,\quad 1.1.2\phi'^6A_2.$$

We have

$$A_0=\frac{1}{\phi'^3},\quad A_1=-\frac{3}{2}\frac{\phi''}{\phi'^4},\quad A_2=-\frac{\phi'''}{\phi'^4}+\frac{3\phi''^2}{\phi'^5},$$

or the identities are

$$\begin{aligned}
2\phi'^3 &= \phi'(\phi^2)''-\phi''(\phi^2)', &&= \phi'(2\phi\phi''+2\phi'^2)-\phi''.2\phi\phi',\\
-6\phi''\phi'^2 &= \phi'''(\phi^2)'-\phi'(\phi^2)''', &&= \phi'''.2\phi\phi'-\phi'(2\phi\phi''+6\phi'\phi''),\\
+6\phi''^2\phi'-2\phi'''\phi'^2 &= \phi''(\phi^2)'''-\phi'''(\phi^2)'', &&= \phi''(2\phi\phi'''+6\phi'\phi'')-\phi'''(2\phi\phi''+2\phi'^2),
\end{aligned}$$

which is right. And in like manner to verify the coefficient of λ^4, we should have to compare the first four terms of the expansion of

$$\frac{1}{\left(\phi'+\frac{\theta}{2}\phi''+\frac{\theta^2}{2.3}\phi'''+\dots\right)^4}.$$

with the determinants formed out of the matrix

$$\left|\begin{matrix} \phi' , & \phi'' , & \phi''' , & \phi'''' \\ (\phi^2)', & (\phi^2)'', & (\phi^2)''', & (\phi^2)'''' \\ (\phi^3)', & (\phi^3)'', & (\phi^3)''', & (\phi^3)'''' \end{matrix}\right| .$$

The series of equalities may be presented as follows, writing as above A to denote the function

$$\phi' + \frac{\theta}{2}\phi'' + \frac{\theta^2}{2.3}\phi''' + \ldots,$$

$$\frac{1}{A} = \frac{1}{\phi'}.1,$$

$$\frac{1}{A^2} = \frac{-1}{\phi'^3}\left|\begin{matrix} \theta, & 1 \\ \phi', & \phi'' \end{matrix}\right| . \frac{1}{1},$$

$$\frac{1}{A^3} = \frac{+1}{\phi'^6}\left|\begin{matrix} \tfrac{1}{2}\theta^2, & \tfrac{1}{2}\theta , & 1 \\ \phi' , & \phi'' , & \phi''' \\ (\phi^2)', & (\phi^2)'', & (\phi^2)''' \end{matrix}\right| . \frac{1}{1.1.2},$$

$$\frac{1}{A^4} = \frac{-1}{\phi'^{10}}\left|\begin{matrix} \tfrac{1}{6}\theta^3, & \tfrac{1}{6}\theta^2 , & \tfrac{1}{3}\theta , & 1 \\ \phi' , & \phi'' , & \phi''' , & \phi'''' \\ (\phi^2)', & (\phi^2)'', & (\phi^2)''', & (\phi^2)'''' \\ (\phi^3)', & (\phi^3)'', & (\phi^3)''', & (\phi^3)'''' \end{matrix}\right| . \frac{1}{1.1.2.1.2.3},$$

&c.,

where in each case the function on the left hand is to be expanded only as far as the power of θ which is contained in the determinant: the numerical coefficients in the top-lines of the several determinants are the reciprocals of

$$n(n-1)\ldots 2.1, \quad n(n-1)\ldots 2, \quad n(n-1), \quad n, \quad 1,$$

where n is the index of the highest power of θ. The demonstration of Wronski's theorem therefore ultimately depends on the establishment of the foregoing equalities As a verification, in the fourth formula, write $\phi = e^a\,(a=0)$, we have

$$\left(\frac{\theta}{e^\theta - 1}\right)^4 \text{ or } \frac{1}{(1+\frac{1}{2}\theta+\frac{1}{6}\theta^2+\frac{1}{24}\theta^3+\ldots)^4} = -\frac{1}{12}\left|\begin{matrix} \tfrac{1}{6}\theta^3, & \tfrac{1}{6}\theta^2, & \tfrac{1}{3}\theta, & 1 \\ 1, & 1, & 1, & 1 \\ 2, & 4, & 8, & 16 \\ 3, & 9, & 27, & 81 \end{matrix}\right| ,$$

where the right hand is

$$= -\tfrac{1}{12}(-1.12 + \tfrac{1}{3}\theta.72 - \tfrac{1}{6}\theta^2.132 + \tfrac{1}{6}\theta^3.72)$$

$$= \quad 1 - 2\theta + \tfrac{11}{6}\theta^2 - \theta^3,$$

and expanding the left hand as far as θ^3, this is

$$\begin{array}{rlr} = & 1 & =1 \\ - & 4\left(\tfrac{1}{2}\theta+\tfrac{1}{6}\theta^2+\tfrac{1}{24}\theta^3\right) & -2\theta-\tfrac{2}{3}\theta^2-\tfrac{1}{6}\theta^3 \\ + & 10\left(\quad\tfrac{1}{4}\theta^2+\tfrac{1}{6}\theta^3\right) & +\tfrac{5}{2}\theta^2+\tfrac{5}{3}\theta^3 \\ - & 20\left(\qquad\tfrac{1}{8}\theta^3\right) & -\tfrac{5}{2}\theta^3 \\ \hline & & 1-2\theta+\tfrac{11}{6}\theta^2-\theta^3, \end{array}$$

which agrees.

Reverting to the above equations, and expanding the several terms $(\phi^2)'=2\phi\phi'$, $(\phi^2)''=2\phi\phi''+2\phi'^2$, &c., then, since in each case the left-hand side contains ϕ', ϕ'', ϕ''', &c. but not ϕ, it is clear that on the right-hand side the terms involving ϕ must disappear of themselves; and assuming that this is so, the equality takes the more simple form obtained by writing in the foregoing expressions $\phi=0$, viz. we thus have $(\phi^2)'=0$, $(\phi^2)''=2\phi'^2$, &c. In order to simplify the formulæ, I replace the series ϕ', $\tfrac{1}{2}\phi''$, $\tfrac{1}{6}\phi'''$, $\tfrac{1}{24}\phi''''$, &c. by b, c, d, e, &c., and I thus find that they assume the following simple form, viz. writing

$$\Theta=b+c\theta+d\theta^2+e\theta^3+\&c.,$$

then we have

$$\frac{1}{\Theta}=\frac{1}{b}\cdot 1,$$

$$\frac{1}{\Theta^2}=-\frac{2}{b^3}\begin{vmatrix} \theta, & \tfrac{1}{2} \\ b, & c \end{vmatrix},$$

$$\frac{1}{\Theta^3}=+\frac{3}{b^6}\begin{vmatrix} \theta^2, & \tfrac{1}{2}\theta, & \tfrac{1}{3} \\ b, & c, & d \\ & b^2, & 2bc \end{vmatrix},$$

$$\frac{1}{\Theta^4}=-\frac{4}{b^{10}}\begin{vmatrix} \theta^3, & \tfrac{1}{2}\theta^2, & \tfrac{1}{3}\theta, & \tfrac{1}{4} \\ b, & c, & d, & e \\ & b^2, & 2bc, & 2bd+c^2 \\ & & b^3, & 3b^2c \end{vmatrix},$$

viz. for Θ^{-n} the right-hand gives the development as far as θ^{n-1}. It will be observed, that in the determinants the several lines are the coefficients in the expansions of Θ, Θ^2, Θ^3, &c. respectively.

The demonstration is very easy; it will be sufficient to take the equation for $\frac{1}{\Theta^4}$. Assume

$$\frac{1}{\Theta^4}=\ldots r\theta^6+q\theta^5+p\theta^4+\beta\theta^3+\tfrac{1}{2}\gamma\theta^2+\tfrac{1}{3}\delta\theta+\tfrac{1}{4}\epsilon,$$

where clearly $\epsilon=\frac{4}{b^4}$, and write also

$$\begin{aligned} \Theta &= B_1+C_1\theta+D_1\theta^2+E_1\theta^3+\ldots, \\ \Theta^2 &= \qquad B_2+C_2\theta+D_2\theta^2+\ldots, \\ \Theta^3 &= \qquad\qquad B_3+C_3\theta+\ldots, \end{aligned}$$

where $B_1 = b$, $B_2 = b^2$, $B_3 = b^3$; we wish to show that

$$\begin{aligned} \beta B_1 + \gamma C_1 + \delta D_1 + \epsilon E_1 &= 0, \\ \gamma B_2 + \delta C_2 + \epsilon D_2 &= 0, \\ \delta B_3 + \epsilon C_3 &= 0, \end{aligned}$$

for this being the case, neglecting the terms in θ^4, θ^5, &c., and writing

$$\beta\theta^3 + \tfrac{1}{2}\gamma\theta^2 + \tfrac{1}{3}\delta\theta + \epsilon\left(\tfrac{1}{4} - \frac{1}{\epsilon\Theta^4}\right) = 0,$$

then eliminating β, γ, δ, ϵ, we have

$$\begin{vmatrix} \theta^3, & \tfrac{1}{2}\theta^2, & \tfrac{1}{3}\theta, & \tfrac{1}{4} - \dfrac{1}{\epsilon\Theta^4} \\ B_1, & C_1, & D_1, & E_1 \\ & B_2, & C_2, & D_2 \\ & & B_3, & C_3 \end{vmatrix} = 0,$$

in which equation the term which contains

$$\frac{1}{\Theta^4} \text{ is } +\frac{1}{\epsilon} B_1 B_2 B_3 \frac{1}{\Theta^4}, \quad = \tfrac{1}{4} b^{10} \frac{1}{\Theta^4};$$

and the equation thus is $\dfrac{1}{\Theta^4} = -\dfrac{4}{b^{10}}$ multiplied by the determinant without the term in question (that is, with $\tfrac{1}{4}$ for its corner term).

To prove the subsidiary theorems, multiply the expression of $\dfrac{1}{\Theta^4}$ by $\dfrac{1}{\theta^4}$, and differentiate in regard to θ, we have

$$\frac{4\,(\theta\Theta)'}{(\theta\Theta)^5} = \dots - 2r\theta - q + \frac{\beta}{\theta^2} + \frac{\gamma}{\theta^3} + \frac{\delta}{\theta^4} + \frac{\epsilon}{\theta^5}.$$

Multiplying by

$$\theta\Theta = B_1\theta + C_1\theta^2 + D_1\theta^3 + E_1\theta^4,$$

we see that $B_1\beta + C_1\gamma + D_1\delta + E_1\epsilon$ is the coefficient of $\dfrac{1}{\theta}$ in $\dfrac{4\,(\theta\Theta)'}{(\theta\Theta)^4}$; and similarly $B_2\gamma + C_2\delta + E_2\epsilon$ is the coefficient of $\dfrac{1}{\theta}$ in $\dfrac{4\,(\theta\Theta)'}{(\theta\Theta)^3}$, and $B_3\delta + C_3\epsilon$ that of $\dfrac{1}{\theta}$ in $\dfrac{4\,(\theta\Theta)'}{(\theta\Theta)^2}$. Now, m being any positive integer, $\dfrac{1}{(\Theta\theta)^m}$ expanded in ascending powers of θ contains negative and positive powers of θ, but of course no logarithmic term; hence differentiating in regard to θ, $\dfrac{(\Theta\theta)'}{(\Theta\theta)^{m+1}}$ contains no term in $\dfrac{1}{\theta}$;* and the expressions in question are thus each $=0$; which completes the demonstration.

The foregoing formulæ giving the expansion of $\dfrac{1}{\Theta^n}$ up to θ^{n-1} in terms of the coefficients in the expansions of Θ, Θ^2, ... Θ^{n-1} are I think interesting.

* This is a well-known method made use of by Jacobi and Murphy.

575.

ON A SPECIAL QUARTIC TRANSFORMATION OF AN ELLIPTIC FUNCTION.

[From the *Quarterly Journal of Pure and Applied Mathematics*, vol. XII. (1873), pp. 266—269.]

It is remarked by Jacobi that a transformation of the order $n'n''$ may lead to a modular equation

$$\frac{\Delta'}{\Delta}=\frac{n'}{n''}\frac{K'}{K},$$

and in particular when $n'=n''$, or the order is square, then the equation may be $\frac{\Delta'}{\Delta}=\frac{K'}{K}$; viz. that instead of a transformation we may have a multiplication. A quartic transformation of the kind in question may be obtained as follows: writing

$$X=(a,\ b,\ c,\ d,\ e\between x,\ 1)^4=a\,(x-\alpha)(x-\beta)(x-\gamma)(x-\delta),$$

H the Hessian, Φ the cubi-covariant, I and J the two invariants, then there is a well known quartic transformation

$$z=\frac{2H}{X},$$

leading to

$$\frac{dz}{\sqrt{(Z)}}=\frac{2\sqrt{(-2)}\,dx}{\sqrt{(X)}},$$

where $Z=z^3-Iz+2J$. In fact we have

$$Z=\frac{2}{X^3}(4H^3-IH^2X+JX^3),\ =\frac{-2}{X^3}\Phi^2,$$

that is,

$$\sqrt{(Z)}=\frac{\sqrt{(-2)}\,\Phi}{X^2}\sqrt{(X)},$$

so that, by Jacobi's general principle, it at once appears that we have a transformation of the form in question.

Now we may establish a linear transformation

$$z=\frac{py+q}{y-\delta},$$

such that to the roots z_1, z_2, z_3 of the equation $z^3-Iz+2J=0$ correspond the values α, β, γ of y; and this being so, we have between y, z the relation

$$\frac{dz}{\sqrt{(Z)}}=\frac{\sqrt{(-2)}\,dy}{\sqrt{(Y)}},$$

where $Y=a\,(y-\alpha)(y-\beta)(y-\gamma)(y-\delta),\ =(a,\ b,\ c,\ d,\ e\between y,\ 1)^4$; that is, we have

$$\frac{py+q}{y-\delta}=\frac{2H}{X},$$

such that

$$\frac{dy}{\sqrt{(Y)}}=\frac{2dx}{\sqrt{(X)}},$$

which is a quartic transformation giving a duplication of the integral. The foundation of the theorem is that we can determine p, q in such wise that the functions

$$\frac{p\alpha+q}{\alpha-\delta},\quad \frac{p\beta+q}{\beta-\delta},\quad \frac{p\gamma+q}{\gamma-\delta}$$

shall be the roots z_1, z_2, z_3 of the equation $z^3-Iz+2J=0$. For writing

$$\begin{aligned}A&=(\beta-\gamma)(\alpha-\delta),\\ B&=(\gamma-\alpha)(\beta-\delta),\\ C&=(\alpha-\beta)(\gamma-\delta),\end{aligned}$$

and observing the equations

$$I=\frac{a^2}{24}(A^2+B^2+C^2),\quad =-\frac{a^2}{12}(BC+CA+AB),$$

(since $A+B+C=0$) and

$$2J=-\frac{a^3}{216}(B-C)(C-A)(A-B),$$

the equation in z is

$$\{z-\tfrac{1}{6}a(B-C)\}\,\{z-\tfrac{1}{6}a(C-A)\}\,\{z-\tfrac{1}{6}a(A-B)\},$$

and the equations for the determination of p, q thus are

$$\begin{aligned}p\alpha+q&=\tfrac{1}{6}a(\alpha-\delta)(B-C),\ =\tfrac{1}{6}a(\alpha-\delta)\{2(\alpha\delta+\beta\gamma)-(\alpha+\delta)(\beta+\gamma)\},\\ p\beta+q&=\tfrac{1}{6}a(\beta-\delta)(C-A),\ =\tfrac{1}{6}a(\beta-\delta)\{2(\beta\delta+\gamma\alpha)-(\beta+\delta)(\gamma+\alpha)\},\\ p\gamma+q&=\tfrac{1}{6}a(\gamma-\delta)(A-B),\ =\tfrac{1}{6}a(\gamma-\delta)\{2(\gamma\delta+\alpha\beta)-(\gamma+\delta)(\alpha+\beta)\},\end{aligned}$$

giving

$$p=\tfrac{1}{6}a\,\{-3\delta^2+2\delta(\alpha+\beta+\gamma)-\beta\gamma-\gamma\alpha-\alpha\beta\},$$
$$q=\tfrac{1}{6}a\,\{\delta^2(\alpha+\beta+\gamma)-2\delta(\beta\gamma+\gamma\alpha+\alpha\beta)+3\alpha\beta\gamma\},$$

or, as these may also be written

$$p=\tfrac{1}{6}a\,\{(\beta-\delta)(\gamma-\delta)+(\gamma-\delta)(\alpha-\delta)+(\alpha-\delta)(\beta-\delta)\},$$
$$q=\tfrac{1}{6}a\,\{\alpha(\beta-\delta)(\gamma-\delta)+\beta(\gamma-\delta)(\alpha-\delta)+\gamma(\alpha-\delta)(\beta-\delta)\}\,;$$

and observe also

$$p\delta+q=\tfrac{1}{2}a\,(\alpha-\delta)(\beta-\delta)(\gamma-\delta).$$

Taking X in the standard form $=(1-x^2)(1-k^2x^2)$, and writing

$$\gamma=-1,\quad \delta=1,\quad \alpha=+\frac{1}{k},\quad \beta=-\frac{1}{k},$$

we have

$$z=\frac{py+q}{y-1}=\frac{-\tfrac{1}{6}\{2k^2(1+k^2)(1+k^2x^4)+(1-10k^2+k^4)x^2\}}{(-x^2)(1-k^2x^2)},$$

$$A=-1+\frac{2}{k}-\frac{1}{k^2},$$

$$B=\ \ 1+\frac{2}{k}+\frac{1}{k^2},$$

$$C=\qquad -\frac{4}{k};$$

$$z_1=\ \ \tfrac{1}{6}(1+6k+k^2),$$
$$z_2=\ \ \tfrac{1}{6}(1-6k+k^2),$$
$$z_3=-\tfrac{1}{3}(1+k^2)\,;$$

$$Z=z^3-\tfrac{1}{12}(1+14k^2+k^4)\,z+\tfrac{1}{108}(1+k^2)(1-34k^2+k^4)$$
$$=(z-z_1)(z-z_2)(z-z_3),$$
$$p=\tfrac{1}{6}(1-5k^2),\quad q=\tfrac{1}{6}(5-k^2),\quad p+q=1-k^2\,;$$

giving as they should do

$$z_1=\frac{\dfrac{p}{k}+q}{\dfrac{1}{k}-1},\quad z_2=\frac{-\dfrac{p}{k}+q}{-\dfrac{1}{k}-1},\quad z_3=\frac{-p+q}{-2}.$$

Write for shortness

$$-\tfrac{1}{6}\{2k^2(1+k^2)(1+x^4)+(1-10k^2+k^4)x^2\}=Q,$$

so that

$$\frac{py+q}{y-1}=\frac{Q}{X},$$

then

$$\frac{Q}{X} - z_1 = \frac{p+q}{k-1} \cdot \frac{ky-1}{y-1},$$

$$\frac{Q}{X} - z_2 = \frac{p+q}{k+1} \cdot \frac{ky+1}{y+1},$$

$$\frac{Q}{X} - z_3 = \frac{p+q}{2} \cdot \frac{y+1}{y-1}.$$

The last of these is

$$\frac{1-k^2}{2} \frac{y+1}{y-1} = \frac{\frac{1}{2}(1-k^2)^2 x^2}{(1-x^2)(1-k^2x^2)},$$

that is,

$$\frac{y+1}{y-1} = \frac{(1-k^2) x^2}{(1+x^2)(1-k^2x^2)},$$

from which the foregoing equation

$$\frac{dy}{\sqrt{(Y)}} = \frac{2dx}{\sqrt{(X)}}$$

may be at once verified.

576.

ADDITION TO MR WALTON'S PAPER "ON THE RAY-PLANES IN BIAXAL CRYSTALS."

[From the *Quarterly Journal of Pure and Applied Mathematics*, vol. XII. (1873), pp. 273—275.]

INSTEAD of Mr Walton's a^2, b^2, c^2 write a, b, c, and assume α, β, $\gamma = b - c$, $c - a$, $a - b$; δ, ϵ, $\zeta = b + c$, $c + a$, $a + b$. Also instead of his x^2, y^2, z^2 write x, y, z. Then instead of the octic cone we have the quartic cone, or say the quartic curve

$$\begin{aligned}
&\alpha^2\beta^2\gamma^2 \sec^2\theta \,.\, xyz\,(x+y+z) \\
&= \alpha^2 yz\,\{(bc-a^2)\,x + a\,(\beta y - \gamma z)\}^2 \\
&+ \beta^2 zx\,\{(ca-b^2)\,y + b\,(\gamma z - \alpha x)\}^2 \\
&+ \gamma^2 xy\,\{(ab-c^2)\,z + c\,(\alpha x - \beta y)\}^2,
\end{aligned}$$

viz. we may herein consider x, y, z as trilinear coordinates, the ratios $x : y : z$ being positive for a point within the fundamental triangle.

The curve passes through the angles of the triangle, and it touches the sides in the points $(x = 0,\ \beta y - \gamma z = 0)$, $(y = 0,\ \gamma z - \alpha x = 0)$, $(z = 0,\ \alpha x - \beta y = 0)$ respectively. Moreover, the tangents at the angles of the triangle lie each of them outside the triangle. Hence, supposing a, b, c each positive, and $a > b > c$, we have α and γ each positive, β negative, and the form of the curve is as shown in the figure, or else the like form with the oval lying outside the triangle. And it is hence clear that, if the side AC $(y = 0)$ instead of touching the curve meets it in a node, this is a conjugate point arising from the evanescence of the oval; and in this case no part of the curve lies within the triangle. Now considering any point $x : y : z = l : m : n$, we obtain a tetrad of points $x : y : z = \pm\sqrt{(l)} : \pm\sqrt{(m)} : \pm\sqrt{(n)}$ on the octic cone or curve; and in order that the point on the octic curve may be real, we must have l, m, n all of the same sign; that is, the point on the quartic curve must lie within the triangle. Hence, when in the quartic curve the oval becomes a conjugate point, the octic curve has no real branch, but it consists wholly of conjugate points; viz. it consists of the points A, B, C as conjugate points; two imaginary conjugate points answering to the

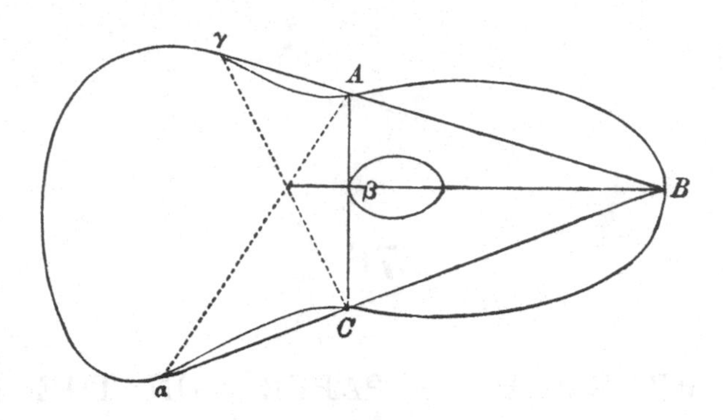

point α of the figure, two other imaginary conjugate points answering to the point γ; and two conjugate points answering to the point β, these last being not ordinary conjugate points, but conjugate tacnodal points, or points of contact of two imaginary branches of the curve.

The case in question, β a conjugate point on the quartic curve, answers to Mr Walton's critical value of $\sec^2\theta$, viz. in the present notation $\sec^2\theta = \dfrac{4b^2+\alpha\gamma}{\alpha\gamma}$. To show this I consider the intersection of the curve by the line $\gamma z - \alpha x = 0$; and I write for convenience $\gamma z = \alpha x = \gamma\alpha u$, that is, $x = \gamma u$, $z = \alpha u$. Substituting these values, the equation divides by yu, or omitting this factor it is

$$\begin{aligned}
&\alpha^3\gamma^3\beta^2 \sec^2\theta \,.\, u\,\{y + (\alpha+\gamma)\,u\} \\
&= \alpha^3\,\{\gamma u\,(bc - a^2 - a\alpha) + a\beta y\}^2 \\
&+ \beta^2\alpha\gamma\,.\,uy\,(ca - b^2)^2 \\
&+ \gamma^3\,\{\alpha u\,(ab - c^2 + c\gamma) - c\beta y\}^2,
\end{aligned}$$

or observing that we have $\alpha+\gamma = -\beta$, $bc - a^2 - a\alpha = \zeta\beta$, $ab - c^2 + c\gamma = -\delta\beta$, this becomes

$$\begin{aligned}
&\alpha^3\gamma^3 \sec^2\theta u\,(y - \beta u) \\
&= \alpha^3\ \ (\gamma\zeta u + ay)^2 \\
&+ \alpha\gamma\,(ca\ \ - b^2)^2\,uy \\
&+ \gamma^3\ \ (\alpha\delta u + cy)^2,
\end{aligned}$$

viz. this is

$$\begin{aligned}
&u^2\,\{\alpha^3\gamma^2\zeta^2 + \alpha^2\gamma^3\delta^2 + \alpha^3\gamma^3\sec^2\theta\,.\,\beta\} \\
&+ uy\,\{2\alpha^3 a\gamma\zeta + 2\gamma^3 c\alpha\delta + \alpha\gamma\,(ca - b^2)^2 - \alpha^3\gamma^3\sec^2\theta\} \\
&+ y^2\,(a^2\alpha^3 + c^2\gamma^3) = 0.
\end{aligned}$$

The required condition is that the coefficient of u^2 shall vanish; viz. we then have

$$\begin{aligned}-\alpha\beta\gamma\sec^2\theta &= \alpha\zeta^2+\gamma\delta^2\\ &= (b-c)(a+b)^2+(a-b)(b+c)^2\\ &= (a-c)\{3b^2+b(a+c)-ac\}\\ &= -\beta(4b^2+\alpha\gamma),\end{aligned}$$

that is,

$$\alpha\gamma\sec^2\theta = 4b^2+\alpha\gamma,$$

agreeing with Mr Walton's value. Giving $\sec^2\theta$ this value, and throwing out the factor u, the equation becomes

$$\begin{aligned}&u\{2a^3\alpha\gamma\zeta+2\gamma^3ca\delta+\alpha\gamma(ca-b^2)^2-\alpha^2\gamma^2(4b^2+\alpha\gamma)\}\\ &+y(a^2\alpha^3+c^2\gamma^3)=0;\end{aligned}$$

or, what is the same thing,

$$\begin{aligned}&\alpha\gamma u\{2a(a+b)(b-c)^2+2c(c+b)(b-a)^2+(ca-b^2)^2-(b-c)(a-b)4b^2-(b-c)^2(b-a)^2\}\\ &+y(a^2\alpha^3+c^2\gamma^3)=0,\end{aligned}$$

say

$$\alpha\gamma Ku+(a^2\alpha^3+c^2\gamma^3)y=0,$$

viz. this equation determines the remaining intersection of the curve by the line $\gamma z-\alpha x=0$; the point in question lies outside the triangle, that is, $u:y$ should be negative; or α, γ, $a^2\alpha^3+c^2\gamma^3$ being each positive, we should have K positive; we in fact find

$$\begin{aligned}K &= 4b^4+b^2(a^2+c^2-6ac)+4a^2c^2\\ &= 4(b^2-ac)^2+b^2(a+c)^2,\end{aligned}$$

which is as it should be.

577.

NOTE IN ILLUSTRATION OF CERTAIN GENERAL THEOREMS OBTAINED BY DR LIPSCHITZ.

[From the *Quarterly Journal of Pure and Applied Mathematics*, vol. XII. (1873), pp. 346—349.]

THE paper by Dr Lipschitz, which follows the present Note [in the *Quarterly Journal, l.c.*], is supplemental to Memoirs by him in *Crelle*, vols. LXX., LXXII., and LXXIV.; and he makes use of certain theorems obtained by him in these memoirs; these theorems may be illustrated by the consideration of a particular example.

Imagine a particle not acted on by any forces, moving in a given surface; and let its position on the surface at the time t be determined by means of the general coordinates x, y. We have then the vis-viva function T, a given function of x, y, x', y'; and the equations of motion are

$$\frac{d}{dt}\frac{dT}{dx'}-\frac{dT}{dx}=0,\quad \frac{d}{dt}\frac{dT}{dy'}-\frac{dT}{dy}=0,$$

which equations serve to determine x, y in terms of t, and of four arbitrary constants; these are taken to be the initial values (or values corresponding to the time $t=t_0$) of x, y, x', y'; say the values are α, β, α', β'.

We have the theorem that x, y are functions of α, β, $\alpha'(t-t_0)$, $\beta'(t-t_0)$.

Suppose for example that x, y, z denote ordinary rectangular coordinates, and that the particle moves on the sphere $x^2+y^2+z^2=c^2$; to fix the ideas, suppose that the coordinates z are measured vertically upwards, and that the particle is on the upper hemisphere; that is, take $z=+\sqrt{(c^2-x^2-y^2)}$, we have

$$T=\tfrac{1}{2}(x'^2+y'^2+z'^2),$$

where z' denotes its value in terms of x, y, x', y'; viz. we have $xx'+yy'+zz'=0$, or

$$z'=-\frac{xx'+yy'}{z},\quad =-\frac{xx'+yy'}{\sqrt{(c^2-x^2-y^2)}};$$

the proper value of T is thus

$$=\tfrac{1}{2}\left\{x'^2+y'^2+\frac{(xx'+yy')^2}{c^2-x^2-y^2}\right\},$$

but it is convenient to retain z, z', taking these to signify throughout their foregoing values in terms of x, y, x', y'.

The constants of integration are, as before, α, β, α', β'; but we use also γ, γ' considered as signifying given functions of these constants, viz. we have

$$\gamma=\sqrt{(c^2-\alpha^2-\beta^2)} \text{ and } \gamma'=-\frac{\alpha\alpha'+\beta\beta'}{\sqrt{(c^2-\alpha^2-\beta^2)}},$$

(in fact, $\alpha^2+\beta^2+\gamma^2=c^2$ and $\alpha\alpha'+\beta\beta'+\gamma\gamma'=0$; γ, γ' being thus the initial values of z, z').

Now, writing

$$\sigma=\frac{(t-t_0)\sqrt{(\alpha'^2+\beta'^2+\gamma'^2)}}{c},$$

the required values of x, y and the corresponding value of z are

$$x=\alpha\cos\sigma+\frac{c\alpha'}{\sqrt{(\alpha'^2+\beta'^2+\gamma'^2)}}\sin\sigma,$$

$$y=\beta\cos\sigma+\frac{c\beta'}{\sqrt{(\alpha'^2+\beta'^2+\gamma'^2)}}\sin\sigma,$$

$$z=\gamma\cos\sigma+\frac{c\gamma'}{\sqrt{(\alpha'^2+\beta'^2+\gamma'^2)}}\sin\sigma.$$

To verify that these are functions of α, β, $\alpha'(t-t_0)$, $\beta'(t-t_0)$, write $\alpha'(t-t_0)=u$, $\beta'(t-t_0)=v$; and take also $\gamma'(t-t_0)=w$; we have $\alpha u+\beta v+\gamma w=0$, viz. w, $=-\frac{1}{\gamma}(\alpha u+\beta v)$, is a function of α, β, u, v; and then

$$\sigma=\frac{\sqrt{(u^2+v^2+w^2)}}{c},$$

and

$$x=\alpha\cos\sigma+\frac{u}{\sigma}\sin\sigma,$$

$$y=\beta\cos\sigma+\frac{v}{\sigma}\sin\sigma,$$

$$z=\gamma\cos\sigma+\frac{w}{\sigma}\sin\sigma;$$

so that x, y (and also z) are each of them a function of α, β, u, v, that is α, β, $\alpha'(t-t_0)$, $\beta'(t-t_0)$, which is the theorem in question.

The original variables are x, y; the quantities $\alpha'(t-t_0)$, $\beta'(t-t_0)$, or u, v are Dr Lipschitz' "Normal-Variables," and the theorem is that the original variables are functions of their initial values, and of the normal-variables.

The vis-viva function T may be expressed in terms of the normal-variables and their derived functions; viz. it is easy to verify that we have

$$T = \tfrac{1}{2}\left(\frac{1}{c^2\sigma^2} - \frac{\sin^2\sigma}{c^2\sigma^4}\right)(uu' + vv' + ww')^2$$

$$- \tfrac{1}{2}\frac{\sin^2\sigma}{\sigma}(u'^2 + v'^2 + w'^2),$$

where w denotes $-\dfrac{1}{\gamma}(\alpha u + \beta v)$ and consequently w' denotes $-\dfrac{1}{\gamma}(\alpha u' + \beta v')$; introducing herein differentials instead of derived functions, or writing

$$\phi(du) = \tfrac{1}{2}\left(\frac{1}{c^2\sigma^2} - \frac{\sin^2\sigma}{c^2\sigma^4}\right)(udu + vdv + wdw)^2$$

$$+ \tfrac{1}{2}\frac{\sin^2\sigma}{\sigma^2}(du^2 + dv^2 + dw^2),$$

where w, dw denote $-\dfrac{1}{\gamma}(\alpha u + \beta v)$, $-\dfrac{1}{\gamma}(\alpha du + \beta dv)$ respectively; then $\phi(du)$ is the function thus denoted by Dr Lipschitz: and writing herein $t - t_0 = 0$, and thence $u = 0$, $v = 0$, $w = 0$, $\sigma = 0$, the resulting value of $\phi(du)$ is

$$f_0(du), \quad = \tfrac{1}{2}(du^2 + dv^2 + dw^2),$$

where $f_0(du)$ is the function thus denoted by him; the corresponding value of $f_0(u)$ is $= \tfrac{1}{2}(u^2 + v^2 + w^2)$. We have thus an illustration of his theorem that the function $\phi(du)$ is such that we have identically

$$\phi(du) - \{d\sqrt{\{f_0(u)\}}\}^2 = \frac{m^2}{2f_0(u)}[f_0(du) - \{d\sqrt{(f_0 u)}\}^2],$$

where m is a function of u, v independent of the differentials du, dv; the value in the present example is in fact $m^2 = c^2\sin^2\sigma$; or the identity is

$$\phi(du) - \{d\sqrt{(f_0 u)}\}^2 = \frac{c^2\sin^2\sigma}{2(f_0 u)}[f_0(du) - \{d\sqrt{(f_0 u)}\}^2],$$

in verification whereof observe that we have

$$d\sqrt{(f_0 u)} = \frac{df_0(u)}{2\sqrt{(f_0 u)}} = \frac{udu + vdv + wdw}{\sqrt{(u^2 + v^2 + w^2)}}$$

$$= \frac{1}{c\sigma}(udu + vdv + wdw)^2.$$

The value of the left-hand side is thus

$$= -\frac{\sin^2\sigma}{c^2\sigma^4}(udu + vdv + wdw)^2 + \tfrac{1}{2}\frac{\sin^2\sigma}{\sigma^2}(du^2 + dv^2 + dw^2),$$

viz. this is

$$= \frac{c^2\sin^2\sigma}{c^2\sigma^2}\left\{\tfrac{1}{2}(du^2 + dv^2 + dw^2) - \frac{1}{c^2\sigma^2}(udu + vdv + wdw)^2\right\};$$

or, finally, it is

$$= \frac{c^2\sigma^2}{2f_0(u)}\left\{f_0(du) - [d\sqrt{\{f_0(u)\}}]^2\right\},$$

which is right.

578.

A MEMOIR ON THE TRANSFORMATION OF ELLIPTIC FUNCTIONS.

[From the *Philosophical Transactions of the Royal Society of London*, vol. CLXIV. (for the year 1874), pp. 397—456. Received November 14, 1873,—Read January 8, 1874.]

THE theory of Transformation in Elliptic Functions was established by Jacobi in the *Fundamenta Nova* (1829); and he has there developed, transcendentally, with an approach to completeness, the general case, n an odd number, but algebraically only the cases $n=3$ and $n=5$; viz. in the general case the formulæ are expressed in terms of the elliptic functions of the nth part of the complete integrals, but in the cases $n=3$ and $n=5$ they are expressed rationally in terms of u and v (the fourth roots of the original and the transformed moduli respectively), these quantities being connected by an equation of the order 4 or 6, the modular equation. The extension of this algebraical theory to any value whatever of n is a problem of great interest and difficulty: such theory should admit of being treated in a purely algebraical manner; but the difficulties are so great that it was found necessary to discuss it by means of the formulæ of the transcendental theory, in particular by means of the expressions involving Jacobi's q $\left(\text{the exponential of } -\frac{\pi K'}{K}\right)$, or say by means of the q-transcendents. Several important contributions to the theory have since been made:—Sohnke, "Equationes modulares pro transformatione functionum ellipticarum," *Crelle*, t. XVI. (1836), pp. 97—130, (where the modular equations are found for the cases $n=3$, 5, 7, 11, 13, 17, and 19); Joubert, "Sur divers équations analogues aux équations modulaires dans la théorie des fonctions elliptiques," *Comptes Rendus*, t. XLVII. (1858), pp. 337—345, (relating among other things to the multiplier equation for the determination of Jacobi's M); and Königsberger, "Algebraische Untersuchungen aus der Theorie der elliptischen Functionen," *Crelle*, t. LXXII. (1870), pp. 176—275; together with other papers by Joubert and by Hermite in later volumes of the *Comptes Rendus*, which need not be more particularly referred to. In the present Memoir I carry on the theory, algebraically, as far as I am able; and I have, it appears to me, put the purely algebraical question in a clearer light than has hitherto been done; but I still find it necessary to resort to the transcendental theory. I remark that the case $n=7$ (next succeeding those of the *Fundamenta Nova*), on account of the peculiarly simple form of the modular equation $(1-u^8)(1-v^8)=(1-uv)^8$, presents but little difficulty; and I give the complete formulæ for this case, obtaining them as well algebraically as transcendentally; I also to a considerable extent discuss algebraically the case of the next succeeding prime value $n=11$. For the sake of completeness I reproduce Sohnke's modular equations, exhibiting them for greater clearness in a square form, and adding to them those for the non-prime cases $n=9$ and $n=15$; also a valuable table given by him for the powers of $f(q)$; and I give other tabular results which are of assistance in the theory.

The General Problem. Article Nos. 1 to 6.

1. Taking n a given odd number, I write

$$\frac{1-y}{1+y}=\frac{1-x}{1+x}\left(\frac{P-Qx}{P+Qx}\right)^2,$$

where P, Q are rational and integral functions of x^2, $P\pm Qx$ being each of them of the order $\frac{1}{2}(n-1)$, or, what is the same thing, $(1\pm x)(P\pm Qx)^2$ being each of them of the order n; that is,

$$\begin{array}{llll} & & n=4p+1, & n=4p+3, \\ \text{Order of } P \text{ in } x^2 \text{ is} & & p \quad , & p, \\ \quad\text{,,} \quad Q & \text{,,} & p-1; & p; \end{array}$$

whence in the first case the number of coefficients in P and Q is $(p+1)+p$, $=\frac{1}{2}(n+1)$, and in the second case the number is $(p+1)+(p+1)$, $=\frac{1}{2}(n+1)$, as before. Taking

$$P=\alpha+\gamma x^2+\epsilon x^4+\ldots,$$

$$Q=\beta+\delta x^2+\zeta x^4+\ldots,$$

the formula is

$$\frac{1-y}{1+y}=\frac{1-x}{1+x}\left(\frac{\alpha-\beta x+\gamma x^2-\ldots}{\alpha+\beta x+\gamma x^2+\ldots}\right)^2,$$

the number of coefficients being as just explained. Starting herefrom I reproduce in a somewhat altered form the investigation in the *Fundamenta Nova*, as follows.

2. If the coefficients are such that the equation remains true when we therein change simultaneously x into $\frac{1}{kx}$ and y into $\frac{1}{\lambda y}$, then the variables x, y will satisfy the differential equation

$$\frac{Mdy}{\sqrt{1-y^2.1-\lambda^2y^2}}=\frac{dx}{\sqrt{1-x^2.1-k^2x^2}},$$

(M a constant, the value of which, as will appear, is given by $\frac{1}{M} = 1 + \frac{2\beta}{\alpha}$); and the problem of transformation is thus to find the coefficients so that the equation may remain true on the above simultaneous change of the values of x, y.

In fact, observing that the original equation and therefore the new equation are each satisfied on changing therein simultaneously x, y into $-x$, $-y$, it follows that the equation may be written in the four forms

$$1 - \ \ y = (1 - \ x) A^2 (\div), \quad 1 + \ \ y = (1 + \ x) B^2 (\div),$$
$$1 - \lambda y = (1 - kx) C^2 (\div), \quad 1 + \lambda y = (1 + kx) D^2 (\div),$$

the common denominator being, say E, where A, B, C, D, E are all of them rational and integral functions of x; and this being so, the differential equation will be satisfied.

3. To develop the condition, observe that the assumed equation gives

$$y = \frac{x(P^2 + 2PQ + Q^2x^2)}{P^2 + 2PQx^2 + Q^2x^2}, \quad = \frac{x\mathfrak{A}}{\mathfrak{B}} \text{ suppose,}$$

where $\mathfrak{A}$, $\mathfrak{B}$ are functions each of them of the degree $\frac{1}{2}(n-1)$ in x^2. (Hence, if with Jacobi $\frac{1}{M}$ denotes the value $(y \div x)_{x=0}$, we have $\frac{1}{M} = \left(1 + \frac{2Q}{P}\right)_{x=0}$, $= 1 + \frac{2\beta}{\alpha}$, as mentioned.)

Suppose in general that U being any integral function $(1, x^2)^p$, we have

$$U^* = (k^2x^2)^p \left(1, \frac{1}{k^2x^2}\right)^p;$$

viz. let U^* be what U becomes when x is changed into $\frac{1}{kx}$ and the whole multiplied by $(k^2x^2)^p$.

Let y^* be the value of y obtained by writing $\frac{1}{kx}$ for x; then, observing that in the expression for y the degree of the numerator exceeds by unity that of the denominator, we have

$$y^* = \frac{1}{kx} \frac{\mathfrak{A}^*}{\mathfrak{B}^*},$$

whence

$$yy^* = \frac{1}{k} \frac{\mathfrak{A}\mathfrak{A}^*}{\mathfrak{B}\mathfrak{B}^*};$$

and the functions $\mathfrak{A}$, $\mathfrak{B}$ may be such that this shall be a constant value, $= \frac{1}{\lambda}$; viz. this will be the case if

$$\frac{\lambda}{k} = \frac{\mathfrak{B}\mathfrak{B}^*}{\mathfrak{A}\mathfrak{A}^*},$$

which being so, the required condition is satisfied.

4. I shall ultimately, instead of k, λ introduce Jacobi's u, v ($u=\sqrt[4]{k}$, $v=\sqrt[4]{\lambda}$); but it is for the present convenient to retain k, and instead of λ to introduce the quantity Ω connected with it by the equation $\lambda = k\Omega^2$; or say the value of Ω is $=v^2 \div u^2$. The modular equation in its standard form is an equation between u, v, which, as will appear, gives rise to an equation of the same order between u^2, v^2; and writing herein $v^2=\Omega u^2$, the resulting equation contains only integer powers of u^4, that is, of k, and we have an Ωk-form of the modular equation, or say an Ωk-modular equation, of the same order in Ω as the standard form is in v; these Ωk-forms for $n=3$, 5, 7, 11 will be given presently.

5. Suppose then, Ω being a constant, that we have identically

$$\mathfrak{A} = \frac{1}{\Omega k^{\frac{1}{2}(n-1)}} \mathfrak{B}^*;$$

this implies

$$\mathfrak{B} = \frac{\Omega}{k^{\frac{1}{2}(n-1)}} \mathfrak{A}^*.$$

(In fact, if

$$\mathfrak{A} = a + cx^2 + \ldots + qx^{n-3} + sx^{n-1},$$
$$\mathfrak{B} = b + dx^2 + \ldots + rx^{n-3} + tx^{n-1},$$

then

$$\mathfrak{A}^* = s + qk^2x^2 + \ldots + ck^{n-3}x^{n-3} + ak^{n-1}x^{n-1},$$
$$\mathfrak{B}^* = t + rk^2x^2 + \ldots + dk^{n-3}x^{n-3} + bk^{n-1}x^{n-1},$$

and the assumed equation gives

$$a = \frac{1}{\Omega k^{\frac{1}{2}(n-1)}} t, \quad c = \frac{k^2}{\Omega k^{\frac{1}{2}(n-1)}} r, \ldots, q = \frac{k^{n-3}}{\Omega k^{\frac{1}{2}(n-1)}} d, \quad s = \frac{k^{n-1}}{\Omega k^{\frac{1}{2}(n-1)}} b,$$

that is,

$$b = \frac{\Omega}{k^{\frac{1}{2}(n-1)}} s, \quad d = \frac{\Omega k^2}{k^{\frac{1}{2}(n-1)}} q, \ldots, r = \frac{\Omega k^{n-3}}{k^{\frac{1}{2}(n-1)}} c, \quad t = \frac{\Omega k^{n-2}}{k^{\frac{1}{2}(n-1)}} a;$$

and therefore $\mathfrak{B} = \frac{\Omega}{k^{\frac{1}{2}(n-1)}} \mathfrak{A}^*$.)

From these equations $\frac{\mathfrak{B}\mathfrak{B}^*}{\mathfrak{A}\mathfrak{A}^*} = \Omega^2$, that is, $=\frac{\lambda}{k}$, as it should be; so that Ω signifying as above, the required condition will be satisfied if only $\mathfrak{A} = \frac{1}{\Omega k^{\frac{1}{2}(n-1)}} \mathfrak{B}^*$; or substituting for $\mathfrak{A}$, $\mathfrak{B}$ their values, if only

$$(P^2 + 2PQx^2 + Q^2x^2)^* = \Omega k^{\frac{1}{2}(n-1)} (P^2 + 2PQ + Q^2x^2),$$

where each side is a function of x^2 of the order $\frac{1}{2}(n-1)$, or the number of terms is $\frac{1}{2}(n+1)$, the several coefficients being obviously homogeneous quadric functions of the $\frac{1}{2}(n+1)$ coefficients of P, Q. We have thus $\frac{1}{2}(n+1)$ equations, each of the form $U = \Omega V$, where U, V are given quadric functions of the coefficients of P, Q, say of the $\frac{1}{2}(n+1)$ coefficients α, β, γ, δ, &c., and where Ω is indeterminate.

6. We may from the $\frac{1}{2}(n+1)$ equations eliminate the $\frac{1}{2}(n-1)$ ratios $\alpha:\beta:\gamma:\ldots$, thus obtaining an equation in Ω (involving of course the parameter k) which is the Ωk-modular equation above referred to; and then Ω denoting any root of this equation, the $\frac{1}{2}(n+1)$ equations give a single value for the set of ratios $\alpha:\beta:\gamma:\delta:\ldots$, so that the ratio of the functions P, Q is determined, and consequently the value of y as given by the equation

$$\frac{1-y}{1+y}=\frac{(1-x)(P-Qx)^2}{(1+x)(P+Qx)^2}, \text{ or } y=\frac{x(P^2+2PQ+Q^2x^2)}{P^2+2PQx^2+Q^2x^2}.$$

The entire problem thus depends on the solution of the system of $\frac{1}{2}(n+1)$ equations,

$$(P^2+2PQx^2+Q^2x^2)^* = \Omega k^{\frac{1}{2}(n-1)}(P^2+2PQ+Q^2x^2).$$

The Ωk-Modular Equations, $n=3, 5, 7, 11$. Article No. 7.

7. For convenience of reference, and to fix the ideas, I give these results, calculated, as above explained, from the standard or uv-forms.

$n=3$:

	k^2	k	1	
Ω^4		$+1$		$=0$
Ω^3	-4			
Ω^2		$+6$		
Ω			-4	
Ω^0		$+1$		
	-4	$+8$	-4	

$\Omega=1$, we have $-4(k-1)^2=0$.

$n=5$:

	k^4	k^3	k^2	k	1	
Ω^6			$+1$			$=0$
Ω^5	-16		$+10$			
Ω^4			$+15$			
Ω^3			-20			
Ω^2			$+15$			
Ω			$+10$		-16	
Ω^0			$+1$			
	-16		$+32$		-16	

$\Omega=1$, we have $-16(k^2-1)^2=0$.

$n=7$:

	k^6	k^5	k^4	k^3	k^2	k	1	
Ω^8				+ 1				= 0
Ω^7	− 64		+ 56					
Ω^6		− 112		+ 140				
Ω^5			− 112		+ 56			
Ω^4				+ 70				
Ω^3			+ 56		− 112			
Ω^2				+ 140		− 112		
Ω					+ 56		− 64	
Ω^0				+ 1				
	− 64	− 112	0	+ 352	0	− 112	− 64	

$\Omega=1$, we have

$$-16(k-1)^2(4k^2+3k+1)(k^2+3k+4)=0.$$

$n=11$:

	k^{10}	k^9	k^8	k^7	k^6	k^5	k^4	k^3	k^2	k^1	k^0	
Ω^{12}						+ 1						+ 1
Ω^{11}	− 1024		+ 1408		− 396							− 12
Ω^{10}		− 5632		+ 4400		+ 1298						+ 66
Ω^9			− 16192		+ 16368		− 396					− 220
Ω^8				− 18656		+ 19151						+ 495
Ω^7			− 16016		− 1144		+ 16368					− 792
Ω^6				+ 4400		− 7876		+ 4400				+ 924
Ω^5					+ 16368		− 1144		− 16016			− 792
Ω^4						+ 19151		− 18656				+ 495
Ω^3					− 396		+ 16368		− 16192			− 220
Ω^2						+ 1298		+ 4400		− 5632		+ 66
Ω^1							− 396		+ 1408		− 1024	− 12
Ω^0						+ 1						+ 1
			− 32208	− 18656	− 1936	− 7876	− 1936	− 18656	− 32208			
			+ 1408	+ 8800	+ 32736	+ 40900	+ 32736	+ 8800	+ 1408			
	− 1024	− 5632	− 30800	− 9856	+ 30800	+ 33024	+ 30800	− 9856	− 30800	− 5632	− 1024	± 94624

Equation-systems for the cases $n = 3, 5, 7, 9, 11$. Art. Nos. 8 to 10.

8. $u = 3$, cubic transformation. $k = u^4$, $\Omega = \dfrac{v^2}{u^2}$ (here and in the other cases).

$P = \alpha$, $Q = \beta$. The condition here is

$$k^2x^2\alpha^2 + (2\alpha\beta + \beta^2) = \Omega k\,\{(\alpha^2 + 2\alpha\beta) + \beta^2x^2\},$$

and the system of equations thus is

$$\begin{aligned} k\alpha^2 &= \Omega\beta^2, \\ 2\alpha\beta + \beta^2 &= \Omega k\,(\alpha^2 + 2\alpha\beta), \end{aligned}$$

and similarly in the other cases; for these it will be enough to write down the equation-systems.

$n = 5$, quintic transformation.

$P = \alpha + \gamma x^2$, $Q = \beta$.

$$\begin{aligned} k^2\alpha^2 &= \Omega\gamma^2, \\ 2\alpha\gamma + 2\alpha\beta + \beta^2 &= \Omega\,(2\alpha\gamma + 2\beta\gamma + \beta^2), \\ \gamma^2 + 2\beta\gamma &= \Omega k^2\,(\alpha^2 + 2\alpha\beta). \end{aligned}$$

$n = 7$, septic transformation.

$P = \alpha + \gamma x^2$, $Q = \beta + \delta x^2$.

$$\begin{aligned} k^3\alpha^2 &= \Omega\delta^2, \\ k\,(2\alpha\gamma + 2\alpha\beta + \beta^2) &= \Omega\,(\gamma^2 + 2\gamma\delta + 2\beta\delta), \\ \gamma^2 + 2\beta\gamma + 2\alpha\delta + 2\beta\delta &= \Omega k\,(2\alpha\gamma + 2\beta\gamma + 2\alpha\delta + \beta^2), \\ \delta^2 + 2\gamma\delta &= \Omega k^3\,(\alpha^2 + 2\alpha\beta). \end{aligned}$$

$n = 9$, enneadic transformation.

$P = \alpha + \gamma x^2 + \epsilon x^4$, $Q = \beta + \delta x^2$.

$$\begin{aligned} k^4\alpha^2 &= \Omega\epsilon^2, \\ k^2\,(2\alpha\gamma + 2\alpha\beta + \beta^2) &= \Omega\,(2\gamma\epsilon + 2\epsilon\delta + \delta^2), \\ 2\alpha\epsilon + \gamma^2 + 2\alpha\delta + 2\gamma\beta + 2\beta\delta &= \Omega\,(2\alpha\epsilon + \gamma^2 + 2\gamma\delta + 2\epsilon\beta + 2\beta\delta), \\ 2\gamma\epsilon + 2\gamma\delta + 2\epsilon\beta + \delta^2 &= \Omega k^2\,(2\alpha\gamma + 2\alpha\delta + 2\gamma\beta + \beta^2), \\ \epsilon^2 + 2\delta\epsilon &= \Omega k^4\,(\alpha^2 + 2\alpha\beta). \end{aligned}$$

$n = 11$, endecadic transformation.

$P = \alpha + \gamma x^2 + \epsilon x^4$, $Q = \beta + \delta x^2 + \zeta x^4$.

$$\begin{aligned} k^5\alpha^2 &= \Omega\zeta^2, \\ k^3\,(2\alpha\gamma + 2\alpha\beta + \beta^2) &= \Omega\,(\epsilon^2 + 2\epsilon\zeta + 2\delta\zeta), \\ k\,(2\alpha\epsilon + \gamma^2 + 2\alpha\delta + 2\gamma\beta + 2\beta\delta) &= \Omega\,(2\gamma\epsilon + 2\gamma\zeta + 2\epsilon\delta + 2\beta\delta + \delta^2), \\ 2\gamma\epsilon + 2\alpha\zeta + 2\gamma\delta + 2\epsilon\beta + 2\beta\zeta + \delta^2 &= \Omega k\,(2\alpha\epsilon + \gamma^2 + 2\alpha\zeta + 2\gamma\delta + 2\epsilon\beta + 2\beta\delta), \\ \epsilon^2 + 2\gamma\zeta + 2\epsilon\delta + 2\delta\zeta &= \Omega k^3\,(2\alpha\gamma + 2\alpha\delta + 2\gamma\beta + \beta^2), \\ 2\epsilon\zeta + \zeta^2 &= \Omega k^5\,(\alpha^2 + 2\alpha\beta). \end{aligned}$$

And so on.

9. It will be noticed that if the coefficients of $P+Qx$ taken in order are

$$\alpha,\ \beta, \ldots,\ \rho,\ \sigma,$$

then in every case the first and last equations are

$$k^{\frac{1}{2}(n-1)}\,\alpha^2 = \Omega\sigma^2,$$

$$2\rho\sigma + \sigma^2 = \Omega k^{\frac{1}{2}(n-1)}\,(\alpha^2 + 2\alpha\beta).$$

Putting in the first of these $k=u^4$, $\Omega = \frac{v^2}{u^2}$, the equation becomes

$$u^{2n}\alpha^2 = v^2\sigma^2,$$

where each side is a perfect square; and in extracting the square root we may without loss of generality take the roots positive, and write $u^n\alpha = v\sigma$.

This speciality, although it renders it proper to employ ultimately u, v in place of k, Ω, produces really no depression of order (viz. the Ωk-form of the modular equation is found to be of the same order in Ω that the standard or uv-form is in v), and is in another point of view a disadvantage, as destroying the uniformity of the several equations: in the discussion of order I consequently retain Ω, k. Ultimately these are to be replaced by u, v; the change in the equation-systems is so easily made that it is not necessary here to write them down in the new form in u, v.

10. The case $\alpha=0$ has to be considered in the discussion of order, but we have thus only solutions which are to be rejected; in the proper solutions α is not $=0$, and it may therefore for convenience be taken to be $=1$. We have then $\sigma = u^n \div v$. The last equation becomes therefore

$$\frac{u^n}{v}\left(2\rho + \frac{u^n}{v}\right) = \frac{v^2}{u^2}\,u^{2n-2}\,(1+2\beta)\,;$$

or recollecting that β is connected with the multiplier M by the relation $\frac{1}{M} = 1+2\beta$, that is,

$$2\beta = \frac{1}{M} - 1,$$

and substituting for $1+2\beta$ its value, the equation becomes

$$2\rho = v^3 u^{n-4}\left(\frac{1}{M} - \frac{u^4}{v^4}\right);$$

that is, the first and the last coefficients are 1, $\frac{u^n}{v}$, and the second and the penultimate coefficients are each expressed in terms of v, M. The cases $n=3$, $n=5$ are so far peculiar, that the only coefficients are α, β, or α, β, γ; in the next case $n=7$, the only coefficients are α, β, γ, δ, and we have in this case all the coefficients expressed as above.

The Ωk-form—Order of the Systems. Art. Nos. 11 to 22.

11. In the general case, n an odd number, we have Ω and $\frac{1}{2}(n+1)$ coefficients connected by a system of $\frac{1}{2}(n+1)$ equations of the form

$$\Omega = \frac{U}{U'} = \frac{V}{V'} = \ldots,$$

where $U, V, \ldots, U', V', \ldots$ are given quadric functions of the coefficients. Omitting the ($\Omega =$), there remains a system of $\frac{1}{2}(n-1)$ equations of the form $\frac{U}{U'} = \frac{V}{V'} = \ldots$, or say

$$\left|\begin{matrix} U, & V, & W, \ldots \\ U', & V', & W', \ldots \end{matrix}\right| = 0,$$

which determine the ratios $\alpha : \beta : \gamma : \ldots$ of the coefficients; and to each set of ratios there corresponds a single value of Ω. The order of the system, or number of sets of ratios, is $= \frac{1}{2}(n+1) \,.\, 2^{\frac{1}{2}(n-1)}$, $= (n+1) \,.\, 2^{\frac{1}{2}(n-3)}$; and this is consequently the number of values of Ω, or the order of the equation for the determination of Ω; viz. but for reduction, the order in Ω of the Ωk-modular equation would be $= (n+1) \,.\, 2^{\frac{1}{2}(n-3)}$. In the case $n = 3$, this is $= 4$, which is right, but for any larger value of n the order is far too high; in fact, assuming (as the case is) that the order is equal to the order in v of the uv-form, the order should for a prime value of n be $= n+1$, and for a composite value not containing any square factor be = the sum of the divisors of n. I do not attempt a general investigation, but confine myself to showing in what manner the reductions arise.

12. I will first consider the cubic transformation; here, writing for convenience $\frac{\alpha}{\beta} = \theta$, the equations give

$$\frac{k\theta^2}{2\theta+1} = \frac{1}{k(\theta^2+2\theta)}, \text{ that is, } k^2\theta^3(\theta+2) - (2\theta+1) = 0,$$

and

$$k\theta^2 = \Omega.$$

The equation in θ gives $(k^2\theta^4 - 1)^2 - 4\theta^2(k^2\theta^2 - 1)^2 = 0$, and we have thence

$$k(\Omega^2 - 1)^2 - 4\Omega(k\Omega - 1)^2 = 0,$$

that is,

$$k\Omega^4 - 4k^2\Omega^3 + 6k\Omega^2 - 4\Omega + k = 0,$$

the modular equation; and then $k^2\theta^4 - 1 + 2\theta(k^2\theta^2 - 1) = 0$, that is, $\Omega^2 - 1 + 2\theta(k\Omega - 1) = 0$, or $2\theta = -\frac{\Omega^2-1}{k\Omega-1}$, which is $= \frac{2\alpha}{\beta}$, say we have $\alpha = \Omega^2 - 1$, $\beta = 2(1 - k\Omega)$; consequently

$$\frac{1-y}{1+y} = \frac{1-x}{1+x}\left\{\frac{\Omega^2 - 1 + 2(k\Omega-1)x}{\Omega^2 - 1 - 2(k\Omega-1)x}\right\}^2,$$

$$\lambda = \Omega^2 k, \text{ and } \frac{1}{M} = \frac{\alpha^2 + 2\alpha\beta}{\alpha^2} = \frac{\theta+2}{\theta} = \frac{\Omega^2 - 4k\Omega + 3}{\Omega^2 - 1},$$

which completes the theory.

13. Reproducing for this case the general theory, it appears *à priori* that Ω is determined by a quartic equation; in fact, from the original equations eliminating Ω, we have an equation

$$\begin{vmatrix} U, & V \\ U', & V' \end{vmatrix} = 0,$$

where U, U', V, V' are quartic functions of α, β; that is, the ratio $\alpha : \beta$ has four values, and to each of these there corresponds a single value of Ω; viz. Ω is determined by a quartic equation.

14. Considering next the case $n=5$, the quintic transformation; the elimination of Ω gives the equations

$$\frac{U}{U'} = \frac{V}{V'} = \frac{W}{W'},$$

where U, U', &c. are all quadric functions of α, β, γ. We have thence $4.4-2.2, =12$ sets of values of $\alpha : \beta : \gamma$; viz. considering α, β, γ as coordinates *in plano*, the curves $UV' - U'V = 0$, $UW' - U'W = 0$ are quartic curves intersecting in 16 points; but among these are included the four points $U=0$, $U'=0$ (in fact, the point $\alpha=0$, $\gamma=0$ four times), which are not points of the curve $VW' - V'W = 0$; there remain therefore $16-4, =12$ intersections, agreeing with the general value $(n+1).2^{\frac{1}{2}(n-3)}$. Hence Ω is in the first instance determined by an equation of the order 12; but the proper order being $=6$, there must be a factor of the order 6 to be rejected. To explain this and to determine the factor, observe that the equations in question are

$$\begin{aligned} k^2\alpha^2(2\alpha\gamma + 2\beta\gamma + \beta^2) - \gamma^2(2\alpha\gamma + 2\alpha\beta + \beta^2) &= 0, \\ k^4\alpha^3(\alpha + 2\beta) \qquad\qquad - \gamma^3(\gamma + 2\beta) \qquad\qquad &= 0; \end{aligned}$$

at the point $\alpha=0$, $\gamma=0$, the first of these has a double point, the second a triple point; or there are at the point in question 6 intersections; but 4 of these are the points which give the foregoing reduction $16-4=12$; we have thus the point $\alpha=0$, $\gamma=0$, counting twice among the twelve points. Writing in the two equations $\beta=0$, the equations become $k^2\alpha^3\gamma - \alpha\gamma^3 = 0$, $k^4\alpha^4 - \gamma^4 = 0$, viz. these will be satisfied if $k^2\alpha^2 - \gamma^2 = 0$, that is, the curves pass through each of the two points $(\beta=0, \gamma=\pm k\alpha)$, and these values satisfy (as in fact they should) the third equation

$$k^2(2\alpha\gamma + 2\alpha\beta + \beta^2)\,\alpha(\alpha + 2\beta) - \gamma(\gamma + 2\beta)(2\alpha\gamma + 2\beta + \beta^2) = 0.$$

It is moreover easily shown that the three curves have at each of the points in question a common tangent; viz. taking A, B, C as current coordinates, the tangent at the point (α, β, γ) of the second curve has for its equation

$$A(2\alpha^3 + 3\alpha^2\beta)k^4 + B(k^4\alpha^3 - \gamma^3) - C(2\gamma^3 + 3\gamma^2\beta) = 0;$$

and for $\beta=0$, $\gamma=\pm k\alpha$, this becomes $2kA + B(k \mp 1) \mp 2C = 0$, viz. this is the line from the point $(\beta=0, \gamma=\pm k\alpha)$ to the point $(1, -2, 1)$. And similarly for the other two curves we find the same equation for the tangent.

Hence among the 12 points are included the point $(\gamma=0,\ \alpha=0)$ twice, and the points $(\beta=0,\ \gamma=\pm k\alpha)$ each twice: we have thus a reduction $=6$.

15. Writing in the equations $\gamma=0$, $\alpha=0$, the first and third are satisfied identically, and the second becomes $\beta^2=\Omega\beta^2$, that is, the equations give $\Omega=1$; writing $\beta=0$, they become

$$k^2\alpha^2=\Omega\gamma^2,\quad \alpha\gamma=\Omega\alpha\gamma,\quad \gamma^2=\Omega k^2\alpha^2,$$

viz. putting herein $\gamma^2=k^2\alpha^2$, the equations again give $\Omega=1$; hence the factor of the order 6 is $(\Omega-1)^6$, and the equation of the twelfth order for the determination of Ω is

$$(\Omega-1)^6\,\{(\Omega,\ 1)^6\}=0,$$

where $(\Omega,\ 1)^6=0$ is the Ωk-modular equation above written down.

16. Reverting to the equation

$$\frac{1-y}{1+y}=\frac{(1-x)(P-Qx)^2}{(1+x)(P+Qx)^2},$$

it is to be observed that for $\alpha=0$, $\gamma=0$, that is, $P=0$, this becomes simply $y=x$, which is the transformation of the order 1; the corresponding value of the modulus λ is $\lambda=k$, and the equation $\lambda=\Omega^2k$ then gives $\Omega^2=1$, which is replaced by $\Omega-1=0$.

If in the same equation we write $\beta=0$, that is, $Q=0$, then (without any use of the equation $\gamma^2=k^2\alpha^2$) we have $y=x$, the transformation of the order 1; but although this is so, the fundamental equation

$$(P^2+2PQx^2+Q^2x^2)^*=\Omega k^2(P^2+2PQ+Q^2x^2),$$

which, putting therein $Q=0$, becomes $(P^2)^*=\Omega k^2P^2$, that is, $(k^2x^2\alpha+\gamma)^2=\Omega k^2(\alpha+\gamma x^2)^2$, is not satisfied by the single relation $\Omega-1=0$, but necessitates the further relation $\gamma^2=k^2\alpha^2$.

The thing to be observed is that the extraneous factor $(\Omega-1)^6$, equated to zero, gives for Ω the value $\Omega=1$ corresponding to the transformation $y=x$ of the order 1.

17. Considering next $n=7$, the septic transformation; we have here between $\alpha,\beta,\gamma,\delta$ a fourfold relation of the form

$$\left(\begin{vmatrix} U, & V, & W, & Z \\ U', & V', & W', & Z' \end{vmatrix}\right)=0,$$

where, as before, U, U', &c. are quadric functions, and the number of solutions is here 8.2^2, $=32$; to each of these corresponds a single value of Ω, or Ω is in the first instance determined by an equation of the order 32. But the order of the modular equation is $=8$; or representing this by $\{(\Omega,\ 1)^8\}=0$, the equation must be

$$(\Omega,\ 1)^{24}\,\{(\Omega,\ 1)^8\}=0,$$

viz. there must be a special factor of the order 24.

18. One way of satisfying the equations is to write therein $\alpha = 0$, $\delta = 0$; the equations thus become

$$k\beta^2 = \Omega\gamma^2,$$
$$\gamma^2 + 2\beta\gamma = \Omega k (2\beta\gamma + \beta^2);$$

or putting $\beta, \gamma = \alpha', \beta'$,

$$k\alpha'^2 = \Omega\beta'^2,$$
$$\beta'^2 + 2\alpha'\beta' = \Omega k (2\alpha'\beta' + \alpha'^2),$$

which (with α', β' instead of α, β) are the very equations which belong to the cubic transformation; hence a factor is $\{(\Omega, 1)^4\}$.

Observe that for the values in question $\alpha = 0$, $\delta = 0$, $P = \beta' x^2$, $Q = \alpha'$,

$$(P \pm Qx)^2 = x^2 (\alpha' \pm \beta' x)^2, \ = x^2 (P' \pm Q'x)^2, \text{ if } P' = \alpha', \ Q' = \beta',$$

and therefore

$$\frac{1-y}{1+y} = \frac{1-x}{1+x}\left(\frac{P' - Q'x}{P' + Q'x}\right)^2,$$

which is the formula for a cubic transformation.

19. The equations may also be satisfied by writing therein $\gamma = k\alpha$, $\delta = k\beta$; in fact, substituting these values, they become

$$k^3\alpha^2 = \Omega k^2\beta^2,$$
$$2k^2\alpha^2 + k(2\alpha\beta + \beta^2) = \Omega k^2 (\alpha^2 + 2\alpha\beta) + 2\Omega k\beta^2,$$
$$k^2\alpha^2 + 2k(\beta^2 + 2\alpha\beta) = 2\Omega k^2 (\alpha^2 + 2\alpha\beta) + \Omega k\beta^2,$$
$$k^2 (\beta^2 + 2\alpha\beta) = \Omega k^3 (\alpha^2 + 2\alpha\beta);$$

the first and last of these are

$$k\alpha^2 = \Omega\beta^2,$$
$$\beta^2 + 2\alpha\beta = \Omega k (\alpha^2 + 2\alpha\beta),$$

which being satisfied the second and third equations are satisfied identically; and these are the formulæ for a cubic transformation; that is, we again have the factor $\{(\Omega, 1)^4\}$.

Observe that for the values in question $\gamma = k\alpha$, $\delta = k\beta$, we have $P = \alpha(1 + kx^2)$, $Q = \beta(1 + kx^2)$; so that, writing $P' = \alpha$, $Q' = \beta$, we have for y the value

$$\frac{1-y}{1+y} = \frac{(1-x)(P' - Q'x)^2}{(1+x)(P' + Q'x)^2},$$

which is the formula for a cubic transformation.

20. It is important to notice that we cannot by writing $\alpha = 0$ or $\delta = 0$ reduce the transformation to a quintic one; in fact, the equation $k^3\alpha^2 = \Omega\delta^2$ shows that if either of these equations is satisfied the other is also satisfied; and we have then the foregoing case $\alpha = 0$, $\delta = 0$, giving not a quintic but a cubic transformation.

And for the same reason we cannot by writing $\alpha = 0$, $\beta = 0$, $\gamma = 0$ or $\beta = 0$, $\gamma = 0$, $\delta = 0$ reduce the transformation to the order 1. There is thus no factor $\Omega - 1$.

21. As regards the non-existence of the factor $\Omega - 1$, I further verify this by writing in the equations $\Omega = 1$; they thus become

$$k^3\alpha^2 = \delta^2,$$
$$k(2\alpha\gamma + 2\alpha\beta + \beta^2) = \gamma^2 + 2\gamma\delta + 2\beta\delta,$$
$$\gamma^2 + 2\beta\gamma + 2\alpha\delta + 2\beta\delta = k(2\alpha\gamma + 2\beta\gamma + 2\alpha\delta + \beta^2),$$
$$\delta^2 + 2\gamma\delta = k^3(\alpha^2 + 2\alpha\beta),$$

which it is to be shown cannot be satisfied in general, but only for certain values of k.

Reducing the last equation, this is $\gamma\delta = k^3\alpha\beta$, which, combined with the first, gives $\alpha\gamma = \beta\delta$; and if for convenience we assume $\alpha = 1$, and write also $\theta = \pm\sqrt{k}$ (that is, $k = \theta^2$), then the values of $\alpha, \beta, \gamma, \delta$ are $\alpha = 1$, $\beta = \gamma\theta^{-3}$, $\gamma = \gamma$, $\delta = \theta^3$; which values, substituted in the second and the third equations, give two equations in γ, θ; and from these, eliminating γ, we obtain an equation for the determination of θ, that is, of k. In fact, the second equation gives

$$\theta^2(2\gamma + 2\gamma\theta^{-3} + \gamma^2\theta^{-6}) = \gamma^2 + 2\gamma\theta^3 + 2\gamma;$$

or, dividing by γ and reducing,

$$\gamma(1 - \theta^4) = 2\theta^3(\theta^2 - 1)(\theta^2 - \theta + 1),$$

that is,

$$\gamma(1 + \theta^2) = -2\theta^3(\theta^2 - \theta + 1),$$

or, as this may also be written,

$$(\gamma + \theta^3)(1 + \theta^2) = -\theta^3(\theta - 1)^2,$$

that is,

$$\gamma + \theta^3 = \frac{-\theta^3(\theta - 1)^2}{\theta^2 + 1}.$$

Moreover the third equation gives

$$\gamma^2 + 2\gamma^2\theta^{-3} + 2\theta^3 + 2\gamma = \theta^2(2\gamma + 2\gamma^2\theta^{-3} + 2\theta^3 + \gamma^2\theta^{-6}),$$

that is,

$$\gamma^2(\theta^4 - 2\theta^3 + 2\theta - 1) - 2(\gamma + \theta^3)\theta^4(\theta^2 - 1) = 0;$$

or dividing by $\theta^2 - 1$, it is

$$\gamma^2(\theta - 1)^2 = 2\theta^4(\gamma + \theta^3);$$

whence also

$$\gamma^2 = \frac{-2\theta^7}{\theta^2 + 1}.$$

Also

$$4\theta^6(\theta^2 - \theta + 1)^2 = \gamma^2(\theta^2 + 1)^2,$$

wherefore

$$2(\theta^2 - \theta + 1)^2 = -\theta(\theta^2 + 1) \text{ or } 2(\theta^2 - \theta + 1)^2 + \theta(\theta^2 + 1) = 0,$$

or

$$\theta(\theta^2 + 1)^2 + 2(\theta^2 - \theta + 1)^2 = 0,$$

that is,

$$2\theta^4 - 3\theta^3 + 6\theta^2 - 3\theta + 2 = 0,$$

or finally

$$(2\theta^2 - \theta + 1)(\theta^2 - \theta + 2) = 0.$$

We have thus $(2\theta^2+1)^2=\theta^2$, that is, $4\theta^4+3\theta^2+1=0$ or $4k^2+3k+1=0$, or else $(\theta^2+2)^2=\theta^3$, that is, $\theta^4+3\theta^2+4=0$ or $k^2+3k+4=0$; viz. the equation in k is

$$(4k^2+3k+1)(k^2+3k+4)=0,$$

these being in fact the values of k given by the modular equation on putting therein $\Omega=1$.

The equation of the order 32 thus contains the factor $\{(\Omega,\ 1)^4\}$ at least twice, and it does not contain either the factor $\Omega-1$, or the factor $\{(\Omega,\ 1)^6\}$ belonging to the quintic transformation; it may be conjectured that the factor $\{(\Omega,\ 1)^4\}$ presents itself six times, and that the form is

$$\{(\Omega,\ 1)^4\}^6(\Omega,\ 1)^8=0;$$

but I am not able to verify this, and I do not pursue the discussion further.

22. The foregoing considerations show the grounds of the difficulty of the purely algebraical solution of the problem; the required results, for instance the modular equation, are obtained not in the simple form, but accompanied with special factors of high order. The transcendental theory affords the means of obtaining the results in the proper form without special factors; and I proceed to develop the theory, reproducing the known results as to the modular and multiplier equations, and extending it to the determination of the transformation-coefficients α, β, ...

The Modular Equation. Art. Nos. 23 to 28.

23. Writing, as usual, $q=e^{-\frac{\pi K'}{K}}$, we have u, a given function of q, viz.

$$\begin{aligned} u &= \sqrt{2}q^{\frac{1}{8}}\frac{1+q^2\,.\,1+q^4\,.\,1+q^6\,..}{1+q\,.\,1+q^3\,.\,1+q^5\,..} \\ &= \sqrt{2}q^{\frac{1}{8}}(1-q+2q^2-3q^3+4q^4-6q^5+9q^6-12q^7+\ldots) \\ &= \sqrt{2}q^{\frac{1}{8}}f(q) \text{ suppose;} \end{aligned}$$

and this being so, the several values of v and of the other quantities in question are all given in terms of q.

The case chiefly considered is that of n an odd prime; and unless the contrary is stated it is assumed that this is so. We have then $n+1$ transformations corresponding to the same number $n+1$ of values of v; these may be distinguished as $v_0, v_1, v_2, \ldots, v_n$; viz. writing α to denote an imaginary nth root of unity, we have

$$v_0=(-)^{\frac{n^2-1}{8}}\sqrt{2}q^{\frac{n}{8}}f(q^n),\quad v_1=\sqrt{2}\,(\alpha q^{\frac{1}{n}})^{\frac{1}{8}}f(\alpha q^{\frac{1}{n}}),\quad v_2=\sqrt{2}\,(\alpha^2 q^{\frac{1}{n}})^{\frac{1}{8}}f(\alpha^2 q^{\frac{1}{n}}),\ \&\text{c.},$$

$$v_n=\sqrt{2}q^{\frac{1}{8n}}f(q^{\frac{1}{n}}).$$

(Observe $(-)^{\frac{n^2-1}{8}}=+$ for $n=8p\pm1$, $-$ for $n=8p\pm3$.)

The occurrence of the fractional exponent $\frac{1}{8}$ is, as will appear, a circumstance of great importance; and it will be convenient to introduce the term "octicity," viz. an expression of the form $q^{\frac{f}{8}}F(q)$ ($f=0$, or a positive integer not exceeding 7, $F(q)$ a rational function of q) may be said to be of the octicity f.

24. The modular equation is of course

$$(v-v_0)(v-v_1)\ldots.(v-v_n)=0;$$

say this is

$$v^{n+1}-Av^n+Bv^{n-1}-\ldots=0,$$

so that $A=\Sigma v_0$, $B=\Sigma v_0v_1$, &c. In the development of these expressions, the terms having a fractional exponent, with denominator n, would disappear of themselves, as involving symmetrically the several nth roots of unity; and each coefficient would be of the form $q^{\frac{g}{8}}F(q)$, F a rational and integral function of q. It is moreover easy to see that, for the several coefficients A, B, C,......, g will denote the positive residue (mod. 8) of n, $2n$, $3n$,... respectively.

Hence assuming, as the fact is, that these coefficients are severally rational and integral functions of q, it follows that the form is

$$au^g+bu^{g+8}+cu^{g+16}+\ldots.,$$

g having the foregoing values for the several coefficients respectively. And it being known that the modular equation is as regards u of the order $=n+1$, there is a known limit to the number of terms in the several coefficients respectively. We have thus for each coefficient an identity of the form

$$A=au^g+bu^{g+8}+\ldots.,$$

where A and u being each of them given in terms of q, the values of the numerical coefficients a, b,.. can be determined; and we thus arrive at the modular equation.

25. It is in effect in this manner that the modular equations are calculated in Sohnke's Memoir. Various relations of symmetry in regard to (u, v) and other known properties of the modular equation are made use of in order to reduce the number of the unknown coefficients to a minimum; and (what in practice is obviously an important simplification) instead of the coefficients Σv_0, Σv_0v_1, &c., it is the sums of powers Σv_0, Σv_0^2, &c., which are compared with their expressions in terms of u, in order to the determination of the unknown numerical coefficients a, b,... The process is a laborious one (although less so than perhaps might beforehand have been imagined), involving very high numbers; it requires the development up to high powers of q, of the high powers of the before-mentioned function $f(q)$; and Sohnke gives a valuable Table, which I reproduce here; adding to it the three columns which relate to ϕq.

q.	ϕq =	$\phi^2 q$ =	$\phi^{-2} q$ =	fq =	$f^2 q$ =	$f^3 q$ =	$f^4 q$ =	$f^5 q$ =	$f^6 q$ =	$f^7 q$ =	$f^8 q$ =	$f^9 q$ =	$f^{10} q$ =	$f^{11} q$ =
0	1	1	1	1	1	1	1	1	1	1	1	1	1	1
1	+ 2	+ 4	− 4	− 1	− 2	− 3	− 4	− 5	− 6	− 7	− 8	− 9	− 10	− 11
2	0	+ 4	+ 12	+ 2	+ 5	+ 9	+ 14	+ 20	+ 27	+ 35	+ 44	+ 54	+ 65	+ 77
3	0	0	− 32	− 3	− 10	− 22	− 40	− 65	− 98	− 140	− 192	− 255	− 330	− 418
4	+ 2	+ 4	+ 76	+ 4	+ 18	+ 48	+ 101	+ 185	+ 309	+ 483	+ 718	+ 1026	+ 1420	+ 1914
5	0	+ 8	− 168	− 6	− 32	− 99	− 236	− 481	− 882	− 1498	− 2400	− 3672	− 5412	− 7733
6	0	0	+ 352	+ 9	+ 55	+ 194	+ 518	+ 1165	+ 2330	+ 4277	+ 7352	+ 11997	+ 18765	+ 28336
7	0	0	− 704	− 12	− 90	− 363	− 1080	− 2665	− 5784	− 11425	− 20992	− 36414	− 60270	− 95931
8	0	+ 4		+ 16	+ 144	+ 657	+ 2162	+ 5820	+ 13644	+ 28889	+ 56549	+ 103977	+ 181645	+ 304062
9	+ 2	+ 4		− 22	− 226	− 1155	− 4180	− 12220	− 30826	− 69734	− 145008	− 281911	− 518660	− 911240
10	0	+ 8		+ 29	+ 346	+ 1977	+ 7840	+ 24802	+ 67107	+ 161735	+ 356388	+ 730953	+ 1413465	+ 2601786
11	0	0		− 38	− 522	− 3312	− 14328	− 48880	− 141444	− 362271	− 844032	− 1822689	− 3697960	− 7120136
12	0	0		+ 50	+ 777	+ 5443	+ 25591	+ 93865	+ 289746	+ 786877	+ 1934534	+ 4390824	+ 9331565	+ 18766759
13	0	+ 8		− 64	− 1138	− 8787	− 44776	− 176125	− 578646	− 1662927	− 4306368	− 10256508	− 22800050	− 47830486
14	0	0		+ 82	+ 1648	+ 13968	+ 76918	+ 323685	+ 1129527	+ 3428770	+ 9337704	+ 23303025	+ 54112825	+ 118270746
15	0	0		− 105	− 2362	− 21894	− 129952	− 583798	− 2159774	− 6913760	− 19771392	− 51631227	− 125090220	− 284527793
16	+ 2	+ 4		+ 132	+ 3348	+ 33873	+ 216240	+ 1035060	+ 4052721	+ 13660346	+ 40965362	+ 111804966	+ 282298020	+ 667553898
17	0	+ 8		− 166	− 4704	− 51795	− 354864	− 1806600	− 7474806	− 26492361	− 83207976	− 237074742	− 623185010	− 1530587256
18	0	+ 4		+ 208	+ 6554	+ 78345	+ 574958	+ 3108085	+ 15063859*	+ 50504755	+ 165944732	+ 493063403	+ 1348033540	+ 3435726536
19	0	0		− 258	− 9056	− 117412								
20	0	+ 8		+ 320	+ 12425	+ 174033								
21	0	0		− 395	− 16932	− 255945								
22	0	0		+ 484										
23	0	0		− 592										
24	0	0		+ 722										
25	+ 2	+ 12		− 876										
26	0	+ 8		+ 1060										

* [Wrongly given by Sohnke as + 3108085.]

ind. of q.	$f^{12}q$ =	$f^{13}q$ =	$f^{14}q$ =	$f^{15}q$ =	$f^{16}q$ =	$f^{17}q$ =	$f^{18}q$ =	$f^{19}q$ =	$f^{20}q$ =
0	1	1	1	1	1	1	1	1	1
1	− 12	− 13	− 14	− 15	− 16	− 17	− 18	− 19	− 20
2	+ 90	+ 104	+ 119	+ 135	+ 152	+ 170	+ 189	+ 209	+ 230
3	− 520	− 637	− 770	− 920	− 1088	− 1275	− 1482	− 1710	− 1960
4	+ 2523	+ 3263	+ 4151	+ 5205	+ 6444	+ 7888	+ 9558	+ 11476	+ 13665
5	− 10764	− 14651	− 19558	− 25668	− 33184	− 42330	− 53352	− 66519	− 82124
6	+ 41534	+ 59345	+ 82936	+ 113675	+ 153152	+ 203201	+ 265923	+ 343710	+ 439270
7	− 147720	− 221091	− 322828	− 461265	− 646528	− 890800	− 1208610	− 1617147	− 2136600
8	+ 490869	+ 768131	+ 1169847	+ 1739710	+ 2533070	+ 3619334	+ 5084478	+ 7034047	+ 9596460
9	− 1539472	− 2514551	− 3988292	− 6164345	− 9311664	− 13780540	− 20021534	− 28607673	− 40260300
10	+ 4592430	+ 7818200	+ 12896562	+ 20690964	+ 32387616	+ 49590581	+ 74438388	+ 109745767	+ 159174524
11	− 13111632	− 23233535	− 39809574	− 66222405	− 107299904	− 169812320	− 263104686		
12	+ 36006362	+ 66328961	+ 117921321	+ 203173760	+ 340436664	+ 556366922	+ 889020813		
13	− 95497116	− 182681916	− 336630840	− 600165795	− 1039026144	− 1752038020	− 2884990266		
14	+ 245457000	+ 487098378	+ 929461993	+ 1713196575	+ 3061896704	+ 5323089708	+ 9026077050		
15	− 613183064	− 1261118313	− 2489690882	− 4740491107	− 8739810688	− 15653783345	− 27314626158		
16	+ 1492474572	+ 3178449222	+ 6486711301	+ 12748926285	+ 24229115109	+ 44679433473	+ 80177033781*		
17	− 3546915228	− 7815313766	− 16475721276	− 33400680615	− 65390485328	− 124069449335	− 228831885054		
18	+ 8245677110	+ 18783535199	+ 40874694490	+ 85415669230	+ 172155210320	+ 335888162944	+ 636376573943		

* [In Sohnke, the figure 1 has] dropped out.

26. I give from Sohnke the series of modular equations, adding those for the composite cases $n=9$ and $n=15$, as to which see the remarks which follow the Table.

$n=3$.

	v^4	v^3	v^2	v	1	
u^4					-1	
u^3		$+2$				
u^2						
u				-2		
1	$+1$					
	1	$+2$	0	-2	-1	$=(v+1)^3(v-1)$.

$n=5$.

	v^6	v^5	v^4	v^3	v^2	v	1	
u^6							-1	
u^5		$+4$						
u^4					-5			
u^3								
u^2			$+5$					
u						-4		
1	$+1$							
	1	$+4$	$+5$	0	-5	-4	-1	$=(v+1)^5(v-1)$.

$n=7$.

	v^8	v^7	v^6	v^5	v^4	v^3	v^2	v	1	
u^8	0								$+1$	
u^7		-8								
u^6			$+28$							
u^5				-56						
u^4					$+70$					
u^3						-56				
u^2							$+28$			
u								-8		
1	$+1$								0	
	1	-8	$+28$	-56	$+70$	-56	$+28$	-8	$+1$	$=(v-1)^8$.

$n = 9.$

	v^{12}	v^{11}	v^{10}	v^9	v^8	v^7	v^6	v^5	v^4	v^3	v^2	v	1	
u^{12}					0								+ 1	
u^{11}				− 16								+ 8		
u^{10}			+ 16								+ 10			
u^9		− 16								− 24				
u^8	0								+ 15					
u^7								+ 48						
u^6							− 84							
u^5						+ 48								
u^4					+ 15								0	
u^3				− 24								− 16		
u^2			+ 10								+ 16			
u		+ 8								− 16				
1	+ 1								0					
	1	− 8	+ 26	− 40	+ 15	+ 48	− 84	+ 48	+ 15	− 40	+ 26	− 8	+ 1	$=(v-1)^{10}(v+1)^2.$

$n = 11.$

	v^{12}	v^{11}	v^{10}	v^9	v^8	v^7	v^6	v^5	v^4	v^3	v^2	v	1	
u^{12}					0								− 1	
u^{11}		+ 32								− 22				
u^{10}							− 44							
u^9				+ 88								+ 22		
u^8	0								− 165					
u^7						+ 132								
u^6			+ 44								− 44			
u^5								− 132						
u^4					+ 165								0	
u^3		− 22								− 88				
u^2							+ 44							
u				+ 22								− 32		
1	+ 1								0					
	1	+ 10	+ 44	+ 110	+ 165	+ 132	0	− 132	− 165	− 110	− 44	− 10	− 1	$=(v+1)^{11}(v-1).$

$n=13.$

	v^{14}	v^{13}	v^{12}	v^{11}	v^{10}	v^9	v^8	v^7	v^6	v^5	v^4	v^3	v^2	v	1
u^{14}							0								− 1
u^{13}		+64								− 52					
u^{12}					0								−65		
u^{11}								+ 208							
u^{10}			0								− 429				
u^9						+ 520								+52	
u^8	0								− 429						
u^7				+ 208								− 208			
u^6							+ 429								0
u^5		−52								− 520					
u^4					+ 429								0		
u^3								− 208							
u^2			+65								0				
u						+ 52								−64	
1	+ 1								0						
	1	+ 12	+ 65	+ 208	+ 429	+ 572	+ 429	0	− 429	− 572	− 429	− 208	− 65	− 12	− 1 $=(v+1)^{13}(v-1)$.

$n=15.$

	v^{24}	v^{23}	v^{22}	v^{21}	v^{20}	v^{19}	v^{18}	v^{17}	v^{16}	v^{15}	v^{14}	v^{13}	v^{12}	v^{11}	v^{10}	v^{9}	v^{8}	v^{7}	v^{6}	v^{5}	v^{4}	v^{3}	v^{2}	v^{1}	v^{0}
u^{24}	0								0								0								+1
u^{23}		0								− 128								+ 120							
u^{22}			0								− 960								+ 980						
u^{21}				0								− 2880								+ 2888					
u^{20}					− 256								− 3920								+ 4050				
u^{19}						− 1920								− 800								+ 2888			
u^{18}							− 5824								+ 5040								+ 980		
u^{17}								− 8960								+ 8160								+ 120	
u^{16}	0								− 26970								+ 27209								0
u^{15}		− 128								− 6960								+ 8160							
u^{14}			− 960								− 5320								+ 5040						
u^{13}				− 2880								+ 3120								− 800					
u^{12}					− 3920								+ 9660								− 3920				
u^{11}						− 800								+ 3120								− 2880			
u^{10}							+ 5040								− 5320								− 960		
u^{9}								+ 8160								− 6960								− 128	
u^{8}	0								+ 27209								− 26970								0
u^{7}		+ 120								+ 8160								− 8960							
u^{6}			+ 980								+ 5040								− 5824						
u^{5}				+ 2888								− 800								− 1920					
u^{4}					+ 4050								− 3920								− 256				
u^{3}						+ 2888								− 2880								0			
u^{2}							+ 980								− 960								0		
u								+ 120								− 128								0	
1	+1								0								0								0
	+1	− 8	+ 20	+ 8	− 126	+ 168	+ 196	− 680	+ 239	+ 1072	− 1240	− 560	+ 1820	− 560	− 1240	+ 1072	+ 239	− 680	+ 196	+ 168	− 126	+ 8	+ 20	− 8	+ 1 $=(v-1)^8(v^2-1)^8$.

$n=17$.

	v^{18}	v^{17}	v^{16}	v^{15}	v^{14}	v^{13}	v^{12}	v^{11}	v^{10}	v^{9}	v^{8}	v^{7}	v^{6}	v^{5}	v^{4}	v^{3}	v^{2}	v	1
u^{18}			0								0								+1
u^{17}		−256								+272								−34	
u^{16}	0								−272								+425		
u^{15}								+1632								−2448			
u^{14}							−4080								+7140				
u^{13}						+4896								−13464					
u^{12}					−4080								+22644						
u^{11}				+1632								−33456							
u^{10}			−272								+44030								0
u^{9}		+272								−49164								−272	
u^{8}	0								+44030								−272		
u^{7}								−33456								+1632			
u^{6}							+22644								−4080				
u^{5}						−13464								+4896					
u^{4}					+7140								−4080						
u^{3}				−2448								+1632							
u^{2}			+425								−272								0
u		−34								+272								−256	
1	+1								0								0		
	1	−18	+153	−816	+3060	−8568	+18564	−31824	+43758	−48520	+43758	−31824	+18564	−8568	+3060	−816	+153	−18	+1 $=(v-1)^{18}$.

	v^{20}	v^{19}	v^{18}	v^{17}	v^{16}	v^{15}	v^{14}	v^{13}	v^{12}	v^{11}	v^{10}	v^{9}	v^{8}	v^{7}	v^{6}	v^{5}	v^{4}	v^{3}	v^{2}	v	1
u^{20}					0								0								−1
u^{19}		+512								−608								+114			
u^{18}							+2432								−2584						
u^{17}				−2432								+3344								−114	
u^{16}	0								+3952								−6859				
u^{15}						+5472								+2280							
u^{14}			−2432								−10488								−2584		
u^{13}								+25536								−2280					
u^{12}					−3952								−21242								0
u^{11}		−608								+20748								−3344			
u^{10}							+10488								−10488						
u^{9}				+3344								−20748								+608	
u^{8}	0								+21242								+3952				
u^{7}						+2280								−25536							
u^{6}			+2584								+10488								+2432		
u^{5}								−2280								−5472					
u^{4}					+6859								−3952								0
u^{3}		+114								−3344								+2432			
u^{2}							+2584								−2432						
u				−114								+608								−512	
1	1								0								0				
	1	+18	+152	+798	+2907	+7752	+15504	+23256	+25194	+16796	0	−16796	−25194	−23256	−15504	−7752	−2907	−798	−152	−18	−1

$=(v+1)^{19}(v-1)$.

$n=19$.

Various remarks arise on the Tables. Attending first to the cases n a prime number; the only terms of the order $n+1$ in v or u are $v^{n+1} \pm u^{n+1}$, viz. $n \equiv 3$ or 5 (mod. 8) the sign is $-$, but $n \equiv 1$ or 7 (mod. 8) the sign is $+$. And there is in every case a pair of terms $v^n u^n$ and vu, having coefficients equal in absolute magnitude, but of opposite signs, or of the same sign, in the two cases respectively.

Each Table is symmetrical in regard to its two diagonals respectively, so that every non-diagonal coefficient occurs (with or without reversal of sign) 4 times; viz. in the case $n \equiv 1$ or 7 (mod. 8) this is a perfect symmetry, without reversal of sign; but in the case $n \equiv 3$ or 5 (mod. 8) it is, as regards the lines parallel to either diagonal, and in regard to the other diagonal, alternately a perfect symmetry without reversal of sign and a skew symmetry with reversal. Thus in the case $n=19$, the lines parallel to the dexter diagonal are -1 (symmetrical), $+114$, -114 (skew), 0, -2584, -6859, -2584, 0 (symmetrical), and so on. The same relation of symmetry is seen in the composite cases $n=9$ and $n=15$, both belonging to $n \equiv 1$ or 7, mod. 8.

If, as before, n is prime, then putting in the modular equation $u=1$, the equation in the case $n \equiv 1$ or 7 (mod. 8) becomes $(v-1)^{n+1}=0$, but in the case $n \equiv 3$ or 5 (mod. 8) it becomes $(v+1)^n (v-1)=0$.

27. In the case n a composite number not containing any square factor, then dividing n in every possible way into two factors $n=ab$ (including the divisions $n.1$ and $1.n$), and denoting by β an imaginary bth root of unity, a value of v is

$$\pm \sqrt{2}\,(\beta q^{\frac{a}{b}})^{\frac{1}{8}} f(\beta q^{\frac{a}{b}});$$

so that the whole number of roots (or order of the modular equation) is $=\nu$, if ν be the sum of the divisors of n. Thus $n=15$, the values are

$$\begin{array}{cccc} \sqrt{2}\,q^{\frac{15}{8}} f(q^{15}), & -\sqrt{2}\,q^{\frac{1}{8}\cdot\frac{5}{3}} f(q^{\frac{5}{3}}), & -\sqrt{2}\,q^{\frac{1}{8}\cdot\frac{3}{5}} f(q^{\frac{3}{5}}), & \sqrt{2}\,q^{\frac{1}{15}\cdot\frac{1}{8}} f(q^{\frac{1}{15}}) \\ 1 \quad , & 3 \quad , & 5 \quad , & 15 \text{ roots}; \end{array}$$

and the order of the modular equation is $=24$. The modular equation might thus be obtained as for a prime number; but it is easier to decompose n into its prime factors, and consider the transformation as compounded of transformations of these prime orders. Thus $n=15$, the transformation is compounded of a cubic and a quintic one. If the v of the cubic transformation be denoted by θ, then we have

$$\theta^4 + 2\theta^3 u^3 - 2\theta u - u^4 = 0;$$

and to each of the four values of θ corresponds the six values of v belonging to the quintic transformation given by

$$v^6 + 4v^5\theta^5 + 5v^4\theta^2 - 5v^2\theta^4 - 4v\theta - \theta^6 = 0.$$

The equation in v is thus

$$(v^6 + 4v^5\theta_1^5 + \ldots - \theta_1^6)(v^6 + \ldots - \theta_2^6)(v^6 + \ldots - \theta_3^6)(v^6 + \ldots - \theta_4^6) = 0,$$

where θ_1, θ_2, θ_3, θ_4 are the roots of the equation in θ, viz. we have

$$\theta^4 + 2\theta^3 u^3 - 2\theta u - u^4 = (\theta - \theta_1)(\theta - \theta_2)(\theta - \theta_3)(\theta - \theta_4);$$

and it was in this way that the equation for the case $n = 15$ was calculated. Observe that writing $u = 1$, we have $(\theta + 1)^3 (\theta - 1) = 0$, or say $\theta_1 = \theta_2 = \theta_3 = -1$, $\theta_4 = +1$. The equation in v thus becomes $\{(v-1)^5 (v+1)\}^3 (v+1)^5 (v-1) = 0$, that is, $(v-1)^{16} (v+1)^8 = 0$.

28. The case where n has a square factor is a little different; thus $n = 9$, the values are

$$\sqrt{2}q^{\frac{9}{8}} f(q^9),\quad -\sqrt{2}q^{\frac{1}{8}\cdot\frac{3}{3}} f(q^{\frac{3}{3}}),\quad \sqrt{2}q^{\frac{1}{8}\cdot\frac{1}{9}} f(q^{\frac{1}{9}}),$$
$$1 \quad , \quad 3 \quad , \quad 9 \quad , \text{ roots};$$

but here ω being an imaginary cube root of unity, the second term denotes the three values,

$$\sqrt{2}q^{\frac{1}{8}} f(q),\quad \sqrt{2}\,(q\omega)^{\frac{1}{8}} f(\omega q),\quad \sqrt{2}\,(\omega^2 q)^{\frac{1}{8}} f(\omega^2 q),$$

the first of which is $= u$, and is to be rejected; there remain $1 + 2 + 9$, $= 12$ roots, or the equation is of the order 12.

Considering the equation as compounded of two cubic transformations, if the value of v for the first of these be θ, then we have

$$\theta^4 + 2\theta^3 u^3 - 2\theta u - u^4 = 0;$$

and to the four values of θ correspond severally the four values of v given by the equation

$$v^4 + 2v^3\theta^3 - 2v\theta - \theta^4 = 0.$$

One of these values is however $v = -u$, since the $v\theta$-equation is satisfied on writing therein $v = -u$; hence, writing

$$\theta^4 + 2\theta^3 u^3 - 2\theta u - u^4 = (\theta - \theta_1)(\theta - \theta_2)(\theta - \theta_3)(\theta - \theta_4),$$

we have an equation

$$(v^4 + 2v^3\theta_1^3 - 2v\theta_1 - \theta_1^4)(v^4 + \ldots - \theta_2^4)(v^4 + \ldots - \theta_3^4)(v^4 + \ldots - \theta_4^4) = 0,$$

which contains the factor $(v + u)^4$ and, divested hereof, gives the required modular equation of the order 12; it was in fact obtained in this manner.

Observe that writing $u = 1$, we have $(\theta + 1)^3 (\theta - 1) = 0$, or say $\theta_1 = \theta_2 = \theta_3 = -1$, $\theta_4 = 1$; the modular equation then becomes

$$\{(v-1)^3 (v+1)\}^3 (v+1)^3 (v-1) \div (v+1)^4 = 0,$$

that is,

$$(v-1)^{10} (v+1)^2 = 0.$$

The Multiplier Equation. Art. No. 29.

29. The theory is in many respects analogous to that of the modular equation. To each value of v there corresponds a single value of M; hence M, or what is the same thing $\frac{1}{M}$, is determined by an equation of the same order as v, viz. n being prime, the order is $=n+1$. The last term of the equation is constant, and the other coefficients are rational and integral functions of u^8, of a degree not exceeding $\frac{1}{2}(n-1)$; and not only so, but they are, $n\equiv 1$ (mod. 4), rational and integral functions of $u^8(1-u^8)$, and $n\equiv 3$ (mod. 4), alternately of this form and of the same form multiplied by the factor $(1-2u^8)$.

The values are in fact given as transcendental functions of q; viz. denoting by $M_0, M_1, M_2, \ldots, M_n$ the values corresponding to $v_0, v_1, v_2, \ldots, v_n$ respectively, and writing

$$\phi(q)=\frac{(1+q)(1+q^3)(1+q^5)\ldots(1-q^2)(1-q^4)(1-q^6)\ldots}{(1-q)(1-q^3)(1-q^5)\ldots(1+q^2)(1+q^4)(1+q^6)\ldots}$$
$$=1+2q+2q^4+2q^9+2q^{16}+\ldots,$$

then we have

$$M_0=\frac{(-)^{\frac{n-1}{2}}}{n}\,\frac{\phi^2(q)}{\phi^2(q^n)},$$

$$M_1=\frac{\phi^2(q)}{\phi^2(\alpha q^{\frac{1}{n}})},\ldots(\alpha \text{ an imaginary } n\text{th root of unity})$$

$$\vdots$$

$$M_n=\frac{\phi^2 q}{\phi^2(q^{\frac{1}{n}})}.$$

Hence, the form of the equation being known, the values of the numerical coefficients may be calculated; and it was in this way that Joubert obtained the following results. I have in some cases changed the sign of Joubert's multiplier, so that in every case the value corresponding to $u=0$ shall be $M=1$.

The equations are:

$$n=3,\quad \frac{1}{M^4}$$
$$+\frac{1}{M^3}\cdot\ 0$$
$$+\frac{1}{M^2}\cdot -6$$
$$+\frac{1}{M}\cdot\ 8(1-2u^8)$$
$$-3=0.$$

$u=0$, this is

$$\left(\frac{1}{M}-1\right)^3\left(\frac{1}{M}+3\right)=0;$$

$u=1$, it is

$$\left(\frac{1}{M}+1\right)^3\left(\frac{1}{M}-3\right)=0.$$

$$
\begin{aligned}
n = 5,\quad & \frac{1}{M^6} \\
& + \frac{1}{M^5} \cdot -10 \\
& + \frac{1}{M^4} \cdot +35 \\
& + \frac{1}{M^3} \cdot -60 \\
& + \frac{1}{M^2} \cdot +55 \\
& \frac{1}{M} \cdot -26 + 256u^8(1-u^8) \\
& \qquad +5 = 0.
\end{aligned}
$$

$u = 0$ or 1, this is

$$\left(\frac{1}{M} - 1\right)^5 \left(\frac{1}{M} - 5\right) = 0.$$

$$
\begin{aligned}
n = 7,\quad & \frac{1}{M^8} \\
& + \frac{1}{M^7} \cdot 0 \\
& + \frac{1}{M^6} \cdot -28 \\
& + \frac{1}{M^5} \cdot +112(1-2u^8) \\
& + \frac{1}{M^4} \cdot -210 \\
& + \frac{1}{M^3} \cdot +224(1-2u^8) \\
& + \frac{1}{M^2} \cdot -140 - 21 \,.\, 256u^8(1-u^8) \\
& + \frac{1}{M} \cdot \{48 + 2048u^8(1-u^8)\}(1-2u^8) \\
& \qquad +7 = 0.
\end{aligned}
$$

$u = 0$, this is

$$\left(\frac{1}{M} - 1\right)^7 \left(\frac{1}{M} - 7\right) = 0;$$

$u = 1$, it is

$$\left(\frac{1}{M} + 1\right)^7 \left(\frac{1}{M} + 7\right) = 0.$$

$$
\begin{aligned}
n = 11,\quad & \frac{1}{M^{12}} \\
& + \frac{1}{M^{11}} \cdot 0 \\
& + \frac{1}{M^{10}} \cdot -66 \\
& + \frac{1}{M^9} \cdot +440(1-2u^8)
\end{aligned}
$$

$u = 0$, this is

$$\left(\frac{1}{M} - 1\right)^{11} \left(\frac{1}{M} + 11\right) = 0;$$

$u = 1$, it is

$$\left(\frac{1}{M} + 1\right)^{11} \left(\frac{1}{M} - 11\right) = 0.$$

$$+\frac{1}{M^8}\,.\,-1485$$

$$+\frac{1}{M^7}\,.\,+3168\,(1-2u^8)$$

$$+\frac{1}{M^6}\,.\,-4620-3\,.\,11^2\,.\,256u^8(1-u^8)$$

$$+\frac{1}{M^5}\,.\,\{+4752+11\,.\,4096u^8(1-u^8)\}\,(1-2u^8)$$

$$+\frac{1}{M^4}\,.\,-3465-3\,.\,7\,.\,11\,.\,512u^8(1-u^8)$$

$$+\frac{1}{M^3}\,.\,\{+1760+11\,.\,83\,.\,2048u^8(1-u^8)\}\,(1-2u^8)$$

$$+\frac{1}{M^2}\,.\,-594-9\,.\,11\,.\,37\,.\,256u^8(1-u^8)-3\,.\,11\,.\,131072\,\{u^8(1-u^8)\}^2$$

$$+\frac{1}{M}\{120+15\,.\,4096u^8(1-u^8)-524288\,\{u^8(1-u^8)\}^2\}\,(1-2u^8)$$

$$-11=0.$$

The Multiplier as a rational function of u, v. Art. Nos. 30 to 36.

30. The multiplier M, as having a single value corresponding to each value of v, is necessarily a rational function of u, v; and such an expression of M can, as remarked by Königsberger, be deduced from the multiplier equation by means of Jacobi's theorem,

$$M^2=\frac{1}{n}\frac{\lambda(1-\lambda^2)}{k(1-k^2)}\frac{dk}{d\lambda};$$

viz. substituting for k, λ their values u^8, v^8, and observing that if the modular equation be $F(u, v)=0$ so that the value of $\frac{du}{dv}$ is $=-F'(v)\div F'(u)$, this is

$$M^2=-\frac{1}{n}\frac{(1-v^8)\,vF'v}{(1-u^8)\,uF'u};$$

and then in the multiplier equation separating the terms which contain the odd and even powers, and writing it in the form $\Phi(M^2)+M\Psi(M^2)=0$, this equation, substituting therein for M^2 its value, gives the value of M rationally.

The rational expression of M in terms of u, v is of course indeterminate, since its form may be modified in any manner by means of the equation $F(u, v)=0$; and in the expression obtained as above, the orders of the numerator and the denominator are far too high. A different form may be obtained as follows: for greater convenience I seek for the value not of M but of $\frac{1}{M}$.

31. Denoting, as above, by $M_0, M_1, \ldots, M_n$ the values which correspond to $v_0, v_1, \ldots, v_n$ respectively, and writing $S\frac{1}{M} = \frac{1}{M_0} + \frac{1}{M_1} + \ldots + \frac{1}{M_n}$, &c., we have $S\frac{1}{M}$, $S\frac{v}{M}$, &c., all of them expressible as determinate functions of u; and we have moreover the theorem that each of these is a rational and integral function of u: we have thus the series of equations

$$S\frac{1}{M} = A,\quad S\frac{v}{M} = B,\ \ldots,\ S\frac{v^n}{M} = H,$$

where $A, B, \ldots, H$ are rational and integral functions of u. These give linearly the different values of $\frac{1}{M}$; in fact, we have

$$(v_0 - v_1)\ldots(v_0 - v_n)\frac{1}{M_0} = H - GSv_1 + FSv_1v_2 - \ldots \pm Av_1v_2\ldots v_n,$$

where Sv_1, Sv_1v_2, &c. denote the combinations formed with the roots $v_1, v_2, \ldots, v_n$ (these can be expressed in terms of the single root v_0); and we have also $(v_0 - v_1)\ldots(v_0 - v_n) = F'(v_0)$: the resulting equation is consequently $F'v_0\frac{1}{M_0} = R(u, v_0)$, R a determinate rational and integral function of (u, v_0); but as the same formula exists for each root of the modular equation, we may herein write M, v in place of M_0, v_0; and the formula thus is

$$F'v \,.\, \frac{1}{M} = R(u, v),$$

viz. we thus obtain the required value of $\frac{1}{M}$ as a rational fraction, the denominator being the determinate function $F'v$, and the numerator being, as is easy to see, a determinate function of the order n as regards v.

32. The method is applicable when M is only known by its expression in terms of q; but if we know for M an expression in terms of v, u, then the method transforms this into a standard form as above. By way of illustration I will consider the case $n=3$, where the modular equation is

$$v^4 + 2v^3u^3 - 2vu - u^4 = 0,$$

and where a known expression of M is $\frac{1}{M} = 1 + \frac{2u^3}{v}$. Here writing S_{-1}, $S_0 (=4)$, S, &c. to denote the sum of the powers -1, 0, 1, &c. of the roots of the equation, we have

$$S\frac{1}{M} = S_0 + 2u^3S_{-1},\ = 0\ ,\ \text{as appears from the values presently given,}$$

$$S\frac{v}{M} = S_1 + 2u^3S_0\ ,\ = 6u^3,$$

$$S\frac{v^2}{M} = S_2 + 2u^3S_1\ ,\ = 0\ ,$$

$$S\frac{v^3}{M} = S_3 + 2u^3S_2\ ,\ = 6u\,;$$

and observing that v_0 being ultimately replaced by v, we have

$$Sv_1 = Sv_0 - v, \qquad Sv_0v_1 = Sv_0v_1 - vSv_0 + v^2, \quad v_1v_2v_3 = Sv_0v_1v_2 - vSv_0v_1 + v_2Sv_0 - v^3,$$

that is,

$$Sv_1 = -2u^3 - v, \quad Sv_1v_2 = 2u^3v + v^2, \quad v_1v_2v_3 = 2u - 2u^3v^2 - v^3,$$

we have

$$\begin{aligned} F'v \,.\, \frac{1}{M} = \; & (S_3 + 2u^3S_2) \\ & + (2u^3 + v)(S_2 + 2u^3S_1) \\ & + (2u^3v + v^2)(S_1 + 2u^3S_0) \\ & + (-2u + 2u^3v^2 + v^3)(S_0 + 2u^3S_{-1}), \end{aligned}$$

viz. this is

$$\begin{aligned} 2(2v^3 + 3v^2u^3 - u)\frac{1}{M} = \; & v^3(S_0 + 2u^3S_{-1}) \\ & + v^2(S_1 + 4u^3S_0 + 4u^6S_{-1}) \\ & + v\,(S_2 + 4u^3S_1 + 4u^6S_0) \\ & + \;\,(S_3 + 4u^3S_2 + 4u^6S_1 - 2uS_0 - 4u^4S_{-1}). \end{aligned}$$

But we have

$$S_{-1} = -\frac{2}{u^3}, \quad S_0 = 4, \quad S_1 = -2u^3, \quad S_2 = 4u^6, \quad S_3 = 6u - 8u^9;$$

and the equation thus is

$$(2v^3 + 3v^2u^3 - u)\frac{1}{M} = 3(v^2u^2 + 2u^5v + 1)\,u;$$

to verify which observe that, substituting herein for $\frac{1}{M}$ its value $1 + \frac{2u^3}{v}$, the equation becomes

$$(2v^3 + 3v^2u^3 - u)(v + 2u^3) - 3vu(v^2u^2 + 2u^5v + 1) = 0;$$

that is,

$$2v^4 + 4v^3u^3 - 4vu - 2u^4 = 0,$$

as it should do.

33. Any expression whatever of M in terms of u, v is in fact one of a system of four expressions; viz. we may simultaneously change

	u	v	$\frac{1}{M}$	that is, signs are $n \equiv 1$	$n \equiv 3$	$n \equiv 5$	$n \equiv 7$ (mod. 8)
into	v,	$(-)^{\frac{n^2-1}{8}}u$,	$(-)^{\frac{n-1}{2}}nM$;	$+ + +$	$+ - -$	$+ - -$	$+ + -$
or	$\frac{1}{u}$,	$\frac{1}{v}$,	$\frac{v^4}{u^4M}$;	$+ + +$	$+ + +$	$+ + +$	$+ + +$
or	$\frac{1}{v}$,	$(-)^{\frac{n^2-1}{8}}\frac{1}{u}$,	$(-)^{\frac{n-1}{2}}\frac{u^4}{v^4}nM$;	$+ + +$	$+ - -$	$+ - +$	$+ + -$

Thus $n=3$, starting from $\frac{1}{M}=1+\frac{2u^3}{v}$, we have

$$\frac{1}{M}=1+\frac{2u^3}{v},\quad -3M=1-\frac{2v^3}{u},\quad \frac{v^4}{u^4M}=1+\frac{2v}{u^3},\quad -\frac{u^4}{v^4}3M=1-\frac{2u}{v^3};$$

and of course if from any two of these we eliminate M, we have either an identity or the modular equation; thus we have the modular equation under the six different forms:

$$\begin{array}{ll}(1,\ 2) & (v+2u^3)(u-2v^3)+3uv\ =0,\\ (1,\ 3) & v^3(v+2u^3)-u(u^3+2v)\ =0,\\ (1,\ 4) & (v+2u^3)(v^3-2u)+3u^4\ =0,\\ (2,\ 3) & (u-2v^3)(u^3+2v)+3v^4\ =0,\\ (2,\ 4) & v(v^3-2u)-u^3(u-2v^3)\ =0,\\ (3,\ 4) & (u^3+2v)(v^3-2u)+3u^3v^3=0.\end{array}$$

34. Next $n=5$. Here, starting from $\frac{1}{M}=\frac{v-u^5}{v(1-uv^3)}$, the changes give

$$\frac{1}{M}=\frac{v-u^5}{v(1-uv^3)},\quad 5M=\frac{u+v^5}{u(1+u^3v)},\quad \frac{v^4}{u^4M}=\frac{v^3(v-u^5)}{u^4(1-uv^3)},\quad \frac{u^4}{v^4}5M=\frac{u^3(u+v^5)}{v^4(1+u^3v)},$$

viz. the third and the fourth forms agree with the first and the second forms respectively; that is, there are only two independent forms, and the elimination of M from these gives

$$5uv(1-uv^3)(1+u^3v)-(v-u^5)(u+v^5)=0,$$

which is a form of the modular equation.

35. In the case $n=7$, starting from $\frac{1}{M}=\frac{-7u(1-uv)(1-uv+u^2v^2)}{u-v^7}$ (as to this see *post*, No. 43), the forms are

$$\frac{1}{M}=\frac{-7u(1-uv)(1-uv+u^2v^2)}{u-v^7}\ \ldots\ldots\ldots\ldots(1),$$

$$-7M=\frac{-7v(1-uv)(1-uv+u^2v^2)}{v-u^7}\ \ldots\ldots\ldots\ldots(2),$$

$$\frac{v^4}{u^4M}=\frac{-7v^4(1-uv)(1-uv+u^2v^2)}{u^3(u-v^7)}\ \ldots\ldots\ldots\ldots(3),$$

$$-\frac{u^4}{v^4}7M=\frac{-7u^4(1-uv)(1-uv+u^2v^2)}{v^3(v-u^7)}\ \ldots\ldots\ldots\ldots(4);$$

so that here again the third and the fourth forms are identical with the second and the third forms respectively; there are thus only two forms, and the elimination of M gives

$$(u-v^7)(v-u^7)+7uv(1-uv)^2(1-uv+u^2v^2)^2=0,$$

which is a form of the modular equation.

36. If in the foregoing equation

$$F'v \cdot \frac{1}{M} = R(u, v),$$

we make the change $u, v, \frac{1}{M}$ into $v, \pm u, \pm nM$, it becomes

$$\pm F'u \cdot nM = R(v, \pm u);$$

combining these equations, we have

$$\pm nM^2 \cdot \frac{F'u}{F'v} = \frac{R(v, \pm u)}{R(u, v)};$$

or, substituting herein the foregoing value

$$M^2 = -\frac{1}{n} \frac{(1-v^8) vF'v}{(1-u^8) uF'u},$$

this becomes

$$\mp \frac{v(1-v^8)}{u(1-u^8)} = \frac{R(v, \pm u)}{R(u, v)} \qquad \begin{array}{l} + \text{ for } n \equiv 3 \text{ or } 5 \text{ (mod. 8)}, \\ - \text{ for } n \equiv 1 \text{ or } 7 \text{ (mod. 8)}, \end{array}$$

which must agree with the modular equation. Thus in the last-mentioned case $n = 3$, we have

$$\tfrac{1}{2}F'v \cdot \frac{1}{M} = 3(v^2u^2 + 2u^5v + 1)u,$$

or, say

$$R(u, v) = \quad (v^2u^2 + 2u^5v + 1)u,$$

and therefore

$$R(v, -u) = \quad (v^2u^2 - 2uv^5 + 1)v;$$

the equation is

$$+\frac{v(1-v^8)}{u(1-u^8)} = \frac{(v^2u^2 - 2uv^5 + 1)v}{(v^2u^2 + 2u^5v + 1)u},$$

which is right; because Jacobi, p. 82, [*Ges. Werke*, t. I., p. 137], for the modular equation, gives

$$1 - u^8 = (1 - u^2v^2)(v^2u^2 + 2u^5v + 1), \quad 1 - v^8 = (1 - u^2v^2)(v^2u^2 - 2uv^5 + 1).$$

Observe that the general equation

$$\mp \frac{v(1-v^8)}{u(1-u^8)} = \frac{R(v, \pm u)}{R(u, v)}$$

no longer contains the functions $F'v$, $F'u$, which enter into Jacobi's expression of M^2.

Theorem in connexion with the multiplication of Elliptic Functions. Art. Nos. 37 to 40.

37. The theory of multiplication gives an important theorem in regard to transformation. Starting with the $n$thic transformation

$$\frac{1-y}{1+y} = \frac{1-x}{1+x} \left(\frac{\alpha - \beta x + \gamma x^2 - \dots}{\alpha + \beta x + \gamma x^2 + \dots}\right)^2, \quad = \frac{1-x}{1+x} \left(\frac{P - Qx}{P + Qx}\right)^2,$$

we may form a like transformation,

$$\frac{1-z}{1+z}=\frac{1-y}{1+y}\left(\frac{\alpha'-\beta'y+\gamma'y^2-\dots}{\alpha'+\beta'y+\gamma'y^2+\dots}\right)^2, \ =\frac{1-y}{1+y}\left(\frac{P'-Q'y}{P'+Q'y}\right)^2,$$

such that the combination of the two gives a multiplication, viz. for the relation between y, z, deriving w from v as v from u, we have $w=u$; and instead of M we have M', $=\pm\frac{1}{nM}$; that is, we have

$$\frac{dx}{\sqrt{1-x^2\,.\,1-u^8x^2}}=\frac{Mdy}{\sqrt{1-y^2\,.\,1-v^8y^2}},$$

$$\frac{dy}{\sqrt{1-y^2\,.\,1-v^8y^2}}=\frac{M'dz}{\sqrt{1-z^2\,.\,1-u^8z^2}},$$

and thence

$$\frac{dx}{\sqrt{1-x^2\,.\,1-u^8x^2}}=\frac{\pm\frac{1}{n}dz}{\sqrt{1-z^2\,.\,1-u^8z^2}};$$

or, writing $x=\text{sn}\,\theta$, we have $z=\pm\,\text{sn}\,n\theta$; $\pm$ is here $(-)^{\frac{n-1}{2}}$, viz. it is $-$ for $n\equiv 3$ or 7 (mod. 8), and $+$ for $n\equiv 1$ or 5 (mod. 8).

Now in part effecting the substitution, we have

$$\frac{1-z}{1+z}=\frac{1-x}{1+x}\left(\frac{P-Qx}{P+Qx}\right)^2\cdot\left(\frac{P'-Q'y}{P'+Q'y}\right)^2,$$

where y denotes its value in terms of x.

And from the theory of elliptic functions, replacing $\text{sn}\,n\theta$, $\text{sn}\,\theta$ by their values $\pm z$, x, we have an equation

$$\frac{1-z}{1+z}=\frac{1-x}{1+x}\left(\frac{A-Bx+Cx^2-\dots}{A+Bx+Cx^2+\dots}\right)^2,$$

where $A-Bx+Cx^2-\dots$, $A+Bx+Cx^2+\dots$ are given functions each of the order $\frac{1}{2}(n^2-1)$; viz. the coefficients are given functions of k, or, what is the same thing, of u^4. Comparing the two results, we see that in the $n$thic transformation the sought-for function, $\alpha+\beta x+\gamma x^2+\dots$ of the order $\frac{1}{2}(n-1)$, is a factor of a given function $A+Bx+Cx^2+\dots$ of the order $\frac{1}{2}(n^2-1)$.

38. Considering the modular equation as known, then by what precedes we have

$$\alpha+\beta x+\gamma x^2+\dots=\alpha\left\{1+\frac{\beta}{\alpha}x+\dots+\frac{u^n}{v}x^{\frac{1}{2}(n-1)}\right\};$$

that is, the given function $A+Bx+Cx^2+\dots$ has a factor $1+\frac{\beta}{\alpha}x+\dots+\frac{u^n}{v}x^{\frac{1}{2}(n-1)}$, of which one (the last) coefficient $\frac{u^n}{v}$ is known, and we are hence able theoretically to determine all the other coefficients rationally in terms of u, v; that is, the modular equation being known, we can theoretically complete the solution of the transformation problem. I do not, however, see the way to obtaining a convenient solution in this manner.

39. The formula in question for $n=3$ is

$$\frac{1+\text{sn}\,3\theta}{1-\text{sn}\,3\theta}=\frac{1-\text{sn}\,\theta}{1+\text{sn}\,\theta}\left(\frac{1+2\,\text{sn}\,\theta-2k^2\,\text{sn}^3\,\theta-k^2\,\text{sn}^4\,\theta}{1-2\,\text{sn}\,\theta+2k^2\,\text{sn}^3\,\theta-k^2\,\text{sn}^4\,\theta}\right)^2,$$

which, putting therein $x=\text{sn}\,\theta$, $z=-\text{sn}\,3\theta$, and replacing k by u^4, may be written

$$1+z\,(\div)=(1+x)\,(1-2x+2u^8x^3-u^8x^4)^2\,(\div),$$

where the signs $(\div)$ indicate denominators which are obtained from the numerators by changing the signs of z, x respectively.

The theorem in regard to $n=3$ thus is, $1+\dfrac{u^3}{v}x$ is a factor of $1-2x+2u^8x^3-u^8x^4$; viz. writing in the last-mentioned function $x=-\dfrac{v}{u^3}$, we ought to have

$$0=1+2\frac{v}{u^3}-2\frac{v^3}{u}-\frac{v^4}{u^4},$$

that is,

$$u^4+2uv-2u^3v^3-v^4=0,$$

which is in fact the modular equation.

40. And so for $n=5$, if $x=\text{sn}\,\theta$, $z=\text{sn}\,5\theta$; and for $n=7$, if $x=\text{sn}\,\theta$, $z=-\text{sn}\,7\theta$; the formulæ are:—

$n=5$,

$$\begin{array}{lll}
1+z=(1+x)\,\{ & 1 & \\
(\div) & +\,2 & x \\
& -\,4 & x^2 \\
& -\,10u^8 & x^3 \\
& +\,5u^8 & x^4 \\
& +\,4u^8\,(3+2u^8) & x^5 \\
& +\,4u^8\,(1-u^8) & x^6 \\
& -\,4u^8\,(2+3u^8) & x^7 \\
& -\,5u^{16} & x^8 \\
& +\,10u^{16} & x^9 \\
& +\,4u^{24} & x^{10} \\
& -\,2u^{24} & x^{11} \\
& -\,u^{24} & x^{12}\}^2\div
\end{array}$$

$n=7$,

$$\begin{array}{lll}
1+z=(1+x)\,\{ & 1 & \\
(\div) & -\,4 & x \\
& -\,4 & x^2 \\
& +\,4\,(2+7u^8) & x^3 \\
& -\,14u^8 & x^4 \\
& -\,28u^8\,(3+2u^8) & x^5 \\
& +\,28u^8\,(4+u^8) & x^6 \\
& +\,4u^8\,(16+51u^8+8u^{16}) & x^7 \\
& -\,u^8\,(144+305u^8+16u^{16}) & x^8 \\
& -\,8u^8\,(4+25u^8+16u^{16}) & x^9 \\
& +\,8u^8\,(8+57u^8+46u^{16}) & x^{10} \\
& +\,56u^{16}\,(2+u^8) & x^{11} \\
& -\,4u^{16}\,(56+161u^8+56u^{16}) & x^{12}
\end{array}$$

Term in { } has factor

$$1+\frac{\beta}{\alpha}x+\frac{u^5}{v}x^2;$$

$u=1$, term in { } is

$$=(1+x)^7(1-x)^5.$$

$$\begin{array}{lr}
+\,56u^{24}(1+2u^8) & x^{13}\\
+\,8u^{24}(46+57u^8+8u^{16}) & x^{14}\\
-\,8u^{24}(16+25u^8+4u^{16}) & x^{15}\\
-\,u^{24}(16+305u^8+144u^{16}) & x^{16}\\
+\,4u^{24}(8+51u^8+16u^{16}) & x^{17}\\
+\,28u^{32}(1+4u^8) & x^{18}\\
-\,28u^{32}(2+3u^8) & x^{19}\\
-\,14u^{40} & x^{20}\\
+\,4u^{40}(7+2u^8) & x^{21}\\
-\,4u^{48} & x^{22}\\
-\,4u^{48} & x^{23}\\
+\,u^{48} & x^{24}\}^2\\
 & (\div)
\end{array}$$

Term in { } has factor

$$1+\frac{\beta}{\alpha}x+\frac{\gamma}{\alpha}x^2+\frac{u^7}{v}x^3;$$

$u=1$, term in { } is $(1+x)^{10}(1-x)^{14}$.

The transformations $n=3, 5, 7, 11$. Art. Nos. 41 to 51.

41. The cubic transformation, $n=3$.

I reproduce the results already obtained; since there are only two coefficients α, β, these are also the last but one and last coefficients ρ, σ. Hence, from the values of α, β, ρ, σ, we have

$$\alpha=1, \qquad 2\alpha=\frac{v^3}{u}\left(\frac{1}{M}-\frac{u^4}{v^4}\right),$$

$$2\beta=\frac{1}{M}-1, \qquad \beta=\frac{u^3}{v};$$

the two values of $\frac{1}{M}$ are thus $\frac{1}{M}=2\frac{u}{v^3}+\frac{u^4}{v^4}$, $=1+\frac{2u^3}{v}$, giving the modular equation

$$v^4+2v^3u^3-2vu-u^4=0;$$

and we then have

$$\frac{1-y}{1+y}=\frac{1-x}{1+x}\left(\frac{v-u^3x}{v+u^3x}\right)^2.$$

42. The quintic transformation, $n=5$.

Here there are the three coefficients α, β, γ, or β, γ are the last but one and last coefficients ρ, σ; we have

$$\alpha = 1, \qquad 2\beta = v^3 u \left(\frac{1}{M} - \frac{u^4}{v^4}\right),$$

$$2\beta = \frac{1}{M} - 1, \qquad \gamma = \frac{u^5}{v}.$$

Comparing the two values of β, we have $\dfrac{1}{M} = \dfrac{v-u^5}{v(1-v^3u)}$, and then

$$\alpha = 1, \quad 2\beta = \frac{u(v^4-u^4)}{v(1-v^3u)}, \quad \gamma = \frac{u^5}{v},$$

so that only the modular equation remains to be determined.

The unused equation is

$$2\alpha\gamma + 2\alpha\beta + \beta^2 = \frac{v^2}{u^2}(2\alpha\gamma + 2\beta\gamma + \beta^2),$$

which, putting therein $\alpha = 1$, may be written

$$(2\gamma + \beta^2)(u^2 - v^2) = 2\beta(\gamma v^2 - u^2);$$

attending to the value of β, this divides by $u^2 - v^2$; in fact the equation may be written

$$2\gamma + \beta^2 = -\frac{u(v^2+u^2)}{v(1-v^3u)}(\gamma v^2 - u^2);$$

and then completing the substitution, and integralizing, this becomes

$$\{8vu^3(1-v^3u)^2 + (v^4-u^4)^2\} = 4uv(u^2+v^2)(1-u^3v)(1-uv^3),$$

viz. this is

$$4(1-v^3u)uv\{2u^2(1-v^3u) - (u^2+v^2)(1-vu^3)\} + (v^4-u^4)^2 = 0;$$

and the term in { } being $= -(v^2-u^2)(1+vu^3)$, the whole again divides by v^2-u^2, and the equation thus becomes

$$(v^2+u^2)(v^4-u^4) - 4uv(1-v^3u)(1+vu^3) = 0,$$

which is the modular equation.

43. The septic transformation, $n = 7$.

I do not propose to complete the solution directly from the fundamental equations for α, β, γ, δ, but resort to the known modular equation, and to an expression of M which I obtain by means thereof.

The modular equation is

$$(1-u^8)(1-v^8) - (1-uv)^8 = 0,$$

which may also be written

$$(v-u^7)(u-v^7)+7uv(1-uv)^2(1-uv+u^2v^2)^2=0,$$

as can be at once verified; but it also follows from Cauchy's identity

$$(x+y)^7-x^7-y^7=7xy(x+y)(x^2+xy+y^2)^2.$$

We then have

$$M^2=-\frac{1}{n}\frac{(1-v^8)\,vF'v}{(1-u^8)uF'u}.$$

Moreover

$$\begin{aligned} uF'u &= -u^8(1-v^8)+uv(1-uv)^7 \\ &= \frac{-u^8}{1-u^8}(1-uv)^8+uv(1-uv)^7 \\ &= \frac{(1-uv)^7}{1-u^8}u(v-u^7); \end{aligned}$$

and similarly

$$F'v=\frac{(1-uv)^7}{1-v^8}v(u-v^7),$$

whence

$$\frac{1}{M^2}=\frac{-7u}{v}\,\frac{(v-u^7)}{u-v^7}.$$

Writing this under the form

$$\frac{1}{M^2}=\frac{-7uv}{v^2}\,\frac{(v-u^7)(u-v^7)}{(u-v^7)^2},\quad =\frac{49u^2(1-uv)^2(1-uv+u^2v^2)^2}{(u-v^7)^2},$$

I find, as will appear, that the root must be taken with the sign $-$, and that we thus have $\frac{1}{M}=-\frac{7u(1-uv)(1-uv+u^2v^2)}{u-v^7}$, whence also $M=\frac{v(1-uv)(1-uv+u^2v^2)}{v-u^7}$.

44. Recurring now to the fundamental equations for the septic transformation, the coefficients are α, β, γ, δ, and we have

$$\alpha=1,\qquad 2\gamma=u^3v^3\left(\frac{1}{M}-\frac{u^4}{v^4}\right),$$

$$2\beta=\frac{1}{M}-1,\quad \delta=\frac{u^7}{v},$$

so that the coefficients are all given in terms of v, M. The unused equations are

$$\begin{aligned} u^6(2\alpha\gamma+2\alpha\beta+\beta^2) &= v^2(\gamma^2+2\gamma\delta+2\beta\delta), \\ u^{-2}(\gamma^2+2\beta\gamma+2\alpha\delta+2\beta\delta) &= v^2(2\alpha\gamma+2\beta\gamma+2\alpha\delta+\beta^2), \end{aligned}$$

which, substituting therein for α, β, γ, δ the foregoing values, give two equations; from these, eliminating M, we should obtain the modular equation, and then M in terms of u, v.

Substituting in the first instance for α, δ their values, the equations are

$$u^6(2\beta+2\gamma+\beta^2)=v^2\left\{\gamma^2+2\frac{u^7}{v}(\beta+\gamma)\right\}$$

$$\gamma^2+2\beta\gamma+(2+2\beta)\frac{u^7}{v}=u^2v^2\left\{2\gamma+2\beta\gamma+2\frac{u^7}{v}+\beta^2\right\}.$$

The first of these is

$$4(1-uv)(2\beta+2\gamma)+4\beta^2-4\frac{v^2}{u^6}\gamma^2=0,$$

viz. this is

$$4(1-uv)\left(\frac{1}{M}-1+\frac{u^3v^3}{M}-\frac{u^7}{v}\right)+\left(\frac{1}{M}-1\right)^2-v^8\left(\frac{1}{M}-\frac{u^4}{v^4}\right)^2=0\,;$$

or observing that in this equation the coefficient of $\dfrac{2}{M}$ is

$$(1-u^2v^2)\{2-2uv+2u^2v^2-1-u^2v^2\},$$

$$=(1-u^2v^2)(1-uv)^2,\ \ =(1-uv)^3(1+uv),$$

the equation becomes

$$(1-v^8)\frac{1}{M^2}+\frac{2}{M}(1-uv)^3(1+uv)+1-u^8-4(1-uv)\left(1+\frac{u^7}{v}\right)=0.$$

45. This should be satisfied identically by the foregoing value of $\dfrac{1}{M}$; viz. it should be satisfied on writing therein

$$\frac{1}{M^2}=-\frac{7u}{v}\,\frac{v-u^7}{u-v^7},$$

$$\frac{1}{M}=-\frac{7u(1-uv)(1-uv+u^2v^2)}{u-v^7}\,;$$

that is, we should have

$$-7\frac{u}{v}(v-u^7)(1-v^8)-14u(1-uv)^4(1+u^3v^3)$$

$$+(u-v^7)\left\{1-u^8-4(1-uv)\left(1+\frac{u^7}{v}\right)\right\}=0,$$

where observe that the $-$ sign of the second term is the sign of the foregoing value of $\dfrac{1}{M}$; so that the identity being verified, it follows that the correct sign has been attributed to the value of $\dfrac{1}{M}$.

46. Multiplying by v, the equation is

$$-7(1-u^8-\overline{1-uv})(1-v^8)-14uv(1-uv)^4(1+u^3v^3)$$

$$+\{1-v^8-\overline{1-uv}\}\{-8(1-uv)+1-u^8\}+4(1-uv)(v-u^7)(u-v^7)=0,$$

viz. this is

$$\begin{aligned}
&-7(1-u^8)(1-v^8)+7(1-uv)(1-v^8)-14uv(1-uv)^4(1+u^3v^3)\\
&+\ (1-u^8)(1-v^8)-8(1-uv)(1-v^8)+\ 8\quad(1-uv)^2\\
&\qquad -1(1-uv)(1-u^8)+\ 4\quad(1-uv)(v-u^7)(u-v^7)=0.
\end{aligned}$$

In the second column the coefficient of $1-uv$ is $2-u^8-v^8$, viz. this is

$$=(1-u^8)(1-v^8)+1-(uv)^8, \text{ or it is } =(1-uv)^8+1-(uv)^8.$$

Reducing also the other two columns by means of the modular equation, the equation thus becomes

$$\begin{aligned}
-6(1-uv)^8-(1-uv)\{(1-uv)^8+1-(uv)^8\}-14uv(1-uv)^4(1+u^3v^3)&\\
+\ 8\quad(1-uv)^2&\\
-28uv(1-uv)^3(1-uv+u^2v^2)^2=0.&
\end{aligned}$$

This is in fact an identity; to show it, writing for convenience θ in place of uv, and observing that the terms

$$\begin{aligned}
&-(1-\theta)(1-\theta^8)+8(1-\theta)^2,\\
&=\quad(1-\theta)^2\{8-(1+\theta+\theta^2+\theta^3+\theta^4+\theta^5+\theta^6+\theta^7)\}
\end{aligned}$$

are

$$=\quad(1-\theta)^3(7+6\theta+5\theta^2+4\theta^3+3\theta^4+2\theta^5+\theta^6),$$

the whole equation divides by $(1-\theta)^3$; or throwing out this factor, it is

$$\begin{aligned}
-6(1-\theta)^5-(1-\theta)^6+7+6\theta+5\theta^2+4\theta^3+3\theta^4+2\theta^5+\theta^6&\\
-14\theta(1-\theta)(1+\theta^3)-28\theta(1-\theta+\theta^2)^2=0.&
\end{aligned}$$

The first line is $=14\theta(3-5\theta+6\theta^2-3\theta^3+\theta^4)$; whence, throwing out the factor 14θ, the equation is

$$3-5\theta+6\theta^2-3\theta^3+\theta^4-(1-\theta)(1+\theta^3)-2(1-\theta+\theta^2)^2,$$

that is,

$$(1-\theta+\theta^2)(3-2\theta+\theta^2)-(1-\theta^2)(1-\theta+\theta^2)-2(1-\theta+\theta^2)^2=0;$$

or throwing out the factor $1-\theta+\theta^2$, the equation is

$$(3-2\theta+\theta^2)-(1-\theta^2)-2(1-\theta+\theta^2)=0,$$

which is an identity.

The other equation is

$$\gamma^2+2\beta\gamma+(2+2\beta)\frac{u^7}{v}=u^2v^2\left(2\gamma+2\beta\gamma+2\frac{u^7}{v}+\beta^2\right);$$

that is,

$$\gamma^2+2\beta\gamma-u^2v^2\beta^2+2(1+\beta)\left(\frac{u^7}{v}-\gamma u^2v^2\right)-2u^9v=0,$$

which might also be verified, but I have not done this.

47. The conclusion is

$$\alpha=1,\quad \beta=\tfrac{1}{2}\left(\frac{1}{M}-1\right),\quad \gamma=\tfrac{1}{2}u^3v^3\left(\frac{1}{M}-\frac{u^4}{v^4}\right),\quad \delta=\frac{u^7}{v},$$

where

$$\frac{1}{M}=\frac{-7u(1-uv)(1-uv+u^2v^2)}{u-v^7},$$

and of course

$$\frac{1-y}{1+y}=\frac{1-x}{1+x}\left(\frac{1-\beta x+\gamma x^2-\delta x^3}{1+\beta x+\gamma x^2+\delta x^3}\right)^2;$$

but the resulting form may admit of simplification.

48. The endecadic transformation, $n=11$.

I have not completed the solution, but the results, so far as I have obtained them, are interesting. The coefficients are $\alpha, \beta, \gamma, \delta, \epsilon, \zeta$; and we have, as in general,

$$\alpha=1\quad,\quad 2\epsilon=u^7v^3\left(\frac{1}{M}-\frac{u^4}{v^4}\right),$$

$$2\beta=\frac{1}{M}-1,\quad \zeta=\frac{u^{11}}{v}.$$

The unused equations then are

$$u^{14}(2\alpha\gamma+2\alpha\beta+\beta^2)=v^2(\epsilon^2+2\epsilon\zeta+2\delta\zeta),$$

$$u^6(\gamma^2+2\alpha\epsilon+2\alpha\delta+2\beta\gamma+2\beta\delta)=v^2(2\gamma\epsilon+2\gamma\zeta+2\delta\epsilon+2\beta\zeta+\delta^2),$$

$$u^{-2}(2\gamma\epsilon+2\alpha\zeta+2\gamma\delta+2\beta\epsilon+2\beta\zeta+\delta^2)=v^2(\gamma^2+2\alpha\epsilon+2\alpha\zeta+2\gamma\delta+2\beta\epsilon+2\beta\delta),$$

$$u^{-10}(\epsilon^2+2\gamma\zeta+2\delta\epsilon+2\delta\zeta)=v^2(2\alpha\gamma+2\alpha\delta+2\beta\gamma+\beta^2);$$

but I attend only to the first and the last, which, it will be observed, contain γ, δ linearly. If in the first instance we substitute only for α, ζ their values, the equations become

$$u^{12}\beta(2+\beta)-\frac{v^2}{u^2}\epsilon\left(\epsilon+2\frac{u^{11}}{v}\right)\qquad+u^{12}.2\gamma\qquad-vu^9.2\delta=0,$$

$$u^{-12}\epsilon^2\qquad-\frac{v^2}{u^2}\beta^2\qquad+\left\{\frac{1}{uv}-\frac{v^2}{u^2}(1+\beta)\right\}.2\gamma+\left\{\frac{1}{uv}-\frac{v^2}{u^2}+u^{-12}\epsilon\right\}.2\delta=0;$$

say, for a moment, these are

$$A+P.2\gamma+Q.2\delta=0,$$

$$B+R.2\gamma+S.2\delta=0,$$

giving

$$1:2\gamma:2\delta=PS-QR:QB-SA:RA-PB.$$

Here

$$PS-QR=\frac{u^{11}}{v}+\epsilon-u^{10}v^2+u^8-u^7v^3(1+\beta)$$

$$=\tfrac{1}{2}\left\{\frac{2u^{11}}{v}+\left(u^7v^3\frac{1}{M}-\frac{u^{11}}{v}\right)-2u^{10}v^2+2u^8-2u^7v^3-\left(u^7v^3\frac{1}{M}-u^7v^3\right)\right\},$$

where the terms containing $\frac{1}{M}$ disappear of themselves, viz. this is

$$= \tfrac{1}{2}\left(\frac{u^{11}}{v} - 2u^{10}v^2 + 2u^8 - u^7v^3\right)$$

$$= -\tfrac{1}{2}\frac{u^7}{v}(v^4 + 2v^3u^3 - 2vu - u^4);$$

observe that the term in ( ), equated to zero, gives the modular equation for the case $n=3$. It thus appears that γ and δ are given as fractions, having in their denominator this function $u^4 + 2uv - 2u^3v^3 - v^4$.

49. To complete the calculation, we have

$$QB - AS = -vu^9\left(\frac{\epsilon^2}{u^{12}} - \frac{v^2}{u^2}\beta^2\right)$$

$$-\left\{u^{12}\beta(2+\beta) - \frac{v^2}{u^2}\epsilon\left(\epsilon + 2\frac{u^{11}}{v}\right)\right\}\left\{\frac{1}{uv} - \frac{v^2}{u^2} + \frac{\epsilon}{u^{12}}\right\};$$

viz. multiplying by 8, and substituting for 2β, 2ϵ their values, this is

$$8(QB - AS) = -2vu^9\left\{u^2v^6\left(\frac{1}{M} - \frac{u^4}{v^4}\right)^2 - \frac{v^2}{u^2}\left(\frac{1}{M} - 1\right)^2\right\}$$

$$-\left\{u^{12}\left(\frac{1}{M} - 1\right)\left(\frac{1}{M} + 3\right) - u^{12}v^8\left(\frac{1}{M} - \frac{u^4}{v^4}\right)\left(\frac{1}{M} + \frac{3u^4}{v^4}\right)\right\}\left(\frac{1}{uv} - \frac{2v^2}{u^2} + \frac{v^3}{u^5}\frac{1}{M}\right),$$

or, what is the same thing,

$$-\frac{8v}{u^7}(QB - AS) = 2\left\{u^4v^8\left(\frac{1}{M} - \frac{u^4}{v^4}\right)^2 - v^4\left(\frac{1}{M} - 1\right)^2\right\}$$

$$+\left\{\left(\frac{1}{M} - 1\right)\left(\frac{1}{M} + 3\right) - v^8\left(\frac{1}{M} - \frac{u^4}{v^4}\right)\left(\frac{1}{M} + 3\frac{u^4}{v^4}\right)\right\}\left(u^4 - 2v^3u^3 + \frac{v^4}{M}\right),$$

viz. the left-hand side is

$$= 2\left\{-v^4(1 - u^4v^4)\frac{1}{M^2} + 2v^4(1 - u^8)\frac{1}{M} + u^{12} - v^4\right\}$$

$$+\left\{\frac{1}{M^2}(1 - v^8) + \frac{2}{M}(1 - u^4v^4) - 3(1 - u^8)\right\}\left\{\frac{1}{M} + u^3(u - 2v^3)\right\};$$

or, say we have $-\frac{8v}{u^7}(QB - SA) = \Pi$, where

$$\begin{aligned}\Pi = &\ \frac{1}{M^3}\cdot v^4(1 - v^8)\\ &+ \frac{1}{M^2}\cdot u^3(u - 2v^3)(1 - v^8)\\ &+ \frac{1}{M}\cdot 4u^7v^7 + v^4(1 - 3u^8) - 4u^3v^3 + 2u^4\\ &+ \quad\cdot -2v^4 + 6v^3u^3(1 - u^8) + u^4(-3 + 5u^8);\end{aligned}$$

wherefore the value of 2γ is $=\frac{1}{4}\Pi \div (v^4+2v^3u^3-2vu-u^4)$. Similarly, writing

$$\begin{aligned}\Pi' = \quad & \frac{1}{M^3} \cdot v^4(1-v^8) \\ & + \frac{1}{M^2} \cdot v(v^3-2u)(1-v^8) \\ & + \frac{1}{M} \cdot 4u^5v^5 + v^4(3-u^8) + 4uv - 2u^4v^8 \\ & + \quad v^4(-5+3u^8) + 6vu(1-u^8) + 2u^{12},\end{aligned}$$

we find

$$2\delta = \quad \tfrac{1}{4}\frac{u^3}{v}\Pi' \div (v^4+2v^3u^3-2vu-u^4);$$

in verification whereof observe that this being so, the first equation gives the identity

$$\left\{\left(\frac{1}{M}-1\right)\left(\frac{1}{M}+3\right) - v^8\left(\frac{1}{M}-\frac{u^4}{v^4}\right)\left(\frac{1}{M}+\frac{3u^4}{v^4}\right)\right\}(v^4+2v^3u^3-2vu-u^4)+\Pi-\Pi'=0.$$

50. The result is that, writing for the moment $v^4+2v^3u^3-2vu-u^4=\Delta$, the values of the coefficients are

$$\begin{array}{cccccc} \alpha, & \beta\;, & \gamma\;, & \delta\;, & \epsilon\;, & \zeta\;, \\ =1, & \frac{1}{2}\left(\frac{1}{M}-1\right), & \frac{1}{8}\frac{\Pi}{\Delta}, & \frac{1}{8}\frac{u^3}{v}\frac{\Pi'}{\Delta}, & \frac{1}{2}u^7v^3\left(\frac{1}{M}-\frac{u^4}{v^4}\right), & \frac{u^{11}}{v},\end{array}$$

and

$$\frac{1-y}{1+y}=\frac{1-x}{1+x}\left(\frac{1-\beta x+\gamma x^2-\delta x^3+\epsilon x^4-\zeta x^5}{1+\beta x+\gamma x^2+\delta x^3+\epsilon x^4+\zeta x^5}\right)^2;$$

the modular equation is known, and to complete the solution we require only an expression for M in terms of u, v.

51. We may herein illustrate the following theorem, viz. we may simultaneously change u, v, $\frac{1}{M}$, $\alpha:\beta:\gamma:\delta:\epsilon:\zeta$; into $\frac{1}{u}$, $\frac{1}{v}$, $\frac{v^4}{u^4}\frac{1}{M}$, $\zeta:\epsilon:\delta:\gamma:\beta:\alpha$.

Thus making the change in the equation

$$\frac{\beta}{\alpha}=\tfrac{1}{2}\left(\frac{1}{M}-1\right),$$

we have

$$\frac{\epsilon}{\zeta}=\tfrac{1}{2}\left(\frac{v^4}{u^4}\frac{1}{M}-1\right);\ \text{that is, } \tfrac{1}{2}\frac{v^4}{u^4}\left(\frac{1}{M}-\frac{u^4}{v^4}\right)=\tfrac{1}{2}\left(\frac{v^4}{u^4}\frac{1}{M}-1\right),$$

which is right.

So in the equation $\frac{\gamma}{\alpha}=\frac{1}{8}\frac{\Pi}{\Delta}$, if for a moment (Π), (Δ) are what Π, Δ become, the equation is $\dfrac{\delta}{\zeta}=\tfrac{1}{8}\dfrac{(\Pi)}{(\Delta)}$, that is, $\dfrac{1}{u^8}\dfrac{\Pi'}{\Delta}=\dfrac{(\Pi)}{(\Delta)}$, or $(\Pi)=\dfrac{1}{u^8}\dfrac{(\Delta)}{\Delta}\Pi'$; but obviously $\dfrac{(\Delta)}{\Delta}=-\dfrac{1}{u^4v^4}$; and the equation thus is $(\Pi)=-\dfrac{1}{u^{12}v^4}\Pi'$, or say $u^{12}v^4(\Pi)=-\Pi'$; that is,

$$\begin{aligned}-\Pi'=u^{12}v^4\Big\{&\frac{v^{12}}{u^{12}}\cdot\frac{1}{M^3}\cdot\frac{1}{v^4}\left(1-\frac{1}{v^8}\right)\\ &+\frac{v^8}{u^8}\cdot\frac{1}{M^2}\cdot\frac{1}{u^3}\left(\frac{1}{u}-\frac{2}{v^3}\right)\left(1-\frac{1}{v^8}\right)\\ &+\frac{v^4}{u^4}\cdot\frac{1}{M}\cdot\frac{4}{u^7v^7}+\frac{1}{v^4}\left(1-\frac{3}{u^8}\right)-\frac{4}{u^3v^3}+\frac{2}{u^4}\\ &-\frac{2}{v^4}+\frac{6}{v^3u^3}\left(1-\frac{1}{u^8}\right)+\frac{1}{u^4}\left(-3+\frac{5}{u^8}\right)\Big\},\end{aligned}$$

which is right.

The general theory by q-transcendents. Art. Nos. 52 to 71.

52. I recur to the formula

$$\frac{1-y}{1+y}=\frac{1-x}{1+x}\left(\frac{\alpha-\beta x+\gamma x^2+\ldots\pm\sigma n^{\frac{1}{2}(n-1)}}{\alpha+\beta x+\gamma x^2+\ldots+\sigma n^{\frac{1}{2}(n-1)}}\right)^2,$$

and seek to express the ratios $\alpha:\beta:\ldots:\sigma$ in terms of q. Writing with Jacobi

$$\omega=\frac{mK+m'iK'}{n},$$

we have in general

$$\alpha+\beta x+\gamma x^2+\ldots+\sigma x^{\frac{1}{2}(n-1)}=\alpha\left(1+\frac{x}{\text{snc }2\omega}\right)\left(1+\frac{x}{\text{snc }4\omega}\right)\ldots\left(1+\frac{x}{\text{snc }(n-1)\,\omega}\right),$$

$$(\text{snc}=\sin\text{co am}\,;\ \text{viz. snc }2\omega=\text{sn}\,(K-2\omega),\ \&\text{c.})\,;$$

and the values of $\alpha, \beta, \ldots, \theta$ which correspond to the moduli $v_0, v_1, \ldots v_n$, or say the values $(\alpha_0, \beta_0, \ldots, \theta_0)$, $(\alpha_1, \beta_1, \ldots, \theta_1)$, ..., $(\alpha_n, \beta_n, \ldots, \theta_n)$, are obtained by giving to ω the values

$$\begin{array}{ccccc}\omega_0\,, & \omega_1\,, & \omega_2 & ,\ldots, & \omega_n\,,\\ =\dfrac{2K}{n}, & \dfrac{2K+iK'}{n}, & \dfrac{4K+iK'}{n} & ,\ldots, & \dfrac{iK'}{n},\end{array}$$

viz. the cases ω_0, ω_n correspond to Jacobi's first and second real transformations, and the others to the imaginary transformations.

I remark that $\omega=\omega_0$ gives for snc $2g\omega$ an expression which is rational as regards q, but $\omega=\omega_n$ gives an expression involving $q^{\frac{1}{n}}$, the real nth root of q; the other values $\omega_1, \omega_2, \ldots$ give the like expressions, involving $\alpha q^{\frac{1}{n}}, \alpha^2q^{\frac{1}{n}}, \ldots$ (α an imaginary nth root of unity), the imaginary nth roots of q.

53. I consider first the expression

$$\frac{1}{\text{snc}\, 2g\omega_0}, \quad = \frac{1}{\text{sn}\,(K - 2g\omega_0)}, \quad = \frac{\text{dn}\, 2g\omega_0}{\text{cn}\, 2g\omega_0}.$$

Here, writing $2g\omega_0 = \frac{2K\xi}{\pi}$ (ξ for Jacobi's x, as x is being used in a different sense), that is,

$$\xi = \frac{\pi}{2K} \cdot 2g \cdot \frac{2K}{n}, \quad = \frac{2g\pi}{n},$$

(and thence $e^{i\xi} = e^{g\frac{2\pi i}{n}} = \alpha^g$, $e^{2i\xi} = \alpha^{2g}$, if $\alpha = e^{\frac{2\pi i}{n}}$, an imaginary nth root of unity), we have (Jacobi, p. 86, [*Ges. Werke*, t. I., p. 143])

$$\frac{1}{\text{snc}\, 2g\omega_0} = \text{dn}\, \frac{2K\xi}{\pi} \div \text{cn}\, \frac{2K\xi}{\pi}$$

$$= \frac{C}{B} \cdot \frac{2e^{i\xi}}{1 + e^{2i\xi}} \cdot \frac{(1 + qe^{2i\xi}) \ldots (1 + qe^{-2i\xi}) \ldots}{(1 + q^2e^{2i\xi}) \ldots (1 + q^2e^{-2i\xi}) \ldots},$$

where

$$\frac{C}{B} = \left\{\frac{(1 + q^2) \ldots}{(1 + q\,) \ldots}\right\}^2 = f^2(q);$$

that is,

$$\frac{1}{\text{snc}\, 2g\omega_0} = \frac{2\alpha^g}{1 + \alpha^{2g}} \cdot f^2(q) \cdot \frac{(1 + \alpha^{2g}\, q) \ldots (1 + \alpha^{n-2g}\, q) \ldots}{(1 + \alpha^{2g} q^2) \ldots (1 + \alpha^{n-2g}\, q) \ldots},$$

where, for shortness, I write $(1 + qe^{2i\xi}) \ldots$ to denote the infinite product

$$(1 + qe^{2i\xi})(1 + q^3e^{2i\xi})(1 + q^5e^{2i\xi}) \ldots,$$

and similarly $(1 + q^2e^{2i\xi}) \ldots$ to denote the infinite product $(1 + q^2e^{2i\xi})(1 + q^4e^{2i\xi})(1 + q^6e^{2i\xi}) \ldots$, and the like for the terms in $e^{-2i\xi}$: the notation, accompanied by its explanation, is quite intelligible, and it would be difficult to make one which would be at the same time complete and not cumbrous. Then attributing to g the values $1, 2, \ldots, \frac{1}{2}(n-1)$, and forming the symmetric functions of these expressions, we have the values of $\frac{\beta}{\alpha}, \frac{\gamma}{\alpha}$, &c., or α being put $= 1$, say the values of $\beta, \gamma, \ldots, \sigma$.

54. I stop to notice a verification afforded by the value of β_0. Putting $u = 0$, that is, $q = 0$, we have

$$\frac{1}{\text{snc}\, 2g\omega_0} = \frac{2\alpha^g}{1 + \alpha^{2g}},$$

and thence

$$\beta_0 = 2\left\{\frac{\alpha}{1 + \alpha^2} + \frac{\alpha^2}{1 + \alpha^4} + \frac{\alpha^3}{1 + \alpha^6} + \ldots + \frac{\alpha^{\frac{1}{2}(n-1)}}{1 + \alpha^{n-1}}\right\},$$

we have $2\beta_0 = \frac{1}{M_0} - 1$; and putting as above $u = 0$, the value of $\frac{1}{M_0}$ is $= (-)^{\frac{n-1}{2}} n$; whence

$$(-)^{\frac{1}{2}(n-1)}\, n - 1 = 4\left\{\frac{\alpha}{1 + \alpha^2} + \frac{\alpha^2}{1 + \alpha^4} + \frac{\alpha^3}{1 + \alpha^6} + \ldots + \frac{\alpha^{\frac{1}{2}(n-1)}}{1 + \alpha^{n-1}}\right\},$$

a theorem relating to the imaginary nth roots of unity, n an odd prime. In particular,

$$n=3, \quad -4=4\left\{\frac{\alpha}{1+\alpha^2}\right\}, \text{ at once verified by } \alpha^2+\alpha+1=0;$$

$$n=5, \quad 4=4\left\{\frac{\alpha}{1+\alpha^2}+\frac{\alpha^2}{1+\alpha^4}\right\}, \text{ verified by } \alpha^5-1=0,$$

viz. the theorem is also true for the real root $\alpha=1$; in fact, the term in $\{\ \}$ is

$$\{\alpha(1+\alpha^4)+\alpha^2(1+\alpha^2)\}\div(1+\alpha^2)(1+\alpha^4), \text{ that is, } (\alpha+1+\alpha^2+\alpha^4)\div(1+\alpha^2+\alpha^4+\alpha),\ =1;$$

$$n=7, \quad -8=4\left\{\frac{\alpha}{1+\alpha^2}+\frac{\alpha^2}{1+\alpha^4}+\frac{\alpha^3}{1+\alpha^6}\right\},$$

which may be verified by means of $\alpha^6+\alpha^5+\alpha^4+\alpha^3+\alpha^2+\alpha+1=0$; and so on.

55. I further remark that we have

$$\frac{1}{M_0}=(-)^{\frac{1}{2}(n-1)}\left\{\frac{\text{sn}\ 2\omega_0\,.\,\text{sn}\ 4\omega_0\,..\,\text{sn}\,(n-1)\,\omega_0}{\text{snc}\ 2\omega_0\,.\,\text{snc}\ 4\omega_0\,..\,\text{snc}\,(n-1)\,\omega_0}\right\}^2.$$

But Jacobi (p. 86, [*l.c.*]),

$$\text{sn}\ 2g\omega_0=\text{sn}\,\frac{2K\xi}{\pi},$$

$$=\frac{AK}{\pi i}\,\frac{e^{2i\xi}-1}{e^{i\xi}}\,\frac{(1-q^2e^{2i\xi})\,..\,(1-q^2e^{-2i\xi})\,..}{(1-qe^{2i\xi})\,..\,(1-qe^{-2i\xi})\,..},$$

where (p. 89, [*l.c.*, p. 146])

$$\frac{AK}{\pi}=\frac{\sqrt[4]{q}}{\sqrt{k}}, \quad =\tfrac{1}{2}\left\{\frac{(1+q)(1+q^3)\,...}{(1+q^2)(1+q^4)\,...}\right\}^2=\tfrac{1}{2}f^{-2}(q),$$

that is,

$$\text{sn}\ 2g\omega=f^{-2}q\,.\,\frac{\alpha^{2g}-1}{2i\alpha^g}\,\frac{(1-\alpha^{2g}q^2)\,..\,(1-\alpha^{n-2g}q^2)\,..}{(1-\alpha^{2g}q)\,..\,(1-\alpha^{n-2g}q)\,..}.$$

Hence

$$\frac{\text{sn}\ 2g\omega_0}{\text{snc}\ 2g\omega_0}=\frac{\alpha^{2g}-1}{i(\alpha^{2g}+1)}\,\frac{1-\alpha^{2g}q^2\,..}{1+\alpha^{2g}q^2\,..}\,\frac{1+\alpha^{2g}q\,..}{1-\alpha^{2g}q\,..}\,\frac{1-\alpha^{n-2g}q^2\,..}{1+\alpha^{n-2g}q^2\,..}\,\frac{1+\alpha^{n-2g}q\,..}{1-\alpha^{n-2g}q\,..};$$

and giving to g the values $1, 2, \ldots, \frac{1}{2}(n-1)$, and multiplying the several expressions, we have the value of $\frac{1}{M_0}$, viz. this is

$$\frac{1}{M_0}=(-)^{\frac{1}{2}(n-1)}\,\Pi\left\{\frac{(\alpha^{2g}+1)^2}{i^2(\alpha^{2g}+1)^2}\right\}R(q),$$

where $R(q)$ denotes the product of the several factors which contain q.

56. The (i^2) of the denominator gives a factor i^{n-1}, $=(-)^{\frac{n-1}{2}}$, which destroys the factor $(-)^{\frac{n-1}{2}}$. We have then a factor

$$\Pi\left(\frac{\alpha^{2g}-1}{\alpha^{2g}+1}\right)^2, \text{ which is } =(-)^{\frac{1}{2}(n-1)}\,n.$$

In fact, $n=3$, this is

$$\left(\frac{\alpha^2-1}{\alpha^2+1}\right)^2=-3,$$

viz. the numerator is $\alpha-2\alpha^2+1$, $=-3\alpha^2$, and the denominator is $(-\alpha)^2$, $=\alpha^2$.

So $n=5$, the formula is

$$\left(\frac{\alpha^2-1}{\alpha^2+1}\cdot\frac{\alpha^4-1}{\alpha^4+1}\right)^2=5, \text{ that is, } \frac{(\alpha^2-1)^4}{(\alpha^4+1)^2}=5;$$

or

$$\frac{\alpha^3-4\alpha^2+6\alpha^4-4\alpha+1}{\alpha^3+2\alpha^4+1}=5,$$

viz. this is $5(1+\alpha^3+2\alpha^4)-(1-4\alpha-4\alpha^2+\alpha^3+6\alpha^4)=0$, which is right; and so in other cases.

We thus have

$$\frac{1}{M_0}=(-)^{\frac{1}{2}(n-1)}\,n\,.\,R(q),$$

which, on putting therein $u=0$, that is, $q=0$, gives, as it should do, $\frac{1}{M_0}=(-)^{\frac{1}{2}(n-1)}\,n$.

57. As regards the expression of $R(q)$, observe that, giving to g its different values, the factors $1-\alpha^{2g}q^2$ and $1-\alpha^{n-2g}q^2$ are all the factors other than $1-q^2$ of $1-q^{2n}$, and so as to the other pairs of factors; viz. we have

$$R(q)=\left(\frac{1-q^{2n}\,..}{1-q^2\;..}\,\frac{1+q^n\,..}{1+q\;..}\,\frac{1+q^2\;..}{1+q^{2n}\,..}\,\frac{1-q\;..}{1-q^n\,..}\right)^2,$$

viz. this is

$$=\left(\frac{1-q^{2n}\,..}{1+q^{2n}\,..}\,\frac{1+q^n\,..}{1-q^n\,..}\right)^2\div\left(\frac{1-q^2\,..}{1+q^2\,..}\,\frac{1+q\,..}{1-q\,..}\right)^2,$$

that is,

$$\frac{1}{M_0}=(-)^{\frac{1}{2}(n-1)}\,n\,\frac{\phi^2(q^n)}{\phi^2(q)},$$

agreeing with a former result.

58. We have of course the identity $2\beta_0=\frac{1}{M_0}-1$; that is,

$$4S\frac{\alpha^g}{1+\alpha^{2g}}f^2(q)\,.\,\frac{(1+\alpha^{2g}q\,)\,..\,(1+\alpha^{n-2g}q\,)\,..}{(1+\alpha^{2g}q^2)\,..\,(1+\alpha^{n-2g}q^2)\,..}=(-)^{\frac{1}{2}(n-1)}\,n\,\frac{\phi^2(q^n)}{\phi^2(q)}-1,$$

$(g=1, 2, ..\frac{1}{2}(n-1))$, which, putting therein $q=0$, is an identity before referred to; a form perhaps more convenient is obtained by dividing each side by $f^2(q)$.

59. I notice further that we have

$$v_0=u^n\,\{\text{snc}\,2\omega_0\,\text{snc}\,4\omega_0\,...\,\text{snc}\,(n-1)\,\omega_0\};$$

the term in { } is

$$\Pi\frac{1+\alpha^{2g}}{2\alpha^g}f^{-g}(q)\frac{(1+\alpha^{2g}q^2)\,..\,(1+\alpha^{n-2g}q)^2\,..}{(1+\alpha^g\,q\,)\,..\,(1+\alpha^{n-2g}q)\;..},$$

where we have $\Pi\frac{1+\alpha^{2g}}{\alpha}=(-)^{\frac{1}{8}(n^2-1)}$. For example, $n=3$, the term is $\frac{1+\alpha^2}{\alpha}=-1$;

$n=5$, it is

$$\frac{(1+\alpha^2)(1+\alpha^4)}{\alpha\,.\,\alpha^2}, \;=\frac{1+\alpha^2+\alpha^4+\alpha}{\alpha^3}, \;=-1;$$

$n=7$, it is

$$\frac{(1+\alpha^2)(1+\alpha^4)(1+\alpha^6)}{\alpha\,.\,\alpha^2\,.\,\alpha^4}, \;=\frac{1+\alpha+\alpha^2+\alpha^3+\alpha^4+\alpha^5+2\alpha^6}{\alpha^6}=1;$$

and so on. The term in question thus is

$$(-)^{\frac{1}{8}(n^2-1)}\cdot\frac{1}{(\sqrt{2})^{n-1}}f^{-n+1}(q)\,\frac{1+q^{2n}\,..\,1+q\,..}{1+q^n\;\,..\,1+q^2\,..},$$

that is,

$$(-)^{\frac{1}{8}(n^2-1)}\,\frac{1}{(\sqrt{2})^{n-1}}f^{-n}(q)\,f(q^n).$$

This has to be multiplied by u^n, $=(\sqrt{2})^n q^{\frac{n}{8}} f^n(q)$, and we thus obtain

$$v_0=(-)^{\frac{1}{8}(n^2-1)}\sqrt{2}q^{\frac{n}{8}}f(q^n),$$

agreeing with a former result.

We have in what precedes a complete q-transcendental solution for the *transformatio prima;* viz. the original modulus $k^2(=u^8)$ being given as a function of q, then, as well the new modulus $\lambda_0^2\,(=v_0^8)$ and the multiplier M_0, as also the several functions which enter into the expression

$$\frac{1-y}{1+y}=\frac{1-x}{1+x}\left\{\frac{\left(1-\dfrac{x}{\operatorname{snc} 2\omega_0}\right)\cdots\left(1-\dfrac{x}{\operatorname{snc}(n-1)\,\omega_0}\right)}{\left(1+\dfrac{x}{\operatorname{snc} 2\omega_0}\right)\cdots\left(1+\dfrac{x}{\operatorname{snc}(n-1)\,\omega_0}\right)}\right\}^2,$$

are all of them expressed as functions of q.

60. I consider in like manner the expression

$$\frac{1}{\operatorname{snc} 2g\omega_n}\;\frac{1}{\operatorname{sn}(K-2g\omega_n)}, \;=\frac{\operatorname{dn} 2g\omega_n}{\operatorname{cn} 2g\omega_n}.$$

Here, writing $2g\omega_n=\dfrac{2K\xi}{\pi}$ (ξ instead of Jacobi's x as before), that is,

$$\xi=\frac{\pi}{2K}\cdot 2g\,.\,\frac{iK'}{n}=\frac{g\pi iK'}{nK},$$

and thence

$$e^{i\xi}=e^{-\frac{g}{n}\frac{\pi K'}{K}}, \;=q^{\frac{g}{n}},$$

we have

$$\frac{1}{\operatorname{snc} 2g\omega_n}=\operatorname{dn}\frac{2K\xi}{\pi}\div\operatorname{cn}\frac{2K\xi}{\pi}$$

$$=f^2(q)\,.\,\frac{2q^{\frac{g}{n}}}{1+q^{\frac{2g}{n}}}\cdot\frac{(1+q^{1+\frac{2g}{n}})\,..\,(1+q^{1-\frac{2g}{n}})\,..}{(1+q^{2+\frac{2g}{n}})\,..\,(1+q^{2-\frac{2g}{n}})\,..},$$

where the notations are as follows:

$$(1+q^{1+\frac{2g}{n}}).. \text{ is the infinite product } (1+q^{1+\frac{2g}{n}})(1+q^{3+\frac{2g}{n}})(1+q^{5+\frac{2g}{n}}).., $$

and

$$(1+q^{2+\frac{2g}{n}}).. \text{ is the infinite product } (1+q^{2+\frac{2g}{n}})(1+q^{4+\frac{2g}{n}})(1+q^{6+\frac{2g}{n}})..;$$

and the like as to the expressions with exponents containing $-\frac{2g}{n}$.

And then attributing to g the values $1, 2, .., \frac{1}{2}(n-1)$, and forming the symmetric functions of these expressions, we have the values of $\frac{\beta}{\alpha}, \frac{\gamma}{\alpha}, .., \frac{\sigma}{\alpha}$; or α being put $=1$, say the values of $\beta, \gamma, \ldots, \sigma$.

It is easy to see, and I do not stop to prove that, if instead of $\omega = \omega_n$ we have $\omega = \omega_1, \omega_2, \ldots$, or ω_{n-1}, we simply multiply $q^{\frac{1}{n}}$ by an imaginary nth root of unity; that is, we replace the real nth root $q^{\frac{1}{n}}$ by an imaginary nth root of q.

In the case $u=0$, that is, $q=0$, we have $\frac{1}{\text{snc}\, 2g\omega_n}=0$, and thence $\beta=0$; and the like for the values $\omega_1, \omega_2, \ldots, \omega_{n-1}$: the equation $2\beta = \frac{1}{M} - 1$ gives consequently for $\frac{1}{M}$, n values each $=1$, agreeing with the multiplier equation.

61. We have for M_n the formula

$$\frac{1}{M_n} = (-)^{\frac{1}{2}(n-1)} \left\{\frac{\text{sn}\, 2\omega_n \, \text{sn}\, 4\omega_n \ldots \text{sn}\,(n-1)\,\omega_n}{\text{snc}\, 2\omega_n \, \text{snc}\, 4\omega_n \ldots \text{snc}\,(n-1)\,\omega_n}\right\}^2,$$

and, as before,

$$\text{sn}\, 2g\omega_n = f^{-2}(q) \, . \, \frac{q^{\frac{2g}{n}}-1}{2iq^{\frac{g}{n}}} \, \frac{(1-q^{2+\frac{2g}{n}}).. \, (1-q^{2-\frac{2g}{n}})..}{(1-q^{1+\frac{2g}{n}}).. \, (1-q^{1-\frac{2g}{n}})..};$$

hence

$$\frac{\text{sn}\, 2g\omega_n}{\text{snc}\, 2g\omega_n} = \frac{q^{\frac{2g}{n}}-1}{i(q^{\frac{2g}{n}}+1)} \, . \, \frac{(1-q^{2+\frac{2g}{n}}).. \, (1+q^{1+\frac{2g}{n}}).. \, (1-q^{2-\frac{2g}{n}}).. \, (1+q^{1-\frac{2g}{n}})..}{(1+q^{2+\frac{2g}{n}}).. \, (1-q^{1+\frac{2g}{n}}).. \, (1+q^{2-\frac{2g}{n}}).. \, (1-q^{1-\frac{2g}{n}})..},$$

and we thence derive the value of $\frac{1}{M_n}$; viz. observing that we have in the denominator $(i^2)^{\frac{1}{2}(n-1)}, =(-)^{\frac{1}{2}(n-1)}$ which destroys this factor in the expression of $\frac{1}{M_n}$, this is

$$\frac{1}{M_n} = \Pi \left\{\frac{1-q^{\frac{2g}{n}}}{1+q^{\frac{2g}{n}}} \, \frac{(1-q^{2+\frac{2g}{n}})..(1+q^{1+\frac{2g}{n}})..(1-q^{2-\frac{2g}{n}})..(1+q^{1-\frac{2g}{n}})..}{(1+q^{2+\frac{2g}{n}})..(1-q^{1+\frac{2g}{n}})..(1+q^{2-\frac{2g}{n}})..(1-q^{1-\frac{2g}{n}})..}\right\}^2.$$

Now, giving to g its values, it is easy to see that we have

$$\Pi (1-q^{\frac{2g}{n}})(1-q^{2+\frac{2g}{n}})\ldots(1-q^{2-\frac{2g}{n}})\ldots = \frac{(1-q^{\frac{2}{n}})\ldots}{(1-q^2)\ldots},$$

where $(1-q^{\frac{2}{n}})\ldots$ denotes $(1-q^{\frac{2}{n}})(1-q^{\frac{4}{n}})(1-q^{\frac{6}{n}})\ldots$, viz. it is the same function of $q^{\frac{1}{n}}$ that $(1-q^2)\ldots$ is of q; also

$$\Pi (1+q^{1+\frac{2g}{n}})\ldots(1+q^{1-\frac{2g}{n}})\ldots = \frac{(1+q^{\frac{1}{n}})\ldots}{(1+q)\ldots},$$

where $(1+q^{\frac{1}{n}})\ldots$ denotes $(1+q^{\frac{1}{n}})(1+q^{\frac{3}{n}})(1+q^{\frac{5}{n}})\ldots$, viz. it is the same function of $q^{\frac{1}{n}}$ that $(1+q)\ldots$ is of q; and the like as to the denominator factors: we thus have

$$\frac{1}{M_n} = \left\{\frac{(1-q^{\frac{2}{n}})\ldots}{(1-q^2)\ldots}\,\frac{(1+q^{\frac{1}{n}})\ldots}{(1+q\ )\ldots}\,\frac{(1+q^2)\ldots}{(1+q^{\frac{2}{n}})\ldots}\,\frac{(1-q\ )\ldots}{(1-q^{\frac{1}{n}})\ldots}\right\}^2,$$

viz. this is

$$= \left\{\frac{(1-q^{\frac{2}{n}})\ldots\,(1+q^{\frac{1}{n}})\ldots}{(1+q^{\frac{2}{n}})\ldots\,(1-q^{\frac{1}{n}})\ldots}\right\}^2 \div \left\{\frac{(1-q^2)\ldots(1+q)\ldots}{(1+q^2)\ldots(1-q)\ldots}\right\}^2;$$

or, we have

$$\frac{1}{M_n} = \phi^2(q^{\frac{1}{n}}) \div \phi^2(q),$$

agreeing with a former result.

We have

$$2\beta_n = \frac{1}{M_n} - 1,$$

that is,

$$2\left\{\frac{1}{\text{snc}\,2\omega_n} + \frac{1}{\text{snc}\,4\omega_n}\ldots + \frac{1}{\text{snc}\,(n-1)\,\omega_n}\right\} = \frac{\phi^2(q^{\frac{1}{n}})}{\phi^2(q)} - 1,$$

a result which, substituting on the left-hand side the foregoing values of the several functions, must be identically true.

62. We have also

$$v_n = u^n\,\{\text{snc}\,2\omega_n\,\text{snc}\,4\omega_n\ldots\text{snc}\,(n-1)\,\omega_n\},$$

where the term in { } is

$$= \Pi f^{-2}(q)\,\frac{(1+q^{\frac{2g}{n}})}{2q^{\frac{g}{n}}}\,\frac{(1+q^{2+\frac{2g}{n}})\ldots(1+q^{2-\frac{2g}{n}})\ldots}{(1+q^{1+\frac{2g}{n}})\ldots(1+q^{1-\frac{2g}{n}})\ldots};$$

or, observing that the sum of the exponents $\frac{g}{n}$ is $\frac{1}{n}\{1+2\ldots+\tfrac{1}{2}(n-1)\} = \frac{n^2-1}{8n}$, this is

$$= f^{-n+1}(q)\,.\,\frac{1}{(\sqrt{2})^{n-1}\,q^{\frac{n^2-1}{8n}}}\,\frac{(1+q^{\frac{2}{n}})\ldots(1+q\ )\ldots}{(1+q^2)\ldots(1+q^{\frac{1}{n}})\ldots};$$

or, the last factor being $f(q^{\frac{1}{n}}) \div f(q)$, the expression is

$$f^{-n}(q) \frac{1}{(\sqrt{2})^{n-1}} q^{-\frac{n}{8}+\frac{1}{8n}} f(q^{\frac{1}{n}});$$

or, multiplying by u^n, $=(\sqrt{2})^n q^{\frac{n}{8}} f^n(q)$, we have

$$v_n = \sqrt{2} q^{\frac{1}{8n}} f(q^{\frac{1}{n}}),$$

agreeing with a former result.

We have in what precedes the complete q-transcendental solution for the *transformatio secunda;* viz. the original modulus $k(=u^4)$ being given as a function of q, then, as well the new modulus $\lambda_n(=v_n^4)$ and the multiplier M_n, as also the several functions which enter into the formula

$$\frac{1-y}{1+y} = \frac{1-x}{1+x} \left\{ \frac{\left(1-\dfrac{x}{\operatorname{snc} 2\omega_n}\right) \ldots \left(1-\dfrac{x}{\operatorname{snc}(n-1)\omega_n}\right)}{\left(1+\dfrac{x}{\operatorname{snc} 2\omega_n}\right) \ldots \left(1+\dfrac{x}{\operatorname{snc}(n-1)\omega_n}\right)} \right\}^2,$$

are all expressed in terms of q. The expressions all contain $q^{\frac{1}{n}}$, and by substituting for this an imaginary nth root of q, we have the formulæ belonging to the several $(n-1)$ imaginary transformations.

63. As an illustration of the formulæ for the *transformatio secunda* I write $n=7$; and putting for greater convenience $q=r^7$, that is, $r=q^{\frac{1}{7}}$, then we have

$$v_7 = \sqrt{2} r^{\frac{1}{8}} f(r), \quad \frac{1}{M_7} = \frac{\phi^2(r)}{\phi^2(r^7)},$$

$$\frac{1}{\operatorname{snc} 2\omega_7} = 2f^2(r^7)\, A, \quad \frac{1}{\operatorname{snc} 4\omega_7} = 2f^2(r^7)\, B, \quad \frac{1}{\operatorname{snc} 6\omega_7} = 2f^2(r^7)\, C,$$

where

$$A = r \cdot \frac{5.19 \ldots \; 9.23 ..}{2.16 \ldots 12.26 ..},$$

$$B = r^2 \cdot \frac{3.17 .. \; 11.25 ..}{4.18 .. \; 10.24 ..},$$

$$C = r^3 \cdot \frac{1.15 .. \; 13.27 ..}{6.20 .. \; 8.22 ..},$$

where the numerator of A denotes $(1+r^5)(1+r^{19}) .. (1+r^9)(1+r^{23}) ..$, and so in other cases, the difference of the exponents being always $=14$. And we have, as mentioned, the identical equation

$$f^2(r^7)(A+B+C) = \tfrac{1}{4}\left\{\frac{\phi^2 r}{\phi^2 r^7} - 1\right\}.$$

The values of the several expressions up to r^{50} are as follows: Mr J. W. L. Glaisher kindly performed for me the greater part of the calculation.

Ind. of r	A	B	C	Sum	Multiplied by $f^2(r^7)$	$\phi^2(r) \div \phi^2(r^7)$
0				0	0	+ 1
1	+ 1			+ 1	+ 1	+ 4
2		+ 1		+ 1	+ 1	+ 4
3	− 1		+ 1	0	0	0
4			+ 1	+ 1	+ 1	+ 4
5	+ 1	+ 1		+ 2	+ 2	+ 8
6	+ 1	− 1		0	0	0
7	− 1			− 1	− 1	− 4
8	− 1			− 1	− 3	− 12
9	+ 1	− 1	− 1	− 1	− 3	− 12
10	+ 2	+ 1	− 1	+ 2	+ 2	+ 8
11	− 1		− 1	− 2	− 4	− 16
12	− 2	− 1	− 1	− 4	− 8	− 32
13		+ 2		+ 2	+ 2	+ 8
14	+ 2	− 1		+ 1	+ 3	+ 12
15	+ 1	− 1	+ 1	+ 1	+ 8	+ 32
16	− 2	+ 2	+ 2	+ 2	+ 9	+ 36
17	− 2	− 2	+ 2	− 2	− 6	− 24
18	+ 1	+ 1	+ 2	+ 4	+ 13	+ 52
19	+ 2	+ 2	+ 2	+ 6	+ 24	+ 96
20		− 3	+ 1	− 2	− 6	− 24
21	− 2	+ 2	− 1	− 1	− 8	− 32
22	− 2	+ 1	− 2	− 3	− 20	− 80
23	+ 2	− 4	− 3	− 5	− 24	− 96
24	+ 3	+ 3	− 4	+ 2	+ 16	+ 64
25		− 1	− 4	− 5	− 33	− 132
26	− 4	− 3	− 3	− 10	− 62	− 248
27	− 2	+ 5	− 1	+ 2	+ 16	+ 64
28	+ 4	− 3	+ 1	+ 2	+ 19	+ 76
29	+ 5	− 1	+ 3	+ 7	+ 46	+ 184
30	− 3	+ 6	+ 5	+ 8	+ 56	+ 224
31	− 7	− 6	+ 7	− 6	− 40	− 160
32	+ 1	+ 1	+ 7	+ 9	+ 77	+ 308
33	+ 9	+ 5	+ 4	+ 18	+ 144	+ 576
34	+ 3	− 8	+ 1	− 4	− 38	− 152
35	− 9	+ 5	− 1	− 5	− 42	− 168
36	− 7	+ 2	− 5	− 10	− 99	− 396
37	+ 7	− 9	− 9	− 11	− 122	− 488
38	+ 11	+ 10	− 11	+ 10	+ 88	+ 352
39	− 4	− 3	− 10	− 17	− 168	− 672
40	− 13	− 8	− 7	− 28	− 310	− 1240
41	− 2	+ 13	− 3	+ 8	+ 82	+ 328
42	+ 13	− 8	+ 3	+ 8	+ 88	+ 352
43	+ 8	− 3	+ 9	+ 14	+ 204	+ 816
44	− 11	+ 14	+ 14	+ 17	+ 252	+ 1008
45	− 14	− 14	+ 16	− 12	− 182	− 728
46	+ 5	+ 4	+ 15	+ 24	+ 344	+ 1376
47	+ 17	+ 11	+ 12	+ 40	+ 632	+ 2528
48	+ 3	− 20	+ 5	− 12	− 168	− 672
49	− 17	+ 13	− 5	− 9	− 175	− 700
50	− 13	+ 5	− 14	− 22	− 401	− 1604

64. As already mentioned, the foregoing expressions of the coefficients in terms of q may be applied to the determination of the coefficients as rational functions of u, v.

Representing by θ any one of the coefficients $\alpha, \beta, \gamma, \ldots, \sigma$, consider the sum

$$S\frac{v^f\theta}{\alpha},$$

f a positive integer, and the summation extending as before to the $n+1$ values of v, and corresponding values of $\frac{\theta}{\alpha}$. This is a rational function of u, and it is also integral. As to this observe that the function, if not integral, must become infinite either for $u=0$ (this would mean that the expression contained a term or terms Au^{-a}) or for some finite value of u. But the function can only become infinite by reason of some term or terms of $Sv^f\frac{\theta}{\alpha}$ becoming infinite; viz. some term $\frac{1}{\text{snc } 2g\omega}$ must become infinite; or attending to the equation

$$v = u^n \{\text{snc } 2\omega \text{ snc } 4\omega \ldots \text{snc } (n-1)\,\omega\},$$

it can only happen if $u=0$, or if $v=\infty$; and from the modular equation it appears that if $v=\infty$, then also $u=\infty$: the expression in question can therefore only become infinite if $u=0$, or if $u=\infty$. Now $u=0$ gives the ratios $\frac{\beta}{\alpha}, \frac{\gamma}{\alpha}, \ldots$, each of them a determinate function of n, that is finite; and gives also $v=0$, so that the expression does not become infinite for $u=0$; hence it does not become infinite either for $u=0$ or for any finite value of u; wherefore it is integral. The like reasoning applies to the sum $Sv^{-f}\frac{\theta}{\alpha}$; viz. this is a rational function of u; and it is quasi-integral, viz. there are no terms having a denominator other than a power of u, the highest denominator being u^{nf}; viz. the expression contains negative and positive integer powers of u, the lowest power (highest negative power) being $\frac{1}{u^{nf}}$.

65. It is to be observed, further, that writing the expression in the form

$$v_0^f\frac{\theta_0}{\alpha_0} + S'v^f\frac{\theta}{\alpha},$$

(where S' refers to the values $v_1, v_2, \ldots, v_n$ of the modulus), and considering the several quantities as expressed in terms of q, then in the sum S' every term involving a fractional power $q^{\frac{h}{n}}$ acquires by the summation the coefficient $(1+\alpha+\alpha^2+\ldots+\alpha^{n-1})$, and therefore disappears; there remains only the radicality $q^{\frac{1}{8}}$ occurring in the expressions of the v's; and if $nf \equiv \mu \pmod{8}$, $\mu=0$, or a positive integer less than 8, then the form of the expression is $q^{\frac{\mu}{8}}$ into a rational function of q. Hence this, being a rational and integral function of u, must be of the form

$$Au^{\mu} + Bu^{\mu+8} + Cu^{\mu+16} + \&c.$$

66. We have thus in general

$$Sv^f\frac{\theta}{\alpha} = Au^\mu \quad + Bu^{\mu+8} \quad + \&c.;$$

and in like manner

$$Sv^{-f}\frac{\theta}{\alpha} = A'u^{-nf} \quad + B'u^{-nf+8} + \&c.$$

We may in these expressions find a limit to the number of terms, by means of the before-mentioned theorem that we may simultaneously interchange $u, v;\ \alpha, \beta, \dots, \rho, \sigma$ into $\frac{1}{u}, \frac{1}{v};\ \sigma, \rho, \dots, \beta, \alpha$. Starting from the expression of $Sv^f\frac{\theta}{\alpha}$, let ϕ be the corresponding coefficient to θ; viz. in the series $\alpha, \beta, .., \theta, .., \phi, .., \rho, \sigma$, let ϕ be as removed from σ as θ is from α; then the equation becomes

$$Sv^{-f}\frac{\phi}{\sigma} = Au^{-\mu} \quad + Bu^{-\mu-8} \quad + \&c.,$$

where $\frac{\phi}{\sigma} = \frac{\phi}{\alpha}\frac{\alpha}{\sigma} = \frac{v}{u^n}\frac{\phi}{\alpha}$; the equation thus is

$$Sv^{1-f}\frac{\phi}{\alpha} = Au^{n-\mu} \quad + Bu^{n-\mu-8} \quad + \&c.;$$

and by what precedes the series on the right-hand side can contain no negative power higher than $\frac{1}{u^{n(f-1)}}$; that is, the series of coefficients $A, B, C, \dots$ goes on to a certain point only, the subsequent coefficients all of them vanishing.

In like manner from the equation for $Sv^{-f}\frac{\theta}{\alpha}$ we have

$$Sv^{f+1}\frac{\phi}{\alpha} = A'u^{(n+1)f} + B'u^{(n+1)f-8} + \&c.,$$

where the indices must be positive; viz. the series of coefficients $A', B', ..$ goes on to a certain point only, the subsequent coefficients all of them vanishing.

67. The like theory applies to the expression $\frac{1}{M}$. We have, putting as before $nf \equiv \mu$ (mod. 8),

$$Sv^f\ \frac{1}{M} = Au^\mu + Bu^{\mu+8} + \dots,$$

$$Sv^{-f}\frac{1}{M} = A'u^{-nf} + B'u^{-nf+8} + \dots$$

and we find a limit to the number of terms by the consideration that we may simultaneously change $u, v, \frac{1}{M}$ into $\frac{1}{u}, \frac{1}{v}, \frac{v^4}{u^4 M}$; the equations thus become

$$Sv^{4-f}\frac{1}{M} = Au^{4-\mu} + Bu^{-4-\mu} + \dots$$

(where, if $f=$ or <4, there must be on the right-hand side no negative power of u; but if $f>4$, then the highest negative power must be $\frac{1}{u^{(f-4)n}}$), and

$$Sv^{4+f}\frac{1}{M}=A'u^{nf+4}+B'u^{nf-4}+\ldots,$$

where on the right-hand side there must be no negative power of u.

68. It is to be remarked that β, ρ being always given linearly in terms of $\frac{1}{M}$ it is the same thing whether we seek in this manner for the values of β, ρ or for that of $\frac{1}{M}$; but the latter course is practically more convenient. Thus in the cases $n=5$, $n=7$ we require only the value of $\frac{1}{M}$.

In the case $n=11$, where the coefficients are α, β, γ, δ, ϵ, ζ, it has been seen that γ, δ are given as cubic functions of $\frac{1}{M}$: seeking for them directly, their values would (if the process be practicable) be obtained in a better form, viz. instead of the denominator $(F'v)^3$ there would be only the denominator $F'(v)$.

69. I consider for $\frac{1}{M}$ the cases $n=3$ and 5:

$$n=3,\ f=0,\ 1,\ 2,\ 3,\ \text{then}\ \mu=0,\ 3,\ 6,\ 1;$$

and we write down the equations

$$S\frac{1}{M}=A,\quad \text{giving}\ S\frac{v^4}{M}=Au^4,$$

$$S\frac{v}{M}=A'u^3,\quad ,,\quad S\frac{v^3}{M}=A'u,$$

$$S\frac{v^2}{M}=0,\quad ,,\quad S\frac{v^2}{M}=0;$$

viz. if we had in the first instance assumed $S\frac{1}{M}=A+Bu^8+\ldots$, this would have given $S\frac{v^4}{M}=Au^4+Bu^{-4}+\ldots$, whence B and the succeeding coefficients all vanish; and so in other cases. We have here only the coefficients A, A'; and these can be obtained without the aid of the q-formulæ by the consideration that for $u=1$ the corresponding values of v, $\frac{1}{M}$ are

$$v=1,\ -1,\ -1,\ -1,$$

$$\frac{1}{M}=3,\ -1,\ -1,\ -1,$$

whence $A=0$, $A'=6$; or we have the equations

$$S\frac{1}{M}=0,\quad S\frac{v}{M}=6u^3,\quad S\frac{v^2}{M}=0,\quad S\frac{v^3}{M}=6u,$$

giving as before

$$(2v^3+3v^2u-u)\frac{1}{M}=3\,(v^2u^2+2u^5v+1)\,u,$$

reducible by means of the modular equation to $\dfrac{1}{M}=1+\dfrac{2u^3}{v}$.

70. $n=5$. Corresponding to $f=0,\ 1,\ 2,\ 3,\ 4,\ 5$, we have $\mu=0,\ 5,\ 2,\ 7,\ 4,\ 1$, and we find

$$\begin{array}{lll}
S\dfrac{1}{M}=A, & \text{giving } S\dfrac{v^4}{M} & =Au^4,\\[2ex]
S\dfrac{v}{M}=0, & \quad\text{,,}\quad\; S\dfrac{v^3}{M} & =0,\\[2ex]
S\dfrac{v^2}{M}=A'u^2, & \quad\text{,,}\quad\; S\dfrac{v^2}{M} & =A'u^2,\\[2ex]
S\dfrac{v^5}{M}=A''u+B''u^9, & \quad\text{,,}\quad\; Sv^{-1}\dfrac{1}{M} & =A''u^3+B''u^{-5}.
\end{array}$$

But for $u=1$ the corresponding values of v, $\dfrac{1}{M}$ are

$$\begin{aligned}
v&=1,\ -1,\ -1,\ -1,\ -1,\ -1,\\
\frac{1}{M}&=5,\quad 1,\quad 1,\quad 1,\quad 1,\quad 1;
\end{aligned}$$

whence $A=A'=10$, $A''+B''=0$, or say the value of $S\dfrac{v^5}{M}$ is $=A''u\,(1-u^8)$.

The value of A'' is found very easily by the q-formulæ, viz. neglecting higher powers of q, we have

$$u=q^{\frac{1}{8}}\sqrt{2},\ v_0=q^{\frac{5}{8}}\sqrt{2},\ \frac{1}{M_0}=5;\ v_5=q^{\frac{1}{40}}\sqrt{2},\ \frac{1}{M_5}=1;$$

hence

$$S\frac{v^5}{M}=\frac{{v_0}^5}{M}+S'\frac{v^5}{M},\ =5q^{\frac{1}{8}}(\sqrt{2})^5=A''q^{\frac{1}{8}}\sqrt{2};$$

that is, $A''=20$, and the equations are

$$S\frac{1}{M}=10,\quad S\frac{v}{M}=0,\quad S\frac{v^2}{M}=10u^2,\quad S\frac{v^3}{M}=0,\quad S\frac{v^4}{M}=10u^4,\quad S\frac{v^5}{M}=20u\,(1-u^8);$$

whence

$$\begin{aligned}
F'v\,.\,\frac{1}{M}=\ &\ 20u\,(1-u^8)\\
&-10u^4\,(Sv_0-v)\\
&-10u^2\,(Sv_0v_1v_2-vSv_0v_1+v^2Sv_0-v^3)\\
&-10\quad(Sv_0v_1v_2v_3v_4-vSv_0v_1v_2v_3+v^2Sv_0v_1v_2-v^3Sv_0v_1+v^4Sv_0-v^5),
\end{aligned}$$

where Sv_0, &c. are the coefficients of the equation

$$v^6 + 4v^5u^5 + 5v^4u^2 - 5v^2u^4 - 4vu - u^6 = 0,$$

viz.

$$Sv_0, \quad v_0v_1, \quad v_0v_1v_2, \quad v_0v_1v_2v_3, \quad v_0v_1v_2v_3v_4$$

are

$$-4u^5, \quad +5u^2, \quad 0\ , \quad -5u^4\ , \quad 4u\ ;$$

or the equation is

$$\begin{aligned} F'v\,.\,\frac{1}{M} = \ & 20u\,(1-u^8) \\ & -10u^4\,(-4u^5 - v) \\ & -10u^2\,(-5u^2v - 4u^5v^2 - v^3) \\ & -10\ (4u + 5u^4v - 5v^3u^2 - 4v^4u^5 - v^5), \end{aligned}$$

or, say

$$\tfrac{1}{2}F'v\,\frac{1}{M} = 5\,\{v^5 + 4v^4u^5 + 6v^3u^2 + 4v^2u^7 + vu^4 - 2u\,(1-u^8)\},$$

where

$$\tfrac{1}{2}F'v \quad = 3\ v^5 + 10v^4u^5 + 10v^3u^2 \qquad - 5vu^4 - 2u.$$

Hence also, reducing by the modular equation,

$$\tfrac{1}{2}vF'v\,\frac{1}{M} = 5u\,\{v^4u + 4v^3u^4 + 6v^2u^3 + 2v\,(1+u^8) + u^5\},$$

the one of which forms is as convenient as the other.

71. Making the change $u,\ v,\ \dfrac{1}{M}$ into $v,\ -u,\ -5M$, we have

$$-\tfrac{1}{2}F'u\,.\,5M = 5\,\{-u^5 + 4v^5u^4 - 6v^2u^3 + 4v^7u^2 - v^4u - 2v\,(1-v^8)\}\ ;$$

and comparing with the equation

$$5M^2 = -\frac{(1-v^8)\,vF'v}{(1-u^8)\,uF'u},$$

we obtain

$$\frac{v\,(1-v^8)}{u\,(1-u^8)} = \frac{-2v\,(1-v^8) - v^4u + 4v^7u^2 - 6v^2u^3 + 4v^5u^4 - u^5}{-2u\,(1-u^8) + u^4v + 4u^7v^2 + 6u^2v^3 + 4u^5v^4 + v^5}.$$

Writing for a moment $M = u^4 + 6u^2v^2 + v^4$, $N = u^2 + v^2$, this is

$$-\frac{v\,(1-v^8)}{u\,(1-u^8)} = \frac{-2v\,(1-v^8) - uM + 4v^5u^2N}{-2u\,(1-u^8) + vM + 4v^2u^5N},$$

that is,

$$-4uv\,(1-u^8)\,(1-v^8) - \{u^2\,(1-u^8) - v^2\,(1-v^8)\}\,M + 4v^3u^3\,\{u^2\,(1-v^8) + v^2\,(1-u^8)\}\,N = 0.$$

But we have

$$\begin{aligned} u^2\,(1-u^8) - v^2\,(1-v^8) &= (u^2 - v^2)\,\{1 - u^8 - u^6v^2 - u^4v^4 - u^2v^6 - v^8\}, \\ u^2\,(1-v^8) + v^2\,(1-u^8) &= (u^2 + v^2)\,\{1 - u^2v^2\,(u^4 - u^2v^2 + v^4)\}. \end{aligned}$$

Hence, replacing M, N by their values, this is

$$\begin{aligned}&-4uv(1-u^8)(1-v^8)\\&-(u^2-v^2)(1-u^8-u^6v^2-u^4v^4-u^2v^6-v^8)(u^4+6u^2v^2+v^4)\\&+4u^3v^3(u^2+v^2)^2\{1-u^2v^2(u^4-u^2v^2+v^4)\}=0;\end{aligned}$$

viz. writing $u^2-v^2=A$, $uv=B$, this is

$$\begin{aligned}&-4B\{1-A^4-4A^2B^2-2B^4+B^8\}\\&-\ A\{1-A^4-5A^2B^2-3B^4\}(A^2+8B^2)\\&+4B^3(A^2+4B^2)\{1-A^2B^2-B^4\}=0,\end{aligned}$$

that is,

$$\begin{aligned}&-4B\{(1-A^4-4A^2B^2-2B^4+B^8)-B^2(A^2+4B^2)(1-A^2B^2-B^4)\}\\&-\ A\ (1-A^4-5A^2B^2-3B^4)(A^2+8B^2)=0;\end{aligned}$$

viz.

$$\begin{aligned}&-4B\ (1-A^4-5A^2B^2-3B^4)(1-B^4)\\&-\ A\ (1-A^4-5A^2B^2-3B^4)(A^2+8B^2)=0;\end{aligned}$$

or throwing out the factor $-(1-A^4-5A^2B^2-3B^4)$, this is

$$A(A^2+8B^2)+4B(1-B^4)=0,$$

the modular equation, which is right.

The four forms of the modular equation, and the curves represented thereby.
Art. Nos. 72 to 79.

72. The modular equation for any value of n has the property that it may be represented as an equation of the same order ($=n+1$, when n is prime) between u, v: or between u^2, v^2: or between u^4, v^4: or between u^8, v^8. As to this, remark that in general an equation $(u, v, 1)^m=0$ of the order m gives rise to an equation $(u^2, v^2, 1)^{2m}=0$ of the order $2m$ between u^2, v^2; viz. the required equation is

$$(u, v, 1)^m(u, -v, 1)^m(-u, v, 1)^m(-u, -v, 1)^m=0,$$

where the left-hand side is a rational function of u^2, v^2 of the form $(u^2, v^2, 1)^{2m}$; or again starting from a given equation $(u, v, w)^m=0$, and transforming by the equations $x:y:z=u^2:v^2:w^2$, the curve in (x, y, z) is of the order $2m$; in fact, the intersections of the curve by the arbitrary line $ax+by+cz=0$ are given by the equations $(u, v, w)^m=0$, $au^2+bv^2+cw^2=0$, and the number of them is thus $=2m$. Moreover, by the general theory of rational transformation, the new curve of the order $2m$ has the same deficiency as the original curve of the order m. The transformed curve in $x, y, z, =u^2, v^2, w^2$ may in particular cases reduce itself to a curve of the order m twice repeated; but it is important to observe that here, taking the single curve of the order m as the transformed curve, this has no longer the same deficiency as the original curve; and in particular the curves represented by the modular equation in its four several forms, writing therein successively u, v; u^2, v^2; u^4, v^4; $u^8, v^8, =x, y$, are not curves of the same deficiency.

73. The question may be looked at as follows: the quantities which enter rationally into the elliptic-function formulæ are k^2, $\lambda^2=u^8$, v^8; if a modular equation $(u, v)^\nu=0$ led to the transformed equation $(u^8, v^8)^{8\nu}=0$, then to a given value of u^8

would correspond 8 values of u, therefore 8ν values of v, giving the same number, 8ν, values of v^8; that is, the values of v^8 corresponding to a given value of u^8 would group themselves in eights corresponding to the 8 values of u. There is, in fact, no such grouping; the equations are $(u, v)^\nu = 0$, $(u^8, v^8)^\nu = 0$; to a given value of u^8 correspond 8 values of u, and therefore 8ν values of v, but these give in eights the same value of v^8, so that the number of values of v^8 is $= \nu$.

74. I consider the case $n = 3$: here, writing x, y for u, v, we have here the sextic curve

I. $y^4 - x^4 + 2xy(x^2y^2 - 1) = 0$;

and it is easy to see that the remaining forms wherein x, y denote u^2, v^2; u^4, v^4; and u^8, v^8 respectively, are derived herefrom as follows; viz.

II. $(y^2 - x^2)^2 - 4xy(xy - 1)^2 = 0$, that is,

$$y^4 + 6x^2y^2 + x^4 - 4xy(x^2y^2 + 1) = 0;$$

III. $(y^2 + 6xy + x^2)^2 - 16xy(xy + 1)^2 = 0$, that is,

$$y^4 + 6x^2y^2 + x^4 - 4xy(4x^2y^2 - 3x^2 - 3y^2 + 4) = 0;$$

IV. $(y^2 + 6xy + x^2)^2 - 16xy(4xy - 3x - 3y + 4)^2 = 0$, that is,

$$y^4 - 762x^2y^2 + x^4 - 4xy\{64x^2y^2 - 96x^2y - 96xy^2 + 33x^2 + 33y^2 - 96x - 96y + 64\} = 0,$$

where it may be noticed that the process is not again repeatable so as to obtain a sextic equation between x, y standing for u^{16}, v^{16} respectively.

The curve I. has a dp (fleflecnode) at the origin, viz. the branches are given by $y^3 - 2x = 0$, $-x^3 - 2y = 0$; and it has 2 cusps at infinity, on the axes $x = 0$, $y = 0$ respectively; viz. the infinite branches are given by $y + 2x^3 = 0$, $-x + 2y^3 = 0$ respectively. These same singularities present themselves in the other curves.

The curve II. has the four dps $(x^2 - y^2 = 0,\ xy - 1 = 0)$, that is,

$$(x = y = 1),\ (x = y = -1),\ (x = i,\ y = -i),\ (x = -i,\ y = i).$$

Corresponding hereto we have in the curve III. the 2 dps $(x = y = 1,\ x = y = -1)$, and in the curve IV. the dp $(x = y = 1)$.

The curve III. has besides the 4 dps $y^2 + 6xy + x^2 = 0$, $xy + 1 = 0$, that is,

$$(1 + \sqrt{2},\ 1 - \sqrt{2}),\ (1 - \sqrt{2},\ 1 + \sqrt{2}),\ (-1 - \sqrt{2},\ -1 + \sqrt{2}),\ (-1 + \sqrt{2},\ -1 - \sqrt{2});$$

and corresponding hereto in the curve IV. we have the 2 dps

$$(3 + 2\sqrt{2},\ 3 - 2\sqrt{2}),\ (3 - 2\sqrt{2},\ 3 + 2\sqrt{2}).$$

The curve IV. has besides the 4 dps $(y^2 + 6xy + x^2 = 0,\ 4xy - 3x - 3y + 4 = 0)$, or say $(2x - \frac{3}{2})(2y - \frac{3}{2}) + \frac{7}{4} = 0$, $2(x + \frac{9}{4})^2 + 2(y + \frac{9}{4})^2 - \frac{177}{8} = 0$. Hence the 4 curves have respectively the dps and deficiency following:—

dps.		dps.	Def.
2, 1	=	3,	7,
2, 1, 4	=	7,	3,
2, 1, 2, 4	=	9,	1,
2, 1, 1, 2, 4	=	10,	0;

viz. the curve IV. representing the equation between u^8 and v^8 is a unicursal sextic.

It may be noticed that, except the fleflecnode at the origin and the cusps at infinity, the dps in question are all acnodes (conjugate points).

75. The foregoing equations may be exhibited in the square diagrams:—

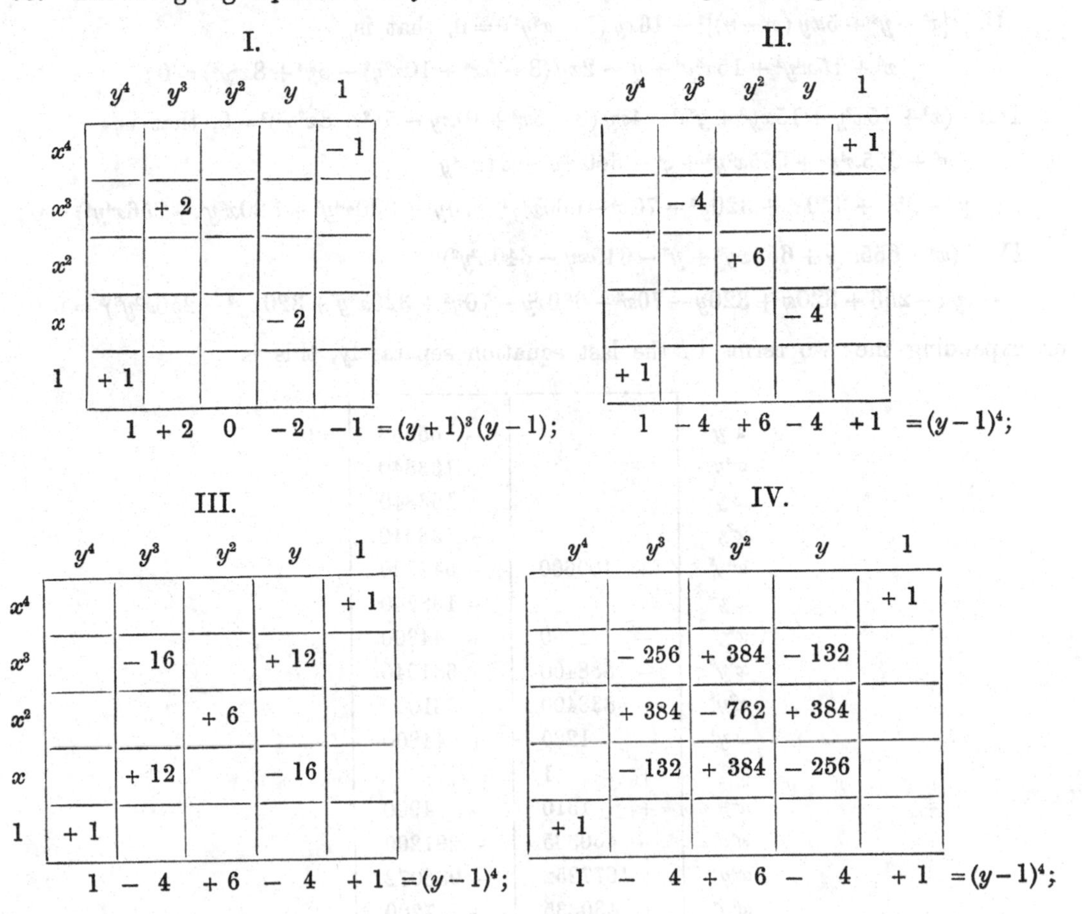

I.

	y^4	y^3	y^2	y	1
x^4					-1
x^3	$+2$				
x^2					
x				-2	
1	$+1$				

$1 \quad +2 \quad 0 \quad -2 \quad -1 = (y+1)^3(y-1);$

II.

	y^4	y^3	y^2	y	1
x^4					$+1$
x^3		-4			
x^2			$+6$		
x				-4	
1	$+1$				

$1 \quad -4 \quad +6 \quad -4 \quad +1 = (y-1)^4;$

III.

	y^4	y^3	y^2	y	1
x^4					$+1$
x^3		-16		$+12$	
x^2			$+6$		
x		$+12$		-16	
1	$+1$				

$1 \quad -4 \quad +6 \quad -4 \quad +1 = (y-1)^4;$

IV.

	y^4	y^3	y^2	y	1
x^4					$+1$
x^3		-256	$+384$	-132	
x^2		$+384$	-762	$+384$	
x		-132	$+384$	-256	
1	$+1$				

$1 \quad - \quad 4 \quad + \quad 6 \quad - \quad 4 \quad +1 = (y-1)^4;$

where the subscript line, showing in each case what the equation becomes on writing therein $x = 1$, serves as a verification of the numerical values.

The curve IV. being unicursal, the coordinates may be expressed rationally in terms of a parameter; in fact, we have

$$x = \frac{\alpha^3(2+\alpha)}{1+2\alpha}, \quad y = \frac{\alpha(2+\alpha)^3}{(1+2\alpha)^3}.$$

These values give

$$16xy = 16\alpha^4(2+\alpha)^4 \qquad \div (1+2\alpha)^4,$$

$$4 + 4xy - 3x - 3y = (4,\ 8,\ 12,\ 32,\ 50,\ 32,\ 12,\ 8,\ 4 \between 1,\ \alpha)^8 \qquad \div (1+2\alpha)^4,$$

$$x^2 + 6xy + y^2 = 4\alpha^2(2+\alpha)^2(4,\ 8,\ 12,\ 32,\ 50,\ 32,\ 12,\ 8,\ 4 \between 1,\ \alpha)^8 \div (1+2\alpha)^6,$$

and the equation of the curve is thus verified.

76. Considering in like manner the modular equation for the quintic transformation, we derive the four forms as follows:—

I. $x^6y^6 + 5x^2y^2(x^2 - y^2) + 4xy(1 - x^4y^4) = 0$;

II. $\{x^3 - y^3 + 5xy(x - y)\}^2 - 16xy(1 - x^2y^2)^2 = 0$, that is,

$$x^6 + 15x^4y^2 + 15x^2y^4 + y^6 - 2xy(8 - 5x^4 + 10x^2y^2 - 5y^4 + 8x^4y^4) = 0;$$

III. $(x^3 + 15x^2y + 15xy^2 + y^3)^2 - 4xy(8 - 5x^2 + 10xy - 5y^2 + 8x^2y^2)^2 = 0$, that is,

$$x^6 + 655x^4y^2 + 655x^2y^4 + y^6 - 640x^2y^2 - 640x^4y^4$$
$$+ xy(-256 + 320x^2 + 320y^2 - 70x^4 - 660x^2y^2 - 70y^4 + 320x^4y^2 + 320x^2y^4 - 256x^4y^4) = 0;$$

IV. $(x^3 + 655x^2y + 655xy^2 + y^3 - 640xy - 640x^2y^2)^2$

$$- xy(-256 + 320x + 320y - 70x^2 - 660xy - 70y^2 + 320x^2y + 320xy^2 - 256x^2y^2)^2 = 0:$$

or, expanding the two terms in the last equation separately, this is

xy		− 65536	= 0.
x^2y		+ 163840	
xy^2		+ 163840	
x^3y		− 138240	
x^2y^2	+ 409600	− 542720	
xy^3		− 138240	
x^4y	− 1280	+ 44800	
x^3y^2	− 838400	+ 631040	
x^2y^3	− 838400	+ 631040	
xy^4	− 1280	+ 44800	
x^6	+ 1		
x^5y	+ 1310	− 4900	
x^4y^2	+ 430335	− 297200	
x^3y^3	+ 1677252	− 986072	
x^2y^4	+ 430335	− 297200	
xy^5	+ 1310	− 4900	
y^6	1		
x^5y^2	− 1280	+ 44800	
x^4y^3	− 838400	+ 631040	
x^3y^4	− 838400	+ 631040	
x^2y^5	− 1280	+ 44800	
x^5y^3		− 138240	
x^4y^4	+ 409600	− 542720	
x^3y^5		− 138240	
x^5y^4		+ 163840	
x^4y^5		+ 163840	
x^5y^5		− 65536	

77. The square diagrams are:—

I.

	y^6	y^5	y^4	y^3	y^2	y	1
x^6							-1
x^5		$+4$					
x^4					-5		
x^3							
x^2			$+5$				
x^1						-4	
1	$+1$						
	1	$+4$	$+5$	0	-5	-4	-1

$$=(y+1)^5(y-1);$$

II.

	y^6	y^5	y^4	y^3	y^2	y	1
x^6							$+1$
x^5		-16				$+10$	
x^4					$+15$		
x^3				-20			
x^2			$+15$				
x^1		$+10$				-16	
1	$+1$						
	1	-6	$+15$	-20	$+15$	-6	$+1$

$$=(y-1)^6;$$

III.

	y^6	y^5	y^4	y^3	y^2	y	1
x^6							$+1$
x^5		-256		$+320$		-70	
x^4			-640		$+655$		
x^3		$+320$		-660		$+320$	
x^2			$+655$		-640		
x		-70		$+320$		-256	
1	$+1$						
	$+1$	-6	$+15$	-20	$+15$	-6	$+1$

$$=(y-1)^6;$$

IV.

	y^6	y^5	y^4	y^3	y^2	y	1
x^6							$+1$
x^5		-65536	$+163840$	-138240	$+43520$	-3590	
x^4		$+163840$	-133120	-207360	$+133135$	$+43520$	
x^3		-138240	-207360	$+691180$	-207360	-138240	
x^2		$+43520$	$+133135$	-207360	-133120	$+163840$	
x		-3590	$+43520$	-138240	$+163840$	-65536	
1	$+1$						
	$+1$	-6	$+15$	-20	$+15$	-6	$+1$

$$=(y-1)^6:$$

where the subscript line, showing in each case what the equation becomes on writing therein $x=1$, serves as a verification of the numerical values.

78. The curve I. has at the origin a dp in the nature of a fleflecnode, viz. the two branches are given by $x^5+4y=0$, $-y^5+4x=0$ respectively; and there are two singular points at infinity on the two axes respectively, viz. the infinite branches are given by $-y-4x^5=0$, $x-4y^5=0$ respectively. Writing the first of these in the form $-yz^4-4x^5=0$, we see that the point at infinity on the axis $x=0$ (i.e. the point $z=0$, $x=0$) is $=6$ dps; and similarly writing for the other branch $xz^4-4y^5=0$, the point at infinity on the axis $y=0$ (i.e. the point $z=0$, $y=0$) is $=6$ dps*.

Moreover, as remarked to me by Professor H. J. S. Smith, the curve has 8 other dps; viz. writing ω to denote an eighth root of -1, $(\omega^8+1=0)$, then a dp is $x=\omega$, $y=\omega^5$. To verify this, observe that these values give

$$\left.\begin{array}{ll|ll} & \omega^5 & & \omega \\ 6x^5 & =\overbrace{+\ 6} & -\ 6y^5 & =\overbrace{+\ 6} \\ +20x^3y^2 & -20 & +10x^4y & -10 \\ -10xy^4 & -10 & -20x^2y^3 & -20 \\ +\ 4y & +\ 4 & +\ 4x & +\ 4 \\ -20x^4y^5 & +20 & -20x^5y^4 & +20 \end{array}\right\|\ ;$$

or the derived functions each vanish. Thus I. has in all $1+12+8, =21$ dps.

In II. we have in like manner $1+12+4, =17$ dps; viz. instead of the 8 dps, we have the 4 dps $x=\omega^2$, $y=\omega^2$, $(\omega^8+1=0)$, or, what is the same thing, $x=\omega$, $y=-\omega$, where $\omega^4+1=0$. But we have besides the 12 dps given by

$$x^3-y^3+5xy(x-y)=0,\ \ 1-x^2y^2=0,$$

viz. we have in all $1+12+4+12, =29$ dps.

In III. we thence have $1+12+2+6, =21$ dps; and, besides, the 12 dps given by

$$x^3+15x^2y+15xy^2+y^3=0,\ \ 8-5x^2+10xy-5y^2+8x^2y^2=0,$$

in all $1+12+2+6+12, =33$ dps.

And in IV. we thence have $1+12+1+3+6, =23$ dps; and, besides, the 12 dps given by

$$x^3+655x^2y+655xy^2+y^3-640xy-640x^2y^2=0,$$

$$-256+320x+320y-70x^2-660xy-70y^2+320x^2y+320xy^2-256x^2y^2=0,$$

(these curves intersect in 16 points, 4 of them at infinity, in pairs on the lines $x=0$, $y=0$ respectively; and the intersections at infinity being excluded, there remain $16-4, =12$ intersections); there are thus in all $1+12+1+3+6+12, =35$ dps.

* These results follow from the general formulæ in the paper "On the Higher Singularities of Plane Curves," *Camb. and Dubl. Math. Journ.* t. VII. (1866), pp. 212—223, [374]; but they are at once seen to be true from the consideration that the curve $yz^4-x^5=0$, which has only the singularity in question, is unicursal; the singularity is thus $=6$ dps.

Arranging the results in a tabular form and adding the values of the deficiency, we have

	dps.	dps.	Def.
I.	$1+12+8$	$=21,$	$=15,$
II.	$1+12+4+12$	$29,$	$7,$
III.	$1+12+2+\ 6+12$	$33,$	$3,$
IV.	$1+12+1+\ 3+\ 6+12$	$35,$	$1,$

so that the curve IV. is a curve of deficiency 1, or bicursal curve. It appears by Jacobi's investigation for the quintic transformation (*Fund. Nov.* pp. 26—28, [*Ges. Werke*, t. I., pp. 77—79]) that we can in fact express x, y, that is, u^8, v^8, rationally in terms of the parameters α, β connected by the equation

$$\alpha^3 = 2\beta(1+\alpha+\beta),$$

which is that of a general cubic (deficiency = 1); in fact, we have

$$\frac{2-\alpha}{\alpha-2\beta} = \frac{v^4}{u^4}, \quad \beta = \frac{u^5}{v},$$

that is,

$$u^8 (=x) = \beta^2 \left(\frac{2-\alpha}{\alpha-2\beta}\right), \quad v^8 (=y) = \beta^2 \left(\frac{2-\alpha}{\alpha-2\beta}\right)^5,$$

where α, β satisfy the relation just referred to. The actual verification of the equation IV. by means of these values would be a work of some labour.

79. In the general case p an odd prime, then in I. we have at the origin one dp (in the nature of a fleflecnode) and at infinity two singular points each $=\frac{1}{2}(p-1)(p-2)$ dps. I infer, from a result obtained by Professor Smith, that there are besides $(p-1)(p-3)$ dps; but I have not investigated the nature of these. And the Table of dps and deficiency then is

			dps.	Def.
I.	$1+(p-1)(p-2)+\ (p-1)(p-3)$	=	$2p^2-7p+6,$	$4p-5,$
II.	$1+(p-1)(p-2)+\frac{1}{2}(p-1)(p-3)+\frac{1}{2}(p^2-1)$		$2p^2-5p+4,$	$2p-3,$
III.	$1+(p-1)(p-2)+\frac{1}{4}(p-1)(p-3)+\frac{1}{4}(p^2-1)+\frac{1}{2}(p^2-1)$		$2p^2-4p+3,$	$p-2,$
IV.	$1+(p-1)(p-2)+\frac{1}{8}(p-1)(p-3)+\frac{1}{8}(p^2-1)+\frac{1}{4}(p^2-1)+\frac{1}{2}(p^2-1)$		$2p^2-\frac{7}{2}p+\frac{5}{2},$	$\frac{1}{2}p-\frac{3}{2};$

viz. his values of the deficiencies being as in the last column, the total number of dps must be as in the last but one column.

579.

ADDRESS DELIVERED BY [PROFESSOR CAYLEY AS] THE PRESIDENT [OF THE ROYAL ASTRONOMICAL SOCIETY] ON PRESENTING THE GOLD MEDAL OF THE SOCIETY TO PROFESSOR SIMON NEWCOMB.

[From the *Monthly Notices of the Royal Astronomical Society*, vol. XXXIV. (1873—1874), pp. 224—233.]

THE Council have awarded the medal to Professor Simon Newcomb for his Researches on the Orbits of *Neptune* and *Uranus*, and for his other contributions to mathematical astronomy. And upon me, as President, the duty has devolved of explaining to you the grounds of their decision.

I think it right to remark that it appears to me that, in the award of their highest honour, the Council of a Society are not bound to institute a comparison between heterogeneous branches of a science, or classes of research—to weigh, for instance, mathematical against observational astronomy or astronomical physics; or, in the several branches respectively, the happy idea which originates a theory against the patience and the skilled labour which develope and carry it out; and still less to decide between the merits of different workers in the science. It is enough that the different branches of a science coming before them in different years, the medal should in every case be bestowed as a recognition of high merit in some important branch of the science.

Before speaking of the Tables, I will notice some of Professor Newcomb's other works.

Memoir "On the secular Variations and mutual Relations of the Orbits of the Asteroids," *Mem. American Academy*, vol. v. (1860), pp. 124—152. The object is to examine those circumstances of the forms, positions, variations, and general relations of the asteroid orbits which may serve as a test, complete or imperfect, of any hypothesis respecting the cause from which they originated, or the reason why they are in a group by themselves. Every *a posteriori* test is founded on the supposition, that the hypothesis necessarily or probably implies that certain conditions must be satisfied by the asteroids or their orbits, viz. in the one case the conditions are those which follow necessarily and immediately from the hypothesis itself, in the other case those which are deducible from it by the principle of random distribution. The two principal hypotheses are that of Olbers, where the asteroids are supposed to be the fragments of a shattered single planet: and the hypothesis that they were formed by the breaking up of a ring of nebulous matter. On the first hypothesis the orbits of all the asteroids once intersected in a common point; the second affords no conclusion equally susceptible of an *a posteriori* test.

But for a rigorous or probable test of either hypothesis, what is needed is rigorous expressions in terms of the time for the eccentricity, inclination, and longitudes of perihelion and node of each of the asteroids considered, or, what is the same thing, the computation of the secular variations of the quantities h, l, p, q, which replace these elements. The investigation is applied to those asteroids the elements of which were determined with sufficient accuracy, and the eccentricities and inclinations of which were sufficiently small (limit taken is 11°). And the backbone of the memoir is the investigation of the h, l, p, q, for twenty-five asteroids included between the numbers (1) and (40). In this calculation, as was clearly necessary, the action of the asteroids on the larger planets and on each other was neglected; the expressions for the h, l, p, q, of the larger planets are regarded as given—they are, in fact, taken from Le Verrier (as calculated by him before the discovery of *Neptune*, but afterwards partially extended to that planet). The effect is that the differential coefficients $\frac{dh}{dt}$, &c. are given each of them as a sum of sines or cosines of arguments varying with the time; and thus, although the calculation is sufficiently laborious, the process is not one of the extreme labour and difficulty which it is in the case of the larger planets. The resulting table of the h, l, p, q, of the twenty-five asteroids has, of course, a value quite independent of the theoretical part of the memoir. Of this it is sufficient to say here that the conclusion is on the whole against Olbers's hypothesis. The subject is resumed, and more fully examined in a paper in the *Astronomische Nachrichten*, t. LVIII.

"Investigation of the Distance of the Sun and of the Elements which depend upon it, from the Observations of *Mars* made during the Opposition of 1862, and from other Sources," *Washington Observations for* 1865, Appendix II., pp. 1—29. The chief part of this valuable Memoir is occupied with a determination of the solar parallax by the discussion of the observations of *Mars* made in 1862 on the plan of Winnecke: three partial discussions had previously appeared, but these having been by comparisons of pairs of observations, one in each hemisphere, many observations in one hemisphere were lost by want of a corresponding observation in the other hemisphere; and out of a total of nearly 300 observations, only 125 were utilised. The idea is, the perturbations of the Earth and *Mars* being perfectly known for the period under consideration, every observation of the planet would lead rigorously to an equation of condition between its parallax, the six elements of its orbit, and the six elements of the Earth's orbit—thus 13 or more observations, when compared with any theory, should suffice to correct the errors of that theory. But the observations extending only over a short interval, say one month, the coefficients would be so minute as to give no trustworthy value of the corrections; the equations only suffice to determine a few functions of the elements which, being determined, the equations will be satisfied by widely differing values of the elements, if only these values are such as to give to the functions their right values. And by fixing *a priori* the entire number of functions in question, and using them in place of the elements of the Earth and *Mars*, the equations will be practically as rigorous as if all the 13 unknown quantities had been introduced. By such considerations as these, each observation is made to give a relation between only 3 unknown quantities, the correction of the Sun's parallax being one of them.

The principle appears to be one of extended application, in regard to the proper mode of dealing with the constantly recurring problem of the determination of a set of corrections from a large number of linear equations; and it is used by the author in regard to the equations which present themselves in his theories of *Neptune* and *Uranus*.

Returning to the *Mars* observations, these were made at six Northern and three Southern Observatories, the total number being 154 Northern, and 143 Southern, together 297 observations. There was the difficulty of reducing to a concordant system the observations at the different Observatories, since (the whole number of comparison stars not being observed on each night) the adopted mean position of each of them was not unimportant. But this being carefully discussed and allowed for, the observations, extending from August 21 to November 3, 1862, are divided into five groups, and from these is deduced a correction to the provisional value 8″·9 of the parallax. The author then reproduces or discusses other determinations, from micrometric observations of *Mars*, the parallactic inequality of the Moon, the lunar equation of the Earth, the transit of 1769, and Foucault's experiment on Light—the last result, as not a strictly astronomical one, and with no means of assigning its probable error, is left out of consideration—and the combination of the remaining ones gives the author's concluded value of the parallax; from which other astronomical constants are deduced.

"On the Right Ascensions of the Equatoreal Fundamental Stars and the Corrections necessary to reduce the Right Ascensions of different Catalogues to a mean homogeneous System," *Washington Observations for* 1870, Appendix III., pp. 1—73.

This important Memoir is referred to in the Council Report for 1873. The object is to do for the right ascensions of the equatoreal and zodiacal Stars what had been done by Auwers for the declinations, namely, to furnish the data necessary to reduce the principal original catalogues of stars to a homogeneous system by freeing them of their systematic differences. The results are contained in two tables of corrections (as depending on the R.A. and N.P.D. respectively) to the several catalogues; and in a table of concluded mean right ascensions for the beginning of each fifth Besselian year,

1750 to 1900, of 32 fundamental Stars, and of periodic terms in the right ascensions of *Sirius* and *Procyon*.

The evil of systematic differences between the observations of different Observatories of course presents itself in every case where such observations have to be combined: for instance, in the just-mentioned determination of the solar parallax by the observations of *Mars*; and in the making of a set of planetary tables: and all that tends to remove or diminish it is most important to the progress of Astronomy. I cannot help thinking that there should be some confederation of Observatories, or Central calculating Board, for publishing the lunar and planetary observations, &c., reduced to a concordant system. It seems hard upon the maker of a set of planetary tables that he should not at least have, ready to hand for comparison with his theory, a single and entire series of the observations of the planet.

"Théorie des Perturbations de la Lune, qui sont dues à l'action des Planètes," *Liouville*, t. XVI. (1871), pp. 1—45. This is a very important theoretical Memoir on the disturbed motion of three bodies: a problem which, so far as I am aware, has not hitherto been considered at all. I have elsewhere remarked that the so-called "Problem of Three Bodies," as usually treated is not really this problem at all, but a different and more simple one—that of disturbed elliptic motion. Thus, in the planetary theory, each planet is considered as moving in an ellipse, and as disturbed by the action of forces represented by means of a disturbing function peculiar to the planet in question. An approach is made to the problem of three bodies when, as in memoirs by Hamilton and Jacobi, the (say) two planets are replaced by two fictitious bodies, and instead of a disturbing function peculiar to each planet, the motion of the system is made to depend on a single disturbing function. And there are memoirs by Jacobi, Bertrand, and Bour, which do relate to the proper problem of three bodies, viz. to their undisturbed motion. But in the present Memoir, Professor Newcomb starts from this problem as if it were actually solved, viz. he takes the coordinates of the three bodies (Sun, Earth, and Moon) as given in terms of the time and of 18 constants of integration *. And then considering the system as acted upon by the attraction of a planet, represented by means of a disturbing function, he applies to the system of the three bodies the method of the variation of the elements. The six elements which determine the motion of the centre of gravity of the system are left out of consideration; there remain to be considered 12 elements only; six of these are $\epsilon_0, \pi_0, \theta_0, \epsilon_0', \pi_0', \theta_0'$ (initial mean longitudes and longitudes of pericentre and node): but the other six k_ϵ, k_π, &c., are functions the invention of which is a leading step in the theory, and it is in fact by means of them that the investigation is brought to a successful conclusion: the expressions of the last-mentioned six functions can, it is stated, be formed with facility by means of the developments (obtainable from the lunar theory) of the rectangular

* Of course the expressions actually used must be approximations: the centre of gravity of the Earth and Moon is regarded as moving round the Sun in an ellipse affected by a secular motion of perihelion (ultimately neglected); and the coordinates of the Moon in regard to the Earth are considered to be given by Delaunay's Lunar Theory. The centre of gravity of the whole system (in the undisturbed motion) moves uniformly in a right line, viz. the coordinates are $a+a't$, $b+b't$, $c+c't$; and we have thus the whole number $6+6+6$, $=18$, of arbitrary constants.

coordinates x, y, z, as periodic functions of the time. With these twelve elements, the expressions for the variations assume the canonical form

$$\frac{dk_\epsilon}{dt}=\frac{dR}{d\epsilon_0},\quad \frac{d\epsilon_0}{dt}=-\frac{dR}{dk_\epsilon},\ \text{\&c.}$$

The concluding part of the Memoir contains approximate calculations which seems to show that the whole process is a very practicable one: but the author remarks that it is only doing justice to Delaunay to say that, starting from his (Delaunay's) final differential equations, and regarding the planet as adding new terms to the disturbing function, there would be obtained equations of the same degree of rigour as those of his own Memoir.

Everything in the Lunar Theory is laborious, and it is impossible to form an opinion as to the comparative facility of methods; but irrespectively of the possible applications of the method, the Memoir is, from the boldness of the conception and beauty of the results, a very remarkable one, and constitutes an important addition to Theoretical Dynamics *.

I come now to the planets *Neptune* and *Uranus*: it is well-known how, historically, the two are connected. The increasing and systematic inaccuracies of Bouvard's Tables of *Uranus* were found to be such as could be accounted for by the existence of an exterior disturbing planet; and it was thus that the planet *Neptune* was discovered by Adams and Le Verrier before it was seen in the telescope, in September 1846. It was afterwards ascertained that the planet had been seen twice by Lalande, in May 1795. The theory of *Neptune* was investigated by Peirce and Walker: viz. Walker, by means of the observations of 1795, and those of 1846—47, and using Peirce's formulæ for the perturbations produced by *Jupiter*, *Saturn*, and *Uranus*, determined successfully two sets of elliptic elements of the planet. The values first obtained showed that it was necessary to revise the perturbation-theory, which Peirce accordingly did, and with the new perturbations and revised normal places, the second set of elements (*Walker's Elliptic Elements II.*) was computed. With these elements and perturbations there was obtained for the planet from the time of its discovery a continuous ephemeris, published in the *Smithsonian Contributions*, *Gould's Astronomical Journal*, and since 1852 in the *American Ephemeris* and the *Nautical Almanac*. The theory was next considered by Kowalski in a work published at Kasan in the year 1855. The long period inequalities are dealt with by him in a manner different from that adopted by Peirce, so that the two theories are not directly comparable, but Professor Newcomb, by a comparison of the ephemerides with observation, arrives at the conclusion that the theory of Kowalski (although derived from observations up to 1853, when the planet had moved through an arc of 16°) was on the whole no nearer the truth than that of Walker;

* Since the above was written, Professor Newcomb has communicated to me some very interesting details as to the extent to which he has carried his computations, and in particular he mentions that, considering the action of each planet from *Mercury* to *Saturn*, he has (in regard to the terms the coefficients of which might become large by integration) estimated the probable limiting value of more than fifty such terms of period from a few years to several thousands without finding any which could become sensible, except the term leading to Hansen's first inequality produced by *Venus*.

he observed, however, that this failure is accounted for by an accidental mistake in the computation of the perturbations of the radius vector by *Jupiter*.

Professor Newcomb's theory of *Neptune* is published in the *Smithsonian Contributions* under the title "An Investigation of the Orbit of *Neptune*, with General Tables of its Motion," (accepted for publication, May 1865). The errors of the published ephemerides were increasing rapidly; in 1863 Walker's was in error by 33″, and Kowalski's by 22″; both might be in error by 5′ before the end of the century. The time was come when (the planet having moved through nearly 40°) the orbit could be determined with some degree of accuracy. The general objects of the work are stated to be:

(1) To determine the elements of the orbit of *Neptune* with as much exactness as a series of observations extending through an arc of 40° would admit of.

(2) To inquire whether the mass of *Uranus* can be concluded from the motion of *Neptune*.

(3) To inquire whether these motions indicate the action of an extra-Neptunian planet, or throw any light on the question of the existence of such planet.

(4) To construct general tables and formulæ, by which the theoretical place of *Neptune* may be found at any time, and more particularly between the years 1600 and 2000.

The formation of the tables of a planet may, I think, be considered as the culminating achievement of Astronomy: the need and possibility of the improvement and approximate perfection of the tables advance simultaneously with the progress of practical astronomy, and the accumulation of accurate observations; and the difficulty and labour increase with the degree of perfection aimed at. The leading steps of the process are in each case the same, and it is well-known what these are; but it will be convenient to speak of them in order, with reference to the present tables: they are *first* to decide on the form of the formulæ, whether the perturbations shall be applied to the elements or the coordinates—or partly to the elements and partly to the coordinates; and as to other collateral matters. These are questions to be decided in each case, in part by reference to the numerical values (in particular, the ratios and approach to commensurability of the mean motions), in part by the degree of accuracy aimed at, or which is attainable—the tables may be intended to hold good for a few centuries, or for a much longer period. The general theory as regards these several forms ought, I think, to be developed to such an extent, that it should be possible to select, according to the circumstances, between two or three ready-made theories; and that the substitution therein of the adopted numerical values should be a mere mechanical operation; but in the planetary theory in its present state, this is very far from being the case, and there is always a large amount of delicate theoretical investigation to be gone through in the selection of the form and development of the algebraical formulæ which serve as the basis of the tables. In Prof. Newcomb's theory the perturbations are applied to the elements; in particular, it was determined that the long inequality arising from the near approach of the mean motion of *Uranus* to twice that of *Neptune* (period about 4,300 years), should be developed as a perturbation, not of the coordinates, but of the elements. And it was best, (as for a theory designed to remain of the highest degree of exactness for only a few centuries) to take not the mean values of the elements, but their values at a particular epoch during the period for which the theory is intended to be used. The adopted provisional elements of *Neptune*, and the elements of the disturbing planets, are accordingly not mean values, but values affected by secular and long inequalities, representing the actual values at the present time. *Secondly*, the form being decided on and the formulæ obtained, the numerical values of the adopted provisional elements of the planet, and of the elements of the disturbing planets and their masses, have to be substituted, so as to obtain the actual formulæ serving for the calculation of a provisional ephemeris; and such ephemeris, first of heliocentric, and then of geocentric positions, has to be computed for the period over which the observations extend. *Thirdly*, the ephemeris, computed as above, has to be compared with the observed positions; viz. in the present case these are, Lalande's two observations of 1795, and the modern observations at the Observatories of Greenwich, Cambridge, Paris, Washington, Hamburg, and Albany, extending over different periods from 1846 to 1864: these are discussed in reference to their systematic differences, and they are then corrected accordingly, so as to reduce the several series of observations to a concordant system. In this way is formed a series of 71 observed longitudes and latitudes (1795, and 1846 to 1864); the comparison of these with the computed values shows the errors of the provisional ephemeris. *Fourthly*, the errors of the provisional elements have to be corrected by means of the last-mentioned series of errors: as regards the longitudes, the comparison gives a series of equations between $\delta\epsilon$, δn, δh, δk, and μ (correction to the assumed mass of *Uranus*). The discussion of the equations shows that no reliable value of μ can be obtained from them; it indeed appears that, if *Uranus* had been unknown, its existence could scarcely have been detected from all the observations hitherto made of *Neptune* (far less is there any indication to be as yet obtained as to the existence of a trans-Neptunian planet): hence, finally, μ is taken $=0$, and the equations used for the determination of the remaining corrections. As regards the latitudes, the comparison gives a series of equations serving for the determination of the values of δp and δq. And applying the corrections to the provisional elements, the author obtains his concluded elements; viz. as already mentioned, these are the values, as affected by the long inequality, belonging to the epoch 1850. *Fifthly*, the tables are computed from the concluded elements, and the perturbations of the provisional theory.

After the elements of *Neptune* were ascertained, the question of its action on *Uranus* was considered by Peirce in a paper in the *Proc. American Acad.*, vol. I. (1848), pp. 334—337. This contains the results of a complete computation of the general perturbations of *Uranus* by *Neptune* in longitude and radius vector, but without any details of the investigation, or statement of the methods employed: it is accompanied by a comparison of the calculated and observed longitudes of *Uranus* (with three different masses of *Neptune*) for years at intervals from 1690 to 1845, and for one of these masses the residuals are so small that it appears that, using these perturbations by *Neptune* and Le Verrier's perturbations by *Jupiter* and *Saturn*, there existed a theory of *Uranus* from which quite accurate tables might have been constructed. But this was never done. The ephemeris of *Uranus* in the *American Ephemeris* was intended to be founded on the theory, but the proper definitive elements do not seem to have been adopted: and in the *Nautical Almanac* for the years up to 1876, Bouvard's Tables of *Uranus* were still employed; for the year 1877 the ephemeris is derived from heliocentric places communicated by Prof. Newcomb.

An extended investigation of the subject was made by Safford, but only a brief general description of his results is published, *Monthly Notices, R.A.S.*, vol. XXII. (1862). The effect of *Neptune* was here computed by mechanical quadratures; and corrections were obtained for the mass of *Neptune* and elements of *Uranus*.

Professor Newcomb's Tables of *Uranus* have only recently appeared. They are published in the *Smithsonian Contributions* under the title "An Investigation of the Orbit of *Uranus*, with General Tables of its Motion," (accepted for publication February, 1873), forming a volume of about 300 pages. The work was undertaken as far back as 1859, but the labour devoted to it at first amounted to little more than tentative efforts to obtain numerical data of sufficient accuracy to serve as a basis of the theory, and to decide on a satisfactory way of computing the general perturbations. First, the elements of *Neptune* had to be corrected, and this led to the foregoing investigation of that planet: it then appeared that the received elements of *Uranus* also differed too widely from the truth to serve as the basis of the work, and they were provisionally corrected by a series of heliocentric longitudes, derived from observations extending from 1781 to 1861. Finally, it was found that the adopted method of computing the perturbations, that of the "variation of the elements," was practically inapplicable to the computation of the more difficult terms, viz. those of the second order in regard to the disturbing force. While entertaining a high opinion of Hansen's method as at once general, practicable, and fully developed, the author conceived that it was on the whole preferable to express the perturbations directly in terms of the time, owing to the ease with which the results of different investigations could be compared, and corrections to the theory introduced; and under these circumstances he worked out the method described in the first chapter of his treatise, not closely examining how much it contained that was essentially new. With these improved elements and methods the work was recommenced in 1868; the investigation has occupied him during the subsequent five years: and, though aided by computers, every part of the work has been done under his immediate direction, and as nearly as possible in the same way as if he had done it himself: a result in some cases obtained only by an amount of labour approximating to that saved by the employment of the computer.

The leading steps of the investigation correspond to those for *Neptune*: there is, *first*, the theoretical investigation already referred to; *secondly*, the formation of the provisional theory with assumed elements; *thirdly*, the comparison with observation; and here the observations are the accidental ones previous to the discovery of *Uranus* as a planet by Herschel in 1781, and the subsequent systematic ones of twelve Observatories, extending over intervals during periods from 1781 to 1872; all which have to be freed from systematic differences, and reduced to a concordant system as before: the operation is facilitated by the existence, since 1830, of ephemerides computed from Bouvard's Tables serving as an intermediate term for the comparison of the observations with the provisional theory. *Fourthly*, the correction of the elements of the provisional theory, viz. the equations for the comparison of the longitudes give $\delta\epsilon$, δn, δh, δk, and a correction to the assumed mass of *Neptune*, which mass is thus brought out $= \frac{1}{19840}$. And the equations for the comparison of latitudes give δp, δq; there is thus obtained a corrected set of elements (Newcomb's Elements IV.), being for the year 1850, the elements as affected with the long inequality; these are the elements upon which the Tables are founded. But it is theoretically interesting to have the absolute mean values of the elements, and the author accordingly obtains these (his Elements V.) together with the corrections corresponding to a varied mass of *Neptune*, $\left(\text{that is, the terms in } \mu \text{ corresponding to a mass } \frac{1+\mu}{19700}\right)$; he remarks that, admitting the mass of *Neptune* to be uncertain by about one-fiftieth of its value, the mean longitude of the perihelion of *Uranus* is from this cause uncertain by more than two minutes, the mean longitude of the planet by nearly a minute, and the mean motion by nearly two seconds in a century. *Fifthly*, the formation of the tables, based on the Elements IV.; the tables calculated with these elements are intended to hold good for the period between the years 1000 and 2200; but by aid of the Elements V. they may be made applicable for a more extended period.

In what precedes I have endeavoured to give you an account of Professor Newcomb's writings: they exhibit all of them a combination, on the one hand, of mathematical skill and power, and on the other hand of good hard work—devoted to the furtherance of Astronomical Science. The Memoir on the Lunar Theory contains the successful development of a highly original idea, and cannot but be regarded as a great step in advance in the method of the variation of the elements and in theoretical dynamics generally; the two sets of planetary tables are works of immense labour, embodying results only attainable by the exercise of such labour under the guidance of profound mathematical skill—and which are needs in the present state of Astronomy. I trust that imperfectly as my task is accomplished, I shall have satisfied you that we have done well in the award of our medal.

The President then, delivering the medal to the Foreign Secretary, addressed him in the following terms:

Mr Huggins—I request that you will have the goodness to transmit to Professor Newcomb this medal, as an expression of the opinion of the Society of the excellence and importance of what he has accomplished; and to assure him at the same time of our best wishes for his health and happiness, and for the long and successful continuation of his career as a worker in our science.

580.

ON THE NUMBER OF DISTINCT TERMS IN A SYMMETRICAL OR PARTIALLY SYMMETRICAL DETERMINANT.

[From the *Monthly Notices of the Royal Astronomical Society*, vol. XXXIV. (1873—1874), pp. 303—307, and p. 335.]

THE determination of a set of unknown quantities by the method of least squares is effected by means of formulæ depending on symmetrical or partially symmetrical determinants; and it is interesting to have an expression for the number of distinct terms in such a determinant.

The terms of a determinant are represented as duads, and the determinant itself as a bicolumn; viz. we write, for instance,

$$\begin{Bmatrix} aa \\ bb \\ pp' \\ qq' \end{Bmatrix} \text{ to represent the determinants } \begin{vmatrix} aa, & ab, & ap', & aq' \\ ba, & bb, & bp', & bq' \\ pa, & pb, & pp', & pq' \\ qa, & qb, & qp', & qq' \end{vmatrix}.$$

This being so if the duads are such that in general $rs = sr$, then the determinant is wholly or partially symmetrical; viz. the determinant just written down, for which the bicolumn contains such symbols as pp' and qq', (each letter $p, q, \ldots$ being distinct from every letter $p', q', \ldots$) is partially symmetrical, but a determinant such as $\begin{Bmatrix} aa \\ bb \\ cc \end{Bmatrix}$ is wholly symmetrical. A determinant for which the bicolumn has m rows aa, bb, &c., and n rows pp', qq', &c. is called a determinant (m, n); and the number of distinct terms in the developed expression of the determinant is taken to be $\phi(m, n)$; the problem is to find the number of distinct terms $\phi(m, n)$.

Consider a determinant (m, n) where n is not $=0$; for instance, the determinant above written down, which is (2, 2); this contains terms multiplied by qa, qb, qp', qq' respectively: where, disregarding signs, the whole factor multiplied by qa is $\left\{\begin{matrix} bb \\ ap' \\ pq' \end{matrix}\right\}$, which is a determinant (1, 2), and similarly the whole factor multiplied by qb is a determinant (1, 2). But the whole factor multiplied by qp' is the determinant $\left\{\begin{matrix} aa \\ bb \\ pq' \end{matrix}\right\}$; which is a determinant (2, 1), and the whole factor multiplied by qq' is also a determinant (2, 1).

Hence, observing that qa, qb, qp', qq' are distinct terms occurring *only* in the last line of the determinant, the number of distinct terms is equal to the sum of the numbers of distinct terms in the several component parts, or we have

$$\phi(2, 2) = 2\phi(1, 2) + 2\phi(2, 1);$$

and so in general:

$$\phi(m, n) = m\phi(m-1, n) + n\phi(m, n-1).$$

Consider next a completely symmetrical determinant $(m, 0)$; for instance (4, 0), the determinant

$$\left\{\begin{matrix} aa \\ bb \\ cc \\ dd \end{matrix}\right\}, = \begin{vmatrix} aa, & ab, & ac, & ad \\ ba, & bb, & bc, & bd \\ ca, & cb, & cc, & cd \\ da, & db, & dc, & dd \end{vmatrix}.$$

We have *first* the terms containing dd; the whole factor is $\left\{\begin{matrix} aa \\ bb \\ cc \end{matrix}\right\}$, which is a determinant (3, 0); *secondly*, the terms containing $ad \,.\, da$, or the like combinations, $bd \,.\, db$ or $cd \,.\, dc$; the whole factor multiplied by $ad \,.\, da$ is $\left\{\begin{matrix} aa \\ bb \end{matrix}\right\}$, which is a determinant (2, 0); *thirdly*, the terms containing $ad \,.\, db + bd \,.\, da$, $= 2ad \,.\, bd$; or the like combinations $2ad \,.\, cd$ or $2bd \,.\, cd$: the whole factor multiplying the term $2ad \,.\, bd$ is $\left\{\begin{matrix} cc \\ ba \end{matrix}\right\}$, which is a determinant (1, 1). Hence observing that ad, bd, cd, $= da$, db, dc, and dd are terms occurring *only* in the last line and column of the original determinant, it is clear that the number of distinct terms in the original determinant is equal to the sum of the numbers of distinct terms in the component parts, or that we have $\phi(4, 0) = \phi(3, 0) + 3\phi(2, 0) + 3\phi(1, 1)$; and so in general:

$$\phi(m, 0) = \phi(m-1, 0) + m\phi(m-2, 0) + \frac{m \,.\, m-1}{2}\phi(m-3, 1).$$

The two equations of differences, together with the initial values $\phi(0,\ 0)=1$, $\phi(1,\ 0)=\phi(0,\ 1)=1$, $\phi(2,\ 0)=\phi(1,\ 1)=\phi(1,\ 2)=2$, enable the calculation of the successive values of $\phi(m,\ n)$: viz. arranging these in the order

$$\begin{array}{c} \phi(0,\ 0), \\ \phi(1,\ 0),\quad \phi(0,\ 1), \\ \phi(2,\ 0),\quad \phi(1,\ 1),\quad \phi(0,\ 2), \\ \phi(3,\ 0),\ \&c.,\ \&c., \end{array}$$

we calculate simultaneously the lines $\phi(m,\ 0)$, $\phi(m,\ 1)$; and thence successively the remaining lines $\phi(m,\ 2)$, $\phi(m,\ 3)$, &c.: the values up to $m+n=6$ being in fact

$$\begin{array}{c} 1, \\ 1,\quad 1, \\ 2,\quad 2,\quad 2, \\ 5,\quad 6,\quad 6,\quad 6, \\ 17,\quad 23,\quad 24,\quad 24,\quad 24, \\ 73,\quad 109,\quad 118,\quad 120,\quad 120,\quad 120, \\ 388,\quad 618,\quad 690,\quad 714,\quad 720,\quad 720,\quad 720: \end{array}$$

where the process for the first two lines is

$$\begin{array}{rlcrl} 5 = & 2 + {}_{2}.\ 1 + \ .\ 1, & \qquad & 6 = & {}_{2}.\ \ 2 + \ 2, \\ 17 = & 5 + {}_{3}.\ 2 + {}_{3}.\ 2, & & 23 = & {}_{3}.\ \ 6 + \ 5, \\ 73 = & 17 + {}_{4}.\ 5 + {}_{6}.\ 6, & & 109 = & {}_{4}.\ \ 23 + 17, \\ 388 = & 73 + {}_{5}.\,17 + \ .\,23, & & 618 = & {}_{5}.\,109 + 23, \\ \vdots & & & \vdots & \end{array}$$

the larger figures being those of the two lines, and the smaller ones numerical multipliers. And then for the third line, fourth line, &c., we have

$$\begin{array}{rlcrl} 6 = & {}_{1}.\ \ 2 + {}_{2}.\ \ 2, & \qquad & 120 = & {}_{2}.\ \ 24 + {}_{3}.\ \ 24, \\ 24 = & {}_{2}.\ \ 6 + {}_{2}.\ \ 6, & & 714 = & {}_{3}.\,120 + {}_{4}.\,118, \\ 118 = & {}_{3}.\ \ 24 + {}_{2}.\ \ 23, & & \vdots & \\ 690 = & {}_{4}.\,118 + {}_{2}.\,109, & & & \end{array}$$

and so on.

This is, in fact, the easiest way of obtaining the actual numerical values; but we may obtain an analytical formula. Considering the two equations

$$\phi(m,\ 1)=m\phi(m-1,\ 1)+\phi(m,\ 0),$$

$$\phi(m,\ 0)=\phi(m-1,\ 0)+m\phi(m-2,\ 0)+\frac{m\,.\,m-1}{2}\,\phi(m-3,\ 1);$$

and using the first of these to eliminate the term $\phi(m-3, 1)$ and resulting terms $\phi(m-4, 1)$, &c. which present themselves in the second equation, this, after a succession of reductions, becomes

$$\begin{aligned}\phi(m, 0) = {} & \phi(m-1, 0) \\ & +(m-1)\phi(m-2, 0) \\ & +\frac{m \, . \, m-1}{2}\{\phi(m-3, 0) \\ & \qquad +(m-3)\phi(m-4, 0) \\ & \qquad \vdots \\ & \qquad +(m-3)\ldots 3 \, . \, 2\phi(1, 0) \\ & \qquad +(m-3)\ldots 3 \, . \, 2 \, . \, 1 \qquad \};\end{aligned}$$

or, observing that the last term $(m-3)\ldots 3 \, . \, 2 \, . \, 1$ is, in fact, $=(m-3)\ldots 3 \, . \, 2 \, . \, 1\phi(0, 0)$, this may be written:

$$\begin{aligned}2\phi(m, 0)-\phi(m-1, 0)-(m-1)\phi(m-2, 0) = {} & \ \phi(m-1, 0) \\ & +(m-1)\phi(m-2, 0) \\ & +(m-1)(m-2)\phi(m-3, 0) \\ & \vdots \\ & +(m-1)\ldots 3 \, . \, 2 \, . \, 1\phi(0, 0).\end{aligned}$$

And hence assuming

$$u=\phi(0, 0)+\frac{x}{1}\phi(1, 0)+\frac{x^2}{1 \, . \, 2}\phi(2, 0)+\ldots+\frac{x^m}{1 \, . \, 2 \ldots m}\phi(m, 0)+\ldots,$$

we find at once

$$2\frac{du}{dx}-u-xu=\frac{u}{1-x},$$

that is,

$$2\frac{du}{u}=dx\left(1+x+\frac{1}{1-x}\right),$$

or integrating and determining the constant so that u shall become $=1$ for $x=0$, we have

$$u=\frac{e^{\frac{1}{2}x+\frac{1}{4}x^2}}{\sqrt{1-x}};$$

wherefore we have

$$\phi(m, 0)=1 \, . \, 2 \ldots m \text{ coefft. } x^m \text{ in } \frac{e^{\frac{1}{2}x+\frac{1}{4}x^2}}{\sqrt{1-x}}.$$

Developing as far as x^6, the numerical process is

	1	$\frac{1}{2}$	$\frac{1}{8}$	$\frac{1}{48}$	$\frac{1}{384}$	$\frac{1}{3840}$	$\frac{1}{46080}$
	1		$\frac{1}{4}$		$\frac{1}{32}$		$\frac{1}{384}$
	1	$\frac{1}{2}$	$\frac{1}{8}$	$\frac{1}{48}$	$\frac{1}{384}$	$\frac{1}{3840}$	$\frac{1}{46080}$
			$\frac{1}{4}$	$\frac{1}{8}$	$\frac{1}{32}$	$\frac{1}{192}$	$\frac{1}{1536}$
					$\frac{1}{32}$	$\frac{1}{64}$	$\frac{1}{256}$
							$\frac{1}{384}$
	1	$\frac{1}{2}$	$\frac{3}{8}$	$\frac{7}{48}$	$\frac{25}{384}$	$\frac{27}{1280}$	$\frac{331}{46080}$
	1	$\frac{1}{2}$	$\frac{3}{8}$	$\frac{5}{16}$	$\frac{35}{128}$	$\frac{63}{256}$	$\frac{231}{1024}$
	1	$\frac{1}{2}$	$\frac{3}{8}$	$\frac{7}{48}$	$\frac{25}{384}$	$\frac{27}{1280}$	$\frac{331}{46080}$
		$\frac{1}{2}$	$\frac{1}{4}$	$\frac{3}{16}$	$\frac{7}{96}$	$\frac{25}{768}$	$\frac{27}{2560}$
			$\frac{3}{8}$	$\frac{3}{16}$	$\frac{9}{64}$	$\frac{7}{128}$	$\frac{25}{1024}$
				$\frac{5}{16}$	$\frac{5}{32}$	$\frac{15}{128}$	$\frac{35}{768}$
					$\frac{35}{128}$	$\frac{35}{256}$	$\frac{105}{1024}$
						$\frac{63}{256}$	$\frac{63}{512}$
							$\frac{231}{1024}$
	1	1	1	$\frac{5}{6}$	$\frac{17}{24}$	$\frac{73}{120}$	$\frac{97}{180}$
× by	1	1	2	6	24	120	720
	1	1	2	5	17	73	388,

agreeing with the former values.

The expression of $\phi(m, 0)$ once found, it is easy thence to obtain

$$\phi(m, 1) = 1 \,.\, 2 \,....\, m \text{ coefft. } x^m \text{ in } \frac{e^{\frac{1}{2}x+\frac{1}{4}x^2}}{(1-x)^{\frac{3}{2}}}$$

$$\phi(m, 2) = 1 \,.\, 2 \,....\, m \text{ coefft. } x^m \text{ in } \frac{2e^{\frac{1}{2}x+\frac{1}{4}x^2}}{(1-x)^{\frac{5}{2}}}$$

$$\phi(m, 3) = 1 \,.\, 2 \,....\, m \text{ coefft. } x^m \text{ in } \frac{2 \,.\, 3e^{\frac{1}{2}x+\frac{1}{4}x^2}}{(1-x)^{\frac{7}{2}}}$$

and so on, the law being obvious.

[*Addition*, p. 335.] The generating function

$$u, \; = 1 + u_1 x + \ldots + u_n \frac{x^n}{1 \,.\, 2 \ldots n} + \ldots, \; = \frac{e^{\frac{1}{2}x+\frac{1}{4}x^2}}{\sqrt{1-x}},$$

was obtained as the solution of the differential equation

$$2\frac{du}{dx} = u\left(1 + x + \frac{1}{1-x}\right).$$

Writing this in the form

$$2(1-x)\frac{du}{dx} = u(2-x^2),$$

we at once obtain for u_n the equation of differences,

$$u_n = nu_{n-1} - \tfrac{1}{2}(n-1)(n-2)u_{n-3};$$

and it thus appears that the values of u_n (number of distinct terms in a symmetrical determination of the order n) can be calculated the one from the other by the process

$$\begin{aligned}
n &= 1, & 1 &= 1\,.\ 1,\\
&= 2, & 2 &= 2\,.\ 1,\\
&= 3, & 5 &= 3\,.\ 2 - 1\,.\,1,\\
&= 4, & 17 &= 4\,.\ 5 - 3\,.\,1,\\
&= 5, & 73 &= 5\,.\,17 - 6\,.\,2,\\
&= 6, & 388 &= 6\,.\,73 - 10\,.\,5,\\
& & &\&c.
\end{aligned}$$

which is one of extreme facility.

581.

ON A THEOREM IN ELLIPTIC MOTION.

[From the *Monthly Notices of the Royal Astronomical Society*, vol. XXXV. (1874—1875), pp. 337—339.]

LET a body move through apocentre between two opposite points of its orbit, say from the point P, eccentric anomaly u, to the point P', eccentric anomaly u', where

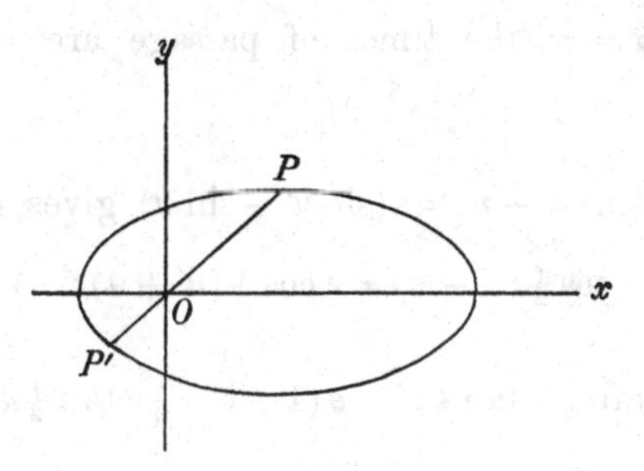

u, u' are each positive, $u < \pi$, $u' > \pi$. Taking the origin at the focus, and the axis of x in the direction through apocentre, then—

$$\text{Coordinates of } P \text{ are } x = a(-\cos u + e), \quad y = a\sqrt{1-e^2}\sin u,$$

$$\text{,, } \quad P' \text{ ,, } \quad x = a(-\cos u' + e), \quad y = a\sqrt{1-e^2}\sin u';$$

whence, expressing that the points P, P' are in a line with the focus,

$$\sin u'(-\cos u + e) - \sin u(-\cos u' + e) = 0,$$

that is,

$$\sin(u' - u) = e(\sin u' - \sin u),$$

which is negative, viz. $u' - u$ is $> \pi$.

The time of passage from P to P' is

$$nt = (u' - e \sin u') - (u - e \sin u),$$
$$= u' - u - e (\sin u' - \sin u),$$
$$= u' - u - \sin (u' - u),$$

which, $u' - u$ being greater than π and $-\sin (u' - u)$ positive, is greater than π; viz. the time of passage is greater than one-half the periodic time. Of course, if P and P' are at pericentre and apocentre, the time of passage is equal one-half the periodic time.

The time of passage from P' to P through the pericentre is

$$nt = 2\pi - (u' - u) + \sin (u' - u),$$

which is

$$= 2\pi - (u' - u) - \sin \{2\pi - (u' - u)\},$$

where $2\pi - (u' - u)$, $= \alpha$ suppose, is an angle $< \pi$. Writing, then

$$nt = \alpha - \sin \alpha,$$

and comparing with the known expression for the time in the case of a body falling directly towards the centre of force, we see that the time of passage from P' to P through the pericentre, is equal to the time of falling directly towards the same centre of force from rest at the distance $2a$ to the distance $a (1 + \cos \alpha)$, where, as above $\alpha = 2\pi - (u' - u)$, $u' - u$ being the difference of the eccentric anomalies at the two opposite points P, P'. If $\alpha = \pi$, the times of passage are each $= \frac{\pi}{n}$, that is, one-half the periodic time.

The foregoing equation $\sin (u' - u) = e (\sin u' - \sin u)$ gives obviously

$$\cos \tfrac{1}{2} (u' - u) = e \cos \tfrac{1}{2} (u' + u);$$

that is,

$$1 + \tan \tfrac{1}{2} u \tan \tfrac{1}{2} u' = e (1 - \tan \tfrac{1}{2} u \tan \tfrac{1}{2} u'),$$

or,

$$-\tan \tfrac{1}{2} u \tan \tfrac{1}{2} u' = \frac{1 - e}{1 + e};$$

(in the figure $\tan \frac{1}{2} u$ is positive, $\tan \frac{1}{2} u'$ negative); and we thence obtain further

$$\sin \tfrac{1}{2} (u' - u) = \cos \tfrac{1}{2} u' \cos \tfrac{1}{2} u (\tan \tfrac{1}{2} u' - \tan \tfrac{1}{2} u),$$
$$\sin \tfrac{1}{2} (u' + u) = \cos \tfrac{1}{2} u' \cos \tfrac{1}{2} u (\tan \tfrac{1}{2} u' + \tan \tfrac{1}{2} u),$$
$$\cos \tfrac{1}{2} (u' - u) = \cos \tfrac{1}{2} u' \cos \tfrac{1}{2} u \,.\, \frac{2e}{1 + e},$$
$$\cos \tfrac{1}{2} (u' + u) = \cos \tfrac{1}{2} u' \cos \tfrac{1}{2} u \,.\, \frac{2}{1 + e};$$

and thence also

$$\cos u + \cos u' = 2 \cos \tfrac{1}{2}(u' + u) \cos \tfrac{1}{2}(u' - u),$$

$$= \cos^2 \tfrac{1}{2} u' \cos^2 \tfrac{1}{2} u \,.\, \frac{8e}{(1+e)^2}.$$

But we have

$$1 + \cos(u' - u) = 2 \cos^2 \tfrac{1}{2}(u' - u) = \cos^2 \tfrac{1}{2} u' \cos^2 \tfrac{1}{2} u \,.\, \frac{8e^2}{(1+e)^2},$$

or, comparing with the last equation,

$$1 + \cos(u' - u) = e(\cos u + \cos u'),$$

or, what is the same thing,

$$1 - \cos(u' - u) = (1 - e\cos u') + (1 - e\cos u);$$

and in like manner,

$$1 + \cos(u' + u) = 2\cos^2 \tfrac{1}{2}(u' + u) = \cos^2 \tfrac{1}{2} u' \,.\, \cos^2 \tfrac{1}{2} u \frac{8}{(1+e)^2};$$

or, comparing with the same equation,

$$1 + \cos(u' + u) = \frac{1}{e}(\cos u + \cos u'):$$

which are formulæ corresponding with the original equation

$$\sin(u' - u) = e(\sin u' - \sin u).$$

582.

NOTE ON THE THEORY OF PRECESSION AND NUTATION.

[From the *Monthly Notices of the Royal Astronomical Society*, vol. XXXV. (1874—1875), pp. 340—343.]

WE have in the dynamical theory of Precession and Nutation (see Bessel's *Fundamenta* (1818), p. 126),

$$C\frac{dp}{dt}+(B-A)\,qr=LS\,(x'y-xy')\,dm'\left(\frac{1}{\Delta^3}-\frac{1}{r^3}\right),$$

$$A\frac{dq}{dt}+(C-B)\,rp=LS\,(y'z-yz')\,dm'\left(\frac{1}{\Delta^3}-\frac{1}{r^3}\right),$$

$$B\frac{dr}{dt}+(A-C)\,pq=LS\,(z'x-zx')\,dm'\left(\frac{1}{\Delta^3}-\frac{1}{r^3}\right),$$

where L is the mass of the Sun or Moon, x, y, z the coordinates of its centre referred to the centre of the Earth as origin,

$$r=\sqrt{x^2+y^2+z^2},$$

the distance of its centre, and

$$\Delta=\sqrt{(x-x')^2+(y-y')^2+(z-z')^2},$$

the distance of its centre from an element dm', coordinates (x', y', z') of the Earth's mass, the sum or integral S being extended to the whole mass of the Earth—I have written dm', r for Bessel's dm, r_1—, we have

$$\Delta^2=r^2-2\,(xx'+yy'+zz')+x'^2+y'^2+z'^2;$$

and thence

$$\frac{1}{\Delta^3}-\frac{1}{r^3}=\frac{3}{r^5}(xx'+yy'+zz')-\tfrac{3}{2}\frac{1}{r^7}\{(x^2+y^2+z^2)(x'^2+y'^2+z'^2)-5\,(xx'+yy'+zz')^2\}+\text{etc.}$$

The principal term is the first one,

$$\frac{3}{r^5}(xx' + yy' + zz');$$

but Bessel takes account also of the second term,

$$-\tfrac{3}{2}\frac{1}{r^7}\{(x^2+y^2+z^2)(x'^2+y'^2+z'^2) - 5(xx'+yy'+zz')^2\},$$

viz. considering the Earth as a solid of revolution (as to density as well as exterior form), he obtains in regard to it the following terms of $\sin\omega\frac{d\psi}{dt}$ and $\frac{d\omega}{dt}$ respectively;

$$\frac{3L}{4r^4}\frac{1}{Cn}\cdot 2(C-A)K(5\sin^2\delta - 1)\cos\delta\sin\alpha,$$

$$-\frac{3L}{4r^4}\frac{1}{Cn}\cdot 2(C-A)K(5\sin^2\delta - 1)\cos\delta\cos\alpha,$$

where

$$2(C-A)K = S(3\mu - 5\mu^3)\,2\pi\rho\,R^5 dR\,d\mu,$$

K being in fact a numerical quantity, relating to the Earth only, and the value of which is by pendulum observations ultimately found to be $=0{\cdot}13603$.

Writing, for shortness,

$$(x^2+y^2+z^2)(x'^2+y'^2+z'^2) - 5(xx'+yy'+zz')^2 = \Omega,$$

then the foregoing terms of $\sin\omega\frac{d\psi}{dt}$ and $\frac{d\omega}{dt}$ depend, as regards their form, on the theorem that for any solid of revolution (about the axis of z) we have

$$S(x'y - xy')\,\Omega dm',\quad S(y'z - yz')\,\Omega dm',\quad S(z'x - zx')\,\Omega dm'$$

$$= 0,$$

$$\tfrac{1}{2}y(x^2+y^2+z^2-5z^2)\,S[3(x'^2+y'^2+z'^2) - 5z'^2]\,z'dm',$$

$$-\tfrac{1}{2}x(x^2+y^2+z^2-5z^2)\,S[3(x'^2+y'^2+z'^2) - 5z'^2]\,z'dm',$$

respectively: viz. writing $x'^2+y'^2+z'^2 = R^2$, and $z' = R\mu$, also $x^2+y^2+z^2 = r^2$ and $x = r\cos\delta\cos\alpha$, $y = r\cos\delta\sin\alpha$, $z = r\sin\delta$, the values would be

$$0,$$

$$\tfrac{1}{2}r^3\cos\delta\sin\alpha(1 - 5\sin^2\delta)\,S(3 - 5\mu^2)\,\mu R^3 dm',$$

$$-\tfrac{1}{2}r^3\cos\delta\sin\alpha(1 - 5\sin^2\delta)\,S(3 - 5\mu^2)\,\mu R^3 dm',$$

which are of the form in question.

The verification is easy: the solid being one of revolution about the axis of z, any integral such as $Sx'z'^2dm'$ or $Sx'y'z'dm'$ which contains an odd power of x' or of y' is $=0$; while such integrals as $Sx'^2z'dm'$, $Sy'^2z'dm'$ are equal to each other, or, what is the same thing, each $=\tfrac{1}{2}S(x'^2+y'^2)\,z'dm'$. That we have $S(x'y - xy')\,\Omega dm' = 0$ is at once seen to be true; considering the next integral $S(y'z - yz')\,\Omega dm'$, the terms of $(y'z - zy')\,\Omega$ which lead to non-evanescent integrals are

$$-yz'\,.\,(x^2+y^2+z^2)(x'^2+y'^2+z'^2),$$
$$-5y'z\,.\,2yzy'z',$$
$$+5yz'\,.\,(x^2x'^2+y^2y'^2+z^2z'^2);$$

giving in the integral the several terms

$$-y(x^2+y^2+z^2)\,S(x'^2+y'^2+z'^2)\,z'dm',$$
$$-10yz^2\,.\,\tfrac{1}{2}S(x'^2+y'^2+z'^2-z'^2)\,z'dm',$$
$$+5y(x^2+y^2+z^2-z^2)\,.\,\tfrac{1}{2}S(x'^2+y'^2+z'^2-z'^2)\,z'dm',$$
$$+yz^2\,Sz'^3\,dm',$$

viz. collecting, the value is

$$(-1+\tfrac{5}{2}=)\quad \tfrac{3}{2}(x^2+y^2+z^2)\,yS(x'^2+y'^2+z'^2)\,z'dm',$$
$$(-\tfrac{5}{2}=)-\tfrac{5}{2}(x^2+y^2+z^2)\,y\Sigma z'^3dm',$$
$$(-\tfrac{5}{2}-5=)-\tfrac{15}{2}yz^2S(x'^2+y'^2+z'^2)\,z'dm',$$
$$(+\tfrac{5}{2}+5+5=)+\tfrac{25}{2}yz^2Sz'^3dm';$$

which is

$$=\ \tfrac{1}{2}y(x^2+y^2+z^2-5z^2)\,S[3(x'^2+y'^2+z'^2)-5z'^2]\,z'dm';$$

and similarly the last term is

$$=-\tfrac{1}{2}x(x^2+y^2+z^2-5z^2)\,S[3(x'^2+y'^2+z'^2)-5z'^2]\,z'dm',$$

which completes the proof.

583.

ON SPHEROIDAL TRIGONOMETRY.

[From the *Monthly Notices of the Royal Astronomical Society*, vol. XXXVII. (1876—1877), p. 92.]

THE fundamental formulæ of Spheroidal Trigonometry are those which belong to a right-angled triangle PSS_0, where P is the pole, PS, PS_0 arcs of meridian, and SS_0 a geodesic line cutting the meridian PS at a given angle, and the meridian PS_0 at right angles. We consider a spherical triangle PSS_0,

$$\begin{array}{llll} \text{Sides} & PS, & PS_0, & SS_0 = \gamma, \quad \gamma_0, \quad s, \\ \text{Angles} & S_0, & S, & P \;= 90^\circ, \quad \theta, \quad l, \end{array}$$

where γ is the reduced colatitude of the point S on the spheroid (and thence also γ_0 the reduced colatitude of S_0) and θ the azimuth of the geodesic SS_0, or angle at which this cuts the meridian SP; and then if S be the length of the geodesic SS_0 measured as a circular arc, radius = Earth's equatoreal radius, and L be the angle SPS_0, S, L differ from the corresponding spherical quantities s, l by terms involving the excentricity of the spheroid, viz. calling this e and writing

$$k = \frac{e \cos \gamma_0}{\sqrt{1 - e^2 \sin^2 \gamma_0}},$$

then (see Hansen's "Geodätische Untersuchungen," *Abh. der K. Sächs. Gesell.*, t. VIII. (1865) pp. 15 and 23, but using the foregoing notation) we have, to terms of the sixth order in e,

$$\begin{aligned} \frac{S}{\sqrt{1-e^2}} = \; & \left(1 + \tfrac{1}{4}k^2 + \tfrac{13}{64}k^4 + \tfrac{45}{256}k^6\right) s \\ & + \left(\tfrac{1}{8}k^2 + \tfrac{3}{32}k^4 + \tfrac{79}{1024}k^6\right) \sin 2s \\ & + \left(\tfrac{1}{256}k^4 + \tfrac{5}{1024}k^6\right) \sin 4s \\ & + \tfrac{1}{3072}k^6 \sin 6s\,; \end{aligned}$$

and

$$\begin{aligned} L = l - \tfrac{1}{2}e^2 \sin \gamma_0 \Big\{ & \left(1 - \tfrac{1}{8}k^2 + \tfrac{1}{4}e^2 - \tfrac{5}{64}k^4 + \tfrac{1}{8}e^4\right) s \\ & - \left(\tfrac{1}{16}k^2 + \tfrac{1}{32}k^4\right) \sin 2s \\ & + \tfrac{1}{256}k^4 \sin 4s \Big\}, \end{aligned}$$

which are the formulæ in question.

584.

ADDITION TO PROF. R. S. BALL'S PAPER, "NOTE ON A TRANSFORMATION OF LAGRANGE'S EQUATIONS OF MOTION IN GENERALISED COORDINATES, WHICH IS CONVENIENT IN PHYSICAL ASTRONOMY."

[From the *Monthly Notices of the Royal Astronomical Society*, vol. XXXVII. (1876—1877), pp. 269—271.]

THE formulæ may be established in a somewhat different way, as follows:—

Consider the masses M_1, M_2, ...

Let X_1, Y_1, Z_1 be the coordinates (in reference to a fixed origin and axes) of the C.G. of M_1;

x_1, y_1, z_1 the coordinates (in reference to a parallel set of axes through the C.G. of M_1) of an element m_1 of the mass M_1, and similarly for the masses M_2, ...; the coordinates (X_1, Y_1, Z_1), (X_2, Y_2, Z_2), ... all belonging to the same origin and axes;

And let $\dot{X}_1$, &c. denote the derived functions $\dfrac{dX_1}{dt}$, &c.

We have

$$\begin{aligned} T = &\quad S\tfrac{1}{2}m_1[(\dot{X}_1+\dot{x}_1)^2+(\dot{Y}_1+\dot{y}_1)^2+(\dot{Z}_1+\dot{z}_1)^2] \\ &+ S\tfrac{1}{2}m_2[(\dot{X}_2+\dot{x}_2)^2+(\dot{Y}_2+\dot{y}_2)^2+(\dot{Z}_2+\dot{z}_2)^2] \\ &\quad \vdots \end{aligned}$$

or since $Sm_1x_1=0$, &c., and therefore also $Sm_1\dot{x}_1=0$, &c., this is

$$\begin{aligned} T = &\quad \tfrac{1}{2}M_1(\dot{X}_1^2+\dot{Y}_1^2+\dot{Z}_1^2) &&+ S\tfrac{1}{2}m_1(\dot{x}_1^2+\dot{y}_1^2+\dot{z}_1^2) \\ &+ \tfrac{1}{2}M_2(\dot{X}_2^2+\dot{Y}_2^2+\dot{Z}_2^2) &&+ S\tfrac{1}{2}m_2(\dot{x}_2^2+\dot{y}_2^2+\dot{z}_2^2) \\ &\quad \vdots && \quad \vdots \end{aligned}$$

Write u, v, w for the coordinates of the C.G. of the whole system: then

$$M_1X_1 + M_2X_2 + \ldots = (M_1 + M_2 \ldots)\, u,$$
$$M_1Y_1 + M_2Y_2 + \ldots = (M_1 + M_2 \ldots)\, v,$$
$$M_1Z_1 + M_2Z_2 + \ldots = (M_1 + M_2 \ldots)\, w;$$

and thence

$$M_1\dot{X}_1 + M_2\dot{X}_2 + \ldots = (M_1 + M_2 \ldots)\, \dot{u},$$
$$M_1\dot{Y}_1 + M_2\dot{Y}_2 + \ldots = (M_1 + M_2 \ldots)\, \dot{v},$$
$$M_1\dot{Z}_1 + M_2\dot{Z}_2 + \ldots = (M_1 + M_2 \ldots)\, \dot{w};$$

and thence

$$T - \tfrac{1}{2}(M_1 + M_2 + \ldots)(\dot{u}^2 + \dot{v}^2 + \dot{w}^2)$$
$$= \frac{1}{M_1 + M_2 \ldots}\{M_1M_2[(\dot{X}_1 - \dot{X}_2)^2 + (\dot{Y}_1 - \dot{Y}_2)^2 + (\dot{Z}_1 - \dot{Z}_2)^2]\}$$
$$\vdots$$
$$+ S\, \tfrac{1}{2} m_1 (\dot{x}_1^2 + \dot{y}_1^2 + \dot{z}_1^2)$$
$$+ S\, \tfrac{1}{2} m_2 (\dot{x}_2^2 + \dot{y}_2^2 + \dot{z}_2^2)$$

or, representing the function on the right-hand side by T', this is

$$T = \tfrac{1}{2}(M_1 + M_2 + \ldots)(\dot{u}_2 + \dot{v}_2 + \dot{w}_2) + T'\ldots, = T_0 + T'.$$

Suppose the positions are determined by means of the $6n$ coordinates $((q))$; the equations of motion are each of them of the form

$$\frac{d}{dt}\cdot\frac{dT_0}{d\dot{q}} - \frac{dT_0}{dq} + \frac{d}{dt}\cdot\frac{dT'}{d\dot{q}} - \frac{dT'}{dq} = -\frac{dV}{dq}.$$

But these admit of further reduction; the part in T_0 depends upon three terms, such as

$$\frac{d}{dt}\left(\dot{u}\frac{d\dot{u}}{d\dot{q}}\right) - \dot{u}\frac{d\dot{u}}{dq}, \quad = \frac{d\dot{u}}{dt}\frac{d\dot{u}}{d\dot{q}} + \dot{u}\left(\frac{d}{dt}\frac{d\dot{u}}{d\dot{q}} - \frac{d\dot{u}}{dq}\right).$$

But we have u a function of $((q))$, and thence

$$\frac{d\dot{u}}{d\dot{q}} = \frac{du}{dq}, \text{ or } \frac{d}{dt}\frac{d\dot{u}}{d\dot{q}} - \frac{d\dot{u}}{dq}, \quad = \frac{d}{dt}\frac{du}{dq} - \frac{d\dot{u}}{dq}, \quad = 0,$$

or the term is simply

$$= \frac{d\dot{u}}{dt}\frac{d\dot{u}}{dq}.$$

The equation thus becomes

$$(M_1 + M_2 \ldots)\left(\frac{d\dot{u}}{dt}\frac{d\dot{u}}{d\dot{q}} + \frac{d\dot{v}}{dt}\frac{d\dot{v}}{d\dot{q}} + \frac{d\dot{w}}{dt}\frac{d\dot{w}}{d\dot{q}}\right) + \frac{d}{dt}\frac{dT'}{d\dot{q}} - \frac{dT'}{dq} = -\frac{dV}{dq}.$$

Suppose now that T', V are functions of $6n-3$ out of the $6n$ coordinates $((q))$, and of the differential coefficients $\dot{q}$ of the same $6n-3$ coordinates, but are independent of the remaining three coordinates and of their differential coefficients; then, first, if q denotes any one of the three coordinates, the equation becomes

$$\frac{d\dot{u}}{dt}\frac{d\dot{u}}{d\dot{q}}+\frac{d\dot{v}}{dt}\frac{d\dot{v}}{d\dot{q}}+\frac{d\dot{w}}{dt}\frac{d\dot{w}}{d\dot{q}}=0;$$

or, better,

$$\frac{d\dot{u}}{dt}\frac{du}{dq}+\frac{d\dot{v}}{dt}\frac{dv}{dq}+\frac{d\dot{w}}{dt}\frac{dw}{dq}=0;$$

and the three equations of this form give

$$\frac{d\dot{u}}{dt}=0,\quad \frac{d\dot{v}}{dt}=0,\quad \frac{d\dot{w}}{dt}=0,$$

viz. these are the equations for the conservation of the motion of the centre of gravity.

And this being so, then, if q now denotes any one of the $6n-3$ coordinates, each of the remaining equations assumes the form

$$\frac{d}{dt}\cdot\frac{dT'}{d\dot{q}}-\frac{dT'}{dq}=-\frac{dV}{dq},$$

viz. we have thus $6n-3$ equations for the relative motion of the bodies of the system.

585.

A NEW THEOREM ON THE EQUILIBRIUM OF FOUR FORCES ACTING ON A SOLID BODY.

[From the *Philosophical Magazine*, vol. XXXI. (1866), pp. 78, 79; *Camb. Phil. Soc. Proc.* vol. I. (1866), p. 235.]

DEFINING the "moment of two lines" as the product of the shortest distance of the two lines into the sine of their inclination, then, if four forces acting along the lines 1, 2, 3, 4 respectively are in equilibrium, the lines must, as is known (Möbius), be four generating lines of an hyperboloid; and if 12 denote the moment of the lines 1 and 2, and similarly 13 the moment of the lines 1 and 3, &c., the forces are as

$$\sqrt{23\,.\,34\,.\,42} : \sqrt{34\,.\,41\,.\,13} : \sqrt{41\,.\,12\,.\,24} : \sqrt{12\,.\,23\,.\,31}.$$

Calling the four forces P_1, P_2, P_3, P_4, it follows as a corollary that we have

$$P_1P_2\,.\,12 = 12\,.\,34\,\sqrt{13\,.\,42}\,.\,\sqrt{14\,.\,23} = P_3P_4\,.\,34\,;$$

viz. the product of any two of the forces into the moment of the lines along which they act is equal to the product of the other two forces into the moment of the lines along which they act,—which is equivalent to Chasles's theorem, that, representing a force by a finite line of proportional magnitude, then in whatever way a system of forces is resolved into two forces, the volume of the tetrahedron formed by joining the extremities of the two representative lines is constant.

586.

ON THE MATHEMATICAL THEORY OF ISOMERS.

[From the *Philosophical Magazine*, vol. XLVII. (1874), pp. 444—446.]

I CONSIDER a "diagram," viz. a set of points H, O, N, C, &c. (any number of each), connected by links into a single assemblage under the condition that through each H there passes not more than one link, through each O not more than two links, through each N not more than three links, through each C not more than four links. Of course through every point there passes at least one link, or the points would not be connected into a single assemblage.

In such a diagram each point having its full number of links is saturate, or nilvalent: in particular, each point H is saturate. A point not having its full number of links is univalent, bivalent, or trivalent, according as it wants one, two, or three of its full number of links. If every point is saturate the diagram is saturate, or nilvalent; or, say, it is a "plerogram"; but if the diagram is susceptible of n more links, then it is n-valent; viz. the valency of the diagram is the sum of the valencies of the component points.

Since each H is connected by a single link (and therefore to a point O, C, &c. as the case may be, but not to another point H), we may without breaking up the diagram remove all the points H with the links belonging to them, and thus obtain a diagram without any points H: such a diagram may be termed a "kenogram": the valency is obviously that of the original diagram *plus* the number of removed H's.

If from a kenogram, we remove every point O, C, &c. connected with the rest of the diagram by a single link only (each with the link belonging to it), and so on indefinitely as long as the process is practicable, we arrive at last at a diagram in which every point O, C, &c. is connected with the rest of the diagram by two links at least: this may be called a "mere kenogram."

Each or any point of a mere kenogram may be made the origin of a "ramification"; viz. we have here links branching out from the original point, and then again from the derived points, and so on any number of times, and never again uniting. We can thus from the mere kenogram obtain (in an infinite variety of ways) a diagram. The diagram completely determines the mere kenogram; and consequently two diagrams cannot be identical unless they have the same mere kenogram. Observe that the mere kenogram may evanesce altogether; viz. this will be the case if the diagram or kenogram is a simple ramification.

A ramification of n points C is $(2n+2)$-valent: in fact, this is so in the most simple case $n=1$; and admitting it to be true for any value of n, it is at once seen to be true for the next succeeding value. But no kenogram of points C is so much as $(2n+2)$-valent; for instance, 3 points C linked into a triangle, instead of being 8-valent are only 6-valent. We have therefore plerograms of n points C and $2n+2$ points H, say plerograms CH^{2n+2}; and in any such plerogram the kenogram is of necessity a ramification of n points C; viz. the different cases of such ramifications are *

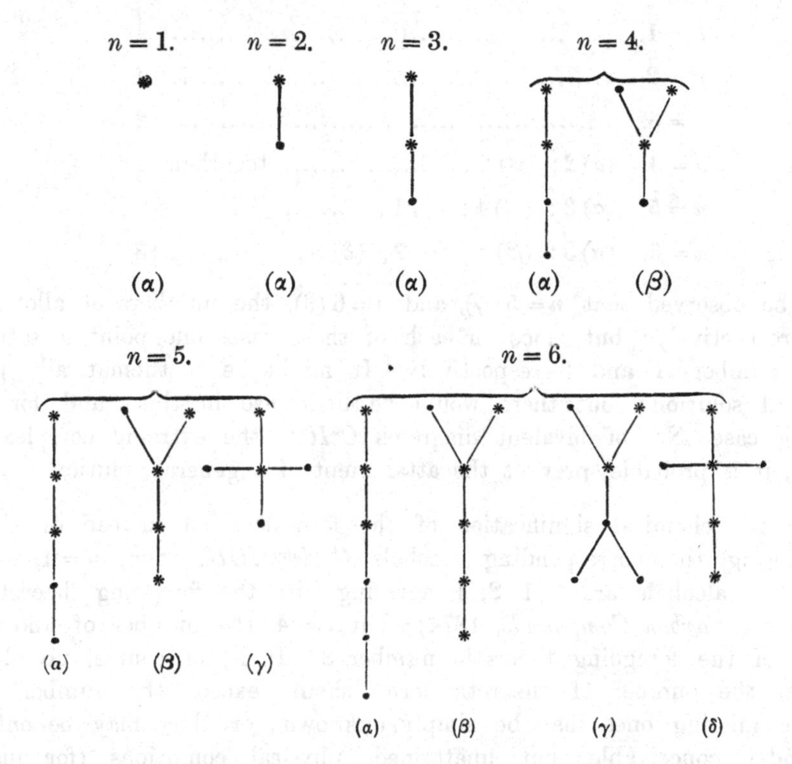

where the mathematical question of the determination of such forms belongs to the class of questions considered in my paper "On the Theory of the Analytical Forms called Trees," *Phil. Mag.* vol. XIII. (1857), [203], and vol. XVIII. (1859), [247], and in some papers on Partitions in the same Journal.

* The distinction in the diagrams of asterisks and dots is to be in the first instance disregarded; it is made in reference to what follows, the explanation as to the allotrious points.

The different forms of univalent diagrams C^nH^{2n+1} are obtained from the same ramifications by adding to each of them all but one of the $2n+2$ points H; that is, by adding to each point C except one its full number of points H, and to the excepted point one less than the full number of points H. The excepted point C must therefore be univalent at least; viz. it cannot be a saturate point, which presents itself for example in the diagrams $n=5\,(\gamma)$ and $n=6\,(\delta)$. And in order to count the number of distinct forms (for the diagrams C^nH^{2n+1}), we must in each of the above ramifications consider what is the number of distinct classes into which the points group themselves, or, say, the number of "allotrious" points. For instance, in the ramification $n=3$ there are two classes only; viz. a point is either terminal or medial; or, say, the number of allotrious points is $=2$: this is shown in the diagrams by means of the asterisks; so that in each case the points which may be considered allotrious are represented by asterisks, and the number of asterisks is equal to the number of allotrious points.

Thus, number of univalent diagrams C^nH^{2n+1}:

$n=1$,	....	1
$n=2$,	....	1
$n=3$,	....	2
$n=4$,	$(\alpha)\,2$; $(\beta)\,2$; .... together	4
$n=5$,	$(\alpha)\,3$; $(\beta)\,4$; $(\gamma)\,1$; .... „	8
$n=6$,	$(\alpha)\,3$; $(\beta)\,5$; $(\gamma)\,2$; $(\delta)\,3$; „	13

where it will be observed that, $n=5\,(\gamma)$, and $n=6\,(\delta)$, the numbers of allotrious points are 2 and 4 respectively; but since in each of these cases one point is saturate, they give only the numbers 1 and 3 respectively. It might be mathematically possible to obtain a general solution; but there would be little use in this; and for even the next succeeding case, No. of bivalent diagrams C^nH^{2n}; the extreme complexity of the question would, it is probable, prevent the attainment of a general solution.

Passing to the chemical signification of the formulæ, and instead of the radicals C^nH^{2n+1} considering the corresponding alcohols $C^nH^{2n+1}.OH$, then, $n=1, 2, 3, 4$, the numbers of known alcohols are 1, 1, 2, 4, agreeing with the foregoing theoretic number (see Schorlemmer's *Carbon Compounds*, 1874); but $n=4$, the number of known alcohols is $=2$, instead of the foregoing theoretic number 8. It is, of course, no objection to the theory that the number of theoretic forms should exceed the number of known compounds; the missing ones may be simply unknown; or they may be only capable of existing under conceivable, but unattained, physical conditions (for instance, of temperature); and if defect from the theoretic number of compounds can be thus accounted for, the theory holds good without modification. But it is also possible that the diagrams, in order that they may represent chemical compounds, may be subject to some as yet undetermined conditions; viz. in this case the theory would stand good as far as it goes, but would require modification.

587.

A SMITH'S PRIZE DISSERTATION.

[From the *Messenger of Mathematics*, vol. III. (1874), pp. 1—4.]

WRITE a dissertation:

On the general equation of virtual velocities.

Discuss the principles of Lagrange's proof of it and employ it [*the general equation*] *to demonstrate the Parallelogram of Forces.*

Imagine a system of particles connected with each other in any manner and subject to any geometrical conditions, for instance, two particles may be such that their distance is invariable, a particle may be restricted to move on a given surface, &c. And let each particle be acted upon by a force [this includes the case of several forces acting on the same particle, since we have only to imagine coincident particles each acted upon by a single force]. Imagine that the system has given to it any indefinitely small displacement consistent with the mutual connexions and geometrical conditions; and suppose that for any particular particle the force acting on it is P, and the displacement in the direction of the force (that is, the actual displacement multiplied into the cosine of the angle included between its direction and that of the force P) is $=\delta p$. Then δp is called the virtual velocity of the particle, and the principle of virtual velocities asserts that the sum of the products $P\delta p$, taken for all the particles of the system, and for any displacement consistent as above, is $=0$; say that we have

$$\Sigma P\delta p=0.$$

This is also the general equation of virtual velocities: as to the mode of using it, observe that the displacements δp are not arbitrary quantities, but are in virtue of the mutual connexions and other geometrical conditions connected together by certain linear relations; or, what is the same thing, they are linear functions of certain independent arbitrary quantities δu. Substituting for δp their expressions in terms of δu

we have $\Sigma P\delta p = \Sigma U\delta u$, where the several expressions U are each of them a linear function of the forces P, and where on the right hand Σ refers to the several quantities δu; and the resulting equation is $\Sigma U\delta u = 0$; viz. since the quantities δu are independent, the equation divides itself into a set of equations $U_1 = 0$, $U_2 = 0, \dots$ which are the equations of equilibrium of the system.

Lagrange imagines the forces produced by means of a weight W at the extremity of a string passing over a set of pulleys, as shown in the figure, viz. assuming the forces commensurable and equal to mW, nW, &c., we must have m strings at A,

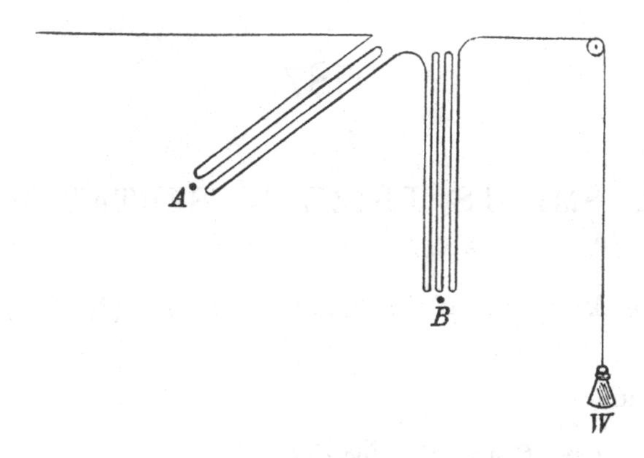

n strings at B, and so on. Suppose any indefinitely small displacement given to the system; each string at A is shortened by δp, or the m strings at A by $m\delta p$; and the like for the other particles at B, &c.; hence, if $m\delta p + n\delta q + \dots, = \frac{1}{W}(P\delta p + Q\delta q + \dots)$, be positive, the weight W will descend through the space

$$\frac{1}{W}(P\delta p + Q\delta q + \dots).$$

Now, in order that the system may be in equilibrium, W must be in its lowest position; or, what is the same thing, if there is any displacement allowing W to descend, W will descend, causing such displacement, and the original position is not a position of equilibrium. That is, if the system be in equilibrium, the sum $\Sigma P\delta p$ cannot be positive.

But it cannot be negative; since, if for any particular values of δp the sum $\Sigma P\delta p$ is negative, then reversing the directions of the several displacements, that is, giving to the several displacements δp the same values with opposite signs, then the sum $\Sigma P\delta p$ will be positive; and we *assume* that it is possible thus to reverse the directions of the several displacements. Hence, if the system be in a position of equilibrium, we cannot have $\Sigma P\delta p$ either positive or negative; that is, we obtain as the condition of equilibrium $\Sigma P\delta p = 0$.

The above is Lagrange's reasoning, and it seems completely unobjectionable. As regards the reversal of the directions of the displacements, observe that we consider such conditions as a condition that the particle shall be always *in* a given plane, but exclude the condition that the particle shall lie *on* a given plane, i.e. that it shall be at liberty to move in one direction (but not in the opposite direction) off from the plane. But the pulley-proof is equally applicable to a case of this kind. Thus, imagine a particle resting on a horizontal plane, and let z be measured vertically downwards, x and y horizontally. Suppose the particle acted on by the forces X, Y, Z, and replacing these by a weight W as above, the condition of equilibrium is, that

$$X\delta x + Y\delta y + Z\delta z$$

shall not be positive. We may have δx and δy, each positive or negative; whence the conditions $X = 0$ and $Y = 0$. But δz is negative; hence the required condition is satisfied if only Z is positive; that is, if the vertical force acts downwards. Clearly this is right, for if it acted upwards it would lift the particle from the plane. The case considered by Lagrange is where the particle is always *in* the plane; here $\delta z = 0$, and there is no condition as to the force Z.

The only omission in Lagrange's proof is, that he does not expressly consider the case of unstable equilibrium, where the weight W is at a position, not of minimum, but of maximum altitude. In such a case, however, the sum $\Sigma P\delta p$ is still $= 0$, taking account (as the proof does) of the displacements considered as infinitesimals of the first order; although taking account of higher powers, the sum $\Sigma P\delta p$ would have a positive value. An explanation as to this point might properly have been added to make the proof "refutation-tight," but the proof is not really in defect.

P.S. Lagrange excludes tacitly, not expressly, the case where the direction of a displacement is not reversible; he observes that the various displacements δp, when not arbitrary, are connected only by linear equations; and "par conséquent les valeurs de toutes ces quantités seront toujours telles qu'elles pourront changer de signe à la fois." The point was brought out more fully by Ostrogradsky, but I think there is no ground for the view that it was not brought out with sufficient clearness by Lagrange himself.

Parallelogram of forces.

Let P, Q, R be the forces, α, β, γ their inclinations to any line; then taking δs the displacement in the direction of this line, the displacements in the directions of the forces are $\delta s \cos \alpha$, $\delta s \cos \beta$, $\delta s \cos \gamma$, and the equation $\Sigma P\delta p = 0$ assumes the form

$$(P \cos \alpha + Q \cos \beta + R \cos \gamma)\, \delta s = 0,$$

that is, we have

$$P \cos \alpha + Q \cos \beta + R \cos \gamma = 0,$$

viz. this equation holds whatever be the fixed line to which the forces are referred. It is easy to see that, supposing it to hold in regard to any two lines, it will hold generally, and that the relation in question is thus equivalent to two independent conditions; and forming these we may obtain from them the theorem of the parallelogram of forces.

But to obtain this more directly, take A, B, C for the angles between the forces Q and R, R and P, P and Q respectively, then $A+B+C=2\pi$, and thence

$$\begin{aligned} \alpha &= \alpha, \\ \beta &= \alpha + C, \\ \gamma &= \alpha + C + A = \alpha + 2\pi - B, \end{aligned}$$

whence writing $\alpha = \frac{1}{2}\pi$, or taking the line of displacement at right angles to the force P, we have

$$\alpha = \tfrac{1}{2}\pi, \quad \beta = \tfrac{1}{2}\pi + C, \quad \gamma = 2\pi + \tfrac{1}{2}\pi - B,$$

and the equation becomes $0P - Q\sin C + R\sin B = 0$, that is, $Q : R = \sin B : \sin C$; and similarly $R : P = \sin C : \sin A$, that is,

$$P : Q : R = \sin A : \sin B : \sin C,$$

equations which in fact express that each force is equal and opposite to the diagonal of the parallelogram formed by the other two forces.

588.

PROBLEM.

[From the *Messenger of Mathematics*, vol. III. (1874), pp. 50—52.]

IT is required to place two given tetrahedra in perspective; or, what is the same thing, the tetrahedra being $ABCD$, $A'B'C'D'$ respectively, to place these so that the lines AA', BB', CC', DD' may meet in a point O.

The following considerations present themselves in regard to the solution of this problem. Take the tetrahedron $ABCD$ to be given in position, and the point O at pleasure; then drawing the lines OA, OB, OC, OD, we may in a determinate number of ways (viz. in 16 different ways) place the tetrahedron $A'B'C'D'$ in such manner that the summits A', B', C' shall be in the lines OA, OB, OC respectively. But the summit D' will then not be in general in the line OD; and in order that it may be so, a two-fold condition must be satisfied by the point O; viz. the locus of this point must be a certain curve in space.

Or again, we may look at the question thus: we have to place a point O in relation to the tetrahedron $ABCD$, and a point O' in relation to the tetrahedron $A'B'C'D'$, in such manner that the edges of the first tetrahedron subtend at O the same angles that the edges of the second tetrahedron subtend at O'; for this being done, then considering O' as rigidly connected with $A'B'C'D'$, we may move the figure $O'A'B'C'D'$ so that O' shall coincide with O, and the lines $O'A'$, $O'B'$, $O'C'$, $O'D'$ with OA, OB, OC, OD respectively. Take a, b, c, f, g, h, for the sides of the tetrahedron $ABCD$ ($BC, CA, AB, AD, BD, CD = a, b, c, f, g, h$ respectively), and take also x, y, z, w for the distances OA, OB, OC, OD respectively; and let $a', b', c', f', g', h', x', y', z', w'$ have the like significations in regard to the tetrahedron $A'B'C'D'$ and the point O', and write

$$\frac{y^2+z^2-a^2}{2yz},\quad \frac{z^2+x^2-b^2}{2zx},\quad \frac{x^2+y^2-c^2}{2xy},\quad \frac{x^2+w^2-f^2}{2xw},\quad \frac{y^2+w^2-g^2}{2yw},\quad \frac{z^2+w^2-h^2}{2zw},$$
$$=A,\qquad B,\qquad C,\qquad F,\qquad G,\qquad H,$$

respectively; and the like as regards the accented letters. Then A, B, C, F, G, H are the cosines of the angles which the edges of the tetrahedron $ABCD$ subtend at O; they are consequently the cosines of the six sides of the spherical quadrangle obtained by the projection of $ABCD$ on a sphere centre O; and they are therefore not independent, but are connected by a single equation; substituting for A, B, C, F, G, H their values, we have a relation between $a, b, c, f, g, h, x, y, z, w$; viz. this is the relation which connects the ten distances of the five points in space O, A, B, C, D (and which relation was originally obtained by Carnot in this very manner). There is of course the like relation between the accented letters.

The conditions as to the two tetrahedra are

$$A = A', \ B = B', \ C = C', \ F = F', \ G = G', \ H = H',$$

which, attending to the relations just referred to and therefore regarding w as a given function of x, y, z, and w' as a given function of x', y', z', are equivalent to five equations (or rather to a five-fold relation); the elimination of x', y', z' from the five-fold relation gives therefore a two-fold relation between x, y, z, that is, between the distances OA, OB, OC; or the locus of O is as before a curve in space.

The conditions may be written:

$$y'^2 + z'^2 - 2Ay'z' = a'^2, \quad x'^2 + w'^2 - 2Fx'w' = f'^2,$$
$$z'^2 + x'^2 - 2Bz'x' = b'^2, \quad y'^2 + w'^2 - 2Gy'w' = g'^2,$$
$$x'^2 + y'^2 - 2Cx'y' = c'^2, \quad z'^2 + w'^2 - 2Hz'w' = h'^2;$$

whence eliminating x', y', z', w', and in the result regarding A, B, C, F, G, H as given functions of x, y, z, w, we have between x, y, z, and w a three-fold relation determining w as a function of x, y, z, and establishing besides a two-fold relation between x, y, z.

As a particular case: One of the tetrahedra may degenerate into a plane quadrangle, and we have then the problem: a given plane quadrangle $ABCD$ being assumed to be the perspective representation of a given tetrahedron $A'B'C'D'$, it is required to determine the positions in space of this tetrahedron and of the point of sight O.

A generalisation of the original problem is as follows: determine the two-fold relation which must subsist between the 4×6, $= 24$ coordinates of four lines, in order that it may be possible to place in the tetrad of lines a given tetrahedron; that is, to place in the four lines respectively the four summits of the given tetrahedron. It may be remarked that considering three of the four lines as given, say these lines are the loci of the summits A, B, C respectively, we can in 16 different ways place in these lines respectively the three summits, and for each of these there are two positions of the summit D; there are consequently 32 positions of D; and the two-fold relation, considered as a relation between the six coordinates of the remaining line, must in effect express that this line passes through some one of the 32 points.

589.

ON RESIDUATION IN REGARD TO A CUBIC CURVE.

[From the *Messenger of Mathematics*, vol. III. (1874), pp. 62—65.]

THE following investigation of Prof. Sylvester's theory of Residuation may be compared with that given in Salmon's *Higher Plane Curves*, 2nd Edition (1873), pp. 133—137:

If the intersections of a cubic curve U_3 with any other curve V_n are divided in any manner into two systems of points, then each of these systems is said to be the residue of the other; and, in like manner, if starting with a given system of points on a cubic curve we draw through them a curve of any order V_n, then the remaining intersections of this curve with the cubic constitute a residue of the original system of points.

If the number of points in the original system is $=3p$, then the number of points in the residual system is $=3q$; and if we again take the residue, and so on indefinitely, the number of points in each residue will be $\equiv 0$ (Mod. 3); viz. we can never in this way arrive at a single point. But if the number of points in the original system be $3p\pm 1$, then that in the residual system will be $3q\mp 1$; and we may in an infinity of different ways arrive at a residue consisting of a single point; or say at a "residual point," viz. after an odd number of steps if the original number of points is $=3p-1$, but after an even number of steps if the original number of points is $=3p+1$. But starting from a given system of points on a given cubic curve, the residual point, however it is arrived at, will be one and the same point; this is Prof. Sylvester's theorem of the residuation of a cubic curve. For instance, starting with two given points on the cubic curve, the line joining these meets the curve in a third point, which is the residual point; any other process leading to a residual point must lead to the same point. Thus if through the 2 points we draw a conic, meeting the cubic besides in 4 points; through these a conic meeting the cubic besides in 2 points; and through this a line meeting the cubic besides in 1 point; this will be the before-mentioned residual point.

The general proof is such as in the following example:

Take on the cubic U_3 a system of $3\kappa - 2$ points, say the points α: through these a curve V_k, besides meeting the cubic in $3k - 3\kappa + 2$ points β: and through these a curve $P_{k-\kappa+1}$, besides meeting the cubic in a point C. And again through the $3\kappa - 2$ points α a curve $W_{k'}$, besides meeting the cubic in $3k' - 3\kappa + 2$ points β': and through these a curve $Q_{k'-\kappa+1}$, besides meeting the cubic in a single point; this will be the point C.

The proof consists in showing that we have a curve $A_{k+k'-\kappa-2}$ such that

$$A_{k+k'-\kappa-2}\, U_3 = Q_{k'-\kappa+1}\, V_k + P_{k-\kappa+1}\, W_{k'}.$$

For this observe that

$Q_{k'-\kappa+1}$ meets $W_{k'}$ in $3k' - 3\kappa + 2$ points β' and besides in $k'^2 - k'(\kappa+2) + 3\kappa - 2$ points ϵ';

$P_{k-\kappa+1}$ meets V_k in $3k - 3\kappa + 2$ points β and besides in $k'^2 - k(\kappa+2) + 3\kappa - 2$ points ϵ;

$P_{k-\kappa+1}$, $Q_{k'-\kappa+1}$ meet in $(k - \kappa + 1)(k' - \kappa + 1)$ points C;

V_k, $W_{k'}$ meet in $3\kappa - 2$ points α and $kk' - 3\kappa + 2$ points a;

$Q_{k'-\kappa+1}\, V_k$ and $P_{k-\kappa+1}\, W_{k'}$ meet in

$$\begin{array}{llllll}
kk' - k(\kappa-1) - k'(\kappa-1) + (\kappa-1)^2 & & & & \text{points} & C \\
 & & 3k' & -3\kappa+2 & \text{,,} & \beta' \\
k'^2 & & -k'(\kappa+2) & +3\kappa-2 & \text{,,} & \epsilon' \\
 & 3k & & -3\kappa+2 & \text{,,} & \beta \\
k^2 & -k(\kappa+2) & & +3\kappa-2 & \text{,,} & \epsilon \\
kk' & & & -3\kappa+2 & \text{,,} & a \\
 & & & 3\kappa-2 & \text{,,} & \alpha \\
\hline
\multicolumn{4}{l}{(k+k')^2 - (2\kappa-2)(k+k') + (\kappa-1)^2} & &
\end{array}$$

$$= (k + k' - \kappa + 1)^2 \text{ points.}$$

Every $(k + k' - \kappa + 1)$thic through

$$\tfrac{1}{2}(k + k' - \kappa + 1)(k + k' - \kappa + 4) - 1$$

of these points passes through all.

Now $A_{k+k'-\kappa-2}$ may be drawn to pass through

$$\tfrac{1}{2}(k + k' - \kappa - 2)(k + k' - \kappa + 1)$$

of the points a.

Hence $A_{k+k'-\kappa-2}\,U_3$ is a $(k+k'-\kappa+1)$thic through

$$\begin{array}{rl}
& \tfrac{1}{2}(k+k'-\kappa-2)(k+k'-\kappa+1) \\
& = \tfrac{1}{2}(k+k')^2-\tfrac{1}{2}(2\kappa+1)(k+k')+\tfrac{1}{2}(\kappa^2+\kappa-2) \text{ points } a \\
& \qquad\qquad 3\kappa-2 \quad ,, \quad \alpha \\
& \qquad 3k \qquad -3\kappa+2 \quad ,, \quad \beta \\
& \qquad 3k' \qquad -3\kappa+2 \quad ,, \quad \beta' \\
\hline
& \tfrac{1}{2}(k+k')^2+(-\kappa+\tfrac{5}{2})(k+k')+\tfrac{1}{2}\kappa^2-\tfrac{5}{2}\kappa+1 \\
& = \tfrac{1}{2}\{(k+k')^2+(k+k')(-2\kappa+5)+(\kappa-1)(\kappa+4)-2\} \\
& = \tfrac{1}{2}(k+k'-\kappa+1)(k+k'-\kappa+4)-1
\end{array}$$

of the points in question; and therefore through all. Whence

$$A_{k+k'-\kappa-2}\,U_3 = Q_{k'-\kappa+1}\,V_k + P_{k-\kappa+1}\,W_{k'}.$$

Also U_3 meets $Q_{k'-\kappa+1}\,V_k$ in $3(k+k'-\kappa+1)$ of the $(k+k'-\kappa+1)^2$ points, viz. these are

$$\begin{array}{rcl}
3\kappa-2 & \text{points} & \alpha, \\
3k-3\kappa+2 & ,, & \beta, \\
3k'-3\kappa+2 & ,, & \beta', \\
1 & ,, & C,
\end{array}$$

and $A_{k+k'-\kappa-2}$ meets $Q_{k'-\kappa+1}\,V_k$ in $(k+k'-\kappa-2)(k+k'-\kappa+1)$,

that is, in

$$(k+k')^2+(k+k')(-2\kappa-1)+\kappa^2+\kappa-2$$

of the

$$(k+k'-\kappa+1)^2 \text{ points,}$$

viz. these are

$$\begin{array}{llrl}
kk' & +(k+k')(-\kappa+1)+\kappa^2-2\kappa & & \text{points } C \\
k'^2 & -k'(\kappa+2) & +3\kappa-2 & ,, \quad \epsilon' \\
k^2 & -k(\kappa+2) & +3\kappa-2 & ,, \quad \epsilon \\
kk' & & -3\kappa+2 & ,, \quad a \\
\hline
\multicolumn{4}{l}{(k+k')^2+(k+k')(-2\kappa-1)+\kappa^2+\kappa-2 \text{ points.}}
\end{array}$$

Hence U_3 passes through 1 of the points C, that is, through an intersection of $Q_{k'-\kappa+1}$ and $P_{k-\kappa+1}$, that is, $Q_{k'-\kappa+1}$ and $P_{k-\kappa+1}$ intersect U_3 in a common point C; which was the theorem to be proved.

In the particular case $3\kappa-2=10$, $k=k'=4$, the theorem is, given on a cubic 10 points, if through these we draw a quartic meeting the cubic besides in 2 points;

and through these a line meeting the cubic besides in a point C; then this is a fixed point, independent of the particular quartic. And the proof is as follows: we have

U a cubic through 10 points α;

V a quartic through the 10 points, and besides meeting the cubic in 2 points β;

W a quartic through the 10 points, and besides meeting the cubic in 2 points β';

P the line joining the two points β, and besides meeting V in two points ϵ;

Q the line joining the two points β', and besides meeting W in two points ϵ'; P, Q meet in the point C;

U, V meet in the 10 points α, and besides in 6 points a;

A a conic through 5 of the points a.

Then quintics QV, PW meet in the 10 points α, 2 points β, 2 points ϵ, 2 points β', 2 points ϵ', 6 points a and 1 point C. Every quintic through 19 of these passes through the 25. But we have AU, a quintic through 5 points a, and the 10 points α, 2 points β and 2 points β'; hence AU passes through all the remaining points, or we have

$$AU = QV + PW,$$

P passes through	$\beta,\ \beta,\ \epsilon,\ \epsilon,\ C,$		
Q ,,	$\beta',\ \beta',\ \epsilon',\ \epsilon',\ C,$		
V ,,	$\epsilon,\ \epsilon,\ \beta,\ \beta,$	6 points a,	10 points α,
W ,,	$\epsilon',\ \epsilon',\ \beta',\ \beta',$	6 points a,	10 points α,
A ,,	$\epsilon,\ \epsilon,\ \epsilon',\ \epsilon',$	6 points a,	
U ,,	$\beta,\ \beta,\ \beta',\ \beta',\ C,$		

or, what is the same thing,

A, P intersect in	$\epsilon,\ \epsilon,$	
A, Q ,,	$\epsilon',\ \epsilon',$	
A, V ,,	$\epsilon,\ \epsilon,$	6 points a,
A, W ,,	$\epsilon',\ \epsilon',$	6 points a,
U, P ,,	$\beta,\ \beta,\ C,$	
U, Q ,,	$\beta',\ \beta',\ C,$	
U, V ,,	$\beta,\ \beta,$	10 points α,
U, W ,,	$\beta',\ \beta',$	10 points α.

In particular U, P, Q intersect in the point C; that is, C as given by the intersection of U by the line P; and as given by the intersection of U by the line Q; is one and the same point.

590.

ADDITION TO PROF. HALL'S PAPER "ON THE MOTION OF A PARTICLE TOWARD AN ATTRACTING CENTRE AT WHICH THE FORCE IS INFINITE."

[From the *Messenger of Mathematics*, vol. III. (1874), pp. 149—152.]

I DO not in the passage referred to* expressly profess to interpret Newton's idea. After referring to his investigation I say, "The method has the advantage of explaining the paradoxical result which presents itself in the case force $\propto$ (dist.)$^{-2}$, and in some other cases where the force becomes infinite. According to theory the velocity becomes infinite at the centre, but the direction of the motion is there abruptly reversed, so that the body in its motion does not pass through the centre, but on arriving there forthwith returns towards its original position; of course such a motion cannot occur in nature, where neither a force nor a velocity is actually infinite;" viz. while assuming that the analysis gives a motion as just described, or in Prof. Hall's figure, a reciprocating motion between A and C, I expressly state that the motion is not one that can occur in nature; in fact, my view is that the question (which, to render it precise, I state as follows: "What happens in nature when the moving point arrives at C") presupposes what is inconceivable. But I consider that the analysis gives a motion as above, viz. that it gives x, t each as a one-valued function of a parameter ϕ, such that this parameter ϕ increasing continuously, we have for the moving point a continuous series of positions corresponding to the motion in question, gives in fact the equations $x = a(1 - \cos\phi)$ and $\dfrac{t\sqrt{(\mu)}}{a^{\frac{3}{2}}} = \phi - \sin\phi$.

In explanation and justification of the assumption, it is interesting to show how the solution just referred to can be obtained from the equation of motion $\dfrac{d^2x}{dt^2} = -\dfrac{\mu}{x^2}$, without (in the process) the extraction of the square root of the two sides of an

[* By Professor Hall in his paper (p. 144, *l.c.*) quoted in the title. The passage is an extract from the British Association Report (1862) *On the progress of the solution of certain special problems of dynamics*, p. 186; [298], *Coll. Math. Papers*, vol. IV. p. 515.]

equation. Taking x as the independent variable and writing for a moment $\frac{dt}{dx} = t'$, $\frac{d^2t}{dx^2} = t''$, the equation is

$$\frac{1}{t'^3} t'' = \frac{\mu}{x^2},$$

and if we herein assume $x = a(1 - \cos\phi)$ and transform to ϕ as the independent variable, it becomes

$$\frac{a^2 \sin^2\phi}{\left(\frac{dt}{d\phi}\right)^3} \left\{\frac{1}{a \sin\phi} \frac{d^2t}{d\phi^2} - \frac{\cos\phi}{a \sin^2\phi} \frac{dt}{d\phi}\right\} = \frac{\mu}{a^2(1 - \cos\phi)^2};$$

or, what is the same thing,

$$\sin\phi \frac{d}{d\phi}\left(\frac{dt}{d\phi}\right) - \cos\phi \left(\frac{dt}{d\phi}\right) = \frac{\mu}{a^3} \frac{1}{(1 - \cos\phi)^2} \left(\frac{dt}{d\phi}\right)^3,$$

a differential equation of the first order for the determination of $\frac{dt}{d\phi}$ as a function of ϕ. Since a is a constant of integration of the original equation, a particular integral only is required, but it is as well to obtain the general integral. For this purpose assume

$$\frac{dt}{d\phi} = \frac{a^{\frac{3}{2}}}{\sqrt{(\mu)}} z(1 - \cos\phi);$$

then, omitting from each side of the equation the factor $\frac{a^{\frac{3}{2}}}{\sqrt{(\mu)}}$, the equation becomes

$$\sin\phi \left\{z \sin\phi + \frac{dz}{d\phi}(1 - \cos\phi)\right\} - \cos\phi \,.\, z(1 - \cos\phi) = (1 - \cos\phi) z^3,$$

viz. the left-hand side being $(1 - \cos\phi)\left(z + \frac{dz}{d\phi}\sin\phi\right)$, the whole equation contains the factor $(1 - \cos\phi)$, and omitting this, the equation becomes

$$z + \frac{dz}{d\phi}\sin\phi = z^3;$$

or, what is the same thing,

$$\frac{dz}{z^3 - z} = \frac{d\phi}{\sin\phi}.$$

The integral of this is

$$\log \frac{z^2 - 1}{z^2} = 2 \log k + 2 \log \tan \tfrac{1}{2}\phi;$$

or, what is the same thing,

$$\frac{z^2 - 1}{z^2} = k^2 \tan^2 \tfrac{1}{2}\phi,$$

where k is the constant of integration.

[In explanation of this constant k, observe that the equation gives

$$z = \frac{1}{\sqrt{(1 - k^2 \tan^2 \frac{1}{2}\phi)}},$$

and that we thence have

$$\frac{dt}{d\phi}=\frac{a^{\frac{3}{2}}}{\sqrt{(\mu)}}\frac{1-\cos\phi}{\sqrt{(1-k^2\tan^2\frac{1}{2}\phi)}};$$

that is,

$$\frac{dt}{dx}=\frac{\sqrt{(\mu)}}{a^{\frac{1}{2}}}\frac{\sin\phi}{1-\cos\phi}\frac{1}{\sqrt{(1-k^2\tan^2\frac{1}{2}\phi)}},\quad =\frac{a^{\frac{1}{2}}}{\sqrt{(\mu)}}\frac{\tan\frac{1}{2}\phi}{\sqrt{(1-k^2\tan^2\frac{1}{2}\phi)}},$$

or, since

$$\tan^2\tfrac{1}{2}\phi=\frac{x}{2a-x},$$

this is

$$\frac{dt}{dx}=\frac{a^{\frac{1}{2}}}{\sqrt{(\mu)}}\frac{\sqrt{(x)}}{\sqrt{(2a-x-k^2x)}};$$

or, what is the same thing,

$$\frac{dt}{dx}=\frac{\sqrt{\left(\frac{a}{1+k^2}\right)}}{\sqrt{(\mu)}}\frac{\sqrt{(x)}}{\sqrt{\left(2\frac{a}{1+k^2}-x\right)}},$$

viz. we in effect have $\frac{a}{1+k^2}$ as a constant of integration in place of the original constant a.]

Recurring to the general solution

$$\frac{z^2-1}{z^2}=k^2\tan^2\tfrac{1}{2}\phi,$$

we may take $z=1$, as a particular solution answering to the value $k=0$ of the constant; and we then have

$$\frac{dt}{d\phi}=\frac{a^{\frac{3}{2}}}{\sqrt{(\mu)}}(1-\cos\phi),$$

viz. reckoning t from the epoch for which ϕ is $=0$, we thus have

$$t=\frac{a^{\frac{3}{2}}}{\sqrt{(\mu)}}(\phi-\sin\phi),$$

which, combined with the assumed equation

$$x=a(1-\cos\phi),$$

gives the foregoing solution.

I quite admit that, considering (with Prof. Hall) the attracted particle as split into two equal particles placed at equal distances above and below the centre C, the motion when the distances become infinitesimal is a motion not as above, but backwards and forwards along the entire line AB; but it remains to be seen whether *at* the limit this can be brought out as an analytical solution of the differential equation $\frac{d^2x}{dt^2}=-\frac{\mu}{x^2}$. Possibly this may be done, and I remark as an objection, not to the foregoing as an admissible solution of the problem but to its generality as the *only* solution, that, in writing $x=a(1-\cos\phi)$ and assuming that ϕ is real, I in effect assume that x is always positive. But the burthen of the proof is with Prof. Hall, to show that there is an analytical solution in which x acquires negative values.

591.

A SMITH'S PRIZE PAPER AND DISSERTATION; SOLUTIONS AND REMARKS.

[From the *Messenger of Mathematics*, vol. III. (1874), pp. 165—183, vol. IV. (1875), pp. 6—8.]

1. *Find the triangular numbers which are also square.*

The "*mise en équation*" is immediate; we have to find n, m such that

$$\tfrac{1}{2}n(n+1)=m^2;$$

or, what is the same thing,

$$(2n+1)^2-8m^2=1.$$

Observing that this is satisfied by $n=m=1$, that is, $2n+1=3$, $2m=2$, we have the general solution given by

$$2n+1+2m\sqrt{(2)}=\{3+2\sqrt{(2)}\}^p,$$

where p is any positive integer; viz. $2n+1$, $2m$ being rational, this implies

$$2n+1-2m\sqrt{(2)}=\{3-2\sqrt{(2)}\}^p,$$

and thence the equation in question. The successive powers

$$3+2\sqrt{(2)},\quad 17+12\sqrt{(2)},\quad 99+70\sqrt{(2)},\ \&c.,$$

give the solutions

$$n,\ m=\ 1,\ 1\ ,\qquad 8,\ 6\ ,\qquad 48,\ 35\ ,\ \&c.;$$

viz. the square triangular numbers are

$$1^2,\ =\tfrac{1}{2}\,1\cdot2\,;\quad 6^2,\ =\tfrac{1}{2}\,8\cdot9\,;\quad 35^2,\ =\tfrac{1}{2}\,49\cdot50,\ \&c.$$

2. *Show how to express any symmetrical function of the roots of an equation in terms of the coefficients. What objection is there to the method which employs the sums of the powers of the roots?*

The ordinary method is that referred to, employing the sums of the powers of the roots; *but it is a very bad one.* In fact, writing

$$x^n - bx^{n-1} + cx^{n-2} - \&c., = (x-\alpha)(x-\beta)(x-\gamma)\ldots = 0,$$

leading to

$$\begin{aligned} S_1 &= b,\\ S_2 &= b^2 - 2c,\\ S_3 &= b^3 - 3bc + 3d,\\ &\vdots \end{aligned}$$

then if the method were employed throughout, we should have for instance to find $S\alpha\beta\gamma$, that is, d, from the formula

$$\begin{aligned} 6S\alpha\beta\gamma = \quad & S_1{}^3 = \ b^3\\ & -3S_1S_2 \quad -3b(b^2-2c)\\ & +2\ S_3 \quad +2(b^3-3bc+3d)\\ & = \quad 6d, \text{ which is right,} \end{aligned}$$

but the process introduces terms b^3 and bc each of a higher order than d (reckoning the order of each coefficient as unity), with numerical coefficients which destroy each other. And, so again, $S\alpha^2\beta$ would be calculated from the formula

$$\begin{aligned} S\alpha^2\beta = \quad & S_1S_2 = \ b(b^2-2c)\\ - & \ S_3 \quad -(b^3-3bc+3d)\\ = & \quad bc - 3d, \text{ which is right,} \end{aligned}$$

but there is here also a term b^3 of a higher order, with numerical coefficients which destroy each other. And the order in which the several expressions are derived the one from the other is a non-natural one; S_3 is required for the determination of $S\alpha^2\beta$, whereas (as will be seen) it is properly $S\alpha^2\beta$ which leads to the value of S_3.

The true method is as follows: we have

$$S\alpha = b,\quad S\alpha\beta = c,\quad S\alpha\beta\gamma = d,\ \&c.,$$

and we thence derive the sets of equations

$$\begin{aligned} b &= S\alpha;\\[4pt] c &= S\alpha\beta,\\ b^2 &= S\alpha^2 + 2S\alpha\beta;\\[4pt] d &= S\alpha\beta\gamma,\\ bc &= S\alpha^2\beta + 3S\alpha\beta\gamma,\\ b^3 &= S\alpha^3 + 3S\alpha^2\beta + 6S\alpha\beta\gamma;\\ &\vdots \end{aligned}$$

viz. we thus have 1 equation to give $S\alpha$; 2 equations to give $S\alpha\beta$ and $S\alpha^2$; 3 equations to give $S\alpha\beta\gamma$, $S\alpha^2\beta$, $S\alpha^3$; and so on. And taking for instance the third set of equations, the first equation gives $S\alpha\beta\gamma$, the second then gives $S\alpha^2\beta$, and the third then gives $S\alpha^3$, viz. we have

$$\begin{aligned} S\alpha\beta\gamma &= d, \\ S\alpha^2\beta &= bc - 3d, \\ S\alpha^3 &= b^3 - 3(bc - 3d) - 6d, \\ &= b^3 - 3bc + 3d. \end{aligned}$$

Of course the process for the formation of the successive sets of equations would require further explanation and development.

3. *Given a point P in the interior of an ellipsoid, show that it is possible to determine an exterior point Q such that for every chord RS through P, the relation $QR : QS = PR : PS$ may hold good.*

There is no difficulty in the analytical solution and in showing thereby that the point Q is determined as the intersection of the polar plane of P by the perpendicular let fall from P on this plane. But a simple and elegant geometrical solution was given in the Examination. Constructing Q as above, let the chord RS meet the polar plane of P in Z; then the polar plane of Z passes through P, that is, the line ZP is harmonically divided in R, S, or we have

$$ZR : ZS = PR : PS.$$

Again ZQP being a right angle, the sphere on ZP as diameter will pass through Q; and R, S being points on the diameter, and Z, Q points on the surface, $ZR : ZS = QR : QS$; whence the required relation $QR : QS = PR : PS$.

4. *Find the number of regions into which infinite space is divided by n planes.*

The number $\frac{1}{6}(n^3 + 5n + 6)$ is a known result, but not a generally known one, and I intended the question as a problem; I do not think it is a difficult one.

Consider the analogous problem for lines in a plane: the first line divides the plane into 2 regions.

The second line is by the first divided into 2 parts, and therefore adds 2 regions.

The third line is by the other two divided into 3 parts, and therefore adds 3 regions; and so on.

That is, the number of regions for

1 line is	$= 2$	$= 2$	regions,
2 lines	$= 2 + 2$	$= 4$	"
3 lines	$= 2 + 2 + 3$	$= 7$	"
⋮			
n lines	$= 2 + 2 + 3 + \ldots + n$	$= \frac{1}{2}(n^2 + n + 2)$	"

In exactly the same way for the problem in space:

The first plane divides space into 2 regions.

The second plane is by the first plane divided into 2 regions, and therefore adds 2 regions.

The third plane is by the other two planes divided into 4 regions, and therefore adds 4 regions.

The fourth plane is by the other three planes divided into 7 regions, and therefore adds 7 regions: and so on.

That is the number of regions for

$$
\begin{array}{lll}
1 \text{ plane is} & = 2 & = 2 \text{ regions} \\
2 \text{ planes} & = 2+2 & = 4 \quad ,, \\
3 \text{ planes} & = 2+2+4 & = 8 \quad ,, \\
4 \text{ planes} & = 2+2+4+7 & = 15 \quad ,, \\
\vdots & & \\
n \text{ planes} & = 2+2+4+7+\ldots+\tfrac{1}{2}(n^2-n+2) = \tfrac{1}{6}(n^3+5n+6), &
\end{array}
$$

where, for effecting the summation, observe that the series is

$$
\begin{aligned}
&= 2+\{1+1+1\ldots(n-1) \text{ terms}\} \\
&\quad +\{1+3+6\ldots+\tfrac{1}{2}n(n-1)\}, \\
&= 2+(n-1)+\tfrac{1}{6}(n+1)n(n-1), = \text{as above.}
\end{aligned}
$$

5. *In the theory of Elliptic Functions, explain and connect together the notations* $F(\theta)$, am u (sinam u, cosam u, Δam u), *illustrating them by reference to the circular functions**.

What is asked for is an explanation of the fundamental notations of Elliptic Functions. To a student acquainted with the subject, the only difficulty is to say enough to bring the meaning fully out, and not to say more than enough.

Defining $F(x)$ by the equation

$$F(x)=\int_0 \frac{dx}{\sqrt{\{(1-x^2)(1-k^2x^2)\}}},$$

(viz. the integral is taken from 0 up to the indefinite value x), then the fundamental property of elliptic functions (derived from consideration of the differential equation

$$\frac{dx}{\sqrt{\{(1-x^2)(1-k^2x^2)\}}}+\frac{dy}{\sqrt{\{(1-y^2)(1-k^2y^2)\}}}=0)$$

consists herein, that the functional relation

$$F(x)+F(y)=F(z)$$

* It would have been better in the question to have written $F(x)$ instead of $F(\theta)$.

is equivalent to an algebraic equation between the arguments x, y, z. $F(x)$ as defined by the foregoing equation is properly an inverse function; this at once appears from a particular case, viz. writing $k=0$, $F(x)=\sin^{-1}x$, and the theory of the function $F(x)$ in the general case corresponds to what the theory of circular functions would be, if writing $F(x)$ to denote $\sin^{-1}x$, we were to work with the equation

$$F(x)+F(y)=F(z)$$

as equivalent to the algebraical equations (one a transformation of the other)

$$z=x\sqrt{(1-y^2)}+y\sqrt{(1-x^2)},$$
$$\sqrt{(1-z^2)}=\sqrt{(1-x^2)}\sqrt{(1-y^2)}-xy.$$

But in the actual theory of circular functions, we introduce the direct symbols sin, cos; writing $F(x)=\theta$, that is, $x=\sin\theta$, $\sqrt{(1-x^2)}=\cos\theta$, and similarly $F(y)=\phi$, that is, $y=\sin\phi$ and $\sqrt{(1-y^2)}=\cos\phi$, then the equation

$$F(x)+F(y)=F(z)$$

becomes $F(z)=\theta+\phi$, that is, $z=\sin(\theta+\phi)$, $\sqrt{(1-z^2)}=\cos(\theta+\phi)$, and the other two equations become

$$\sin(\theta+\phi)=\sin\theta\cos\phi+\sin\phi\cos\theta,$$
$$\cos(\theta+\phi)=\cos\theta\cos\phi-\sin\theta\sin\phi,$$

viz. these are the addition-equations for the functions sin and cos.

In passing from the original notation $F(x)$ to the notation am u, we make the like step of passing from an inverse to a set of direct functions; first modifying the meaning of F, so as to denote by $F(\theta)$ what was originally $F(\sin\theta)$, we have as the new definition

$$F(\theta)=\int_0\frac{d\theta}{\sqrt{(1-k^2\sin^2\theta)}}=\int_0\frac{d\theta}{\Delta(\theta)},$$

(if as usual $\Delta\theta$ denotes $\sqrt{(1-k^2\sin^2\theta)}$), and this being so, the relation $F(\theta)+F(\phi)=F(\mu)$ is equivalent to a relation between the sine, cosine, and Δ of θ, ϕ, μ. Writing then $F(\theta)=u$, and considering this equation as determining θ as a function of u, $\theta=\text{am}\,u$, we have $\sin\theta=\sin.\text{am}\,u$, $\cos\theta=\cos.\text{am}\,u$, and $\Delta\theta=\Delta.\text{am}\,u$, and similarly $F(\phi)=v$, $\phi=\text{am}\,v$, &c., then the equation $F(\theta)+F(\phi)=F(\mu)$ becomes $F(\mu)=u+v$, that is, $\mu=\text{am}(u+v)$; and the algebraic relation in its various forms gives the values of $\sin.\text{am}(u+v)$, $\cos.\text{am}(u+v)$, $\Delta.\text{am}(u+v)$ in terms of the like functions of u, v respectively, viz. it is the addition-theorem for the function am.

Observe that am u is considered as a certain function of u, sin . am u is its sine, cos . am u its cosine, and

$$\Delta.\text{am}\,u=\sqrt{(1-k^2\sin^2.\text{am}\,u)},$$

a function analogous to a cosine. But making only a slight change in the point of view, we have sinam u, a certain function of u, and

$$\text{cosam}\,u\,\{=\sqrt{(1-\text{sinam}^2u)}\},\ \Delta\text{am}\,u\,\{=\sqrt{(1-k^2\,\text{sinam}^2u)}\},$$

two allied functions, viz. sinam u is analogous to a sine, and the other two functions to cosines; the algebraical equations give the sinam, cosam, and Δam of $u+v$ in terms of the like functions of u and v respectively, viz. they constitute the addition-theorem for these functions.

6. *Find the differential equation satisfied by a hypergeometric series, and express by means of such series the coefficients of the expansion of* $(1-2a\cos\theta+a^2)^{-n}$ *according to multiple cosines of* θ.

I understand the expression "hypergeometric series" in the restricted sense in which it signifies the series

$$F(\alpha,\ \beta,\ \gamma,\ x)=1+\frac{\alpha\,.\,\beta}{1\,.\,\gamma}x+\frac{\alpha(\alpha+1)\beta(\beta+1)}{1\,.\,2\gamma(\gamma+1)}x^2+\&\text{c.}$$

I find it was understood in the more general sense of a series

$$u=a_0+a_1x+a_2x^2+\ldots+a_nx^n+\ldots,$$

where the coefficient a_{n+1} is given in terms of the preceding one a_n by an equation of the form $a_{n+1}=\phi(n)\,.\,a_n$. In this latter sense, but supposing for greater simplicity, that $\phi(n)$ is a rational and integral function of n, the solution is as follows: we operate on the series with the symbol $\phi\left(x\dfrac{d}{dx}\right)$; viz. $x\dfrac{d}{dx}$ is regarded as a single symbol of operation; $x\dfrac{d}{dx}\,.\,x^n=nx^n$, $\left(x\dfrac{d}{dx}\right)^2x^n=n^2x^n$, &c.; thus $x\dfrac{d}{dx}$ is, as regards x^n, $=n$, and therefore $\phi\left(x\dfrac{d}{dx}\right)=\phi(n)$. We thence have

$$\begin{aligned}\phi\left(x\frac{d}{dx}\right)u&=\phi(0)\,a_0+\phi(1)\,a_1x+\phi(2)\,a_2x^2\ldots+\phi(n)\,a_nx^n+\ldots\\&=\quad a_1\quad+\quad a_2x\quad+\quad a_3x^2\quad\ldots+\quad a_{n+1}\,x^n\quad+\ldots,\end{aligned}$$

and consequently

$$x\phi\left(x\frac{d}{dx}\right)u=u-a_0,$$

which is the required differential equation. This is equivalent to the process given in Boole, only he writes $x=e^\theta$, in order to reduce $x\dfrac{d}{dx}$ to a mere differentiation $\dfrac{d}{d\theta}$. I regard this introduction of a new variable θ *as most unfortunate*; the effect is entirely to conceal the real nature of the operation; the notion of $x\dfrac{d}{dx}$ as a single symbol of operation is quite as simple as that of $\dfrac{d}{d\theta}$; and by means of it we retain the original variable.

The process is substantially the same when $\phi(n)$ is a rational fraction, but I give the investigation directly for the hypergeometric series in the restricted sense, viz. writing u for the series $F(\alpha,\ \beta,\ \gamma,\ x)$, we find

$$x\left(x\frac{d}{dx}+\alpha\right)\left(x\frac{d}{dx}+\beta\right)u=x\frac{d}{dx}\left(x\frac{d}{dx}+\gamma-1\right)u;$$

or, what is the same thing,

$$\left\{\left(x\frac{d}{dx}+\alpha\right)\left(x\frac{d}{dx}+\beta\right)-\frac{d}{dx}\left(x\frac{d}{dx}+\gamma-1\right)\right\}u=0,$$

as at once appears by writing the general term successively under the two forms

$$\frac{\alpha\,.\,\alpha+1\ldots\alpha+n-1\,.\,\beta\,.\,\beta+1\ldots\beta+n-1}{1\,.\,2\;\ldots\quad n\qquad\gamma\,.\,\gamma+1\ldots\gamma+n-1}\,x^n,$$

and

$$\frac{\alpha\,.\,\alpha+1\ldots\alpha+n\,.\,\beta\,.\,\beta+1\ldots\beta+n}{1\,.\,2\;\ldots n+1\,.\,\gamma\,.\,\gamma+1\ldots\gamma+n}\,x^{n+1}.$$

The differential equation may also be written

$$\left[(x^2-x)\frac{d^2}{dx^2}+\{(1+\alpha+\beta)\,x-\gamma\}\frac{d}{dx}+\alpha\beta\right]u=0.$$

Take next the function

$$(1-2a\cos\theta+a^2)^{-n},$$

$$=\left\{1-a\left(x+\frac{1}{x}\right)+a^2\right\}^{-n}$$

$$=\left\{(1-ax)\left(1-a\frac{1}{x}\right)\right\}^{-n},\text{ if } x+\frac{1}{x}=2\cos\theta,$$

$$=\left(1+\frac{n}{1}ax+\frac{n\,.\,n+1}{1\,.\,2}a^2x^2+\ldots\right)\left(1+\frac{n}{1}\frac{a}{x}+\frac{n\,.\,n+1}{1\,.\,2}\frac{a^2}{x^2}+\ldots\right)$$

$$=\left\{1^2+\left(\frac{n}{1}\right)^2a^2+\left(\frac{n\,.\,n+1}{1\,.\,2}\right)^2a^4\ldots\right\}$$

$$+\left\{1\,.\,\frac{n}{1}+\frac{n}{1}\,\frac{n\,.\,n+1}{1\,.\,2}a^2+\frac{n\,.\,n+1}{1\,.\,2}\,\frac{n\,.\,n+1\,.\,n+2}{1\,.\,2\,.\,3}a^4+\ldots\right\}a\left(x+\frac{1}{x}\right)(=2a\cos\theta)$$

$$+\left\{1\,.\,\frac{n\,.\,n+1}{1\,.\,2}+\&\text{c.}\qquad\qquad\right\}a^2\left(x^2+\frac{1}{x^2}\right)(=2a^2\cos 2\theta),$$

&c. &c.

where the second term contains the factor $\frac{n}{1}a$, the third the factor $\frac{n\,.\,n+1}{1\,.\,2}a^2$, and so on. Throwing these out, the remaining factors are each of them a hypergeometric series, viz. representing the whole expression by

$$A_0+2A_1\cos\theta+2A_2\cos 2\theta+\&\text{c.},$$

we have

$$A_0=F(n,\ n,\ 1,\ a^2),$$

$$A_1=\frac{n}{1}aF(n,\ n+1,\ 2,\ a^2),$$

$$\vdots$$

and generally

$$A_r=\frac{n\,.\,n+1\ldots n+r-1}{1\,.\,2\ldots r}\,a^rF(n,\ n+r,\ r+1,\ a^2).$$

7. *The function* $e^{-\frac{1}{(x-a)^2}}$ *has been suggested as an exception to the theorem that if a function and all its differential coefficients vanish for a given value of the variable, then the function is identically* $=0$; *discuss the question as regards the precise meaning of the theorem, and validity of the exception.*

The suggestion was made by Sir W. R. Hamilton; the following remarks arise in regard to it:

The function $e^{-\frac{1}{(x-a)^2}}$ is a function which *in a certain sense* satisfies the condition that for a given value ($=a$) of the variable, the function and all its differential coefficients vanish; viz. each differential coefficient is of the form $Xe^{-\frac{1}{(x-a)^2}}$, where X is a finite series of negative powers of $x-a$; if then $x=a\pm r$, where r is real and positive, and if r continually diminishes to zero, then $(x-a)^2$, remaining always real and positive, continually diminishes to zero, that is, $-\frac{1}{(x-a)^2}$ remaining always real and negative continually increases to $-\infty$, and $e^{-\frac{1}{(x-a)^2}}$ remaining always real and positive continually diminishes to zero. And, moreover, (X containing only a finite series of negative powers of $x-a$) the expression $Xe^{-\frac{1}{(x-a)^2}}$ will in like manner, remaining always real, continually approximate to zero. But assume $x=a+r(\cos\theta+i\sin\theta)$, r real and positive, θ real; then $(x-a)^2=r^2(\cos 2\theta+i\sin 2\theta)$, and if $\cos 2\theta$ be positive, then the real part of $(x-a)^2$, being always positive, continually diminishes to zero, and the like conclusions follow. If however $\cos 2\theta$ be negative, then the real part of $(x-a)^2$ is negative, and the real part of $-\frac{1}{(x-a)^2}$ is positive, and as r diminishes continually approximates to $+\infty$; so far from $e^{-\frac{1}{(x-a)^2}}$ continually approximating to zero, it is in general an imaginary quantity continually approximating to infinity; and the like is the case with its successive differential coefficients; the conclusion is, it is not true *simpliciter* that the function $e^{-\frac{1}{(x-a)^2}}$, or any one of its successive differential coefficients, vanishes for the value a of the variable.

Generally, if a real or imaginary quantity $\alpha+\beta i$ is represented by the point whose rectangular coordinates are α, β; say if the value a of the variable x is represented by the point P, and any other value $a+h+ki$, by the point Q (h, k being therefore the coordinates of Q measured from the origin P), then a function $F(x)$ which as Q (no matter in what direction) approaches and ultimately coincides with P, tends to become and becomes ultimately $=0$, may be said to vanish *simpliciter* for the value a of the variable; but if this is only the case when Q approaches P in a certain direction or within certain limits of direction, the function not becoming zero when Q approaches in a different direction, then the function may be said to vanish *sub modo* for the value a of the variable.

Taking the theorem to mean "If for a given value a of the variable, a function and its differential coefficients vanish *sub modo*, the function is identically $=0$," the instance of the function $e^{-\frac{1}{(x-a)^2}}$ shows that the theorem is certainly not true; but taking the theorem to mean "If for a given value a of the variable, the function and its differential coefficients vanish *simpliciter*, then the function is identically $=0$"; the instance does not apply to it, and the truth of the theorem remains an open question.

The above view is consistent with a theorem obtained by Cauchy and others, defining within what limits of h the expansion by Taylor's theorem of the function $F(a+h)$ is applicable, viz. a and h being in general imaginary as above, if the function (or ? the function and its successive differential coefficients) is (or are) finite and continuous so long as the distance PQ does not exceed a certain real and positive value ρ, then the expansion is applicable for any point Q, whose distance PQ does not exceed this value ρ: but it ceases to be applicable for a point Q, the distance of which is equal to or exceeds ρ. In the case of a function such as $e^{-\frac{1}{(x-a)^2}}$, discontinuity arises *at* the point P, that is, for the value $\rho=0$, and according to the theorem in question, the expansion is not applicable for any value of ρ however small.

I wish to remark on a view which appears to me to be founded on a radical misconception of the notion of convergence. Writing $F(x)=e^{-\frac{1}{(x-a)^2}}$, consider the series

$$F(a)+F'(a)\frac{h}{1}+F''(a)\frac{h^2}{1.2}+\&c. \ldots$$

Then admitting that the exponential $e^{-\frac{1}{(x-a)^2}}$ becomes $=0$ for $x=a$, the successive functions $F(a)$, $F'(a)$, $F''(a)$, ... are each $=0$ as containing this exponential: but inasmuch as the successive differentiations introduce negative powers of $x-a$, each successive function is regarded as an infinitesimal of a lower order than those which precede it; say $F(a)$ being $=0^\mu$, the successive terms are multiples 0^μ, $0^{\mu-3}$, $0^{\mu-6}$, $0^{\mu-9}$, &c. respectively; where however μ is infinite, so that the several exponents μ, $\mu-3$, $\mu-6$, &c., however far the series is continued, remain all of them positive. This being so, it is said that the series $F(a)+F'(a)\frac{h}{1}+\&c.$, as being really of the form $0^\mu+0^{\mu-3}+0^{\mu-6}+\ldots$ is divergent, and for this reason fails to give a correct value of $F(a+h)$. I apprehend that the notion of divergence is a strictly numerical one; a series of numbers $a+b+c+d+\ldots$ is divergent when the successive sums a, $a+b$, $a+b+c$, $a+b+c+d$, &c., are numbers not continually tending to a determinate limit. In the actual case the series is $0+0+0+0+\ldots$, viz. each term is by hypothesis an absolute zero; the successive sums 0, $0+0$, $0+0+0$, ... are each $=0$, and we cannot, by the process of numerical summation, make the sum of the series to be anything else than 0. If it could, there would be an end of all numerical equality between infinite series; for taking any convergent series $a+b+c+d+\ldots$, if 0 means 0, this is the same thing as the series, also a convergent one,

$$(a+0)+(b+0)+(c+0)+\&c.,$$

and their difference $0+0+0+\ldots$ must be $=0$. I regard the view as a mere failure to reconcile the equation

$$F(a+h)=Fa+\frac{h}{1}F'(a)+\&c.,$$

with the *supposed* fact in regard to the function $e^{-\frac{1}{(x-a)^2}}$.

8. *Find the value of the definite integrals*

$$\int e^{-x^2}dx, \int \sin x^2 dx, \int \cos x^2 dx,$$

the limits being in each case ∞, $-\infty$. *Examine whether the last two integrals can be found by a process such as Laplace's (depending on a double integral) for the first integral.*

Laplace's process for the integral $\int e^{-x^2}dx$ is as follows: write $u=\int e^{-x^2}dx$, then also $u=\int e^{-y^2}dy$, and thence

$$u^2=\iint e^{-(x^2+y^2)}\,dx\,dy,$$

which, considering x, y as rectangular coordinates and substituting for them the polar coordinates r, θ, becomes

$$u^2=\iint e^{-r^2}\,r\,dr\,d\theta;$$

and then considering the double integral as extending over the infinite plane, and taking the limits to be $r=0$ to $r=\infty$, $\theta=0$ to $\theta=2\pi$, we obtain

$$u^2=(-\tfrac{1}{2}e^{-r^2})_0^\infty\,2\pi, =\tfrac{1}{2}\,.\,2\pi, =\pi,$$

that is,

$$u=\int e^{-x^2}dx=\sqrt{(\pi)}.$$

There is an assumption the validity of which requires examination. We have u the limit of the integral $\int_{-a}^{a}e^{-x^2}dx$, as a approaches to ∞; and this being so, we have u^2 the limit of

$$\int_{-a}^{a}\int_{-a}^{a}e^{-(x^2+y^2)}\,dx\,dy,$$

viz. u^2 is the integral of $e^{-(x^2+y^2)}$ taken over a square, the side of which is $2a$, a being ultimately infinite. But making the transformation to polar coordinates, and integrating as above, we in fact take the integral over a circle radius $=\beta$, β being ultimately infinite. And we assume that the two values are equal; or, generally, that taking the integral over an area bounded by a curve which is such that the distance of every point from the origin is ultimately infinite, the value of the integral is independent of the form of the curve.

This is really the case under the following conditions: 1°. For a curve of a given form, the integral tends to a fixed limit, as the size is continually increased. 2°. The quantity under the integral sign is always of the same sign (say always positive); (the last condition is sufficient, but not necessary). For, to fix the ideas, let the curves be as before the square and the circle: take a square; surrounding this, a circle; and surrounding the circle, a square. Imagine the two squares and the circle continually to increase in magnitude; the integral over the smaller square and that over the larger square, each tend to the same fixed limit; consequently the integral over the area enclosed between the two squares tends to the limit zero; and *à fortiori* the integral over the area enclosed between the circle and either of the two squares tends to the limit zero; that is, the integral over the square, and that over the circle, tend to the same limit. In the case under consideration, the function $e^{-(x^2+y^2)}$ is always positive; and the integral $\iint e^{-(x^2+y^2)}\,dx\,dy$, taken over the circle, tends (as in effect shown above) to the limit π: hence the process is a legitimate one.

But endeavour to apply it to the other two integrals; write

$$u = \int \sin x^2 dx = \int \sin y^2 dy, \qquad v = \int \cos x^2 dx = \int \cos y^2 dy,$$

then

$$\iint \sin (x^2 + y^2)\, dx\, dy = 2uv, \quad \iint \cos (x^2 + y^2)\, dx\, dy = v^2 - u^2,$$

where the double integrals on the left-hand side really denote integrals taken over a square and are not equal to the like integrals taken over a circle. This appears *à posteriori* if we only assume that the integrals u, v have determinate values; for taking the integrals over a circle they would be

$$\iint \begin{matrix}\sin\\ \cos\end{matrix}\, r^2 . r\, dr\, d\theta,$$

and would involve the indeterminate functions $\begin{matrix}\sin\\ \cos\end{matrix}\,\infty$; that is, if it were allowable to take the integrals over a circle, we should have $2uv$ and $v^2 - u^2$ indeterminate instead of determinate.

A process of finding them is as follows: in the equation $\int e^{-x^2} dx = \sqrt{(\pi)}$, substituting in the first instance $x\sqrt{(a)}$ for x, a real and positive, we have

$$\int e^{-ax^2} dx = \frac{\sqrt{(\pi)}}{\sqrt{(a)}},$$

and if it be assumed that this equation extends to the case where $a = \alpha + \beta i$, *the* *real part* a *real and positive**; or, what is the same thing, $a=\rho(\cos\theta+i\sin\theta)$, ρ real and positive, θ between the limits 0 and $\frac{1}{2}\pi$, then we have

$$\int e^{-\rho(\cos\theta+i\sin\theta)x^2}\,dx=\frac{\surd(\pi)}{\surd(\rho)}(\cos\tfrac{1}{2}\theta-i\sin\tfrac{1}{2}\theta),$$

or, separating the real and imaginary parts and taking $\rho=1$, we have

$$\int e^{-x^2\cos\theta}\cos(x^2\sin\theta)\,dx=\surd(\pi)\cos\tfrac{1}{2}\theta,$$

$$\int e^{-x^2\cos\theta}\sin(x^2\sin\theta)\,dx=\surd(\pi)\sin\tfrac{1}{2}\theta.$$

Admitting these formulæ to be true in general, there is still considerable difficulty in seeing that they hold good in the limiting case $\theta=\frac{1}{2}\pi$. But assuming that they do, the formulæ then become

$$\int\cos x^2\,dx=\frac{\surd(\pi)}{\surd(2)},\quad\int\sin x^2\,dx=\frac{\surd(\pi)}{\surd(2)},$$

which are the values of the integrals in question.

9. *Considering in a solid body a system of two, three, four, five, or six lines, determine in each case the relations between the lines in order that it may be possible to find along them forces to hold the body in equilibrium.*

If there are two lines, the condition obviously is that these must be one and the same line.

If three lines, then these must lie in a plane, and meet in a point.

The conditions in the other cases ought to be in the text-books; they in fact are not, and I assumed that they would not be known, and considered the question as a problem; it is, in regard to the cases of four and five lines, *a very easy problem when the solution is seen.*

In the case of four lines; imagine in the solid body an axis meeting any three of the lines, and let this axis be fixed; the condition of equilibrium about this axis is that the fourth line shall meet the axis. The required condition therefore is that every line meeting three of the four lines shall meet the fourth line; or, what is the same thing, the four lines must be generators (of the same kind) of a skew hyperboloid.

In the case of five lines, taking any four of them, we have two lines (tractors) each meeting the four lines; and taking either of the two lines as an axis, then for equilibrium the fifth line must also meet this axis; the required relations therefore are that the fifth line shall meet each of the two lines which meet the other four lines; or, what is the same thing, that there shall be two lines each meeting the five given lines.

* The equation is clearly *not true* unless this is so: for a being negative, then in virtue of the factor e^{-ax^2}, the exponential, instead of decreasing will increase, and ultimately become infinite as x increases to $\pm\infty$

The case of six lines is one the answer to which could not have been discovered in an examination; the relations in fact are that the six lines shall form an involution; viz. this is a system such that taking five of the lines as given, then if the sixth line is taken to pass through a given point it may be any line whatever in a determinate plane through this point; or, what is the same thing, if the sixth line is taken to be in a given plane, it may be any line whatever through a determinate point in this plane. But in a particular case, the answer is easy; suppose five of the six given lines to be met by a single line, then the sixth line may be any line whatever meeting this single line.

10. *If* $X, Y, Z, \ldots$ *are the roots of the equation*

$$(1, P, Q, \ldots)(c, 1)^n = 0,$$

show that the differential equation obtained by the elimination of c *is* $\zeta X'Y'Z' = 0$, *where* ζ *denotes the product of the squared differences of the roots* $X, Y, Z, \ldots$, *and* $X', Y', Z', \ldots$ *are the derived functions of these roots; and connect this result with the theory of singular solutions.*

We have identically

$$(1, P, Q \ldots)(c, 1)^n = (c-X)(c-Y)(c-Z)\ldots;$$

the original equation and its derived equation

$$(0, P', Q', \ldots)(c, 1)^n = 0$$

(the latter of them of degree $n-1$) may therefore be written

$$(c-X)(c-Y)(c-Z)\ldots = 0,$$

$$X'(c-Y)(c-Z)\ldots + Y'(c-X)(c-Z)\ldots + \&c. = 0.$$

To eliminate c, we have in the *nilfactum* of the second equation to substitute successively the values $c = X$, $c = Y$, &c., multiply the several functions together and equate the result to zero; the factors are evidently

$$X'(X-Y)(X-Z)\ldots,\quad Y'(Y-X)(Y-Z)\ldots,\ \&c.,$$

where each difference occurs twice, e.g. $X-Y$ under the two forms $X-Y$ and $Y-X$ respectively; the result thus is

$$X'Y'Z'\ldots(X-Y)^2(X-Z)^2(Y-Z)^2\ldots = 0;$$

that is,

$$\zeta . X'Y'Z'\ldots = 0.$$

Thus in particular in the case of a quadric equation

$$(1, P, Q)(c, 1)^2, = (c-X)(c-Y), = 0,$$

the differential equation is

$$(X-Y)^2 X'Y' = 0;$$

viz. since $X+Y=-P$, and $XY=Q$, this is

$$(P^2-4Q)\,X'Y'=0,$$

and writing also

$$X=-\tfrac{1}{2}\{P+\sqrt{(P^2-4Q)}\},\quad Y=-\tfrac{1}{2}\{P-\sqrt{(P^2-4Q)}\},$$

we find

$$X'Y'=\tfrac{1}{4}\left\{P'^2-\frac{(PP'-2Q')^2}{P^2-4Q}\right\}:$$

the differential equation thus is

$$(P^2-4Q)\left\{P'^2-\frac{(PP'-2Q')^2}{P^2-4Q}\right\}=0.$$

The application to the theory of singular solutions is that, in the case where the function $(1, P, Q\ldots)(c, 1)^n$ breaks up into rational factors $c-X$, $c-Y,\ldots$, the factor $\zeta=(X-Y)^2(X-Z)^2\ldots$ divides out and should be rejected from the differential equation, which in its true form is $X'Y'Z'\ldots=0$; viz. this is what we obtain immediately, considering the given integral equation as meaning the system of curves $c-X=0$, $c-Y=0,\ldots$, and there is not really any singular solution; whereas in the case where the factors are not rational, the factor in question, when the product $X'Y'Z'\ldots$ is expressed in terms of the coefficients $P, Q,\ldots$, and their derived coefficients does not divide out from the equation; and in this case, equated to zero, it gives a proper singular solution of the equation.

11. *In the theory of elliptic motion, v denoting the mean anomaly and e the eccentricity, if m' be an angle such that $\tan\frac{1}{2}v=\frac{1+e}{1-e}\tan\frac{1}{2}m'$, find in terms of e, m' the mean anomaly m.*

Taking as usual u for the eccentric anomaly, to commence the solution write down

$$\tan\tfrac{1}{2}v=\sqrt{\left(\frac{1+e}{1-e}\right)}\tan\tfrac{1}{2}u$$

$$=\frac{1+e}{1-e}\tan\tfrac{1}{2}m',$$

that is,

$$\tan\tfrac{1}{2}u=\sqrt{\left(\frac{1+e}{1-e}\right)}\tan\tfrac{1}{2}m',$$

and u being given hereby as a function of m', we have by substitution in the equation $m=u-e\sin u$, to find m as a function of m'.

A creditable approximate solution would be $m=m'+0\,.\,e$, viz. this would be to show that neglecting terms in e^2, &c., we have $m=m'$. In fact, taking e small, we have

$$\tan\tfrac{1}{2}u=(1+e)\tan\tfrac{1}{2}m',$$

and thence if $u=m'+x$, we have

$$\tan\tfrac{1}{2}m'+\tfrac{1}{2}x\sec^2\tfrac{1}{2}m'=(1+e)\tan\tfrac{1}{2}m',$$

that is,

$$x = 2e \cos^2 \tfrac{1}{2}m' \tan \tfrac{1}{2}m' = e \sin m' ; \quad u = m' + e \sin m',$$

and

$$\begin{aligned} m &= m' + e \sin m' \\ &\quad - e \sin (m' + \ldots) \\ &= m' + 0 \,.\, e. \end{aligned}$$

The complete solution would be obtained by expanding u in terms of e, m' from the equation $\tan \frac{1}{2}u = \sqrt{\left(\frac{1+e}{1-e}\right)} \tan \frac{1}{2}m'$ (which is of the form $\tan \frac{1}{2}u = n \tan \frac{1}{2}m'$, giving for u a known series $= m' +$ multiple sines of m'), and then observing that the same equation leads to

$$\sin u = \frac{\sqrt{(1-e^2)} \sin m'}{1 - e \cos m'},$$

we have

$$m = \text{series} - \frac{e \sqrt{(1-e^2)} \sin m'}{1 - e \cos m'},$$

where the second term has also to be expanded in a series of multiple sines of m'; which can be done without difficulty.

12. *If* (u, v) *are given functions of the coordinates* (x, y), *neither of them a maximum or a minimum at a given point* O; *and if through* O *we draw* Ox' *in the direction in which* v *is constant and* u *increases, and* Oy' *in the direction in which* u *is constant and* v *increases; then the rotation (through an angle not greater than* π), *from* Ox' *to* Oy' *is in the same direction with that from* Ox *to* Oy, *or in the contrary direction, according as* $\frac{du}{dx}\frac{dv}{dy} - \frac{du}{dy}\frac{dv}{dx}$ *is positive or negative.*

The theorem has not, so far as I am aware, been noticed, and it seems to be one of some importance; there is no difficulty in it, but the answer requires some care in writing out; of course where the whole question is one of sign and direction, the omission to state that a subsidiary quantity is positive may render an answer worthless.

It depends on the following lemma: Consider the triangle $OX'Y'$, such that Ox, Oy being any rectangular axes through the origin O, the coordinates of X' are h, k, and those of Y' are h_1, k_1: then considering the area as positive, the double area is $= \pm (hk_1 - h_1k)$, viz. the sign is $+$ or $-$ according as the rotation from OX' to OY' (through an angle less than π) is in the same direction with that from Ox to Oy, or in the contrary direction; or, what is the same thing, $hk_1 - h_1k$ is in the first case positive and in the second case negative.

To show this, suppose for a moment that the lines OX', OY' are each of them in the quadrant xOy, say in the first quadrant, the inclination of OY' to Ox exceeding that of OX' to Ox; then h, k, h_1, k_1 are all positive, and $\frac{k_1}{h_1} > \frac{k}{h}$, that is, $hk_1 - h_1k$ is $+$, and the rotation from OX' to OY' is in the same direction as that from Ox to Oy; or the lemma holds good. Now OX' remaining fixed, let OY' revolve in the direction Ox to Oy; so long as OY' remains in the first quadrant, $\frac{k_1}{h_1}$ continues to increase, and we have always $\frac{k_1}{h_1} > \frac{k}{h}$, and $hk_1 - h_1k = +$; when OY' comes into the second quadrant (h, k being always positive), h_1 is negative and k_1 positive, consequently $hk_1 - h_1k$ is the sum of two positive terms, and therefore $= +$; as OY' continues to revolve and passes into the third quadrant, we have h_1, k_1 each negative, but $\frac{k_1}{h_1} < \frac{k}{h}$, and therefore $hk_1 - h_1k$ still $= +$; when, however, OY' comes into the position opposite to OX', then $\frac{k_1}{h_1} = \frac{k}{h}$, and $hk_1 - h_1k$ is $= 0$; and when OY', continuing in the third quadrant, has passed the position in question, we have $\frac{k_1}{h_1} > \frac{k}{h}$, and therefore $hk_1 - h_1k = -$, but now the angle $X'OY'$ measured in the original direction has become $> \pi$, and the rotation OX' to OY' through an angle less than π will be in the opposite direction, that is, in the direction opposite to that from Ox to Oy; and, similarly, when OY' passes into the fourth quadrant, and until, passing into the first quadrant, it approaches the position OX', the sign of $hk_1 - h_1k$ will be $-$, and the rotation will be in the direction contrary to that from Ox to Oy. The lemma is thus true for any position of OX' in the first quadrant; and the like reasoning would show that it is true for any position of OX' in the second, the third, or the fourth quadrant; hence the lemma is true generally.

This being so, taking a new origin, let the coordinates of O be x, y; and drawing through O the axes Ox', Oy' as directed, let X' be the point belonging to the values $u + \delta u$, v of (u, v), and Y' the point belonging to the values u, $v + \delta v$ of (u, v); taking δu positive, X' will be on Ox' in the direction O to x', and similarly taking δv positive, Y' will be on Oy' in the direction O to y'. Taking as before (h, k) for the coordinates of X', and (h_1, k_1) for the coordinates of Y', these coordinates being measured from the point O as origin, we have

$$\delta u = \frac{du}{dx} h + \frac{du}{dy} k,$$

$$0 = \frac{dv}{dx} h + \frac{dv}{dy} k,$$

whence, writing for a moment $J = \frac{du}{dx}\frac{dv}{dy} - \frac{du}{dy}\frac{dv}{dx}$, we have $Jh = +\frac{dv}{dy}\delta u$, $Jk = -\frac{dv}{dx}\delta u$. And in like manner

$$0 = \frac{du}{dx} h_1 + \frac{du}{dy} k_1,$$

$$\delta v = \frac{dv}{dx} h_1 + \frac{dv}{dy} k_1,$$

whence

$$Jh_1 = -\frac{du}{dy}\delta v, \quad Jk_1 = \frac{du}{dx}\delta v;$$

and hence

$$J^2(hk_1 - h_1k) = \left(\frac{du}{dx}\frac{dv}{dy} - \frac{du}{dy}\frac{dv}{dx}\right)\delta u\,\delta v, \;= J\delta u\delta v,$$

that is,

$$hk_1 - h_1k = \frac{1}{J}\delta u\delta v,$$

and δu, δv being as above each of them positive, J has the same sign as $hk_1 - h_1k$. But the rotation from OX' to OY' is in the same direction as that from Ox to Oy, or in the contrary direction, according as $hk_1 - h_1k$ is + or −, that is, according as $J, = \frac{du}{dx}\frac{dv}{dy} - \frac{du}{dy}\frac{dv}{dx}$, is + or −; which is the theorem in question.

13. Write a dissertation on:

The theory and constructions of Perspective.

In Perspective we represent an object in space by means of its central projection upon a plane: viz. any point P_1* of the object is represented by P', the intersection with the plane of projection of the line D_1P_1 from the centre of projection (or say the eye) D_1 to the point P_1; and considering any line or curve in the object, this is represented by the line or curve which is the locus of the points P', the projections of the corresponding points P_1 of the line or the curve in the object.

The fundamental construction in perspective is derived from the following considerations: viz. considering through P_1 (fig. 1) a line meeting the plane of projection in Q, and drawing parallel thereto through D_1 a line to meet the plane of projection in M and joining the points M, Q, then the lines D_1M, MQ, QP_1 are in a plane; that is, the plane through D_1 and the line P_1Q meets the plane of projection in MQ;

Fig. 1.

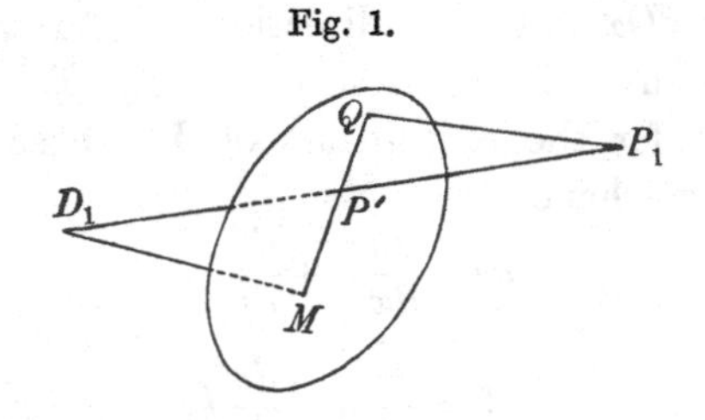

and consequently the projection P' of any point P_1 in the line P_1Q lies in the line QM; and not only so, but considering only the points P_1 of this line which lie behind the plane of projection (D_1 being considered as in front of it), the projections of all these points lie on the terminated line MQ; viz. Q is the projection of the point Q, and M the projection of the point at infinity on the line QP_1; or, if we please, the finite line QM is the projection of the line $QP_1\infty$.

* The subscript unity is used to denote a point not in the plane of projection, considered as a point out of this plane; a point in the plane of projection, used in the constructions of perspective as a conventional representation of a point P_1, will be denoted by the same letter P without the subscript unity. And the like as regards D_1 and D.

If we consider a set of lines parallel to P_1Q, these all give rise to the same point M, and thus their projections MQ all pass through this point M, which is said to be the "vanishing point" of the system of parallel lines. Again, if we consider any two or more lines through P_1, to each of these there correspond different points M and Q, and, therefore, a different line MQ, but these all intersect in a common point P' which is the projection of P_1. If the lines are all in one and the same plane through P_1, then the locus of the points Q is a line, the intersection of this plane with the plane of projection, say the "trace" line; and the locus of the points M is a parallel line, the intersection of the parallel plane through D_1 with the plane of projection; say this is the "vanishing line" for the plane in question.

A construction in perspective presupposes a conventional representation on the plane of projection (or say on the paper) as well of the position of the eye as of the object to be projected. If for simplicity we suppose the object to be a figure in one plane, then this plane intersects the paper in a trace line, and we may imagine the plane made to rotate about the trace line until it comes to coincide with the paper, and we have thus the plane object conventionally represented on the paper. Similarly considering the parallel plane through the eye D_1, and regarding D_1 as a point of this plane, the plane meets the paper in the vanishing line, and we may imagine the plane made to rotate (in the direction opposite to that of the first rotation) until it comes to coincide with the paper, bringing the point D_1 to coincide with a point D of the paper. We have thus the "point of distance" D, being a conventional representation on the paper of the position of the eye D_1; but which point D has, observe, a different position for different directions of the plane of the object.

To fix the ideas, suppose the plane of projection to be vertical, and the plane of the object to be a horizontal plane situate below the eye. The trace line will be represented by a horizontal line HH' (fig. 2), and the object by a figure in the plane

Fig. 2.

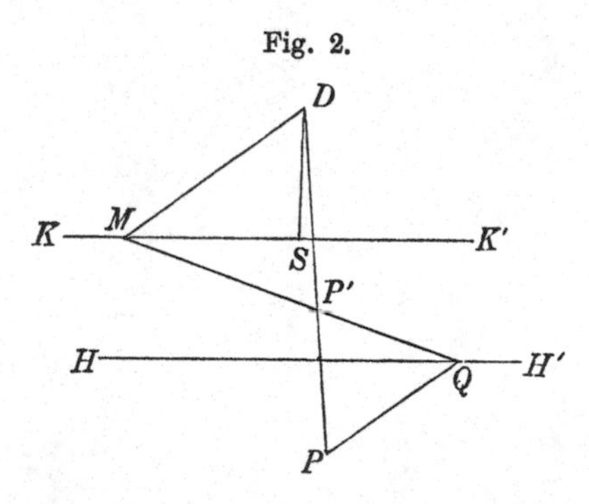

of the paper below the line HH' such that, bending this portion of the paper backwards through a right angle round HH', the figure would be brought to coincide with the object*. The vanishing line will be a horizontal line KK' above HH', and the

* It is assumed in the text, that the figure on the paper is equal in magnitude to the object; but practically the figure is drawn on a reduced scale, the distance between the lines KK', HH', and the distance DS (representing respectively the distance between the parallel planes, and the distance of the eye from the plane of projection) being drawn on the same reduced scale.

eye will be represented by a point D above KK', in suchwise that, bending the upper part of the paper round KK' forwards through a right angle, the point D would come to coincide with the position D_1 of the eye. This being so, taking any line PQ in the representation of the object, we draw through D the parallel line DM, and then joining the points M and Q, we have MQ as the perspective representation of the line $QP\infty$, which represents a line $QP_1\infty$ of the object. And drawing through P any number of lines, each of these gives a point Q and a point M, but the lines MQ all meet in a common point P', which is the perspective representation of the point P; which point P' may, it is clear, be obtained as the intersection with any one line MQ of the line DP drawn to join P with the point of distance D. The plane of the object has for convenience been taken to be horizontal; but its position may be any whatever, and in particular the construction is equally applicable in the case where the plane is vertical.

In the case of an object not in one plane, any point Q_1 of the object may be determined by means of its projection by a vertical line upon a given horizontal plane, say this is P_1, and of its altitude Q_1P_1 above this plane. We in fact determine the object by means of its groundplan, and of the altitudes of the several points thereof. It is easy, from the foregoing principles, to see that, drawing through P the vertical line PQ equal to the altitude, and joining the points Q, D, then the vertical line through P' meets this line QD in a point Q', which will be the perspective representation of Q_1. We have thus a construction applicable to any solid figure whatever.

592.

ON THE MERCATOR'S-PROJECTION OF A SKEW HYPERBOLOID OF REVOLUTION.

[From the *Messenger of Mathematics*, vol. IV. (1875), pp. 17—20.]

IN a note "On the Mercator's-projection of a surface of revolution" read before the British Association, [555, (5)], I remarked that the surface might be, by its meridians and parallels, divided into infinitesimal squares; and that these would be on the map represented by two systems of parallel lines at right angles to each other, dividing the map into infinitesimal squares; and that, by taking the squares not infinitesimal but small, for instance, by considering the meridians at intervals of 10° or 5°, we might approximately construct a Mercator's-projection of the surface. But it is worth while, for the skew hyperboloid of revolution, to develop analytically the ordinary accurate solution.

Taking the equation of the surface to be

$$\frac{x^2+y^2}{a^2}-\frac{z^2}{c^2}=1,$$

{or, if as usual $a^2+c^2=a^2e^2$, then $x^2+y^2-(e^2-1)z^2=a^2$}, and writing $x=r\cos\theta$, $y=r\sin\theta$, the meridians corresponding to the several longitudes θ are in the map represented by the parallel lines $X=a\theta$, and the parallels corresponding to the several values of z are in the map represented by a set of parallel lines $Z=f(z)$, the form of the function being so determined that the infinitesimal rectangles on the map are similar to those on the surface. The required relation is readily found to be

$$Z=\int\frac{\sqrt{\{(a^2+c^2)z^2+c^4\}}}{z^2+c^2}dz,$$

where the integral is taken from the value $z=0$.

The substitution which first presents itself is to write herein $z = \frac{c^2}{\sqrt{(a^2+c^2)}} \tan\phi$; or, what is the same thing,

$$z = a\left(e - \frac{1}{e}\right)\tan\phi,$$

where observe that $a\left(e-\frac{1}{e}\right)$ is the distance between a focus and its corresponding directrix. The equation of the surface is satisfied by writing therein $\sqrt{(x^2+y^2)} = a\sec\psi$, $z = c\tan\psi$, and ψ as thus defined is the "parametric latitude"; hence the foregoing angle ϕ is a deduced latitude connected with the parametric latitude ψ by the equation

$$\tan\phi = \frac{c}{a\left(e-\frac{1}{e}\right)}\tan\psi, \quad = \frac{e}{\sqrt{(e^2-1)}}\tan\psi.$$

The resulting formula in terms of ϕ is

$$Z = \int_0 \frac{c^2\sqrt{(a^2+c^2)}\,d\phi}{\cos\phi\,(c^2+a^2\cos^2\phi)},$$

or, if we write herein $\zeta = \tan\frac{1}{2}\phi$, the formula becomes

$$Z = 2c^2\sqrt{(a^2+c^2)}\int_0 \frac{(1+\zeta^2)^2}{1-\zeta^2}\,\frac{d\zeta}{c^2(1+\zeta^2)^2+a^2(1-\zeta^2)^2},$$

viz. the function under the integral sign is rational. The expression is, however, complicated, and a more simple formula is obtained by using instead of ϕ the parametric latitude ψ; viz. we have $z = c\tan\psi$, and thence

$$Z = \int \frac{\sqrt{(a^2\sin^2\psi+c^2)}}{\cos\psi}\,d\psi,$$

or, putting herein

$$\sin\psi = \frac{c}{a}\,\frac{u}{\sqrt{(1-u^2)}},$$

and therefore

$$\cos^2\psi = \frac{a^2-(a^2+c^2)\,u^2}{a^2(1-u^2)},$$

and

$$a^2\sin^2\psi + c^2 = \frac{c^2}{1-u^2}, \quad \cos\psi\,d\psi = \frac{c}{a}\,\frac{du}{(1-u^2)^{\frac{3}{2}}},$$

the formula becomes

$$Z = c^2 a\int_0 \frac{du}{(1-u^2)\,\{a^2-(a^2+c^2)\,u^2\}},$$

or, what is the same thing,

$$= a\,(e^2-1)\int_0 \frac{du}{(1-u^2)(1-e^2u^2)};$$

viz. we thus have

$$Z = \tfrac{1}{2}a\left\{\log\frac{1-u}{1+u} - e\log\frac{1-eu}{1+eu}\right\},$$

the logarithms being hyperbolic.

As already mentioned, u is connected with the parametric latitude ψ by the equation

$$\sin\psi = \frac{c}{a}\frac{u}{\sqrt{(1-u^2)}}, \quad = \frac{u\sqrt{(e^2-1)}}{\sqrt{(1-u^2)}},$$

that is,

$$\sin\psi = \sqrt{(e^2-1)}\tan p, \text{ if } u = \sin p,$$

or conversely

$$u = \frac{\sin\psi}{\sqrt{(e^2-1+\sin^2\psi)}},$$

so that the point passing to infinity along the branch of the hyperbola, or ψ passing from 0 to 90°, u passes from 0 to $\frac{1}{e}$; and for $u=\frac{1}{e}$ the value of Z becomes, as it should do, infinite. The value of z in terms of u is

$$z = \frac{(e^2-1)u}{\sqrt{(1-e^2u^2)}}, \text{ or conversely } u = \frac{z}{\sqrt{(c^2z^2+e^2-1)}},$$

and we have, moreover,

$$u = \frac{1}{e}\frac{2\zeta}{1+\zeta^2}, \quad = \frac{1}{e}\sin\phi, \quad = (\text{as before}) \ \frac{\sin\psi}{\sqrt{(e^2-1+\sin^2\psi)}}.$$

It will be recollected that, in the Mercator's-projection of the sphere, the longitude and latitude being θ, ϕ, the values of X, Z are

$$X = a\theta, \quad Z = \log\tan\left(\frac{\pi}{4}+\tfrac{1}{2}\phi\right),$$

the logarithm being hyperbolic.

In the case of the rectangular hyperbola $a=c$, $=1$ suppose,

$$e=\sqrt{(2)}, \quad z=\tan\psi, \quad u=\frac{\sin\psi}{\sqrt{(1+\sin^2\psi)}}, \quad =\sin p, \text{ if } \sin\psi=\tan p;$$

whence

$$Z = \tfrac{1}{2}h\,.\,l\,\frac{\tan(45°-\frac{1}{2}p)}{\tan(45°+\frac{1}{2}p)} - \tfrac{1}{2}\sqrt{(2)}\,h\,.\,l\,\frac{\tan(22°30'-\frac{1}{2}p)}{\tan(22°30'+\frac{1}{2}p)},$$

the first term being of course

$$= h\,.\,l\tan(45°-\tfrac{1}{2}p), \text{ or } -h\,.\,l\tan(45°+\tfrac{1}{2}p).$$

Transforming to ordinary logarithms, this is

$$Z = \frac{1}{\sqrt{(2)}\log e}\left[-\sqrt{(2)}\log\tan(45°+\tfrac{1}{2}p) + \{\log\tan(22°30'+\tfrac{1}{2}p) - \log\tan(22°30'-\tfrac{1}{2}p)\}\right],$$

say this is

$$Z = \frac{1}{\sqrt{(2)}\log e}(-A+B),$$

where

$$A = \sqrt{(2)} \log \tan (45^\circ + \tfrac{1}{2}p),$$

$$B = \log \tan (22^\circ 30' + \tfrac{1}{2}p) - \log \tan (22^\circ 30' - \tfrac{1}{2}p).$$

Taking ψ as the argument, I tabulate $z, = \tan \psi$, and $Z . \sqrt{(2)} \log e, = - A + B$, as shown in the annexed table: the last column of which gives, therefore, the positions of the several parallels of 5°, 10°, ..., 85°; the interval of 5° between two meridians is, on the same scale,

$$\sqrt{(2)} \log e . \frac{\pi}{36} = (1{\cdot}4142136)({\cdot}4342945)({\cdot}0872665), = {\cdot}05360;$$

viz. this is nearly equal to the arc of meridian 0° to 5°, and the table shows that the arcs 0°—5°, 5°—10°, &c. continually increase as in a Mercator's-projection of the sphere, but more rapidly; there is, however, nothing in this comparison, since the determination of latitude on the hyperboloid by the equation $z = \tan \psi$ is altogether arbitrary.

ψ	$z = \tan \psi$	p	A	B	$-A+B$
0°	0·	0°	0·	0·	0·
5	·08749	4° 58′ 51″	·05348	·10719	·05370
10	·17632	9 51 0	·10611	·21439	·10827
15	·26795	14 30 40	·15724	·32174	·16450
20	·36397	18 53 0	·20619	·42943	·22324
25	·46631	22 54 30	·25342	·53780	·28539
30	·57735	26 34 0	·29557	·64758	·35201
35	·70020	29 50 20	·33538	·75959	·42421
40	·83910	32 44 0	·37170	·87506	·50337
45	1·00000	35 15 50	·40446	·99554	·59108
50	1·19175	37 27 20	·43360	1·12355	·68995
55	1·42815	39 19 20	·45913	1·26151	·80238
60	1·73205	40 53 40	·48117	1·41445	·93328
65	2·14450	42 11 10	·49971	1·58840	1·08869
70	2·74747	43 13 10	·51478	1·79504	1·28026
75	3·73205	44 0 30	·52646	2·05524	1·52877
80	5·67128	44 33 40	·53469	2·41347	1·87877
85	11·43005	44 53 30	·53973	3·02355	2·48381
90	∞	45° 0′ 0″	·54133	∞	∞

593.

A SHEEPSHANKS' PROBLEM (1866).

[From the *Messenger of Mathematics*, vol. IV. (1875), pp. 34—36.]

Apply the formulæ of elliptic motion to determine the motion of a body let fall from the top of a tower at the equator.

The earth is regarded as rotating with the angular velocity ω round a fixed axis, so that the body is in fact projected from the apocentre with an angular velocity $=\omega$; and we write α for the equatorial radius, β for the height of the tower; then g denoting the force of gravity, and μ, h, n, a, e, θ, as in the theory of elliptic motion, we have

$$\mu = \quad n^2a^3 \quad = g\alpha^2,$$

$$h = (\alpha+\beta)^2\,\omega = na^2\sqrt{(1-e^2)},$$

$$\alpha+\beta \quad = a\,(1+e);$$

whence

$$(\alpha+\beta)^4\,\omega^2 = g\alpha^2 a\,(1-e^2),$$

$$(\alpha+\beta) \quad = \quad a\,(1+e),$$

$$\frac{(\alpha+\beta)^3\,\omega^2}{g\alpha^2} = \frac{\alpha\omega^2}{g}\left(1+\frac{\beta}{\alpha}\right)^3 = 1-e,$$

where $\dfrac{\alpha\omega^2}{g}$ = ratio of centrifugal force to gravity,

$$= \tfrac{1}{289},$$

so that $1-e$ is small;

$$r = \frac{a\,(1-e^2)}{1-e\cos\theta}, \quad = \frac{(\alpha+\beta)(1-e)}{1-e\cos\theta},$$

whence

$$1-e\cos\theta \quad = \frac{(\alpha+\beta)(1-e)}{r}.$$

Suppose

$$r = \alpha,$$

$$1 - e\cos\theta = (1-e)\left(1+\frac{\beta}{\alpha}\right) = 1 - e + \frac{\beta}{\alpha}(1-e),$$

which is nearly

$$= 1 - e,$$

that is, θ is small, and therefore approximately

$$1 - e + \tfrac{1}{2}e\theta^2 = 1 - e + \frac{\beta}{\alpha}(1-e),$$

or

$$\theta^2 = \frac{2\beta}{\alpha}\,\frac{1-e}{e};$$

we then have

$$r^2 d\theta = h dt, \text{ or } dt = \frac{r^2 d\theta}{h} = \frac{(\alpha+\beta)^2(1-e)^2}{(\alpha+\beta)^2\,\omega\,(1-e\cos\theta)^2}\,d\theta$$

$$= \frac{(1-e)^2}{\omega^2(1-e\cos\theta)^2}\,d\theta$$

$$= \frac{(1-e)^2}{\omega\,(1-e+\tfrac{1}{2}e\theta^2)^2}\,d\theta$$

$$= \frac{1}{\omega\left(1+\frac{\tfrac{1}{2}e}{1-e}\,\theta^2\right)^2}\,d\theta$$

$$= \frac{1}{\omega}\left(1 - \frac{e}{1-e}\,\theta^2\right)d\theta;$$

that is,

$$\omega\,dt = \left(1 - \frac{e}{1-e}\,\theta^2\right)d\theta.$$

Integrating, we have

$$\theta\left(1 - \frac{\tfrac{1}{3}e\theta^2}{1-e}\right) = \omega t,$$

where $\omega t =$ earth's rotation in time t, $= \phi$ suppose; therefore

$$\theta\left(1 - \frac{\tfrac{1}{3}e\theta^2}{1-e}\right) = \phi,$$

hence, if θ be as above, the angle described in falling to the surface,

$$\frac{e\theta^2}{1-e} = \frac{2\beta}{\alpha},$$

$$\theta\left(1 - \tfrac{2}{3}\frac{\beta}{\alpha}\right) = \phi,$$

$$\theta - \phi = \tfrac{2}{3}\frac{\beta}{\alpha}\,\theta = \tfrac{2}{3}\frac{\beta}{\alpha}\sqrt{\left(\frac{2\beta}{\alpha}\,\frac{1-e}{e}\right)}.$$

Writing herein

$$\frac{1-e}{e}, \;=1-e,\; =\frac{\alpha\omega^2}{g},$$

this is

$$=\tfrac{2}{3}\frac{\beta}{\alpha}\sqrt{\left(\frac{2\beta}{\alpha}\,\frac{\alpha\omega^2}{g}\right)}=\frac{(2\beta)^{\frac{3}{2}}}{3\alpha^{\frac{3}{2}}}\sqrt{\left(\frac{\alpha\omega^2}{g}\right)},$$

viz.

$$\theta-\phi=\frac{2^{\frac{3}{2}}\,.\,\beta^{\frac{3}{2}}}{3\alpha^{\frac{3}{2}}}\sqrt{\left(\frac{\alpha\omega^2}{g}\right)},$$

whence

$$\alpha\,(\theta-\phi)=\frac{2\sqrt{(2)}}{3}\sqrt{\left(\frac{\beta}{\alpha}\right)}\sqrt{\left(\frac{\alpha\omega^2}{g}\right)}\,\beta,$$

or say

$$\frac{\alpha\,(\theta-\phi)}{\beta}=\frac{2\sqrt{(2)}}{3}\sqrt{\left(\frac{\beta}{\alpha}\right)}\sqrt{\left(\frac{\alpha\omega^2}{g}\right)},$$

where $\alpha\,(\theta-\phi)$ is the distance at which the body falls from the foot of the tower. Substituting for $\sqrt{\left(\frac{\alpha\omega^2}{g}\right)}$ its value, $=\frac{1}{17}$, we have

$$\frac{\alpha\,(\theta-\phi)}{\beta}=\frac{2\sqrt{(2)}}{51}\sqrt{\left(\frac{\beta}{\alpha}\right)},\;=\cdot 056\sqrt{\left(\frac{\beta}{\alpha}\right)}.$$

594.

ON A DIFFERENTIAL EQUATION IN THE THEORY OF ELLIPTIC FUNCTIONS.

[From the *Messenger of Mathematics*, vol. IV. (1875), pp. 69, 70.]

THE following equation presented itself to me in connexion with the cubic transformation:

$$Q^2 - Q\left(k+\frac{1}{k}\right) - 3 = 3(1-k^2)\frac{dQ}{dk}.$$

Writing as usual $k = u^4$, I was aware that a solution was

$$Q = \frac{v^2}{u^2} + 2uv,$$

where u, v are connected by the modular equation

$$u^4 - v^4 + 2uv(1-u^2v^2) = 0;$$

but it was no easy matter to verify that the differential equation was satisfied. After a different solution, it occurred to me to obtain the relation between (Q, u); or, what is the same thing, (Q, k), viz. eliminating v, we find

$$Q^4 - 6Q^2 - 4\left(u^4 + \frac{1}{u^4}\right)Q - 3 = 0,$$

or say

$$\frac{1}{Q}(Q^4 - 6Q^2 - 3) = 4\left(k + \frac{1}{k}\right),$$

whence also

$$\frac{1}{Q}(Q^4 - 6Q^2 \pm 8Q - 3) = 4\left(k \pm 2 + \frac{1}{k}\right),$$

that is,

$$\frac{1}{Q}(Q-1)^3(Q+3)=4\left\{\sqrt{k}+\frac{1}{\sqrt{k}}\right\}^2,$$

and

$$\frac{1}{Q}(Q+1)^3(Q-3)=4\left\{\sqrt{k}-\frac{1}{\sqrt{k}}\right\}^2,$$

and thence

$$\frac{(Q+1)^3(Q-3)}{(Q-1)^3(Q+3)}=\left(\frac{k-1}{k+1}\right)^2;$$

viz. the value of Q thus determined must satisfy the differential equation. This is easily verified, for, in virtue of the assumed integral, we have

$$Q^2-3-\tfrac{1}{4}(Q^4-6Q^2-3)=3(1-k^2)\frac{dQ}{dk};$$

that is,

$$Q^4-10Q^2+9=-12(1-k^2)\frac{dQ}{dk},$$

or finally

$$(Q^2-1)(Q^2-9)=-12(1-k^2)\frac{dQ}{dk},$$

an equation which is at once obtained by differentiating logarithmically the former result, and we have thus the verification of the solution. This is, however, a particular integral only; and it appears doubtful whether there exists a general integral of an algebraical form.

595.

ON A SENATE-HOUSE PROBLEM.

[From the *Messenger of Mathematics*, vol. IV. (1875), pp. 75—78.]

THE following was given [5 Jan., 1874,] as a problem of elementary algebra:

"Solve the equations

$$u(2a-x)=x(2a-y)=y(2a-z)=z(2a-u)=b^2,$$

and prove that unless $b^2=2a^2$, $x=y=z=u$, but that if $b^2=2a^2$, the equations are not independent."

This is really a very remarkable theorem in regard to the intersections of a certain set of four quadric surfaces in four-dimensional space; viz. slightly altering the notation, we may write the equations in the form

$$x(2\theta-y)=m\theta^2 \ldots (12),$$
$$y(2\theta-z)=m\theta^2 \ldots (23),$$
$$z(2\theta-w)=m\theta^2 \ldots (34),$$
$$w(2\theta-x)=m\theta^2 \ldots (41),$$

where, regarding (x, y, z, w, θ) as coordinates in four-dimensional space, each equation represents a quadric surface. I remark that in such a space we have the notions, point-system, curve, subsurface, surface, according as the number of equations is 4, 3, 2, or 1.

Four quadric surfaces intersect in general in 16 points. But for the system in question (m being arbitrary), the common intersection consists of two lines and the two points

$$x=y=z=w=\theta\{1\pm\surd(1-m)\};$$

and in the case where $m=2$, then the intersection consists of two lines and a certain unicursal quartic curve.

To obtain these results, I consider the four points

$$\theta=0,\quad x=0,\quad y=0,\quad z=0, \ldots 123,$$
$$\theta=0,\quad y=0,\quad z=0,\quad w=0, \ldots 234,$$
$$\theta=0,\quad z=0,\quad w=0,\quad x=0, \ldots 341,$$
$$\theta=0,\quad w=0,\quad x=0,\quad y=0, \ldots 412:$$

the two points

$$x=y=z=w=\theta\{1 \pm \sqrt{(1-m)}\}, \ldots PQ:$$

and the six lines

$$\theta=0,\quad x=0,\quad y=0, \ldots 12,$$
$$\theta=0,\quad y=0,\quad z=0, \ldots 23,$$
$$\theta=0,\quad z=0,\quad w=0, \ldots 34,$$
$$\theta=0,\quad w=0,\quad x=0, \ldots 41,$$
$$\theta=0,\quad x=0,\quad z=0, \ldots 13,$$
$$\theta=0,\quad y=0,\quad w=0, \ldots 24,$$

being the edges of a tetrahedron, the vertices of which are the four points, viz. the point 123 is the intersection of the lines 12, 13, 23, and so for the other points.

The surfaces contain the several lines, viz.

the surface 12 contains $(12)^2$, 13, 14, 23, 24,
" 23 " $(23)^2$, 12, 24, 13, 34,
" 34 " $(34)^2$, 13, 23, 14, 24,
" 41 " $(41)^2$, 24, 34, 12, 13,

where $(12)^2$ denotes that 12 is a double line on the surface, and so in other cases. And it thus appears that the surfaces pass all four of them through the lines 13, 24, so that these lines are a part of the common intersection. To obtain the residual intersection, observe that the equations give

$$x=2\theta-m\frac{\theta^2}{w}=\frac{m\theta^2}{2\theta-y},$$

$$z=2\theta-m\frac{\theta^2}{y}=\frac{m\theta^2}{2\theta-w},$$

whence

$$(2\theta-y)\left(2\theta-\frac{m\theta^2}{w}\right)=m\theta^2,$$

$$(2\theta-w)\left(2\theta-\frac{m\theta^2}{y}\right)=m\theta^2,$$

or omitting from each equation the factor θ, the equations become

$$(2\theta - y)(2w - m\theta) = m\theta w,$$

$$(2\theta - w)(2y - m\theta) = m\theta y,$$

that is,

$$(4 - 2m)\,\theta w - 2m\theta^2 - 2yw + m\theta\,(y + w) = 0,$$

$$(4 - 2m)\,\theta y - 2m\theta^2 - 2yw + m\theta\,(y + w) = 0.$$

Whence, m not being $= 2$, we have $y = w$, and then

$$w^2 - 2\theta w + m\theta^2 = 0,$$

or, what is the same thing,

$$2\theta - w = \frac{m\theta^2}{w},$$

giving $x = y = z = w = \theta\,\{1 \pm \sqrt{(1 - m)}\}$, viz. the surfaces each pass through the points P, Q. As regards the omitted factor θ, it is to be observed that, writing in the equations of the four surfaces $\theta = 0$, the equations become $xy = 0$, $yz = 0$, $zw = 0$, $wx = 0$, satisfied by $x = 0$, $z = 0$, or by $y = 0$, $w = 0$, we have thus $(\theta = 0,\ x = 0,\ z = 0)$ and $(\theta = 0,\ y = 0,\ w = 0)$, viz. the before-mentioned lines 13 and 24.

In the case $m = 2$, we have between y, w the single equation

$$yw - \theta\,(y + w) + 2\theta^2 = 0,$$

giving

$$y = \frac{\theta\,(w - 2\theta)}{w - \theta},$$

and thence

$$x = \frac{2\theta\,(w - \theta)}{w},$$

$$z = \frac{-2\theta^2}{w - \theta};$$

or, writing for convenience $\alpha = \dfrac{w}{\theta}$, then the equations are

$$\frac{w}{\theta} = \alpha,$$

$$\frac{y}{\theta} = \frac{\alpha - 2}{\alpha - 1},$$

$$\frac{z}{\theta} = \frac{-2}{\alpha - 2},$$

$$\frac{x}{\theta} = \frac{2\,(\alpha - 1)}{\alpha};$$

or, what is the same thing,

$$\begin{array}{lll} x = & 2(\alpha-1)^2(\alpha-2) & \left(1-\frac{\alpha}{\infty}\right) \\ : y \;:& \alpha \;\ldots\; (\alpha-2)^2 & \left(1-\frac{\alpha}{\infty}\right) \\ : z \;:& -2\alpha(\alpha-1) \;\ldots & \left(1-\frac{\alpha}{\infty}\right)^2 \\ : w \;:& \alpha^2(\alpha-1)(\alpha-2) & \ldots \\ : \theta \;:& \alpha(\alpha-1)(\alpha-2) & \left(1-\frac{\alpha}{\infty}\right), \end{array}$$

where, for the sake of homogeneity, I have introduced the factors $\left(1-\frac{\alpha}{\infty}\right)$ and $\left(1-\frac{\alpha}{\infty}\right)^2$; viz. we have x, y, z, w, θ proportional to quartic functions of the arbitrary parameter α, or the curve is a unicursal quartic. Writing in the equations $\alpha = 0, 1, 2, \infty$ successively, we see that this quartic curve passes through the four points 123, 234, 341, 412 (intersecting at these points the lines 13 and 24 respectively); and writing also $\alpha = 1 \pm i$ we see that the curve passes through the points P, Q, the coordinates of which now are

$$x = y = z = w = (1 \pm i)\,\theta.$$

It should admit of being proved by general considerations that, in 4-dimensional geometry when 4 quadric surfaces partially intersect in two lines, the residual intersection consists of 2 points; and that, when they intersect in the two lines and in a unicursal quartic met twice by each of the lines, there is no residual intersection—but this theory has not yet been developed.

596.

NOTE ON A THEOREM OF JACOBI'S FOR THE TRANSFORMATION OF A DOUBLE INTEGRAL.

[From the *Messenger of Mathematics*, vol. IV. (1875), pp. 92—94.]

JACOBI, in the Memoir "De Transformatione Integralis Duplicis..." &c., *Crelle*, t. VIII. (1832) pp. 253—279 and 321—357, [*Ges. Werke*, t. III., pp. 91—158], after establishing a theorem which includes the addition-theorem of elliptic functions, viz. this last is "the differential equation

$$\frac{d\eta}{\sqrt{(G'^2\cos^2\eta+G''^2\sin^2\eta-G^2)}}+\frac{d\theta}{\sqrt{(G'^2\cos^2\theta+G''^2\sin^2\theta-G^2)}},$$

has for its *complete* integral

$$G+G'\cos\eta\cos\theta+G''\sin\eta\sin\theta=0,"$$

{observe, as to the integral being complete, that the differential equation contains only the constant $G^2-G'^2\div(G^2-G''^2)$, whereas the integral equation contains the two constants $G'\div G$ and $G''\div G$}, obtains a corresponding theorem for double integrals; viz. this, in the corresponding special case, is as follows: If the variables $(\phi,\ \psi)$ and $(\eta,\ \theta)$ are connected by the two equations

$$\left.\begin{array}{ll}\alpha & =0,\\ +\alpha'\cos\phi & .\cos\eta\\ +\alpha''\sin\phi\cos\psi & .\sin\eta\cos\theta\\ +\alpha'''\sin\phi\sin\psi & .\sin\eta\sin\theta\end{array}\right| \begin{array}{ll}\beta & =0,\\ +\beta'\cos\phi & .\cos\eta\\ +\beta''\sin\phi\cos\psi & .\sin\eta\cos\theta\\ +\beta'''\sin\phi\sin\psi & .\sin\eta\sin\theta\end{array}$$

and if putting for shortness

$$\begin{array}{ll}\alpha''\beta'''-\alpha'''\beta''=f, & \alpha\beta'-\alpha'\beta=a,\\ \alpha'''\beta'-\alpha'\beta'''=g, & \alpha\beta''-\alpha''\beta=b,\\ \alpha'\beta''-\alpha''\beta'=h, & \alpha\beta'''-\alpha'''\beta=c,\end{array}$$

(whence $af+bg+ch=0$);

$$\begin{aligned}R^2 = \ & f^2 (\sin\phi\cos\psi)^2 (\sin\phi\sin\psi)^2 \\ & + g^2 (\sin\phi\cos\psi)^2 (\cos\phi)^2 \\ & + h^2 (\cos\phi)^2 (\sin\phi\cos\psi)^2 \\ & - a^2 (\cos\phi)^2 \\ & - b^2 (\sin\phi\cos\psi)^2 \\ & - c^2 (\sin\phi\sin\psi)^2,\end{aligned}$$

$$\begin{aligned}S^2 = \ & f^2 (\sin\eta\cos\theta)^2 (\sin\eta\sin\theta)^2 \\ & + g^2 (\sin\eta\sin\theta)^2 (\cos\eta)^2 \\ & + h^2 (\cos\eta)^2 (\sin\eta\cos\theta)^2 \\ & - a^2 (\cos\eta)^2 \\ & - b^2 (\sin\eta\cos\theta)^2 \\ & - c^2 (\sin\eta\sin\theta)^2,\end{aligned}$$

then we have

$$\frac{\sin\phi\, d\phi\, d\psi}{R} = \frac{\sin\eta\, d\eta\, d\theta}{S}.$$

And it may be added that the integral equations are, so to speak, a complete integral of the differential relation; viz. in virtue of the identity $af + bg + ch = 0$, the differential relation contains really only four constants; the integral relations contain the six constants $\alpha : \alpha' : \alpha'' : \alpha'''$ and $\beta : \beta' : \beta'' : \beta'''$, or we have *two* constants introduced by the integration.

The best form of statement is, in the first theorem, to write x, y for $\cos\eta, \sin\eta, (x^2+y^2=1)$, ξ, η for $\cos\theta, \sin\theta, (\xi^2+\eta^2=1)$, and similarly in the second theorem to introduce the variables x, y, z connected by $x^2+y^2+z^2=1$, and ξ, η, ζ connected by $\xi^2+\eta^2+\zeta^2=1$; then in the first theorem $d\eta, d\theta$ represent elements of circular arc, and in the second theorem $\sin\phi\, d\phi\, d\psi$ and $\sin\eta\, d\eta\, d\theta$ represent elements of spherical surface, and the theorems are:

I. If (x, y) are coordinates of a point on the circle $x^2+y^2=1$, and (ξ, η) coordinates of a point on the circle $\xi^2+\eta^2=1$, and if $ds, d\sigma$ are the corresponding circular elements, then

$$\frac{ds}{\sqrt{(ax^2+by^2-c)}} = \frac{d\sigma}{\sqrt{(a\xi^2+b\eta^2-c)}},$$

has for its complete integral

$$ax\xi + by\eta - c = 0.$$

II. If (x, y, z) are coordinates of a point on the sphere $x^2+y^2+z^2=1$, and (ξ, η, ζ) coordinates of a point on the sphere $\xi^2+\eta^2+\zeta^2=1$; and if ds, $d\sigma$ are the corresponding spherical elements, and

$$\begin{aligned} \beta\gamma'-\beta'\gamma&=f, & \alpha\delta'-\alpha'\delta&=a,\\ \gamma\alpha'-\gamma'\alpha&=g, & \beta\delta'-\beta'\delta&=b,\\ \alpha\beta'-\alpha'\beta&=h, & \gamma\delta'-\gamma'\delta&=c, \end{aligned}$$
$$(\text{whence } af+bg+ch=0);$$

and for shortness

$$\begin{aligned} S^2&=f^2y^2z^2+g^2z^2x^2+h^2x^2y^2-a^2x^2-b^2y^2-c^2z^2,\\ \Sigma^2&=f^2\eta^2\zeta^2+g^2\zeta^2\xi^2+h^2\xi^2\eta^2-a^2\xi^2-b^2\eta^2-c^2\zeta^2, \end{aligned}$$

then the differential relation

$$\frac{ds}{\sqrt{(S)}}=\frac{d\sigma}{\sqrt{(\Sigma)}},$$

has for its complete integral the system

$$\begin{aligned} \alpha x\xi+\beta y\eta+\gamma z\zeta+\delta&=0,\\ \alpha' x\xi+\beta' y\eta+\gamma' z\zeta+\delta'&=0, \end{aligned}$$

where by complete integral is meant a system of two equations containing two arbitrary constants.

597.

ON A DIFFERENTIAL EQUATION IN THE THEORY OF ELLIPTIC FUNCTIONS.

[From the *Messenger of Mathematics*, vol. IV. (1875), pp. 110—113.]

THE differential equation

$$Q^2 - Q\left(k + \frac{1}{k}\right) - 3 = 3(1-k^2)\frac{dQ}{dk},$$

considered *ante*, p. 69, [594, this volume, p. 244], belongs to a class of equations transformable into linear equations of the second order, and consequently is such that, knowing a particular solution, we can obtain the general solution.

In fact, assuming

$$Q = -3(1-k^2)\frac{1}{z}\frac{dz}{dk},$$

the equation becomes

$$9(1-k^2)^2\frac{1}{z^2}\left(\frac{dz}{dk}\right)^2 + 3(1-k^2)\left(k+\frac{1}{k}\right)\frac{1}{z}\frac{dz}{dk} - 3$$
$$= 3(1-k^2)\left\{3(1-k^2)\frac{1}{z^2}\frac{dz}{dk^2} + 6k\frac{1}{z}\frac{dz}{dk} - 3(1-k^2)\frac{1}{z}\frac{d^2z}{dk^2}\right\},$$

viz. omitting the terms in $\frac{1}{z^2}\left(\frac{dz}{dk}\right)^2$ which destroy each other, and dividing by $3(1-k^2)$, this is

$$\left(k+\frac{1}{k}\right)\frac{1}{z}\frac{dz}{dk} - \frac{1}{1-k^2} = 6k\frac{1}{z}\frac{dz}{dk} - 3(1-k^2)\frac{1}{z}\frac{d^2z}{dk^2},$$

or finally

$$3(1-k^2)\frac{d^2z}{dk^2} + \frac{1-5k^2}{k}\frac{dz}{dk} - \frac{1}{1-k^2}z = 0.$$

But knowing a particular value of Q we have

$$z = \exp. \left\{ -\tfrac{1}{8} \int \frac{Q dz}{1-k^2} \right\},$$

a particular value of z, and thence in the ordinary manner the general value of z, giving the general value of Q.

The solution given in my former paper may be exhibited in a more simple form by introducing, instead of k, the variable α connected with it by the equation $k^2 = \frac{\alpha^3(2+\alpha)}{1+2\alpha}$. We have in fact, *Fundamenta Nova*, p. 25, [Jacobi's *Ges. Werke*, t. I., p. 76],

$$u^8 = \alpha^3 \frac{2+\alpha}{1+2\alpha}, \quad = k^2,$$

$$v^8 = \alpha \left(\frac{2+\alpha}{1+2\alpha}\right)^3, \quad = \lambda^2,$$

viz. these expressions of u, v in terms of the parameter α, are equivalent to, and replace, the modular equation $u^4 - v^4 + 2uv(1 - u^2v^2) = 0$. We thence obtain

$$u^8 v^8 = \frac{\alpha^4 (2+\alpha)^4}{(1+2\alpha)^4}, \quad \frac{v^8}{u^8} = \frac{(2+\alpha)^2}{\alpha^2 (1+2\alpha)^2},$$

that is,

$$uv = \surd(\alpha) \sqrt{\left(\frac{2+\alpha}{1+2\alpha}\right)}, \quad \frac{v^2}{u^2} = \frac{1}{\surd(\alpha)} \sqrt{\left(\frac{2+\alpha}{1+2\alpha}\right)},$$

and the particular solution, $Q = \frac{v^2}{u^2} + 2uv$, becomes

$$Q = \frac{1}{\surd(\alpha)} \surd(1+2\alpha \,.\, 2+\alpha), \quad = \sqrt{\left\{5 + 2\left(\alpha + \frac{1}{\alpha}\right)\right\}}.$$

Introducing into the differential equation α in place of k, this is found to be

$$Q^2 - Q \frac{\frac{1}{\alpha^2} + \alpha^2 + 2\left(\frac{1}{\alpha} + \alpha\right)}{\sqrt{\left\{5 + 2\left(\alpha + \frac{1}{\alpha}\right)\right\}}} - 3 = (1-\alpha^2) \sqrt{\left\{5 + 2\left(\alpha + \frac{1}{\alpha}\right)\right\}} \frac{dQ}{d\alpha}.$$

But from this form it at once appears that it is convenient in place of α to introduce the new variable β, $= \alpha + \frac{1}{\alpha}$; the equation thus becomes

$$Q^2 + Q \frac{2 - 2\beta - \beta^2}{\surd(5+2\beta)} - 3 = (4-\beta^2) \surd(5+2\beta) \frac{dQ}{d\beta},$$

satisfied by $Q = \surd(5+2\beta)$; or, what is the same thing, writing $5 + 2\beta = \gamma^2$, that is, $\beta = -\frac{5}{2} + \gamma^2$, the equation becomes

$$4Q^2 + \frac{Q}{\gamma}(3 + 6\gamma^2 - \gamma^4) - 12 = -(\gamma^2 - 1)(\gamma^2 - 9) \frac{dQ}{d\gamma},$$

satisfied by $Q = \gamma$.

Writing here

$$Q = \tfrac{1}{4}(\gamma^2-1)(\gamma^2-9)\frac{1}{z}\frac{dz}{d\gamma},$$

we have for z the equation

$$(\gamma^2-1)(\gamma^2-9)\frac{d^2z}{d\gamma^2} + (3\gamma^4-14\gamma^2+3)\frac{dz}{d\gamma} - \frac{48}{(\gamma^2-1)(\gamma^2-9)}z = 0,$$

satisfied by

$$z = \left(\frac{\gamma^2-9}{\gamma^2-1}\right)^{\frac{1}{4}}.$$

[In fact, this value gives

$$z = (\gamma^2-9)^{\frac{1}{4}}(\gamma^2-1)^{-\frac{1}{4}},$$

$$\frac{dz}{d\gamma} = 4\gamma(\gamma^2-9)^{-\frac{3}{4}}(\gamma^2-1)^{-\frac{5}{4}},$$

$$\frac{d^2z}{d\gamma^2} = (-12\gamma^4+57\gamma^2+36)(\gamma^2-9)^{-\frac{7}{4}}(\gamma^2-1)^{-\frac{9}{4}},$$

which verify the equation as they should do.]

Representing for a moment the differential equation by $A\dfrac{d^2z}{d\gamma^2} + B\dfrac{dz}{d\gamma} + Cz = 0$, and putting $z_1 = \left(\dfrac{\gamma^2-9}{\gamma^2-1}\right)^{\frac{1}{4}}$, then assuming $z = z_1\int y\,d\gamma$, we find

$$A\left(z_1\frac{dy}{d\gamma} + 2y\frac{dz_1}{d\gamma}\right) + Byz_1 = 0,$$

that is,

$$\frac{1}{y}\frac{dy}{d\gamma} + \frac{2}{z_1}\frac{dz_1}{d\gamma} + \frac{B}{A} = 0,$$

viz.

$$\frac{1}{y}\frac{dy}{d\gamma} + \frac{2}{z_1}\frac{dz_1}{d\gamma} + \frac{3\gamma^4-14\gamma^2+3}{(\gamma^2-1)(\gamma^2-9)} = 0,$$

or

$$\frac{1}{y}\frac{dy}{d\gamma} + \frac{2}{z_1}\frac{dz_1}{d\gamma} + 3 + \frac{1}{\gamma^2-1} + \frac{15}{\gamma^2-9} = 0\,;$$

whence, integrating

$$\log yz_1^2 + 3\gamma - \tfrac{1}{2}\log\frac{\gamma+1}{\gamma-1} - \tfrac{5}{2}\log\frac{\gamma+3}{\gamma-3} = 0,$$

that is,

$$\begin{aligned} y &= e^{-3\gamma}\frac{1}{z_1^2}\left(\frac{\gamma+1}{\gamma-1}\right)^{\frac{1}{2}}\left(\frac{\gamma+3}{\gamma-3}\right)^{\frac{5}{2}} \\ &= e^{-3\gamma}\left(\frac{\gamma-1\,.\,\gamma+1}{\gamma-3\,.\,\gamma+3}\right)^{\frac{1}{2}}\left(\frac{\gamma+1}{\gamma-1}\right)^{\frac{1}{2}}\left(\frac{\gamma+3}{\gamma-3}\right)^{\frac{5}{2}} \\ &= \frac{(\gamma+1)(\gamma+3)^2}{(\gamma-3)^3}e^{-3\gamma}. \end{aligned}$$

Hence, the general value of z is

$$z = K\left(\frac{\gamma^2-9}{\gamma^2-1}\right)^{\frac{1}{4}} \int_{\gamma_0} \frac{(\gamma+1)(\gamma+3)^2}{(\gamma-3)^3} e^{-3\gamma} d\gamma,$$

the constants of integration being K and γ_0, or, what is the same thing,

$$z = \left(\frac{\gamma^2-9}{\gamma^2-1}\right)^{\frac{1}{4}} \left\{C + D\int_{\infty} \frac{(\gamma+1)(\gamma+3)^2}{(\gamma-3)^3} e^{-3\gamma} d\gamma\right\},$$

the corresponding value of Q being

$$Q = \tfrac{1}{4}(\gamma^2-1)(\gamma^2-9)\frac{1}{z}\frac{dz}{d\gamma},$$

which contains the single arbitrary constant $\frac{D}{C}$; when this vanishes, we have the foregoing particular solution $Q=\gamma$.

I recall that the expression of γ is

$$\gamma = \sqrt{(5+2\beta)}, \quad = \sqrt{\left\{5+2\left(\alpha+\frac{1}{\alpha}\right)\right\}}, \quad = \frac{1}{\sqrt{(\alpha)}}\sqrt{\{(2+\alpha)(1+2\alpha)\}},$$

where α is connected with k by the relation

$$k^2 = \frac{\alpha^3(2+\alpha)}{1+2\alpha}.$$

598.

NOTE ON A PROCESS OF INTEGRATION.

[From the *Messenger of Mathematics*, vol. IV. (1875), pp. 149, 150.]

I HAD occasion to consider the integral

$$\int_0^R \frac{r^{s-1}\,dr}{\{r^2+e^2\}^{\frac{1}{2}s+q}},$$

where e is small in regard to R and q is negative. The integral is finite when $e=0$, and it might be imagined that it could be expanded in positive powers of e; and, assuming it to be thus expansible, that the process would simply be to expand under the integral sign in ascending powers of e, and integrate each term separately, so that the series would be in integer powers of e^2.

Take two particular cases. First, let

$$s=2,\quad q=-\tfrac{3}{2};$$

the integral is

$$\int_0^R r\sqrt{(r^2+e^2)}\,dr = \int_0^R dr\,(r^2+\tfrac{1}{2}e^2r^0-\tfrac{1}{8}e^4r^{-2}+\ldots)$$

$$=\tfrac{1}{3}R^3+\tfrac{1}{2}e^2R+\infty\,e^4+\ldots,$$

viz. the integral is not thus obtainable: the series is right as far as it goes, but the true expansion contains a term in e^3; and the failure of the series to give the true expansion is indicated by the appearance of infinite coefficients. In fact, the indefinite integral is $\frac{1}{3}(r^2+e^2)^{\frac{3}{2}}$; taking this between the limits, it is

$$\tfrac{1}{3}(R^2+e^2)^{\frac{3}{2}}-\tfrac{1}{3}e^3,\quad =\tfrac{1}{3}R^3+\tfrac{1}{2}e^2R+\ldots-\tfrac{1}{3}e^3.$$

Again, let $s=1$, $q=-2$; the integral is

$$\int_0^R (r^2+e^2)^{\frac{3}{2}}\,dr=\int_0^R (r^3+\tfrac{3}{2}e^2r+\tfrac{3}{8}e^4r^{-1}+\ldots)$$

$$=\tfrac{1}{4}R^4+\tfrac{3}{4}e^2R^2+\infty\,e^4+\ldots,$$

viz. the integral is not thus obtainable: the series is right as far as it goes, but the true expansion contains a term as $e^4 \log e$, and the failure is indicated by the infinite coefficients. In fact, the indefinite integral is

$$(\tfrac{1}{4}r^3 + \tfrac{5}{8}e^2 r)\sqrt{(r^2+e^2)} + \tfrac{3}{8}e^4 \log\{r + \sqrt{(r^2+e^2)}\},$$

which between the limits is

$$(\tfrac{1}{4}R^3 + \tfrac{5}{8}e^2 R)\sqrt{(R^2+e^2)} + \tfrac{3}{8}e^4 \log \frac{R+\sqrt{(R^2+e^2)}}{e},$$

$$= \tfrac{1}{4}R^4 + \tfrac{3}{4}e^2R^2 + \ldots - \tfrac{3}{8}e^4 \log e.$$

In the general case, the term causing the failure is Ke^{-2q} when q is fractional, and $Ke^{-2q}\log e$ when q is integral. As a step towards determining the entire expansion, I notice that, writing $x = \dfrac{e^2}{e^2+r^2}$ or $r = ex^{-\frac{1}{2}}(1-x)^{\frac{1}{2}}$, the value of the integral is

$$= \tfrac{1}{2}e^{-2q}\int_X^1 x^{q-1}(1-x)^{\frac{1}{2}s-1}\,dx,$$

where

$$X = \frac{e^2}{e^2+R^2}.$$

599.

A SMITH'S PRIZE DISSERTATION.

[From the *Messenger of Mathematics*, vol. IV. (1875), pp. 157—160.]

Write a dissertation on Bernoulli's Numbers and their use in Analysis.

The function $\frac{t}{e^t-1}+\frac{1}{2}t$ is an *even* function of t, as appears by expressing it in the form

$$\tfrac{1}{2}t\frac{e^t+1}{e^t-1},\quad =\tfrac{1}{2}t\frac{e^{\frac{1}{2}t}+e^{-\frac{1}{2}t}}{e^{\frac{1}{2}t}-e^{-\frac{1}{2}t}},$$

and its value for $t=0$ being obviously $=1$, we may write

$$\frac{t}{e^t-1}+\tfrac{1}{2}t=1+B_1\frac{t^2}{1.2}-B_2\frac{t^4}{1.2.3.4}+\&c.;$$

or, what is the same thing,

$$\frac{t}{e^t-1}=1-\tfrac{1}{2}t+B_1\frac{t^2}{1.2}-B_2\frac{t^4}{1.2.3.4}+\ldots+(-)^{n-1}B_n\frac{t^{2n}}{1.2\ldots 2n}+\ldots,$$

where the several coefficients B_1, B_2, B_3, &c., are, as is at once seen, rational fractions, and, as it may be shown, are all of them positive. These numerical coefficients B_1, B_2, B_3, &c., are called Bernoulli's numbers.

There is no difficulty in calculating directly the first few terms; viz. we have

$$\begin{aligned}\frac{t}{e^t-1}=\frac{1}{1+(\frac{1}{2}t+\frac{1}{6}t^2+\frac{1}{24}t^3+\ldots)}=1&-t\,(\tfrac{1}{2}+\tfrac{1}{6}t+\tfrac{1}{24}t^2+\tfrac{1}{120}t^3+\ldots)\\ &+t^2(\tfrac{1}{4}+\tfrac{1}{6}t+\tfrac{5}{72}t^2+\ldots)\\ &-t^3(\tfrac{1}{8}+\tfrac{1}{8}t+\ldots)\\ &+t^4(\tfrac{1}{16}+\ldots)\end{aligned}$$

$$\begin{aligned}=1+t(-\tfrac{1}{2})+t^2(-\tfrac{1}{6}+\tfrac{1}{4},\ =+\tfrac{1}{12})&+t^3(-\tfrac{1}{24}+\tfrac{1}{6}-\tfrac{1}{8}=0)\\ &+t^4(-\tfrac{1}{120}+\tfrac{5}{72}-\tfrac{1}{8}+\tfrac{1}{16}=-\tfrac{1}{720})+\ldots,\end{aligned}$$

viz.

$$=1-\tfrac{1}{2}t+\tfrac{1}{12}t^2-\tfrac{1}{720}t^4+\ldots,$$

which is therefore

$$=1-\tfrac{1}{2}t+\tfrac{1}{2}B_1t^2-\tfrac{1}{24}B_2t^4+\ldots,$$

and consequently

$$B_1=\tfrac{1}{6},\quad B_2=\tfrac{1}{30},$$

and so a few more terms might have been found.

But a more convenient method is to express the numbers in terms of the differences of 0^m by means of a general formula for the expansion of a function of e^t, viz. this is

$$\phi(e^t)=\phi(1+\Delta)\,e^{t.0},$$

where

$$e^{t.0}=0^0+\frac{t}{1}0^1+\frac{t^2}{1.2}0^2+\frac{t^3}{1.2.3}0^3+\&c.,$$

and the $\phi(1+\Delta)$ is to be applied to the terms 0^0, 0^1, 0^2, 0^3, &c. We have thus

$$\begin{aligned}\frac{t}{e^t-1}&=\frac{\log(e^t)}{e^t-1}\\&=\frac{\log(1+\Delta)}{\Delta}e^{t.0}\\&=\frac{\log(1+\Delta)}{\Delta}\left\{0^0+\frac{t}{1}0^1+\frac{t^2}{1.2}0^2+\&c.\ldots+\frac{t^{2n-1}}{1.2\ldots 2n-1}0^{2n-1}+\frac{t^{2n}}{1.2\ldots 2n}0^{2n}+\ldots\right\}.\end{aligned}$$

We have, as may be at once verified,

$$\frac{\log(1+\Delta)}{\Delta}0^0=1,\quad \frac{\log(1+\Delta)}{\Delta}0^1=-\tfrac{1}{2},$$

and by what precedes, since the coefficient of every higher odd power of t vanishes,

$$\frac{\log(1+\Delta)}{\Delta}0^{2n-1}=0;$$

and then, by comparing the even powers of t,

$$(-)^{n-1}B_n=\frac{\log(1+\Delta)}{\Delta}0^{2n},$$

that is,

$$(-)^{n-1}B_n=\left(1-\tfrac{1}{2}\Delta+\tfrac{1}{3}\Delta^2\ldots+\frac{1}{2n+1}\Delta^{2n}\right)0^{2n},$$

the series for $\dfrac{\log(1-\Delta)}{\Delta}$ being stopped at this point since $\Delta^{2n+1}0^{2n}=0$, &c. For instance, in the case $n=1$, we have

$$\begin{aligned}B_1=(1-\tfrac{1}{2}\Delta+\tfrac{1}{3}\Delta^2)\,0^2=\;&0^2\\&-\tfrac{1}{2}(1^2-0^2)\\&+\tfrac{1}{3}(2^2-2.1^2+0^2)\\=\;&-\tfrac{1}{2}+\tfrac{2}{3},\;=\tfrac{1}{6}\text{ as above.}\end{aligned}$$

The formula shows, not only that B_n is a rational fraction but that its denominator is at most = least common multiple of the numbers 2, 3, ..., $2n+1$; the actual denominator of the fraction in its least terms is, however, much less than this, there being as to its value a theorem known as Staudt's theorem. It does not obviously show that the Numbers are positive, or afford any indication of the rate of increase of the successive terms of the series.

These last requirements are satisfied by an expression for B_n as the sum of an infinite numerical series, which expression is obtained by means of the function $\cot\theta$, as follows:

We have

$$\frac{t}{e^t-1}+\tfrac{1}{2}t, \ =\tfrac{1}{2}t\frac{e^{\frac{1}{2}t}+e^{-\frac{1}{2}t}}{e^{\frac{1}{2}t}-e^{-\frac{1}{2}t}}=1+B_1\frac{t^2}{1.2}-B_2\frac{t^4}{1.2.3.4}+\&\text{c.},$$

or, writing herein $t=2i\theta$ $\{i=\sqrt{(-1)}$ as usual$\}$, this is

$$\theta\cot\theta=1-B_1\frac{2^2\theta^2}{1.2}-B_2\frac{2^4\theta^4}{1.2.3.4}-\&\text{c.}$$

But we have

$$\log\sin\theta=\log\theta+\log\left(1-\frac{\theta^2}{\pi^2}\right)+\log\left(1-\frac{\theta^2}{2^2\pi^2}\right)+\dots,$$

and thence, by differentiation,

$$\theta\cot\theta=1-\frac{2\theta^2}{\pi^2}\left\{\frac{1}{1^2\left(1-\frac{\theta^2}{1^2\pi^2}\right)}+\frac{1}{2^2\left(1-\frac{\theta^2}{2^2\pi^2}\right)}+..\right\}$$

$$=1-\frac{2\theta^2}{\pi^2}\left\{\frac{1}{1^2}+\frac{1}{2^2}+\&\text{c.}\right\}$$
$$-\frac{2\theta^4}{\pi^4}\left\{\frac{1}{1^4}+\frac{1}{2^4}+\&\text{c.}\right\}$$
$$-\&\text{c.}$$

Hence

$$B_n\frac{2^{2n}}{1.2\dots 2n}=\frac{2}{\pi^{2n}}\left\{\frac{1}{1^{2n}}+\frac{1}{2^{2n}}+\dots\right\},$$

that is,

$$B_n=\frac{2\,(1.2..2n)}{(2\pi)^{2n}}\left(\frac{1}{1^{2n}}+\frac{1}{2^{2n}}+\dots\right),$$

showing first, that B_n is positive, and next, that it rapidly increases with n, viz. n being large, we have

$$B_n=\frac{2\,(1.2\dots 2n)}{(2\pi)^{2n}},$$

or, instead of $1.2\dots 2n$ writing its approximate value $\sqrt{(2\pi)}\,.\,(2n)^{2n+\frac{1}{2}}e^{-2n}$, this is

$$B_n=4\sqrt{(n\pi)}\left(\frac{n}{\pi e}\right)^{2n}.$$

The result may of course be considered from the opposite point of view, as giving a determination of the sum $\frac{1}{1^{2n}}+\frac{1}{2^{2n}}+\&c. \ldots$ in terms of Bernoulli's Numbers, assumed to be known, viz. we thus have

$$\frac{1}{1^{2n}}+\frac{1}{2^{2n}}+\ldots=\frac{(2\pi)^{2n}}{2(1.2\ldots 2n)}B_n.$$

For instance, $n=1$,

$$\frac{1}{1^2}+\frac{1}{2^2}+\ldots=\frac{(2\pi)^2}{2.1.2}.\tfrac{1}{6}, \ =\frac{\pi^2}{6},$$

and this is one and a good instance of the use of Bernoulli's Numbers in Analysis.

Another and very important one is in the summation of a series, or say in the determination of $\Sigma u_x, =u_0+u_1+\ldots+u_{x-1}$; viz. starting from

$$\frac{1}{e^t-1}=\frac{1}{t}-\tfrac{1}{2}+\frac{B_1}{1.2}t-\frac{B_2}{1.2.3.4}t^3+\&c.,$$

and writing herein $t=d_x$, and therefore

$$\frac{1}{e^t-1}=\frac{1}{e^{d_x}-1}, \ =\frac{1}{\Delta} \text{ or } \Sigma,$$

and applying each side to a function u_x of x, we have

$$\Sigma u_x=C+\int dx\, u_x-\tfrac{1}{2}u_x+\frac{B_1}{1.2}d_xu_x-\frac{B_2}{1.2.3.4}d_x{}^3u_x+\ldots,$$

or taking the two sides each between the integer limits a, x,

$$u_a+u_{a+1}\ldots+u_{x-1}=\int_a^x dxu_x-\tfrac{1}{2}(u_x-u_a)+\frac{B_1}{1.2}(d_xu_x)_a^x-\frac{B_2}{1.2.3.4}(d_x{}^3u_x)_a^x+\ldots,$$

where if u_x is a rational and integral function the series on the right-hand side is finite. If for instance $u_x=x$, the equation is

$$a+(a+1)\ldots+(x-1)=\tfrac{1}{2}(x^2-a^2)-\tfrac{1}{2}(x-a),$$

viz.

$$\{1+2\ldots+(x-1)\}-\{1+2\ldots+(a-1)\}=\tfrac{1}{2}(x^2-n)-\tfrac{1}{2}(a^2-a),$$

which is right.

Applying the formula to the function $\log x$, we deduce theorems as to the Γ-function; and it is also interesting to apply it to $\frac{1}{x}$.

The above is given as a specimen of what might be expected in an examination: I remark as faults the omission to make it clear that B_n is a rational fraction; and the giving the series-formula as a formula for the convenient calculation of B_n. The omission to give the first-mentioned straightforward process of development strikes me as curious.

600.

THEOREM ON THE nth ROOTS OF UNITY.

[From the *Messenger of Mathematics*, vol. IV. (1875), p. 171.]

IF n be an odd prime, and α an imaginary nth root of unity, then

$$(-)^{\frac{1}{2}(n-1)}\, n - 1 = 4\left\{\frac{\alpha}{1+\alpha^2} + \frac{\alpha^2}{1+\alpha^4} + \frac{\alpha^3}{1+\alpha^6} \dots + \frac{\alpha^{\frac{1}{2}(n-1)}}{1+\alpha^{n-1}}\right\};$$

for instance,

$$n = 3, \quad -4 = 4\,\frac{\alpha}{1+\alpha^2},$$

verified at once by means of the equation $1+\alpha+\alpha^2=0$:

$$n = 5, \quad 4 = 4\left(\frac{\alpha}{1+\alpha^2} + \frac{\alpha^2}{1+\alpha^4}\right),$$

where the term in ( ) is

$$\frac{\alpha(1+\alpha^4) + \alpha^2(1+\alpha^2)}{(1+\alpha^2)(1+\alpha^4)},$$

that is,

$$= \frac{\alpha + 1 + \alpha^2 + \alpha^4}{1+\alpha^2+\alpha^4+\alpha}, \quad = 1:$$

and so in other cases.

601.

NOTE ON THE CASSINIAN.

[From the *Messenger of Mathematics*, vol. IV. (1875), pp. 187, 188.]

A SYMMETRICAL bicircular quartic has in general on the axis two nodofoci and four ordinary foci; viz. joining a nodofocus with either of the circular points at infinity, the joining line is a tangent to the curve at the circular point (and, this being a node of the curve, the tangent has there a three-pointic intersection): and joining an ordinary focus with either of the circular points at infinity, the joining line is at some other point a tangent to the curve, viz. an ordinary tangent of two-pointic intersection. In the case of the Cassinian, each circular point at infinity is a fleflecnode (node with an inflexion on each branch); of the four ordinary foci on the axis, one coincides with one nodofocus, another with the other nodofocus, and there remain only two ordinary foci on the axis; the so-called foci of the Cassinian are in fact the nodofoci, viz. each of these points is by what precedes a nodofocus *plus* an ordinary focus, and the line from either of these points to a circular point at infinity, *quà* tangent at a fleflecnode, has there a four-pointic intersection with the curve.

The analytical proof is very easy; writing the equation under the homogeneous form

$$\{(x-az)^2+y^2\}\{(x+az)^2+y^2\}-c^4z^4=0,$$

then the so-called foci are the points $(x=az,\ y=0)$, $(x=-az,\ y=0)$; at either of these, say the first of them, the line drawn to one of the circular points at infinity is $x=az+iy$, and substituting this value in the equation of the curve we obtain $z^4=0$, viz. the line is a tangent of four-pointic intersection; this implies that there is an inflexion at the point of contact on the branch touched by the line $x=az+iy$; and there is similarly an inflexion at the point of contact on the branch touched by the line $x=-az+iy$; viz. the circular point $x=iy$, $z=0$ is a fleflecnode; and similarly the circular point $x=-iy$, $z=0$, is also a fleflecnode.

To verify that there are on the axis only two ordinary foci, we write in the equation $x = \alpha z + iy$, and determine α by the condition that the resulting equation for y (which equation, by reason that the circular point $z = 0$, $x = iy$, is a node, will be a quadric equation only) shall have two equal roots; the equation is in fact

$$\{(\alpha - a)^2 z^2 + 2(\alpha - a) iyz\} \{(\alpha + a)^2 z^2 - 2(\alpha + a) iyz\} - c^2 z^4 = 0,$$

viz. throwing out the factor z^2, this is

$$(\alpha^2 - a^2) \{(\alpha - a) z + 2iy\} \{(\alpha + a) z + 2iy\} - c^4 z^2 = 0,$$

or, what is the same thing, it is

$$(\alpha^2 - a^2) \{(\alpha z + 2iy)^2 - a^2 z^2\} - c^4 z^2 = 0,$$

viz. it is

$$(2iy + \alpha z)^2 - \left(a^2 + \frac{c^4}{\alpha^2 - a^2}\right) z^2 = 0.$$

The condition in order that this may have equal roots is

$$a^2 + \frac{c^4}{\alpha^2 - a^2} = 0, \text{ that is, } \alpha^2 = a^2 - \frac{c^4}{a^2};$$

hence α has only the two values $\pm\sqrt{\left(a^2 - \frac{c^4}{a^2}\right)}$, viz. there are only two ordinary foci.

602.

ON THE POTENTIALS OF POLYGONS AND POLYHEDRA.

[From the *Proceedings of the London Mathematical Society*, vol. VI. (1874—1875), pp. 20—34. Read December 10, 1874.]

THE problem of the attraction of polyhedra is treated of by Mehler, *Crelle*, t. LXVI. pp. 375—381 (1866); but the results here obtained are exhibited under forms, which are very different from his and which give rise to further developments of the theory.

General Formulæ for the Potentials of a Cone and a Shell.

1. The law of attraction is taken to be according to the inverse square of the distance; and I commence with the general case of a cone standing upon any portion of a surface Σ as its base, and attracting a point at its vertex, the cone being considered as a mass of density unity.

2. Considering, in the first instance, an element of mass, the position of which is determined by its distance r from the vertex (or origin) and by two angular coordinates defining the position of the radius vector r, then the element is $= r^2\, dr\, d\omega$ (where $d\omega$ is the element of solid angle, or surface of the unit-sphere), and the corresponding element of potential is $\dfrac{1}{r} r^2\, dr\, d\omega, = r\, dr\, d\omega$; whence

$$V = \int r\, dr\, d\omega,$$

which, integrating from $r = 0$ to $r =$ its value at the surface, is

$$= \tfrac{1}{2}\int r^2\, d\omega,$$

where r now denotes the radius vector at a point of the surface, being, therefore, a given function of the two angular coordinates: and the remaining (double) integration is to be extended to all values of the angular coordinates belonging to a position of r within the conical surface which is the other boundary of the attracting mass, or say over the spherical aperture of the cone.

3. If the value of the radius vector at the surface is taken to be mr (m a constant), then we have obviously

$$V = \tfrac{1}{2}m^2 \int r^2\, d\omega\,;$$

and hence also, writing $m + dm$ instead of m, we obtain, for the potential of the portion of the shell lying between the similar and similarly situated surfaces Σ and Σ', belonging to the parameters m and $m + dm$ respectively, the value

$$V = m\, dm \int r^2\, d\omega\,;$$

this is $= 2\dfrac{dm}{m}$ into the potential of the cone; and we thus see that it is the same problem to determine the potential of the cone, and that of the subtended portion of the indefinitely thin shell included between the two surfaces.

4. The same result may be arrived at as follows: the element of solid angle $d\omega$ determines on the surface an element of surface $d\Sigma$, and if $d\nu$ be the corresponding normal thickness of the shell, then the element of mass is $= d\nu\, d\Sigma$, and the element of potential is $= \dfrac{1}{mr}\, d\nu\, d\Sigma$ (mr being, as before, the radius vector at the surface). Take α the complement of the inclination of the radius vector to the tangent plane—that is, α the inclination of the radius vector to the normal, or, what is the same thing, to the perpendicular from the origin on the tangent plane (whence, also, if mp be the length of this perpendicular, then $p = r\cos\alpha$). The shell-thickness in the direction of the radius vector is $= r\, dm$, or we have $d\nu = r\, dm \cos\alpha$; the element of potential is therefore $= \dfrac{dm}{m}\cos\alpha\, d\Sigma$. But $d\omega$ being the spherical aperture of the cone subtending the element $d\Sigma$, the perpendicular section at the distance mr is $= m^2 r^2 d\omega$; we have therefore $d\Sigma = \dfrac{1}{\cos\alpha}\, m^2 r^2\, d\omega$; and hence the element of potential is $= m\, dm\,.\, r^2\, d\omega$, or the potential of the subtended portion of the shell is as before, $= m\, dm \int r^2\, d\omega$.

5. It may be added that, integrating between the values m, n ($m > n$), we obtain $\tfrac{1}{2}(m^2 - n^2) \int r^2\, d\omega$ for the potential of the shell-portion included between the surfaces mr, nr; and if $n = 0$, then, as before, the potential of the cone is $= \tfrac{1}{2}m^2 \int r^2\, d\omega$.

Cone on a plane base, and plane figure.

6. Suppose that the surface Σ is a plane; the surface Σ' is, of course, a parallel plane. Taking here mp for the perpendicular distance of the plane Σ from the origin, then, if δ be the infinitesimal distance of the two planes from each other, we have $\delta = p\,dm$, that is, $dm = \frac{\delta}{p}$; the potential of the cone is, as before, $= \frac{1}{2}m^2 \int r^2\,d\omega$, and that of the plane figure, thickness δ, is $= \frac{m\delta}{p}\int r^2 d\omega$.

7. Taking, for greater convenience, $m = 1$, we have

$$\text{Potential of cone} \qquad = \tfrac{1}{2}\int r^2\,d\omega,$$

$$\text{Do.} \quad \text{of plane figure} = \frac{\delta}{p}\int r^2\,d\omega,$$

where p is now the perpendicular distance of the plane from the vertex; or if, as regards the plane figure, the infinitesimal thickness δ is taken as unity, then

$$\text{Potential of plane figure} = \frac{1}{p}\int r^2\,d\omega.$$

In each case r is the value of the radius vector corresponding to a point of the plane figure which is the base of the cone, and the integration extends over the spherical aperture of the cone.

8. If the position of the radius vector is determined by the usual angular coordinates, θ its inclination to the axis of z, and ϕ its azimuth from the plane of zx—viz. if we have

$$x = r \sin\theta \cos\phi,$$

$$y = r \sin\theta \sin\phi,$$

$$z = r \cos\theta;$$

then, as is well-known, $d\omega = \sin\theta\,d\theta\,d\phi$, and the integral $\int r^2\,d\omega$ is $= \int r^2 \sin\theta\,d\theta\,d\phi$.

Taking the inclination of p to the axes to be α, β, γ respectively, the equation of the plane which is the base of the cone is

$$x\cos\alpha + y\cos\beta + z\cos\gamma = p;$$

viz. we have

$$r\,[(\cos\alpha\cos\phi + \cos\beta\sin\phi)\sin\theta + \cos\gamma\cos\theta] = p;$$

that is,

$$r = \frac{p}{(\cos\alpha\cos\phi + \cos\beta\sin\phi)\sin\theta + \cos\gamma\cos\theta},$$

and the integral $\int r^2\,d\omega$ is therefore

$$=p^2\int\frac{\sin\theta\,d\theta\,d\phi}{[(\cos\alpha\cos\phi+\cos\beta\sin\phi)\sin\theta+\cos\gamma\cos\theta]^2};$$

and, in particular, if p coincide with the axis of z, so that the equation of the plane is $z=p$, then the integral is

$$=p^2\int\frac{\sin\theta\,d\theta\,d\phi}{\cos^2\theta}.$$

9. The integration in regard to θ can be at once performed; viz. in the latter case we have $\int\frac{\sin\theta\,d\theta}{\cos^2\theta}=\sec\theta$; and in the former case, writing, as we may do,

$$(\cos\alpha\cos\phi+\cos\beta\sin\phi)\sin\theta+\cos\gamma\cos\theta=M\cos(\theta-N),$$

then

$$\int\frac{\sin\theta\,d\theta}{[(\cos\alpha\cos\phi+\cos\beta\sin\phi)\sin\theta+\cos\gamma\cos\theta]^2}=\frac{1}{M^2}\int\frac{\sin(\theta-N+N)\,d\theta}{\cos^2(\theta-N)}$$

$$=\frac{1}{M^2}\left[\cos N\int\frac{\sin(\theta-N)\,d\theta}{\cos^2(\theta-N)}+\sin N\int\frac{d\theta}{\cos(\theta-N)}\right]$$

$$=\frac{1}{M^2}\left[\cos N\sec(\theta-N)+\sin N\log\tan\{\tfrac{1}{4}\pi+\tfrac{1}{2}(\theta-N)\}\right].$$

Case of a Polyhedron or a Polygon.

10. Consider now the pyramid, vertex the origin O, standing on a polygonal base. Letting fall from the vertex a perpendicular OM on the base of the pyramid, and drawing planes through OM and the several vertices of the polygon, we thus divide the pyramid into triangular pyramids; viz. AB being any side of the polygon, a component pyramid (or tetrahedron) will be $OMAB$, vertex O and base MAB, where MO is a perpendicular at M to the triangular base MAB. And drawing through MO a plane at right angles to AB, meeting it in D (viz. MD is the perpendicular from M on the base AB of the triangle), we divide the triangular pyramid into two pyramids $OMAD$, $OMBD$, each having for its base a right-angled triangle; viz. the vertex is O, the base is the triangle ADM (or, as the case may be, BDM) right-angled at D, and OM is a perpendicular at the vertex M to the plane of the triangle. It is to be observed that, in speaking of the original pyramid as thus divided, we mean that the pyramid is the sum of the component pyramids taken each with the proper sign, + or −, as the case may be.

11. In the case of a polyhedron, this is in the like sense divisible into pyramids having for the common vertex the origin or point O, and standing on the several faces respectively; hence the polyhedron is ultimately divisible into triangular pyramids such as $OADM$, where ADM is a triangle right-angled at D, and where OM is a perpendicular at M to the plane of the triangle. Hence the potential of the polyhedron

in regard to the point O depends upon that of the pyramid $OADM$; and (what is the same thing) the potential of any plane polygon in regard to the point O depends upon that of the right-angled triangle ADM, situate as above in regard to the point O. I

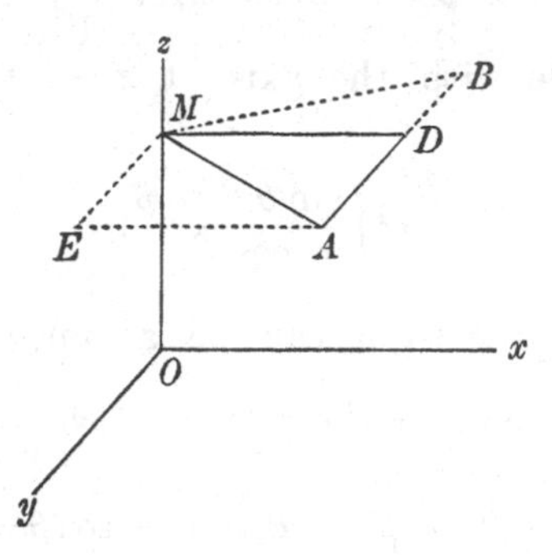

take $OM = h$, $MD = f$, $DA = g$; viz. supposing, as we may do, that the plane of the triangle is parallel to that of xy, the point M on the axis of z, and the side MD parallel to the axis of x, then f, g, h will be the coordinates of the point A.

Formulæ for component triangular Pyramid, and Triangle.

12. Writing, as above, $x = r \sin\theta \cos\phi$, $y = r\sin\theta\sin\phi$, $z = r\cos\theta$, and observing that h is the perpendicular distance originally called p, we have, for the potential of the pyramid,

$$V = \tfrac{1}{2}\int r^2\, d\omega = \tfrac{1}{2}h^2 \int \frac{\sin\theta\, d\theta\, d\phi}{\cos^2\theta}$$

$$= \tfrac{1}{2}h^2 \int d\phi\, (\sec\theta),$$

where, ϕ being regarded as a given angle, the integral expression $\sec\theta$ must be taken from $\theta = 0$ to the value of θ corresponding to a point in the side AD. For any such point we have $f = r\sin\theta\cos\phi$, $h = r\cos\theta$, that is, $\frac{f}{h} = \tan\theta\cos\phi$, or the required value of θ is $= \tan^{-1}\frac{f}{h\cos\phi}$, and consequently that of $\sec\theta$ is

$$\sqrt{1 + \frac{f^2}{h^2\cos^2\phi}},\quad = \frac{1}{h\cos\phi}\sqrt{f^2 + h^2\cos^2\phi},$$

or, as this may also be written,

$$= \frac{1}{h}\sqrt{f^2 + h^2 + f^2\tan^2\phi};$$

hence

$$V = \tfrac{1}{2}h\int \left(\sqrt{f^2 + h^2 + f^2\tan^2\phi} - h\right) d\phi.$$

13. The first term of the integral, writing therein for a moment $\tan\phi = x$, is

$$\int \frac{(f^2+h^2+f^2x^2)\,dx}{(1+x^2)\sqrt{f^2+h^2+f^2x^2}},$$

$$= f^2\int\frac{dx}{\sqrt{f^2+h^2+f^2x^2}} + h^2\int\frac{dx}{(1+x^2)\sqrt{f^2+h^2+f^2x^2}}$$

$$= f\log\left(fx+\sqrt{f^2+h^2+f^2x^2}\right) + h\tan^{-1}\frac{hx}{\sqrt{f^2+h^2+f^2x^2}}.$$

Hence, replacing x by its value, we have

$$V = \tfrac{1}{2}h\left\{h\tan^{-1}\frac{h\tan\phi}{\sqrt{f^2+h^2+f^2\tan^2\phi}} + f\log\left(f\tan\phi+\sqrt{f^2+h^2+f^2\tan^2\phi}\right) - h\phi\right\},$$

to be taken from $\phi = 0$ to the value of ϕ corresponding to the point A; viz. we have here $f = r\sin\theta\cos\phi$, $g = r\sin\theta\sin\phi$, $h = r\cos\theta$, and thence $\tan\phi = \frac{g}{f}$ or $f\tan\phi = g$; whence, writing for shortness, $s = \sqrt{f^2+g^2+h^2}$ (viz. s denotes the distance OA), we have

$$V = \tfrac{1}{2}h\left\{h\tan^{-1}\frac{gh}{fs} + f\log\frac{s+g}{\sqrt{f^2+h^2}} - h\tan^{-1}\frac{g}{f}\right\};$$

or, observing that

$$\frac{s+g}{s-g} = \frac{(s+g)^2}{\sqrt{f^2+h^2}},$$

this is

$$V = \tfrac{1}{2}h\left\{h\tan^{-1}\frac{gh}{fs} + \tfrac{1}{2}f\log\frac{s+g}{s-g} - h\tan^{-1}\frac{g}{f}\right\},$$

for the potential of the pyramid $OMDA$ in regard to the point O; by omitting the factor $\frac{1}{2}h$, we have

$$V = h\tan^{-1}\frac{gh}{fs} + \tfrac{1}{2}f\log\frac{s+g}{s-g} - h\tan^{-1}\frac{g}{f}$$

for the potential of the triangle MDA. The expression $\tan^{-1}$ denotes, here and elsewhere, an arc included between the limits $-\frac{\pi}{2}$, $+\frac{\pi}{2}$: it is therefore + or − according as the tangent is + or −.

Formulæ for rectangular Pyramid, and Rectangle.

14. Completing the rectangle $MDAE$, the potential of the triangle AME is obtained by interchanging the letters g and f; viz. we have

$$V = h\tan^{-1}\frac{fh}{gs} + \tfrac{1}{2}g\log\frac{s+f}{s-f} - h\tan^{-1}\frac{f}{g}$$

for the potential of the triangle MEA.

The sum of the two gives the potential of the rectangle $MDAE$; viz. for this rectangle, we have

$$V = h\left(\tan^{-1}\frac{gh}{fs} + \tan^{-1}\frac{fh}{gs} - \frac{\pi}{2}\right) + \tfrac{1}{2}f\log\frac{s+g}{s-g} + \tfrac{1}{2}g\log\frac{s+f}{s-f}.$$

But we have

$$\tan^{-1}\frac{gh}{fs} + \tan^{-1}\frac{hf}{gs} + \tan^{-1}\frac{fg}{hs} = \frac{\pi}{2};$$

for the function on the left hand is

$$= \tan^{-1}\frac{\dfrac{gh}{fs} + \dfrac{hf}{gs} + \dfrac{fg}{hs} - \dfrac{fgh}{s^3}}{1 - \dfrac{f^2}{s^2} - \dfrac{g^2}{s^2} - \dfrac{h^2}{s^2}};$$

viz. the denominator being $1 - \dfrac{f^2+g^2+h^2}{s^2}$, $=0$, the tangent of the arc is ∞, and the component arcs being each positive and less than $\dfrac{\pi}{2}$, the arc in question can only be $=\dfrac{\pi}{2}$. We have consequently

$$V = -h\tan^{-1}\frac{fg}{hs} + \tfrac{1}{2}f\log\frac{s+g}{s-g} + \tfrac{1}{2}g\log\frac{s+f}{s-f}$$

for the potential of the rectangle $MDAE$. And, multiplying this by $\tfrac{1}{2}h$, we have

$$V = -\tfrac{1}{2}h^2\tan^{-1}\frac{fg}{hs} + \tfrac{1}{4}hf\log\frac{s+g}{s-g} + \tfrac{1}{4}gh\log\frac{s+f}{s-f}$$

for the potential of the rectangular pyramid, vertex O and base $MDAE$.

Formula for the Cuboid.

15. Completing the rectangular parallelopiped, or, say for shortness, the "cuboid," the sides whereof are (f, g, h); this breaks up into three pyramids, standing on the rectangles fg, gh, and hf respectively; and the potentials for the last two pyramids are at once obtained from the last-mentioned expression of V by mere cyclical interchanges of the letters. Adding the three expressions, we obtain

$$V = \tfrac{1}{2}gh\log\frac{s+f}{s-f} + \tfrac{1}{2}hf\log\frac{s+g}{s-g} + \tfrac{1}{2}fg\log\frac{s+h}{s-h} - \tfrac{1}{2}f^2\tan^{-1}\frac{gh}{fs} - \tfrac{1}{2}g^2\tan^{-1}\frac{hf}{gs} - \tfrac{1}{2}h^2\tan^{-1}\frac{fg}{hs}$$

for the potential of the cuboid.

Group of Results, for Point, Line, Rectangle, and Cuboid.

16. It is convenient to prefix two results, that for the potential of the point A (mass taken to be unity), and that for the potential of the line AE (density taken to

be unity, or mass of an element of length dx, taken to be $=dx$). We have, the attracted point being always at O,

$$\text{Potential of point } A \qquad = \frac{1}{s}, \qquad (s=\sqrt{f^2+g^2+h^2}, \text{ as before}),$$

$$\text{Potential of line } AE \qquad = \tfrac{1}{2}\log\frac{s+f}{s-f},$$

$$\text{Potential of rectangle } MDAE = \tfrac{1}{2}g\log\frac{s+f}{s-f}+\tfrac{1}{2}f\log\frac{s+g}{s-g}-h\tan^{-1}\frac{fg}{hs},$$

$$\text{Potential of cuboid} \qquad = \tfrac{1}{2}gh\log\frac{s+f}{s-f}+\tfrac{1}{2}hf\log\frac{s+g}{s-g}+\tfrac{1}{2}fg\log\frac{s+h}{s-h}$$

$$-\tfrac{1}{2}f^2\tan^{-1}\frac{gh}{fs}-\tfrac{1}{2}g^2\tan^{-1}\frac{hf}{gs}-\tfrac{1}{2}h^2\tan^{-1}\frac{fg}{hs},$$

which functions may be called $A(f, g, h)$, $B(f, g, h)$, $C(f, g, h)$, and $D(f, g, h)$ respectively. It is to be observed that f, g, h are taken to be each of them positive, and that s denotes in every case the positive value of $\sqrt{f^2+g^2+h^2}$; for a symmetrically situated body, corresponding to negative values of each or any of these quantities, the potential has in each case its original value, without change of sign. But B is an odd function as regards f, C an odd function as regards f or g, D an odd function as regards f, g, or h; for example, $C(-f, g, \pm h)$ and $C(f, -g, \pm h)$ are each $=-C(f, g, h)$, and therefore of course $C(-f, -g, \pm h)=C(f, g, h)$.

Extension to case where the attracted point has an arbitrary position.

17. The attracted point has thus far been considered as in a definite position in regard to the attracting mass; but it is easy to pass to the general case of any relative position whatever. Thus, for a line AB, if M be the foot of the perpendicular let fall from the point O, and if, to fix the ideas, the order of succession of the three points is A, B, M, then, with respect to the point O,

$$\text{line } AB = \text{line } AM - \text{line } BM.$$

A B M

Taking the y- and z-coordinates to be b, c, the x-coordinates for the points A, B, M to be x_0, x_1, a respectively, and in the figure $a>x_1$, $x_1>x_0$, then $a-x_0$, $a-x_1$ are each of them positive, $a-x_0$ being the greater, the potential of the line AM is $=B(a-x_0, b, c)$, that of BM is $=B(a-x_1, b, c)$, and the potential of the whole line is

$$=B(a-x_0, b, c)-B(a-x_1, b, c);$$

viz. this formula is proved for the case where M is situate as in the figure. But supposing that A and B retain their relative position (viz. $x_1>x_0$), then the formula holds good for any other position of M; thus, if M be between the points A, B—viz. if the order is A, M, B—then

$$\text{line } AB = \text{line } AM + \text{line } BM,$$

and potential is

$$= B(a - x_0,\ b,\ c) + B(x_1 - a,\ b,\ c),$$

where the second term is $= - B(a - x_1,\ b,\ c)$; and so, if the order is M, A, B, then

$$\text{line } AB = \text{line } BM - \text{line } AM,$$

and the potential is $B(x_1 - a,\ b,\ c) - B(x_0 - a,\ b,\ c)$, which is

$$= - B(a - x_1,\ b,\ c) + B(a - x_0,\ b,\ c).$$

18. Similarly for a rectangle $ABCD$, if M, the foot of the perpendicular from the point O, has the position shown in the figure, then

$$\begin{aligned} \text{rectangle } AD = &\ \ \text{rectangle } MC \\ &- \text{rectangle } MA \\ &- \text{rectangle } MD \\ &+ \text{rectangle } MB, \end{aligned}$$

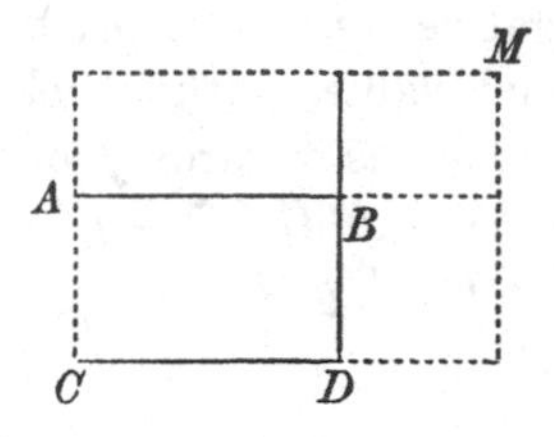

where O is a point on the perpendicular at the common vertex M of the four rectangles; and the resulting expression for the rectangle AD will apply to any position of the point M.

19. And in like manner for a cuboid; taking the point O in any determinate position, the cuboid may be decomposed into eight cuboids (each with the sign + or − as the case may be) having the point O for a common vertex; and the resulting expression for the potential will apply to any position whatever of the point M.

20. The results may be collected and exhibited as follows:—the coordinates of the attracted point are a, b, c; and it is assumed that $x_1 > x_0$, $y_1 > y_0$, $z_1 > z_0$, (viz. for x the order is $+\infty$, x_1, x_0, $-\infty$, and so for y and z respectively).

Potential of point (x, y, z) is $= A(a - x,\ b - y,\ c - z)$;

Potential of line (x_1, y, z), (x_0, y, z) is

$$= B(a - x_0,\ b - y,\ c - z) - B(a - x_1,\ b - y,\ c - z);$$

Potential of rectangle (x_1, y_1, z), (x_0, y_1, z), (x_1, y_0, z), (x_0, y_0, z) is

$$\begin{aligned} = &\ \ C(a - x_0,\ b - y_0,\ c - z) \\ &- C(a - x_0,\ b - y_1,\ c - z) \\ &- C(a - x_1,\ b - y_0,\ c - z) \\ &+ C(a - x_1,\ b - y_1,\ c - z); \end{aligned}$$

$$\begin{array}{lll}\text{Potential of cuboid} & (x_1, y_1, z_1), (x_0, y_1, z_1) \text{ is } = & D(a-x_0, b-y_0, c-z_0)\\ & (x_1, y_0, z_1), (x_0, y_0, z_1) & -D(a-x_1, b-y_0, c-z_0)\\ & (x_1, y_1, z_0), (x_0, y_1, z_0) & -D(a-x_0, b-y_1, c-z_0)\\ & (x_1, y_0, z_0), (x_0, y_0, z_0) & +D(a-x_1, b-y_1, c-z_0)\\ & & -D(a-x_0, b-y_0, c-z_1)\\ & & +D(a-x_1, b-y_0, c-z_1)\\ & & +D(a-x_0, b-y_1, c-z_1)\\ & & -D(a-x_1, b-y_1, c-z_1).\end{array}$$

21. These are connected together as follows, viz.:—

$$\text{Potential of line} \quad = \int_{x_0}^{x_1} dx \text{ Potential of point,}$$

$$\text{Potential of rectangle} = \int_{y_0}^{y_1} dy \text{ Potential of line,}$$

$$\text{Potential of cuboid} \quad = \int_{z_0}^{z_1} dz \text{ Potential of rectangle,}$$

equations which are in fact of the form

$$B(x, y, z) = \int dx\, A(x, y, z),$$

$$C(x, y, z) = \int dy\, B(x, y, z),$$

$$D(x, y, z) = \int dz\, C(x, y, z).$$

Differential properties of the functions A, B, C, D.

22. These relations, with other allied ones, may be verified as follows. Writing $r = \sqrt{x^2+y^2+z^2}$, the fundamental forms are

$$\log\frac{r+x}{r-x}, \quad \text{and} \quad \tan^{-1}\frac{yz}{rx}.$$

We have $d_x r = \frac{x}{r}$, &c., and thence

$$d_x \log\frac{r+x}{r-x} = \frac{1+\frac{x}{r}}{r+x} - \frac{-1+\frac{x}{r}}{r-x}, \ = \frac{1}{r}+\frac{1}{r}, \ = \frac{2}{r};$$

$$d_y \log\frac{r+x}{r-x} = \frac{\frac{y}{r}}{r+x} - \frac{\frac{y}{r}}{r-x}, \ = \frac{y}{r}\left(\frac{1}{r+x} - \frac{1}{r-x}\right), \ = \frac{-2xy}{r(r^2-x^2)};$$

&c. &c. &c.

$$d_x \tan^{-1}\frac{yz}{rx} = \frac{-yz\left(r+\frac{x^2}{r}\right)}{r^2x^2+y^2z^2} = \frac{-yz}{r}\,\frac{r^2+x^2}{r^2x^2+y^2z^2},$$

or, since

$$r^2+x^2=(r^2-y^2)+(r^2-z^2) \text{ and } r^2x^2+y^2z^2=(r^2-y^2)(r^2-z^2),$$

this is

$$=-\frac{yz}{r}\left(\frac{1}{r^2-y^2}+\frac{1}{r^2-z^2}\right);$$

$$d_y \tan^{-1}\frac{yz}{rx}=\frac{rx\,.\,z-yz\,.\,\dfrac{xy}{r}}{r^2x^2+y^2z^2}=\frac{xz}{r}\,\frac{r^2-y^2}{r^2x^2+y^2z^2},$$

which, the denominator being, as before, $(r^2-y^2)(r^2-z^2)$, is

$$=\frac{xz}{r}\,\frac{1}{r^2-z^2}.$$

It is now easy to form the following results:—

23. First,

$$u=A(x,\ y,\ z)=\frac{1}{r} \text{ (symmetrical)},$$

$$d_xu=-\frac{x}{r^3},\ \text{\&c.},$$

$$d_x{}^2u=\frac{3x^2}{r^5}-\frac{1}{r^3},\ \text{\&c.},\ d_xd_yu=\frac{3xy}{r^5},\ \text{\&c.},$$

and thence

$$(d_x{}^2+d_y{}^2+d_z{}^2)\,u=0.$$

24. Secondly,

$$u=B(x,\ y,\ z)=\tfrac{1}{2}\log\frac{r+x}{r-x} \text{ (symmetrical as to } y,\ z);$$

then

$$d_xu=\frac{1}{r}\left(=A(x,\ y,\ z)\right),\quad d_yu=\frac{-xy}{r(r^2-x^2)},\ \text{\&c.},$$

$$d_x{}^2u=-\frac{x}{r^3},\quad d_xd_yu=-\frac{y}{r^3},\quad d_yd_zu=\frac{xyz}{r^3(r^2-x^2)}+\frac{2xyz}{r(r^2-x^2)^2},\ \text{\&c.},$$

$$d_y{}^2u=\frac{-x}{r(r^2-x^2)}+\frac{2xy^2}{r(r^2-x^2)^2}+\frac{xy^2}{r^3(r^2-x^2)},\ \text{\&c.};$$

and thence

$$(d_x{}^2+d_y{}^2+d_z{}^2)\,u=-\frac{x}{r^3}-\frac{2x}{r(r^2-x^2)}+\frac{2x}{r(r^2-x^2)}+\frac{x}{r^3},\ =0.$$

25. Thirdly,

$$u=C(x,\ y,\ z)=\tfrac{1}{2}y\log\frac{r+x}{r-x}+\tfrac{1}{2}x\log\frac{r+y}{r-y}-z\tan^{-1}\frac{xy}{zr} \text{ (symmetrical as to } x,\ y)$$

$$d_yu=\tfrac{1}{2}\log\frac{r+x}{r-x}\left(=B(x,\ y,\ z)\right):$$

in verification whereof, observe that the remaining terms are

$$= -\frac{xy^2}{r \,.\, r^2 - x^2} + \frac{x}{r} - \frac{xz^2}{r} \frac{r^2 - y^2}{(r^2 - x^2)(r^2 - y^2)},$$

$$= \frac{x}{r}\left(-\frac{y^2}{r^2 - x^2} + 1 - \frac{z^2}{r^2 - x^2}\right),$$

$$= \frac{x}{r(r^2 - x^2)}(-y^2 + r^2 - x^2 - z^2),$$

which is $=0$;

$$d_z u = -\frac{xyz}{r(r^2 - x^2)} - \frac{xyz}{r(r^2 - y^2)} + \frac{xyz}{r} \frac{r^2 - x^2 + r^2 - y^2}{(r^2 - x^2)(r^2 - y^2)} - \tan^{-1}\frac{xy}{zr},$$

$$= -\tan^{-1}\frac{xy}{zr},$$

$$d_x^2 u = -\frac{xy}{r(r^2 - y^2)},$$

$$d_z^2 u = \frac{xy}{r} \frac{r^2 - x^2 + r^2 - y^2}{(r^2 - x^2)(r^2 - y^2)}, \quad = \frac{xy}{r}\left(\frac{1}{r^2 - x^2} + \frac{1}{r^2 - y^2}\right),$$

and thence

$$(d_x^2 + d_y^2 + d_z^2)\, u = 0.$$

26. Fourthly,

$$u = D(x, y, z) = \tfrac{1}{2} yz \log\frac{r+x}{r-x} + \tfrac{1}{2} zx \log\frac{r+y}{r-y} + \tfrac{1}{2} xy \log\frac{r+z}{r-z}$$

$$- \tfrac{1}{2} x^2 \tan^{-1}\frac{yz}{xr} - \tfrac{1}{2} y^2 \tan^{-1}\frac{zx}{yr} - \tfrac{1}{2} z^2 \tan^{-1}\frac{xy}{zr} \text{ (symmetrical)},$$

$$d_z u = \tfrac{1}{2} y \log\frac{r+x}{r-x} + \tfrac{1}{2} x \log\frac{r+y}{r-y} - z \tan^{-1}\frac{xy}{zr} = C(x, y, z),$$

$$d_z^2 u = -\tan^{-1}\frac{xy}{zr};$$

and thence

$$(d_x^2 + d_y^2 + d_z^2)\, u = -\tan^{-1}\frac{yz}{xr} - \tan^{-1}\frac{zx}{yr} - \tan^{-1}\frac{xy}{zr}$$

$$= -\tan^{-1}\frac{\dfrac{yz}{xr} + \dfrac{zx}{yr} + \dfrac{xy}{zr} - \dfrac{xyz}{r^3}}{1 - \dfrac{x^2}{r^2} - \dfrac{y^2}{r^2} - \dfrac{z^2}{r^2}};$$

viz. the denominator being $=0$, the arc is $\pm\frac{\pi}{2}$, or we have

$$(d_x^2 + d_y^2 + d_z^2)\, u = \mp\frac{\pi}{2},$$

the value being $-\frac{\pi}{2}$ if x, y, z are all three of them, or only one, positive; but $+\frac{\pi}{2}$ if they are all three of them, or only one, negative.

Application to the Potentials of the Point, the Line, the Rectangle, and the Cuboid.

27. Take now V to denote in succession the foregoing expressions of the potential of a point, a line, a rectangle, or a cuboid, at the point (a, b, c). In the first three cases respectively, each of the component terms is reduced to zero by the operator $d_a^2 + d_b^2 + d_c^2$; and we have, therefore,

$$(d_a^2 + d_b^2 + d_c^2)\, V = 0,$$

which is as it should be. But in the case of the cuboid, each of the eight component terms is by the operator reduced to $\mp \frac{\pi}{2}$, and we have therefore

$$(d_a^2 + d_b^2 + d_c^2)\, V = \Sigma \left\{ \pm \left(\mp \frac{\pi}{2} \right) \right\};$$

Σ denoting the sum of eight terms, the $\pm$ denoting $+$ or $-$, according to the sign of the term in the formula (viz. in four cases this is $+$, and in four cases it is $-$), and the $\mp \frac{\pi}{2}$ denoting the value $\frac{\pi}{2}$ with its proper sign depending on the signs of the quantities $(a - x_0,\ b - y_0,\ c - z_0)$, &c., as explained in the preceding Number.

Suppose for a moment $a > x_1,\ b > y_1,\ c > z_1$, or the attracted point in one of the regions exterior to the cuboid; then $\mp \frac{\pi}{2}$ will in each case be $= -\frac{\pi}{2}$, and the sign $\mp$, being $+$ for four of the terms and $-$ for the four remaining terms, the sum is $= 0$. And similarly, in all cases where the attracted point is exterior to the cuboid, the sum of the eight terms is $= 0$. But when the attracted point is interior, that is, when $a > x_0 < x_1,\ b > y_0 < y_1,\ c > z_0 < z_1$, then it is found that, for the four terms which have the sign $+$, the value of $\mp \frac{\pi}{2}$ is $= -\frac{\pi}{2}$; and for the four terms which have the sign $-$, its value is $= +\frac{\pi}{2}$; whence, in the sum, each term is $= -\frac{\pi}{2}$, or the value is $= -4\pi$. Hence, in the case of the cuboid, we have

$$(d_a^2 + d_b^2 + d_c^2)\, V = 0 \text{ or } -4\pi,$$

according as the attracted point is external or internal.

Verification in regard to the Rectangle.

28. I start from the formula

$$\begin{aligned} V = {} & \ \ C(a - x_0,\ b - y_0,\ c) \\ & - C(a - x_1,\ b - y_0,\ c) \\ & - C(a - x_0,\ b - y_1,\ c) \\ & + C(a - x_1,\ b - y_1,\ c), \end{aligned}$$

where, as before, $x_1 > x_0$, $y_1 > y_0$. V is here a function of (a, b, c), satisfying the partial differential equation

$$(d_a^2 + d_b^2 + d_c^2)\, V = 0,$$

and (as is easily verified) vanishing when any one of the variables a, b, c becomes infinite; it does not become infinite for any finite values of a or b, or any positive value of c. Hence, by a theorem of Green's*, there exists on the plane $z=0$ a distribution of matter giving rise to the potential V; and not only so, but the density at any point (x, y) of the plane is given by the formula

$$\rho = -\frac{1}{2\pi}\left(\frac{dW}{dc}\right)_{c=0},$$

where W is what V becomes on writing therein x, y in place of a, b, and $c=0$ is regarded as an indefinitely small positive quantity.

We have

$$d_c C(x, y, c) = -\tan^{-1}\frac{xy}{cr}, \text{ where } r = \sqrt{x^2+y^2+c^2}.$$

And hence

$$\begin{aligned} d_c W = &-\tan^{-1}\frac{(x-x_0)(y-y_0)}{c\sqrt{(x-x_0)^2+(y-y_0)^2+c^2}} \\ &+\tan^{-1}\frac{(x-x_1)(y-y_0)}{c\sqrt{(x-x_1)^2+(y-y_0)^2+c^2}} \\ &+\tan^{-1}\frac{(x-x_0)(y-y_1)}{c\sqrt{(x-x_0)^2+(y-y_1)^2+c^2}} \\ &-\tan^{-1}\frac{(x-x_1)(y-y_1)}{c\sqrt{(x-x_1)^2+(y-y_1)^2+c^2}}. \end{aligned}$$

Putting $c=0$, as above, each arc is $=\frac{\pi}{2}$ or $-\frac{\pi}{2}$, according as the fraction under the $\tan^{-1}$ is positive or negative—that is, according as the numerator is positive or negative. Suppose for a moment $x > x_1$, $y > y_1$, viz. the point (x, y) is here in a region exterior to the rectangle (x_1, y_1), (x_1, y_0), (x_0, y_1), (x_0, y_0): the value of $d_c W$ is $=-\frac{\pi}{2}+\frac{\pi}{2}+\frac{\pi}{2}-\frac{\pi}{2}, =0$; and similarly, for every other position of the point (x, y)

* The theorem in question is a particular case of Green's, $4\pi\rho = -\frac{dV}{dw} - \frac{dV'}{dw}$ ("Essay on the Application of Mathematical Analysis to the Theories of Electricity and Magnetism" (1828), see p. 31 of the *Collected Works*); viz. the surface is here a plane, and $V=V'$. And it is also a particular case of the formula $\rho' = \frac{-\Gamma\frac{1}{2}(n-1)}{2\pi^{\frac{1}{2}s}\,\Gamma\frac{1}{2}(n-s+1)}P'$ ("Memoir on the Determination of the Exterior and Interior Attraction of Ellipsoids of Variable Densities" (1835), see p. 199 of the *Collected Works*); viz. s is taken $=2$; and Green's extra-spatial coordinate u then becomes the coordinate z of ordinary tri-dimensional space.

exterior to the rectangle, the value is $=0$. But for a point interior to the rectangle, we have $x < x_1 > x_0$, $y < y_1 > y_0$, and in this case the value is

$$-\frac{\pi}{2}+\left(-\frac{\pi}{2}\right)+\left(-\frac{\pi}{2}\right)-\frac{\pi}{2}, \quad =-2\pi.$$

Hence

$$\rho, \;=-\frac{1}{2\pi}(d_c W)_{c=0}, \text{ is } =0 \text{ or } 1,$$

according as the point is exterior or interior to the rectangle, viz. the distribution producing the potential in question is a uniform distribution (density unity) over the rectangle, which is as it should be.

Potential of a Cuboidal Surface.

29. The preceding formulæ lead to the expression of the potential of a cuboidal surface (viz. the surface composed of the six faces of a cuboid, each of them being considered as a plate of the same uniform density) upon a point a, b, c. Writing, for convenience,

$$E(f,\, g,\, h)=\tfrac{1}{2}(g+h)\log\frac{s+f}{s-f}+\tfrac{1}{2}(h+f)\log\frac{s+g}{s-g}+\tfrac{1}{2}(f+g)\log\left(\frac{s+h}{s-h}\right)$$

$$-f\tan^{-1}\frac{gh}{fs}-g\tan^{-1}\frac{hf}{gs}-h\tan^{-1}\frac{fg}{hs},$$

where each term is supposed to have (compounded with its expressed sign) a sign $\pm$, as follows: viz. in any fg term $\left(\tfrac{1}{2}f\log\frac{s+g}{s-g},\; \tfrac{1}{2}g\log\frac{s+f}{s-f},\text{ or } h\tan^{-1}\frac{fg}{hs}\right)$, this sign $\pm$ is $+$ if f and g are both positive or both negative, but is $-$ if f and g are the one of them positive and the other negative; and the like as to the gh terms and the hf terms respectively. And this being so, the expression for the potential (applying as well to an interior as to an exterior point) is

$$\begin{aligned} V = \;& E(a-x_0,\, b-y_0,\, c-z_0)\\ &+E(a-x_1,\, b-y_0,\, c-z_0)\\ &+E(a-x_0,\, b-y_1,\, c-z_0)\\ &+E(a-x_1,\, b-y_1,\, c-z_0)\\ &+E(a-x_0,\, b-y_0,\, c-z_1)\\ &+E(a-x_1,\, b-y_0,\, c-z_1)\\ &+E(a-x_0,\, b-y_1,\, c-z_1)\\ &+E(a-x_1,\, b-y_1,\, c-z_1). \end{aligned}$$

It is, in fact, easy to verify that the final result, interpreted as above, represents the sum of the six positive values, which are the values of the potential for the six faces of the cuboid respectively.

603.

ON THE POTENTIAL OF THE ELLIPSE AND THE CIRCLE.

[From the *Proceedings of the London Mathematical Society*, vol. VI. (1874—1875), pp. 38—58. Read January 14, 1875.]

The Potential of the Ellipse.

1. I CONSIDER the potential of an ellipse (or say an elliptic plate of uniform density); viz. this is

$$V = \int \frac{dx\,dy}{\sqrt{(a-x)^2 + (b-y)^2 + c^2}},$$

the limits being given by the equation $\frac{x^2}{f^2} + \frac{y^2}{g^2} = 1$.

Writing herein $x = mf\cos u$, $y = mg\sin u$, we have $dxdy = fg\, mdmdu$; and consequently

$$V = fg \int \frac{m\,dm\,du}{\sqrt{(a - mf\cos u)^2 + (b - mg\sin u)^2 + c^2}},$$

where the integrations are to be taken from $m = 0$ to $m = 1$, and from $u = 0$ to $u = 2\pi$.

2. It is to be remarked that, by first performing the integration in regard to m, we may reduce the potential to the form $\int du\,.\,F$, where F is an *algebraic* function of $\cos u$, $\sin u$; and that the result so obtained, although in the general case too complex to be manageable, is a useful one in the case $f = g$, where the ellipse becomes a circle. The case of the circle will be treated of separately, but in the general case it will be sufficient to show that the integral is of the form in question.

3. To accomplish this, writing

$$A = a^2 + b^2 + c^2,$$
$$B = af\cos u + bg\sin u,$$
$$C = f^2\cos^2 u + g^2\sin^2 u,$$

then the integral in regard to m is

$$\int \frac{m\,dm}{\sqrt{A - 2Bm + Cm^2}},$$

which is

$$= \frac{1}{C}\sqrt{A - 2Bm + Cm^2} + \frac{B}{C\sqrt{C}}\log\left\{Cm - B + \sqrt{C}\sqrt{A - 2Bm + Cm^2}\right\}.$$

Taken between the limits 0 and 1, this is

$$= \frac{1}{C}(\sqrt{A - 2B + C} - \sqrt{A}) + \frac{B}{C\sqrt{C}}\log\left\{\frac{C - B + \sqrt{C}\sqrt{A - 2B + C}}{-B + \sqrt{C}\sqrt{A}}\right\};$$

and we have therefore

$$V = fg\int du\,\frac{1}{C}(\sqrt{A - 2B + C} - \sqrt{A}) + fg\int du\,\frac{(af\cos u + bg\sin u)}{(f^2\cos^2 u + g^2\sin^2 u)^{\frac{3}{2}}}\log\Upsilon,$$

where, for greater clearness, the value of the coefficient $\dfrac{B}{C\sqrt{C}}$ of the logarithmic term has been written at full length.

4. But this coefficient admits of algebraic integration, viz. we have

$$fg\int du\,\frac{af\cos u + bg\sin u}{(f^2\cos^2 u + g^2\sin^2 u)^{\frac{3}{2}}} = \frac{ag\sin u - bf\cos u}{(f^2\cos^2 u + g^2\sin^2 u)^{\frac{1}{2}}};$$

hence, integrating the second term by parts, we have

$$V = fg\int du\,\frac{1}{C}\{\sqrt{A - 2B + C} - \sqrt{A}\}$$
$$+ \frac{ag\sin u - bf\cos u}{(f^2\cos^2 u + g^2\sin^2 u)^{\frac{1}{2}}}\log\Upsilon$$
$$- \int du\,\frac{ag\sin u - bf\cos u}{(f^2\cos^2 u + g^2\sin^2 u)^{\frac{1}{2}}}\cdot\frac{\Upsilon'}{\Upsilon},$$

where the second term, taken between the limits $u = 0$, $u = 2\pi$, is $= 0$; and $\dfrac{\Upsilon'}{\Upsilon}$ being an algebraic function of $\sin u$, $\cos u$, the potential is expressed in the form in question.

5. But we may, by means of a transformation upon u (that made use of in Gauss' Memoir* on the attraction of an elliptic ring), transform the expression so as

* [*Ges. Werke*, t. III., pp. 333—355; in particular, *l.c.*, p. 338].

to obtain the integral in regard to m under a much more simple form. We, in fact, assume

$$\cos u = \frac{\alpha + \alpha' \cos T + \alpha'' \sin T}{\gamma + \gamma' \cos T + \gamma'' \sin T},$$

$$\sin u = \frac{\beta + \beta' \cos T + \beta'' \sin T}{\gamma + \gamma' \cos T + \gamma'' \sin T},$$

where the nine coefficients are such that identically

$$(\alpha + \alpha' \cos T + \alpha'' \sin T)^2 + (\beta + \beta' \cos T + \beta'' \sin T)^2 - (\gamma + \gamma' \cos T + \gamma'' \sin T)^2 = \cos^2 T + \sin^2 T - 1,$$

(this of course renders the two equations consistent); and also that

$$(a - mf \cos u)^2 + (b - mg \sin u)^2 + c^2 = \frac{1}{(\gamma + \gamma' \cos T + \gamma'' \sin T)^2} (G + G' \cos^2 T + G'' \sin^2 T).$$

This last condition gives, for the determination of the coefficients G, G', G'', the identity

$$(\theta - G)(\theta + G')(\theta + G'') = -(\theta + m^2 f^2)(\theta + m^2 g^2)\, \theta \left\{ \frac{a^2}{\theta + m^2 f^2} + \frac{b^2}{\theta + m^2 g^2} + \frac{c^2}{\theta} - 1 \right\};$$

or, what is the same thing, G, $-G'$, $-G''$ are the roots of the equation

$$\frac{a^2}{\theta + m^2 f^2} + \frac{b^2}{\theta + m^2 g^2} + \frac{c^2}{\theta} - 1 = 0.$$

This equation has one positive root, which may be taken to be G, and two negative roots, which will then be $-G'$, $-G''$; viz. G, G', G'' are thus all positive; and G denotes the positive root of the last-mentioned equation.

6. We have

$$du = \frac{dT}{(G + G' \cos T + G'' \sin T)^2},$$

and thence

$$V = fg \int m\, dm \int \frac{dT}{(G + G' \cos^2 T + G'' \sin^2 T)^{\frac{1}{2}}},$$

the integral in regard to T being taken from 0 to 2π; or, what is the same thing, we may multiply by 4 and take the integral only from 0 to $\frac{\pi}{2}$; viz. we thus have

$$V = 4fg \int m\, dm \int_0^{\frac{1}{2}\pi} \frac{dT}{(G + G' \cos^2 T + G'' \sin^2 T)^{\frac{1}{2}}},$$

where the integral in regard to T can be at once reduced to the standard form of an elliptic function, or it might be calculated by Gauss' method of the arithmetico-geometrical mean.

7. But, for the present purpose, a further reduction is required. Writing

$$t = G + (G + G') \cot^2 T,$$

we have

$$t - G \ = (G + G')\frac{\cos^2 T}{\sin^2 T},$$

$$t + G' = (G + G')\frac{1}{\sin^2 T},$$

$$t + G'' = (G + G'\cos^2 T + G''\sin^2 T)\frac{1}{\sin^2 T}:$$

whence

$$\sqrt{t - G\,.\,t + G'\,.\,t + G''} = (G + G')(G + G'\cos^2 T + G''\sin^2 T)^{\frac{1}{2}}\frac{\cos T}{\sin^3 T};$$

moreover

$$dt = -\,2\,(G + G')\frac{\cos T}{\sin^3 T}\,dT.$$

Hence

$$\frac{dt}{\sqrt{t - G\,.\,t + G'\,.\,t + G''}} = \frac{-\,2dT}{\{G + G'\cos^2 T + G''\sin^2 T\}^{\frac{1}{2}}};$$

and, observing that to the limits 0, $\frac{\pi}{2}$ of T correspond the limits ∞, G of t, we thence obtain

$$V = 2fg\int m\,dm\int_G^\infty \frac{dt}{\sqrt{t - G\,.\,t + G'\,.\,t + G''}};$$

or, what is the same thing,

$$V = 2fg\int m\,dm\int_G^\infty \frac{dt}{\sqrt{t\,(t + m^2f^2)\,(t + m^2g^2)\left(1 - \dfrac{a^2}{t + m^2f^2} - \dfrac{b^2}{t + m^2g^2} - \dfrac{c^2}{t}\right)}},$$

where G denotes, as before, the positive root of the equation

$$\frac{a^2}{\theta + m^2f^2} + \frac{b^2}{\theta + m^2g^2} + \frac{c^2}{\theta} - 1 = 0.$$

8. Writing for t, m^2t, and for G, m^2G, the formula becomes

$$V = 2fg\int m\,dm\int_G^\infty \frac{dt}{\sqrt{t\,.\,t + f^2\,.\,t + g^2\left(m^2 - \dfrac{a^2}{t + f^2} - \dfrac{b^2}{t + g^2} - \dfrac{c^2}{t}\right)}},$$

where G now denotes the positive root of the equation

$$\frac{a^2}{\theta + f^2} + \frac{b^2}{\theta + g^2} + \frac{c^2}{\theta} - m^2 = 0.$$

Thus G is a function of m; but it is to be remarked that the integration in respect to m can be performed through the integral sign $\int_G^\infty dt$ in precisely the same way as if G were constant, and that we, in fact, have

$$V = 2fg\left[\int_G^\infty dt \sqrt{m^2 - \frac{a^2}{t+f^2} - \frac{b^2}{t+g^2} - \frac{c^2}{t}}\, \frac{1}{\sqrt{t \,.\, t+f^2 \,.\, t+g^2}}\right],$$

where the function of m is to be taken between the limits 0 and 1. The reason is that, differentiating this last integral in respect to m, the term depending on the variation of the limit G is

$$-\sqrt{m^2 - \frac{a^2}{G+f^2} - \frac{b^2}{G+g^2} - \frac{c^2}{G}}\, \frac{1}{\sqrt{G \,.\, G+f^2 \,.\, G+g^2}}\, \frac{dG}{dm},$$

which is $=0$ in virtue of the equation which defines G; hence the whole result is the term arising from the variation of m in so far as it appears explicitly.

9. Proceeding next to take the function of m between the two limits: for $m=0$ we have $G=\infty$, and the integral vanishes; for $m=1$ we have G the positive root of the equation

$$\frac{a^2}{\theta+f^2} + \frac{b^2}{\theta+g^2} + \frac{c^2}{\theta} - 1 = 0,$$

or, using θ to denote the positive root of this equation, the value is $G=\theta$; we thus finally obtain

$$V = 2fg\int_\theta^\infty dt \sqrt{1 - \frac{a^2}{t+f^2} - \frac{b^2}{t+g^2} - \frac{c^2}{t}}\, \frac{1}{\sqrt{t \,.\, t+f^2 \,.\, t+g^2}}$$

as the expression for the potential of the ellipse semiaxes (f, g) on the point (a, b, c).

Case where the Attracted Point is on the Focal Hyperbola.

10. The result becomes very simple when the attracted point is in the focal hyperbola of the ellipse, viz. when we have $b=0$ and $\dfrac{a^2}{f^2-g^2} - \dfrac{c^2}{g^2} = 1$. The function $1 - \dfrac{a^2}{t+f^2} - \dfrac{b^2}{t+g^2} - \dfrac{c^2}{t}$ is here

$$= \frac{a^2}{f^2-g^2} - \frac{c^2}{g^2} - \frac{a^2}{f^2+t} - \frac{c^2}{t}$$

$$= (t+g^2)\left\{\frac{a^2}{(t+f^2)(f^2-g^2)} - \frac{c^2}{g^2 t}\right\}$$

$$= (t+g^2)\left\{\left(1+\frac{c^2}{g^2}\right)\frac{1}{t+f^2} - \frac{c^2}{g^2 t}\right\}$$

$$= \frac{t+g^2}{t(t+f^2)}\left(t - \frac{c^2 f^2}{g^2}\right).$$

Hence also $\theta=\dfrac{c^2f^2}{g^2}$; introducing this value, the function in question becomes

$$=\frac{(t+g^2)(t-\theta)}{t(t+f^2)},$$

and we have

$$V=2fg\int_\theta^\infty dt\frac{\sqrt{t+g^2 . t-\theta}}{\sqrt{t . t+f^2}}\frac{1}{\sqrt{t . t+f^2 . t+g^2}}$$

$$=2fg\int_\theta^\infty\frac{dt\sqrt{t-\theta}}{t . t+f^2},$$

which, writing $t=x^2+\theta$, becomes

$$V=4fg\int_0^\infty\frac{x^2\,dx}{x^2+\theta . x^2+\theta+f^2}$$

$$=\frac{4fg}{f^2}\int_0^\infty\left(\frac{\theta+f^2}{x^2+\theta+f^2}-\frac{\theta}{x^2+\theta}\right)dx$$

$$=\frac{4g}{f}\left(\sqrt{\theta+f^2}\tan^{-1}\frac{x}{\sqrt{\theta+f^2}}-\sqrt{\theta}\tan^{-1}\frac{x}{\sqrt{\theta}}\right)_0^\infty$$

$$=2\pi\frac{g}{f}(\sqrt{\theta+f^2}-\sqrt{\theta});$$

or, substituting for θ its value $\dfrac{c^2f^2}{g^2}$, this is

$$V=2\pi(\sqrt{c^2+g^2}-c),$$

which is, in fact, the potential of the circle $x^2+y^2=g^2$ on the axial point $(0, 0, c)$; and, observing that the value is independent of f, we have at once the theorem that, considering f as variable, and taking the attracted point at the constant altitude c in the focal hyperbola $\dfrac{x^2}{f^2-g^2}-\dfrac{z^2}{g^2}=1$, the potential is the same, whatever is the value of the semi-axis major f of the ellipse.

11. A point in the focal hyperbola determines, with the ellipse, a right circular cone having for its axis the tangent to the hyperbola; viz. the tangent in question is equally inclined to the two lines joining the point with the foci of the hyperbola, or with the extremities of the major axis of the ellipse. Taking θ for the inclination of the tangent to either of these lines, viz. θ is the semi-aperture of the cone, and γ for the inclination of the tangent to the axis of z, then it is easy to show that

$$\sqrt{c^2+g^2}=c\frac{\cos\gamma}{\sqrt{\cos^2\gamma-\sin^2\theta}};$$

and we thence have

$$V=2\pi c\left(\frac{\cos\gamma}{\sqrt{\cos^2\gamma-\sin^2\theta}}-1\right),$$

viz. the ellipse is here considered as the section of a right cone of semi-aperture θ, the perpendicular distance from the vertex being $=c$, and the inclination of this distance to the axis of the cone being $=\gamma$; and this being so, the potential is then expressed by the last preceding equation. It will be observed that, when $\gamma=\frac{\pi}{2}-\theta$, the section becomes a parabola, and the potential is infinite; for any larger value of γ, the section is a hyperbola, and the formula ceases to be applicable.

12. I originally obtained the result by thus considering the ellipse as the section of a right cone. Consider for a moment, in the case of any cone whatever, the plate included between the plane, perpendicular distance from the vertex $=c$, and the consecutive parallel plane, distance $=c+dc$. Let $d\Sigma$ denote an element of the first plane, r its distance from the vertex, and $r+dr$ the distance produced to meet the second plane; also let $d\omega$ denote the subtended solid angle. We have $d\Sigma\, dc = r^2\, dr\, d\omega$, or, since $\frac{dc}{c}=\frac{dr}{r}$, we obtain $d\Sigma=\frac{1}{c}r^3\, d\omega$, or $\frac{1}{r}\, d\Sigma=\frac{1}{c}r^2\, d\omega$; wherefore the potential of the plane section is $V=\frac{1}{c}\int r^2\, d\omega$, where r denotes the value at a point of the plane section, and the integration extends over the spherical aperture of the cone.

13. Let the position of r be determined by means of its inclination θ to the axis of the cone, and the azimuth ϕ of the plane through r and the axis of the cone; viz. taking the axis of the cone for the axis of z, suppose, as usual, $x=r\sin\theta\cos\phi$, $y=r\sin\theta\sin\phi$, $z=r\cos\theta$. We have then, as usual, $d\omega=\sin\theta\, d\theta\, d\phi$; and if the equation of the plane be $x\cos\alpha+y\cos\beta+z\cos\gamma=c$, then the value of r is obtained from the equation

$$r\{(\cos\alpha\cos\phi+\cos\beta\sin\phi)\sin\theta+\cos\gamma\cos\theta\}=c;$$

so that we have for the potential

$$V=c\int\frac{\sin\theta\, d\theta\, d\phi}{\{(\cos\alpha\cos\phi+\cos\beta\sin\phi)\sin\theta+\cos\gamma\cos\theta\}^2},$$

where the integration is extended over the whole spherical aperture of the cone; viz. in the case of a right cone of semi-aperture θ, the limits are from $\theta=0$ to $\theta=\theta$ and from $\phi=0$ to $\phi=2\pi$.

14. Write

$$(\cos\alpha\cos\phi+\cos\beta\sin\phi)\sin\theta+\cos\gamma\cos\theta=M\cos(\theta-N),$$

where M, N are given functions of ϕ; then we have

$$V=c\int\frac{d\phi}{M^2}\int\frac{\sin\theta\, d\theta}{\cos^2(\theta-N)}$$

and the θ-integral is

$$\int\frac{[\sin(\theta-N)\cos N+\cos(\theta-N)\sin N]\, d\theta}{\cos^2(\theta-N)},$$

$$=\cos N\sec(\theta-N)+\sin N\log\tan\{\tfrac{1}{4}\pi+\tfrac{1}{2}(\theta-N)\},$$

which between the limits is

$$= \cos N \{\sec(\theta - N) - \sec N\} + \sin N \{\log\tan[\tfrac{1}{4}\pi + \tfrac{1}{2}(\theta - N)] - \log\tan(\tfrac{1}{4}\pi - \tfrac{1}{2}N)\},$$

θ now denoting the semi-aperture of the right cone. And we have

$$V = c\int \frac{d\phi}{M^2}\left\{\cos N\left(\frac{1}{\cos(N-\theta)} - \frac{1}{\cos N}\right) + \sin N\left[\log\tan\{\tfrac{1}{4}\pi + \tfrac{1}{2}(\theta - N)\} - \log\tan(\tfrac{1}{4}\pi - \tfrac{1}{2}N)\right]\right\}.$$

We may without loss of generality write $\cos\beta = 0$, and therefore $\cos\alpha = \sin\gamma$, where γ now is the inclination of the perpendicular on the plane to the axis of the cone. We thus have

$$\cos\gamma\cos\theta + \sin\gamma\cos\phi\sin\theta = M\cos(\theta - N),$$

that is,

$$\cos\gamma = M\cos N,$$

$$\sin\gamma\cos\phi = M\sin N;$$

whence

$$\tan N = \tan\gamma\cos\phi \quad\text{or}\quad N = \tan^{-1}(\tan\gamma\cos\phi),$$

$$M^2 = \cos^2\gamma + \sin^2\gamma\cos^2\phi = 1 - \sin^2\gamma\sin^2\phi,$$

and

$$\frac{\cos N}{\cos(N-\theta)} = \frac{1}{\cos\theta + \sin\theta\tan\gamma\cos\phi}.$$

15. We have, therefore,

$$V = c\int \frac{d\phi}{1 - \sin^2\gamma\sin^2\phi}\left(\frac{1}{\cos\theta + \sin\theta\tan\gamma\cos\phi} - 1\right)$$

$$+ c\int \frac{d\phi\sin\gamma\cos\phi}{(1-\sin^2\gamma\sin^2\phi)^{\frac{3}{2}}}\{\log\tan[\tfrac{1}{4}\pi + \tfrac{1}{2}\theta - \tfrac{1}{2}\tan^{-1}(\tan\gamma\cos\phi)] - \log\tan[\tfrac{1}{4}\pi - \tfrac{1}{2}\tan^{-1}(\tan\gamma\cos\phi)]\}.$$

But

$$\int \frac{d\phi\cos\phi}{(1 - \sin^2\gamma\sin^2\phi)^{\frac{3}{2}}} = \frac{\sin\phi}{(1 - \sin^2\gamma\sin^2\phi)^{\frac{1}{2}}};$$

hence the second line is

$$c\sin\gamma\frac{\sin\phi}{(1 - \sin^2\gamma\sin^2\phi)^{\frac{1}{2}}}\{\log\tan[\tfrac{1}{4}\pi + \tfrac{1}{2}\theta - \tfrac{1}{2}\tan^{-1}(\tan\gamma\cos\phi)] - \log\tan[\tfrac{1}{4}\pi - \tfrac{1}{2}\tan^{-1}(\tan\gamma\cos\phi)]\}$$

$$- c\sin\gamma\int d\phi\frac{\sin\phi}{(1 - \sin^2\gamma\sin^2\phi)^{\frac{1}{2}}}\frac{d}{d\phi}\{\log\tan[\tfrac{1}{4}\pi + \tfrac{1}{2}\theta - \tfrac{1}{2}\tan^{-1}(\tan\gamma\cos\phi)] - \log\tan[\tfrac{1}{4}\pi - \tfrac{1}{2}\tan^{-1}(\tan\gamma\cos\phi)]\}.$$

But, restoring for a moment N in place of $\tan^{-1}(\tan\gamma\cos\phi)$, we have

$$\frac{d}{d\phi}\log\tan(\tfrac{1}{4}\pi + \tfrac{1}{2}\theta - N) = -\frac{dN}{d\phi}\frac{1}{\cos(N-\theta)} = \frac{\sin\gamma\cos\gamma\sin\phi}{1 - \sin^2\gamma\sin^2\phi}\frac{1}{\cos(N-\theta)},$$

$$\frac{d}{d\phi}\log\tan(\tfrac{1}{4}\pi - N) = -\frac{dN}{d\phi}\frac{1}{\cos N} = \frac{\sin\gamma\cos\gamma\sin\phi}{1 - \sin^2\gamma\sin^2\phi}\frac{1}{\cos N}.$$

And then, in place of $\frac{1}{\cos(N-\theta)}-\frac{1}{\cos N}$, writing

$$\frac{1}{\cos\gamma\sqrt{1-\sin^2\gamma\sin^2\phi}}\left(\frac{1}{\cos\theta+\sin\theta\tan\gamma\cos\phi}-1\right),$$

the expression in question becomes

$$c\sin\gamma\frac{\sin\phi}{(1-\sin^2\gamma\sin^2\phi)^{\frac{1}{2}}}\{\log\tan[\tfrac{1}{4}\pi+\tfrac{1}{2}\theta-\tfrac{1}{2}\tan^{-1}(\tan\gamma\cos\phi)]$$
$$-\log\tan[\tfrac{1}{4}\pi-\tfrac{1}{2}\tan^{-1}(\tan\gamma\cos\phi)]\}$$
$$-c\int d\phi\frac{\sin^2\gamma\sin^2\phi}{1-\sin^2\gamma\sin^2\phi}\left(\frac{1}{\cos\theta+\sin\theta\tan\gamma\cos\phi}-1\right).$$

And we have

$$V=\frac{c\sin\gamma\sin\phi}{(1-\sin^2\gamma\sin^2\phi)^{\frac{1}{2}}}\{\log\tan[\tfrac{1}{4}\pi+\tfrac{1}{2}\theta-\tfrac{1}{2}\tan^{-1}(\tan\gamma\cos\phi)]$$
$$-\log\tan[\tfrac{1}{4}\pi-\tfrac{1}{2}\tan^{-1}(\tan\gamma\cos\phi)]\}+c\int d\phi\left(\frac{1}{\cos\theta+\sin\theta\tan\gamma\cos\phi}-1\right).$$

16. The integral is here

$$=\int d\phi\left\{\frac{\cos\gamma(\cos\theta\cos\gamma-\sin\theta\sin\gamma\cos\phi)}{\cos^2\theta\cos^2\gamma-\sin^2\theta\sin^2\gamma\cos^2\phi}-1\right\}$$
$$=\cos^2\gamma\cos\theta\int\frac{d\phi}{\cos^2\theta\cos^2\gamma-\sin^2\theta\sin^2\gamma\cos^2\phi}$$
$$-\cos\gamma\sin\gamma\sin\theta\int\frac{\cos\phi\,d\phi}{\cos^2\theta\cos^2\gamma-\sin^2\theta\sin^2\gamma\cos^2\phi}-\int d\phi$$
$$=\frac{\cos\gamma}{\sqrt{\cos^2\gamma-\sin^2\theta}}\tan^{-1}\frac{\cos\theta\cos\gamma\tan\phi}{\sqrt{\cos^2\gamma-\sin^2\theta}}$$
$$-\frac{\cos\gamma}{\sqrt{\cos^2\gamma-\sin^2\theta}}\tan^{-1}\frac{\sin\theta\sin\gamma\sin\phi}{\sqrt{\cos^2\gamma-\sin^2\theta}}-\phi,$$

as may be immediately verified.

Hence

$$V=\frac{c\sin\gamma\sin\phi}{\sqrt{1-\sin^2\gamma\sin^2\phi}}\{\log\tan[\tfrac{1}{4}\pi+\tfrac{1}{2}\theta-\tfrac{1}{2}\tan^{-1}(\tan\gamma\cos\phi)]$$
$$-\log\tan[\tfrac{1}{4}\pi-\tfrac{1}{2}\tan^{-1}(\tan\gamma\cos\phi)]\}$$
$$+\frac{c\cos\gamma}{\sqrt{\cos^2\theta-\sin^2\gamma}}\tan^{-1}\frac{\cos\theta\cos\gamma\tan\phi}{\sqrt{\cos^2\theta-\sin^2\gamma}}$$
$$-\frac{c\cos\gamma}{\sqrt{\cos^2\theta-\sin^2\gamma}}\tan^{-1}\frac{\sin\theta\sin\gamma\sin\phi}{\sqrt{\cos^2\theta-\sin^2\gamma}}$$
$$-c\phi,$$

which is to be taken between the limits 0 and 2π; or, what is the same thing, the integral may be taken between the limits 0, π, and multiplied by 2. But as ϕ passes from 0 to π, the arc of the form $\tan^{-1}(A\tan\phi)$ passes through the values 0, $\frac{\pi}{2}$, $-\frac{\pi}{2}$, 0, but the other arc of the form $\tan^{-1}(B\sin\phi)$ through the values 0, $\frac{\pi}{2}$, $\frac{\pi}{2}$, 0; the first arc gives therefore a term π, the second arc a term 0, and the final result is

$$V = 2c\pi\left(\frac{\cos\gamma}{\sqrt{\cos^2\gamma-\sin^2\theta}}-1\right),$$

which is right.

The Potential of the Circle.

17. In the case of the circle we have $g=f$; the terms containing a^2, b^2 unite throughout into a single term containing a^2+b^2, and there is obviously no loss of generality in assuming $b=0$, and so reducing this to a^2; viz. we take the axis of x to pass through the projection of the attracted point, the coordinates of this point being therefore $(a, 0, c)$. We in fact consider the potential

$$V=\int\frac{dx\,dy}{\sqrt{(a-x)^2+y^2+c^2}}$$

over the circle $x^2+y^2=f^2$; or, writing $x=mf\cos\phi$, $y=mf\sin\phi$, we have $dxdy=f^2mdmd\phi$, and therefore

$$V=f^2\int\frac{m\,dm\,d\phi}{\sqrt{a^2+c^2+m^2f^2-2maf\cos\phi}},$$

the integral being taken from $m=0$ to $m=1$, and $\phi=0$ to $\phi=2\pi$.

Writing in the general formula $g=f$ and $b=0$, we have

$$V=2f^2\int_\theta^\infty\frac{dt\sqrt{1-\dfrac{a^2}{t+f^2}-\dfrac{c^2}{t}}}{(t+f^2)\sqrt{t}},$$

where θ denotes the positive root of the equation

$$1-\frac{a^2}{\theta+f^2}-\frac{c^2}{\theta}=0;$$

or, observing that

$$\begin{aligned}1-\frac{a^2}{t+f^2}-\frac{c^2}{t}&=a^2\left(\frac{1}{\theta+f^2}-\frac{1}{t+f^2}\right)+c^2\left(\frac{1}{\theta}-\frac{1}{t}\right)\\&=(t-\theta)\left\{\frac{a^2}{(\theta+f^2)(t+f^2)}+\frac{c^2}{\theta t}\right\}\\&=\frac{t-\theta}{t\,.\,t+f^2}\left\{\left(1-\frac{c^2}{\theta}\right)t+\frac{c^2}{\theta}(t+f^2)\right\}\\&=\frac{(t-\theta)\left(t+\dfrac{c^2f^2}{\theta}\right)}{t\,.\,t+f^2},\end{aligned}$$

we have also

$$V = 2f^2 \int_\theta^\infty \frac{\sqrt{\left(t - \theta \, . \, t + \dfrac{c^2 f^2}{\theta}\right)} \, dt}{t \, (t + f^2) \sqrt{t + f^2}} .$$

18. The present particular case gives rise to some interesting investigations. We may, in the first place, complete the process of first integrating directly in regard to m. Writing

$$V = f \iint \frac{[(mf - a\cos\phi) + a\cos\phi] \, dm \, d\phi}{\{(mf - a\cos\phi)^2 + a^2\sin^2\phi + c^2\}^{\frac{3}{2}}},$$

the integral in regard to m is

$$= \frac{1}{f} \{\sqrt{(mf - a\cos\phi)^2 + a^2\sin^2\phi + c^2} + a\cos\phi \log\{mf - a\cos\phi + \sqrt{(mf - a\cos\phi)^2 + a^2\sin^2\phi + c^2}\}\}$$

to be taken from $m = 0$ to $m = 1$; and we thus obtain

$$V = \int d\phi \, \{\sqrt{a^2 + c^2 + f^2 - 2af\cos\phi} - \sqrt{a^2 + c^2}$$

$$+ a\cos\phi \, [\log(f - a\cos\phi + \sqrt{a^2 + c^2 + f^2 - 2af\cos\phi}) - \log(-a\cos\phi + \sqrt{a^2 + c^2})]\}.$$

Writing for shortness $\sqrt{a^2 + c^2 + f^2 - 2af\cos\phi} = \Delta$, the second line of this is

$$a\sin\phi \, [\log(f - a\cos\phi + \Delta) - \log(-a\cos\phi + \sqrt{a^2 + c^2})]$$

$$- \int d\phi \, a^2\sin^2\phi \left\{\frac{f + \Delta}{\Delta \, (f - a\cos\phi + \Delta)} - \frac{1}{-a\cos\phi + \sqrt{a^2 + c^2}}\right\},$$

and we thus have

$$V = a\sin\phi \, \{\log(f - a\cos\phi + \Delta) - \log(-a\cos\phi + \sqrt{a^2 + c^2})\}$$

$$+ \int d\phi \left\{\Delta - \sqrt{a^2 + c^2} - \frac{a^2\sin^2\phi \, (f + \Delta)}{\Delta \, (f - a\cos\phi + \Delta)} + \frac{a^2\sin^2\phi}{-a\cos\phi + \sqrt{a^2 + c^2}}\right\}.$$

19. We have

$$\frac{f + \Delta}{\Delta \, (f - a\cos\phi + \Delta)} = \frac{(f + \Delta)(f - a\cos\phi - \Delta)}{\Delta \, \{(f - a\cos\phi)^2 - \Delta^2\}},$$

the numerator of which is $f^2 - \Delta^2 - a\cos\phi \, (f + \Delta)$,

$$= f^2 + \Delta^2 + a\cos\phi \, (f - a\cos\phi - \Delta) - 2af\cos\phi + a^2\cos^2\phi,$$

$$= -c^2 - a^2\sin^2\phi + a\cos\phi \, (f - a\cos\phi - \Delta),$$

and the denominator is $= -\Delta \, (c^2 + a^2\sin^2\phi)$. The second line of V is thus

$$= \int d\phi \left\{\Delta - \sqrt{a^2 + c^2} - \frac{a^2\sin^2\phi}{\Delta} + \frac{a^2\sin^2\phi\cos\phi}{\Delta} \frac{f - a\cos\phi - \Delta}{c^2 + a^2\sin^2\phi} + \frac{a^2\sin^2\phi \, (\sqrt{a^2 + c^2} + a\cos\phi)}{c^2 + a^2\sin^2\phi}\right\},$$

which is easily reduced to

$$\int d\phi \left\{\frac{c^2+f^2-af\cos\phi}{\Delta} - \frac{c^2 a\cos\phi\,(f-a\cos\phi)}{(c^2+a^2\sin^2\phi)\,\Delta} - \frac{c^2\sqrt{a^2+c^2}}{c^2+a^2\sin^2\phi}\right\},$$

the last term of which is $= -c\tan^{-1}\dfrac{\sqrt{a^2+c^2}\tan\phi}{c}$; and we thus have

$$V = a\sin\phi\,\{\log(f-a\cos\phi+\Delta) - \log(-a\cos\phi+\sqrt{a^2+c^2})\} - c\tan^{-1}\frac{\sqrt{a^2+c^2}\tan\phi}{c}$$

$$+\int d\phi\left\{\frac{c^2+f^2-af\cos\phi}{\Delta} - \frac{c^2 a\cos\phi\,(f-a\cos\phi)}{(c^2+a^2\sin^2\phi)\,\Delta}\right\}$$

between the limits 0, 2π; or, finally,

$$V = -2c\pi + 2\int_0^\pi d\phi\left\{\frac{c^2+f^2-af\cos\phi}{\Delta} - \frac{c^2 a\cos\phi\,(f-a\cos\phi)}{(c^2+a^2\sin^2\phi)\,\Delta}\right\};$$

in partial verification whereof observe that for $a=0$ we have $\Delta=\sqrt{c^2+f^2}$, and the value becomes

$$V = 2\pi(\sqrt{c^2+f^2}-c),$$

which, writing therein g in place of f, agrees with a foregoing result.

20. The process applied to finding the Potential of the Ellipse is really applicable step by step to the Circle; but if we begin by assuming $g=f$, it presents itself under a different and simplified form. Starting from

$$V = f^2\int m\,dm\int\frac{d\phi}{\sqrt{a^2+c^2+m^2f^2-2maf\cos\phi}},$$

for convenience we assume

$$P^2+Q^2 = a^2+c^2+m^2f^2,$$
$$PQ = maf,$$

thereby converting the radical into $\sqrt{P^2+Q^2-2PQ\cos\phi}$. Writing also

$$\Omega = a^4+c^4+m^4f^4+2a^2c^2+2m^2c^2f^2-2m^2a^2f^2, \quad = (P^2-Q^2)^2,$$

and hence assuming $P^2-Q^2=\sqrt{\Omega}$, and combining with the foregoing equation

$$P^2+Q^2 = a^2+c^2+m^2f^2,$$

we have

$$P^2 = \tfrac{1}{2}(a^2+c^2+m^2f^2+\sqrt{\Omega}),$$
$$Q^2 = \tfrac{1}{2}(a^2+c^2+m^2f^2-\sqrt{\Omega}).$$

21. This being so, the transformation-equations to the new variable T are

$$\cos\phi = \frac{P\cos T+Q}{P+Q\cos T}, \quad \text{whence} \quad \cos T = \frac{P\cos\phi-Q}{P-Q\cos\phi},$$

$$\sin\phi = \frac{\sqrt[4]{\Omega}\sin T}{P+Q\cos T}, \qquad \sin T = \frac{\sqrt[4]{\Omega}\sin\phi}{P-Q\cos\phi};$$

and also

$$\sqrt{\Omega} = (P + Q\cos T)(P - Q\cos\phi), \quad = P^2 - Q^2.$$

We find moreover

$$d\phi = \frac{\sqrt[4]{\Omega}\, dT}{P + Q\cos T}, \qquad dT = \frac{\sqrt[4]{\Omega}\, d\phi}{P - Q\cos\phi},$$

and

$$P^2 + Q^2 - 2PQ\cos\phi = \frac{\sqrt{\Omega}\,(P - Q\cos T)}{P + Q\cos T},$$

whence

$$\frac{d\phi}{\sqrt{P^2 + Q^2 - 2PQ\cos\phi}} = \frac{dT}{\sqrt{P^2 - Q^2\cos^2 T}};$$

and hence

$$V = f^2 \int m\, dm \int \frac{dT}{\sqrt{P^2 - Q^2\cos^2 T}},$$

where the limits of T are from 0 to 2π, or, what is the same thing, we may multiply by 4, and take them to be 0, $\frac{1}{2}\pi$.

22. Assuming next

$$t = P^2 - m^2f^2 + (P^2 - Q^2)\cot^2 T,$$

we have

$$t - P^2 + m^2f^2 = (P^2 - Q^2)\frac{\cos^2 T}{\sin^2 T},$$

$$t - Q^2 + m^2f^2 = (P^2 - Q^2)\frac{1}{\sin^2 T},$$

$$t \qquad + m^2f^2 = (P^2 - Q^2\cos^2 T)\frac{1}{\sin^2 T},$$

and thence

$$\sqrt{t - P^2 + m^2f^2 \,.\, t - Q^2 + m^2f^2 \,.\, t + m^2f^2} = (P^2 - Q^2)\frac{\cos T}{\sin^3 T}\sqrt{P^2 - Q^2\cos^2 T};$$

also

$$dt = -2(P^2 - Q^2)\frac{\cos T}{\sin^3 T}\, dT;$$

and consequently

$$\frac{dt}{\sqrt{t - P^2 + m^2f^2 \,.\, t - Q^2 + m^2f^2 \,.\, t + m^2f^2}} = \frac{-2dT}{\sqrt{P^2 - Q^2\cos^2 T}}.$$

$T = 0$ gives $t = \infty$, and $T = \frac{1}{2}\pi$ gives $t = P^2 - m^2f^2$, $= G$ suppose; and we thus have

$$V = 2f^2 \int m\, dm \int_G^\infty \frac{dt}{\sqrt{t - P^2 + m^2f^2 \,.\, t - Q^2 + m^2f^2 \,.\, t + m^2f^2}}.$$

23. We have

$$(t - P^2 + m^2f^2)(t - Q^2 + m^2f^2) = t^2 + (m^2f^2 - a^2 - c^2)\, t - m^2c^2f^2,$$

or, putting m^2t in the place of t, this is

$$= m^2\{m^2t^2 + (m^2f^2 - a^2 - c^2)\, t - c^2f^2\},$$

or, what is the same thing,

$$= m^2 t\,(t+f^2)\left\{m^2 - \frac{a^2}{t+f^2} - \frac{c^2}{t}\right\};$$

whence, completing the substitution, we have

$$V = 2f^2 \int m\,dm \int_\theta^\infty \frac{1}{\sqrt{m^2 - \frac{a^2}{t+f^2} - \frac{c^2}{t}}}\,\frac{dt}{\sqrt{t}\,(t+f^2)},$$

where the inferior limit θ, $=\frac{1}{m^2}G$, $=\frac{1}{m^2}P^2 - f^2$ is, in fact, the positive root of the equation

$$m^2 - \frac{a^2}{\theta+f^2} - \frac{c^2}{\theta} = 0.$$

24. We may hence integrate in regard to m, through the sign $\int dt$, in the same way as if θ were constant; viz. we have

$$V = 2f^2 \left[\int_\theta^\infty \sqrt{m^2 - \frac{a^2}{t+f^2} - \frac{c^2}{t}}\,\frac{dt}{\sqrt{t}\,(t+f^2)}\right],$$

where the function of m is to be taken between the limits 0, 1: for $m=0$, we have $\theta=\infty$, and the function vanishes; hence, writing $m=1$, we obtain

$$V = 2f^2 \int_\theta^\infty \sqrt{1 - \frac{a^2}{t+f^2} - \frac{c^2}{t}}\,\frac{dt}{\sqrt{t}\,(t+f^2)},$$

where θ now denotes the positive root of

$$1 - \frac{a^2}{\theta+f^2} - \frac{c^2}{\theta} = 0.$$

25. But it is interesting to reverse the transformation, so as to bring the radical back into its original form. For this purpose, taking now

$$P^2 + Q^2 = a^2 + c^2 + f^2,$$

$$PQ = af,$$

and consequently

$$P^2 = \tfrac{1}{2}\,(a^2 + c^2 + f^2 + \sqrt{\Omega}),$$

$$Q^2 = \tfrac{1}{2}\,(a^2 + c^2 + f^2 - \sqrt{\Omega}),$$

where

$$\Omega = a^4 + c^4 + f^4 + 2a^2c^2 + 2c^2f^2 - 2a^2f^2,$$

and writing

$$t = P^2 - f^2 + (P^2 - Q^2)\cot^2 T,$$

we first obtain

$$V = f^2 \int_0^{2\pi} \frac{\Omega\,\cos^2 T\,dT}{(P^2 - Q^2\cos^2 T - f^2\sin^2 T)\,(P^2 - Q^2\cos^2 T)^{\frac{3}{2}}};$$

and then, writing

$$\cos T = \frac{P\cos\phi - Q}{P - Q\cos\phi},$$

$$\sin T = \frac{\sqrt[4]{\Omega}\sin\phi}{P - Q\cos\phi},$$

we bring in the variable ϕ. But it is important to remark that this is not the quantity which was, at the beginning of the investigation, represented by this letter, and that it is not easy to see the connexion between the two quantities ϕ. We find

$$V = f^2\int_0^{2\pi}\frac{(P - Q\cos\phi)^2(P\cos\phi - Q)^2\,d\phi}{(a^2+c^2+f^2\cos^2\phi - 2af\cos\phi)(a^2+c^2+f^2-2af\cos\phi)^{\frac{3}{2}}}.$$

26. To reduce this, write as before

$$\Delta = \sqrt{a^2+c^2+f^2-2af\cos\phi},$$

and also

$$\Phi = a^2+c^2-2af\cos\phi+f^2\cos^2\phi,$$

so that the denominator in the integral is $=\Phi\Delta^3$.

We have

$$\begin{aligned}(P - Q\cos\phi)^2(P\cos\phi - Q)^2 &= (\Delta^2 - Q^2\sin^2\phi)(\Delta^2 - P^2\sin^2\phi),\\ &= \Delta^4 - (a^2+c^2+f^2)\,\Delta^2\sin^2\phi + a^2f^2\sin^4\phi,\\ &= \Delta^2\{\Delta^2 - (c^2+f^2)\sin^2\phi\} - a^2\sin^2\phi\,(\Delta^2 - f^2\sin^2\phi),\\ &= \Delta^2\{\Delta^2 - (c^2+f^2)\sin^2\phi\} - a^2\sin^2\phi\,.\,\Phi,\end{aligned}$$

and hence

$$V = \int\frac{f^2[\Delta^2 - (c^2+f^2)\sin^2\phi]\,d\phi}{\Phi\Delta} - a^2f^2\int\frac{\sin^2\phi\,d\phi}{\Delta^3},$$

the limits being always 0, 2π. But we have identically

$$\frac{d}{d\phi}\,\frac{\sin\phi}{\Delta} = \frac{\cos\phi}{\Delta} - \frac{af\sin^2\phi}{\Delta^3},$$

and thence

$$\int\frac{\sin^2\phi\,d\phi}{\Delta^3} = -\frac{1}{af}\left(\frac{\sin\phi}{\Delta}\right) - \frac{1}{af}\int\frac{\cos\phi\,d\phi}{\Delta},$$

where the term $\left(\frac{\sin\phi}{\Delta}\right)$ is to be taken between the limits, but for the present I retain it as it stands. Moreover, $\Delta^2 = \Phi + f^2\sin^2\phi$, and consequently

$$\Delta^2 - (c^2+f^2)\sin^2\phi = \Phi - c^2\sin^2\phi,$$

and we thus obtain the result

$$V = af\left(\frac{\sin\phi}{\Delta}\right) - af\int\frac{\cos\phi\,d\phi}{\Delta} + f^2\int\frac{d\phi}{\Delta} - c^2f^2\int\frac{\sin^2\phi\,d\phi}{\Phi\Delta},$$

where the denominators under the integral signs are

$$\Delta, = \sqrt{a^2+c^2+f^2-2af\cos\phi}, \text{ and } \Phi\Delta, = (a^2+c^2-2af\cos\phi+f^2\cos^2\phi)\,\Delta.$$

27. We may, by a transformation such as that for the change of parameter in an elliptic integral of the third kind, make the denominators to be Δ and $(c^2+a^2\sin^2\phi)\Delta$; viz. for this purpose we assume $\Lambda=\tan^{-1}\dfrac{B\Delta}{A}$, where B and A are functions of ϕ such that we have identically $A^2+B^2\Delta^2=(c^2+a^2\sin^2\phi)(a^2+c^2-2af\cos\phi+f^2\cos^2\phi)$; the values of B, A are found to be $c\cos\phi$ and $\sin\phi\,(a^2+c^2-af\cos\phi)$, whence, dividing each of these for greater convenience by $\sin\phi$, we have

$$\Lambda=\tan^{-1}\left(\frac{c\cot\phi\,\Delta}{a^2+c^2-af\cos\phi}\right),$$

so that, writing now B, $A=c\cot\phi$ and $a^2+c^2-af\cos\phi$ respectively, the value is

$$\Lambda=\tan^{-1}\left(\frac{B\Delta}{A}\right),$$

where

$$A^2+B^2\Delta^2=\frac{1}{\sin^2\phi}\,\Pi\Phi;$$

and, as before, $\Phi=a^2+c^2-2af\cos\phi+f^2\cos^2\phi$, and also $\Pi=c^2+a^2\sin^2\phi$. We have

$$\frac{d\Lambda}{d\phi}=\frac{(AB'-A'B)\,\Delta^2+\tfrac{1}{2}AB\,(\Delta^2)'}{(A^2+B^2\Delta^2)\,\Delta},\quad\left(A'=\frac{dA}{d\phi},\ \&\text{c.}\right);$$

and then

$$AB'-A'B=\frac{a}{\sin^2\phi}(-a^2-c^2+af\cos^3\phi),$$

$$\tfrac{1}{2}AB\,(\Delta^2)'=\frac{c\sin\phi\cos\phi}{\sin^2\phi}(a^2+c^2-af\cos\phi)\,af\sin\phi,$$

and the numerator thus is

$$\frac{c}{\sin^2\phi}\{(-a^2-c^2+af\cos^3\phi)(a^2+c^2+f^2-2af\cos\phi)$$
$$+af\cos\phi\,(1-\cos^2\phi)(a^2+c^2-af\cos\phi)\},$$

which is in fact

$$=\frac{c}{\sin^2\phi}\{-(c^2+a^2\sin^2\phi)(a^2+c^2+f^2-2af\cos\phi)$$
$$+(af\cos\phi-a^2\cos^2\phi)(a^2+c^2-2af\cos\phi+f^2\cos^2\phi)\},$$

$$=\frac{c}{\sin^2\phi}\{-\Pi\Delta^2+(af\cos\phi-a^2\cos^2\phi)\,\Phi\};$$

or, what is the same thing,

$$=\frac{c}{\sin^2\phi}\{-\Pi\phi-\Pi f^2\sin^2\phi+(af\cos\phi-a^2\cos^2\phi)\,\Phi\},$$

and the denominator, by what precedes, is

$$=\frac{1}{\sin^2\phi}\,.\,\Pi\Phi\Delta.$$

We thus have

$$\frac{1}{c}\frac{d\Lambda}{d\phi} = -\frac{1}{\Delta} - \frac{f^2\sin^2\phi}{\Phi\Delta} + \frac{af\cos\phi - a^2\cos^2\phi}{\Pi\Delta},$$

whence, by integration,

$$\frac{1}{c}\tan^{-1}\left(\frac{c\cot\phi\,\Delta}{a^2+c^2-af\cos\phi}\right) = -\int\frac{d\phi}{\Delta} + \int\frac{(af\cos\phi - a^2\cos^2\phi)\,d\phi}{\Pi\Delta} - f^2\int\frac{\sin^2\phi\,d\phi}{\Phi\Delta},$$

which is the required formula of transformation.

28. Multiplying by c^2, and subtracting from the value of V, we find

$$V = c\tan^{-1}\left(\frac{c\cot\phi\,\Delta}{a^2+c^2-af\cos\phi}\right) + af\left(\frac{\sin\phi}{\Delta}\right)$$

$$+\int\frac{(c^2+f^2-af\cos\phi)d\phi}{\Delta} - c^2a\int\frac{\cos\phi\,(f-a\cos\phi)\,d\phi}{(c^2+a^2\sin^2\phi)\,\Delta},$$

which is to be taken between the limits 0 and 2π; viz. we thus have

$$V = -2c\pi + 2\int_0^\pi\frac{(c^2+f^2-af\cos\phi)\,d\phi}{\Delta} - 2c^2a\int_0^\pi\frac{\cos\phi\,(f-a\cos\phi)\,d\phi}{(c^2+a^2\sin^2\phi)\,\Delta},$$

agreeing with a former result.

29. But this former result, previous to the final step of taking the integrals between the limits, was

$$V = 2a\sin\phi\log\left(\frac{f-a\cos\phi+\Delta}{-a\cos\phi+\sqrt{a^2+c^2}}\right) - c\tan^{-1}\left(\frac{\sqrt{a^2+c^2}\tan\phi}{c}\right)$$

$$+\int\frac{(c^2+f^2-af\cos\phi)\,d\phi}{\Delta} - c^2a\int\frac{\cos\phi\,(f-a\cos\phi)\,d\phi}{(c^2+a^2\sin^2\phi)\,\Delta};$$

viz. the integrals are the same, but the integrated terms are altogether different; the explanation of course is that the ϕ's are different in the two formulæ, which therefore do not correspond element by element but only in their ultimate value between the limits.

30. In order to discuss numerically the Potential of the Circle,

$$V = 2f^2\int_\theta^\infty\frac{\sqrt{\left(t-\theta\,.\,t+\dfrac{c^2f^2}{\theta}\right)}\,dt}{t\,(t+f^2)\,\sqrt{t+f^2}},$$

this must be reduced to elliptic functions. Writing $t=\theta+x^2$, we have

$$V = 4f^2\int_0^\infty\frac{x^2\sqrt{x^2+\beta^2}\,dx}{(x^2+\theta)\,(x^2+\alpha^2)^{\frac{3}{2}}};$$

if for shortness

$$\theta + f^2 \quad = \alpha^2,$$

$$\theta + \frac{c^2 f^2}{\theta} = \beta^2.$$

The constants α, β, θ may be considered as replacing the original constants a, c, f; viz. from the last two equations and the equation

$$\frac{a^2}{\theta + f^2} + \frac{c^2}{\theta} = 1,$$

we deduce

$$a^2 = \frac{\alpha^2(\alpha^2 - \beta^2)}{\alpha^2 - \theta}, \quad c^2 = \frac{\theta(\beta^2 - \theta)}{\alpha^2 - \theta}, \quad f^2 = \alpha^2 - \theta;$$

showing that α^2, β^2, θ are in order of decreasing magnitude; viz. $\alpha^2 - \beta^2$, $\beta^2 - \theta$, $\alpha^2 - \theta$ are all positive. The formula may be written

$$\tfrac{1}{4}V = (\alpha^2 - \theta)\int_0^\infty \frac{x^2(x^2+\beta^2)\,dx}{(x^2+\theta)(x^2+\alpha^2)\sqrt{x^2+\alpha^2 \,.\, x^2+\beta^2}};$$

which, in virtue of the identity

$$(\alpha^2 - \theta)\,x^2(x^2+\beta^2) = (\alpha^2-\theta)(x^2+\theta)(x^2+\alpha^2) - \alpha^2(\alpha^2-\beta^2)(x^2+\theta) - \theta(\beta^2-\theta)(x^2+\alpha^2),$$

becomes

$$\begin{aligned}\tfrac{1}{4}V = &\quad (\alpha^2-\theta)\int_0^\infty \frac{dx}{\sqrt{x^2+\alpha^2 \,.\, x^2+\beta^2}}\\ &- \alpha^2(\alpha^2-\beta^2)\int_0^\infty \frac{dx}{(x^2+\alpha^2)\sqrt{x^2+\alpha^2 \,.\, x^2+\beta^2}}\\ &- \theta(\beta^2-\theta)\int_0^\infty \frac{dx}{(x^2+\theta)\sqrt{x^2+\alpha^2 \,.\, x^2+\beta^2}}.\end{aligned}$$

31. Writing here $x = \alpha \cot u$, and therefore $dx = -\alpha \operatorname{cosec}^2 u\, du$, to the values $x = \infty, 0$ correspond $u = 0, \tfrac{1}{2}\pi$, and we have

$$\tfrac{1}{4}V = \int_0^{\frac{1}{2}\pi} \frac{du}{\sqrt{\alpha^2\cos^2 u + \beta^2 \sin^2 u}}\left\{\alpha^2 - \theta - (\alpha^2-\beta^2)\sin^2 u - \frac{\theta(\beta^2-\theta)\sin^2 u}{\alpha^2\cos^2 u + \theta\sin^2 u}\right\}$$

$$= \int_0^{\frac{1}{2}\pi} \frac{du}{\sqrt{\alpha^2\cos^2 u + \beta^2\sin^2 u}}\left\{\alpha^2 - \theta + \frac{\theta(\beta^2-\theta)}{\alpha^2-\theta} - (\alpha^2-\beta^2)\sin^2 u - \frac{\alpha^2\theta(\beta^2-\theta)}{\alpha^2-\theta}\,\frac{1}{\alpha^2\cos^2 u + \theta\sin^2 u}\right\}.$$

Writing $k^2 = 1 - \dfrac{\beta^2}{\alpha^2}$, we have

$$\sqrt{\alpha^2\cos^2 u + \beta^2\sin^2 u} = \alpha\sqrt{1 - k^2\sin^2 u},$$

and thence

$$\tfrac{1}{4}V = \int_0^{\frac{1}{2}\pi} \frac{du}{\sqrt{1-k^2\sin^2 u}} \times \left\{\alpha(1-k^2\sin^2 u) - \frac{k^2\theta}{1-\dfrac{\theta}{\alpha^2}} - \frac{\dfrac{\theta}{\alpha}\,\dfrac{\beta^2-\theta}{\alpha^2}}{1-\dfrac{\theta}{\alpha^2}}\,\frac{1}{1-\left(1-\dfrac{\theta}{\alpha^2}\right)\sin^2 u}\right\};$$

viz. writing $n=-1+\frac{\theta}{\alpha^2}$ (so that n is negative and in absolute magnitude <1), and moreover $\beta^2=\alpha^2k'^2$ and $\theta=(n+1)\alpha^2$, this is

$$\tfrac{1}{4}V=\int_0^{\frac{1}{2}\pi}\frac{du}{\sqrt{1-k^2\sin^2 u}}\times\left\{(1-k^2\sin^2 u)\,\alpha+\frac{k^2}{n}(n+1)\,\alpha-\frac{n+1\,.\,n+k^2}{n}\,\alpha\,\frac{1}{1+n\sin^2 u}\right\};$$

viz. this is

$$=\alpha\left\{E_1k+k^2\frac{n+1}{n}F_1k-\frac{n+1\,.\,n+k^2}{n}\Pi_1(n,k)\right\}.$$

32. This may be further reduced by substituting for the complete function $\Pi_1(n,k)$, its value; viz. writing

$$n=\left(-1+\frac{\theta}{\alpha^2}\right)=-1+k'^2\sin^2\lambda,$$

that is, $\sin^2\lambda=\frac{\theta}{\beta^2}$; then, writing the value first in the form

$$\alpha\left\{E_1k-(n+1)F_1k-\frac{n+1\,.\,n+k^2}{n}[\Pi_1(n,k)-F_1k]\right\},$$

and observing that

$$\frac{n+1\,.\,n+k^2}{n}[\Pi_1(n,k)-F_1k]=\frac{k'^4\sin^2\lambda\cos^2\lambda}{1-k'^2\sin^2\lambda}[\Pi_1(n,k)-F_1k]$$

$$=\frac{k'^2\sin\lambda\cos\lambda}{\sqrt{1-k'^2\sin^2\lambda}}\left\{\tfrac{1}{2}\pi+(F_1k-E_1k)\,F(k',\lambda)-F_1k\,.\,E(k',\lambda)\right\},$$

we have

$$\tfrac{1}{4}V=\alpha\left\{E_1k-k'^2\sin^2\lambda\,F_1k-\frac{k'^2\sin\lambda\cos\lambda}{\sqrt{1-k'^2\sin^2\lambda}}[\tfrac{1}{2}\pi+(F_1k-E_1k)\,F(k',\lambda)-F_1k\,.\,E(k',\lambda)]\right\},$$

where

$$\alpha^2=\theta+f^2,\quad k^2=1-\frac{\beta^2}{\alpha^2},\quad =1-\frac{\theta+\frac{c^2f^2}{\theta}}{\theta+f^2},\quad =\frac{f^2\left(1-\frac{c^2}{\theta}\right)}{\theta+f^2},$$

or, what is the same thing,

$$k=\frac{af}{f^2+\theta},\quad \sin^2\lambda=\frac{\theta}{\beta^2},\quad =\frac{1}{1+\frac{c^2f^2}{\theta^2}},$$

θ being, it will be recollected, the positive root of

$$\frac{a^2}{f^2+\theta}+\frac{c^2}{\theta}=1.$$

33. Thus when in particular $a=0$, we have $\theta=c^2$, and thence

$$\alpha=\sqrt{c^2+f^2},\quad k=0,\quad k'=1,\quad \sin\lambda=\frac{c}{\sqrt{c^2+f^2}};$$

whence

$$\tfrac{1}{4}V = \tfrac{1}{2}\pi\sqrt{c^2+f^2}\,\{1-\sin^2\lambda - \sin\lambda\,(1-\sin\lambda)\},$$

$$= \tfrac{1}{2}\pi\sqrt{c^2+f^2}\,(1-\sin\lambda), \quad = \tfrac{1}{2}\pi\,(\sqrt{c^2+f^2}-c),$$

or

$$V = 2\pi\,(\sqrt{c^2+f^2}-c),$$

which is right.

34. If $c=0$, a being $>f$, then $\theta = a^2-f^2$, $k=\dfrac{f}{a}$, $\lambda=\tfrac{1}{2}\pi$, $\alpha = a$; so that, retaining k as standing for its value $\dfrac{f}{a}$, we have

$$\tfrac{1}{4}V = a\,(E_1k - k'^2F_1k), \text{ or } V = 4a\,(E_1k-k'^2\,F_1k),$$

which may easily be verified.

If $c=0$, a being $<f$, then, recurring to the original equation for the determination of θ, viz. $(\theta+f^2)^2\,\theta\left(\dfrac{a^2}{\theta+f^2}+\dfrac{c^2}{\theta}-1\right)=0$, which for $c=0$ becomes $\theta\,(\theta+f^2)(\theta-a^2+f^2)=0$, we have (as the positive root of this equation) $\theta=0$; whence $\alpha=f$; also, observing that $1-\dfrac{c^2}{\theta}=\dfrac{a^2}{f^2}$, $k=\dfrac{a}{f}$, and $\sin^2\lambda = \dfrac{\theta}{\theta+\dfrac{c^2}{\theta}f^2}$ $\left(\text{where } \dfrac{c^2}{\theta} \text{ is finite}\right)$, $=0$, and retaining k to denote its value $=\dfrac{a}{f}$, we obtain $\tfrac{1}{4}V=fE_1k$, or $V=4fE_1k$.

If $a=f$, then in each of the formulæ $k=1$; and since in the first formula k'^2F_1k, k nearly $=1$, is $=k'^2\log\dfrac{4}{k'}$, vanishing for $k=1$ or $k'=0$, we have $V=4fE_11$, $=4f$.

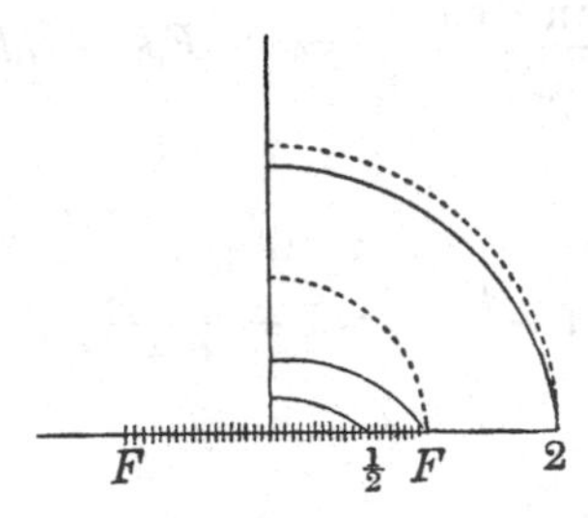

Section of Equipotential surfaces of a Circle.

It would be interesting to consider the value of the potential at different points of the ellipse $\dfrac{a^2}{f^2+\theta}+\dfrac{c^2}{\theta}=1$ (θ constant, a, c current coordinates). For this purpose writing $a=\sqrt{f^2+\theta}\cos q$, $c=\sqrt{\theta}\sin q$, we should have $\alpha=\sqrt{f^2+\theta}$ (a constant), and

$$k=\frac{f\cos q}{\sqrt{f^2+\theta}}, \quad k'=\frac{\sqrt{\theta+f^2\sin^2 q}}{\sqrt{f^2+\theta}},$$

$$\sin\lambda = \frac{\sqrt{\theta}}{\sqrt{\theta+f^2\sin^2 q}}, \quad \cos\lambda=\frac{f\sin q}{\sqrt{\theta+f^2\sin^2 q}};$$

and then V through k, k', λ, is a given function of q.

35. Suppose, to fix the ideas, $f=1$, and consider the points $(0, c)$ and $(a, 0)$, which have equal potentials. First, if $a > f$ (that is, $a > 1$), then writing $k=\frac{1}{a}$, the relation is

$$2\pi\left(\sqrt{1+c^2}-c\right)=\frac{4}{k}\left(E_1k-k'^2F_1k\right);$$

and we have

$$F_1\,30^\circ=1{\cdot}68575,\quad E_1\,30^\circ=1{\cdot}46746,\quad \frac{4}{\pi}=1{\cdot}27324.$$

Secondly, if $a < f$ (that is, $a < 1$), then writing $k=a$, the relation is

$$2\pi\left(\sqrt{1+c^2}-c\right)=4E_1k.$$

(1) In particular $a=\frac{1}{2}$, $=\sin 30^\circ$, this is

$$\sqrt{1+c^2}-c=\frac{2}{\pi}E_1 30^\circ \qquad = {\cdot}93421.$$

(2) $a=1$, then

$$\sqrt{1+c^2}-c=\frac{2}{\pi} \qquad = {\cdot}63662.$$

(3) $a=2$, $k=\frac{1}{2}$, $=\sin 30^\circ$,

$$\sqrt{1+c^2}-c=\frac{4}{\pi}\left\{E_1(30^\circ)-\tfrac{3}{4}F_1(30^\circ)\right\} = {\cdot}25866.$$

But if $\sqrt{1+c^2}-c=m$, then $c=\frac{1}{2}\left(\frac{1}{m}-m\right)$; whence

a	c
0	·0
$\frac{1}{2}$	·06810
1	·46709
2	1·80376

for the values of c, corresponding to the foregoing values of a.

604.

DETERMINATION OF THE ATTRACTION OF AN ELLIPSOIDAL SHELL ON AN EXTERIOR POINT.

[From the *Proceedings of the London Mathematical Society*, vol. VI. (1874—1875), pp. 58—67. Read January 14, 1875.]

THE shell in question is the indefinitely thin shell included between two concentric, similar, and similarly situated ellipsoidal surfaces, the density being uniform and the attraction varying as the inverse square of the distance.

It was shown by Poisson that the attraction was in the direction of the axis of the circumscribed cone, and expressible in finite terms; the theorem as to the direction of the attraction was afterwards demonstrated geometrically by Steiner, *Crelle*, t. XII. (1834), his method being to divide the shell into elements by means of conical surfaces having their vertices at an interior point Q; and the investigation was about two years ago completed by Prof. Adams, so as to obtain from it the finite expression for the attraction of the shell. The process was explained in a lecture at which I was present: I did not particularly attend to the details of it; and I now reproduce the solution in my own form, stating, in the first place, the geometrical theorems on which it depends.

Statement of the Geometrical Theorems.

1. We consider (see figure, p. 305) an ellipsoid, and two corresponding points, an external point P, and an internal point Q; as will appear, the correspondence is not a reciprocal one. The points are such that each of them is, in regard to the ellipsoid, in the polar plane of the other; moreover PQ is the perpendicular at P to the polar plane of Q; that is, Q being regarded as given, then P is determined as the foot of the perpendicular let fall from Q upon its polar plane; to a given position of Q there corresponds thus a single position of P. It follows that PQ is the normal at

P to the confocal ellipsoid through this point; that is, given the position of P, then Q is the intersection of the polar plane of P by the normal at P to the confocal ellipsoid. Analytically, to a given position of P, there correspond three positions of Q, viz. these are the intersections of the polar plane of P by the normals at P to the three confocal surfaces through this point, and the correspondence of the points P, Q is a (1, 3) correspondence: but the other two positions of Q are external to the ellipsoid, and we are not concerned with them; we determine Q as above by means of the normal to the confocal ellipsoid.

2. If through the point Q we draw at pleasure a chord $R'QR''$, and join the extremities R', R'' with P, then the line PQ bisects the angle $R'PR''$; whence also $PR' : QR' = PR'' : QR''$, or writing $QR', QR'' = r', r''$ and $PR', PR'' = \rho', \rho''$, then $\frac{\rho'}{r'} = \frac{\rho''}{r''}$. Putting each of these equal ratios $= \frac{\Omega}{R}$, where Ω is a length depending on the position of Q but independent of the direction of the chord $R'QR''$, then R will be a length depending on the direction of the chord, and if along the chord (say in the sense Q to R') we measure off from Q a length $QT, = R$, thence the locus of the extremity T of this line will be an ellipsoid, centre Q, similarly situate to the given ellipsoid, say this is the "auxiliary ellipsoid."

Consider now the given ellipsoid and a concentric and similarly situated similar ellipsoid, exterior to and indefinitely near it. To fix the ideas, let the semi-axes of the given ellipsoid be mf, mg, mh, and those of the consecutive ellipsoid be $(m+dm)f$, $(m+dm)g$, $(m+dm)h$. Producing the chord $R'R''$ to meet the consecutive ellipsoid in S', S'', then the radial thicknesses $R'S'$, $R''S''$ of the included shell will be equal to each other, or say each $= \Lambda dm$, where Λ is a quantity dependent as well on the position of the point Q as on the direction of the chord $R'R''$ through this point.

3. Let 2ϕ denote the angle $R'PR''$, or, what is the same thing, let ϕ denote either of the equal angles $R'PQ$, $R''PQ$; then, R, Λ being as above, it is found that $\cos\phi = \frac{mR}{\Lambda}$.

Determination of the Attraction of the Shell.

4. We may now solve the attraction-problem. We consider the indefinitely thin shell (density unity) included between the given ellipsoid and the consecutive ellipsoid, and attracting the exterior point P. We determine the corresponding interior point Q, and then dividing the shell into elements by means of indefinitely thin cones having their vertices at Q, we consider in conjunction the elements determined by any two opposite cones, say the two opposite cones, having for their axis the chord $R'QR''$ and a spherical aperture $= d\omega$. The shell-element at R' is

$$r'^2 d\omega \,.\, R'S' = r'^2 \Lambda d\omega\, dm;$$

its attraction on P is therefore

$$\frac{r'^2}{\rho'^2}\Lambda\, d\omega\, dm, = \frac{R^2}{\Omega^2}\Lambda\, dm\, d\omega,$$

and the attractions in the directions QR' and PQ are this quantity multiplied by $\sin\phi$ and $\cos\phi$ respectively.

5. But the shell-element at R'' exerts upon P the same attraction $\frac{R^2}{\Omega^2}\Lambda\, dm\, d\omega$, and the attractions in the directions QR'' and PQ are this quantity multiplied by $\sin\phi$ and $\cos\phi$ respectively: hence the attractions in the directions QR', QR'' exactly counterbalance each other, and there remain only the two equal attractions in the direction PQ; viz. this, for either of the elements in question, say for the element at R', is

$$= \frac{R^2}{\Omega^2}\Lambda \cos\phi\, dm\, d\omega,$$

or, substituting for $\cos\phi$ its value, $= \frac{mR}{\Lambda}$, this is

$$= \frac{m\, dm}{\Omega^2} R^3\, d\omega.$$

Hence the whole attraction of the shell is in the direction PQ, its value being

$$\frac{m\, dm}{\Omega^2}\iint R^3\, d\omega,$$

over the whole solid angle at Q; and recollecting that R denotes the radius vector in the auxiliary ellipsoid, we have the volume of this ellipsoid

$$= \iiint r^2\, dr\, d\omega = \tfrac{1}{3}\iint R^3\, d\omega,$$

that is, $\iint R^3\, d\omega =$ thrice the volume of the auxiliary ellipsoid, $= 4\pi FGH$, if F, G, H are the semiaxes of the auxiliary ellipsoid. That is,

$$\text{Attraction of shell} = \frac{m\, dm}{\Omega^2} 4\pi FGH.$$

The problem is now solved; but it remains to prove the geometrical theorems, and to determine the values of the quantities Ω, F, G, H, which enter into the expression for the attraction; and we may also deduce the formula for the attractions of a solid ellipsoid.

Proof of the Geometrical Theorems.

6. I take

$$\frac{x^2}{f^2} + \frac{y^2}{g^2} + \frac{z^2}{h^2} = m^2$$

for the equation of the ellipsoid; a, b, c for the coordinates of P; ξ, η, ζ for those of Q; α, β, γ for the cosine-inclinations of the radius QR' to the axes. Hence, in the equation of the ellipsoid, substituting for x, y, z the values $\xi + r\alpha,\ \eta + r\beta,\ \zeta + r\gamma$, and writing for shortness

$$A = \frac{\alpha^2}{f^2} + \frac{\beta^2}{g^2} + \frac{\gamma^2}{h^2},$$

$$B = \frac{\alpha\xi}{f^2} + \frac{\beta\eta}{g^2} + \frac{\gamma\zeta}{h^2},$$

$$C = \frac{\xi^2}{f^2} + \frac{\eta^2}{g^2} + \frac{\zeta^2}{h^2} - m^2, \text{ (C being therefore negative)},$$

we have $r', -r''$ as the roots of the equation

$$Ar^2 + 2Br + C = 0;$$

viz.

$$\frac{2B}{A} = -r' + r'',\ \frac{C}{A} = -r'r'',$$

and thence

$$r' = \frac{-B + \sqrt{B^2 - AC}}{A},\ r'' = \frac{B + \sqrt{B^2 - AC}}{A},\ r' + r'' = \frac{2\sqrt{B^2 - AC}}{A}.$$

7. Suppose for a moment that the semidiameter parallel to $R'R''$ is $= mv$; we have evidently $v^2 = \frac{1}{A}$. And then, if in the central section through $R'R''$ the conjugate semidiameter is mu, the equation of the section referred to these conjugate axes will be $\frac{x^2}{m^2u^2} + \frac{y^2}{m^2v^2} = 1$, or say, $y^2 = m^2v^2 - \frac{v^2}{u^2}x^2$, where y is the coordinate parallel to $R'R''$, so that, taking the coordinate to belong to the point R', we have

$$y = \tfrac{1}{2}(r' + r'') = \frac{\sqrt{B^2 - AC}}{A}.$$

For the exterior surface of the shell, m is to be changed into $m + dm$; hence, y and m alone varying, we have

$$y\,dy = mv^2\,dm,\ = m\,dm\,\frac{1}{A},$$

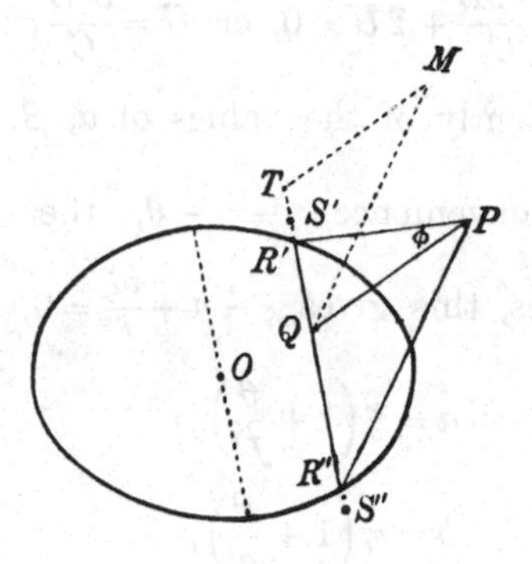

that is,

$$dy = m\,dm\,\frac{1}{\sqrt{B^2 - AC}};$$

viz. this is the value of the radial thickness $R'S'$ of the shell; or, since the same process applies to the point R'', we have

$$R'S' = R''S'' = m\,dm\,\frac{1}{\sqrt{B^2 - AC}},$$

or, calling this, as above, $\Lambda\,dm$, the value of Λ is $= \dfrac{m}{\sqrt{B^2 - AC}}$.

8. The points P and Q are connected by the condition that, for every direction whatever of the chord $R'R''$, we have

$$PR' : PR'' = QR' : QR'',$$

or, what is the same thing, that the line QP bisects the angle $R'PR''$. Taking $PR' = \rho'$, $PR'' = \rho''$, the condition is $\rho' : r' = \rho'' : r''$; and taking (a, b, c) as the coordinates of the point P, we have

$$\begin{aligned}\rho'^2 &= (\xi + r'\alpha - a)^2 + (\eta + r'\beta - b)^2 + (\zeta + r'\gamma - c)^2\\ &= \sigma^2 + 2r'U + r'^2,\end{aligned}$$

if, for shortness,

$$\begin{aligned}\sigma^2 &= (\xi - a)^2 + (\eta - b)^2 + (\zeta - c)^2,\ (= \overline{QP}^2),\\ U &= \alpha(\xi - a) + \beta(\eta - b) + \gamma(\zeta - c);\end{aligned}$$

and similarly

$$\rho''^2 = \sigma^2 - 2r''U + r''^2.$$

The required condition therefore is

$$\frac{\sigma^2}{r'^2} + \frac{2}{r'}U = \frac{\sigma^2}{r''^2} - \frac{2}{r''}U,$$

viz. this is

$$\sigma^2\left(\frac{1}{r'^2} - \frac{1}{r''^2}\right) + 2U\left(\frac{1}{r'} + \frac{1}{r''}\right) = 0,$$

so that, omitting a factor, it becomes

$$\sigma^2\left(\frac{1}{r'} - \frac{1}{r''}\right) + 2U = 0,$$

that is,

$$\sigma^2 . \frac{-2B}{C} + 2U = 0, \text{ or } U = \frac{\sigma^2 B}{C},$$

which must be satisfied independently of the values of α, β, γ.

9. Writing, for greater convenience, $\dfrac{\sigma^2}{C} = -\theta$, the equation is $U = -\theta B$, viz. substituting for U, B their values, this gives $\xi - a + \dfrac{\theta\xi}{f^2} = 0$, &c., or say,

$$\begin{aligned}a &= \xi\left(1 + \frac{\theta}{f^2}\right),\\ b &= \eta\left(1 + \frac{\theta}{g^2}\right),\\ c &= \zeta\left(1 + \frac{\theta}{h^2}\right);\end{aligned}$$

and the assumed relation $\frac{\sigma^2}{C} = -\theta$ is

$$(\xi - a)^2 + (\eta - b)^2 + (\zeta - c)^2 = -\theta\left(\frac{\xi^2}{f^2} + \frac{\eta^2}{g^2} + \frac{\zeta^2}{h^2} - m^2\right);$$

viz. substituting for a, b, c the foregoing values, and omitting a factor θ, this is

$$\theta\left(\frac{\xi^2}{f^4} + \frac{\eta^2}{g^4} + \frac{\zeta^2}{h^4}\right) = -\left(\frac{\xi^2}{f^2} + \frac{\eta^2}{g^2} + \frac{\zeta^2}{h^2} - m^2\right);$$

or, writing for shortness

$$\frac{1}{\Omega^2} = \left(\frac{\xi^2}{f^4} + \frac{\eta^2}{g^4} + \frac{\zeta^2}{h^4} - m^2\right),$$

the equation is

$$\theta = -\Omega^2 C.$$

We thus see that, (ξ, η, ζ) being given, θ, and therefore also (a, b, c), are uniquely determined. It may be added that, writing $C = -\frac{\sigma^2}{\theta}$, we have $\theta^2 = \Omega^2\sigma^2$, or say $\Omega\sigma = \theta$.

10. We have, moreover,

$$\frac{a^2}{\theta + f^2} = \frac{\xi^2}{f^4}(\theta + f^2), \text{ \&c.,}$$

and

$$\frac{\xi^2}{f^4}(\theta + f^2) + \frac{\eta^2}{g^4}(\theta + g^2) + \frac{\zeta^2}{h^4}(\theta + h^2) = \frac{1}{\Omega^2}\theta + \frac{\xi^2}{f^2} + \frac{\eta^2}{g^2} + \frac{\zeta^2}{h^2}$$

$$= -C + m^2 + C, \; = m^2;$$

whence

$$\frac{a^2}{\theta + f^2} + \frac{b^2}{\theta + g^2} + \frac{c^2}{\theta + h^2} = m^2,$$

or, regarding (a, b, c) as given, θ is determined as a function of (a, b, c) by this cubic equation; and θ being (in accordance with the foregoing equation $\theta = -\Omega^2 C$) assumed to be positive, we have θ the positive root of this equation, and $m^2(\theta + f^2)$, $m^2(\theta + g^2)$, $m^2(\theta + h^2)$ as the squared semiaxes of the confocal ellipsoid through the point P. And θ being known, ξ, η, ζ are, by the foregoing equations $a = \xi\left(1 + \frac{\theta}{f^2}\right)$, &c., determined in terms of ξ, η, ζ; that is, starting from the given external point P, we have the internal point Q. And it appears that PQ is the normal at P to the confocal ellipsoid, or, what is the same thing, the axis of the circumscribed cone, vertex P.

11. The foregoing equation

$$\frac{\xi^2}{f^4}(\theta + f^2) + \frac{\eta^2}{g^4}(\theta + g^2) + \frac{\zeta^2}{h^4}(\theta + h^2) = m^2,$$

considering a, b, c, and therefore θ, as given, shows further that the point Q is situate on an ellipsoid which is the inverse of the confocal ellipsoid $\frac{x^2}{\theta + f^2} + \frac{y^2}{\theta + g^2} + \frac{z^2}{\theta + h^2} = m^2$ in regard to the given ellipsoid $\frac{x^2}{f^2} + \frac{y^2}{g^2} + \frac{z^2}{h^2} = m^2$.

12. Expressing Ω in terms of a, b, c, we have

$$\frac{1}{\Omega^2} = \frac{a^2}{(\theta+f^2)^2} + \frac{b^2}{(\theta+g^2)^2} + \frac{c^2}{(\theta+h^2)^2}.$$

We have $\sigma^2 = \frac{\theta^2}{\Omega^2}, = C^2\Omega^2$, and

$$U = \alpha(\xi - a) + \beta(\eta - b) + \gamma(\zeta - c),$$

$$= -\theta\left(\frac{\alpha\xi}{f^2} + \frac{\beta\eta}{g^2} + \frac{\gamma\zeta}{h^2}\right), \quad = -\theta B, \quad = BC\Omega^2;$$

whence

$$\frac{\rho'^2}{r'^2} = \frac{\sigma^2}{r'^2} + 2U\frac{1}{r'} + 1, \quad = \frac{C^2\Omega^2}{r'^2} + \frac{2BC\Omega^2}{r'} + 1,$$

$$= C\Omega^2\left(\frac{C}{r'^2} + \frac{2B}{r'}\right) + 1;$$

or, since

$$A + 2B\frac{1}{r'} + C\frac{1}{r'^2} = 0,$$

this is

$$\frac{\rho'^2}{r'^2} = -AC\Omega^2 + 1 = \Omega^2\left(\frac{1}{\Omega^2} - AC\right)$$

$$= \frac{\Omega^2}{R^2}, \text{ if } \frac{1}{R^2} = \frac{1}{\Omega^2} - AC.$$

This last equation may also be written

$$\frac{1}{R^2} = \frac{1}{\Omega^2}(\alpha^2 + \beta^2 + \gamma^2) - C\left(\frac{\alpha^2}{f^2} + \frac{\beta^2}{g^2} + \frac{\gamma^2}{h^2}\right);$$

or, what is the same thing,

$$\frac{1}{R^2} = \frac{\alpha^2}{F^2} + \frac{\beta^2}{G^2} + \frac{\gamma^2}{H^2};$$

if for shortness

$$\frac{1}{F^2} = \frac{1}{\Omega^2} - \frac{C}{f^2},$$

$$\frac{1}{G^2} = \frac{1}{\Omega^2} - \frac{C}{g^2},$$

$$\frac{1}{H^2} = \frac{1}{\Omega^2} - \frac{C}{h^2};$$

viz. substituting herein for C its value $-\frac{\theta}{\Omega^2}$, these equations give

$$F = \frac{\Omega f}{\sqrt{\theta + f^2}}, \quad G = \frac{\Omega g}{\sqrt{\theta + g^2}}, \quad H = \frac{\Omega h}{\sqrt{\theta + h^2}};$$

where Ω stands for its expression in terms of a, b, c.

13. The expression for $\frac{1}{R^2}$ shows that R is the radius vector, cosine-inclinations α, β, γ, in an ellipsoid semi-axes F, G, H, which may be regarded as having its centre at Q; viz. this is the "auxiliary ellipsoid." And this being so, we have

$$\frac{\rho'}{r'} = \frac{\rho''}{r''} = \frac{\Omega}{R}.$$

It appears from these equations that, drawing from Q parallel to PR'' a line QM, $=\Omega$, and from its extremity M parallel to PQ a line to meet QR' in T, the locus of T is the auxiliary ellipsoid.

14. By what precedes, the angles $R'PQ$, $R''PQ$ are equal to each other, say each is $=\phi$; the triangle $R'PR''$ gives

$$\cos 2\phi = \frac{\rho'^2 + \rho''^2 - (r' + r'')^2}{2\rho'\rho''},$$

that is,

$$\cos^2\phi = \frac{(\rho' + \rho'')^2 - (r' + r'')^2}{4\rho'\rho''};$$

viz. this is

$$= \left(\frac{\Omega^2}{R^2} - 1\right)(r' + r'')^2 \div 4\frac{\Omega^2}{R^2} r'r'',$$

$$= R^2\left(\frac{1}{R^2} - \frac{1}{\Omega^2}\right)\frac{(r' + r'')^2}{4r'r''}$$

$$= -ACR^2.\frac{4(B^2 - AC)}{A^2}.\frac{-A}{4C}$$

$$= R^2(B^2 - AC);$$

or say

$$\cos\phi = R\sqrt{B^2 - AC};$$

a remarkable equation which may also be written

$$\cos\phi = \frac{R}{v^2}.\tfrac{1}{2}(r' + r''),$$

if, as before, v is the semi-diameter parallel to $R'R''$.

In virtue of the equation $\Lambda = \frac{m}{\sqrt{B^2 - AC}}$ which defines Λ, the equation becomes

$$\cos\phi = \frac{mR}{\Lambda};$$

and we thus complete the demonstration of the several geometrical theorems upon which the investigation was founded.

Analytical Expressions for the Attraction of the Shell, and for the Resolved Attractions.

15. The attraction of the shell was shown to be

$$= \frac{m\,dm}{\Omega^2} 4\pi FGH;$$

or, since the mass of the shell, the density being unity, is

$$\frac{4\pi}{3} fgh \,.\, 3m^2 dm = 4m^2\, dm\, \pi\, fgh,$$

we have

$$\text{Attraction} \div \text{Mass} = \frac{1}{m\Omega^2} \frac{FGH}{fgh};$$

which, by what precedes, is

$$= \frac{\Omega}{m\sqrt{(f^2+\theta)(g^2+\theta)(h^2+\theta)}},$$

where

$$\frac{1}{\Omega^2} = \frac{a^2}{(f^2+\theta)^2} + \frac{b^2}{(g^2+\theta)^2} + \frac{c^2}{(h^2+\theta)^2},$$

θ being the positive root of

$$\frac{a^2}{f^2+\theta} + \frac{b^2}{g^2+\theta} + \frac{c^2}{h^2+\theta} = m^2.$$

16. It is to be observed that the cosine-inclinations of the line PQ to the axes are

$$\frac{a\Omega}{f^2+\theta},\ \frac{b\Omega}{g^2+\theta},\ \frac{c\Omega}{h^2+\theta},$$

respectively; so that, considering, for instance, the attraction parallel to the axis of x, we have

$$\text{Resolved Attraction} \div \text{Mass} = \frac{a\Omega^2}{m(f^2+\theta)\sqrt{(f^2+\theta)(g^2+\theta)(h^2+\theta)}}.$$

Resolved Attractions of the Ellipsoid $\frac{x^2}{f^2} + \frac{y^2}{g^2} + \frac{z^2}{h^2} = 1.$

17. We may find the attraction of the solid ellipsoid

$$\frac{x^2}{f^2} + \frac{y^2}{g^2} + \frac{z^2}{h^2} = 1.$$

For this purpose, dividing it into shells, semi-axes mf, mg, mh, and $(m+dm)f$, $(m+dm)\,g$, $(m+dm)\,h$ respectively, we have for the shell in question

$$\text{Resolved Attraction} \div \text{Mass} = \frac{a\Omega^2}{m(f^2+\theta)\sqrt{(f^2+\theta)(g^2+\theta)(h^2+\theta)}},$$

and the mass of the shell is $\frac{4\pi}{3} fgh \,.\, 3m^2 dm$, where the first factor is the mass of the ellipsoid; whence

$$\text{Resolved Attraction} \div \text{Mass of Ellipsoid} = \frac{a \,.\, 3m\Omega^2\, dm}{(f^2+\theta)\sqrt{(f^2+\theta)(g^2+\theta)(h^2+\theta)}},$$

θ being here a function of m, and m extending from 0 to 1. But taking θ as the variable in place of m, the equation

$$\frac{a^2}{f^2+\theta} + \frac{b^2}{g^2+\theta} + \frac{c^2}{h^2+\theta} = m^2$$

gives

$$-\frac{1}{\Omega^2} d\theta = 2m\, dm;\ \text{that is,}\ 3m\Omega^2\, dm = -\tfrac{3}{2} d\theta.$$

Moreover $m=0$ gives $\theta=\infty$, and $m=1$ gives $\theta=$ its value as defined by the equation

$$\frac{a^2}{f^2+\theta} + \frac{b^2}{g^2+\theta} + \frac{c^2}{h^2+\theta} = 1,$$

so that, reversing the sign, the limits are ∞, θ; or, finally, writing under the integral sign ϕ in place of θ, the formula is

$$\text{Resolved Attraction} \div \text{Mass of Ellipsoid} = \tfrac{3}{2} a \int_\theta^\infty \frac{d\phi}{(f^2+\phi)\sqrt{(f^2+\phi)(g^2+\phi)(h^2+\phi)}},$$

which is a known formula.

605.

NOTE ON A POINT IN THE THEORY OF ATTRACTION.

[From the *Proceedings of the London Mathematical Society*, vol. VI. (1874—1875), pp. 79—81. Read February 11, 1875.]

CONSIDER a mass of matter distributed in any manner on a surface, and attracting points P, Q not on the surface. Consider a point Q accessible from P, viz. such that we can pass continuously from P to Q without passing through the surface. (It is hardly necessary to remark that, if for example the matter is distributed over a hemisphere or segment of a closed surface, then by the surface we mean the hemisphere or segment, not the whole closed surface.) The potential and its differential coefficients *ad infinitum*, in regard to the coordinates of the attracted point, all vary continuously as we pass from P to Q; and it follows that the potential is one and the same analytical function of (a, b, c), the coordinates of the attracted point, for the whole series of points accessible from the original point P; in particular, if the surface be an unclosed surface, for instance a hemisphere or segment of a sphere, then every point Q whatever not on the surface is accessible from P; and the theorem is that the potential is one and the same analytical function of (a, b, c), the coordinates of the attracted point, for any position whatever of this point (not being a point on the surface). But this seems to give rise to a difficulty. Consider the matter as uniformly distributed over a closed surface, and divide the closed surface into two segments: the potential of the whole shell is the sum of the potentials of the two segments; and the potential of the first segment being always one and the same function of (a, b, c), whatever may be the position of the attracted point, and similarly the potential of the second segment being always one and the same function of (a, b, c), whatever may be the position of the attracted point; then the potential of the whole shell is one and the same function of (a, b, c), whatever may be the position of the attracted point. This we know is not the case for a uniform spherical shell; for the potential is a different function for external and interior points, viz. for internal points it is a constant, $= M \div \text{radius}$; for external points it is $= \dfrac{M}{\sqrt{a^2+b^2+c^2}}$, if a, b, c are the coordinates measured from the centre of the sphere.

The difficulty is rather apparent than real. Reverting to the case of an unclosed surface or segment, and considering the continuous curve from P to Q, let this be completed by a curve from Q to P through the segment; viz. we thus have P, Q points on a closed curve or circuit meeting the segment in a single point. To fix the ideas, the circuit may be taken to be a plane curve, and the position of a point on the circuit may be determined by means of its distance s from a fixed point on the circuit. Considering this circuit as drawn on a cylinder, we may at each point of the circuit measure off, say upwards, along the generating line of the cylinder, a length or ordinate z, proportional to the potential of the point on the circuit, the extremities of these distances forming a curve on the cylinder, say the potential curve. We may draw a figure representing this curve only; the points P, Q being marked

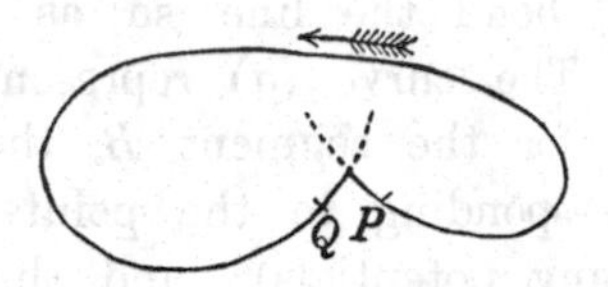

as if they were points on the curve (viz. at the upper instead of the lower extremities of the corresponding ordinates z): the generating lines of the cylinder, and the plane section which is the circuit, not being shown in the figure. The potential curve is then, as shown in the figure, a continuous curve, viz. we pass from P to Q in the direction of the arrow, or along that part of the circuit which does not meet the segment, a curve without any abrupt change in the value of the ordinate z or of any of its differential coefficients, $\dfrac{dz}{ds}$, $\dfrac{d^2z}{ds^2}$, &c.; but there is, corresponding to the point where the circuit meets the surface, an abrupt change in the direction of the potential curve or value of the differential coefficient $\dfrac{dz}{ds}$, viz. the point on the curve is really a node, the two branches crossing at an angle, as shown by the dotted lines, but without any potentials corresponding to these dotted lines.

In the case of two segments forming a closed surface, or say two segments forming a complete spherical shell; then, if the points P, Q are one of them internal, the other external, the circuit, assuming it to meet the first segment in one point only, will meet the second segment in at least one point; the potential curves corresponding to the two segments respectively will have each of them, at the point corresponding to the intersection of the circuit with the segment, a node; and it hence appears how, in the potential curve corresponding to the whole shell (for which curve the ordinate z is the sum of the ordinates belonging to the two segments respectively), there will be a discontinuity of form corresponding to the passage from an exterior to an interior point.

This is best shown by the annexed figure, which represents a uniform spherical shell made up of two segments, one of which is taken to be a small segment or disc having the point A for its centre, the other the large segment B, which is the remainder of the shell; the circuit is taken to be the right line $..PAQB..$ through

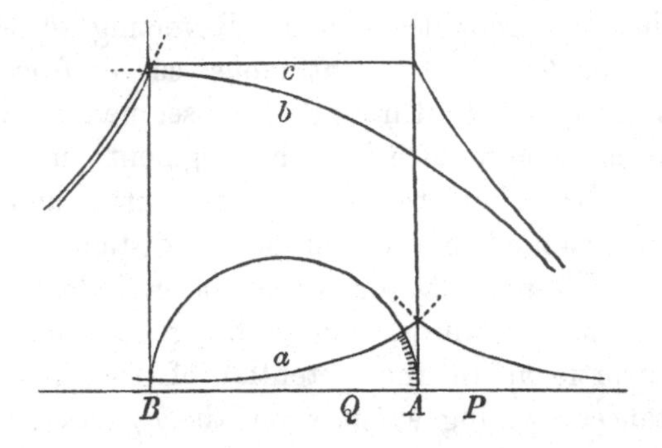

the centre of the sphere (viz. we may imagine the two extremities meeting at infinity, or we may, outside the sphere, bend the line so as to unite the two extremities, thus forming a closed curve). The curve (a) represents the potential curve for the segment A, the curve (b) that for the segment B, these two curves having, as shown by the dotted lines, nodes corresponding to the points A, B respectively (but these dotted portions not indicating any potentials); and then, drawing at each point the ordinate which is the sum of those for the curves (a), (b) respectively, we have the discontinuous curve (c), composed of a horizontal portion and two hyperbolic branches, which is the potential curve for the whole spherical shell.

Practically the figure is constructed by drawing the curves (c), (a), and from them deducing the curve (b). As regards the curve (a) it may be noticed that, treating the segment (a) as a plane disc, the curve (a) is made up of portions of two hyperbolas; viz. it breaks up into two curves, instead of being, as assumed in the discussion, a single curve; this is a mere accident, not affecting the theory; and, in fact, taking the segment to be what it really is, the segment of a sphere the potential curve does not thus break up.

606.

ON THE EXPRESSION OF THE COORDINATES OF A POINT OF A QUARTIC CURVE AS FUNCTIONS OF A PARAMETER.

[From the *Proceedings of the London Mathematical Society*, vol. VI. (1874—1875), pp. 81—83. Read February 11, 1875.]

THE present short Note is merely the development of a process of Prof. Sylvester's. It will be recollected that the general quartic curve has the deficiency 3 (or it is 4-cursal); the question is therefore that of the determination of the subrational* functions of a parameter which have to be considered in the theory of curves of the deficiency 3.

Taking the origin at a point of the curve, the equation is

$$(x,\ y)^4 + (x,\ y)^3 + (x,\ y)^2 + (x,\ y) = 0\,;$$

and writing herein $y = \lambda x$, the equation, after throwing out the factor x, becomes

$$(1,\ \lambda)^4 x^3 + (1,\ \lambda)^3 x^2 + (1,\ \lambda)^2 x + (1,\ \lambda) = 0\,;$$

or, say

$$ax^3 + 3bx^2 + 3cx + d = 0,$$

where we write for shortness

$$a,\ b,\ c,\ d = (1,\ \lambda)^4,\ \tfrac{1}{3}(1,\ \lambda)^3,\ \tfrac{1}{3}(1,\ \lambda)^2,\ (1,\ \lambda)\,;$$

viz. a, b, c, d stand for functions of λ of the degrees 4, 3, 2, and 1 respectively.

The equation may be written

$$(ax + b)^3 - 3(b^2 - ac)(ax + b) + a^2 d - 3abc + 2b^3 = 0\,;$$

* The expression "subrational" includes irrational, but it is more extensive; if Y, X are rational functions, the same or different, of y, x respectively and Y is determined as a function of x by an equation of the form $Y = X$, then y is a subrational function of x. The notion is due to Prof. Sylvester.

viz. writing for a moment $ax+b=2\sqrt{b^2-ac}\,.\,u$, this is

$$4u^3-3u+\frac{a^2d-3abc+2b^3}{2(b^2-ac)\sqrt{b^2-ac}}=0.$$

Hence, assuming

$$-\cos\phi=\frac{a^2d-3abc+2b^3}{2(b^2-ac)\sqrt{b^2-ac}},$$

then we have $4u^3-3u-\cos\phi=0$; consequently u has the three values $\cos\frac{1}{3}\phi$, $\cos\frac{1}{3}(\phi+2\pi)$, $\cos\frac{1}{3}(\phi-2\pi)$, and we may regard $\cos\frac{1}{3}\phi$ as representing any one of these values.

We have thus $ax+b=2\sqrt{b^2-ac}\cos\frac{1}{3}\phi$, and $y=\lambda x$, giving x and y as functions of λ and ϕ, that is, of λ. But for their expression in this manner we introduce the irrationality $\sqrt{b^2-ac}$, which is of the form $\sqrt{(1,\ \lambda)^6}$, and the trisection or derivation of $\cos\frac{1}{3}\phi$ from a given value of $\cos\phi$; viz. we have, as above, $-\cos\phi$, a function of λ of the form

$$(1,\ \lambda)^9\div(1,\ \lambda)^6\sqrt{(1,\ \lambda)^6}.$$

The equation for ϕ may be expressed in the equivalent forms

$$\sin\phi=\frac{a\sqrt{-(a^2d^2+4ac^3+4b^3d-6abcd-3b^2c^2)}}{(b^2-ac)\sqrt{b^2-ac}},$$

$$-\tan\phi=\frac{a\sqrt{-(a^2d^2+4ac^3+4b^3d-6abcd-3b^2c^2)}}{a^2d-3abc+2b^3};$$

and inasmuch as we have

$$2\sqrt{b^2-ac}=-\frac{a^2d-3abc+2b^3}{(b^2-ac)\cos\phi},$$

we may, instead of

$$ax+b=2\sqrt{b^2-ac}\cos\tfrac{1}{3}\phi,$$

write

$$ax+b=-\frac{(a^2d-3abc+2b^3)\cos\frac{1}{3}\phi}{(b^2-ac)\cos\phi};$$

or, what is the same thing,

$$=\frac{-(a^2d-3abc+2b^3)}{(b^2-ac)(4\cos^2\frac{1}{3}\phi-3)}.$$

The formulæ may be simplified by introducing μ, a function of λ, determined by the equation

$$c\mu^2-2b\mu+a=0;$$

viz. this equation is

$$\tfrac{1}{3}(1,\ \lambda)^2\mu^2-\tfrac{2}{3}(1,\ \lambda)^3\mu+(1,\ \lambda)^4=0,$$

so that $(\lambda,\ \mu)$ may be regarded as coordinates of a point on a nodal quartic curve, or a quartic curve of the next inferior deficiency 2. And we then have

$$(c\mu-b)=\sqrt{b^2-ac},$$

and consequently

$$-\cos\phi = \frac{a^2d - 3abc + 2b^3}{2(c\mu - b)^3};$$

viz. $\cos\phi$ is given as a rational function of the coordinates (λ, μ); there is, as before, the trisection; and we then have

$$ax + b = 2(c\mu - b)\cos\tfrac{1}{3}\phi, \quad y = \lambda x,$$

giving x and y as functions of λ, μ, ϕ; that is, ultimately, as functions of λ. I have not succeeded in obtaining in a good geometrical form the relation between the point (x, y) on the given quartic and the point (λ, μ) on the nodal quartic.

Reverting to the expression of $\tan\phi$, it may be remarked that $a = 0$ gives the values of λ which correspond to the four points at infinity on the given quartic curve; $a^2d^2 + 4ac^3 + 4b^3d - 6abcd - 3b^2c^2 = 0$, the values corresponding to the ten tangents from the origin; and $a^2d - 3abc + 2b^3 = 0$, the values corresponding to the nine lines through the origin, which are each such that the origin is the centre of gravity of the other three points on the line.

I take the opportunity of mentioning a mechanical construction of the Cartesian. The equation $r' = -A\cos\theta - N$ represents a limaçon (which is derivable mechanically from the circle $r' = -A\cos\theta$), and if we effect the transformation $r' = r + \frac{B}{r}$, the new curve is $r + \frac{B}{r} + A\cos\theta + N = 0$; that is, $r^2 + r(A\cos\theta + N) + B = 0$, which is, in fact, the equation of a Cartesian. The assumed transformation $r' = r + \frac{B}{r}$ can be effected immediately by a Peaucellier cell.

607.

A MEMOIR ON PREPOTENTIALS.

[From the *Philosophical Transactions of the Royal Society of London,* vol. CLXV. Part II. (1875), pp. 675—774. Received April 8,—Read June 10, 1875.]

THE present Memoir relates to multiple integrals expressed in terms of the $(s+1)$ ultimately disappearing variables $(x, .., z, w)$, and the same number of parameters $(a, .., c, e)$; they are of the form

$$\int \frac{\rho\, d\varpi}{\{(a-x)^2+..+(c-z)^2+(e-w)^2\}^{\frac{1}{2}s+q}},$$

where ρ and $d\varpi$ depend only on the variables $(x, .., z, w)$. Such an integral, in regard to the index $\frac{1}{2}s+q$, is said to be "prepotential," and in the particular case $q=-\frac{1}{2}$ to be "potential."

I use throughout the language of hyper-tridimensional geometry: $(x, .., z, w)$ and $(a, .., c, e)$ are regarded as coordinates of points in $(s+1)$-dimensional space, the former of them determining the position of an element $\rho\, d\varpi$ of attracting matter, the latter being the attracted point; viz. we have a mass of matter $=\int \rho\, d\varpi$ distributed in such manner that, $d\varpi$ being the element of $(s+1)$- or lower-dimensional volume at the point $(x, .., z, w)$, the corresponding density is ρ, a given function of $(x, .., z, w)$, and that the element of mass $\rho\, d\varpi$ exerts on the attracted point $(a, .., c, e)$ a force inversely proportional to the $(s+2q+1)$th power of the distance $\{(a-x)^2+..+(c-z)^2+(e-w)^2\}^{\frac{1}{2}}$. The integration is extended so as to include the whole attracting mass $\int \rho\, d\varpi$; and the integral is then said to represent the Prepotential of the mass in regard to the point $(a, .., c, e)$. In the particular case $s=2$, $q=-\frac{1}{2}$, the force is as the inverse square of the distance, and the integral represents the Potential in the ordinary sense of the word.

The element of volume $d\varpi$ is usually either the element of solid (spatial or $(s+1)$-dimensional) volume $dx\, ..\, dz\, dw$, or else the element of superficial (s-dimensional) volume dS. In particular, when the surface (s-dimensional locus) is the (s-dimensional) plane $w=0$, the superficial element dS is $=dx \ldots dz$. The cases of a less-than-s-dimensional volume are in the present memoir considered only incidentally. It is scarcely necessary to remark that the notion of density is dependent on the dimensionality of the element of volume $d\varpi$: in passing from a spatial distribution, $\rho\, dx \ldots dz\, dw$, to a superficial distribution, $\rho\, dS$, we alter the signification of ρ. In fact, if, in order to connect the two, we imagine the spatial distribution as made over an indefinitely thin layer or stratum bounded by the surface, so that at any element dS of the surface the normal thickness is $d\nu$, where $d\nu$ is a function of the coordinates $(x, \ldots, z, w)$ of the element dS, the spatial element is $=d\nu\, dS$, and the element of mass $\rho\, dx \ldots dz\, dw$ is $=\rho\, d\nu\, dS$; and then changing the signification of ρ, so as to denote by it the product $\rho\, d\nu$, the expression for the element of mass becomes $\rho\, dS$, which is the formula in the case of the superficial distribution.

The space or surface over which the distribution extends may be spoken of as the material space or surface; so that the density ρ is not $=0$ for any finite portion of the material space or surface; and if the distribution be such that the density becomes $=0$ for any point or locus of the material space or surface, then such point or locus, considered as an infinitesimal portion of space or surface, may be excluded from and regarded as not belonging to the material space or surface. It is allowable, and frequently convenient, to regard ρ as a discontinuous function, having its proper value within the material space or surface, and having its value $=0$ beyond these limits; and this being so, the integrations may be regarded as extending as far as we please beyond the material space or surface (but so always as to include the whole of the material space or surface)—for instance, in the case of a spatial distribution, over the whole $(s+1)$-dimensional space; and in the case of a superficial distribution, over the whole of the s-dimensional surface of which the material surface is a part.

In all cases of surface-integrals it is, unless the contrary is expressly stated, assumed that the attracted point does not lie on the material surface; to make it do so is, in fact, a particular supposition. As to solid integrals, the cases where the attracted point is not, and is, in the material space may be regarded as cases of coordinate generality; or we may regard the latter one as the general case, deducing the former one from it by supposing the density at the attracted point to become $=0$.

The present memoir has chiefly reference to three principal cases, which I call A, C, D, and a special case, B, included both under A and C: viz. these are:—

A. The prepotential-plane case; q general, but the surface is here the plane $w=0$, so that the integral is

$$\int \frac{\rho\, dx \ldots dz}{\{(a-x)^2+\ldots+(c-z)^2+e^2\}^{\frac{1}{2}s+q}}.$$

B. The potential-plane case; $q=-\frac{1}{2}$, and the surface the plane $w=0$, so that the integral is

$$\int \frac{\rho\, dx \ldots dz}{\{(a-x)^2+\ldots+(c-z)^2+e^2\}^{\frac{1}{2}s-\frac{1}{2}}}.$$

C. The potential-surface case; $q=-\frac{1}{2}$, the surface arbitrary, so that the integral is

$$\int \frac{\rho\, dS}{\{(a-x)^2+\ldots+(c-z)^2+(e-w)^2\}^{\frac{1}{2}s-\frac{1}{2}}}.$$

D. The potential-solid case; $q=-\frac{1}{2}$, and the integral is

$$\int \frac{\rho\, dx \ldots dz\, dw}{\{(a-x)^2+\ldots+(c-z)^2+(e-w)^2\}^{\frac{1}{2}s-\frac{1}{2}}}.$$

It is, in fact, only the prepotential-plane case which is connected with the partial differential equation

$$\left(\frac{d^2}{da^2}+\ldots+\frac{d^2}{dc^2}+\frac{d^2}{de^2}+\frac{2q+1}{e}\frac{d}{de}\right)V=0,$$

considered in Green's memoir* "On the Attractions of Ellipsoids" (1835), and called here "the prepotential equation." For this equation is satisfied by the function

$$\frac{1}{\{a^2+\ldots+c^2+e^2\}^{\frac{1}{2}s+q}},$$

and therefore also by

$$\frac{1}{\{(a-x)^2+\ldots+(c-z)^2+e^2\}^{\frac{1}{2}s+q}},$$

and consequently by the integral

$$\int \frac{\rho\, dx \ldots dz}{\{(a-x)^2+\ldots+(c-z)^2+e^2\}^{\frac{1}{2}s+q}} \quad \ldots\ldots\ldots\ldots\ldots\ldots\text{(A)},$$

that is, by the prepotential-plane integral; but the equation is *not* satisfied by the value

$$\frac{1}{\{(a-x)^2+\ldots+(c-z)^2+(e-w)^2\}^{\frac{1}{2}s+q}},$$

nor, therefore, by the prepotential-solid, or general superficial, integral.

But if $q=-\frac{1}{2}$, then, instead of the prepotential equation, we have "the potential equation"

$$\left(\frac{d^2}{da^2}+\ldots+\frac{d^2}{dc^2}+\frac{d^2}{de^2}\right)V=0;$$

and this is satisfied by

$$\frac{1}{\{a^2+\ldots+c^2+e^2\}^{\frac{1}{2}s-\frac{1}{2}}},$$

and therefore also by

$$\frac{1}{\{(a-x)^2+\ldots+(c-z)^2+(e-w)^2\}^{\frac{1}{2}s-\frac{1}{2}}}.$$

Hence it is satisfied by

$$\int \frac{\rho\, dx \ldots dz\, dw}{\{(a-x)^2+\ldots+(c-z)^2+(e-w)^2\}^{\frac{1}{2}s-\frac{1}{2}}} \quad \ldots\ldots\ldots\ldots\ldots\text{(D)},$$

* [*Green's Mathematical Papers*, pp. 185—222.]

the potential-solid integral, *provided that the point* $(a, .., c, e)$ *does not lie within the material space:* I would rather say that the integral does *not* satisfy the equation, but of this more hereafter; and it is satisfied by

$$\int \frac{\rho\, dS}{\{(a-x)^2 + \ldots + (c-z)^2 + (e-w)^2\}^{\frac{1}{2}s-\frac{1}{2}}} \quad \ldots\ldots\ldots\ldots\ldots\ldots\ldots\ldots(\mathrm{C}),$$

the potential-surface integral. The potential-plane integral (B), as a particular case of (C), of course also satisfies the equation.

Each of the four cases give rise to what may be called a distribution-theorem; viz. given V a function of $(a, .., c, e)$ satisfying certain prescribed conditions, but otherwise arbitrary, then the form of the theorem is that there exists and that we can find an expression for ρ, the density or distribution of matter over the space or surface to which the theorem relates, such that the corresponding integral V has its given value: viz. in A and B there exists such a distribution over the plane $w = 0$, in C such a distribution over a given surface, and in D such a distribution in space. The establishment, and exhibition in connexion with each other, of these four distribution-theorems is the principal object of the present memoir; but the memoir contains other investigations which have presented themselves to me in treating the question. It is to be noticed that the theorem A belongs to Green, being in fact the fundamental theorem of his memoir of 1835, already referred to. Theorem C, in the particular case of tridimensional space, belongs also to him, being given in his "Essay on the Application of Mathematical Analysis to the theories of Electricity and Magnetism" (Nottingham, 1828*), being partially rediscovered by Gauss** in the year 1840; and theorem D, in the same case of tridimensional space, to Lejeune-Dirichlet: see his memoir "Sur un moyen général de vérifier l'expression du potentiel relatif à une masse quelconque homogène ou hétérogène," *Crelle*, t. XXXII. pp. 80—84 (1840). I refer more particularly to these and other researches by Gauss, Jacobi, and others in an Annex to the present memoir.

On the Prepotential Surface-integral. Art. Nos. 1 to 18.

1. In what immediately follows we require

$$V = \int \frac{dx \ldots dz}{(x^2 + \ldots + z^2 + e^2)^{\frac{1}{2}s+q}},$$

limiting condition $x^2 + \ldots + z^2 = R^2$, the prepotential of a uniform (s-coordinal) circular disk†, radius R, in regard to a point $(0, .., 0, e)$ on the axis; and in particular the

* [Also *Crelle*, t. XXXIX., pp. 73—89, t. XLIV., pp. 356—374, t. XLVII., pp. 161—221; *Green's Mathematical Papers*, pp. 1—115.]

** ["Allgemeine Lehrsätze in Beziehung auf die im verkehrten Verhältnisse des Quadrats der Entfernung wirkenden Anziehungs- und Abstossungskräfte," *Ges. Werke*, t. V., pp. 195—242.]

† It is to be throughout borne in mind that $x, .., z$ denotes a set of s coordinates, $x, .., z, w$ a set of $s+1$ coordinates; the adjective coordinal refers to the number of coordinates which enter into the equation; thus, $x^2+\ldots+z^2+w^2=f^2$ is an $(s+1)$-coordinal sphere (observe that the surface of such a sphere is s-dimensional); $x^2+\ldots+z^2=f^2$, according as we tacitly associate with it the condition $w=0$, or w arbitrary, is an s-coordinal circle, or cylinder, the surface of such circle or cylinder being s-dimensional, but the circumference of the circle $(s-1)$-dimensional; or if we attend only to the s-dimensional space constituted by the plane $w=0$, the locus may be considered as an s-coordinal sphere, its surface being $(s-1)$-dimensional.

value is required in the case where the distance e (taken to be always positive) is indefinitely small in regard to the radius R.

Writing $x = r\xi, .., z = r\zeta$, where the s new variables $\xi, .., \zeta$ are such that $\xi^2 + \ldots + \zeta^2 = 1$, the integral becomes

$$\int \frac{r^{s-1}\,dr\,dS}{(r^2+e^2)^{\frac{1}{2}s+q}}, \quad = \int dS \int_0^R \frac{r^{s-1}\,dr}{(r^2+e^2)^{\frac{1}{2}s+q}},$$

where dS is the element of surface of the s-dimensional unit-sphere $\xi^2 + \ldots + \zeta^2 = 1$; the integral $\int dS$ denotes the entire surface of this sphere, which (see Annex I.) is $= \dfrac{2\,(\Gamma\frac{1}{2})^s}{\Gamma\frac{1}{2}s}$. The other factor,

$$\int_0^R \frac{r^{s-1}\,dr}{(r^2+e^2)^{\frac{1}{2}s+q}},$$

is the r-integral of Annex II.

2. We now consider the prepotential-surface integral

$$V = \int \frac{\rho\,dS}{\{(a-x)^2 + \ldots + (c-z)^2 + (e-w)^2\}^{\frac{1}{2}s+q}}.$$

As already mentioned, it is only a particular case of this, the prepotential-plane integral, which is specially discussed; but at present I consider the general case, for the purpose of establishing a theorem in relation thereto. The surface (s-dimensional surface) S is any given surface whatever.

Let the attracted point P be situate indefinitely near to the surface, on the normal thereto at a point N, say the normal distance NP is $= \varepsilon$*; and let this point N be taken at the centre of an indefinitely small circular (s-dimensional) disk or segment (of the surface), the radius of which R, although indefinitely small, is indefinitely large in comparison with the normal distance ε. I proceed to determine the prepotential of the disk; for this purpose, transforming to new axes, the origin being at N and the axes of $x, .., z$ in the tangent-plane at N, then the coordinates of the attracted point P will be $(0, .., 0, \varepsilon)$, and the expression for the prepotential of the disk will be

$$V = \int \frac{\rho\,dx \ldots dz}{\{x^2 + \ldots + z^2 + \varepsilon^2\}^{\frac{1}{2}s+q}},$$

where the limits are given by $x^2 + \ldots + z^2 < R^2$.

Suppose for a moment that the density at the point N is $= \rho'$, then the density throughout the disk may be taken $= \rho'$, and the integral becomes

$$V = \rho' \int \frac{dx \ldots dz}{\{x^2 + \ldots + z^2 + \varepsilon^2\}^{\frac{1}{2}s+q}},$$

where instead of ρ' I write ρ; viz. ρ now denotes the density at the point N. Making this change, then (by what precedes) the value is

$$= \rho\,\frac{2\,(\Gamma\frac{1}{2})^s}{\Gamma(\frac{1}{2}s)} \int_0^R \frac{r^{s-1}\,dr}{\{r^2+\varepsilon^2\}^{\frac{1}{2}s+q}}.$$

* ε is positive; in afterwards writing $\varepsilon = 0$, we mean by 0 the limit of an indefinitely small positive quantity.

$q = Positive.$ Art. Nos. 3 to 7.

3. I consider first the case where q is positive. The value is here

$$= \rho \frac{2(\Gamma\tfrac{1}{2})^s}{\Gamma(\tfrac{1}{2}s)} \frac{1}{2s^{2q}} \left\{ \frac{\Gamma\tfrac{1}{2}s\,\Gamma q}{\Gamma(\tfrac{1}{2}s+q)} - \int_0^{\frac{s^2}{R^2}} \frac{x^{q-1}\,dx}{(1+x)^{\frac{1}{2}s+q}} \right\};$$

or, since $\frac{s}{R}$ is indefinitely small, the x-integral may be neglected, and the value is

$$= \frac{1}{s^{2q}} \rho \frac{(\Gamma\tfrac{1}{2})^s\,\Gamma q}{\Gamma(\tfrac{1}{2}s+q)}.$$

Observe that this value is independent of R, and that the expression is thus the same as if (instead of the disk) we had taken the whole of the infinite tangent-plane, the density at every point thereof being $= \rho$. It is proper to remark that the neglected terms are of the orders

$$\frac{1}{s^{2q}} \left\{ \left(\frac{s}{R}\right)^{2q},\ \left(\frac{s}{R}\right)^{2q+2},\ \&c. \right\};$$

so that the complete value multiplied by s^{2q} is equal to the constant $\rho \frac{(\Gamma\frac{1}{2})^s\,\Gamma q}{\Gamma(\frac{1}{2}s+q)}$ + terms of the orders $\left(\frac{s}{R}\right)^{2q}$, $\left(\frac{s}{R}\right)^{2q+2}$, &c.

4. Let us now consider the prepotential of the remaining portion of the surface; every part thereof is at a distance from P exceeding, in fact far exceeding, R; so that imagining the whole mass $\int \rho\,dS$ to be collected at the distance R, the prepotential of the remaining portion of the surface is less than

$$\frac{\int \rho\,dS}{R^{s+2q}};$$

viz. we have thus, in the case where the mass $\int \rho\,dS$ is finite, a superior limit to the prepotential of the remaining portion of the surface. This will be indefinitely small in comparison with the prepotential of the disk, provided only s^{2q} is indefinitely small compared with R^{s+2q}, that is, s indefinitely small in comparison with $R^{1+\frac{s}{2q}}$. The proof assumes that the mass $\int \rho\,dS$ is finite; but considering the very rough manner in which the limit $\frac{\int \rho\,dS}{R^{s+2q}}$ was obtained, it can scarcely be doubted that, if not universally, at least for very general laws of distribution, even when $\int \rho\,dS$ is infinite, the same thing is true; viz. that by taking s sufficiently small in regard to R, we can make the prepotential of the remaining portion of the surface vanish in comparison with that of the disk. But without entering into the question I assume that the prepotential of the remaining portion does thus vanish; the prepotential of the whole surface in regard to the indefinitely near point P is thus equal to the prepotential of the disk; viz. its value is

$$= \frac{1}{\varepsilon^{2q}} \rho \frac{(\Gamma\frac{1}{2})^s \Gamma q}{\Gamma(\frac{1}{2}s+q)},$$

which, observe, is infinite for a point P on the surface.

5. Considering the prepotential V at an arbitrary point $(a, .., c, e)$ as a given function of $(a, .., c, e)$ the coordinates of this point, and taking $(x, .., z, w)$ for the coordinates of the point N, which is, in fact, an arbitrary point on the surface, then the value of V at the point P indefinitely near to N will be $= W$, if W denote the same function of $(x, .., z, w)$ that V is of $(a, .., c, e)$. The result just obtained is therefore

$$W = \frac{1}{\varepsilon^{2q}} \rho \frac{(\Gamma\frac{1}{2})^s \Gamma q}{\Gamma(\frac{1}{2}s+q)}, \ (\varepsilon = 0),$$

or, what is the same thing,

$$\rho = \frac{\Gamma(\frac{1}{2}s+q)}{(\Gamma\frac{1}{2})^s \Gamma q} (\varepsilon^{2q} W)_{\varepsilon=0}.$$

As to this, remark that V is not an arbitrary function of $(a, .., c, e)$: *non constat* that there is any distribution of matter, and still less that there is any distribution of matter on the surface, which will produce at the point $(a, .., c, e)$, that is, at every point whatever, a prepotential the value of which shall be a function assumed at pleasure of the coordinates $(a, .., c, e)$. But suppose that V, the given function of $(a, .., c, e)$, is such that there does exist a corresponding distribution of matter on the surface, (viz. that V satisfies the conditions, whatever they are, required in order that this may be the case), then the foregoing formula determines the distribution, viz. it gives the expression of ρ, that is, the density at any point of the surface.

6. The theorem may be presented in a somewhat different form; regarding the prepotential as a function of the normal distance ε, its derived function in regard to ε is

$$= -\frac{2q}{\varepsilon^{2q+1}} \rho \frac{(\Gamma\frac{1}{2})^s \Gamma q}{\Gamma(\frac{1}{2}s+q)},$$

that is,

$$= -\frac{1}{\varepsilon^{2q+1}} \rho \frac{2(\Gamma\frac{1}{2})^s \Gamma(q+1)}{\Gamma(\frac{1}{2}s+q)};$$

and we thus have

$$\frac{dW}{d\varepsilon} = -\frac{1}{\varepsilon^{2q+1}} \rho \frac{2(\Gamma\frac{1}{2})^s \Gamma(q+1)}{\Gamma(\frac{1}{2}s+q)}, \ (\varepsilon = 0),$$

or, what is the same thing,

$$\rho = -\frac{\Gamma(\frac{1}{2}s+q)}{2(\Gamma\frac{1}{2})^s \Gamma(q+1)} \left(\varepsilon^{2q+1} \frac{dW}{d\varepsilon}\right)_{\varepsilon=0},$$

where, however, W being given as a function of $(x, .., z, w)$, the notation $\frac{dW}{d\mathfrak{s}}$ requires explanation. Taking $\cos\alpha, .., \cos\gamma$ to be the inclinations of the normal at N, in the direction NP in which the distance $\mathfrak{s}$ is measured, to the positive parts of the axes of $(x, .., z)$, viz. these cosines denote the values of

$$\frac{dS}{dx}, \ldots, \frac{dS}{dz},$$

each taken with the same sign + or −, and divided by the square root of the sum of the squares of the last-mentioned quantities, then the meaning is

$$\frac{dW}{d\mathfrak{s}} = \frac{dW}{dx}\cos\alpha + \ldots + \frac{dW}{dz}\cos\gamma.$$

7. The surface S may be the plane $w = 0$, viz. we have then the prepotential-plane integral

$$V = \int \frac{\rho\, dx \ldots dz}{\{(a-x)^2 + \ldots + (c-z)^2 + e^2\}^{\frac{1}{2}s-q}} \quad \ldots\ldots\ldots\ldots\ldots\ldots\ldots\ldots \text{(A)},$$

where e (like $\mathfrak{s}$) is positive. In afterwards writing $e = 0$, we mean by 0 the limit of an indefinitely small positive quantity.

The foregoing distribution-formulæ then become

$$\rho = \frac{\Gamma(\frac{1}{2}s+q)}{(\Gamma\frac{1}{2})^s\,\Gamma q}\,(e^{2q}\,W)_{e=0} \quad \ldots\ldots\ldots\ldots\ldots\ldots\ldots\ldots\ldots\ldots \text{(A)},$$

and

$$\rho = -\frac{\Gamma(\frac{1}{2}s+q)}{2\,(\Gamma\frac{1}{2})^s\,\Gamma(q+1)}\left(e^{2q+1}\frac{dW}{de}\right)_{e=0} \quad \ldots\ldots\ldots\ldots\ldots\ldots\ldots \text{(A*)},$$

which will be used in the sequel.

It will be remembered that in the preceding investigation it has been assumed that q is positive, the limiting case $q = 0$ being excluded†.

$q = -\frac{1}{2}$. Art. Nos. 8 to 13.

8. I pass to the case $q = -\frac{1}{2}$, viz. we here have the potential-surface integral

$$V = \int \frac{\rho\, dS}{\{(a-x)^2 + \ldots + (c-z)^2 + (e-w)^2\}^{\frac{1}{2}s-\frac{1}{2}}} \quad \ldots\ldots\ldots\ldots\ldots\ldots \text{(C)}:$$

it will be seen that the results present themselves under a remarkably different form.

The potential of the disk is, as before,

$$\rho\,\frac{2\,(\Gamma\frac{1}{2})^s}{\Gamma\frac{1}{2}s}\int \frac{r^{s-1}\,dr}{(r^2+\mathfrak{s}^2)^{\frac{1}{2}s-\frac{1}{2}}},$$

† This is, as regards q, the case throughout; a limiting value, if not expressly stated to be included, is always excluded.

where ρ here denotes the density at the point N; and the value of the r-integral

$$= R\left(1 + \text{terms in } \frac{s^2}{R^2}, \frac{s^4}{R^4}, \ldots\right) - s\,\frac{\Gamma\frac{1}{2}s\,\Gamma\frac{1}{2}}{\Gamma(\frac{1}{2}s - \frac{1}{2})}.$$

Observe that this is indefinitely small, and remains so for a point P on the surface; the potential of the remaining portion of the surface (for a point P near to or on the surface) is finite, that is, neither indefinitely large nor indefinitely small, and it varies continuously as the attracted point passes through the disk (or aperture in the material surface now under consideration); hence the potential of the whole surface is finite for an attracted point P on the surface, and it varies continuously as P passes through the surface.

It will be noticed that there is in this case a term in V independent of s; and it is on this account necessary, instead of the potential, to consider its derived function in regard to s; viz. neglecting the indefinitely small terms which contain powers of $\frac{s}{R}$, I write

$$\frac{dV}{ds} = -\frac{2(\Gamma\frac{1}{2})^{s+1}}{\Gamma(\frac{1}{2}s - \frac{1}{2})}\rho.$$

The corresponding term arising from the potential of the other portion of the surface, viz. the derived function of the potential in regard to s, is not indefinitely small; and calling it Q, the formula for the whole surface becomes

$$\frac{dV}{ds} = Q - \frac{2(\Gamma\frac{1}{2})^{s+1}}{\Gamma(\frac{1}{2}s - \frac{1}{2})}\rho.$$

9. I consider positions of the point P on the two opposite sides of the point N, say at the normal distances s', s'', these being positive distances measured in opposite directions from the point N. The function V, which represents the potential of the surface in regard to the point P, is or may be a different function of the coordinates $(a, .., c, e)$ of the point P, according as the point is situate on the one side or the other of the surface (as to this more presently). I represent it in the one case by V', and in the other case by V''; and in further explanation state that s' is measured *into* the space to which V' refers, s'' *into* that to which V'' refers; and I say that the formulæ belonging to the two positions of the point P are

$$\frac{dW'}{ds'} = Q' - \frac{2(\Gamma\frac{1}{2})^{s+1}}{\Gamma(\frac{1}{2}s - \frac{1}{2})}\rho,$$

$$\frac{dW''}{ds''} = Q'' - \frac{2(\Gamma\frac{1}{2})^{s+1}}{\Gamma(\frac{1}{2}s - \frac{1}{2})}\rho,$$

where, instead of V', V'', I have written W', W'', to denote that the coordinates, as well of P' as of P'', are taken to be the values $(x, .., z, w)$ which belong to the point N. The symbols denote

$$\frac{dW'}{ds'} = \frac{dW'}{dx}\cos\alpha' + \ldots + \frac{dW'}{dz}\cos\gamma',$$

$$\frac{dW''}{ds''} = \frac{dW''}{dx}\cos\alpha'' + \ldots + \frac{dW''}{dz}\cos\gamma'',$$

where $(\cos\alpha', .., \cos\gamma')$ and $(\cos\alpha'', .., \cos\gamma'')$ are the cosine-inclinations of the normal distances $\mathfrak{s}'$, $\mathfrak{s}''$ to the positive parts of the axes of $(x, .., z)$; since these distances are measured in opposite directions, we have $\cos\alpha'' = -\cos\alpha', .., \cos\gamma'' = -\cos\gamma'$. If we imagine a curve through N cutting the surface at right angles, or, what is the same thing, an element of the curve coinciding in direction with the normal element $P'NP''$, and if s denote the distance of N from a fixed point of the curve, and for the point P' if s become $s + \delta' s$, while for the point P'' it becomes $s - \delta'' s$, or, what is the same thing, if s increase in the direction of NP' and decrease in that of NP'', then if any function Θ of the coordinates $(x, .., z, w)$ of N be regarded as a function of s, we have

$$\frac{d\Theta}{ds} = \frac{d\Theta}{d\mathfrak{s}'}, \quad \frac{d\Theta}{ds} = -\frac{d\Theta}{d\mathfrak{s}''}.$$

10. In particular, let Θ denote the potential of the remaining portion of the surface, that is, of the whole surface exclusive of the disk; the curve last spoken of is a curve which does not pass through the material surface, viz. the portion to which Θ has reference: and there is no discontinuity in the value of Θ as we pass along this curve through the point N. We have $Q' =$ value of $\frac{d\Theta}{d\mathfrak{s}'}$ at the point P', and $Q'' =$ value of $\frac{d\Theta}{d\mathfrak{s}''}$ at the point P''; and the two points P', P'' coming to coincide together at the point N, we have then

$$Q' = \frac{d\Theta}{d\mathfrak{s}'}, \quad = \quad \frac{d\Theta}{ds},$$

$$Q'' = \frac{d\Theta}{d\mathfrak{s}''}, \quad = -\frac{d\Theta}{ds}.$$

We have in like manner $\frac{dW'}{d\mathfrak{s}'} = \frac{dW'}{ds}$, $\frac{dW''}{d\mathfrak{s}''} = -\frac{dW''}{ds}$; and the equation obtained above may be written

$$\frac{dW'}{ds} = \frac{d\Theta}{ds} - \frac{2(\Gamma\frac{1}{2})^{s+1}}{\Gamma(\frac{1}{2}s - \frac{1}{2})}\rho,$$

$$\frac{dW''}{ds} = \frac{d\Theta}{ds} + \frac{2(\Gamma\frac{1}{2})^{s+1}}{\Gamma(\frac{1}{2}s - \frac{1}{2})}\rho;$$

in which form they show that as the attracted point passes through the surface from the position P' on the one side to P'' on the other, there is an abrupt change in the value of $\frac{dW}{ds}$, or say of $\frac{dV}{ds}$, the first derived function of the potential in regard to the orthotomic arc s, that is, in the rate of increase of V in the passage of the attracted point normally to the surface. It is obvious that, if the attracted point traverses the surface obliquely instead of normally, viz. if the arc s cuts the surface obliquely, there is the like abrupt change in the value of $\frac{dV}{ds}$.

Reverting to the original form of the two equations, and attending to the relation $Q' + Q'' = 0$, we obtain

$$\frac{dW'}{d\mathfrak{s}'} + \frac{dW''}{d\mathfrak{s}''} = \frac{-4(\Gamma\tfrac{1}{2})^{s+1}}{\Gamma(\tfrac{1}{2}s - \tfrac{1}{2})}\rho,$$

or, what is the same thing,

$$\rho = -\frac{\Gamma(\tfrac{1}{2}s-\tfrac{1}{2})}{4(\Gamma\tfrac{1}{2})^{s+1}}\left(\frac{dW'}{d\mathfrak{s}'} + \frac{dW''}{d\mathfrak{s}''}\right) \text{..............................(C)}.$$

11. I recall the signification of the symbols:—V', V'' are the potentials, it may be different functions of the coordinates $(a, .., c, e)$ of the attracted point, for positions of this point on the two sides of the surface (as to this more presently): and W', W'' are what V', V'' respectively become when the coordinates $(a, .., c, e)$ are replaced by $(x, .., z, w)$, the coordinates of a point N on the surface. The explanation of the symbols $\frac{dW'}{d\mathfrak{s}'}$, $\frac{dW''}{d\mathfrak{s}''}$ is given a little above; ρ denotes the density at the point $(x, .., z, w)$.

12. The like remarks arise as with regard to the former distribution theorem (A); the functions V', V'' cannot be assumed at pleasure; *non constat* that there is any distribution in space, and still less any distribution on the surface, which would give such values to the potential of a point $(a, .., c, e)$ on the two sides of the surface respectively; but assuming that the functions V', V'' are such that they do arise from a distribution on the surface, or say that they satisfy all the conditions, whatever they are, required in order that this may be so, then the formula determines the distribution, viz. it gives the value of ρ, the density at a point $(x, .., z, w)$ of the surface.

13. In the case where the surface is the plane $w = 0$, viz. in the case of the potential-plane integral,

$$V = \int \frac{\rho\, dx \ldots dz}{\{(a-x)^2 + \ldots + (c-z)^2 + e^2\}^{\frac{1}{2}s-\frac{1}{2}}} \text{........................(B)},$$

(e assumed to be positive); then, since the conformation is symmetrical on the two sides of the plane, V' and V'' are the same functions of $(a, .., c, e)$, say they are each $= V$; W', W'' are each of them the same function, say they are each $= W$, of $(x, .., z, e)$ that V is of $(a, .., c, e)$; the distribution-formula becomes

$$\rho = -\frac{\Gamma(\tfrac{1}{2}s-\tfrac{1}{2})}{2(\Gamma\tfrac{1}{2})^{s+1}}\left(\frac{dW}{de}\right)_{e=0} \text{.............................. (B)},$$

viz. this is also what one of the prepotential-plane formulæ becomes on writing therein $q = -\tfrac{1}{2}$.

$q = 0$, *or Negative.* Art. Nos. 14 to 18.

14. Consider the case $q = 0$. The prepotential of the disk is

$$\rho\frac{2(\Gamma\tfrac{1}{2})^s}{\Gamma\tfrac{1}{2}s}(\log R + N - \log \mathfrak{s} \ldots);$$

to get rid of the constant term we must consider the derived function in regard to ς, viz. this is

$$= -\rho\,\frac{2\,(\Gamma\tfrac{1}{2})^s}{\Gamma\tfrac{1}{2}s}\cdot\frac{1}{\varsigma},$$

and we have thus for the whole surface

$$\frac{dV}{d\varsigma} = Q - \rho\,\frac{2\,(\Gamma\tfrac{1}{2})^s}{\Gamma\tfrac{1}{2}s}\,\frac{1}{\varsigma},$$

where Q, which relates to the remaining portion of the surface, is finite; we have thence, writing, as before, W in place of V,

$$\varsigma\,\frac{dW}{d\varsigma} = -\rho\,\frac{2\,(\Gamma\tfrac{1}{2})^s}{\Gamma\tfrac{1}{2}s},$$

or say

$$\rho = -\frac{\Gamma\tfrac{1}{2}s}{2\,(\Gamma\tfrac{1}{2})^s}\left(\varsigma\,\frac{dW}{d\varsigma}\right)_{\varsigma=0}.$$

15. Consider the case q negative, but $-q < \tfrac{1}{2}$. The prepotential of the disk is here

$$= \rho\,\frac{2\,(\Gamma\tfrac{1}{2})^s}{\Gamma\tfrac{1}{2}s}\left\{\frac{R^{-2q}}{-2q} + \tfrac{1}{2}\varsigma^{-2q}\,\frac{\Gamma\tfrac{1}{2}s\,\Gamma q}{\Gamma(\tfrac{1}{2}s+q)} + \ldots\right\};$$

to get rid of the first term we must consider the derived function in regard to ς, viz. this is

$$-\,\varsigma^{-2q-1}\,\rho\,\frac{2\,(\Gamma\tfrac{1}{2})^s\,\Gamma(q+1)}{\Gamma(\tfrac{1}{2}s+q)};$$

whence, for the potential of the whole surface,

$$\frac{dV}{d\varsigma} = Q - \varsigma^{-2q-1}\,\rho\,\frac{2\,(\Gamma\tfrac{1}{2})^s\,\Gamma(q+1)}{\Gamma(\tfrac{1}{2}s+q)},$$

where Q, the part relating to the remaining portion of the surface, is finite. Multiplying by ς^{2q+1} (where the index $2q+1$ is positive), the term in Q disappears; and writing, as before, W in place of V, this is

$$\varsigma^{2q+1}\,\frac{dW}{d\varsigma} = -\rho\,\frac{2\,(\Gamma\tfrac{1}{2})^s\,\Gamma(q+1)}{\Gamma\tfrac{1}{2}s+q},$$

or, say

$$\rho = -\frac{\Gamma(\tfrac{1}{2}s+q)}{2\,(\Gamma\tfrac{1}{2})^s\,\Gamma(q+1)}\left(\varsigma^{2q+1}\,\frac{dW}{d\varsigma}\right)_{\varsigma=0};$$

viz. we thus see that the formula (A*) originally obtained for the case q positive extends to the case $q=0$, and $q=-$ but $-q<\tfrac{1}{2}$; but, as already seen, it does not extend to the limiting case $q=-\tfrac{1}{2}$.

16. If q be negative and between $-\tfrac{1}{2}$ and -1, we have in like manner a formula

$$\frac{dV}{d\varsigma} = Q - \rho\,\frac{2\,(\Gamma\tfrac{1}{2})^s\,\Gamma(q+1)}{\Gamma(\tfrac{1}{2}s+q)}\,\varsigma^{-2q-1};$$

but here, $2q+1$ being negative, the term $s^{2q+1}Q$ does not disappear: the formula has to be treated in the same way as for $q=-\frac{1}{2}$, and we arrive at

$$\left\{s'^{2q+1}\frac{dW'}{ds'}+s''^{2q+1}\frac{dW''}{ds''}\right\}=-\frac{4(\Gamma\frac{1}{2})^s\,\Gamma(q+1)}{\Gamma(\frac{1}{2}s+q)}\rho\,;$$

viz. the formula is of the same form as for the potential case $q=-\frac{1}{2}$. Observe that the formula does not hold good in the limiting case $q=-1$.

17. We have, in fact, for $q=-1$, the potential of the disk

$$=\frac{2(\Gamma\frac{1}{2})^s}{\Gamma(\frac{1}{2}s)}\rho\left\{\frac{R^2}{2}-s^2\log s\,\frac{\Gamma\frac{1}{2}s}{\Gamma(\frac{1}{2}s-1)}\right\};$$

whence

$$\frac{dV}{ds}=Q-\frac{2(\Gamma\frac{1}{2})^s}{\Gamma(\frac{1}{2}s-1)}\rho\,(2s\log s),$$

since, in the complete differential coefficient $s+2s\log s$, the term s vanishes in comparison with $2s\log s$. Then, proceeding as before, we find

$$\frac{1}{s'\log s'}\,\frac{dW'}{ds'}+\frac{1}{s''\log s''}\,\frac{dW''}{ds''}=\frac{-8(\Gamma\frac{1}{2})^s}{\Gamma(\frac{1}{2}s-1)}\rho\,;$$

but I have not particularly examined this formula.

18. If q be negative and >-1 (that is, $-q>1$), then the prepotential for the disk is

$$=\rho\frac{(\Gamma\frac{1}{2})^s}{\Gamma\frac{1}{2}s}\left\{\frac{R^{-2q}}{-2q}+\frac{\frac{1}{2}s+q}{1}\,\frac{R^{-2q-2}}{-2q-2}\cdot s^2+\ldots+Ks^{-2q}\right\};$$

and it would seem that, in order to obtain a result, it would be necessary to proceed to a derived function higher than the first; but I have not examined the case.

Continuity of the Prepotential-surface Integral. Art. Nos. 19 to 25.

19. I again consider the prepotential-surface integral

$$\int\frac{\rho\,dS}{\{(a-x)^2+\ldots+(c-z)^2+(e-w)^2\}^{\frac{1}{2}s-q}}$$

in regard to a point $(a,..,c,e)$ not on the surface; q is either positive or negative, as afterwards mentioned.

The integral or prepotential and all its derived functions, first, second, &c. *ad infinitum*, in regard to each or all or any of the coordinates $(a,..,c,e)$, are all finite. This is certainly the case when the mass $\int\rho\,dS$ is finite, and possibly in other cases also; but to fix the ideas we may assume that the mass is finite. And the prepotential and its derived functions vary continuously with the position of the attracted point $(a,.., c, e)$, so long as this point in its course does not traverse the material surface. For greater clearness we may consider the point as moving along a continuous curve (one-dimensional locus), which curve, or the part of it under consideration, does not meet the surface; and the meaning is that the prepotential and each of its derived functions vary continuously as the point $(a, .., c, e)$ passes continuously along the curve.

20. Consider a "region," that is, a portion of space any point of which can be, by a continuous curve not meeting the material surface, connected with any other point of the region. It is a legitimate inference, from what just precedes, that the prepotential is, for any point $(a,.., c, e)$ whatever within the region, one and the same function of the coordinates $(a,.., c, e)$, viz. the theorem, rightly understood, is true; but the theorem gives rise to a difficulty, and needs explanation.

Consider, for instance, a closed surface made up of two segments, the attracting matter being distributed in any manner over the whole surface (as a particular case $s+1=3$, a uniform spherical shell made up of two hemispheres); then, as regards the first segment (now taken as the material surface), there is no division into regions, but the whole of the $(s+1)$-dimensional space is one region; wherefore the prepotential of the first segment is one and the same function of the coordinates $(a,.., c, e)$ of the attracted point for any position whatever of this point. But in like manner the prepotential of the second segment is one and the same function of the coordinates $(a,.., c, e)$ for any position whatever of the attracted point. And the prepotential of the whole surface, being the sum of the prepotentials of the two segments, is consequently one and the same function of the coordinates $(a,.., c, e)$ of the attracted point for any position whatever of this point; viz. it is the same function for a point in the region inside the closed surface and for a point in the outside region. That this is not in general the case we know from the particular case, $s+1=3$, of a uniform spherical shell referred to above.

21. Consider in general an unclosed surface or segment, with matter distributed over it in any manner; and imagine a closed curve or circuit cutting the segment once; and let the attracted point $(a,.., c, e)$ move continuously along the circuit. We may consider the circuit as corresponding to (in ordinary tridimensional space) a plane curve of equal periphery, the corresponding points on the circuit and the plane curve being points at equal distances s along the curves from fixed points on the two curves respectively; and then treating the plane curve as the base of a cylinder, we may represent the potential as a length or ordinate, $V=y$, measured upwards from the point on the plane curve along the generating line of the cylinder, in such wise that the upper extremity of the length or ordinate y traces out on the cylinder a curve, say the prepotential curve, which represents the march of the prepotential. The attracted point may, for greater convenience, be represented as a point *on* the prepotential curve, viz. by the upper instead of the lower extremity of the length or ordinate y; and the ordinate, or height of this point above the base of the cylinder, then represents the value of the prepotential. The before-mentioned continuity-theorem is that the prepotential curve, corresponding to any portion (of the circuit) which

does not meet the material surface, is a continuous curve: viz. that there is no abrupt change of value either in the ordinate y $(= V)$ of the prepotential curve, or in the first or any other of the derived functions $\frac{dy}{ds}$, $\frac{d^2y}{ds^2}$, &c. We have thus (in each of the two figures) a continuous curve as we pass (in the direction of the arrow) from

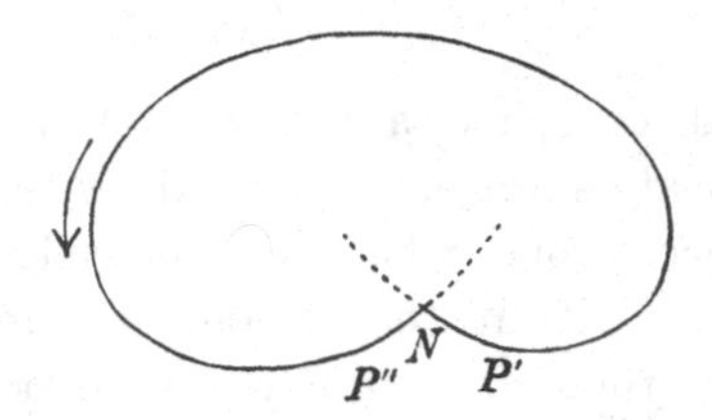

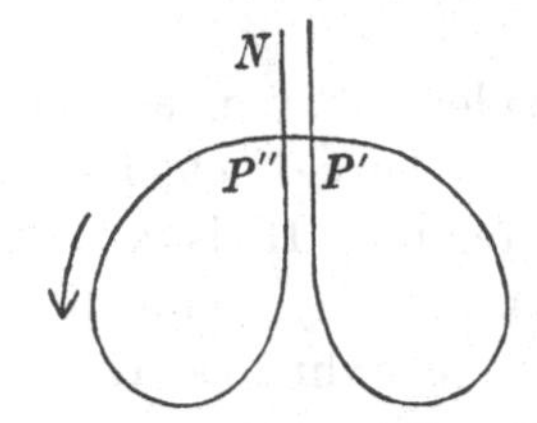

a point P' on one side of the segment to a point P'' on the other side of the segment; but this continuity does not exist in regard to the remaining part, from P'' to P', of the prepotential curve corresponding to the portion (of the circuit) which traverses the material surface.

22. I consider first the case $q = -\frac{1}{2}$ (see the left-hand figure): the prepotential is here a potential. At the point N, which corresponds to the passage through the material surface, then, as was seen, the ordinate y (= the Potential V) remains finite and continuous; but there is an abrupt change in the value of $\frac{dy}{ds}$, that is, in the direction of the curve: the point N is really a node with two branches crossing at this point, as shown in the figure; but the dotted continuations have only an analytical existence, and do not represent values of the potential. And by means of this branch-to-branch discontinuity at the point N, we escape from the foregoing conclusion as to the continuity of the potential on the passage of the attracted point through a closed surface.

23. To show how this is, I will for greater clearness examine the case $(s+1)=3$, in ordinary tridimensional space, of the uniform spherical shell attracting according to the inverse square of the distance; instead of dividing the shell into hemispheres, I divide it by a plane into any two segments (see the figure, wherein A, B represent the centres of the two segments respectively, and where for graphical convenience the segment A is taken to be small).

We may consider the attracted point as moving along the axis xx', viz. the two extremities may be regarded as meeting at infinity, or we may outside the sphere bend the line round, so as to produce a closed circuit. We are only concerned with what happens at the intersections with the spherical surface. The ordinates represent the potentials, viz. the curves are a, b, c for the segments A, B, and the whole spherical surface respectively. Practically, we construct the curves c, a, and deduce the curve b by taking for its ordinate the difference of the other two ordinates. The curve c is, as we know, a discontinuous curve, composed of a horizontal line and two hyperbolic branches; the curve a can be laid down approximately by treating the segment A as a plane circular disk; it is of the form shown in the figure, having a node at the point corresponding to A. (In the case where the segment A is actually a plane disk, the curve is made up of portions of branches of two hyperbolas; but taking the segment A as being what it is, the segment of a spherical surface, the curve is a single curve, having a node as mentioned above.) And from the

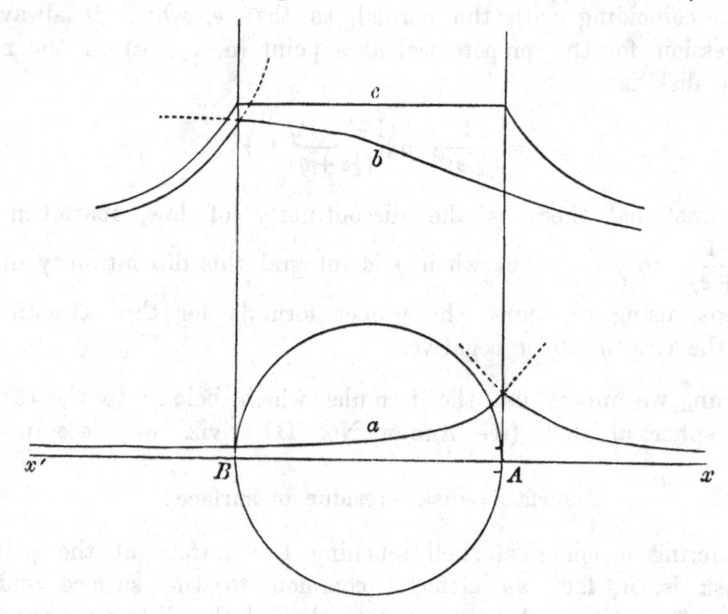

curves c and a, deducing the curve b, we see that this is a curve without any discontinuity corresponding to the passage of the attracted point through A (but with an abrupt change of direction or node corresponding to the passage through B). And conversely, using the curves a, b to determine the curve c, we see how, on the passage of the attracted point at A into the interior of the sphere, in consequence of the branch-to-branch discontinuity of the curve a, the curve c, obtained by combination of the two curves, undergoes a change of law, passing abruptly from a hyperbolic to a rectilinear form, and how similarly on the passage of the attracted point at B from the interior to the exterior of the sphere, in consequence of the branch-to-branch discontinuity of the curve b, the curve c again undergoes a change of law, abruptly reverting to the hyperbolic form.

24. In the case q positive, the prepotential curve is as shown by the right-hand figure on p. 332, viz. the ordinate is here infinite at the point N corresponding to the passage through the surface; the value of the derived function changes between + infinity and − infinity; and there is thus a discontinuity of value in the derived function. It would seem that, when q is fractional, this occasions a change of law on passage through the surface: but that there is no change of law when q is integral.

In illustration, consider the closed surface as made up of an infinitesimal circular disk, as before, and of a residual portion; the potential of the disk at an indefinitely near point is found as before, and the prepotential of the whole surface is

$$=\frac{1}{s^{2q}}\rho\frac{(\Gamma\frac{1}{2})^s\,\Gamma q}{\Gamma(\frac{1}{2}s+q)}+V_1,$$

where V_1, the prepotential of the remaining portion of the surface, is a function which varies (and its derived functions vary) continuously as the attracted point traverses the disk. To fix the ideas, we may take the origin at the centre of the disk, and the axis of e as coinciding with the normal, so that $\mathfrak{s}$, which is always positive, is $= \pm e$; the expression for the prepotential at a point $(a, .., c, e)$ on the normal through the centre of the disk is

$$= \frac{1}{(\pm e)^{2q}} \cdot \rho \frac{(\Gamma\tfrac{1}{2})^s \cdot \Gamma q}{\Gamma(\tfrac{1}{2}s + q)} + V_1;$$

viz. when q is fractional there is the discontinuity of law, inasmuch as the term changes from $\frac{1}{(+e)^{2q}}$ to $\frac{1}{(-e)^{2q}}$: but when q is integral this discontinuity disappears. The like considerations, using of course the proper formula for the attraction of the disk, would apply to the case $q = 0$ or negative.

25. Or again, we might use the formulæ which belong to the case of a uniform $(s+1)$-coordinal spherical shell (see Annex No. III.), viz. we decompose the surface as follows,

surface = disk + residue of surface;

and then, considering a spherical shell touching the surface at the point in question (so that the disk is, in fact, an element common to the surface and the spherical shell), and being of a uniform density equal to that of the disk, we have

disk = spherical shell − residue of spherical shell;

and consequently

surface = spherical shell − residue of spherical shell + residue of surface;

and then, considering the attracted point as passing through the disk, it does not pass through either of the two residues, and there is not any discontinuity, as regards the prepotentials of these residues respectively; there is consequently, as regards the prepotential of the surface, the same discontinuity that there is as regards the prepotential of the spherical shell. But I do not further consider the question from this point of view.

The Potential Solid Integral. Art. No. 26.

26. We have further to consider the prepotential (and in particular the potential) of a material space; to fix the ideas, consider for the moment the case of a distribution over the space included within a closed surface, the exterior density being zero, and the interior density being, supposed for the moment, constant; we consider the discontinuity which takes place as the attracting point passes from the exterior space through the bounding surface into the interior material space. We may imagine the interior space divided into indefinitely thin shells by a series of closed surfaces similar, if we please, to the bounding surface; and we may conceive the matter included between any two consecutive surfaces as concentrated on the exterior of the

two surfaces, so as to give rise to a series of consecutive material surfaces; the quantity of such matter is infinitesimal, and the density of each of the material surfaces is therefore also infinitesimal. As the attracted point comes from the external space to pass through the first of the material surfaces—suppose, to fix the ideas, it moves continuously along a curve the arc of which measured from a fixed point is $=s$—there is in the value of V (or, as the case may be, in the values of its derived functions $\frac{dV}{ds}$, &c.) the discontinuity due to the passage through the material surface; and the like as the attracted point passes through the different material surfaces respectively. Take the case of a potential, $q=-\tfrac{1}{2}$; then, if the surface-density were finite, there would be no finite change in the value of V, but there would be a finite change in the value of $\frac{dV}{ds}$; as it is, the changes are to be multiplied by the infinitesimal density, say ρ, of the material surface; there is consequently no finite change in the value of the first derived function; but there is, or may be, a finite change in the value of $\frac{d^2V}{ds^2}$ and the higher derived functions. But there is in V an infinitesimal change corresponding to the passage through the successive material surfaces respectively; that is, as the attracted point enters into the material space, there is a change in the law of V considered as a function of the coordinates $(a,.., c, e)$ of the attracted point; but by what precedes this change of law takes place without any abrupt change of value either of V or of its first derived function; which derived function may be considered as representing the derived function in regard to any one of the coordinates $a,.., c, e$. The suppositions, that the density outside the bounding surface was zero and inside it constant, were made for simplicity only, and were not essential; it is enough if the density, changing abruptly at the bounding surface, varies continuously in the material space within the bounding surface*. The conclusion is that V', V'' being the values at points within and without the bounding surface, V' and V'' are in general different functions of the coordinates $(a,.., c, e)$ of the attracted point; but that *at* the surface we have not only $V'=V''$, but that the first derived functions are also equal, viz. that we have

$$\frac{dV'}{da}=\frac{dV''}{da},\ldots,\ \frac{dV'}{dc}=\frac{dV''}{dc},\ \frac{dV'}{de}=\frac{dV''}{de}.$$

27. In the general case of a Potential, we have

$$V=\int\frac{\rho\,dx\ldots dz\,dw}{\{(a-x)^2+\ldots+(c-z)^2+(e-w)^2\}^{\frac{1}{2}s-\frac{1}{2}}}.$$

If ρ does not vanish at the attracted point $(a,.., c, e)$, but has there a value ρ' different from zero, we may consider the attracting $(s+1)$-dimensional mass as made

* It is, indeed, enough if the density varies continuously within the bounding surface in the neighbourhood of the point of passage through the surface; but the condition may without loss of generality be stated as in the text, it being understood that for each abrupt change of density within the bounding surface we must consider the attracted point as passing through a new bounding surface, and have regard to the resulting discontinuity.

up of an indefinitely small sphere, radius ϵ and density ρ', which includes within it the attracted point, and of a remaining portion external to the attracted point. Writing ∇ to denote $\frac{d^2}{da^2}+\ldots+\frac{d^2}{dc^2}+\frac{d^2}{de^2}$, then, as regards the potential of the sphere, we have $\nabla V = -\frac{4(\Gamma\frac{1}{2})^{s+1}}{\Gamma(\frac{1}{2}s-\frac{1}{2})}\rho'$ (see Annex III. No. 67), and as regards the remaining portion $\nabla V=0$; hence, as regards the whole attracting mass, ∇V has the first-mentioned value, that is, we have

$$\left(\frac{d^2}{da^2}+\ldots+\frac{d^2}{dc^2}+\frac{d^2}{de^2}\right)V = -\frac{4(\Gamma\frac{1}{2})^{s+1}}{\Gamma(\frac{1}{2}s-\frac{1}{2})}\rho',$$

where ρ' is the same function of the coordinates $(a,\ldots,c,\ e)$ that ρ is of $(x,\ldots,z,\ w)$; viz. the potential of an attracting mass distributed not on a surface, but over a portion of space, *does not satisfy the potential equation*

$$\left(\frac{d^2}{da^2}+\ldots+\frac{d^2}{dc^2}+\frac{d^2}{de^2}\right)V=0,$$

but it satisfies the foregoing equation, which only agrees with the potential equation in regard to a point $(a,\ldots,c,\ e)$ outside the material space, and for which, therefore, ρ' is $=0$.

The equation may be written

$$\rho' = -\frac{\Gamma(\frac{1}{2}s-\frac{1}{2})}{4(\Gamma\frac{1}{2})^{s+1}}\left(\frac{d^2}{da^2}+\ldots+\frac{d^2}{dc^2}+\frac{d^2}{de^2}\right)V;$$

or, considering V as a given function of $(a,\ldots,c,\ e)$, in general a discontinuous function but subject to certain conditions as afterwards mentioned, and taking W the same function of $(x,\ldots,z,\ w)$ that V is of $(a,\ldots,c,\ e)$, then we have

$$\rho = -\frac{\Gamma(\frac{1}{2}s-\frac{1}{2})}{4(\Gamma\frac{1}{2})^{s+1}}\left(\frac{d^2}{dx^2}+\ldots+\frac{d^2}{dz^2}+\frac{d^2}{dw^2}\right)W \quad \ldots\ldots\ldots\ldots\ldots\ldots\text{(D)},$$

viz. this equation determines ρ as a function, in general a discontinuous function, of $(x,\ldots,z,\ w)$ such that the corresponding integral

$$V=\int\frac{\rho\, dx\ldots dz\, dw}{\{(a-x)^2+\ldots+(c-z)^2+(e-w)^2\}^{\frac{1}{2}s+q}}$$

may be the given function of the coordinates $(a,\ldots,c,\ e)$. The equation is, in fact, the distribution-theorem D.

28. It is to be observed that the given function of $(a,\ldots,c,\ e)$ must satisfy certain conditions as to value at infinity and continuity, but it is not (as in the distribution-theorems A, B, and C it is) required to satisfy a partial differential equation; the function, except as regards the conditions as to value at infinity and continuity, is absolutely arbitrary.

The potential (assuming that the matter which gives rise to it lies wholly within a finite closed surface) must vanish for points at an infinite distance: or, more accurately, it must for indefinitely large values of $a^2 + \ldots + c^2 + e^2$ be of the form, Constant $\div (a^2 + \ldots + c^2 + e^2)^{\frac{1}{2}s-\frac{1}{2}}$. It may be a discontinuous function; for instance, outside a given closed surface it may be one function, and inside the same surface a different function of the coordinates $(a, .., c, e)$; viz. this may happen in consequence of an abrupt change of the density of the attracting matter on the one and the other side of the given closed surface, but not in any other manner; and, happening in this manner, then V' and V'' being the values for points within and without the surface respectively, it has been seen to be necessary that, at the surface, not only $V' = V''$, but also $\frac{dV'}{da} = \frac{dV''}{da}, .., \frac{dV'}{dc} = \frac{dV''}{dc}, \frac{dV'}{de} = \frac{dV''}{de}$. Subject to these conditions as to value at infinity and continuity, V may be any function whatever of the coordinates $(a, .., c, e)$; and then taking W, the same function of $(x, .., z, w)$, the foregoing equation determines ρ, viz. determines it to be $= 0$ for those parts of space which do not belong to the material space, and to have its proper value as a function of $(x, .., z, w)$ for the remaining or material space.

The Prepotential-Plane Theorem A. Art. Nos. 29 to 36.

29. We have seen that, if there exists on the plane $w = 0$ a distribution of matter producing at the point $(a, .., c, e)$ a given prepotential V—viz. V is to be regarded as a given function of $(a, .., c, e)$—, then the distribution or density ρ is given by a determinate formula; but it was remarked that the prepotential V cannot be a function assumed at pleasure: it must be a function satisfying certain conditions. One of these is the condition of continuity; the function V and all its derived functions must vary continuously as we pass, without traversing the material plane, from any given point to any other given point. But it is sufficient to attend to points on one side of the plane, say the upperside, or that for which e is positive; and since any such point is accessible from any other such point by a path which does not meet the plane, it is sufficient to say that the function V must vary continuously for a passage by such path from any such point to any such point; the function V must therefore be one and the same function (and that a continuous one in value) for all values of the coordinates $(a, .., c)$ and positive values of the coordinate e.

If, moreover, we assume that the distribution which corresponds to the given potential V is a distribution of a finite mass $\int \rho dx \ldots dz$ over a finite portion of the plane $w = 0$, viz. over a portion or area such that the distance of a point within the area from a fixed point, or say from the origin $(a, .., c) = (0, .., 0)$, is always finite; this being so, we have the further condition that the prepotential V must, for indefinitely large values of all or any of the coordinates $(a, .., c, e)$, reduce itself to the form

$$\left(\int \rho \, dx \ldots dz\right) \div (a^2 + \ldots + c^2 + e^2)^{\frac{1}{2}s+q}.$$

The assumptions upon which this last condition is obtained are perhaps unnecessary; instead of the condition in the foregoing form we, in fact, use only the condition that the prepotential vanishes for a point at infinity, that is, when all or any one or more of the coordinates $(a, .., c, e)$ are or is infinite.

Again, as we have seen, the prepotential V must satisfy the prepotential equation

$$\left(\frac{d^2}{da^2}+\ldots+\frac{d^2}{dc^2}+\frac{d^2}{de^2}+\frac{2q+1}{e}\frac{d}{de}\right)V=0.$$

These conditions satisfied, to the given prepotential V there corresponds, on the plane $w=0$, a distribution given by the foregoing formula; it will be a distribution over a finite portion of the plane, as already mentioned.

30. The proof depends upon properties of the prepotential equation

$$\left(\frac{d^2}{dx^2}+\ldots+\frac{d^2}{dz^2}+\frac{d^2}{de^2}+\frac{2q+1}{e}\frac{d}{de}\right)W=0,$$

or, what is the same thing,

$$\frac{d}{dx}\left(e^{2q+1}\frac{dW}{dx}\right)+\ldots+\frac{d}{dz}\left(e^{2q+1}\frac{dW}{dz}\right)+\frac{d}{de}\left(e^{2q+1}\frac{dW}{de}\right)=0,$$

say, for shortness, $\square W=0$.

Consider, in general, the integral

$$\int dx\ldots dz\, de\, .\, e^{2q+1}\left\{\left(\frac{dW}{dx}\right)^2+\ldots+\left(\frac{dW}{dz}\right)^2+\left(\frac{dW}{de}\right)^2\right\},$$

taken over a closed surface S lying altogether on the positive side of the plane $e=0$, the function W being in the first instance arbitrary.

Writing the integral under the form

$$\int dx\ldots dz\, de\left(e^{2q+1}\frac{dW}{dx}\cdot\frac{dW}{dx}+\ldots+e^{2q+1}\frac{dW}{dz}\cdot\frac{dW}{dz}+e^{2q+1}\frac{dW}{de}\cdot\frac{dW}{de}\right),$$

we reduce the several terms by an integration by parts as follows:—

The term in $\dfrac{dW}{dx}$ is $=\int dy\ldots dz\, de\, We^{2q+1}\dfrac{dW}{dx}-\int dx\ldots dz\, de\, W\dfrac{d}{dx}\left(e^{2q+1}\dfrac{dW}{dx}\right)$,

$\vdots$

$\dfrac{dW}{dz}$ is $=\int dx\ldots\ldots de\, We^{2q+1}\dfrac{dW}{dz}-\int dx\ldots dz\, de\, W\dfrac{d}{dz}\left(e^{2q+1}\dfrac{dW}{dz}\right)$,

$\dfrac{dW}{de}$ is $=\int dx\ldots\ldots dz\, We^{2q+1}\dfrac{dW}{de}-\int dx\ldots dz\, de\, W\dfrac{d}{de}\left(e^{2q+1}\dfrac{dW}{de}\right)$.

Write dS to denote an element of surface at the point $(x, .., z, e)$. Then taking $\alpha, .., \gamma, \delta$ to denote the inclinations of the interior normal at that point to the positive axes of coordinates, we have

$$\begin{aligned} dy \ldots dz\, de &= - dS \cos \alpha, \\ &\vdots \\ dx \ldots\ldots de &= - dS \cos \gamma, \\ dx \ldots\ldots dz &= - dS \cos \delta; \end{aligned}$$

and the first terms are together

$$= - \int e^{2q+1}\, W \left(\frac{dW}{dx} \cos \alpha + \ldots + \frac{dW}{dz} \cos \gamma + \frac{dW}{de} \cos \delta \right) dS,$$

W here denoting the value at the surface, and the integration being extended over the whole of the closed surface: this may also be written

$$= - \int e^{2q+1}\, W \frac{dW}{d\mathfrak{s}}\, dS,$$

where $\mathfrak{s}$ denotes an element of the internal normal.

The second terms are together

$$= - \int dx \,..\, dz\, de\, W \left\{ \frac{d}{dx} \left(e^{2q+1} \frac{dW}{dx} \right) + \ldots + \frac{d}{dz} \left(e^{2q+1} \frac{dW}{dz} \right) + \frac{d}{de} \left(e^{2q+1} \frac{dW}{de} \right) \right\} = - \int dx \,..\, dz\, de\, W \square W.$$

We have consequently

$$\int dx \ldots dz\, de \,.\, e^{2q+1} \left\{ \left(\frac{dW}{dx} \right)^2 + \ldots + \left(\frac{dW}{dz} \right)^2 + \left(\frac{dW}{de} \right)^2 \right\}$$

$$= - \int e^{2q+1}\, W \frac{dW}{d\mathfrak{s}}\, dS - \int dx \ldots dz\, de\, e^{2q+1}\, W \square W.$$

31. The second term vanishes if W satisfies the prepotential equation $\square W = 0$; and this being so, if also $W = 0$ for all points of the closed surface S, then the first term also vanishes, and we therefore have

$$\int dx \ldots dz\, de \,.\, e^{2q+1} \left\{ \left(\frac{dW}{dx} \right)^2 + \ldots + \left(\frac{dW}{dz} \right)^2 + \left(\frac{dW}{de} \right)^2 \right\} = 0,$$

where the integration extends over the whole space included within the closed surface; whence, W being a real function,

$$\frac{dW}{dx} = 0, \,..,\, \frac{dW}{dz} = 0, \quad \frac{dW}{de} = 0,$$

for all points within the closed surface; consequently, since W vanishes at the surface, $W = 0$ for all points within the closed surface.

32. Considering W as satisfying the equation $\square W = 0$, we may imagine the closed surface to become larger and larger, and ultimately infinite, at the same time flattening

itself out into coincidence with the plane $e=0$, so that it comes to include the whole space above the plane $e=0$: say the surface breaks up into the surface positive infinity and the infinite plane $e=0$.

The integral $\int e^{2q+1}\, W \frac{dW}{d\mathfrak{s}}\, dS$ separates itself into two parts, the first relating to the surface positive infinity, and vanishing if $W=0$ at infinity (that is, if all or any of the coordinates $x, .., z, e$ are infinite); the second, relating to the plane $e=0$, is $\int W \left(e^{2q+1} \frac{dW}{de}\right) dx \ldots dz$, W here denoting its value at the plane, that is, when $e=0$, and the integral being extended over the whole plane. The theorem thus becomes

$$\int dx \ldots dz\, de\,.\, e^{2q+1} \left\{ \left(\frac{dW}{dx}\right)^2 + \ldots + \left(\frac{dW}{dz}\right)^2 + \left(\frac{dW}{de}\right)^2 \right\} = -\int W \left(e^{2q+1} \frac{dW}{de}\right) dx \ldots dz.$$

Hence also, if $W=0$ at all points of the plane $e=0$, the right-hand side vanishes, and we have

$$\int dx \ldots dz\, de\,.\, e^{2q+1} \left\{ \left(\frac{dW}{dx}\right)^2 + \ldots + \left(\frac{dW}{dz}\right)^2 + \left(\frac{dW}{de}\right)^2 \right\} = 0.$$

Consequently $\frac{dW}{dx} = 0, \ldots, \frac{dW}{dz} = 0$, $\frac{dW}{de} = 0$, for all points whatever of positive space; and therefore also $W=0$ for all points whatever of positive space.

33. Take next U, W, each of them a function of $(x, .., z, e)$, and consider the integral

$$\int dx \ldots dz\, de\,.\, e^{2q+1} \left(\frac{dU}{dx}\frac{dW}{dx} + \ldots + \frac{dU}{dz}\frac{dW}{dz} + \frac{dU}{de}\frac{dW}{de} \right),$$

taken over the space within a closed surface S; treating this in a similar manner, we find it to be

$$= -\int e^{2q+1}\, W \frac{dU}{d\mathfrak{s}}\, dS - \int dx \ldots dz\, de\,.\, e^{2q+1}\, W \square U,$$

where the integration extends over the whole of the closed surface S; and by parity of reasoning it is also

$$= -\int e^{2q+1}\, U \frac{dW}{d\mathfrak{s}}\, dS - \int dx \ldots dz\, de\,.\, e^{2q+1}\, U \square W,$$

with the same limits of integration; that is, we have

$$\int e^{2q+1}\, W \frac{dU}{d\mathfrak{s}}\, dS + \int dx \ldots dz\, de\,.\, e^{2q+1}\, W \square U = \int e^{2q+1}\, U \frac{dW}{d\mathfrak{s}}\, dS + \int dx \ldots dz\, de\,.\, e^{2q+1}\, U \square W,$$

which, if U and W each satisfy the prepotential equation, becomes

$$\int e^{2q+1}\, W \frac{dU}{d\mathfrak{s}}\, dS = \int e^{2q+1}\, U \frac{dW}{d\mathfrak{s}}\, dS.$$

And if we now take the closed surface S to be the surface positive infinity, together with the plane $e=0$, then, provided only U and V vanish at infinity, for each integral the portion belonging to the surface positive infinity vanishes, and there remains only the portion belonging to the plane $e=0$; we have therefore

$$\int e^{2q+1}\, W \frac{dU}{de}\, dx \ldots dz = \int e^{2q+1}\, U \frac{dW}{de}\, dx \ldots dz,$$

where the functions U, W have each of them the value belonging to the plane $e=0$: viz. in U, W considered as given functions of $(x, .., z, e)$ we regard e as a positive quantity ultimately put $=0$; and where the integrations extend each of them over the whole infinite plane.

34. Assume

$$U = \frac{1}{\{(a-x)^2 + \ldots + (c-z)^2 + e^2\}^{\frac{1}{2}s+q}},$$

an expression which, regarded as a function of $(x, .., z, e)$, satisfies the prepotential equation in regard to these variables, and which vanishes at infinity when all or any of these coordinates $(x, .., z, e)$ are infinite.

We have

$$\frac{dU}{de} = \frac{-2\,(\tfrac{1}{2}s+q)\, e}{\{(a-x)^2 + \ldots + (c-z)^2 + e^2\}^{\frac{1}{2}s+q+1}};$$

and we have consequently

$$\int W \frac{-2\,(\tfrac{1}{2}s+q)\, e^{2q+2}}{\{(a-x)^2 + \ldots + (c-z)^2 + e^2\}^{\frac{1}{2}s+q+1}}\, dx \ldots dz = \int \left(e^{2q+1} \frac{dW}{de}\right) \frac{dx \ldots dz}{\{(a-x)^2 + \ldots + (c-z)^2 + e^2\}^{\frac{1}{2}s+q}},$$

where it will be recollected that e is ultimately $=0$; to mark this, we may for W write W_0.

Attend to the left-hand side; take V_0 the same function of $a, .., c, e=0$, that W_0 is of $x, .., z, e=0$; then, first writing the expression in the form

$$V_0 \int \frac{-2\,(\tfrac{1}{2}s+q)\, e^{2q+2}\, dx \ldots dz}{\{(a-x)^2 + \ldots + (c-z)^2 + e^2\}^{\frac{1}{2}s+q+1}},$$

write $x = a + e\xi, .., z = c + e\zeta$, the expression becomes

$$= V_0 \int \frac{-2\,(\tfrac{1}{2}s+q)\, e^{2q+2} \,.\, e^s d\xi \ldots d\zeta}{\{e^2\,(1 + \xi^2 + \ldots + \zeta^2)\}^{\frac{1}{2}s+q+1}}, \quad = -2\,(\tfrac{1}{2}s+q)\, V_0 \int \frac{d\xi \ldots d\zeta}{\{1 + \xi^2 + \ldots + \zeta^2\}^{\frac{1}{2}s+q+1}},$$

where the integral is to be taken from $-\infty$ to $+\infty$ for each of the new variables $\xi, .., \zeta$.

Writing $\xi = r\alpha, .., \zeta = r\gamma$, where $\alpha^2 + \ldots + \gamma^2 = 1$, we have $d\xi \ldots d\zeta = r^{s-1}\, dr\, dS$: also $\xi^2 + \ldots + \zeta^2 = r^2$, and the integral is

$$= \int \frac{r^{s-1}\, dr\, dS}{(1+r^2)^{\frac{1}{2}s+q+1}}, \quad = \int dS \int_0^\infty \frac{r^{s-1}\, dr}{(1+r^2)^{\frac{1}{2}s+q+1}},$$

where $\int dS$ denotes the surface of the s-coordinal unit sphere $\alpha^2 + \ldots + \gamma^2 = 1$, and the r-integral is to be taken from $r = 0$ to $r = \infty$; the values of the two factors thus are

$$\int dS = \frac{2 (\Gamma \tfrac{1}{2})^s}{\Gamma (\tfrac{1}{2} s)}, \text{ and } \int \frac{r^{s-1}\, dr}{(1 + r^2)^{\frac{1}{2}s+q+1}} = \frac{\tfrac{1}{2}\Gamma \tfrac{1}{2} s\, \Gamma (q+1)}{\Gamma (\tfrac{1}{2} s + q + 1)}.$$

Hence the expression in question is

$$-2 (\tfrac{1}{2} s + q)\, V_0 \frac{2 (\Gamma \tfrac{1}{2})^s}{\Gamma \tfrac{1}{2} s} \frac{\tfrac{1}{2}\Gamma \tfrac{1}{2} s\, \Gamma (q+1)}{\Gamma (\tfrac{1}{2} s + q + 1)}, \quad = \frac{-2 (\Gamma \tfrac{1}{2})^s\, \Gamma (q+1)}{\Gamma (\tfrac{1}{2} s + q)} V_0,$$

and we have

$$\int \left(e^{2q+1} \frac{dW}{de} \right)_0 \frac{dx \ldots dz}{\{(a-x)^2 + \ldots + (c-z)^2 + e^2\}^{\frac{1}{2}s+q}} = \frac{-2 (\Gamma \tfrac{1}{2})^s\, \Gamma (q+1)}{\Gamma (\tfrac{1}{2} s + q)} V_0;$$

or, what is the same thing,

$$V_0 = \int \frac{\dfrac{-\Gamma (\tfrac{1}{2} s + q)}{2 (\Gamma \tfrac{1}{2})^s\, \Gamma (q+1)} \left(e^{2q+1} \dfrac{dW}{de} \right)_0 dx \ldots dz}{\{(a-x)^2 + \ldots + (c-z)^2 + e^2\}^{\frac{1}{2}s+q}}.$$

35. Take now V a function of $(a, .., c, e)$ satisfying the prepotential equation in regard to these variables, always finite, and vanishing at infinity; and let W be the same function of $(x, .., z, e)$, W therefore satisfying the prepotential equation in regard to the last-mentioned variables. Consider the function

$$V - \int \frac{-\dfrac{\Gamma (\tfrac{1}{2} s + q)}{2 (\Gamma \tfrac{1}{2})^s\, \Gamma (q+1)} \left(e^{2q+1} \dfrac{dW}{de} \right)_0 dx \ldots dz}{\{(a-x)^2 + \ldots + (c-z)^2 + e^2\}^{\frac{1}{2}s+q}},$$

where the integral is taken over the infinite plane $e = 0$; then this function (V – the integral) satisfies the prepotential equation (for each term separately satisfies it), is always finite, and it vanishes at infinity. It also, as has just been seen, vanishes for any point whatever of the plane $e = 0$. Consequently it vanishes for all points whatever of positive space. Or, what is the same thing, if we write

$$V = \int \frac{\rho\, dx \ldots dz}{\{(a-x)^2 + \ldots + (c-z)^2 + e^2\}^{\frac{1}{2}s+q}} \quad \text{.......................(A)},$$

where ρ is a function of $(x, .., z)$, and the integral is taken over the whole infinite plane, then if V is a function of $(a, .., c, e)$ satisfying the above conditions, there exists a corresponding value of ρ; viz. taking W the same function of $(x, .., z, e)$ which V is of $(a, .., c, e)$, the value of ρ is

$$\rho = -\frac{\Gamma (\tfrac{1}{2} s + q)}{2 (\Gamma \tfrac{1}{2})^s\, \Gamma (q+1)} \left(e^{2q+1} \frac{dW}{de} \right)_0 \quad \text{....................... (A)},$$

where e is to be put $= 0$ in the function $e^{2q+1} \dfrac{dW}{de}$. This is the prepotential-plane theorem; viz. taking for the prepotential in regard to a given point $(a, .., c, e)$ a function of $(a, .., c, e)$ satisfying the prescribed conditions, but otherwise arbitrary, there exists on the plane $e = 0$ a distribution ρ given by the last-mentioned formula.

36. It is assumed in the proof that $2q+1$ is positive or zero; viz. q is positive, or if negative then $-q \not> \frac{1}{2}$; the limiting case $q=-\frac{1}{2}$ is included.

It is to be remarked that, by what precedes, if q be positive (but excluding the case $q=0$), the density ρ is given by the equivalent more simple formula

$$\rho = \frac{\Gamma(\frac{1}{2}s+q)}{(\Gamma\frac{1}{2})^s \Gamma q} (e^{2q} W)_0.$$

The foregoing proof is substantially that given in Green's memoir on the Attraction of Ellipsoids; it will be observed that the proof only imposes upon V the condition of vanishing at infinity, without obliging it to assume for large values of $(a, .., c, e)$ the form

$$\frac{M}{\{a^2+\ldots+c^2+e^2\}^{\frac{1}{2}s+q}}.$$

The Potential-surface Theorem C. Art. Nos. 37 to 42.

37. In the case $q=-\frac{1}{2}$, writing here $\nabla = \frac{d^2}{dx^2}+\ldots+\frac{d^2}{dz^2}+\frac{d^2}{de^2}$, we have, precisely as in the general case,

$$\int W \frac{dU}{ds} dS + \int dx \ldots dz\, de\, W \nabla U = \int U \frac{dW}{ds} dS + \int dx \ldots dz\, de\, U \nabla W;$$

and if the functions U, W satisfy the equations $\nabla U=0$, $\nabla W=0$, then (subject to the exception presently referred to) the second terms on the two sides respectively each of them vanish.

But, instead of taking the surface to be the surface positive infinity together with the plane $e=0$, we now leave it an arbitrary closed surface, and for greater symmetry of notation write w in place of e; and we suppose that the functions U and W, or one of them, may become infinite at points within the closed surface; then, on this last account, the second terms do not in every case vanish.

38. Suppose, for instance, that U at a point indefinitely near the point $(a, .., c, e)$ within the surface becomes

$$= \frac{1}{\{(x-a)^2+\ldots+(z-c)^2+(w-e)^2\}^{\frac{1}{2}s-\frac{1}{2}}};$$

then if V be the value of W at the point $(a, .., c, e)$, we have

$$\int dx \ldots dz\, dw\, W \nabla U = V \int dx \ldots dz\, dw\, \nabla U;$$

and since $\nabla U=0$, except at the point in question, the integral may be taken over any portion of space surrounding this point, for instance, over the space included within the sphere, radius R, having the point $(a,.., c, e)$ for its centre; or taking the origin at this point, we have to find $\int dx \ldots dz\, dw\, \nabla U$, where

$$U = \frac{1}{\{x^2 + \ldots + z^2 + w^2\}^{\frac{1}{2}s-\frac{1}{2}}},$$

and the integration extends over the space within the sphere $x^2 + \ldots + z^2 + w^2 = R^2$.

39. This may be accomplished most easily by means of a particular case of the last-mentioned theorem; viz. writing $W = 1$, we have

$$\int \frac{dU}{d\mathfrak{s}} dS + \int dx \ldots dz\, dw\, \nabla U = 0,$$

or the required value is $= -\int \frac{dU}{d\mathfrak{s}} dS$ over the surface of the last-mentioned sphere. We have, if for a moment $r^2 = x^2 + \ldots + z^2 + w^2$,

$$\frac{dU}{d\mathfrak{s}} = -\left(\frac{x}{r}\frac{d}{dx} + \ldots + \frac{z}{r}\frac{d}{dz} + \frac{w}{r}\frac{d}{dw}\right) U, \quad = -\frac{dU}{dr}\left\{\left(\frac{x}{r}\frac{d}{dx} + \ldots + \frac{z}{r}\frac{d}{dz} + \frac{w}{r}\frac{d}{dw}\right) r\right\}, \quad = -\frac{dU}{dr},$$

that is, $\frac{dU}{d\mathfrak{s}} = \frac{s-1}{r^s}$, $= \frac{s-1}{R^s}$ at the surface; and hence

$$\int \frac{dU}{d\mathfrak{s}} dS = \frac{s-1}{R^s} \int dS,$$

where $\int dS$ is the whole surface of the sphere $x^2 + \ldots + z^2 + w^2 = R^2$, viz. it is $= R^s$, multiplied by the surface of the unit-sphere $x^2 + \ldots + z^2 + w^2 = 1$. This spherical surface, say $\int d\Sigma$, is

$$= \frac{2(\Gamma\tfrac{1}{2})^{s+1}}{\Gamma\tfrac{1}{2}(s+1)}, \quad = \frac{4(\Gamma\tfrac{1}{2})^{s+1}}{(s-1)\,\Gamma\tfrac{1}{2}(s-1)},$$

and we have thus $\int \frac{dU}{d\mathfrak{s}} dS = \frac{4(\Gamma\tfrac{1}{2})^{s+1}}{\Gamma\tfrac{1}{2}(s-1)}$, and consequently

$$\int dx \ldots dz\, dw\, \nabla U = -\frac{4(\Gamma\tfrac{1}{2})^{s+1}}{\Gamma(\tfrac{1}{2}s - \tfrac{1}{2})}.$$

40. Treating in like manner the case, where W at a point indefinitely near the point (a,.., c, e) within the surface becomes

$$= \frac{1}{\{(x-\mathrm{a})^2 + \ldots + (z-\mathrm{c})^2 + (w-\mathrm{e})^2\}^{\frac{1}{2}s-\frac{1}{2}}},$$

and writing T to denote the same function of (a,.., c, e) that U is of $(x,.., z, w)$, we have, instead of the foregoing, the more general theorem

$$\int W \frac{dU}{d\mathfrak{s}} dS + \int dx \ldots dz\, dw\, W \nabla U - \frac{4(\Gamma\tfrac{1}{2})^{s+1}}{\Gamma(\tfrac{1}{2}s - \tfrac{1}{2})} V$$

$$= \int U \frac{dW}{d\mathfrak{s}} dS + \int dx \ldots dz\, dw\, U \nabla W - \frac{4(\Gamma\tfrac{1}{2})^{s+1}}{\Gamma(\tfrac{1}{2}s - \tfrac{1}{2})} T,$$

where, in the two solid integrals, we exclude from consideration the space in the immediate neighbourhood of the two critical points $(a, .., c, e)$ and (a, .., c, e) respectively.

Suppose that W is always finite within the surface, and that U is finite except at the point $(a, .., c, e)$: and moreover that U, W are such that $\nabla U = 0$, $\nabla W = 0$; then the equation becomes

$$\int W \frac{dU}{d\mathfrak{s}} dS - \frac{4(\Gamma\tfrac{1}{2})^{s+1}}{\Gamma\tfrac{1}{2}s - \tfrac{1}{2}} V = \int U \frac{dW}{d\mathfrak{s}} dS.$$

In particular, this equation holds good if U is

$$= \frac{1}{\{(a-x)^2 + \ldots + (e-w)^2\}^{\frac{1}{2}s-\frac{1}{2}}}.$$

41. Imagine now on the surface S a distribution $\rho\, dS$ producing at a point $(a', .., c', e')$ within the surface a potential V', and at a point $(a'', .., c'', e'')$ without the surface a potential V''; where, by what precedes, V'' is in general not the same function of $(a'', .., c'', e'')$ that V' is of $(a', .., c', e')$.

It is further assumed that at a point $(a, .., c, e)$ on the surface we have $V' = V''$:

that V', or any of its derived functions, are not infinite for any point $(a', .., c', e')$ within the surface:

that V'', or any of its derived functions, are not infinite for any point $(a'', .., c'', e'')$ without the surface:

and that $V'' = 0$ for any point at infinity.

Consider V' as a given function of $(a, .., c, e)$; and take W' the same function of $(x, .., z, w)$. Then if, as before,

$$U = \frac{1}{\{(a-x)^2 + \ldots + (c-z)^2 + (e-w)^2\}^{\frac{1}{2}s-\frac{1}{2}}},$$

we have

$$\left(\frac{d^2}{dx^2} + \ldots + \frac{d^2}{dz^2} + \frac{d^2}{dw^2}\right) U = 0,$$

$$\int U \frac{dW'}{d\mathfrak{s}'} dS = \int W' \frac{dU}{d\mathfrak{s}'} dS - \frac{4(\Gamma\tfrac{1}{2})^{s+1}}{\Gamma(\tfrac{1}{2}s - \tfrac{1}{2})} V'.$$

Similarly, considering V'' as a given function of $(a, .., c, e)$, take W'' the same function of $(x, .., z, e)$. Then, by considering the space outside the surface S, or say between this surface and infinity, and observing that U does not become infinite for any point in this space, we have

$$\int U \frac{dW''}{d\mathfrak{s}''} dS = \int W'' \frac{dU}{d\mathfrak{s}''} dS;$$

adding these two equations, we have

$$\int U \left(\frac{dW'}{d\mathfrak{s}'} + \frac{dW''}{d\mathfrak{s}''}\right) dS = \int \left(W' \frac{dU}{d\mathfrak{s}'} + W'' \frac{dU}{d\mathfrak{s}''}\right) dS - \frac{4(\Gamma\tfrac{1}{2})^{s+1}}{\Gamma(\tfrac{1}{2}s - \tfrac{1}{2})} V'.$$

But in this equation the functions W' and W'' each of them belong to a point $(x, .., z, w)$ on the surface, and we have at the surface $W' = W''$, $= W$ suppose; the term on the right-hand side thus is $\int W\left(\frac{dU}{d\mathfrak{s}'} + \frac{dU}{d\mathfrak{s}''}\right) dS$, which vanishes in virtue of $\frac{dU}{d\mathfrak{s}'} + \frac{dU}{d\mathfrak{s}''} = 0$; and the equation thus becomes

$$\int U\left(\frac{dW'}{d\mathfrak{s}'} + \frac{dW''}{d\mathfrak{s}''}\right) dS = -\frac{4(\Gamma\frac{1}{2})^{s+1}}{\Gamma(\frac{1}{2}s - \frac{1}{2})} V':$$

that is, the point $(a, .., c, e)$ being interior, we have

$$V' = \int \frac{-\Gamma(\frac{1}{2}s - \frac{1}{2})}{4(\Gamma\frac{1}{2})^{s+1}} \left(\frac{dW'}{d\mathfrak{s}'} + \frac{dW''}{d\mathfrak{s}''}\right) \frac{dS}{\{(a-x)^2 + \ldots + (c-z)^2 + (e-w)^2\}^{\frac{1}{2}s-\frac{1}{2}}}.$$

In exactly the same way, if $(a, .., c, e)$ be an exterior point, then we have

$$\int U \frac{dW'}{d\mathfrak{s}'} dS = \int W' \frac{dU}{d\mathfrak{s}'} dS,$$

$$\int U \frac{dW''}{d\mathfrak{s}''} dS = \int W'' \frac{dU}{d\mathfrak{s}''} dS - \frac{4(\Gamma\frac{1}{2})^{s+1}}{\Gamma(\frac{1}{2}s - \frac{1}{2})} V'';$$

adding, and omitting the terms which vanish,

$$\int U\left(\frac{dW'}{d\mathfrak{s}'} + \frac{dW''}{d\mathfrak{s}''}\right) dS = -\frac{4(\Gamma\frac{1}{2})^{s+1}}{\Gamma(\frac{1}{2}s - \frac{1}{2})} V'',$$

that is,

$$V'' = \int \frac{-\Gamma(\frac{1}{2}s - \frac{1}{2})}{4(\Gamma\frac{1}{2})^{s+1}} \left(\frac{dW'}{d\mathfrak{s}'} + \frac{dW''}{d\mathfrak{s}''}\right) \frac{dS}{\{(a-x)^2 + \ldots + (c-z)^2 + (e-w)^2\}^{\frac{1}{2}s-\frac{1}{2}}}.$$

42. Comparing the two results with

$$V = \int \frac{\rho \, dS}{\{(a-x)^2 + \ldots + (c-z)^2 + (e-w)^2\}^{\frac{1}{2}s-\frac{1}{2}}},$$

we see that, V' and V'' satisfying the foregoing conditions, there exists a distribution ρ on the surface, producing the potentials V' and V'' at an interior point and an exterior point respectively; the value of ρ in fact being

$$\rho = -\frac{\Gamma(\frac{1}{2}s - \frac{1}{2})}{4(\Gamma\frac{1}{2})^{s+1}} \left(\frac{dW'}{d\mathfrak{s}'} + \frac{dW''}{d\mathfrak{s}''}\right) \quad \ldots\ldots\ldots\ldots\ldots\ldots\ldots\ldots\text{(C)},$$

where W', W'' are respectively the same functions of $(x, .., z, w)$ that V', V'' are of $(a, .., c, e)$.

The Potential-solid Theorem D. Art. No. 43.

43. We have as before (No. 40),

$$\int W \frac{dU}{d\mathfrak{s}} dS + \int dx \ldots dz\, dw\, W \nabla U - \frac{4(\Gamma\frac{1}{2})^{s+1}}{\Gamma(\frac{1}{2}s - \frac{1}{2})} V$$
$$= \int U \frac{dW}{d\mathfrak{s}} dS + \int dx \ldots dz\, dw\, U \nabla W - \frac{4(\Gamma\frac{1}{2})^{s+1}}{\Gamma(\frac{1}{2}s - \frac{1}{2})} T,$$

where, assuming first that W is not infinite for any point $(x, .., z, w)$ whatever, we have no term in T; and taking next $U = \dfrac{1}{\{(a-x)^2 + \ldots + (c-z)^2 + (e-w)^2\}^{\frac{1}{2}s-\frac{1}{2}}}$ as before, we have $\nabla U = 0$; the equation thus becomes

$$\int W \frac{dU}{d\mathfrak{s}} dS - \int U \frac{dW}{d\mathfrak{s}} dS - \frac{4(\Gamma\frac{1}{2})^{s+1}}{\Gamma(\frac{1}{2}s - \frac{1}{2})} V = \int dx \ldots dz\, dw\, U \nabla W,$$

where W may be a discontinuous function of the coordinates $(x, .., z, w)$, provided only there is no abrupt change in the value either of W or of any of its first derived functions $\dfrac{dW}{dx}, .., \dfrac{dW}{dz}, \dfrac{dW}{dw}$, viz. it may be any function which can represent the potential of a solid mass on an attracted point $(x, .., z, w)$; the resulting value of ∇W is of course discontinuous. Taking, then, for the closed surface S the boundary of infinite space, U and W each vanish at this boundary, and the equation becomes

$$-\frac{(\Gamma\frac{1}{2})^{s+1}}{\Gamma(\frac{1}{2}s - \frac{1}{2})} V = \int dx \ldots dz\, dw\, U \nabla W;$$

viz. substituting for U its value, and comparing with

$$V = \int \frac{\rho\, dx \ldots dz\, dw}{\{(a-x)^2 + \ldots + (c-z)^2 + (e-w)^2\}^{\frac{1}{2}s-\frac{1}{2}}},$$

where the integral in the first instance extends to the whole of infinite space, but the limits may be ultimately restricted by ρ being $= 0$, we see that the value of ρ is

$$\rho = -\frac{\Gamma(\frac{1}{2}s - \frac{1}{2})}{(\Gamma\frac{1}{2})^{s+1}} \left(\frac{d^2}{dx^2} + \ldots + \frac{d^2}{dz^2} + \frac{d^2}{dw^2}\right) W,$$

W being the same function of $(x, .., z, w)$ that V is of $(a, .., c, e)$: which is the theorem D.

Examples of the foregoing Theorems. Art. Nos. 44 to 50.

44. It will be remarked, as regards all the theorems, that we do not start with known limits; we start with V a function of $(a, .., c, e)$, the coordinates of the attracted point, satisfying certain prescribed conditions, and we thence find ρ, a function of the coordinates $(x, .., z)$ or $(x, .., z, w)$, as the case may be, which function is found to be $= 0$ for values of $(x, .., z)$ or $(x, .., z, w)$ lying beyond certain limits, and to have a determinate non-evanescent value for values of $(x, .., z)$ or $(x, .., z, w)$ lying within these limits; and we thus, as a result, obtain these limits for the limits of the multiple integral V.

45. Thus in theorem A, in the example where the limiting equation is ultimately found to be $x^2 + \ldots + z^2 = f^2$, we start with V a certain function of $a^2 + \ldots + c^2$ ($= \kappa^2$ suppose) and e^2, viz. V is a function of these quantities through θ, which denotes the positive root of the equation

$$\frac{\kappa^2}{f^2 + \theta} + \frac{e^2}{\theta} = 1,$$

the value in fact being $V = \int_\theta^\infty t^{-q-1}(t+f^2)^{-\frac{1}{2}s}\,dt$, and the resulting value of ρ is found to be $=0$ for values of $(x,..,z)$ for which $x^2+\ldots+z^2>f^2$. Hence V denotes an integral

$$\int \frac{\rho\,dx\ldots dz}{\{(a-x)^2+\ldots+(c-z)^2+e^2\}^{\frac{1}{2}s+q}},$$

the limiting equation being $x^2+\ldots+z^2=f^2$: say this is the s-coordinal sphere.

And similarly, in the examples where the limiting equation is ultimately found to be $\frac{x^2}{f^2}+\ldots+\frac{z^2}{h^2}=1$, we start with V a certain function of $a,..,c,\ e$ through θ (or directly and through θ), where θ denotes the positive root of the equation

$$\frac{a^2}{f^2+\theta}+\ldots+\frac{c^2}{h^2+\theta}+\frac{e^2}{\theta}=1,$$

and the resulting value of ρ is found to be $=0$ for values of $(x,..,z)$ for which

$$\frac{x^2}{f^2}+\ldots+\frac{z^2}{h^2}>1.$$

Hence V denotes an integral

$$\int \frac{\rho\,dx\ldots dz}{\{(a-x)^2+\ldots+(c-z)^2+e^2\}^{\frac{1}{2}s+q}},$$

the limiting equation being $\frac{x^2}{f^2}+\ldots+\frac{z^2}{h^2}=1$: say this is the s-coordinal ellipsoid. It is clear that this includes the before-mentioned case of the s-coordinal sphere; but, on account of the more simple form of the θ-equation, it is worth while to work out directly an example for the sphere.

46. Three examples are worked out in Annex IV.; the results are as follows:—

First, θ defined for the sphere as above; $q+1$ positive;

$$V = \int \frac{\left(1-\frac{x^2+\ldots+z^2}{f^2}\right)^q dx\ldots dz}{\{(a-x)^2+\ldots+(c-z)^2+e^2\}^{\frac{1}{2}s+q}}$$

over the sphere $x^2+\ldots+y^2=f^2$,

$$= \frac{(\Gamma\tfrac{1}{2})^s\,\Gamma(q+1)}{\Gamma(\tfrac{1}{2}s+q)}\,f^s\int_\theta^\infty t^{-q-1}(t+f^2)^{-\frac{1}{2}s}\,dt.$$

This is included in the next-mentioned example for the ellipsoid.

Secondly, θ defined for the ellipsoid as above; $q+1$ positive;

$$V = \int \frac{\left(1-\frac{x^2}{f^2}-\ldots-\frac{z^2}{h^2}\right)^q dx\ldots dz}{\{(a-x)^2+\ldots+(c-z)^2+e^2\}^{\frac{1}{2}s+q}}$$

over the ellipsoid $\frac{x^2}{f^2}+\ldots+\frac{z^2}{h^2}=1$,

$$=\frac{(\Gamma\tfrac{1}{2})^s\Gamma(q+1)}{\Gamma(\tfrac{1}{2}s+q)}\cdot f\ldots h\int_\theta^\infty t^{-q-1}\{(t+f^2)\ldots(t+h^2)\}^{-\frac{1}{2}}\,dt.$$

This result is included in the next-mentioned example; but the proof for the general value of m is not directly applicable to the value $m=0$ for the case in question.

Thirdly, θ defined for the ellipsoid as above; $q+1$ positive; $m=0$ or positive, and apparently in other cases,

$$V=\int\frac{\left(1+\frac{x^2}{f^2}-\ldots-\frac{z^2}{h^2}\right)^{q+m}dx\ldots dz}{\{(a-x)^2+\ldots+(c-z)^2+e^2\}^{\frac{1}{2}s+q}}$$

over the ellipsoid as above,

$$=\frac{(\Gamma\tfrac{1}{2})^s\Gamma(1+q+m)}{\Gamma(\tfrac{1}{2}s+q)\Gamma(1+m)}\cdot f\ldots h\int_0^\infty\left(1-\frac{a^2}{f^2+\theta}-\ldots-\frac{c^2}{h^2+\theta}-\frac{e^2}{\theta}\right)^m t^{-q-1}\{(t+f^2)\ldots(t+h^2)\}^{-\frac{1}{2}}\,dt.$$

And we have in Annex V. a fourth example; here θ and the ellipsoid are as above: the result involves the Greenian functions.

47. We may in the foregoing results write $e=0$; the results,—writing therein $s+1$ for s, and in the new forms taking $(a,\ldots,c,e)$ and $(x,\ldots,z,w)$ for the two sets of coordinates respectively, also writing $q-\frac{1}{2}$ for q—, would give integrals of the form

$$\int\frac{\rho\,dx\ldots dz\,dw}{\{(a-x)^2+\ldots+(c-z)^2+(e-w)^2\}^{\frac{1}{2}s+q}}$$

for the $(s+1)$-coordinal sphere and ellipsoid $x^2+\ldots+z^2+w^2=f^2$ and $\frac{x^2}{f^2}+\ldots+\frac{z^2}{h^2}+\frac{w^2}{k^2}=1$: say these are prepotential-solid integrals; and then, writing $q=-\frac{1}{2}$, we should obtain potential-solid integrals, such as are also given by the theorem D. The change can be made if necessary; but it is more convenient to retain the results in their original forms, as relating to the s-coordinal sphere and ellipsoid.

There are two cases, according as the attracted point $(a,\ldots,c)$ is external or internal.

For the sphere:—For an external point $\kappa^2>f^2$; writing $e=0$, the equation $\frac{\kappa^2}{f^2+\theta}=1$ has a positive root, viz. this is $\theta=\kappa^2-f^2$; and θ will have, or it may be replaced by, this value κ^2-f^2: for an internal point $\kappa^2<f^2$; as e approaches zero, the positive root of the original equation gradually diminishes and becomes ultimately $=0$, viz. in the formulæ θ is to be replaced by this value 0.

For the ellipsoid:—For an external point $\frac{a^2}{f^2}+\ldots+\frac{c^2}{h^2}>1$; writing $e=0$, the equation $\frac{a^2}{\theta+f^2}+\ldots+\frac{c^2}{\theta+h^2}=1$ has a positive root, and θ will denote this positive root: for an internal point $\frac{a^2}{f^2}+\ldots+\frac{c^2}{h^2}<1$; as e approaches zero the positive root of the original equation gradually diminishes and becomes ultimately $=0$, viz. in the formulæ θ is to be replaced by this value 0.

The resulting formulæ for the sphere $x^2+\ldots+z^2=f^2$ may be compared with formulæ for the spherical shell, Annex VI., and each set with formulæ obtained by direct integration in Annex III.

We may in any of the formulæ write $q=-\frac{1}{2}$, and so obtain examples of theorem B.

48. As regards theorem C, we might in like manner obtain examples of potentials relating to the surfaces of the $(s+1)$-coordinal sphere $x^2+\ldots+z^2+w^2=f^2$, and ellipsoid $\frac{x^2}{f^2}+\ldots+\frac{z^2}{h^2}+\frac{w^2}{k^2}=1$, or say to spherical and ellipsoidal shells; but I have confined myself to the sphere. We have to assume values V' and V'' belonging to the cases of an internal and an external point respectively, and thence to obtain a value ρ, or distribution over the spherical surface, which shall produce these potentials respectively. The result (see Annex VI.) is

$$\int\frac{dS}{\{(a-x)^2+\ldots+(c-z)^2+(e-w)^2\}^{\frac{1}{2}s-\frac{1}{2}}}$$

over the surface of the $(s+1)$-coordinal sphere $x^2+\ldots+z^2+w^2=f^2$,

$$=\frac{2\,(\Gamma\tfrac{1}{2})^{s+1}f^s}{\Gamma\,(\tfrac{1}{2}s+\tfrac{1}{2})}\,\frac{1}{\kappa^{s-1}}\text{ for exterior point }\kappa>f,$$

and

$$=\frac{2\,(\Gamma\tfrac{1}{2})^{s+1}f^s}{\Gamma\,(\tfrac{1}{2}s+\tfrac{1}{2})}\,\frac{1}{f^{s-1}}\text{ for interior point }\kappa<f,$$

where $\kappa^2=a^2+\ldots+c^2+e^2$. Observe that for the interior point the potential is a mere constant multiple of f.

The same Annex VI. contains the case of the s-coordinal cylinder $x^2+\ldots+z^2=f^2$, which is peculiar in that the cylinder is not a finite closed surface; but the theorem C is found to extend to it.

49. As regards theorem D, we might in like manner obtain potentials relating to the $(s+1)$-coordinal sphere $x^2+\ldots+z^2+w^2=f^2$ and ellipsoid $\frac{x^2}{f^2}+\ldots+\frac{z^2}{h^2}+\frac{w^2}{k^2}=1$; but I confine myself to the case of the sphere (see Annex VII.). We here assume values V' and V'' belonging to an internal and an external point respectively, and thence obtain a value ρ, or distribution over the whole $(s+1)$-dimensional space, which density is found to be $=0$ for points outside the sphere. The result obtained is

$$V=\int\frac{dx\ldots dz\,dw}{\{(a-x)^2+\ldots+(c-z)^2+(e-w)^2\}^{\frac{1}{2}s-\frac{1}{2}}}$$

over the $(s+1)$-coordinal sphere $x^2+\ldots+z^2+w^2=f^2$,

$$=\frac{(\Gamma\tfrac{1}{2}s)^{s+1}}{\Gamma(\tfrac{1}{2}s+\tfrac{3}{2})}\cdot\frac{f^{s+1}}{\kappa^{s-1}} \text{ for an exterior point } \kappa>f,$$

$$=\frac{(\Gamma\tfrac{1}{2}s)^{s+1}}{\Gamma(\tfrac{1}{2}s+\tfrac{3}{2})}\{(\tfrac{1}{2}s+\tfrac{1}{2})f^2-(\tfrac{1}{2}s-\tfrac{1}{2})\kappa^2\} \text{ for an interior point } \kappa<f,$$

where $\kappa^2=a^2+\ldots+c^2+e^2$.

50. The remaining Annexes VIII. and IX. have no immediate reference to the theorems A, B, C, D, which are the principal objects of the memoir. The subjects to which they relate will be seen from the headings and introductory paragraphs.

ANNEX I. *Surface and Volume of Sphere* $x^2+\ldots+z^2+w^2=f^2$. Art. Nos. 51 and 52.

51. We require in $(s+1)$-dimensional space, $\int dx\ldots dz\,dw$, the volume of the sphere $x^2+\ldots+z^2+w^2=f^2$, and $\int dS$, the surface of the same sphere.

Writing $x=f\sqrt{\xi},\ldots,z=f\sqrt{\zeta}$, $w=f\sqrt{\omega}$, we have

$$dx\ldots dz\,dw=\frac{1}{2^{s+1}}f^{s+1}\,\xi^{-\frac{1}{2}}\ldots\zeta^{-\frac{1}{2}}\,\omega^{-\frac{1}{2}}\,d\xi\ldots d\zeta\,d\omega,$$

with the limiting condition $\xi+\ldots+\zeta+\omega=1$; but in order to take account as well of the negative as the positive values of $x,\ldots,z,w$, we must multiply by 2^{s+1}. The value is therefore

$$=f^{s+1}\int\xi^{-\frac{1}{2}}\ldots\zeta^{-\frac{1}{2}}\,\omega^{-\frac{1}{2}}\,d\xi\ldots d\zeta\,d\omega,$$

extended to all positive values of $\xi,\ldots,\zeta,\omega$, such that $\xi+\ldots+\zeta+\omega<1$; and we obtain this by a known theorem, viz.

$$\text{Volume of } (s+1)\text{-dimensional sphere}=f^{s+1}\frac{(\Gamma\tfrac{1}{2})^{s+1}}{\Gamma(\tfrac{1}{2}s+\tfrac{3}{2})}.$$

Writing $x=f\xi,\ldots,z=f\zeta$, $w=f\omega$, we obtain $dS=f^s\,d\Sigma$, where $d\Sigma$ is the element of surface of the unit-sphere $\xi^2+\ldots+\zeta^2+\omega^2=1$; we have element of volume $d\xi\ldots d\zeta\,d\omega$ $=r^s\,dr\,d\Sigma$, where r is to be taken from 0 to 1, and thence

$$\int d\xi\ldots d\zeta\,d\omega=\int_0^1 r^s\,dr\int d\Sigma=\frac{1}{s+1}\int d\Sigma,$$

that is,

$$\int d\Sigma=(s+1)\int d\xi\ldots d\zeta\,d\omega,\ =2(\tfrac{1}{2}s+\tfrac{1}{2})\frac{(\Gamma\tfrac{1}{2})^{s+1}}{\Gamma(\tfrac{1}{2}s+\tfrac{3}{2})}=\frac{2(\Gamma\tfrac{1}{2})^{s+1}}{\Gamma(\tfrac{1}{2}s+\tfrac{1}{2})};$$

consequently $\int dS=\text{surface of } (s+1)\text{-dimensional sphere}=f^s\dfrac{2(\Gamma\tfrac{1}{2})^{s+1}}{\Gamma(\tfrac{1}{2}s+\tfrac{1}{2})}$.

52. Writing $s-1$ for s, we have

$$\text{Volume of } (s-1)\text{-dimensional sphere} = f^s \frac{(\Gamma \tfrac{1}{2})^s}{\Gamma(\tfrac{1}{2}s+1)},$$

$$\text{Surface of} \qquad \text{do.} \qquad = f^{s-1} \frac{2(\Gamma \tfrac{1}{2})^s}{\Gamma(\tfrac{1}{2}s)},$$

which forms are sometimes convenient.

Writing in the first forms $s+1=3$, or in the second forms $s=3$, we find in ordinary space

$$\text{Volume of sphere} = f^3 \frac{(\Gamma \tfrac{1}{2})^3}{\Gamma(\tfrac{5}{2})} = f^3 \frac{\pi^{\frac{3}{2}}}{\tfrac{3}{2}\cdot\tfrac{1}{2}\cdot\sqrt{\pi}}, \quad = \frac{4\pi f^3}{3},$$

and

$$\text{Surface of sphere} = f^2 \frac{2(\Gamma \tfrac{1}{2})^3}{\Gamma \tfrac{3}{2}} = f^2 \frac{2\pi^{\frac{3}{2}}}{\tfrac{1}{2}\sqrt{\pi}}, \quad = 4\pi f^2,$$

as they should be.

ANNEX II. *The Integral* $\int_0^R \frac{r^{s-1}\,dr}{(r^2+e^2)^{\frac{1}{2}s+q}}$. Art. Nos. 53 to 63.

53. The integral in question (which occurs *antè*, No. 2) may also be considered as arising from a prepotential integral in tridimensional space; the prepotential of an element of mass dm is taken to be $= \frac{dm}{d^{s+2q}}$, where d is the distance of the element from the attracted point P. Hence if the element of mass be an element of the plane $z=0$, coordinates (x, y), ρ being the density, and if the attracted point be situate in the axis of z at a distance e from the origin, the prepotential is

$$V = \int \frac{\rho\, dx\, dy}{(x^2+y^2+e^2)^{\frac{1}{2}s+q}}.$$

For convenience, it is assumed throughout that e is positive.

Suppose that the attracting body is a circular disk, radius R, having the origin for its centre (viz. that bounded by the curve $x^2+y^2=R^2$); then writing $x=r\cos\theta$, $y=r\sin\theta$, we have

$$V = \int \frac{\rho r\, dr\, d\theta}{(r^2+e^2)^{\frac{1}{2}s+q}},$$

which, if ρ is a function of r only, is

$$= 2\pi \int \frac{\rho r\, dr}{(r^2+e^2)^{\frac{1}{2}s+q}}$$

and in particular, if $\rho = r^{s-2}$, then the value is

$$= 2\pi \int \frac{r^{s-1}\, dr}{(r^2+e^2)^{\frac{1}{2}s+q}},$$

the integral in regard to r being taken from $r=0$ to $r=R$. It is assumed that $s-1$ is not negative, viz. it is positive or (it may be) zero. I consider the integral

$$\int_0^R \frac{r^{s-1}\,dr}{(r^2+e^2)^{\frac{1}{2}s+q}},$$

which I call the r-integral, more particularly in the case where e is small in comparison with R. It is to be observed that e not being $=0$, and R being finite, the integral contains no infinite element, and is therefore finite, whether q is positive, negative, or zero.

54. Writing $r=e\sqrt{v}$, the integral is

$$=\tfrac{1}{2}e^{-2q}\int \frac{v^{\frac{1}{2}s-1}\,dv}{(1+v)^{\frac{1}{2}s+q}},$$

the limits being $\dfrac{R^2}{e^2}$ and 0.

In the case where q is positive, this is

$$=\tfrac{1}{2}e^{-2q}\left(\int_0^\infty - \int_{\frac{R^2}{e^2}}^\infty\right)\frac{v^{\frac{1}{2}s-1}\,dv}{(1+v)^{\frac{1}{2}s+q}};$$

viz. the first term of this is

$$=\tfrac{1}{2}e^{-2q}\,\frac{\Gamma\tfrac{1}{2}s\,.\,\Gamma q}{\Gamma(\tfrac{1}{2}s+q)},$$

and the second term is a term expansible in a series containing the powers $2q$, $2q+2$, &c. of the small quantity $\dfrac{e^2}{R^2}$, as appears by effecting therein the substitution $v=\dfrac{1}{x}$; viz. the value of the entire integral is by this means found to be

$$=\tfrac{1}{2}e^{-2q}\left\{\frac{\Gamma\tfrac{1}{2}s\,.\,\Gamma q}{\Gamma(\tfrac{1}{2}s+q)} - \int_0^{\frac{e^2}{R^2}} \frac{x^{q-1}\,dx}{(1+x)^{\frac{1}{2}s+q}}\right\}.$$

55. In the case where q is $=0$, or negative, the formula fails by reason that the element $\dfrac{v^{\frac{1}{2}s-1}\,dv}{(1+v)^{\frac{1}{2}s+q}}$ of the integrals $\displaystyle\int_0^\infty$, $\displaystyle\int_{\frac{R^2}{e^2}}^\infty$ becomes infinite for indefinitely large values of v. Recurring to the original form $\displaystyle\int_0^R \frac{r^{s-1}\,dr}{(r^2+e^2)^{\frac{1}{2}s+q}}$, it is to be observed that the integral has a finite value when $e=0$; and it might therefore at first sight be imagined that the factor $(r^2+e^2)^{-\frac{1}{2}s-q}$ might be expanded in ascending powers of e^2, and the value of the integral consequently obtained as a series of positive powers of e^2. But the series thus obtained is of the form $e^{2k}\displaystyle\int_0^R r^{-2q-2k-1}\,dr$, where $2q$ being positive, the exponent $-2q-2k-1$ is for a sufficiently small value of k at first positive, or if negative less than -1, and the value of the integral is finite; but as k increases the exponent becomes negative, and equal or greater than -1, and the value of the

integral is then infinite. The inference is that the series *commences* in the form $A + Be^2 + Ce^4 \ldots$: but that we come at last when q is fractional to a term of the form Ke^{-2q}, and when q is $=0$ or is integral, to a term of the form $Ke^{-2q}\log e$; the process giving the coefficients $A, B, C, \ldots$, so long as the exponent of the corresponding term $e^0, e^2, e^4, \ldots$ is less than $-2q$ (in particular $q=0$, there is a term $k \log e$, *and the expansion-process does not give any term of the result*), and the failure of the series after this point being indicated by the values of the subsequent coefficients coming out $=\infty$.

56. In illustration, we may consider any of the cases in which the integral can be obtained in finite terms. For instance,

$s=2,\ q=-\tfrac{3}{2}$,

$$\text{Integral is } \int r(r^2+e^2)^{\frac{1}{2}}\,dr,\ = \tfrac{1}{3}(r^2+e^2)^{\frac{3}{2}}, \text{ from } 0 \text{ to } R,$$

$$= \tfrac{1}{3}(R^2+e^2)^{\frac{3}{2}} - \tfrac{1}{3}e^3;$$

viz. expanding in ascending powers of e, this is

$$= \tfrac{1}{3}R^3 + \tfrac{1}{2}Re^2 - \ldots - \tfrac{1}{3}e^3,$$

or we have here a term in e^3. And so,

$s=1,\ q=-2$,

$$\text{Integral is } \int (r^2+e^2)^{\frac{3}{2}}\,dr,\ = (\tfrac{1}{4}r^2 + \tfrac{5}{8}e^2)\,r\sqrt{r^2+e^2} + \tfrac{3}{8}e^4\log(r+\sqrt{r^2+e^2}), \text{ from } 0 \text{ to } R,$$

$$= (\tfrac{1}{4}R^2 + \tfrac{5}{8}e^2)\,R\sqrt{R^2+e^2} + \tfrac{3}{8}e^4 \log\frac{R+\sqrt{R^2+e^2}}{e};$$

viz. expanding in ascending powers of e, this is

$$= \tfrac{1}{4}R^4 + \tfrac{3}{4}R^2e^2 + \ldots + \tfrac{3}{8}e^4\log\frac{R}{e}\,*,$$

or we have here a term in $e^4 \log e$.

57. Returning to the form

$$\tfrac{1}{2}e^{-2q}\int_0^{\frac{R^2}{e^2}} \frac{v^{\frac{1}{2}s-1}\,dv}{(1+v)^{\frac{1}{2}s+q}},$$

and writing herein $v=\dfrac{1-x}{x}$, or, what is the same thing, $x=\dfrac{1}{1+v}$, and for shortness $X=\dfrac{e^2}{e^2+R^2}$, $=\dfrac{1}{1+\dfrac{R^2}{e^2}}$, the value is

$$= \tfrac{1}{2}e^{-2q}\int_X^1 x^{q-1}(1-x)^{\frac{1}{2}s-1}\,dx,$$

where observe that $q-1$ is 0 or negative, but X being a positive quantity less than 1, the function $x^{q-1}(1-x)^{\frac{1}{2}s-1}$ is finite for the whole extent of the integration.

* Term is $\tfrac{3}{8}e^4\log\dfrac{2R}{e}$, $=\tfrac{3}{8}e^4\left(\log\dfrac{R}{e}+\log 2\right)$, which, $\dfrac{R}{e}$ being large, is reduced to $\tfrac{3}{8}e^4\log\dfrac{R}{e}$.

58. If $q=0$, this is

$$=\tfrac{1}{2}\int_X^1 \frac{1-\{1-(1-x)^{\frac{1}{2}s-1}\}\,dx}{x}$$

$$=\tfrac{1}{2}\log X-\tfrac{1}{2}\int_X^1 \frac{\{1-(1-x)^{\frac{1}{2}s-1}\}\,dx}{x}$$

$$=\tfrac{1}{2}\log\sqrt{1+\frac{R^2}{e^2}}-\tfrac{1}{2}\int_0^1 \frac{\{1-(1-x)^{\frac{1}{2}s-1}\}\,dx}{x}+\tfrac{1}{2}\int_0^X \frac{\{1-(1-x)^{\frac{1}{2}s-1}\}\,dx}{x}:$$

where observe that, in virtue of the change made from $\frac{1}{x}(1-x)^{\frac{1}{2}s-1}$ to $\frac{1}{x}\{1-(1-x)^{\frac{1}{2}s-1}\}$ (a function which becomes infinite, to one which does not become infinite, for $x=0$), it has become allowable in place of $\int_X^1$ to write $\int_0^1-\int_0^X$.

When e is small, the integral which is the third term of the foregoing expression is obviously a quantity of the order e^2; the first term is $\tfrac{1}{2}\left(\log\frac{R}{e}+\log\sqrt{1+\frac{e^2}{R^2}}\right)$, which, neglecting terms in e^2, is $=\tfrac{1}{2}\log\frac{R}{e}$, and hence the approximate value of the r-integral $\int_0^R \frac{r^{s-1}\,dr}{(r^2+e^2)^{\frac{1}{2}s}}$ is

$$=\log\frac{R}{e}-\tfrac{1}{2}\int_0^1 dx\,\frac{1-(1-x)^{\frac{1}{2}s-1}}{x},$$

or, what is the same thing, it is

$$=\log\frac{R}{e}-\tfrac{1}{2}\int_0^1 dy\,\frac{1-y^{\frac{1}{2}s-1}}{1-y},$$

where the integral in this expression is a mere numerical constant, which, when $\tfrac{1}{2}s-1$ is a positive integer, has the value

$$\tfrac{1}{1}+\tfrac{1}{2}+\ldots+\frac{1}{\tfrac{1}{2}s-1};$$

neglecting this in comparison with the logarithmic term, the approximate value is

$$=\log\frac{R}{e}.$$

59. I consider also the case $q=-\tfrac{1}{2}$; the integral is here

$$\tfrac{1}{2}e\int_X^1 x^{-\frac{3}{2}}(1-x)^{\frac{1}{2}s-1}\,dx$$

$$=\tfrac{1}{2}e\int_X^1 x^{-\frac{3}{2}}(1-\{1-(1-x)^{\frac{1}{2}s-1}\})\,dx$$

$$=e\,(X^{-\frac{1}{2}}-1)+\tfrac{1}{2}e\int_X^1 x^{-\frac{3}{2}}\{1-(1-x)^{\frac{1}{2}s-1}\}\,dx;$$

and the first term of this being $=\sqrt{e^2+R^2}-e$, this is consequently

$$=\sqrt{R^2+e^2}+\tfrac{1}{2}e\int_0^X x^{-\frac{3}{2}}\{1-(1-x)^{\frac{1}{2}s-1}\}\,dx-e\left(1+\tfrac{1}{2}\int_0^1 x^{-\frac{3}{2}}\{1-(1-x)^{\frac{1}{2}s-1}\}\,dx\right).$$

As regards the second term of this, we have

$$-2x^{-\frac{1}{2}}\{1-(1-x)^{\frac{1}{2}s-1}\}+2(\tfrac{1}{2}s-1)\int x^{-\frac{1}{2}}(1-x)^{\frac{1}{2}s-2}\,dx=\int x^{-\frac{3}{2}}\{1-(1-x)^{\frac{1}{2}s-1}\}\,dx;$$

or, taking each term between the limits 1, 0,

$$-2+2(\tfrac{1}{2}s-1)\frac{\Gamma\tfrac{1}{2}\Gamma(\tfrac{1}{2}s-1)}{\Gamma(\tfrac{1}{2}s-\tfrac{1}{2})}=\int_0^1 x^{-\frac{3}{2}}\{1-(1-x)^{\frac{1}{2}s-1}\}\,dx;$$

viz. this integral has the value

$$-2+\frac{2\Gamma\tfrac{1}{2}s\,\Gamma\tfrac{1}{2}}{\Gamma(\tfrac{1}{2}s-\tfrac{1}{2})};$$

and the value of the r-integral $\int_0^R \frac{r^{s-1}\,dr}{(r^2+e^2)^{\frac{1}{2}s-\frac{1}{2}}}$ is consequently

$$=\sqrt{R^2+e^2}+\tfrac{1}{2}e\int_0^X x^{-\frac{3}{2}}\{1-(1-x)^{\frac{1}{2}s-1}\}\,dx-e\frac{\Gamma\tfrac{1}{2}s\,\Gamma\tfrac{1}{2}}{\Gamma(\tfrac{1}{2}s-\tfrac{1}{2})},$$

which is of the form

$$R\left\{1+\text{terms in }\frac{e^2}{R^2},\ \frac{e^4}{R^4},\ldots\right\}-e\frac{\Gamma\tfrac{1}{2}s\,\Gamma\tfrac{1}{2}}{\Gamma(\tfrac{1}{2}s-\tfrac{1}{2})};$$

say the approximate value is

$$R-e\frac{\Gamma\tfrac{1}{2}s\,\Gamma\tfrac{1}{2}}{\Gamma(\tfrac{1}{2}s-\tfrac{1}{2})},$$

where the first term R is the term $\int_0^R dr$, given by the expansion in ascending powers of e^2; the second term is the term in e^{-2q}. And observe that the term is the value of

$$\tfrac{1}{2}e\int_0^1 x^{-\frac{3}{2}}(1-x)^{\frac{1}{2}s-1}\,dx,$$

calculated by means of the ordinary formula for a Eulerian integral (which formula, on account of the negative exponent $-\frac{3}{2}$, is not really applicable, the value of the integral being $=\infty$) on the assumption that the Γ of a negative q is interpreted in accordance with the equation $\Gamma(q+1)=q\Gamma q$; viz. the value thus calculated is

$$=\tfrac{1}{2}e\frac{\Gamma(-\tfrac{1}{2})\Gamma(\tfrac{1}{2}s)}{\Gamma(\tfrac{1}{2}s-\tfrac{1}{2})},\ =-e\frac{\Gamma\tfrac{1}{2}\,\Gamma\tfrac{1}{2}s}{\Gamma(\tfrac{1}{2}s-\tfrac{1}{2})}$$

on the assumption $\Gamma\tfrac{1}{2}=-\tfrac{1}{2}\Gamma(-\tfrac{1}{2})$; and this agrees with the foregoing value.

60. It is now easy to see in general how the foregoing transformed value $\tfrac{1}{2}e^{-2q}\int_X^1 x^{q-1}(1-x)^{\frac{1}{2}s-1}\,dx$, where q is negative and fractional, gives at once the value of

the term in e^{-2q}. Observe that in the integral x is always between 1 and X $\left(=\frac{e^2}{e^2+R^2},\right.$ a positive quantity less than $\left.1\right)$; the function to be integrated never becomes infinite. Imagine for a moment an integral $\int_X^1 x^\alpha\,dx$, where α is positive or negative. We may conventionally write this $=\int_0^1 x^\alpha\,dx - \int_0^X x^\alpha\,dx$, understanding the first symbol to mean $\frac{1^{1+\alpha}}{1+\alpha}$, and the second to mean $\frac{X^{1+\alpha}}{1+\alpha}$; they of course properly mean $\frac{1^{1+\alpha}-0^{1+\alpha}}{1+\alpha}$ and $\frac{X^{1+\alpha}-0^{1+\alpha}}{1+\alpha}$; but the terms in $0^{1+\alpha}$, whether zero or infinite, destroy each other, the original form $\int_X^1 x^\alpha\,dx$, in fact, showing that no such terms can appear in the result.

In accordance with the convention, we write

$$\int_X^1 x^{q-1}(1-x)^{\frac{1}{2}s-1}\,dx = \int_0^1 x^{q-1}(1-x)^{\frac{1}{2}s-1}\,dx - \int_0^X x^{q-1}(1-x)^{\frac{1}{2}s-1}\,dx\,;$$

and it follows that the term in e^{-2q} is

$$\tfrac{1}{2}e^{-2q}\int_0^1 x^{q-1}(1-x)^{\frac{1}{2}s-1}\,dx,$$

this last expression (wherein q, it will be remembered, is a negative fraction) being understood according to the convention; and so understanding it, the value of the term is

$$=\tfrac{1}{2}e^{-2q}\frac{\Gamma\frac{1}{2}s\,\Gamma q}{\Gamma(\frac{1}{2}s+q)},$$

where the Γ of the negative q is to be interpreted in accordance with the equation $\Gamma(q+1)=q\Gamma q$; viz. we have $\Gamma q=\frac{1}{q}\Gamma(q+1),=\frac{1}{q(q+1)}\Gamma(q+2)$, &c., so as to make the argument of the Γ positive. Observe that under this convention we have

$$\Gamma q\Gamma(1-q)=\frac{\Gamma^2(\frac{1}{2})}{\sin q\pi}:\text{ or the term is }\tfrac{1}{2}e^{-2q}.\frac{\Gamma^2(\frac{1}{2})}{\sin q\pi}\frac{\Gamma\frac{1}{2}s}{\Gamma(\frac{1}{2}s+q)\,\Gamma(1-q)}.$$

61. An example in which $\frac{1}{2}s-1$ is integral will make the process clearer, and will serve instead of a general proof. Suppose $q=-\frac{1}{7}$, $\frac{1}{2}s-1=4$, the expression

$$\int_0^1 x^{-\frac{8}{7}}(1-x)^4\,dx = \int_0^1 (x^{-\frac{8}{7}}-4x^{-\frac{1}{7}}+6x^{\frac{6}{7}}-4x^{\frac{13}{7}}+x^{\frac{20}{7}})\,dx$$

is used, in accordance with the convention, to denote the value

$$\begin{aligned}&-7-\tfrac{14}{3}+\tfrac{42}{13}-\tfrac{7}{5}+\tfrac{7}{27}\\&=7(-1-\tfrac{2}{3}+\tfrac{6}{13}-\tfrac{1}{5}+\tfrac{1}{27}),=7(-\tfrac{44}{27}-\tfrac{1}{5}+\tfrac{6}{13}),=\frac{-7.2401}{5.13.27},=\frac{-7^5}{5.13.27}.\end{aligned}$$

But we have

$$\frac{\Gamma\frac{1}{2}s\,\Gamma q}{\Gamma(\frac{1}{2}s+q)}=\frac{\Gamma 5\,\Gamma(-\frac{1}{7})}{\Gamma(5-\frac{1}{7})}=\frac{24\,\Gamma(-\frac{1}{7})}{\frac{27}{7}\cdot\frac{20}{7}\cdot\frac{13}{7}\cdot\frac{6}{7}\cdot\frac{-1}{7}\,\Gamma(-\frac{1}{7})}=\frac{-7^5}{5\,.\,13\,.\,27},$$

agreeing with the former value.

62. The case of a negative integer is more simple. To find the logarithmic term of

$$\tfrac{1}{2}e^{-2q}\int_X^1 x^{q-1}(1-x)^{\frac{1}{2}s-1}\,dx,$$

we have only to expand the factor $(1-x)^{\frac{1}{2}s-1}$ so as to obtain the term involving x^{-q}. We have thus the term

$$\tfrac{1}{2}\,e^{-2q}\int_X^1 x^{q-1}(-)^q\frac{\Gamma\frac{1}{2}s}{\Gamma(1-q)\,\Gamma(\frac{1}{2}s+q)}\,x^{-q}\,dx$$

$$=\tfrac{1}{2}(-)^q\,e^{-2q}\frac{\Gamma\frac{1}{2}s}{\Gamma(1-q)\,\Gamma(\frac{1}{2}s+q)}\log\frac{1}{X},$$

where $\log\frac{1}{X}=\log\left(1+\frac{R^2}{e^2}\right)$, $=2\log\frac{R}{e}+2\log\sqrt{1+\frac{e^2}{R^2}}$; so that, neglecting the terms in $\frac{e^2}{R^2}$, &c., this is $=2\log\frac{R}{e}$, and the term in question is

$$=(-)^q\,e^{-2q}\frac{\Gamma\frac{1}{2}s}{\Gamma(1-q)\,\Gamma(\frac{1}{2}s+q)}\log\frac{R}{e}.$$

The general conclusion is that q being negative, the r-integral

$$\int_0^R\frac{r^{s-1}\,dr}{(r^2+e^2)^{\frac{1}{2}s+q}}$$

has for its value a series proceeding in powers of e^2, which series up to a certain point is equal to the series obtained by expanding in ascending powers of e^2 and integrating each term separately; viz. the series to the point in question is

$$\frac{R^{-2q}}{-2q}-\frac{\frac{1}{2}s+q}{1}\,\frac{R^{-2q-2}}{-2q-2}\,e^2+\frac{\frac{1}{2}s+q\,.\,\frac{1}{2}s+q+1}{1\,.\,2}\,\frac{R^{-2q-4}}{-2q-4}\,e^4+\ldots,$$

continued so long as the exponent of e is less than $-2q$; together with a term Ke^{-2q} when q is fractional, and $Ke^{-2q}\log\frac{R}{e}$ when q is integral; viz. q fractional, this term is

$$=\tfrac{1}{2}e^{-2q}\frac{\Gamma\frac{1}{2}s\,\Gamma q}{\Gamma(\frac{1}{2}s+q)},\quad=\tfrac{1}{2}e^{-2q}\frac{\Gamma^2\frac{1}{2}}{\sin q\pi}\,\frac{\Gamma\frac{1}{2}s}{\Gamma(\frac{1}{2}s+q)\,\Gamma(1-q)},$$

and q integral, it is

$$=(-)^q\,e^{-2q}\frac{\Gamma\frac{1}{2}s}{\Gamma(1-q)\,\Gamma(\frac{1}{2}s+q)}\log\frac{R}{e}.$$

63. It has been tacitly assumed that $\frac{1}{2}s+q$ is positive; but the formulæ hold good if $\frac{1}{2}s+q$ is $=0$ or negative. Suppose $\frac{1}{2}s+q$ is 0 or a negative integer, then $\Gamma(\frac{1}{2}s+q)=\infty$, and the special term involving e^{-2q} or $e^{-2q}\log e$ vanishes; in fact, in this case the r-integral is

$$=\int_0^R r^{\frac{1}{2}s-1}(r^2+e^2)^{-(\frac{1}{2}s+q)}\,dr,$$

where $(r^2+e^2)^{-(\frac{1}{2}s+q)}$ has for its value a finite series, and the integral is therefore equal to a finite series $A+Be^2+Ce^4+\&c.$ If $\frac{1}{2}s+q$ be fractional, then the Γ of the negative quantity $\frac{1}{2}s+q$ must be understood as above, or, what is the same thing, we may, instead of $\Gamma(\frac{1}{2}s+q)$, write

$$\frac{(\Gamma\frac{1}{2})^2}{\sin(\frac{1}{2}s+q)\pi\,\Gamma(1-q-\frac{1}{2}s)};$$

thus, q being integral, the exceptional term is

$$=(-)^q\,e^{-2q}\,\frac{\Gamma\frac{1}{2}s\sin(\frac{1}{2}s+q)\pi\,.\,\Gamma(1-q-\frac{1}{2}s)}{(\Gamma\frac{1}{2})^2\,\Gamma(1-q)}\log\frac{R}{e}.$$

For instance, $s=1$, $q=-2$, the term is

$$\tfrac{1}{2}e^4\,\frac{\Gamma\frac{1}{2}\sin(-\frac{3}{2}\pi)\,\Gamma\frac{5}{2}}{(\Gamma\frac{1}{2})^2\,.\,\Gamma 3}\log\frac{R}{e};$$

or, since $\Gamma\frac{5}{2}=\frac{3}{2}\,.\,\frac{1}{2}\Gamma\frac{1}{2}$, and $\Gamma 3=2$, the term is $+\frac{3}{8}e^4\log\frac{R}{e}$, agreeing with a preceding result.

ANNEX III. *Prepotentials of Uniform Spherical Shell and Solid Sphere.* Art. Nos. 64 to 92.

64. The prepotentials in question depend ultimately upon two integrals, which also arise, as will presently appear, from prepotential problems in two-dimensional space, and which are for convenience termed the ring-integral and the disk-integral respectively. The analytical investigation in regard to these, depending as it does on a transformation of a function allied with the hypergeometric series, is I think interesting.

65. Consider first the prepotential of a uniform $(s+1)$-dimensional spherical shell. This is

$$V=\int\frac{dS}{\{(a-x)^2+\ldots+(c-z)^2+(e-w)^2\}^{\frac{1}{2}s+q}},$$

the equation of the surface being $x^2+\ldots+z^2+w^2=f^2$; and there are the two cases of an internal point, $a^2+\ldots+c^2+e^2<f^2$, and an external point, $a^2+\ldots+c^2+e^2>f^2$.

The value is a function of $a^2+\ldots+c^2+e^2$, say this is $=\kappa^2$. Taking the axes so that the coordinates of the attracted point are $(0,\ldots,0,\kappa)$, the integral is

$$=\int\frac{dS}{\{x^2+\ldots+z^2+(\kappa-w)^2\}^{\frac{1}{2}s+q}},$$

where the equation of the surface is still $x^2 + \ldots + z^2 + w^2 = f^2$. Writing $x = f\xi, \ldots, z = f\zeta$, $w = f\omega$, where $\xi^2 + \ldots + \zeta^2 + \omega^2 = 1$, we have $dS = \dfrac{f^s\, d\xi \ldots d\zeta}{\omega}$, or the integral is

$$= f^s \int \frac{d\xi \ldots d\zeta}{\omega\,(f^2 - 2\kappa f\omega + \kappa^2)^{\frac{1}{2}s+q}}.$$

Assume $\xi = px, \ldots, \zeta = pz$, where $x^2 + \ldots + z^2 = 1$; then $p^2 + \omega^2 = 1$. Moreover, $d\xi \ldots d\zeta$, $= p^{s-1}\, dp\, d\Sigma$, where $d\Sigma$ is the element of surface of the s-dimensional unit-sphere $x^2 + \ldots + z^2 = 1$; or for p, substituting its value $\sqrt{1-\omega^2}$, we have $dp = \dfrac{-\omega d\omega}{\sqrt{1-\omega^2}}$; and thence $d\xi \ldots d\zeta = -(1-\omega^2)^{\frac{1}{2}s-1}\, \omega\, d\omega\, d\Sigma$. The integral as regards p is from $p = -1$ to $+1$, or as regards ω from 1 to -1; whence reversing the sign, the integral will be from $\omega = -1$ to $+1$; and the required integral is thus

$$= f^s \int_{-1}^{1} \frac{(1-\omega^2)^{\frac{1}{2}s-1}\, d\omega\, d\Sigma}{(f^2 - 2\kappa f\omega + \kappa^2)^{\frac{1}{2}s+q}}, \quad = f^s \int d\Sigma \int_{-1}^{1} \frac{(1-\omega^2)^{\frac{1}{2}s-1}\, d\omega}{(f^2 - 2\kappa f\omega + \kappa^2)^{\frac{1}{2}s+q}},$$

where $\int d\Sigma$ is the surface of the s-dimensional unit-sphere (see Annex I.), $= \dfrac{2\,(\Gamma\frac{1}{2})^s}{\Gamma\frac{1}{2}s}$; and for greater convenience transforming the second factor by writing therein $\omega = \cos\theta$, the required integral is $= \dfrac{(\Gamma\frac{1}{2})^s}{\Gamma(\frac{1}{2}s)}$ multiplied by

$$2f^s \int_0^{\pi} \frac{\sin^{s-1}\theta\, d\theta}{(f^2 - 2\kappa f\cos\theta + \kappa^2)^{\frac{1}{2}s+q}},$$

which last expression—including the factor $2f^s$, but without the factor $\dfrac{(\Gamma\frac{1}{2})^s}{\Gamma\frac{1}{2}s}$—is the ring-integral discussed in the present Annex. It may be remarked that the value can be at once obtained in the particular case $s = 2$, which belongs to tridimensional space: viz. we then have

$$V = 2\pi f^2 \int_0^{\pi} \frac{\sin\theta\, d\theta}{(f^2 - 2\kappa f\cos\theta + \kappa^2)^{q+1}}$$

$$= \frac{2\pi f^2}{2\kappa f q}(f^2 - 2\kappa f\cos\theta + \kappa^2)^{-q}$$

$$= \frac{\pi f}{\kappa q}\{(f-\kappa)^{-2q} - (f+\kappa)^{-2q}\},$$

which agrees with a result given, *Mécanique Céleste*, Book XII. Chap. II.

66. Consider next the prepotential of the uniform solid $(s+1)$-dimensional sphere,

$$V = \int \frac{dx \ldots dz dw}{\{(a-x)^2 + \ldots + (c-z)^2 + (e-w)^2\}^{\frac{1}{2}s+q}},$$

the equation of the surface being $x^2 + \ldots + z^2 + w^2 = f^2$; there are the two cases of an internal point $\kappa < f$, and an external point $\kappa > f$ $(a^2 + \ldots + c^2 + e^2 = \kappa^2$ as before).

Transforming so that the coordinates of the attracted point are $0, \ldots, 0, \kappa$, the integral is

$$=\int \frac{dx \ldots dz\, dw}{\{x^2 + \ldots + z^2 + (\kappa - w)^2\}^{\frac{1}{2}s+q}},$$

where the equation is still $x^2 + \ldots + z^2 + w^2 = f^2$. Writing here $x = r\xi, \ldots, z = r\zeta$, where $\xi^2 + \ldots + \zeta^2 = 1$, we have $dx \ldots dz = r^{s-1}\, dr\, d\Sigma$, where $d\Sigma$ is an element of surface of the s-dimensional unit-sphere $\xi^2 + \ldots + \zeta^2 = 1$; the integral is therefore

$$=\int \frac{r^{s-1}\, dr\, d\Sigma\, dw}{\{r^2 + (\kappa - w)^2\}^{\frac{1}{2}s+q}}$$

$$=\int d\Sigma \int \frac{r^{s-1}\, dr\, dw}{\{r^2 + (\kappa - w)^2\}^{\frac{1}{2}s+q}},$$

where, as regards r and w, the integration extends over the circle $r^2 + w^2 = f^2$. The value of the first factor (see Annex I.) is $= \frac{2 (\Gamma\frac{1}{2})^s}{\Gamma\frac{1}{2}s}$; writing y and x in place of r and w respectively, the integral is $= \frac{2 (\Gamma\frac{1}{2})^s}{\Gamma\frac{1}{2}s}$ multiplied by

$$\int \frac{y^{s-1}\, dx\, dy}{\{(x-\kappa)^2 + y^2\}^{\frac{1}{2}s+q}}$$

over the circle $x^2 + y^2 = f^2$; viz. this last expression $\left(\text{without the factor } \frac{2 (\Gamma\frac{1}{2})^s}{\Gamma(\frac{1}{2}s)}\right)$ is the disk-integral discussed in the present Annex.

67. We find, for the value in regard to an internal point $\kappa < f$,

$$V = \frac{(\Gamma\frac{1}{2})^{s+1}}{\Gamma(\frac{1}{2}s + q)\, \Gamma(\frac{1}{2} - q)} f^{s+1} \int_0^\infty (t + f^2 - \kappa^2)^{\frac{1}{2}-q}\, t^{-\frac{1}{2}-q}\, (t + f^2)^{-\frac{1}{2}s+q-1}\, dt,$$

which, in the particular case $q = -\frac{1}{2}$, is

$$= \frac{(\Gamma\frac{1}{2})^{s+1}}{\Gamma(\frac{1}{2}s - \frac{1}{2})} f^{s+1} \int_0^\infty (t + f^2 - \kappa^2)\, (t + f^2)^{-\frac{1}{2}s-\frac{3}{2}}\, dt;$$

viz. the integral in t is here

$$= \int_0^\infty \{(t + f^2)^{-\frac{1}{2}s-\frac{1}{2}} - \kappa^2 (t + f^2)^{-\frac{1}{2}s-\frac{3}{2}}\}\, dt, \quad = \frac{1}{f^{s+1}} \left(\frac{f^2}{\frac{1}{2}s - \frac{1}{2}} - \frac{\kappa^2}{\frac{1}{2}s + \frac{1}{2}} \right),$$

or we have

$$V = \frac{(\Gamma\frac{1}{2})^{s+1}}{\Gamma(\frac{1}{2}s - \frac{1}{2})} \left(\frac{f^2}{\frac{1}{2}s - \frac{1}{2}} - \frac{\kappa^2}{\frac{1}{2}s + \frac{1}{2}} \right).$$

It may be added that, in regard to an external point $\kappa > f$, the value is

$$V = \frac{(\Gamma\frac{1}{2})^{s+1}}{\Gamma(\frac{1}{2}s + q)\, \Gamma(\frac{1}{2} - q)} f^{s+1} \text{ and } \int_{\kappa^2 - f^2}^\infty (t + f^2 - \kappa^2)^{\frac{1}{2}-q}\, t^{-\frac{1}{2}-q}\, (t + f^2)^{-\frac{1}{2}s+q-1}\, dt,$$

which, in the same case $q=-\frac{1}{2}$, is

$$=\frac{(\Gamma\frac{1}{2})^{s+1}}{\Gamma(\frac{1}{2}s-\frac{1}{2})}f^{s+1} \text{ and } \int_{\kappa^2-f^2}^{\infty}(t+f^2-\kappa^2)(t+f^2)^{-\frac{1}{2}s-\frac{3}{2}}\,dt,$$

where the t-integral is

$$=\int_{\kappa^2-f^2}^{\infty}\{(t+f^2)^{-\frac{1}{2}s-\frac{1}{2}}-\kappa^2(t+f^2)^{-\frac{1}{2}s-\frac{3}{2}}\}\,dt,\quad =\frac{\kappa^{-s+1}}{\frac{1}{2}s-\frac{1}{2}}-\frac{\kappa^2.\kappa^{-s-1}}{\frac{1}{2}s+\frac{1}{2}},\quad =\frac{\kappa^{-s+1}}{\frac{1}{2}s-\frac{1}{2}.\frac{1}{2}s+\frac{1}{2}};$$

and the value of V is therefore

$$=\frac{(\Gamma\frac{1}{2})^{s+1}}{\Gamma(\frac{1}{2}s+\frac{1}{2})}\frac{f^{s+1}}{\kappa^{s-1}}.$$

Recurring to the case of the internal point; then, writing $\nabla=\frac{d^2}{da^2}+\ldots+\frac{d^2}{dc^2}+\frac{d^2}{de^2}$, and observing that $\nabla(\kappa^2)=4(\frac{1}{2}s+\frac{1}{2})$, we have

$$\nabla V=-\frac{4(\Gamma\frac{1}{2})^{s+1}}{\Gamma(\frac{1}{2}s-\frac{1}{2})}:$$

(in particular, for ordinary space $s+1=3$, or the value is $\frac{-4\pi^{\frac{3}{2}}}{\sqrt{\pi}}$, $=-4\pi$, which is right).

68. The integrals referred to as the ring-integral and the disk-integral arise also from the following integrals in two-dimensional space, viz. these are

$$\int\frac{y^{s-1}\,dS}{\{(x-\kappa)^2+y^2\}^{\frac{1}{2}s+q}},\quad \int\frac{y^{s-1}\,dx\,dy}{\{(x-\kappa)^2+y^2\}^{\frac{1}{2}s+q}},$$

in the first of which dS denotes an element of arc of the circle $x^2+y^2=f^2$, the integration being extended over the whole circumference, and in the second the integration extends over the circle $x^2+y^2=f^2$; y^{s-1} is written for shortness instead of $(y^2)^{\frac{1}{2}(s-1)}$, viz. this is considered as always positive, whether y is positive or negative; it is moreover assumed that $s-1$ is zero or positive.

Writing in the first integral $x=f\cos\theta$, $y=f\sin\theta$, the value is

$$=f^s\int\frac{(\sin\theta)^{s-1}\,d\theta}{(f^2-2\kappa f\cos\theta+\kappa^2)^{\frac{1}{2}s+q}};$$

viz. this represents the prepotential of the circumference of the circle, density varying as $(\sin\theta)^{s-1}$, in regard to a point $x=\kappa$, $y=0$ in the plane of the circle; and similarly the second integral represents the prepotential of the circular disk, density of the element at the point $(x, y)=y^{s-1}$, in regard to the same point $x=\kappa$, $y=0$; it being in each case assumed that the prepotential of an element of mass $\rho\,d\varpi$ at a point at distance d is $=\frac{\rho\,d\varpi}{d^{s+2q}}$.

69. In the case of the circumference, it is assumed that the attracted point is not on the circumference, κ not $=f$; and the function under the integral sign, and therefore the integral itself, is in every case finite. In the case of the circle, if κ be an interior point, then if $2q-1$ be $=0$ or positive, the element at the attracted point becomes infinite; but to avoid this we consider, not the potential of the whole circle, but the potential of the circle *less* an indefinitely small circle radius ϵ having the attracted point for its centre; which being so, the element under the integral sign, and consequently the integral itself, remains finite.

It is to be remarked that the two integrals are connected with each other; viz. the circle of the second integral being divided into rings by means of a system of circles concentric with the bounding circle $x^2+y^2=f^2$, then the prepotential of each ring or annulus is determined by an integral such as the first integral; or, analytically, writing in the second integral $x=r\cos\theta$, $y=r\sin\theta$, and therefore $dx\,dy=r\,dr\,d\theta$, the second integral is

$$=\int r^s\,dr\int\frac{(\sin\theta)^{s-1}\,d\theta}{(r^2+\kappa^2-2\kappa r\cos\theta)^{\frac{1}{2}s+q}},$$

viz. the integral in regard to θ is here the same function of r, κ that the first integral is of f, κ; and the integration in regard to r is of course to be taken from $r=0$ to $r=f$. But the θ-integral is not, in its original form, such a function of r as to render possible the integration in regard to r; and I, in fact, obtain the second integral by a different and in some respects a better process.

70. Consider first the ring-integral which, writing therein as above $x=f\cos\theta$, $y=f\sin\theta$, and multiplying by 2 in order that the integral, instead of being taken from 0 to 2π, may be taken from 0 to π, becomes

$$=2f^s\int\frac{(\sin\theta)^{s-1}\,d\theta}{(f^2-2\kappa f\cos\theta+\kappa^2)^{\frac{1}{2}s+q}}.$$

Write $\cos\frac{1}{2}\theta=\sqrt{x}$; then $\sin\frac{1}{2}\theta=\sqrt{1-x}$, $\sin\theta=2x^{\frac{1}{2}}(1-x)^{\frac{1}{2}}$; $d\theta=-x^{-\frac{1}{2}}(1-x)^{-\frac{1}{2}}\,dx$; $\cos\theta=-1+2x$; $\theta=0$ gives $x=1$, $\theta=\pi$ gives $x=0$, and the integral is

$$=2^{s-1}f^s\int_0^1\frac{x^{\frac{1}{2}s-1}(1-x)^{\frac{1}{2}s-1}\,dx}{\{(f+\kappa)^2-4\kappa fx\}^{\frac{1}{2}s+q}},$$

$$=\frac{2^{s-1}f^s}{(f+\kappa)^{s+2q}}\int_0^1\frac{x^{\frac{1}{2}s-1}(1-x)^{\frac{1}{2}s-1}\,dx}{(1-ux)^{\frac{1}{2}s+q}},$$

if for shortness $u=\dfrac{4\kappa f}{(\kappa+f)^2}$, so that obviously $u<1$.

The integral in x is here an integral belonging to the general form

$$\Pi(\alpha,\beta,\gamma,u)=\int_0^1 x^{\alpha-1}(1-x)^{\beta-1}(1-ux)^{-\gamma}\,dx,$$

viz. we have

$$\text{Ring-integral}=\frac{2^{s-1}f^s}{(f+\kappa)^{s+2q}}\,\Pi(\tfrac{1}{2}s,\tfrac{1}{2}s,\tfrac{1}{2}s+q,u).$$

71. The general function $\Pi(\alpha, \beta, \gamma, u)$ is

$$\Pi(\alpha, \beta, \gamma, u) = \frac{\Gamma\alpha\,\Gamma\beta}{\Gamma(\alpha+\beta)} F(\alpha, \gamma, \alpha+\beta, u),$$

or, what is the same thing,

$$F(\alpha, \beta, \gamma, u) = \frac{\Gamma\gamma}{\Gamma\alpha\,\Gamma(\gamma-\alpha)} \Pi(\alpha, \gamma-\alpha, \beta, u),$$

and consequently transformable by means of various theorems for the transformation of the hypergeometric series, in particular, by the theorems

$$F(\alpha, \beta, \gamma, u) = F(\beta, \alpha, \gamma, u),$$
$$F(\alpha, \beta, \gamma, u) = (1-u)^{\gamma-\alpha-\beta} F(\gamma-\alpha, \gamma-\beta, \gamma, u);$$

and if $v = \left(\frac{1-\sqrt{1-u}}{1+\sqrt{1-u}}\right)^2$, or, what is the same thing, $u = \frac{4\sqrt{v}}{(1+\sqrt{v})^2}$, then

$$F(\alpha, \beta, 2\beta, u) = (1+\sqrt{v})^{2\alpha} F(\alpha, \alpha-\beta+\tfrac{1}{2}, v).$$

In verification, observe that if $u=1$ then also $v=1$, and that with these values, calculating each side by means of the formulæ

$$F(\alpha, \beta, \gamma, 1) = \frac{\Gamma\gamma\,\Gamma(\gamma-\alpha-\beta)}{\Gamma(\gamma-\alpha)\,\Gamma(\gamma-\beta)}, \qquad \Pi(\alpha, \beta, \gamma, 1) = \frac{\Gamma\alpha\,\Gamma(\beta-\gamma)}{\Gamma(\alpha+\beta-\gamma)},$$

the resulting equation, $F(\alpha, \beta, 2\beta, 1) = 2^{2\alpha} F(\alpha, \alpha-\beta+\frac{1}{2}, \beta+\frac{1}{2}, 1)$, becomes

$$\frac{\Gamma 2\beta\,\Gamma(\beta-\alpha)}{\Gamma(2\beta-\alpha)\,\Gamma\beta} = 2^{2\alpha} \frac{\Gamma(\beta+\frac{1}{2})\,\Gamma(2\beta-2\alpha)}{\Gamma(2\beta-\alpha)\,\Gamma(\beta-\alpha+\frac{1}{2})},$$

that is,

$$\frac{\Gamma 2\beta}{\Gamma\beta\,\Gamma(\beta+\frac{1}{2})} = 2^{2\alpha} \frac{\Gamma(2\beta-2\alpha)}{\Gamma(\beta-\alpha)\,\Gamma(\beta-\alpha+\frac{1}{2})},$$

which is true, in virtue of the relation $\frac{\Gamma 2x\,\Gamma\frac{1}{2}}{\Gamma x\,\Gamma(x+\frac{1}{2})} = 2^{2x-1}$.

72. The foregoing formulæ, and in particular the formula which I have written $F(\alpha, \beta, 2\beta, u) = (1+\sqrt{v})^{2\alpha} F(\alpha, \alpha-\beta+\frac{1}{2}, \beta+\frac{1}{2}, v)$, are taken from Kummer's Memoir, "Ueber die hypergeometrische Reihe," *Crelle*, t. XV. (1836), viz. the formula in question is, under a slightly different form, his formula (41), p. 76; the formula (43), p. 77, is intended to be equivalent thereto; but there is an error of transcription, $2\alpha-2\beta+1$, in place of $\beta+\frac{1}{2}$, which makes the formula (43) erroneous.

It may be remarked as to the formulæ generally that, although very probably $\Pi(\alpha, \beta, \gamma, u)$ may denote a proper function of u, whatever be the values of the indices (α, β, γ), and the various transformation-theorems hold good accordingly (the Γ-function of a negative argument being interpreted in the usual manner by means of the equation $\Gamma x = \frac{1}{x}\Gamma(1+x)$, $= \frac{1}{x(x+1)}\Gamma(2+x)$ &c.), yet that the function $\Pi(\alpha, \beta, \gamma, u)$, used as denoting the definite integral $\int_0^1 x^{\alpha-1}(1-x)^{\beta-1}(1-ux)^{-\gamma}\,dx$, has no meaning except in the case where α and β are each of them positive.

In what follows we obtain for the ring-integral and the disk-integral various expressions in terms of Π-functions, which are afterwards transformed into t-integrals with a superior limit ∞ and inferior limit 0, or κ^2-f^2; but for values of the variable index, q lying beyond certain limits, the indices α and β, or one of them, of the Π-function will become negative, viz. the integral represented by the Π-function, or, what is the same thing, the t-integral, will cease to have a determinate value, and at the same time, or usually so, the argument or arguments of one or more of the Γ-functions will become negative. It is quite possible that in such cases the results are not without meaning, and that an interpretation for them might be found; but they have not any obvious interpretation, and we must in the first instance consider them as inapplicable.

73. We require further properties of the Π-functions. Starting with the foregoing equation

$$F(\alpha,\ \beta,\ 2\beta,\ u)=(1+\sqrt{v})^{2\alpha}\,F(\alpha,\ \alpha-\beta+\tfrac{1}{2},\ \beta+\tfrac{1}{2},\ v),$$

each side may be expressed in a fourfold form:—

$F(\alpha,\ \beta,\ 2\beta,\ u)$	$(1+\sqrt{v})^{2\alpha}F(\alpha,\ \alpha-\beta+\tfrac{1}{2},\ \beta+\tfrac{1}{2},\ v)$
$=F(\beta,\ \alpha,\ 2\beta,\ u)$	$=(1+\sqrt{v})^{2\alpha}F(\alpha-\beta+\tfrac{1}{2},\ \alpha,\ \beta+\tfrac{1}{2},\ v)$
$=(1-u)^{\beta-\alpha}F(2\beta-\alpha,\ \beta,\ 2\beta,\ u)$	$=(1+\sqrt{v})^{2\alpha}(1-v)^{2\beta-2\alpha}F(\beta-\alpha+\tfrac{1}{2},\ 2\beta-\alpha,\ \beta+\tfrac{1}{2},\ v)$
$=(1-u)^{\beta-\alpha}F(\alpha,\ 2\beta-\alpha,\ 2\beta,\ u)$	$=(1+\sqrt{v})^{2\alpha}(1-v)^{2\beta-2\alpha}F(2\beta-\alpha,\ \beta-\alpha+\tfrac{1}{2},\ \beta+\tfrac{1}{2},\ v)$,

where, instead of $(1+\sqrt{v})^{2\alpha}(1-v)^{2\beta-2\alpha}$, it is proper to write $(1+\sqrt{v})^{2\beta}(1-\sqrt{v})^{2\beta-2\alpha}$; and then to each form applying the transformation

$$F(\alpha,\ \beta,\ \gamma,\ u)=\frac{\Gamma\gamma}{\Gamma\alpha\,\Gamma(\gamma-\alpha)}\Pi(\alpha,\ \gamma-\alpha,\ \beta,\ u),$$

we have

$$\frac{\Gamma\,2\beta}{\Gamma\alpha\,\Gamma(2\beta-\alpha)}\Pi(\alpha,\ 2\beta-\alpha,\ \beta,\ u)$$

$$=\frac{\Gamma\,2\beta}{\Gamma\beta\,\Gamma\beta}\Pi(\beta,\ \beta,\ \alpha,\ u)$$

$$=(1-u)^{\beta-\alpha}\frac{\Gamma\,2\beta}{\Gamma(2\beta-\alpha)\,\Gamma\alpha}\Pi(2\beta-\alpha,\ \alpha,\ \beta,\ u)$$

$$=(1-u)^{\beta-\alpha}\frac{\Gamma\,2\beta}{\Gamma\alpha\,\Gamma(2\beta-\alpha)}\Pi(\alpha,\ 2\beta-\alpha,\ 2\beta-\alpha,\ u);$$

$$= (1+\sqrt{v})^{2\alpha} \frac{\Gamma(\beta+\tfrac{1}{2})}{\Gamma\alpha\,\Gamma(\beta-\alpha+\tfrac{1}{2})} \Pi(\alpha,\ \beta-\alpha+\tfrac{1}{2},\ \alpha-\beta+\tfrac{1}{2},\ v)$$

$$= (1+\sqrt{v})^{2\alpha} \frac{\Gamma(\beta+\tfrac{1}{2})\,1}{\Gamma(\alpha-\beta+\tfrac{1}{2})\,\Gamma(2\beta-\alpha)} \Pi(\alpha-\beta+\tfrac{1}{2},\ 2\beta-\alpha,\ \alpha,\ v)$$

$$= (1+\sqrt{v})^{2\beta}(1-\sqrt{v})^{2\beta-2\alpha} \frac{\Gamma(\beta+\tfrac{1}{2})}{\Gamma(\beta-\alpha+\tfrac{1}{2})\,\Gamma\alpha} \Pi(\beta-\alpha+\tfrac{1}{2},\ \alpha,\ 2\beta-\alpha,\ v)$$

$$= (1+\sqrt{v})^{2\beta}(1-\sqrt{v})^{2\beta-2\alpha} \frac{\Gamma(\beta+\tfrac{1}{2})}{\Gamma(2\beta-\alpha)\,\Gamma(\alpha-\beta+\tfrac{1}{2})} \Pi(2\beta-\alpha,\ \alpha-\beta+\tfrac{1}{2},\ \beta-\alpha+\tfrac{1}{2},\ v).$$

I select the second of the first four forms; equating it successively to each of the second four forms, and attending to the relation $\dfrac{\Gamma\beta\,\Gamma(\beta+\tfrac{1}{2})}{\Gamma 2\beta} = 2^{1-2\beta}\,\Gamma\tfrac{1}{2}$, we find

$$\Pi(\beta,\ \beta,\ \alpha,\ u) = (1+\sqrt{v})^{2\alpha}\,2^{1-2\beta} \frac{\Gamma\beta\,\Gamma\tfrac{1}{2}}{\Gamma\alpha\,\Gamma(\beta-\alpha+\tfrac{1}{2})} \Pi(\alpha,\ \beta-\alpha+\tfrac{1}{2},\ \alpha-\beta+\tfrac{1}{2},\ v)$$

$$= (1+\sqrt{v})^{2\alpha}\,2^{1-2\beta} \frac{\Gamma\beta\,\Gamma\tfrac{1}{2}}{\Gamma(\alpha-\beta+\tfrac{1}{2})\,\Gamma(2\beta-\alpha)} \Pi(\alpha-\beta+\tfrac{1}{2},\ 2\beta-\alpha,\ \alpha,\ v)$$

$$= (1+\sqrt{v})^{2\beta}(1-\sqrt{v})^{2\beta-2\alpha}\,2^{1-2\beta} \frac{\Gamma\beta\,\Gamma\tfrac{1}{2}}{\Gamma(\beta-\alpha+\tfrac{1}{2})\,\Gamma\alpha} \Pi(\beta-\alpha+\tfrac{1}{2},\ \alpha,\ 2\beta-\alpha,\ v)$$

$$= (1+\sqrt{v})^{2\beta}(1-\sqrt{v})^{2\beta-2\alpha}\,2^{1-2\beta} \frac{\Gamma\beta\,\Gamma\tfrac{1}{2}}{\Gamma(2\beta-\alpha)\,\Gamma(\alpha-\beta+\tfrac{1}{2})} \Pi(2\beta-\alpha,\ \alpha-\beta+\tfrac{1}{2},\ \beta-\alpha+\tfrac{1}{2},\ v).$$

Putting herein $\beta=\tfrac{1}{2}s,\ \alpha=\tfrac{1}{2}s+q$, the formulæ become

$$\Pi(\tfrac{1}{2}s,\ \tfrac{1}{2}s,\ \tfrac{1}{2}s+q,\ u) = (1+\sqrt{v})^{s+2q}\,2^{1-s} \frac{\Gamma\tfrac{1}{2}s\,\Gamma\tfrac{1}{2}}{\Gamma(\tfrac{1}{2}s+q)\,\Gamma(\tfrac{1}{2}-q)} \Pi(\tfrac{1}{2}s+q,\ \tfrac{1}{2}-q,\ \tfrac{1}{2}+q,\ v) \quad \text{.........(I.)}$$

$$= (1+\sqrt{v})^{s+2q}\,2^{1-s} \frac{\Gamma\tfrac{1}{2}s\,\Gamma\tfrac{1}{2}}{\Gamma(\tfrac{1}{2}+q)\,\Gamma(\tfrac{1}{2}s-q)} \Pi(\tfrac{1}{2}+q,\ \tfrac{1}{2}s-q,\ \tfrac{1}{2}s+q,\ v) \quad \text{......(II.)}$$

$$= (1+\sqrt{v})^{s}(1-\sqrt{v})^{-2q}\,2^{1-s} \frac{\Gamma\tfrac{1}{2}s\,\Gamma\tfrac{1}{2}}{\Gamma(\tfrac{1}{2}-q)\,\Gamma(\tfrac{1}{2}s+q)} \Pi(\tfrac{1}{2}-q,\ \tfrac{1}{2}s+q,\ \tfrac{1}{2}s-q,\ v) \quad \text{......(III.)}$$

$$= (1+\sqrt{v})^{s}(1-\sqrt{v})^{-2q}\,2^{1-s} \frac{\Gamma\tfrac{1}{2}s\,\Gamma\tfrac{1}{2}}{\Gamma(\tfrac{1}{2}s-q)\,\Gamma(\tfrac{1}{2}+q)} \Pi(\tfrac{1}{2}s-q,\ \tfrac{1}{2}+q,\ \tfrac{1}{2}-q,\ v) \quad \text{......(IV.)},$$

where observe that on the right-hand side the Π-functions in I. and IV. only differ by the sign of q, and so also the Π-functions in II. and III. only differ by the sign of q. We hence have

$$\Pi(\tfrac{1}{2}s,\ \tfrac{1}{2}s,\ \tfrac{1}{2}s-q,\ u) = (1+\sqrt{v})^{s-2q}\,2^{1-s}\,.\,\frac{\Gamma\tfrac{1}{2}s\,\Gamma\tfrac{1}{2}}{\Gamma(\tfrac{1}{2}s-q)\,\Gamma(\tfrac{1}{2}+q)} \Pi(\tfrac{1}{2}s-q,\ \tfrac{1}{2}+q,\ \tfrac{1}{2}-q,\ v);$$

and comparing with (IV.),

$$\Pi\left(\tfrac{1}{2}s,\ \tfrac{1}{2}s,\ \tfrac{1}{2}s+q,\ u\right)=\left(\frac{1+\sqrt{v}}{1-\sqrt{v}}\right)^{2q}\Pi\left(\tfrac{1}{2}s,\ \tfrac{1}{2}s,\ \tfrac{1}{2}s-q,\ u\right).$$

74. The foregoing formula,

$$\text{Ring-integral}=\frac{2^{s-1}f^s}{(f+\kappa)^{s+2q}}\,\Pi\left(\tfrac{1}{2}s,\ \tfrac{1}{2}s,\ \tfrac{1}{2}s+q,\ u\right),$$

where $u=\dfrac{4\kappa f}{(f+\kappa)^2}$, gives, as well in the case of an exterior as an interior point, a convergent series for the integral; but this series proceeds according to the powers of $\dfrac{4\kappa f}{(f+\kappa)^2}$. We may obtain more convenient formulæ applying to the cases of an internal and an external point respectively.

75. For an internal point $\kappa<f$, $\sqrt{1-u}=\dfrac{f-\kappa}{f+\kappa}$, and therefore $v=\dfrac{\kappa^2}{f^2}$.

$$\begin{aligned}
\Pi\left(\tfrac{1}{2}s,\ \tfrac{1}{2}s,\ \tfrac{1}{2}s+q,\ u\right)&=\left(\frac{f+\kappa}{f}\right)^{s+2q}2^{1-s}\frac{\Gamma\tfrac{1}{2}s\,\Gamma\tfrac{1}{2}}{\Gamma(\tfrac{1}{2}s+q)\,\Gamma(\tfrac{1}{2}-q)}\,\Pi\left(\tfrac{1}{2}s+q,\ \tfrac{1}{2}-q,\ \tfrac{1}{2}+q,\ \frac{\kappa^2}{f^2}\right)\\
&=\left(\frac{f+\kappa}{f}\right)^{s+2q}2^{1-s}\frac{\Gamma\tfrac{1}{2}s\,\Gamma\tfrac{1}{2}}{\Gamma(\tfrac{1}{2}+q)\,\Gamma(\tfrac{1}{2}s-q)}\,\Pi\left(\tfrac{1}{2}+q,\ \tfrac{1}{2}s-q,\ \tfrac{1}{2}s+q,\ \frac{\kappa^2}{f^2}\right)\\
&=\left(\frac{f+\kappa}{f}\right)^{s}\left(\frac{f-\kappa}{f}\right)^{-2q}2^{1-s}\frac{\Gamma\tfrac{1}{2}s\,\Gamma\tfrac{1}{2}}{\Gamma(\tfrac{1}{2}-q)\,\Gamma(\tfrac{1}{2}s+q)}\,\Pi\left(\tfrac{1}{2}-q,\ \tfrac{1}{2}s+q,\ \tfrac{1}{2}s-q,\ \frac{\kappa^2}{f^2}\right)\\
&=\left(\frac{f+\kappa}{f}\right)^{s}\left(\frac{f-\kappa}{f}\right)^{-2q}2^{1-s}\frac{\Gamma\tfrac{1}{2}s\,\Gamma\tfrac{1}{2}}{\Gamma(\tfrac{1}{2}s-q)\,\Gamma(\tfrac{1}{2}+q)}\,\Pi\left(\tfrac{1}{2}s-q,\ \tfrac{1}{2}+q,\ \tfrac{1}{2}-q,\ \frac{\kappa^2}{f^2}\right):
\end{aligned}$$

where the Π-functions on the right-hand side are respectively

$$\begin{aligned}
&=f^{2q+1}\int_0^1\frac{x^{\frac{1}{2}s+q-1}(1-x)^{-q-\frac{1}{2}}\,dx}{(f^2-\kappa^2x)^{q+\frac{1}{2}}} &&=\frac{f^{2q+1}}{(f^2-\kappa^2)^{2q}}\int_0^\infty t^{\frac{1}{2}s+q-1}(t+f^2-\kappa^2)^{-\frac{1}{2}s+q}(t+f^2)^{-q-\frac{1}{2}}\,dt\\
&=f^{s+2q}\int_0^1\frac{x^{q-\frac{1}{2}}(1-x)^{\frac{1}{2}s-q-1}\,dx}{(f^2-\kappa^2x)^{\frac{1}{2}s+q}} &&=\frac{f^{s+2q}}{(f^2-\kappa^2)^{2q}}\int_0^\infty t^{q-\frac{1}{2}}(t+f^2-\kappa^2)^{q-\frac{1}{2}}(t+f^2)^{-\frac{1}{2}s-q}\,dt\\
&=f^{s-2q}\int_0^1\frac{x^{-q-\frac{1}{2}}(1-x)^{\frac{1}{2}s+q-1}\,dx}{(f^2-\kappa^2x)^{\frac{1}{2}s-q}} &&=\frac{f^{s-2q}}{(f^2-\kappa^2)^{-2q}}\int_0^\infty t^{-q-\frac{1}{2}}(t+f^2-\kappa^2)^{-q-\frac{1}{2}}(t+f^2)^{-\frac{1}{2}s+q}\,dt\\
&=f^{-2q+1}\int_0^1\frac{x^{\frac{1}{2}s-q+1}(1-x)^{q-\frac{1}{2}}\,dx}{(f^2-\kappa^2x)^{-q+\frac{1}{2}}} &&=\frac{f^{-2q+1}}{(f^2-\kappa^2)^{-2q}}\int_0^\infty t^{\frac{1}{2}s-q-1}(t+f^2-\kappa^2)^{-\frac{1}{2}s-q}(t+f^2)^{q-\frac{1}{2}}\,dt,
\end{aligned}$$

the t-forms being obtained by means of the transformation $x=\dfrac{t}{t+f^2-\kappa^2}$; viz. this gives

$$1-x=\frac{f^2-\kappa^2}{t+f^2-\kappa^2},\quad f^2-\kappa^2x=\frac{(f^2-\kappa^2)(t+f^2)}{t+f^2-\kappa^2},\quad dx=\frac{(f^2-\kappa^2)\,dt}{(t+f^2-\kappa^2)^2},$$

whence the results just written down.

We hence have

$$\text{Ring-integral} = \frac{f}{(f^2-\kappa^2)^{2q}} \frac{\Gamma\tfrac{1}{2}s\,\Gamma\tfrac{1}{2}}{\Gamma(\tfrac{1}{2}s+q)\,\Gamma(\tfrac{1}{2}-q)} \int_0^\infty t^{\frac{1}{2}s+q-1}(t+f^2-\kappa^2)^{-\frac{1}{2}s+q}(t+f^2)^{-q-\frac{1}{2}}\,dt$$

$$= \frac{f^s}{(f^2-\kappa^2)^{2q}} \frac{\Gamma\tfrac{1}{2}s\,\Gamma\tfrac{1}{2}}{\Gamma(\tfrac{1}{2}+q)\,\Gamma(\tfrac{1}{2}s-q)} \int_0^\infty t^{q-\frac{1}{2}}\quad(t+f^2-\kappa^2)^{q-\frac{1}{2}}\quad(t+f^2)^{-\frac{1}{2}s-q}\,dt$$

$$= \quad f^s \quad \frac{\Gamma\tfrac{1}{2}s\,\Gamma\tfrac{1}{2}}{\Gamma(\tfrac{1}{2}-q)\,\Gamma(\tfrac{1}{2}s+q)} \int_0^\infty t^{-q-\frac{1}{2}}\quad(t+f^2-\kappa^2)^{-q-\frac{1}{2}}\quad(t+f^2)^{-\frac{1}{2}s+q}\,dt$$

$$= \quad f \quad \frac{\Gamma\tfrac{1}{2}s\,\Gamma\tfrac{1}{2}}{\Gamma(\tfrac{1}{2}s-q)\,\Gamma(\tfrac{1}{2}+q)} \int_0^\infty t^{\frac{1}{2}s-q-1}(t+f^2-\kappa^2)^{-\frac{1}{2}s-q}(t+f^2)^{q-\frac{1}{2}}\,dt.$$

As a verification write $\kappa=0$, the four integrals are

$$\int_0^\infty \frac{t^{\frac{1}{2}s+q-1}\,dt}{(t+f^2)^{\frac{1}{2}s+\frac{1}{2}}},\quad = f^{2q-1}\,\frac{\Gamma(\tfrac{1}{2}s+q)\,\Gamma(\tfrac{1}{2}-q)}{\Gamma(\tfrac{1}{2}s+\tfrac{1}{2})},$$

$$\int_0^\infty \frac{t^{\frac{1}{2}+q-1}\,dt}{(t+f^2)^{\frac{1}{2}s+\frac{1}{2}}},\quad = f^{2q-s}\,\frac{\Gamma(\tfrac{1}{2}+q)\,\Gamma(\tfrac{1}{2}s-q)}{\Gamma(\tfrac{1}{2}s+\tfrac{1}{2})},$$

$$\int_0^\infty \frac{t^{\frac{1}{2}-q-1}\,dt}{(t+f^2)^{\frac{1}{2}s+\frac{1}{2}}},\quad = f^{-2q-s}\,\frac{\Gamma(\tfrac{1}{2}-q)\,\Gamma(\tfrac{1}{2}s+q)}{\Gamma(\tfrac{1}{2}s+\tfrac{1}{2})},$$

$$\int_0^\infty \frac{t^{\frac{1}{2}s-q-1}\,dt}{(t+f^2)^{\frac{1}{2}s+\frac{1}{2}}},\quad = f^{-2q-1}\,\frac{\Gamma(\tfrac{1}{2}s-q)\,\Gamma(\tfrac{1}{2}+q)}{\Gamma(\tfrac{1}{2}s+\tfrac{1}{2})};$$

hence from each of them

$$\text{Ring-integral} = \frac{1}{f^{2q}}\,\frac{\Gamma\tfrac{1}{2}s\,\Gamma\tfrac{1}{2}}{\Gamma(\tfrac{1}{2}s+\tfrac{1}{2})},$$

which is, in fact, the value obtained from

$$\text{Ring-integral} = \frac{2^{s-1}f^s}{(f+\kappa)^{s+2q}}\,\Pi\left(\tfrac{1}{2}s,\ \tfrac{1}{2}s,\ \tfrac{1}{2}s+q,\ \frac{4\kappa f}{(\kappa+f)^2}\right)$$

on putting therein $\kappa=0$; viz. the value is

$$= \frac{2^{s-1}}{f^{2q}}\int_0^1 x^{\frac{1}{2}s-1}(1-x)^{\frac{1}{2}s-1}\,dx,\quad = \frac{1}{f^{2q}}\,\frac{2^{s-1}\,\Gamma\tfrac{1}{2}s\,.\,\Gamma\tfrac{1}{2}s}{\Gamma s}.$$

76. For an external point $\kappa > f$, $\sqrt{1-u} = \dfrac{\kappa-f}{\kappa+f}$, and therefore $v = \dfrac{f^2}{\kappa^2}$.

$$\Pi(\tfrac{1}{2}s,\tfrac{1}{2}s,\tfrac{1}{2}s+q,u) = \left(\frac{\kappa+f}{\kappa}\right)^{s+2q} \qquad 2^{1-s}\frac{\Gamma\tfrac{1}{2}s\,\Gamma\tfrac{1}{2}}{\Gamma(\tfrac{1}{2}s+q)\,\Gamma(\tfrac{1}{2}-q)}\,\Pi\left(\tfrac{1}{2}s+q,\tfrac{1}{2}\ -q,\tfrac{1}{2}\ +q,\frac{f^2}{\kappa^2}\right)$$

$$= \left(\frac{\kappa+f}{\kappa}\right)^{s+2q} \qquad 2^{1-s}\frac{\Gamma\tfrac{1}{2}s\,\Gamma\tfrac{1}{2}}{\Gamma(\tfrac{1}{2}+q)\,\Gamma(\tfrac{1}{2}s-q)}\,\Pi\left(\tfrac{1}{2}\ +q,\tfrac{1}{2}s-q,\tfrac{1}{2}s+q,\frac{f^2}{\kappa^2}\right)$$

$$= \left(\frac{\kappa+f}{\kappa}\right)^{s}\left(\frac{\kappa-f}{\kappa}\right)^{-2q} 2^{1-s}\frac{\Gamma\tfrac{1}{2}s\,\Gamma\tfrac{1}{2}}{\Gamma(\tfrac{1}{2}-q)\,\Gamma(\tfrac{1}{2}s+q)}\,\Pi\left(\tfrac{1}{2}\ -q,\tfrac{1}{2}s+q,\tfrac{1}{2}s-q,\frac{f^2}{\kappa^2}\right)$$

$$= \left(\frac{\kappa+f}{\kappa}\right)^{s}\left(\frac{\kappa-f}{\kappa}\right)^{-2q} 2^{1-s}\frac{\Gamma\tfrac{1}{2}s\,\Gamma\tfrac{1}{2}}{\Gamma(\tfrac{1}{2}s-q)\,\Gamma(\tfrac{1}{2}+q)}\,\Pi\left(\tfrac{1}{2}s-q,\tfrac{1}{2}\ +q,\tfrac{1}{2}\ -q,\frac{f^2}{\kappa^2}\right):$$

where the Π-functions on the right hand are respectively

$$= \kappa^{2q+1}\int_0^1 \frac{x^{\frac{1}{2}s+q-1}(1-x)^{-q-\frac{1}{2}}\,dx}{(\kappa^2-f^2x)^{q+\frac{1}{2}}} \quad\Big|\quad = \frac{\kappa^{2q+1}}{(\kappa^2-f^2)^{2q}}\int_{\kappa^2-f^2}^{\infty} t^{-\frac{1}{2}s+q}(t+f^2-\kappa^2)^{\frac{1}{2}s+q-1}(t+f^2)^{-q-\frac{1}{2}}\,dt,$$

$$= \kappa^{s+2q}\int_0^1 \frac{x^{q-\frac{1}{2}}(1-x)^{\frac{1}{2}s-q-1}\,dx}{(\kappa^2-f^2x)^{\frac{1}{2}s+q}} \quad\Big|\quad = \frac{\kappa^{s+2q}}{(\kappa^2-f^2)^{2q}}\int_{\kappa^2-f^2}^{\infty} t^{q-\frac{1}{2}}\;(t+f^2-\kappa^2)^{q-\frac{1}{2}}\;(t+f^2)^{-\frac{1}{2}s-q}\,dt,$$

$$= \kappa^{s-2q}\int_0^1 \frac{x^{-q-\frac{1}{2}}(1-x)^{\frac{1}{2}s+q-1}\,dx}{(\kappa^2-f^2x)^{\frac{1}{2}s-q}} \quad\Big|\quad = \frac{\kappa^{s-2q}}{(\kappa^2-f^2)^{-2q}}\int_{\kappa^2-f^2}^{\infty} t^{-q-\frac{1}{2}}(t+f^2-\kappa^2)^{-q-\frac{1}{2}}\;(t+f^2)^{-\frac{1}{2}s+q}\,dt,$$

$$= \kappa^{-2q+1}\int_0^1 \frac{x^{\frac{1}{2}s-q+1}(1-x)^{q-\frac{1}{2}}\,dx}{(\kappa^2-f^2x)^{-q+\frac{1}{2}}} \quad\Big|\quad = \frac{\kappa^{-2q+1}}{(\kappa^2-f^2)^{-2q}}\int_{\kappa^2-f^2}^{\infty} t^{-\frac{1}{2}s-q}(t+f^2-\kappa^2)^{\frac{1}{2}s-q-1}(t+f^2)^{q-\frac{1}{2}}\;dt.$$

We have then

$$\text{Ring-integral} = \frac{f^s\,\kappa^{1-s}}{(\kappa^2-f^2)^{2q}}\,\frac{\Gamma\frac{1}{2}s\,\Gamma\frac{1}{2}}{\Gamma(\frac{1}{2}s+q)\,\Gamma(\frac{1}{2}-q)}\int_{\kappa^2-f^2}^{\infty} t^{-\frac{1}{2}s+q}(t+f^2-\kappa^2)^{\frac{1}{2}s+q-1}(t+f^2)^{-q-\frac{1}{2}}\,dt$$

$$= \frac{f^s}{(\kappa^2-f^2)^{2q}}\,\frac{\Gamma\frac{1}{2}s\,\Gamma\frac{1}{2}}{\Gamma(\frac{1}{2}+q)\,\Gamma(\frac{1}{2}s-q)}\int_{\kappa^2-f^2}^{\infty} t^{q-\frac{1}{2}}\;(t+f^2-\kappa^2)^{q-\frac{1}{2}}\;(t+f^2)^{-\frac{1}{2}s-q}\,dt$$

$$= \quad f^s\quad \frac{\Gamma\frac{1}{2}s\,\Gamma\frac{1}{2}}{\Gamma(\frac{1}{2}-q)\,\Gamma(\frac{1}{2}s+q)}\int_{\kappa^2-f^2}^{\infty} t^{-q-\frac{1}{2}}(t+f^2-\kappa^2)^{-q-\frac{1}{2}}\;(t+f^2)^{-\frac{1}{2}s+q}\,dt$$

$$= \frac{f^s\,\kappa^{1-s}}{(\kappa^2-f^2)^{2q}}\,\frac{\Gamma\frac{1}{2}s\,\Gamma\frac{1}{2}}{\Gamma(\frac{1}{2}s-q)\,\Gamma(\frac{1}{2}+q)}\int_{\kappa^2-f^2}^{\infty} t^{-\frac{1}{2}s-q}(t+f^2-\kappa^2)^{\frac{1}{2}s-q-1}(t+f^2)^{q-\frac{1}{2}}\,dt.$$

Observe that in II. and III. the integrals, except as to the limits, are the same as in the corresponding formulæ for the interior point.

If in the t-integrals we put $t+\kappa^2-f^2$ in place of t, and ultimately suppose κ indefinitely large in comparison with f, they severally become

$$\int_0^\infty (t+\kappa^2-f^2)^{-\frac{1}{2}s+q}\,t^{\frac{1}{2}s+q-1}(t+\kappa^2)^{-q-\frac{1}{2}}\,dt = \int_0^\infty \frac{t^{\frac{1}{2}s+q-1}\,dt}{(t+\kappa^2)^{\frac{1}{2}s+\frac{1}{2}}} = \kappa^{2q-1}\;\frac{\Gamma(\frac{1}{2}s+q)\,\Gamma(\frac{1}{2}-q)}{\Gamma(\frac{1}{2}s+\frac{1}{2})},$$

$$\int_0^\infty (t+\kappa^2-f^2)^{q-\frac{1}{2}}\quad t^{q+\frac{1}{2}}\quad (t+\kappa^2)^{-\frac{1}{2}s-q}\,dt = \int_0^\infty \frac{t^{\frac{1}{2}+q-1}\,dt}{(t+\kappa^2)^{\frac{1}{2}s+\frac{1}{2}}} = \kappa^{2q-s}\;\frac{\Gamma(\frac{1}{2}+q)\,\Gamma(\frac{1}{2}s-q)}{\Gamma(\frac{1}{2}s+\frac{1}{2})},$$

$$\int_0^\infty (t+\kappa^2-f^2)^{-q-\frac{1}{2}}\,t^{-q-\frac{1}{2}}\;(t+\kappa^2)^{-\frac{1}{2}s+q}\,dt = \int_0^\infty \frac{t^{\frac{1}{2}-q-1}\,dt}{(t+\kappa^2)^{\frac{1}{2}s+\frac{1}{2}}} = \kappa^{-2q-s}\,\frac{\Gamma(\frac{1}{2}-q)\,\Gamma(\frac{1}{2}s+q)}{\Gamma(\frac{1}{2}s+\frac{1}{2})},$$

$$\int_0^\infty (t+\kappa^2-f^2)^{-\frac{1}{2}s-q}\,t^{\frac{1}{2}s-q-1}(t+\kappa^2)^{q-\frac{1}{2}}\;\;dt = \int_0^\infty \frac{t^{\frac{1}{2}s-q-1}\,dt}{(t+\kappa^2)^{\frac{1}{2}s+\frac{1}{2}}} = \kappa^{-2q-1}\,\frac{\Gamma(\frac{1}{2}s-q)\,\Gamma(\frac{1}{2}+q)}{\Gamma(\frac{1}{2}s+\frac{1}{2})};$$

and they all four give

$$\text{Ring-integral} = \frac{f^s}{\kappa^{s+2q}}\cdot\frac{\Gamma\frac{1}{2}s\,\Gamma\frac{1}{2}}{\Gamma(\frac{1}{2}s+\frac{1}{2})},$$

which agrees with the value

$$\frac{2^{s-1}f^s}{(\kappa+f)^{s+2q}}\,\Pi\left(\tfrac{1}{2}s,\ \tfrac{1}{2}s,\ \tfrac{1}{2}s+q,\ \frac{4\kappa f}{(\kappa+f)^2}\right),\quad = \frac{2^{s-1}f^s}{\kappa^{s+2q}}\,\Pi\left(\tfrac{1}{2}s,\ \tfrac{1}{2}s,\ \tfrac{1}{2}s+q,\ 0\right),$$

when $\dfrac{\kappa}{f}$ is indefinitely large.

77. We come now to the disk-integral,

$$\int \frac{y^{s-1}\,dx\,dy}{\{(x-\kappa)^2+y^2\}^{\frac{1}{2}s+q}},$$

over the circle $x^2+y^2=f^2$. Writing $x=\kappa+\rho\cos\phi$, $y=\rho\sin\phi$, we have $dx\,dy=\rho\,d\rho\,d\phi$, and the integral therefore is

$$\iint \frac{\sin^{s-1}\phi\,d\rho\,d\phi}{\rho^{2q}},$$

where the integration in regard to ρ is performed at once; viz. the integral is

$$=\frac{1}{1-2q}\int(\rho^{1-2q})\sin^{s-1}\phi\,d\phi\,;$$

or multiplying by 2, in order that the integration may be taken only over the semicircle, $y=$ positive, this is

$$=\frac{1}{\frac{1}{2}-q}\int(\rho^{1-2q})\sin^{s-1}\phi\,d\phi,$$

the term (ρ^{1-2q}) being taken between the proper limits.

78. Consider first an interior point $\kappa<f$. As already mentioned, we exclude an indefinitely small circle radius ϵ, and the limits for ρ are from $\rho=\epsilon$ to $\rho=$ its value at

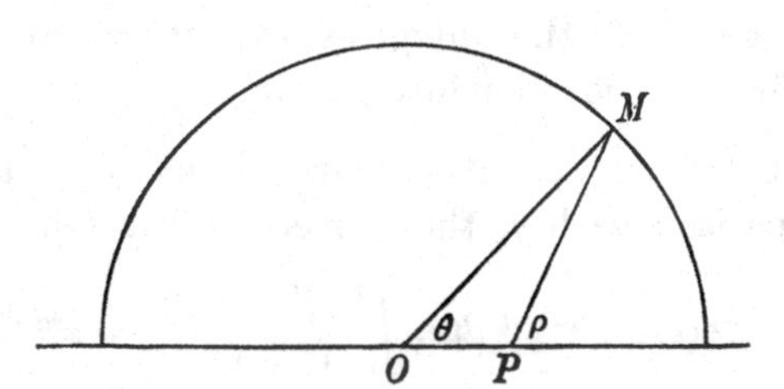

the circumference; viz. if here $x=f\cos\theta$, $y=f\sin\theta$, then we have $f\cos\theta=\kappa+\rho\cos\phi$, $f\sin\theta=\rho\sin\phi$, and consequently

$$\rho^2=\kappa^2+f^2-2\kappa f\cos\theta,$$

$$\sin\phi=\frac{f}{\rho}\sin\theta,\ =\frac{f\sin\theta}{\sqrt{\kappa^2+f^2-2\kappa f\cos\theta}},$$

and the integral therefore is

$$=\frac{1}{\frac{1}{2}-q}\left(\frac{f^{s-1}\sin^{s-1}\theta}{\{\kappa^2+f^2-2\kappa f\cos\theta\}^{\frac{1}{2}s+q-1}}-\epsilon^{1-2q}\sin^{s-1}\phi\right)d\phi.$$

As regards the second term, this is $=-\dfrac{\epsilon^{1-2q}}{\frac{1}{2}-q}\displaystyle\int\sin^{s-1}\phi\,d\phi$, from $\phi=0$ to $\phi=\pi$; or, what is the same thing, we may multiply by 2 and take the integral from $\phi=0$ to $\phi=\dfrac{\pi}{2}$.

Writing then $\sin\phi = \sqrt{x}$, and consequently $\sin^{s-1}\phi\, d\phi = \frac{1}{2}x^{\frac{1}{2}s-1}(1-x)^{-\frac{1}{2}}\,dx$, the term is $= -\dfrac{\epsilon^{1-2q}}{\frac{1}{2}-q}\,\dfrac{\Gamma\frac{1}{2}s\,\Gamma\frac{1}{2}}{\Gamma(\frac{1}{2}s+\frac{1}{2})}$: the value of the disk-integral is

$$= \frac{f^{s-1}}{\frac{1}{2}-q}\int \frac{\sin^{s-1}\theta\, d\phi}{(\kappa^2+f^2-2\kappa f\cos\theta)^{\frac{1}{2}s+q-1}} - \frac{\epsilon^{1-2q}}{\frac{1}{2}-q}\,\frac{\Gamma\frac{1}{2}s\,\Gamma\frac{1}{2}}{\Gamma(\frac{1}{2}s+\frac{1}{2})}.$$

But we have

$$\sin\phi = \frac{f\sin\theta}{\rho},\quad \cos\phi = \frac{f\cos\theta-\kappa}{\rho},$$

and thence

$$\tan\phi = \frac{f\sin\theta}{f\cos\theta-\kappa},\quad \sec^2\phi\, d\phi = \frac{f(f-\kappa\cos\theta)\,d\theta}{(f\cos\theta-\kappa)^2};$$

that is,

$$d\phi = \frac{f(f-\kappa\cos\theta)\,d\theta}{\rho^2},\quad = \frac{f(f-\kappa\cos\theta)\,d\theta}{f^2+\kappa^2-2\kappa f\cos\theta},$$

or, what is the same thing,

$$= \frac{\frac{1}{2}\{(f^2-\kappa^2)+(f^2+\kappa^2-2\kappa f\cos\theta)\}}{f^2+\kappa^2-2\kappa f\cos\theta};$$

the expression for the disk-integral is therefore

$$= \frac{\frac{1}{2}f^{s-1}}{\frac{1}{2}-q}\int_0^\pi \frac{\sin^{s-1}\theta\,\{(f^2-\kappa^2)+(f^2+\kappa^2-2\kappa f\cos\theta)\}\,d\theta}{\{f^2+\kappa^2-2\kappa f\cos\theta\}^{\frac{1}{2}s+q}} - \frac{\epsilon^{1-2q}}{\frac{1}{2}-q}\,\frac{\Gamma\frac{1}{2}s\,\Gamma\frac{1}{2}}{\Gamma(\frac{1}{2}s+\frac{1}{2})}.$$

79. Writing as before $\cos\frac{1}{2}\theta = \sqrt{x}$, $\sin\frac{1}{2}\theta = \sqrt{1+x}$, &c., and $u = \dfrac{4\kappa f}{(\kappa+f)^2}$, this is

$$= \frac{2^{s-2}f^{s-1}}{(\frac{1}{2}-q)(\kappa+f)^{s+2q-2}}\left\{\frac{(f^2-\kappa^2)}{(\kappa+f)^2}\,\Pi(\tfrac{1}{2}s,\tfrac{1}{2}s,\tfrac{1}{2}s+q,u)+\Pi(\tfrac{1}{2}s,\tfrac{1}{2}s,\tfrac{1}{2}s+q-1,u)\right\} - \frac{\epsilon^{1-2q}}{\frac{1}{2}-q}\frac{\Gamma\frac{1}{2}s\,\Gamma\frac{1}{2}}{\Gamma(\frac{1}{2}s+\frac{1}{2})}.$$

As a verification, observe that, if $\kappa=0$, each of the Π-functions becomes

$$= \int_0^1 x^{\frac{1}{2}s-1}(1-x)^{\frac{1}{2}s-1}\,dx,\quad = \frac{\Gamma\frac{1}{2}s\,\Gamma\frac{1}{2}s}{\Gamma s};$$

hence the whole first term is $= \dfrac{2\,.\,2^{s-2}\,.\,f^{1-2q}}{\frac{1}{2}-q}\,.\,\dfrac{\Gamma\frac{1}{2}s\,\Gamma\frac{1}{2}s}{\Gamma s}$, viz. this is $= \dfrac{f^{1-2q}}{\frac{1}{2}-q}\,\dfrac{\Gamma\frac{1}{2}s\,\Gamma\frac{1}{2}}{\Gamma(\frac{1}{2}s+\frac{1}{2})}$, and the complete value is

$$= \frac{1}{\frac{1}{2}-q}\,\frac{\Gamma\frac{1}{2}s\,\Gamma\frac{1}{2}}{\Gamma(\frac{1}{2}s+\frac{1}{2})}\{f^{1-2q}-\epsilon^{1-2q}\},$$

vanishing, as it should do, if $f=\epsilon$.

80. In the case of an exterior point $\kappa > f$, the process is somewhat different; but the result is of a like form. We have

$$\text{Disk-integral} = \frac{1}{\frac{1}{2}-q}\int(\rho_1^{1-2q}-\rho^{1-2q})\sin^{s-1}\phi\, d\phi,$$

where ρ_1 refers to the point M' and ρ to the point M. Attending first to the integral $\int \rho^{1-2q} \sin^{s-1}\phi\, d\phi$, and writing as before $f\cos\theta = \kappa + \rho\cos\phi$, $f\sin\theta = \rho\sin\phi$, this is

$$= f^{s-1}\int \frac{\sin^{s-1}\theta\, d\phi}{\{\kappa^2 + f^2 - 2\kappa f\cos\theta\}^{\frac{1}{2}s+q}}$$

$$= \tfrac{1}{2} f^{s-1}\int \frac{\sin^{s-1}\theta\,\{(f^2-\kappa^2) + (f^2+\kappa^2-2\kappa f\cos\theta)\}\, d\theta}{(f^2+\kappa^2-2f\kappa\cos\theta)^{\frac{1}{2}s+q}},$$

the inferior and the superior limits being here the values of θ which correspond to the points N, A respectively, say $\theta + \alpha$, and $\theta = 0$; hence, reversing the sign and inter-

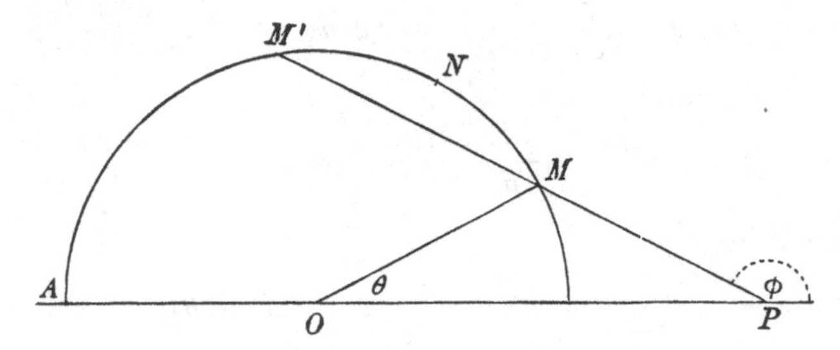

changing the two limits, the value of $-\int \rho^{1-2q}\sin^{s-1}\theta\, d\phi$ is the above integral taken from 0 to α. But similarly the value of $+\int \rho_1^{1-2q}\sin^{s-1}\theta\, d\phi$ is the same integral taken from α to π. For the two terms together, the value is the same integral from 0 to π; viz. we thus find

$$\text{Disk-integral} = \frac{\frac{1}{2}f^{s-1}}{\frac{1}{2}-q}\int_0^\pi \frac{\sin^{s-1}\theta\,\{-(\kappa^2-f^2) + (f^2+\kappa^2-2\kappa f\cos\theta)\}\, d\theta}{(f^2+\kappa^2-2f\kappa\cos\theta)^{\frac{1}{2}s+q}};$$

or, writing as before $\cos\frac{1}{2}\theta = \sqrt{x}$, &c., and $u = \dfrac{4\kappa f}{(\kappa+f)^2}$, this is

$$= \frac{2^{s-2} f^{s-1}}{(\frac{1}{2}-q)(\kappa+f)^{s+2q-2}}\left\{-\frac{\kappa^2-f^2}{(\kappa+f)^2}\cdot \Pi(\tfrac{1}{2}s,\ \tfrac{1}{2}s,\ \tfrac{1}{2}s+q,\ u) + \Pi(\tfrac{1}{2}s,\ \tfrac{1}{2}s,\ \tfrac{1}{2}s+q-1)\right\}.$$

81. As a verification, suppose that κ is indefinitely large: we must recur to the last preceding formula; the value is thus

$$= \frac{f^s}{(\frac{1}{2}-q)\,\kappa^{s+2q-1}}\int_0^\pi \frac{\sin^{s-1}\theta\left(-\cos\theta + \dfrac{f}{\kappa}\right)}{\left(1-\dfrac{2f}{\kappa}\cos\theta\right)^{\frac{1}{2}s+q}}\, d\theta;$$

viz. this is

$$= \frac{f^s}{(\frac{1}{2}-q)\,\kappa^{s+2q-1}}\int_0^\pi \sin^{s-1}\theta\left\{-\cos\theta + [1-(s+2q)\cos^2\theta]\frac{f}{\kappa}\right\} d\theta,$$

where the integral of the first term vanishes; the value is thus

$$= \frac{f^{s+1}}{(\frac{1}{2}-q)\,\kappa^{s+2q}}\int_0^\pi \sin^{s-1}\theta\,[1-(s+2q)\cos^2\theta]\, d\theta,$$

where we may multiply by 2 and take the integral from 0 to $\frac{\pi}{2}$. Writing then $\sin\theta = \sqrt{x}$, the value is

$$= \frac{f^{s+1}}{(\frac{1}{2}-q)\,\kappa^{s+2q}} \int_0^1 x^{\frac{1}{2}s-\frac{1}{2}} \{1-(s+2q)(1-x)\}(1-x)^{-\frac{1}{2}}\,dx,$$

where the integral is

$$= \frac{\Gamma\frac{1}{2}s\,\Gamma\frac{1}{2}}{\Gamma(\frac{1}{2}s+\frac{1}{2})}\left\{1-\frac{\frac{1}{2}(s+2q)}{\frac{1}{2}s+\frac{1}{2}}\right\},\ = \frac{\Gamma\frac{1}{2}s\,\Gamma\frac{1}{2}}{\Gamma(\frac{1}{2}s+\frac{1}{2})}\cdot\frac{\frac{1}{2}-q}{\frac{1}{2}s+\frac{1}{2}},$$

and hence the value is

$$= \frac{f^{s+1}}{\kappa^{s+2q}}\cdot\frac{\Gamma\frac{1}{2}s\,\Gamma\frac{1}{2}}{\Gamma(\frac{1}{2}s+\frac{3}{2})};$$

viz. this is $=\frac{1}{\kappa^{s+2q}}\int y^{s-1}\,dx\,dy$, over the circle $x^2+y^2=f^2$, as is easily verified.

82. Reverting to the interior point $\kappa < f$,

Disk-integral

$$= \frac{2^{s-2} f^{s-1}}{(\frac{1}{2}-q)(\kappa+f)^{s+2q-2}}\left\{\frac{f-\kappa}{f+\kappa}\,\Pi(\tfrac{1}{2}s,\tfrac{1}{2}s,\tfrac{1}{2}s+q,u)+\Pi(\tfrac{1}{2}s,\tfrac{1}{2}s,\tfrac{1}{2}s+q-1,u)\right\} - \frac{\epsilon^{1-2q}}{\frac{1}{2}-q}\,\frac{\Gamma\frac{1}{2}s\,\Gamma\frac{1}{2}}{\Gamma(\frac{1}{2}s+\frac{1}{2})};$$

then reducing the expression in { } by the transformations for $\Pi(\frac{1}{2}s, \frac{1}{2}s, \frac{1}{2}s+q, u)$ and the like transformations for $\Pi(\frac{1}{2}s, \frac{1}{2}s, \frac{1}{2}s+q-1, u)$, the term in { } may be expressed in the four forms:—

$$2^{1-s}\frac{\Gamma\frac{1}{2}s\,\Gamma\frac{1}{2}}{\Gamma(\frac{1}{2}s+q)\,\Gamma(\frac{1}{2}-q)}\,\frac{(f+\kappa)^{s+2q-2}}{f^{s+2q-2}}\ \text{multiplied by}$$

$$\left[\left(1-\frac{\kappa^2}{f^2}\right)\Pi\left(\tfrac{1}{2}s+q,\ \tfrac{1}{2}-q,\ \tfrac{1}{2}+q,\ \frac{\kappa^2}{f^2}\right)+\frac{\frac{1}{2}s+q-1}{\frac{1}{2}-q}\,\Pi\left(\tfrac{1}{2}s+q-1,\ \tfrac{3}{2}-q,\ -\tfrac{1}{2}+q,\ \frac{\kappa^2}{f^2}\right)\right],$$

$$2^{1-s}\frac{\Gamma\frac{1}{2}s\,\Gamma\frac{1}{2}}{\Gamma(\frac{1}{2}+q)\,\Gamma(\frac{1}{2}s-q)}\,\frac{(f+\kappa)^{s+2q-2}}{f^{s+2q-2}}\ \text{multiplied by}$$

$$\left[\left(1-\frac{\kappa^2}{f^2}\right)\Pi\left(\tfrac{1}{2}+q,\ \tfrac{1}{2}s-q,\ \tfrac{1}{2}s+q,\ \frac{\kappa^2}{f^2}\right)+\frac{-\frac{1}{2}+q}{\frac{1}{2}s+q}\,\Pi\left(-\tfrac{1}{2}+q,\ \tfrac{1}{2}s-q+1,\ \tfrac{1}{2}s+q-1,\ \frac{\kappa^2}{f^2}\right)\right],$$

$$2^{1-s}\frac{\Gamma\frac{1}{2}s\,\Gamma\frac{1}{2}}{\Gamma(\frac{1}{2}-q)\,\Gamma(\frac{1}{2}s+q)}\,\frac{(f+\kappa)^{s-1}(f-\kappa)^{1-2q}}{f^{s-2q}}\ \text{multiplied by}$$

$$\left[\Pi\left(\tfrac{1}{2}-q,\ \tfrac{1}{2}s+q,\ \tfrac{1}{2}s-q,\ \frac{\kappa^2}{f^2}\right)+\left(1-\frac{\kappa^2}{f^2}\right)\frac{\frac{1}{2}s+q-1}{\frac{1}{2}-q}\,\Pi\left(\tfrac{3}{2}-q,\ \tfrac{1}{2}s+q-1,\ \tfrac{1}{2}s-q+1,\ \frac{\kappa^2}{f^2}\right)\right],$$

$$2^{1-s}\frac{\Gamma\frac{1}{2}s\,\Gamma\frac{1}{2}}{\Gamma(\frac{1}{2}s-q)\,\Gamma(\frac{1}{2}+q)}\,\frac{(f+\kappa)^{s-1}(f-\kappa)^{1-2q}}{f^{s-2q}}\ \text{multiplied by}$$

$$\left[\Pi\left(\tfrac{1}{2}s-q,\ \tfrac{1}{2}+q,\ \tfrac{1}{2}-q,\ \frac{\kappa^2}{f^2}\right)+\left(1-\frac{\kappa^2}{f^2}\right)\frac{-\frac{1}{2}+q}{\frac{1}{2}s-q}\,\Pi\left(\tfrac{1}{2}s-q+1,\ -\tfrac{1}{2}+q,\ \tfrac{3}{2}-q,\ \frac{\kappa^2}{f^2}\right)\right].$$

83. The first and the fourth of these are susceptible of a reduction which does not appear to be applicable to the second and the third. Consider in general the function

$$(1-v)\,\Pi(\alpha,\ \beta,\ 1-\beta,\ v)+\frac{\alpha-1}{\beta}\,\Pi(\alpha-1,\ \beta+1,\ -\beta,\ v);$$

the second Π-function is here

$$\int_0^1 x^{\alpha-2}(1-x\,.\,1-vx)^{\beta}\,dx;$$

viz. this is

$$=\frac{x^{\alpha-1}}{\alpha-1}(1-x\,.\,1-vx)^{\beta}-\frac{1}{\alpha-1}\int_0^1 x^{\alpha-1}\frac{d}{dx}(1-x\,.\,1-vx)^{\beta}\,dx,$$

or, since the first term vanishes between the limits, this is

$$=\frac{\beta}{\alpha-1}\int_0^1 x^{\alpha-1}\,.\,(1-x\,.\,1-vx)^{\beta-1}(1+v-2vx)\,dx,$$

$$=\frac{\beta}{\alpha-1}\{(1+v)\,\Pi(\alpha,\ \beta,\ 1-\beta,\ v)-2v\int_0^1 x^{\alpha}(1-x\,.\,1-vx)^{\beta-1}\,dx\}.$$

Hence the two Π-functions together are

$$=(1-v+1+v)\int_0^1 x^{\alpha-1}(1-x\,.\,1-vx)^{\beta-1}\,dx-2\int_0^1 vx\,.\,x^{\alpha-1}(1-x\,.\,1-vx)^{\beta-1}\,dx,$$

$$=2\int_0^1 x^{\alpha-1}(1-x)^{\beta-1}(1-vx)^{\beta}\,dx,$$

that is,

$$(1-v)\,\Pi(\alpha,\ \beta,\ 1-\beta,\ v)+\frac{\alpha-1}{\beta}\,\Pi(\alpha-1,\ \beta+1,\ -\beta,\ v)=2\Pi(\alpha,\ \beta,\ -\beta,\ v).$$

We have therefore

$$\left(1-\frac{\kappa^2}{f^2}\right)\Pi\left(\tfrac{1}{2}s+q,\ \tfrac{1}{2}-q,\ \tfrac{1}{2}+q,\ \frac{\kappa^2}{f^2}\right)+\frac{\tfrac{1}{2}s+q-1}{\tfrac{1}{2}-q}\,\Pi\left(\tfrac{1}{2}s+q-1,\ \tfrac{3}{2}-q,\ -\tfrac{1}{2}+q,\ \frac{\kappa^2}{f^2}\right)$$

$$=2\Pi\left(\tfrac{1}{2}s+q,\ \tfrac{1}{2}-q,\ -\tfrac{1}{2}+q,\ \frac{\kappa^2}{f^2}\right);$$

and from the same equation, written in the form

$$\Pi(\alpha-1,\ \beta+1,\ -\beta,\ v)+\frac{\beta}{\alpha-1}(1-v)\,\Pi(\alpha,\ \beta,\ 1-\beta,\ v)=2\,\frac{\beta}{\alpha-1}\,\Pi(\alpha,\ \beta,\ -\beta,\ v),$$

we obtain

$$\Pi\left(\tfrac{1}{2}s-q,\ \tfrac{1}{2}+q,\ \tfrac{1}{2}-q,\ \frac{\kappa^2}{f^2}\right)+\frac{-\tfrac{1}{2}+q}{\tfrac{1}{2}s-q}\left(1-\frac{\kappa^2}{f^2}\right)\Pi\left(\tfrac{1}{2}s-q+1,\ -\tfrac{1}{2}+q,\ \tfrac{3}{2}-q,\ \frac{\kappa^2}{f^2}\right)$$

$$=\frac{2(-\tfrac{1}{2}+q)}{\tfrac{1}{2}s-q}\,\Pi\left(\tfrac{1}{2}s-q+1,\ -\tfrac{1}{2}+q,\ \tfrac{1}{2}-q,\ \frac{\kappa^2}{f^2}\right).$$

84. Hence the terms in [ ] in the first and the fourth expressions in No. 82 are

$$=\frac{2^{2-s}\,\Gamma\tfrac{1}{2}s\,\Gamma\tfrac{1}{2}}{\Gamma(\tfrac{1}{2}s+q)\,\Gamma(\tfrac{1}{2}-q)}\cdot\frac{(f+\kappa)^{s+2q-2}}{f^{s+2q-2}}\,.\,\Pi\left(\tfrac{1}{2}s+q,\ \tfrac{1}{2}-q,\ -\tfrac{1}{2}+q,\ \frac{\kappa^2}{f^2}\right),$$

$$=\frac{2^{2-s}(-\tfrac{1}{2}+q)\,.\,\Gamma\tfrac{1}{2}s\,\Gamma\tfrac{1}{2}}{\Gamma(\tfrac{1}{2}s-q+1)\,\Gamma(\tfrac{1}{2}+q)}\,\frac{(f+\kappa)^{s-1}(f-\kappa)^{1-2q}}{f^{s-2q}}\,\Pi\left(\tfrac{1}{2}s-q+1,\ -\tfrac{1}{2}+q,\ \tfrac{1}{2}-q,\ \frac{\kappa^2}{f^2}\right),$$

respectively; the corresponding values of the disk-integral are

$$\frac{\Gamma\frac{1}{2}s\,\Gamma\frac{1}{2}}{\Gamma(\frac{3}{2}-q)\,\Gamma(\frac{1}{2}s+q)}\,f^{1-2q}\,.\,\Pi\left(\tfrac{1}{2}s+q,\ \tfrac{1}{2}-q,\ -\tfrac{1}{2}+q,\ \frac{\kappa^2}{f^2}\right)\qquad-\frac{\epsilon^{1-2q}}{\frac{1}{2}-q}\,\frac{\Gamma\frac{1}{2}s\,\Gamma\frac{1}{2}}{\Gamma(\frac{1}{2}s+\frac{1}{2})},$$

$$\frac{-\Gamma\frac{1}{2}s\,\Gamma\frac{1}{2}}{\Gamma(\frac{1}{2}s-q+1)\,\Gamma(\frac{1}{2}+q)}\left(\frac{f^2-\kappa^2}{f}\right)^{1-2q}.\,\Pi\left(\tfrac{1}{2}s-q+1,\ -\tfrac{1}{2}+q,\ \tfrac{1}{2}-q,\ \frac{\kappa^2}{f^2}\right)-\frac{\epsilon^{1-2q}}{\frac{1}{2}-q}\,\frac{\Gamma\frac{1}{2}s\,\Gamma\frac{1}{2}}{\Gamma(\frac{1}{2}s+\frac{1}{2})},$$

which we may again verify by writing therein $\kappa=0$, viz. the Π-functions thus become

$$\frac{\Gamma(\frac{1}{2}s+q)\,.\,\Gamma(\frac{1}{2}-q)}{\Gamma(\frac{1}{2}s+\frac{1}{2})}\ \text{ and }\ \frac{\Gamma(\frac{1}{2}s-q+1)\,\Gamma(-\frac{1}{2}+q)}{\Gamma(\frac{1}{2}s+\frac{1}{2})},$$

and consequently the integral is

$$=\frac{1}{\frac{1}{2}-q}\cdot\frac{\Gamma\frac{1}{2}s\,\Gamma\frac{1}{2}}{\Gamma(\frac{1}{2}s+\frac{1}{2})}\,(f^{1-2q}-\epsilon^{1-2q}).$$

85. But the forms nevertheless belong to a system of four. In the formulæ

$$\begin{aligned}
&\Pi(\alpha,\ \beta,\ \gamma,\ v)\\
=\ &\frac{\Gamma\alpha\,\Gamma\beta}{\Gamma\gamma\,\Gamma(\alpha+\beta-\gamma)}\,\Pi(\gamma,\ \alpha+\beta-\gamma,\ \alpha,\ v)\\
=\ &(1-v)^{\beta-\gamma}\,\Pi(\beta,\ \alpha,\ \alpha+\beta-\gamma,\ v)\\
=\ &(1-v)^{\beta-\gamma}\,\frac{\Gamma\alpha\,\Gamma\beta}{\Gamma(\alpha+\beta-\gamma)\,\Gamma\gamma}\,\Pi(\alpha+\beta-\gamma,\ \gamma,\ \beta,\ v),
\end{aligned}$$

writing $\alpha=\frac{1}{2}s+q$, $\beta=\frac{1}{2}-q$, $\gamma=-\frac{1}{2}+q$, we deduce

$$\begin{aligned}
&\Pi(\tfrac{1}{2}s+q,\ \tfrac{1}{2}-q,\ -\tfrac{1}{2}+q,\ v)\\
=\ &\frac{\Gamma(\frac{1}{2}s+q)\,\Gamma(\frac{1}{2}-q)}{\Gamma(-\frac{1}{2}+q)\,\Gamma(\frac{1}{2}s-q+1)}\,\Pi(-\tfrac{1}{2}+q,\ \tfrac{1}{2}s-q+1,\ \tfrac{1}{2}s+q,\ v)\\
=\ &(1-v)^{1-2q}\,\Pi(\tfrac{1}{2}-q,\ \tfrac{1}{2}s+q,\ \tfrac{1}{2}s-q+1,\ v)\\
=\ &(1-v)^{1-2q}\,\frac{\Gamma(\frac{1}{2}s+q)\,\Gamma(\frac{1}{2}-q)}{\Gamma(\frac{1}{2}s-q+1)\,\Gamma(-\frac{1}{2}+q)}\,\Pi(\tfrac{1}{2}s-q+1,\ -\tfrac{1}{2}+q,\ \tfrac{1}{2}-q,\ v);
\end{aligned}$$

and the last-mentioned values of the disk-integral may thus be written in the four forms:

$$\begin{aligned}
&\frac{\Gamma\frac{1}{2}s\,\Gamma\frac{1}{2}}{\Gamma(\frac{3}{2}-q)\,\Gamma(\frac{1}{2}s+q)}\,f^{1-2q}\,\Pi\left(\tfrac{1}{2}s+q,\ \tfrac{1}{2}-q,\ -\tfrac{1}{2}+q,\ \frac{\kappa^2}{f^2}\right)-\text{term in }\epsilon,\\
&\frac{-\Gamma\frac{1}{2}s\,\Gamma\frac{1}{2}}{\Gamma(\frac{1}{2}+q)\,\Gamma(\frac{1}{2}s-q+1)}\,f^{1-2q}\,\Pi\left(-\tfrac{1}{2}+q,\ \tfrac{1}{2}s-q+1,\ \tfrac{1}{2}s+q,\ \frac{\kappa^2}{f^2}\right)-\quad\text{,,}\quad,\\
&\frac{\Gamma\frac{1}{2}s\,\Gamma\frac{1}{2}}{\Gamma(\frac{3}{2}-q)\,\Gamma(\frac{1}{2}s+q)}\left(f-\frac{\kappa^2}{f}\right)^{1-2q}\Pi\left(\tfrac{1}{2}-q,\ \tfrac{1}{2}s+q,\ \tfrac{1}{2}s-q+1,\ \frac{\kappa^2}{f^2}\right)-\quad\text{,,}\quad,\\
&\frac{-\Gamma\frac{1}{2}s\,\Gamma\frac{1}{2}}{\Gamma(\frac{1}{2}+q)\,\Gamma(\frac{1}{2}s-q+1)}\left(f-\frac{\kappa^2}{f}\right)^{1-2q}\Pi\left(\tfrac{1}{2}s-q+1,\ -\tfrac{1}{2}+q,\ \tfrac{1}{2}-q,\ \frac{\kappa^2}{f^2}\right)-\quad\text{,,}\quad;
\end{aligned}$$

and since the last of these is in fact the second of the original forms, it is clear that, if instead of the first we had taken the second of the original forms, we should have obtained again the same system of four forms.

86. Writing as before $x = \frac{t}{t + f^2 - \kappa^2}$, &c., the forms are

$$\frac{\Gamma\frac{1}{2}s\,\Gamma\frac{1}{2}}{\Gamma(\frac{3}{2}-q)\,\Gamma(\frac{1}{2}s+q)}(f^2-\kappa^2)^{1-2q}\int_0^\infty t^{\frac{1}{2}s+q-1}(t+f^2-\kappa^2)^{-\frac{1}{2}s+q-1}(t+f^2)^{-\frac{1}{2}s-q}\,dt - \text{term in } \epsilon,$$

$$\frac{-\Gamma\frac{1}{2}s\,\Gamma\frac{1}{2}}{\Gamma(\frac{1}{2}+q)\,\Gamma(\frac{1}{2}s-q+1)}f^{s+1}(f^2-\kappa^2)^{1-2q}\int_0^\infty t^{-\frac{3}{2}+q}(t+f^2-\kappa^2)^{q-\frac{1}{2}}(t+f^2)^{-\frac{1}{2}s-q}\,dt - \quad ,, \quad ,$$

$$\frac{\Gamma\frac{1}{2}s\,\Gamma\frac{1}{2}}{\Gamma(\frac{3}{2}-q)\,\Gamma(\frac{1}{2}s+q)}f^{s+1}\int_0^\infty t^{-q-\frac{1}{2}}(t+f^2-\kappa^2)^{-q+\frac{1}{2}}(t+f^2)^{-\frac{1}{2}s+q-1}\,dt - \quad ,, \quad ,$$

$$\frac{-\Gamma\frac{1}{2}s\,\Gamma\frac{1}{2}}{\Gamma(\frac{1}{2}s-q+1)\,\Gamma(\frac{1}{2}+q)}\int_0^\infty t^{\frac{1}{2}s-q}(t+f^2-\kappa^2)^{-\frac{1}{2}s-q}(t+f^2)^{-\frac{1}{2}+q}\,dt - \quad ,, \quad .$$

87. The third of these possesses a remarkable property. Write mf instead of f, and at the same time change t into m^2t: the integral becomes

$$\frac{\Gamma\frac{1}{2}s\,\Gamma\frac{1}{2}}{\Gamma(\frac{3}{2}-q)\,\Gamma(\frac{1}{2}s+q)}f^{s+1}\int_0^\infty t^{-q-\frac{1}{2}}\{m^2(t+f^2)-\kappa^2\}^{-q+\frac{1}{2}}(t+f^2)^{-\frac{1}{2}s+q-1}\,dt - \text{term in } \epsilon\,;$$

and hence, writing $mf = f + \delta f$ or $m = 1 + \frac{\delta f}{f}$, and therefore $m^2 = 1 + 2\frac{\delta f}{f}$, the value is

$$\frac{\Gamma\frac{1}{2}s\,\Gamma\frac{1}{2}}{\Gamma(\frac{3}{2}-q)\,\Gamma(\frac{1}{2}s+q)}f^{s+1}\int_0^\infty t^{-q-\frac{1}{2}}\left\{t+f^2-\kappa^2+\frac{2\delta f}{f}(t+f^2)\right\}^{-q+\frac{1}{2}}.(t+f^2)^{-\frac{1}{2}s+q-1}\,dt - \text{term in } \epsilon.$$

Hence the term in δf is

$$= 2(-q+\tfrac{1}{2})\frac{\delta f}{f}\cdot\frac{\Gamma\frac{1}{2}s\,\Gamma\frac{1}{2}}{\Gamma(\frac{3}{2}-q)\,\Gamma(\frac{1}{2}s+q)}f^{s+1}\int_0^\infty t^{-q-\frac{1}{2}}(t+f^2-\kappa^2)^{-q-\frac{1}{2}}(t+f^2)^{-\frac{1}{2}s+q}\,dt,$$

$$= \delta f \text{ into expression } \frac{2\Gamma\frac{1}{2}s\,.\,\Gamma\frac{1}{2}}{\Gamma(\frac{1}{2}-q)\,\Gamma(\frac{1}{2}s+q)}f^{s}\int_0^\infty t^{-q-\frac{1}{2}}(t+f^2-\kappa^2)^{-q-\frac{1}{2}}(t+f^2)^{-\frac{1}{2}s+q}\,dt,$$

where the factor which multiplies δf is, as it should be, the ring-integral; it in fact agrees with one of the expressions previously obtained for this integral.

88. Similarly for an exterior point $\kappa > f$; starting in like manner from, Disk-integral

$$= \frac{2^{s-2}f^{s-1}}{(\frac{1}{2}-q)(\kappa+f)^{s+2q-2}}\left\{-\frac{\kappa-f}{\kappa+f}\Pi(\tfrac{1}{2}s,\ \tfrac{1}{2}s,\ \tfrac{1}{2}s+q,\ u)+\Pi(\tfrac{1}{2}s,\ \tfrac{1}{2}s,\ \tfrac{1}{2}s+q-1,\ u)\right\},$$

and reducing in like manner, the term in { } may be expressed in the four forms

$$2^{1-s}\frac{\Gamma\frac{1}{2}s\,\Gamma\frac{1}{2}}{\Gamma(\frac{1}{2}s+q)\,\Gamma(\frac{1}{2}-q)}\frac{(\kappa+f)^{s+2q-2}}{\kappa^{s+2q-2}} \text{ multiplied by}$$

$$\left[-\left(1-\frac{f^2}{\kappa^2}\right)\Pi\left(\tfrac{1}{2}s+q,\ \tfrac{1}{2}-q,\ \tfrac{1}{2}+q,\ \frac{f^2}{\kappa^2}\right)+\frac{\frac{1}{2}s+q-1}{\frac{1}{2}-q}\Pi\left(\tfrac{1}{2}s+q-1,\ \tfrac{3}{2}-q,\ -\tfrac{1}{2}+q,\ \frac{f^2}{\kappa^2}\right)\right],$$

$$2^{1-s}\frac{\Gamma\tfrac{1}{2}s\,\Gamma\tfrac{1}{2}}{\Gamma(\tfrac{1}{2}+q)\,\Gamma(\tfrac{1}{2}s-q)}\,\frac{(\kappa+f)^{s+2q-2}}{\kappa^{s+2q-2}}\text{ multiplied by}$$

$$\left[-\left(1-\frac{f^2}{\kappa^2}\right)\Pi\left(\tfrac{1}{2}+q,\ \tfrac{1}{2}s-q,\ \tfrac{1}{2}s+q,\ \frac{f^2}{\kappa^2}\right)+\frac{-\tfrac{1}{2}+q}{\tfrac{1}{2}s-q}\ \Pi\left(-\tfrac{1}{2}+q,\ \tfrac{1}{2}s-q+1,\ \tfrac{1}{2}s+q-1,\ \frac{f^2}{\kappa^2}\right)\right],$$

$$2^{1-s}\frac{\Gamma\tfrac{1}{2}s\,\Gamma\tfrac{1}{2}}{\Gamma(\tfrac{1}{2}-q)\,\Gamma(\tfrac{1}{2}s+q)}\left(\frac{\kappa+f}{\kappa}\right)^{s-1}\left(\frac{\kappa-f}{\kappa}\right)^{-2q+1}\text{ multiplied by}$$

$$\left[-\Pi\left(\tfrac{1}{2}-q,\ \tfrac{1}{2}s+q,\ \tfrac{1}{2}s-q,\ \frac{f^2}{\kappa^2}\right)+\left(1-\frac{f^2}{\kappa^2}\right)\frac{\tfrac{1}{2}s+q-1}{\tfrac{1}{2}-q}\ \Pi\left(\tfrac{1}{2}-q,\ \tfrac{1}{2}s+q,\ \tfrac{1}{2}s-q,\ \frac{f^2}{\kappa^2}\right)\right],$$

$$2^{1-s}\frac{\Gamma\tfrac{1}{2}s\,\Gamma\tfrac{1}{2}}{\Gamma(\tfrac{1}{2}s-q)\,\Gamma(\tfrac{1}{2}+q)}\left(\frac{\kappa+f}{\kappa}\right)^{s-1}\left(\frac{\kappa-f}{\kappa}\right)^{-2q+1}\text{ multiplied by}$$

$$\left[-\Pi\left(\tfrac{1}{2}s-q,\ \tfrac{1}{2}+q,\ \tfrac{1}{2}-q,\ \frac{f^2}{\kappa^2}\right)+\left(1-\frac{f^2}{\kappa^2}\right)\frac{-\tfrac{1}{2}+q}{\tfrac{1}{2}s-q}\ \Pi\left(\tfrac{1}{2}s-q+1,\ -\tfrac{1}{2}+q,\ \tfrac{3}{2}-q,\ \frac{f^2}{\kappa^2}\right)\right].$$

89. For the reduction of the first and the fourth of these, we have to consider

$$-(1-v)\,\Pi(\alpha,\ \beta,\ 1-\beta,\ v)+\frac{\alpha-1}{\beta}\,\Pi(\alpha-1,\ \beta+1,\ -\beta,\ v);$$

viz. this is

$$(-1+v+1+v)\int_0^1 x^{\alpha-1}(1-x\,.\,1-vx)^{\beta-1}\,dx-2\int_0^1 vx\,.\,x^{\alpha-1}(1-x\,.\,1-vx)^{\beta-1}\,dx,$$

$$=2v\int_0^1 x^{\alpha-1}(1-x)(1-x\,.\,1-vx)^{\beta-1}\,dx,$$

$$=2v\,\Pi(\alpha,\ \beta+1,\ -\beta+1,\ v);$$

that is,

$$-(1-v)\,\Pi(\alpha,\ \beta,\ 1-\beta,\ v)+\frac{\alpha-1}{\beta}\,\Pi(\alpha-1,\ \beta+1,\ -\beta,\ v)=2v\Pi(\alpha,\ \beta+1,\ -\beta+1,\ v).$$

I repeat, for comparison, the foregoing equation

$$+(1-v)\,\Pi(\alpha,\ \beta,\ 1-\beta,\ v)+\frac{\alpha-1}{\beta}\,\Pi(\alpha-1,\ \beta+1,\ -\beta,\ v)=2\Pi(\alpha,\ \beta,\ -\beta,\ v);$$

by adding and subtracting these we obtain two new formulæ; for reduction of the fourth formula, the equation may be written

$$-\Pi(\alpha-1,\ \beta+1,\ -\beta,\ v)+(1-v)\,\frac{\beta}{\alpha-1}\,\Pi(\alpha,\ \beta,\ 1-\beta,\ v)=-2\,\frac{\beta}{\alpha-1}\,v\Pi(\alpha,\ \beta+1-\beta+1,\ v).$$

90. But it is sufficient to consider the first formula; the term in [ ] is

$$=\frac{2^{2-s}\,\Gamma\tfrac{1}{2}s\,\Gamma\tfrac{1}{2}}{\Gamma(\tfrac{1}{2}s+q)\,\Gamma(\tfrac{1}{2}-q)}\quad\left(\frac{\kappa+f}{\kappa}\right)^{s+2q-2}\frac{f^2}{\kappa^2}\ \Pi\left(\tfrac{1}{2}s+q,\ \tfrac{3}{2}-q,\ \tfrac{1}{2}+q,\ \frac{f^2}{\kappa^2}\right),$$

and the corresponding value of the disk-integral is

$$=\frac{\Gamma\tfrac{1}{2}s\,\Gamma\tfrac{1}{2}}{\Gamma(\tfrac{1}{2}s+q)\,\Gamma(\tfrac{3}{2}-q)}\quad\frac{f^{s+1}}{\kappa^{s+2q}}\qquad\Pi\left(\tfrac{1}{2}s+q,\ \tfrac{3}{2}-q,\ \tfrac{1}{2}+q,\ \frac{f^2}{\kappa^2}\right),$$

which we may again verify by taking therein κ indefinitely large; viz. the value is then $=\dfrac{\Gamma\tfrac{1}{2}s\,\Gamma\tfrac{1}{2}}{\Gamma(\tfrac{1}{2}s+\tfrac{3}{2})}\dfrac{f^{s+1}}{\kappa^{s+2q}}$, as above. It is the first of a system of four forms, the others of which are

$$=\frac{\Gamma\tfrac{1}{2}s\,\Gamma\tfrac{1}{2}}{\Gamma(\tfrac{1}{2}+q)\,\Gamma(\tfrac{1}{2}s-q+1)}\frac{f^{s+1}}{\kappa^{s+2q}}\qquad\Pi\left(\tfrac{1}{2}+q,\ \tfrac{1}{2}s-q+1,\ \tfrac{1}{2}s+q,\ \frac{f^2}{\kappa^2}\right),$$

$$=\frac{\Gamma\tfrac{1}{2}s\,\Gamma\tfrac{1}{2}}{\Gamma(\tfrac{1}{2}s+q)\,\Gamma(\tfrac{3}{2}-q)}\quad\frac{f^{s+1}}{\kappa^{s+2q}}\left(1-\frac{f^2}{\kappa^2}\right)^{1-2q}\Pi\left(\tfrac{3}{2}-q,\ \tfrac{1}{2}s+q,\ \tfrac{1}{2}s-q+1,\ \frac{f^2}{\kappa^2}\right),$$

$$=\frac{\Gamma\tfrac{1}{2}s\,\Gamma\tfrac{1}{2}}{\Gamma(\tfrac{1}{2}s-q+1)\,\Gamma(\tfrac{1}{2}+q)}\frac{f^{s+1}}{\kappa^{s+2q}}\left(1-\frac{f^2}{\kappa^2}\right)^{1-2q}\Pi\left(\tfrac{1}{2}s-q+1,\ \tfrac{1}{2}+q,\ \tfrac{3}{2}-q,\ \frac{f^2}{\kappa^2}\right).$$

And hence, writing as before $x=\dfrac{t+f^2-\kappa^2}{t}$, &c., the four values are

$$=\frac{\Gamma\tfrac{1}{2}s\,\Gamma\tfrac{1}{2}}{\Gamma(\tfrac{1}{2}s+q)\,\Gamma(\tfrac{3}{2}-q)}\quad\frac{f^{s+1}}{\kappa^{s-1}}(\kappa^2-f^2)^{1-2q}\int_{\kappa^2-f^2}^{\infty}t^{-\frac{1}{2}s+q-1}(t+f^2-\kappa^2)^{\frac{1}{2}s+q-1}(t+f^2)^{-\frac{1}{2}-q}\,dt,$$

$$=\frac{\Gamma\tfrac{1}{2}s\,\Gamma\tfrac{1}{2}}{\Gamma(\tfrac{1}{2}+q)\,\Gamma(\tfrac{1}{2}s-q+1)}f^{s+1}(\kappa^2-f^2)^{1-2q}\int_{\kappa^2-f^2}^{\infty}t^{q-\frac{3}{2}}\quad(t+f^2-\kappa^2)^{q-\frac{1}{2}}\quad(t+f^2)^{-\frac{1}{2}s-q}\,dt,$$

$$=\frac{\Gamma\tfrac{1}{2}s\,\Gamma\tfrac{1}{2}}{\Gamma(\tfrac{1}{2}s+q)\,\Gamma(\tfrac{3}{2}-q)}\quad f^{s+1}\qquad\int_{\kappa^2-f^2}^{\infty}t^{-q-\frac{1}{2}}\quad(t+f^2-\kappa^2)^{-q+\frac{1}{2}}\quad(t+f^2)^{-\frac{1}{2}s+q-1}\,dt,$$

$$=\frac{\Gamma\tfrac{1}{2}s\,\Gamma\tfrac{1}{2}}{\Gamma(\tfrac{1}{2}s-q+1)\,\Gamma(\tfrac{1}{2}+q)}\frac{f^{s+1}}{\kappa^{s-1}}\qquad\int_{\kappa^2-f^2}^{\infty}t^{-\frac{1}{2}s-q}\quad(t+f^2-\kappa^2)^{\frac{1}{2}s-q}\quad(t+f^2)^{q-\frac{3}{2}}\quad dt,$$

where we may in the integrals write $t+\kappa^2-f^2$ in place of t, making the limits ∞, 0; but the actual form is preferable.

91. In the third form, for f write mf, at the same time changing t into mt; the new value of the disk-integral is

$$=\frac{\Gamma\tfrac{1}{2}s\,\Gamma\tfrac{1}{2}}{\Gamma(\tfrac{1}{2}s+q)\,\Gamma(\tfrac{3}{2}-q)}f^{s+1}\int_{\frac{\kappa^2}{m^2}-f^2}^{\infty}t^{-q-\frac{1}{2}}\{m^2(t+f^2)-\kappa^2\}^{-q+\frac{1}{2}}(t+f^2)^{-\frac{1}{2}s+q-1}\,dt.$$

Writing here $mf=f+\delta f$, that is, $m=1+\dfrac{\delta f}{f}$, $m^2=1+\dfrac{2\delta f}{f}$, and observing that, if $-q+\tfrac{1}{2}$ be positive, the factor $\{m^2(t+f^2)-\kappa^2\}^{-q+\frac{1}{2}}$ vanishes for the value $t=\dfrac{\kappa^2}{m^2}-f^2$ at the lower limit, we see that on this supposition, $-q+\tfrac{1}{2}$ positive, the value is

$$=\frac{\Gamma\tfrac{1}{2}s\,\Gamma\tfrac{1}{2}}{\Gamma(\tfrac{1}{2}s+q)\,\Gamma(\tfrac{3}{2}-q)}f^{s+1}\int_{\kappa^2-f^2}^{\infty}t^{-q-\frac{1}{2}}\left\{t+f^2-\kappa^2+\frac{2\delta f}{f}(t+f^2)\right\}^{-q+\frac{1}{2}}(t+f^2)^{-\frac{1}{2}s+q-1}\,dt;$$

viz. the term in δf is $=\delta f$ multiplied by the expression

$$2(\tfrac{1}{2}-q)\frac{\Gamma\tfrac{1}{2}s\,\Gamma\tfrac{1}{2}}{\Gamma(\tfrac{1}{2}s+q)\,\Gamma(\tfrac{3}{2}-q)}f^{s}\int_{\kappa^2-f^2}^{\infty}t^{-q-\frac{1}{2}}(t+f^2-\kappa^2)^{-q-\frac{1}{2}}(t+f^2)\,(t+f^2)^{-\frac{1}{2}s+q-1}\,dt,$$

that is, multiplied by

$$2\frac{\Gamma\tfrac{1}{2}s\,\Gamma\tfrac{1}{2}}{\Gamma(\tfrac{1}{2}s+q)\,\Gamma(\tfrac{1}{2}-q)}f^{s}\int_{\kappa^2-f^2}^{\infty}t^{-q-\frac{1}{2}}(t+f^2-\kappa^2)^{-q-\frac{1}{2}}(t+f^2)^{-\frac{1}{2}s+q}\,dt,$$

which is in fact $=\delta f$ multiplied by the value of the ring-integral.

92. Comparing for the cases of an interior point $\kappa<f$ and an exterior point $\kappa>f$, the four expressions for the disk-integral, it will be noticed that only the third expressions correspond precisely to each other; viz. these are: interior point, $\kappa<f$; the value is

$$\frac{\Gamma\tfrac{1}{2}s\,\Gamma\tfrac{1}{2}}{\Gamma(\tfrac{1}{2}s+q)\,\Gamma(\tfrac{3}{2}-q)}f^{s+1}\int_{0}^{\infty}t^{-q-\frac{1}{2}}(t+f^2-\kappa^2)^{-q+\frac{1}{2}}(t+f^2)^{-\frac{1}{2}s+q-1}\,dt-\frac{\epsilon^{1-2q}}{\tfrac{1}{2}-q}\,\frac{\Gamma\tfrac{1}{2}s\,\Gamma\tfrac{1}{2}s}{\Gamma(\tfrac{1}{2}s+q)},$$

where, if $\tfrac{1}{2}-q$ be positive (which is, in fact, a necessary condition in order to the applicability of the formula), the term in ϵ vanishes, and may therefore be omitted: and exterior point, $\kappa>f$; the value is

$$=\frac{\Gamma\tfrac{1}{2}s\,\Gamma\tfrac{1}{2}}{\Gamma(\tfrac{1}{2}s+q)\,\Gamma(\tfrac{3}{2}-q)}f^{s+1}\int_{\kappa^2-f^2}^{\infty}t^{-q-\frac{1}{2}}(t+f^2-\kappa^2)^{-q+\frac{1}{2}}(t+f^2)^{-\frac{1}{2}s+q-1}\,dt,$$

differing only from the preceding one in the inferior limit κ^2-f^2 in place of 0 of the integral. We have $\tfrac{1}{2}-q$ positive, and also $\tfrac{1}{2}s+q$ positive; viz. q may have any value diminishing from $\tfrac{1}{2}$ to $-\tfrac{1}{2}s$, the extreme values *not* admissible.

ANNEX IV. *Examples of Theorem A.* Art. Nos. 93 to 112.

93. It is remarked in the text that, in the examples which relate to the s-coordinal sphere and ellipsoid respectively, we have a quantity θ, a function of the coordinates $(a,\ldots,c,e)$ of the attracted point; viz. in the case of the sphere, writing $a^2+\ldots+c^2=\kappa^2$, we have

$$\frac{\kappa^2}{f^2+\theta}+\frac{e^2}{\theta}=1:$$

in the case of the ellipsoid, we have

$$\frac{a^2}{f^2+\theta}+\ldots+\frac{c^2}{h^2+\theta}+\frac{e^2}{\theta}=1,$$

the equations having in each case a positive root which is called θ. The properties of the equations are the same in each case; but for the sphere, the equation being a quadric one, can be solved. The equation in fact is

$$\theta^2-\theta(e^2+\kappa^2-f^2)-e^2f^2=0,$$

and the positive root is therefore

$$\theta=\tfrac{1}{2}\{e^2+\kappa^2-f^2+\sqrt{(e^2+\kappa^2-f^2)^2+4e^2f^2}\}.$$

Suppose e to diminish gradually and become $=0$; for an exterior point, $\kappa>f$, the value of the radical is $=\kappa^2-f^2$, and we have $\theta=\kappa^2-f^2$; for an interior point, $\kappa<f$,

the value of the radical, supposing e only indefinitely small, is $=f^2-\kappa^2+\frac{f^2+\kappa^2}{f^2-\kappa^2}e^2$, and we have $\theta=\frac{1}{2}e^2\left(1+\frac{f^2+\kappa^2}{f^2-\kappa^2}\right), =\frac{e^2f^2}{f^2-\kappa^2}$, or, what is the same thing, $\frac{e^2}{\theta}=\left(1-\frac{\kappa^2}{f^2}\right)$; viz. the positive root of the equation continually diminishes with e, and becomes ultimately $=0$.

If κ or e be indefinitely large, then the radical may be taken $=e^2+\kappa^2$, and we have θ indefinitely large, $=e^2+\kappa^2$.

94. The result is similar for the general equation

$$\frac{a^2}{f^2+\theta}+\ldots+\frac{c^2}{h^2+\theta}+\frac{e^2}{\theta}=1;$$

the left-hand side is $=0$ for $\theta=\infty$, and (as θ decreases) continually increases, becoming infinite for $\theta=0$; there is consequently a single positive value of θ for which the value is $=1$; viz. the equation has a single positive root, and θ is taken to denote this root.

In the last-mentioned equation, let e gradually diminish and become $=0$; then for an exterior point, viz. if

$$\frac{a^2}{f^2}+\ldots+\frac{c^2}{h^2}>1, \text{ the equation } \frac{a^2}{f^2+\theta}+\ldots+\frac{c^2}{h^2+\theta}=1$$

has (as is at once seen) a single positive root, and θ becomes equal to the positive root of this equation; but for an interior point, or $\frac{a^2}{f^2}+\ldots+\frac{c^2}{h^2}<1$, the equation just written down has no positive root, and θ becomes $=0$, that is, the positive root of the original equation continually diminishes with e, and for $e=0$ becomes ultimately $=0$; its value for e small is, in fact, given by $\frac{e^2}{\theta}=\left(1-\frac{a^2}{f^2}-\ldots-\frac{c^2}{h^2}\right)$. Also $a,\ldots, c, e$ (or any of them) indefinitely large, θ is indefinitely large, $=a^2+\ldots+c^2+e^2$.

95. We have an interesting geometrical illustration in the case $s+1=2$; θ is here determined by the equation

$$\frac{a^2}{f^2+\theta}+\frac{b^2}{g^2+\theta}+\frac{e^2}{\theta}=1;$$

viz. θ is the squared z-semiaxis of the ellipsoid, confocal with the conic $\frac{x^2}{f^2}+\frac{y^2}{g^2}=1$, which passes through the point (a, b, e). Taking $e=0$, the point in question, if $\frac{a^2}{f^2}+\frac{b^2}{g^2}>1$, is a point in the plane of xy, outside the ellipse, and we have through the point a proper confocal ellipsoid, whose squared z-semiaxis does not vanish; but if $\frac{a^2}{f^2}+\frac{b^2}{g^2}<1$, then the point is within the ellipse, and the only confocal ellipsoid through the point is the indefinitely thin ellipsoid, squared semiaxes $(f^2, g^2, 0)$, which in fact coincides with the ellipse.

96. The positive root θ of the equation

$$J, = 1 - \frac{a^2}{f^2+\theta} - \dots - \frac{c^2}{h^2+\theta} - \frac{e^2}{\theta}, = 0$$

has certain properties which connect themselves with the function

$$\Theta, = \theta^{-q-1}\{(\theta+f^2)\dots(\theta+h^2)\}^{-\frac{1}{2}}.$$

We have, the accents denoting differentiations in regard to θ,

$$J'\frac{d\theta}{da} - \frac{2a}{\theta+f^2} = 0, \quad \text{or} \quad \frac{d\theta}{da} = \frac{1}{J'}\frac{2a}{\theta+f^2},$$

where

$$J' = \frac{a^2}{(f^2+\theta)^2} + \dots + \frac{c^2}{(h^2+\theta)^2} + \frac{e^2}{\theta^2};$$

and we have the like formulæ for .., $\dfrac{d\theta}{dc}, \dfrac{d\theta}{de}$.

We deduce

$$\frac{a}{\theta+f^2}\frac{d\theta}{da} + \dots + \frac{c}{\theta+h^2}\frac{d\theta}{dc} + \frac{e}{\theta}\frac{d\theta}{de} = \frac{2}{J'}\left\{\frac{a^2}{(\theta+f^2)^2} + \dots + \frac{c^2}{(\theta+h^2)^2} + \frac{e^2}{\theta}\right\}, = 2;$$

and to this we may join, η being arbitrary,

$$\frac{a}{\theta+\eta+f^2}\frac{d\theta}{da} + \dots + \frac{c}{\theta+\eta+h^2}\frac{d\theta}{dc} + \frac{e}{\theta+\eta}\frac{d\theta}{de} = \frac{2}{J'}\left\{\frac{a^2}{\theta+f^2\,.\,\theta+\eta+f^2} + \dots \right.$$
$$\left. + \frac{c^2}{\theta+h^2\,.\,\theta+\eta+h^2} + \frac{e^2}{\theta\,.\,\theta+\eta}\right\}.$$

Again, defining $\nabla_1\theta$ and $\square\theta$ as immediately appears, we have

$$\nabla_1\theta, = \left(\frac{d\theta}{da}\right)^2 + \dots + \left(\frac{d\theta}{dc}\right)^2, \quad = \frac{1}{J'^2}\,.\,4J', \quad = \frac{4}{J'};$$

and passing to the second differential coefficients, we have

$$\frac{d^2\theta}{da^2} = \frac{2}{J'(\theta+f^2)} - \frac{8a^2}{J'^2(\theta+f^2)^3} - \frac{4a^2J''}{J'^3(\theta+f^2)^2},$$

where

$$J'' = -2\left\{\frac{a^2}{(\theta+f^2)^3} + \dots + \frac{c^2}{(\theta+h^2)^3} + \frac{e^2}{\theta^3}\right\},$$

and the like formulæ for .., $\dfrac{d^2\theta}{dc^2}, \dfrac{d^2\theta}{de^2}$. Joining to these $\dfrac{2q+1}{e}\dfrac{d\theta}{de} = \dfrac{4q+2}{J'\theta}$, we obtain

$$\square\theta, = \left(\frac{d^2\theta}{da^2} + \dots + \frac{d^2\theta}{dc^2} + \frac{d^2\theta}{de^2} + \frac{2q+1}{e}\frac{d\theta}{de}\right),$$
$$= \frac{2}{J'}\left\{\frac{1}{\theta+f^2} + \dots + \frac{1}{\theta+h^2} + \frac{1+(2q+1)}{\theta}\right\}$$
$$- \frac{8}{J'^2}\left(-\tfrac{1}{2}J''\right) - \frac{4J''}{J'^3}(J'),$$

where the last two terms destroy each other; observing that we have

$$\frac{\Theta'}{\Theta} = -\tfrac{1}{2}\left(\frac{1}{\theta+f^2} + \dots + \frac{1}{\theta+h^2} + \frac{2q+2}{\theta}\right),$$

the result is

$$\Box\theta = \frac{2}{J'}\left(-\frac{2\Theta'}{\Theta}\right), \quad = -\frac{4\Theta'}{J'\Theta}.$$

97. First example. $\kappa^2 = a^2 + \dots + c^2$, and θ the positive root of $\dfrac{\kappa^2}{f^2+\theta} + \dfrac{e^2}{\theta} = 1$.

V is assumed $= \displaystyle\int_\theta^\infty t^{-q-1}(t+f^2)^{-\frac{1}{2}s}\,dt$, where $q+1$ is positive.

I do not work the example out; it corresponds step by step with, and is hardly more simple than, the next example, which relates to the ellipsoid. The result is

$$\rho = 0, \text{ if } x^2 + \dots + z^2 > f^2,$$

$$\rho = \frac{\Gamma(\tfrac{1}{2}s+q)}{(\Gamma\tfrac{1}{2})^s\,\Gamma(q+1)} f^{-s}\left(1 - \frac{x^2+\dots+z^2}{f^2}\right)^q, \text{ if } x^2+\dots+z^2 < f^2;$$

hence the integral

$$\int \frac{\left(1 - \dfrac{x^2+\dots+z^2}{f^2}\right)^q dx \dots dz}{\{(a-x)^2 + \dots + (c-z)^2 + e^2\}^{\frac{1}{2}s+q}},$$

taken over the sphere $x^2 + \dots + z^2 = f^2$,

$$= \frac{(\Gamma\tfrac{1}{2})^s\,\Gamma(q+1)}{\Gamma(\tfrac{1}{2}s+q)} \int_\theta^\infty t^{-q-1}(t+f^2)^{-\frac{1}{2}s}\,dt.$$

98. Second example. θ the positive root of $\dfrac{a^2}{f^2+\theta} + \dots + \dfrac{c^2}{h^2+\theta} + \dfrac{e^2}{\theta} = 1$; $q+1$ positive.

Consider here the function

$$V = \int_\theta^\infty t^{-q-1}\{(t+f^2)\dots(t+h^2)\}^{-\frac{1}{2}}\,dt;$$

this satisfies the prepotential equation. We have in fact

$$\frac{dV}{da} = -\Theta\frac{d\theta}{da};\quad \frac{d^2V}{da^2} = -\Theta\frac{d^2\theta}{da^2} - \Theta'\left(\frac{d\theta}{da}\right)^2,$$

with the like expressions for $\dots, \dfrac{d^2V}{dc^2}, \dfrac{d^2V}{de^2}$; also

$$\frac{2q+1}{e}\frac{dV}{de} = -\Theta\frac{2q+1}{e}\frac{d\theta}{de}.$$

Hence

$$\Box V = -\Theta\Box\theta - \Theta'\nabla_1\theta,$$

or, substituting for $\square\theta$ and $\nabla_1\theta$ their values, this is

$$=-\Theta\left(-\frac{4\Theta'}{J\Theta}\right)-\Theta'.4J,\ =0.$$

Moreover V does not become infinite for any values of $(a,..,c,e)$, e not $=0$; and it vanishes for points at ∞. And not only so, but for indefinitely large values of any of the coordinates $(a,..,c,e)$ it reduces itself to a numerical multiple of $(a^2+\ldots+c^2+e^2)^{-\frac{1}{2}s+q}$; in fact, in this case θ is indefinitely large, $=a^2+\ldots+c^2+e^2$. Consequently throughout the integral, t is indefinitely large, and we may therefore write

$$V=\int_\theta^\infty t^{-q-1}.t^{-\frac{1}{2}s}\,dt,\ =-\frac{1}{\frac{1}{2}s+q}(t^{-\frac{1}{2}s-q})_\theta^\infty,\ =\frac{1}{\frac{1}{2}s+q}\theta^{-\frac{1}{2}s-q},$$

that is,

$$V=\frac{1}{\frac{1}{2}s+q}(a^2+\ldots+c^2+e^2)^{-\frac{1}{2}s-q}.$$

The conditions of the theorem are thus satisfied, and we have for ρ either of the formulæ

$$\rho=\frac{\Gamma(\frac{1}{2}s+q)}{(\Gamma\frac{1}{2})^s\,\Gamma q}(e^{2q}W)_0,\quad \rho=\frac{-\Gamma(\frac{1}{2}s+q)}{2(\Gamma\frac{1}{2})^s\,\Gamma(q+1)}\left(e^{2q+1}\frac{dW}{de}\right)_0:$$

in the former of them q must be positive; in the latter it is sufficient if $q+1$ be positive.

99. We have W the same function of $(x,..,z,e)$ that V is of $(a,..,c,e)$; viz. writing λ for the positive root of

$$\frac{x^2}{f^2+\lambda}+\ldots+\frac{z^2}{h^2+\lambda}+\frac{e^2}{\lambda}=1,$$

the value of W is

$$=\int_\lambda^\infty t^{-q-1}\{(t+f^2)\ldots(t+h^2)\}^{-\frac{1}{2}}\,dt.$$

Considering the formula which involves $e^{2q}W$,—first, if $\frac{x^2}{f^2}+\ldots+\frac{z^2}{h^2}>1$, then, when e is $=0$ the value of λ is not $=0$; the integral W is therefore finite (not indefinitely large), and we have $e^{2q}W=0$, consequently $\rho=0$.

But if $\frac{x^2}{f^2}+\ldots+\frac{z^2}{h^2}<1$, then, when e is indefinitely small λ is also indefinitely small; viz. we then have $\frac{e^2}{\lambda}=1-\frac{x^2}{f^2}-\ldots-\frac{z^2}{h^2}$; the value of W is

$$W=(f\ldots h)^{-1}\int_\lambda^\infty t^{-q-1}\,dt,\ =(f\ldots h)^{-1}\frac{1}{q}\lambda^{-q},$$

and hence

$$\rho=\frac{\Gamma(\frac{1}{2}s+q)}{(\Gamma\frac{1}{2})^s\,\Gamma q}\,\frac{1}{q}\left(\frac{e^2}{\lambda}\right)^q(f\ldots h)^{-1},\ =\frac{\Gamma(\frac{1}{2}s+q)}{(\Gamma\frac{1}{2})^s\,\Gamma(q+1)}(f\ldots h)^{-1}\left(1-\frac{x^2}{f^2}-\ldots-\frac{z^2}{h^2}\right)^q.$$

100. Again, using the formula which involves $\left(e^{2q+1}\frac{dW}{de}\right)$; we have here $\frac{dV}{de}=-\Theta\frac{d\theta}{de}$, or substituting for Θ and $\frac{d\theta}{de}$ their values and multiplying by e^{2q+1}, we find

$$e^{2q+1}\frac{dV}{de}=2e^{2q+2}\,\theta^{-1}\,J'^{-1}\,\Theta,$$

$$=2e^{2q+2}\,\theta^{-q-2}\left[\frac{a^2}{(f^2+\theta)^2}+\ldots+\frac{c^2}{(h^2+\theta)^2}+\frac{e^2}{\theta^2}\right]^{-1}\{(\theta+f^2)\ldots(\theta+h^2)\}^{-\frac{1}{2}},$$

and therefore

$$e^{2q+1}\frac{dW}{de}=2e^{2q+2}\lambda^{-q-2}\left[\frac{x^2}{(f^2+\lambda)^2}+\ldots+\frac{z^2}{(h^2+\lambda)^2}+\frac{e^2}{\lambda^2}\right]^{-1}\{(\lambda+f^2)\ldots(\lambda+h^2)\}^{-\frac{1}{2}}.$$

Hence, writing $e=0$: first, for an exterior point or $\frac{x^2}{f^2}+\ldots+\frac{z^2}{h^2}>1$, λ is not $=0$, and the expression vanishes in virtue of the factor e^{2q+2}, whence also $\rho=0$; next, for an interior point or $\frac{x^2}{f^2}+\ldots+\frac{z^2}{h^2}<1$, λ is $=0$, hence also $\frac{e^2}{\lambda^2}=\frac{1}{\lambda}\left(1-\frac{x^2}{f^2}-\ldots-\frac{z^2}{h^2}\right)$ is infinite; neglecting in comparison with it the other terms $\frac{x^2}{(f^2+\lambda)^2}+\ldots$, the value is

$$2\left(\frac{e^2}{\lambda}\right)^q(f\ldots h)^{-1},\ =2\left(1-\frac{x^2}{f^2}-\ldots-\frac{z^2}{h^2}\right)^q(f\ldots h)^{-1},$$

and we have, as before,

$$\rho=\frac{\Gamma(\frac{1}{2}s+q)}{(\Gamma\frac{1}{2})^s\,\Gamma(q+1)}(f\ldots h)^{-1}\left(1-\frac{x^2}{f^2}-\ldots-\frac{z^2}{h^2}\right)^q.$$

101. Hence in the formula

$$V=\int\frac{\rho\,dx\ldots dz}{\{(a-x)^2+\ldots+(c-z)^2+e^2\}^{\frac{1}{2}s+q}}$$

$$=\int_\theta^\infty t^{-q-1}\{(t+f^2)\ldots(t+h^2)\}^{-\frac{1}{2}}\,dt,$$

ρ has the value just found, or, what is the same thing, we have

$$\int\frac{\left(1-\frac{x^2}{f^2}-\ldots-\frac{z^2}{h^2}\right)^q dx\ldots dz}{\{(a-x)^2+\ldots+(c-z)^2+e^2\}^{\frac{1}{2}s+q}},$$

taken over ellipsoid $\frac{x^2}{f^2}+\ldots+\frac{z^2}{h^2}=1$,

$$=\frac{(\Gamma\frac{1}{2})^s\,\Gamma(q+1)}{\Gamma(\frac{1}{2}s+q)}(f\ldots h)\int_\theta^\infty t^{-q-1}\{(t+f^2)\ldots(t+h^2)\}^{-\frac{1}{2}}\,dt.$$

102. We may in this result write $e=0$. There are two cases, according as the attracted point is exterior or interior: if it is exterior, $\frac{a^2}{f^2}+\ldots+\frac{c^2}{h^2}>1$, θ will denote the positive root of the equation $\frac{a^2}{f^2+\theta}+\ldots+\frac{c^2}{h^2+\theta}=1$; if it be interior, $\frac{a^2}{f^2}+\ldots+\frac{c^2}{h^2}<1$, θ will be $=0$; and we thus have

$$\int\frac{\left(1-\frac{x^2}{f^2}-\ldots-\frac{z^2}{h^2}\right)^q dx\ldots dz}{\{(a-x)^2+\ldots+(c-z)^2\}^{\frac{1}{2}s+q}}$$

$$=\frac{(\Gamma\tfrac{1}{2})^s\,\Gamma(q+1)}{\Gamma(\tfrac{1}{2}s+q)}(f\ldots h)\int_\theta^\infty t^{-q-1}\{(t+f^2)\ldots(t+h^2)\}^{-\frac{1}{2}}dt, \text{ for exterior point } \frac{a^2}{f^2}+\ldots+\frac{c^2}{h^2}>1:$$

$$=\frac{(\Gamma\tfrac{1}{2})^s\,\Gamma(q+1)}{\Gamma(\tfrac{1}{2}s+q)}(f\ldots h)\int_0^\infty t^{-q-1}\{(t+f^2)\ldots(t+h^2)\}^{-\frac{1}{2}}dt, \text{ for interior point } \frac{a^2}{f^2}+\ldots+\frac{c^2}{h^2}<1;$$

but as regards the value for an interior point it is to be observed that, unless q be negative (between 0 and -1, since $1+q$ is positive by hypothesis), the two sides of the equation will be each of them infinite.

103. Third example. We assume here

$$V=\int_\theta^\infty dt\, I^m\, T,$$

where

$$I=1-\frac{a^2}{f^2+t}-\ldots-\frac{c^2}{h^2+t}-\frac{e^2}{t},$$

$$T=t^{-q-1}\{(t+f^2)\ldots(t+h^2)\}^{-\frac{1}{2}};$$

as before, θ is the positive root of the equation

$$J=1-\frac{a^2}{f^2+\theta}-\ldots-\frac{c^2}{h^2+\theta}-\frac{e^2}{\theta},\ =0,$$

and $\frac{1}{2}s+q$ is positive in order that the integral may be finite; also m is positive.

104. In order to show that V satisfies the prepotential equation $\square V=0$, I shall, in the first place, consider the more general expression,

$$V=\int_{\theta+\eta}^\infty dt\, I^m T,$$

where η is a constant positive quantity which will be ultimately put $=0$. The functions previously called J and Θ will be written J_0 and Θ_0, and J, Θ will now denote

$$J,\ =1-\frac{a^2}{\theta+\eta+f^2}-\ldots-\frac{c^2}{\theta+\eta+h^2}-\frac{e^2}{\theta+\eta},$$

$$\Theta,\ =(\theta+\eta)^{-q-1}\{(\theta+\eta+f^2)\ldots(\theta+\eta+h^2)\}^{-\frac{1}{2}};$$

whence also, subtracting from J the evanescent function J_0, we have

$$J = \eta\left(\frac{a^2}{\theta+f^2 . \theta+\eta+f^2} + \ldots + \frac{c^2}{\theta+h^2 . \theta+\eta+h^2} + \frac{e^2}{\theta . \theta+\eta}\right),$$

say this is

$$J = \eta P;$$

and we have thence, by former equations and in the present notation,

$$\frac{a}{\theta+\eta+f^2}\frac{d\theta}{da} + \ldots + \frac{c}{\theta+\eta+h^2}\frac{d\theta}{dc} + \frac{e}{\theta+\eta}\frac{d\theta}{de} = \frac{2}{J_0}P,$$

$$\nabla_1\theta = \frac{4}{J_0},$$

$$\square\theta = \frac{-4\Theta_0'}{J_0'\Theta_0}.$$

In virtue of the equation which determines θ, we have

$$\frac{dV}{da} = \int_{\theta+\eta}^{\infty} dt\, mI^{m-1}\frac{-2a}{t+f^2}T - J^m\Theta\frac{d\theta}{da};$$

and thence

$$\begin{aligned}\frac{d^2V}{da^2} = \int_{\theta+\eta}^{\infty} dt &\left\{mI^{m-1}\frac{-2}{t+f^2} + m(m-1)I^{m-2}\frac{4a^2}{(t+f^2)^2}\right\}T \\ &- mJ^{m-1}\left(-\frac{2a}{\theta+\eta+f^2}\right)\Theta\frac{d\theta}{da} \\ &- mJ^{m-1}\left(\frac{-2a}{\theta+\eta+f^2}\right)\Theta\frac{d\theta}{da} \\ &- \frac{d}{d\theta}(J^m\Theta)\left(\frac{d\theta}{da}\right)^2 \\ &- J^m\Theta\frac{d^2\theta}{da^2};\end{aligned}$$

with like expressions for .., $\frac{d^2V}{dc^2}$, $\frac{d^2V}{de^2}$. Also

$$\frac{2q+1}{e}\frac{dV}{de} = \int_{\theta+\eta}^{\infty} dt\, mI^{m-1}\frac{-4q-2}{t}T - \frac{2q+1}{e}J^m\Theta\frac{d\theta}{de};$$

and hence

$$\begin{aligned}\square V = \int_{\theta+\eta}^{\infty} dt &\left[-2m\,I^{m-1}\left\{\frac{1}{t+f^2} - \ldots - \frac{1}{t+h^2} + \frac{1+(2q+1)}{t}\right\}T\right. \\ &\left. + m(m-1)\,I^{m-2}.\,4\left\{\frac{a^2}{(t+f^2)^2} + \ldots + \frac{c^2}{(t+h^2)^2} + \frac{e^2}{t^2}\right\}T\right] \\ &+ 4m\,J^{m-1}\Theta\left(\frac{a}{\theta+\eta+f^2}\frac{d\theta}{da} + \ldots + \frac{c}{\theta+\eta+h^2}\frac{d\theta}{dc} + \frac{e}{\theta+\eta}\frac{d\theta}{de}\right) \\ &- \frac{d}{d\theta}(J^m\Theta)\left\{\left(\frac{d\theta}{da}\right)^2 + \ldots + \left(\frac{d\theta}{dc}\right)^2 + \left(\frac{d\theta}{de}\right)^2\right\} \\ &- J^m\Theta\left(\frac{d^2\theta}{da^2} + \ldots + \frac{d^2\theta}{dc^2} + \frac{d^2\theta}{de^2} + \frac{2q+1}{e}\frac{d\theta}{de}\right).\end{aligned}$$

105. Writing I', T' for the first derived coefficients of I, T in regard to t, we have

$$I' = \frac{a^2}{(t+f^2)^2} + \dots + \frac{c^2}{(t+h^2)^2} + \frac{e^2}{t^2}, \qquad \frac{T'}{T} = -\tfrac{1}{2}\left(\frac{1}{t+f^2} + \dots + \frac{1}{t+h^2} + \frac{2q+2}{t}\right).$$

The integral is therefore

$$\int_{\theta+\eta}^{\infty} dt \left\{2m\, I^{m-1} \frac{2T'}{T}\, T + m(m-1)\, I^{m-2} \,.\, 4I'T\right\},$$

$$= \int_{\theta+\eta}^{\infty} dt\, \{4m\, I^{m-1}\, T' + 4m(m-1)\, I^{m-2} I'T\},$$

$$= \int_{\theta+\eta}^{\infty} dt\; 4m \frac{d}{dt}(I^{m-1}\, T);$$

viz. $I^{m-1}\, T$ vanishing for $t = \infty$, this is

$$= -4m\, J^{m-1}\, \Theta.$$

Hence, writing $(J^m \Theta)'$ instead of $\dfrac{d}{d\theta}(J^m \Theta)$, we have

$$\begin{aligned}
\square V = &- 4m\, J^{m-1}\, \Theta \\
&+ 4m\, J^{m-1}\, \Theta \left(\frac{a}{\theta+\eta+f^2}\frac{d\theta}{da} + \dots + \frac{c}{\theta+\eta+f^2}\frac{d\theta}{dc} + \frac{e}{\theta+\eta}\frac{d\theta}{de}\right) \\
&- (J^m \Theta)'\, \nabla_1 \theta \\
&- J^m\, \Theta\, \square\theta\,;
\end{aligned}$$

viz. this is

$$\begin{aligned}
\square V = &- 4m\, J^{m-1}\, \Theta \\
&+ 8m\, J^{m-1}\, \Theta \,.\, \frac{P}{J_0'} \\
&- 4\,(J^m \Theta)' \frac{1}{J_0'} \\
&+ 4J^m\, \Theta\, \frac{\Theta_0'}{J_0'\,\Theta_0}\,;
\end{aligned}$$

or, writing $m\, J^{m-1}\, J'\Theta + J^m \Theta'$ instead of $(J^m\Theta)'$, this is

$$\square V = -\frac{4m\, J^{m-1}\Theta}{J_0'}(J' - 2P + J) - \frac{4J^m}{J_0'\Theta_0}(\Theta'\Theta_0 - \Theta\Theta_0').$$

We have here

$$J' - 2P + J = a^2\left\{\frac{1}{(\theta+\eta+f^2)^2} - \frac{2}{(\theta+\eta+f^2)(\theta+f^2)} + \frac{1}{(\theta+f^2)^2}\right\} + \dots + e^2\left\{\frac{1}{(\theta+\eta)^2} - \frac{2}{(\theta+\eta)\theta} + \frac{1}{\theta^2}\right\}$$

$$= \eta^2\left\{\frac{a^2}{(\theta+f^2)^2(\theta+\eta+f^2)^2} + \dots + \frac{c^2}{(\theta+h^2)^2(\theta+\eta+h^2)^2} + \frac{e^2}{\theta^2(\theta+\eta)^2}\right\}$$

$= \eta^2 Q$, suppose.

Also $\Theta'\Theta_0 - \Theta\Theta_0'$ contains the factor η, is $= \eta M$ suppose.

106. Substituting for J, $J'-2P+J$, and $\Theta'\Theta_0 - \Theta\Theta_0'$ their values ηP, ηQ, and ηM, the whole result contains the factor η^{m+1}, viz. we have

$$\square V = -\frac{4\eta^{m+1} P^{m-1}}{J_0'}\left(Q\Theta + \frac{PM}{\Theta_0}\right).$$

If here, except in the term η^{m+1}, we write $\eta = 0$, we have

$$P = \frac{a^2}{(\theta+f^2)^2} + \ldots + \frac{c^2}{(\theta+h^2)^2} + \frac{e^2}{\theta^2},\ = J_0,$$

$$Q = \frac{a^2}{(\theta+f^2)^4} + \ldots + \frac{c^2}{(\theta+h^2)^4} + \frac{e^2}{\theta^4},\ = \tfrac{1}{6} J_0''',$$

$$M = \Theta_0\Theta_0'' - \Theta_0'^2;$$

the formula becomes

$$\square V = -4\eta^{m+1} J_0'^{m-2}\left\{\tfrac{1}{6}J_0'''\Theta_0 + J_0'\left(\Theta_0'' - \frac{\Theta_0'^2}{\Theta_0}\right)\right\};$$

or (instead of J_0, Θ_0) using now J, Θ in their original significations

$$J = 1 - \frac{a^2}{\theta+f^2} - \ldots - \frac{c^2}{\theta+h^2} - \frac{e^2}{\theta},\ \text{and}\ \Theta = \theta^{-q-1}\{(\theta+f^2)\ldots(\theta+h^2)\}^{-\frac{1}{2}},$$

this is

$$\square V = -4\eta^{m+1} J'^{m-2}\left\{\tfrac{1}{6}J'''\Theta + J'\left(\Theta'' - \frac{\Theta'^2}{\Theta}\right)\right\},$$

or, what is the same thing,

$$= -4\eta^{m+1} J'^{m-2}\Theta\left\{\tfrac{1}{6}J''' + J'\left(\frac{\Theta'}{\Theta}\right)'\right\};$$

viz. the expression in $\{\ \}$ is

$$= \left[\frac{a^2}{(\theta+f^2)^4} + \ldots + \frac{c^2}{(\theta+h^2)^4} + \frac{e^2}{\theta^4}\right]$$

$$-\left[\frac{a^2}{(\theta+f^2)^2} + \ldots + \frac{c^2}{(\theta+h^2)^2} + \frac{e^2}{\theta^2}\right]\left[\frac{1}{(\theta+f^2)^2} + \ldots + \frac{1}{(\theta+h^2)^2} + \frac{2q+2}{\theta^2}\right].$$

We thus see that, η being infinitesimal, $\square V$ is infinitesimal of the order η^{m+1}; and hence, η being $=0$, we have

$$\square V = 0;$$

viz. the prepotential equation is satisfied by the value

$$V = \int_\theta^\infty dt\, I^m\, T,$$

where $m+1$ is positive.

107. We have consequently a value of ρ corresponding to the foregoing value of V; and this value is

$$\rho = -\frac{\Gamma(\frac{1}{2}s+q)}{2\pi^{\frac{1}{2}s}\Gamma(q+1)}\left(e^{2q+1}\frac{dW}{de}\right)_{e=0},$$

where, writing λ for the positive root of

$$1-\frac{x^2}{\lambda+f^2}-\ldots-\frac{z^2}{\lambda+h^2}-\frac{e^2}{\lambda}=0,$$

we have

$$W=\int_\lambda^\infty dt\left(1-\frac{x^2}{t+f^2}-\ldots-\frac{z^2}{t+h^2}-\frac{e^2}{t}\right)^m t^{-q-1}\{(t+f^2)\ldots(t+h^2)\}^{-\frac{1}{2}};$$

we thence obtain

$$\frac{dW}{de}=\int_\lambda^\infty dt\,.-\frac{2me}{t}\left(1-\frac{x^2}{t+f^2}-\ldots-\frac{z^2}{t+h^2}-\frac{e^2}{t}\right)^{m-1} t^{-q-1}\{(t+f^2)\ldots(t+h^2)\}^{-\frac{1}{2}}$$

$$-\left(1-\frac{x^2}{\lambda+f^2}-\ldots-\frac{z^2}{\lambda+h^2}-\frac{e^2}{\lambda}\right)^m \lambda^{-q-1}\{(\lambda+f^2)\ldots(\lambda+h^2)\}^{-\frac{1}{2}}\frac{d\lambda}{de};$$

or, multiplying by e^{2q+1} and substituting for $\frac{d\lambda}{de}$ its value

$$=\frac{\dfrac{2e}{\lambda}}{\left\{\dfrac{x^2}{(\lambda+f^2)^2}+\ldots+\dfrac{z^2}{(\lambda+h^2)^2}+\dfrac{e^2}{\lambda^2}\right\}},$$

we have

$$e^{2q+1}\frac{dW}{de}=\int_\lambda^\infty dt\,.-\frac{2me^{2q+2}}{t^{q+2}}\left(1-\frac{x^2}{t+f^2}-\ldots-\frac{z^2}{t+h^2}-\frac{e^2}{t}\right)^{m+1}\{(t+f^2)\ldots(t+h^2)\}^{-\frac{1}{2}}$$

$$-\frac{\dfrac{2e^{2q+2}}{\lambda^{q+2}}}{\left\{\dfrac{x^2}{(\lambda+f^2)^2}+\ldots+\dfrac{z^2}{(\lambda+h^2)^2}+\dfrac{e^2}{\lambda^2}\right\}}\left(1-\frac{x^2}{\lambda+f^2}\ldots-\frac{z^2}{\lambda+h^2}-\frac{e^2}{\lambda}\right)^m\{(\lambda+f^2)\ldots(\lambda+h^2)\}^{-\frac{1}{2}},$$

where the second term, although containing the evanescent factor

$$\left(1-\frac{x^2}{\lambda+f^2}-\ldots-\frac{z^2}{\lambda+h^2}-\frac{e^2}{\lambda}\right)^m,$$

is for the present retained.

108. I attend to the second term.

1°. Suppose $\frac{x^2}{f^2}+\ldots+\frac{z^2}{h^2}>1$; then, as e diminishes and becomes $=0$, λ does not become zero, but it becomes the positive root of the equation

$$1-\frac{x^2}{\lambda+f^2}-\ldots-\frac{z^2}{\lambda+h^2}=0;$$

hence the term, containing as well the evanescent factor e^{2q+2} as the other evanescent factor $\left(1-\frac{x^2}{\lambda+f^2}-\ldots-\frac{z^2}{\lambda+h^2}-\frac{e^2}{\lambda}\right)^m$, is $=0$.

2°. Suppose $\frac{x^2}{f^2}+\ldots+\frac{z^2}{h^2}<1$; then, as e diminishes to zero, λ tends to become $=0$, but $\frac{e^2}{\lambda}$ is finite and $=1-\frac{x^2}{f^2}-\ldots-\frac{z^2}{h^2}$, whence $\frac{e^2}{\lambda^2}$ is indefinitely large; and since $\frac{x^2}{(\lambda+f^2)^2}+\ldots+\frac{z^2}{(\lambda+h^2)^2}$ becomes $=\frac{x^2}{f^4}+\ldots+\frac{z^2}{h^4}$, which is finite, the denominator may be reduced to $\frac{e^2}{\lambda^2}$, and the term therefore is

$$=-2\left(\frac{e^2}{\lambda}\right)^q\left(1-\frac{x^2}{\lambda+f^2}-\ldots-\frac{z^2}{\lambda+h^2}-\frac{e^2}{\lambda}\right)^m\{(\lambda+f^2)\ldots(\lambda+h^2)\}^{-\frac{1}{2}},$$

$$=-2\left(1-\frac{x^2}{f^2}-\ldots-\frac{z^2}{h^2}\right)^q\left(1-\frac{x^2}{\lambda+f^2}-\ldots-\frac{z^2}{\lambda+h^2}-\frac{e^2}{\lambda}\right)^m(f\ldots h)^{-1},$$

which, the other factor being finite, vanishes in virtue of the evanescent factor

$$\left(1-\frac{x^2}{\lambda+f^2}-\ldots-\frac{z^2}{\lambda+h^2}-\frac{e^2}{\lambda}\right)^m.$$

Hence the second term always vanishes, and we have (e being $=0$)

$$e^{2q+1}\frac{dW}{de}=\int_\lambda^\infty dt\,.-\frac{2me^{2q+2}}{t^{q+2}}\left(1-\frac{x^2}{t+f^2}-\ldots-\frac{z^2}{t+h^2}-\frac{e^2}{t}\right)^m\{(t+f^2)\ldots(t+h^2)\}^{-\frac{1}{2}}.$$

109. Considering first the case $\frac{x^2}{f^2}+\ldots+\frac{z^2}{h^2}>1$: then, as e diminishes to zero, λ does not become $=0$; the integral contains no infinite element, and it consequently vanishes in virtue of the factor e^{2q+2}.

But if $\frac{x^2}{f^2}+\ldots+\frac{z^2}{h^2}<1$, then, introducing instead of t the new variable ξ, $=\frac{e^2}{t}$, that is, $t=\frac{e^2}{\xi}$, $dt=\frac{-e^2\,d\xi}{\xi^2}$, and writing for shortness

$$R=1-\frac{x^2}{f^2+\frac{e^2}{\xi}}-\ldots-\frac{z^2}{h^2+\frac{e^2}{\xi}},$$

the term becomes

$$=\int d\xi\,.\,2m\,(R-\xi)^{m-1}\,\xi^q\left\{\left(f^2+\frac{e^2}{\xi}\right)\ldots\left(h^2+\frac{e^2}{\xi}\right)\right\}^{-\frac{1}{2}},$$

where, as regards the limits, corresponding to $t=\infty$ we have $\xi=0$, and corresponding to $t=\lambda$ we have ξ the positive root of $R-\xi=0$. But e is indefinitely small; except for indefinitely small values of ξ, we have

$$R=1-\frac{x^2}{f^2}-\ldots-\frac{z^2}{h^2},\ \text{and}\ \left\{\left(f^2+\frac{e^2}{\xi}\right)\ldots\left(h^2+\frac{e^2}{\xi}\right)\right\}^{-\frac{1}{2}}=(f\ldots h)^{-1};$$

and if ξ be indefinitely small, then, whether we take the accurate or the reduced expressions, the elements are finite, and the corresponding portion of the integral is indefinitely small. We may consequently reduce as above; viz. writing now

$$R = 1 - \frac{x^2}{f^2} - \dots - \frac{z^2}{h^2},$$

the formula is

$$e^{2q+1}\frac{dW}{de} = \int_R^0 d\xi \,.\, 2m\,(R-\xi)^{m-1}\xi^q\,(f\dots h)^{-1},$$

$$= -\,2m\,(f\dots h)^{-1}\int_0^R d\xi \,.\, \xi^q\,(R-\xi)^{m-1};$$

or writing $\xi = Ru$, the integral becomes $= R^{q+m}\int_0^1 du\,.\,u^q\,(1-u)^{m-1}$, which is

$$= \frac{\Gamma(1+q)\,\Gamma(m)}{\Gamma(1+q+m)}\,R^{q+m};$$

that is, we have

$$e^{2q+1}\frac{dW}{de} = -\,2\,(f\dots h)^{-1}\frac{\Gamma(1+q)\,\Gamma(1+m)}{\Gamma(1+q+m)},$$

and consequently

$$\rho = \frac{\Gamma(\tfrac{1}{2}s+q)}{2\,(\Gamma\tfrac{1}{2})^s\,\Gamma(1+q)}\,2\,(f\dots h)^{-1}\frac{\Gamma(1+q)\,\Gamma(1+m)}{\Gamma(1+q+m)}\,R^{q+m},$$

that is,

$$\rho = (f\dots h)^{-1}\frac{\Gamma(\tfrac{1}{2}s+q)\,\Gamma(1+m)}{(\Gamma\tfrac{1}{2})^s\,\Gamma(1+q+m)}\,R^{q+m},$$

viz. ρ has this value for values of $(x,..,z)$ such that $\frac{x^2}{f^2}+\dots+\frac{z^2}{h^2} < 1$, but is $=0$ if $\frac{x^2}{f^2}+\dots+\frac{z^2}{h^2} > 1$.

110. Multiplying by a constant factor so as to reduce ρ to the value R^{q+m}, the final result is that the integral

$$V = \int\frac{\left(1-\frac{x^2}{f^2}-\dots-\frac{z^2}{h^2}\right)^{q+m} dx\dots dz}{[(a-x)^2+\dots+(c-z)^2+\dots+e^2]^{\frac{1}{2}s+q}},$$

the limits being given by the equation

$$\frac{x^2}{f^2}+\dots+\frac{z^2}{h^2} = 1,$$

is equal to

$$\frac{\Gamma(\tfrac{1}{2})^s\,\Gamma(1+q+m)}{\Gamma(\tfrac{1}{2}s+q)\Gamma(1+m)}(f\dots h)\int_\theta^\infty dt\,.\,t^{-q-1}\left(1-\frac{a^2}{t+f^2}-\dots-\frac{c^2}{t+h^2}\right)^m\{(t+f^2)\dots(t+h^2)\}^{-\frac{1}{2}},$$

where θ is the positive root of

$$1-\frac{a^2}{\theta+f^2}-\dots-\frac{c^2}{\theta+h^2}-\frac{e^2}{\theta} = 0.$$

In particular, if $e=0$, or

$$V=\int\frac{\left(1-\frac{x^2}{f^2}-\ldots-\frac{z^2}{h^2}\right)^{q+m}dx\ldots dz}{\{(a-x)^2+\ldots+(c-z)^2\}^{\frac{1}{2}s+q}},$$

there are two cases:

exterior, $\frac{a^2}{f^2}+\ldots+\frac{c^2}{h^2}>1$, θ is positive root of $1-\frac{a^2}{f^2}-\ldots-\frac{c^2}{h^2}=0$,

interior, $\frac{a^2}{f^2}+\ldots+\frac{c^2}{h^2}<1$, θ vanishes, viz. the limits in the integral are ∞, 0;

q must be *negative*, $1+q$ positive as before, in order that the t-integral may not be infinite in regard to the element $t=0$.

It is assumed in the proof that m and $1+q$ are each of them positive; but, as appears by the second example, the theorem is true for the extreme value $m=0$; it does not, however, appear that the proof can be extended to include the extreme value $q=-1$. The formula seems, however, to hold good for values of m, q beyond the foregoing limits; and it would seem that the only necessary conditions are $\frac{1}{2}s+q$, $1+m$, and $1+q+m$, each of them positive. The theorem is, in fact, a particular case of the following one, proved Annex X. No. 162, viz.

$$V=\int\frac{\phi\left(1-\frac{x^2}{f^2}-\ldots-\frac{z^2}{h^2}\right)dx\ldots dz}{\{(a-x)^2+\ldots+(c-z)^2+e^2\}^{\frac{1}{2}s+q}},$$

taken over the ellipsoid $\frac{x^2}{f^2}+\ldots+\frac{z^2}{h^2}=1$, is equal to

$$\frac{(\Gamma\frac{1}{2})^s(f\ldots h)}{\Gamma(-q)\,\Gamma(\frac{1}{2}s+q)}\int_\theta^\infty dt\,t^{-q-1}\{(t+f^2)\ldots(t+h^2)\}^{-\frac{1}{2}}(1-\sigma)^{-q}\int_0^1 x^{-q-1}\phi\{\sigma+(1-\sigma)x\}\,dx,$$

where σ denotes $\frac{a^2}{f^2+t}+\ldots+\frac{c^2}{h^2+t}+\frac{e^2}{t}$: assuming $\phi u=(1-u)^{q+m}$, we have

$$\phi\{\sigma+(1-\sigma)x\}=(1-\sigma)^{q+m}(1-x)^{q+m},$$

and the theorem is thus proved.

111. Particular cases: $m=0$;

$$\int\frac{\left(1-\frac{x^2}{f^2}-\ldots-\frac{z^2}{h^2}\right)^q dx\ldots dz}{[(a-x)^2+\ldots+(c-z)^2+e^2]^{\frac{1}{2}s+q}}=\frac{(\Gamma\frac{1}{2})^s\,\Gamma(1+q)}{\Gamma(\frac{1}{2}s+q)}(f\ldots h)\int_\theta^\infty dt\,t^{-q-1}\{(t+f^2)\ldots(t+h^2)\}^{-\frac{1}{2}}$$

Cor. In a somewhat similar manner it may be shown that

$$\int\frac{\left(1-\frac{x^2}{f^2}-\ldots-\frac{z^2}{h^2}\right)^q x\,dx\ldots dz}{\{(a-x)^2+\ldots+(c-z)^2+e^2\}^{\frac{1}{2}s+q}}=\frac{(\Gamma\frac{1}{2})^s\,\Gamma(1+q)}{\Gamma(\frac{1}{2}s+q)}(f\ldots h)\int_\theta^\infty dt\,\frac{af^2}{t+f^2}t^{-q-1}\{(t+f^2)\ldots(t+h^2)\}^{-\frac{1}{2}}.$$

Multiplying the first by a and subtracting it from the second, we have

$$\int \frac{\left(1-\frac{x^2}{f^2}-\ldots-\frac{z^2}{h^2}\right)^q (a-x)\,dx\ldots dz}{\{(a-x)^2+\ldots+(c-z)^2+e^2\}^{\frac{1}{2}s+q}} = \frac{(\Gamma\tfrac{1}{2})^s\,\Gamma(1+q)}{\Gamma(\tfrac{1}{2}s+q)}(f\ldots h)\int_\theta^\infty dt\,\frac{a}{t+f^2}t^{-q}\{(t+f^2)\ldots(t+h^2)\}^{-\frac{1}{2}};$$

or, writing $q+1$ for q, this is

$$\int \frac{\left(1-\frac{x^2}{f^2}-\ldots-\frac{z^2}{h^2}\right)^{q+1} (a-x)\,dx\ldots dz}{\{(a-x)^2+\ldots+(c-z)^2+e^2\}^{\frac{1}{2}s+q+1}} = \frac{(\Gamma\tfrac{1}{2})^s\,\Gamma(2+q)}{\Gamma(\tfrac{1}{2}s+q+1)}(f\ldots h)\int_\theta^\infty dt\,\frac{a}{t+f^2}t^{-q-1}\{(t+f^2)\ldots(t+h^2)\}^{-\frac{1}{2}};$$

and we have similar formulæ with .., $(c-z)$, e, instead of $(a-x)$, in the numerator.

112. If $m=1$, we have

$$\int \frac{\left(1-\frac{x^2}{f^2}-\ldots-\frac{z^2}{h^2}\right)^{q+1} dx\ldots dz}{\{(a-x)^2+\ldots+(c-z)^2+e^2\}^{\frac{1}{2}s+q}}$$

$$= \frac{(\Gamma\tfrac{1}{2})^s\,\Gamma(2+q)}{\Gamma(\tfrac{1}{2}s+q)}(f\ldots h)\int_\theta^\infty dt\left\{1-\frac{a^2}{t+f^2}-\ldots-\frac{c^2}{t+h^2}-\frac{e^2}{t}\right\}t^{-q-1}\{(t+f^2)\ldots(t+h^2)\}^{-\frac{1}{2}},$$

which, differentiated in respect to a, gives the $(a-x)$-formula; hence conversely, assuming the $a-x, .., c-z$, e-formulæ, we obtain by integration the last preceding formula to a constant *près*, viz. we thereby obtain the multiple integral $=C+$ right-hand function, where C is independent of $(a, .., c, e)$; by taking these all infinite, and observing that then $\theta=\infty$, the two integrals each vanish, and we obtain $C=0$.

In particular, when $s=3$, $q=-1$, then

$$\int\frac{dx\,dy\,dz}{\{(a-x)^2+(b-y)^2+(c-z)^2+e^2\}^{\frac{1}{2}}}=\pi fgh\int_\theta^\infty dt\left\{1-\frac{a^2}{t+f^2}-\frac{b^2}{t+g^2}-\frac{c^2}{t+h^2}-\frac{e^2}{t}\right\}(t+f^2.t+g^2.t+h^2)^{-\frac{1}{2}},$$

which, putting therein $e=0$, gives the potential of an ellipsoid for the cases of an exterior point and an interior point respectively.

ANNEX V. GREEN'S *Integration of the Prepotential Equation* $\left(\frac{d^2}{da^2}+\ldots+\frac{d^2}{dc^2}+\frac{d^2}{de^2}+\frac{2q+1}{e}\frac{d}{de}\right)V=0$. Art. Nos. 113 to 128.

113. In the present Annex, I in part reproduce Green's process for the integration of this equation by means of a series of functions, which are analogous to Laplace's Functions and may be termed "Greenians" (see his Memoir on the Attraction of Ellipsoids, referred to above, p. 320); each such function gives rise to a Prepotential Integral.

Green shows, by a complicated and difficult piece of general reasoning, that there exist solutions of the form $V=\Theta\phi$ (see *post*, No. 116), where ϕ is a function of the

s new variables $\alpha, \beta, .., \gamma$ without θ, such that $\nabla\phi = \kappa\phi$, κ being a function of θ only; these functions ϕ of the variables $\alpha, \beta, .., \gamma$ are in fact the Greenian Functions in question. The function of the order 0 is $\phi = 1$; those of the order 1 are $\phi = \alpha$, $\phi = \beta, .., \phi = \gamma$; those of the order 2 are $\phi = \alpha\beta$, &c., and s functions each of the form

$$\tfrac{1}{2}\{A\alpha^2 + B\beta^2 + \ldots + C\gamma^2\} + D.$$

The existence of the functions just referred to other than the s functions involving the squares of the variables is obvious enough; the difficulty first arises in regard to these s functions; and the actual development of them appears to me important by reason of the light which is thereby thrown upon the general theory. This I accomplish in the present Annex; and I determine by Green's process the corresponding prepotential integrals. I do not go into the question of the Greenian Functions of orders superior to the second.

114. I write for greater clearness $(a, b, .., c, e)$ instead of $(a, .., c, e)$ to denote the series of $(s+1)$ variables; viz. $(a, b, .., c)$ will denote a series of s variables; corresponding to these we have the semiaxes $(f, g, .., h)$, and the new variables $(\alpha, \beta, .., \gamma)$; these last, with the before-mentioned function θ, are the $s+1$ new variables of the problem; and, for convenience, there is introduced also a quantity ϵ; viz. we have

$$\begin{aligned} a &= \sqrt{f^2+\theta}\ \alpha, \\ b &= \sqrt{g^2+\theta}\ \beta, \\ &\vdots \\ c &= \sqrt{h^2+\theta}\ \gamma, \\ e &= \sqrt{\theta}\quad\ \ \epsilon, \end{aligned}$$

where $1 = \alpha^2 + \beta^2 + \ldots + \gamma^2 + \epsilon^2$.

That is, we have θ a function of $a, b, .., c, e$, determined by

$$\frac{a^2}{f^2+\theta} + \frac{b^2}{g^2+\theta} + \ldots + \frac{c^2}{h^2+\theta} + \frac{e^2}{\theta} = 1;$$

and then $\alpha, \beta, .., \gamma$ are given as functions of the same quantities $a, b, .., c, e$ by the equations

$$\alpha^2 = \frac{a^2}{f^2+\theta},\quad \beta^2 = \frac{b^2}{g^2+\theta}, \ldots, \gamma^2 = \frac{c^2}{h^2+\theta};$$

also ϵ, considered as a function of the same quantities, is

$$= \sqrt{1 - \frac{a^2}{f^2+\theta} - \frac{b^2}{g^2+\theta} - \ldots - \frac{c^2}{h^2+\theta}}.$$

115. Introducing instead of $a, b, .., c, e$ the new variables $\alpha, \beta, .., \gamma, \theta$, the transformed differential equation is

$$4\theta\frac{d^2V}{d\theta^2} + 2\frac{dV}{d\theta}\left(s + 2q + 2 - \frac{f^2}{f^2+\theta} - \ldots - \frac{h^2}{h^2+\theta}\right) + \nabla V = 0$$

where for shortness

$$\nabla V = \frac{1}{f^2+\theta}\left\{-\alpha^2 - \frac{g^2}{g^2+\theta}\beta^2 - \ldots - \frac{h^2}{h^2+\theta}\gamma^2+1\right\}\frac{d^2V}{d\alpha^2}$$

$$+\frac{1}{g^2+\theta}\left\{-\frac{f^2}{f^2+\theta}\alpha^2 - \beta^2 - \ldots - \frac{h^2}{h^2+\theta}\gamma^2+1\right\}\frac{d^2V}{d\beta^2}$$

$$\vdots$$

$$+\frac{1}{h^2+\theta}\left\{-\frac{f^2}{f^2+\theta}\alpha^2 - \frac{g^2}{g^2+\theta}\beta^2 - \ldots - \gamma^2+1\right\}\frac{d^2V}{d\gamma^2}$$

$$-\frac{2\theta}{f^2+\theta\,.\,g^2+\theta}\,\frac{d^2V}{d\alpha\, d\beta} - \&\text{c.}$$

$$+\frac{1}{f^2+\theta}\left\{-2q-2-\theta\left(\frac{1}{g^2+\theta}+\ldots+\frac{1}{h^2+\theta}\right)\right\}\alpha\,\frac{dV}{d\alpha}$$

$$+\frac{1}{g^2+\theta}\left\{-2q-2-\theta\left(\frac{1}{f^2+\theta}+\ldots+\frac{1}{h^2+\theta}\right)\right\}\beta\,\frac{dV}{d\beta}$$

$$\vdots$$

$$+\frac{1}{h^2+\theta}\left\{-2q-2-\theta\left(\frac{1}{f^2+\theta}+\frac{1}{g^2+\theta}+\ldots\right)\right\}\gamma\,\frac{dV}{d\gamma}.$$

Also

$$\epsilon^{2q+1}\frac{dV}{de} = -\,\theta^{q+1}\,\epsilon^{2q+2}\left\{1-\frac{f^2\alpha^2}{f^2+\theta}-\ldots-\frac{h^2\gamma^2}{\gamma^2+\theta}\right\}^{-1}\left(\frac{1}{f^2+\theta}\alpha\frac{dV}{d\alpha}+\ldots+\frac{1}{h^2+\theta}\gamma\frac{dV}{d\gamma}-2\frac{dV}{d\theta}\right).$$

116. To integrate the equation for V, we assume

$$V = \Theta\phi,$$

where Θ is a function of θ only, and ϕ a function of $\alpha, \beta, .., \gamma$ (without θ), such that

$$\nabla\phi = \kappa\phi,$$

κ being a function of θ only. Assuming that this is possible, the remaining equation to be satisfied is obviously

$$4\theta\frac{d^2\Theta}{d\theta^2}+2\frac{d\Theta}{d\theta}\left\{2q+2+\theta\left(\frac{1}{f^2+\theta}+\ldots+\frac{1}{h^2+\theta}\right)\right\}+\kappa\Theta=0.$$

Solutions of the form in question are

$$\phi = 1\ ,\quad \kappa = 0,$$

$$\phi = \alpha\ ,\quad \kappa = \frac{1}{f^2+\theta}\left\{-2q-2-\frac{\theta}{g^2+\theta}-\ldots-\frac{\theta}{h^2+\theta}\right\},$$

$$\phi = \beta\ ,\quad \kappa = \quad ,, \quad\quad ,,$$

$$\vdots$$

$$\phi = \alpha\beta,\quad \kappa = \frac{-2\theta}{f^2+\theta\,.\,g^2+\theta}+\frac{1}{f^2+\theta}\left\{-2q-2-\frac{\theta}{g^2+\theta}-\ldots-\frac{\theta}{h^2+\theta}\right\}$$

$$+\frac{1}{g^2+\theta}\left\{-2q-2-\frac{\theta}{f^2+\theta}-\ldots-\frac{\theta}{h^2+\theta}\right\};$$

and it can be shown next that there is a solution of the form

$$\phi = \tfrac{1}{2}(A\alpha^2 + B\beta^2 + \ldots + C\gamma^2) + D.$$

117. In fact, assuming that this satisfies $\nabla\phi - \kappa\phi = 0$, we must have identically

$$\begin{aligned}
&\frac{A}{f^2+\theta}\left\{-\alpha^2 - \frac{g^2}{g^2+\theta}\beta^2 - \ldots - \frac{h^2}{h^2+\theta}\gamma^2 + 1\right\}\\
&+\frac{B}{g^2+\theta}\left\{-\frac{f^2}{f^2+\theta}\alpha^2 - \beta^2 - \ldots - \frac{h^2}{h^2+\theta}\gamma^2 + 1\right\}\\
&\vdots\\
&+\frac{C}{h^2+\theta}\left\{-\frac{f^2}{f^2+\theta}\alpha^2 - \frac{g^2}{g^2+\theta}\beta^2 - \ldots - \gamma^2 + 1\right\}\\
&+\frac{A}{f^2+\theta}\left\{-s - 2q - 1 + \frac{g^2}{g^2+\theta} + \ldots + \frac{h^2}{h^2+\theta}\right\}\\
&+\frac{B}{g^2+\theta}\left\{-s - 2q - 1 + \frac{f^2}{f^2+\theta} + \ldots + \frac{h^2}{h^2+\theta}\right\}\\
&\vdots\\
&+\frac{C}{h^2+\theta}\left\{-s - 2q - 1 + \frac{f^2}{f^2+\theta} + \frac{g^2}{g^2+\theta} + \ldots\right\}\\
&+\kappa\left\{\tfrac{1}{2}(A\alpha^2 + B\beta^2 + \ldots + C\gamma^2) + D\right\};
\end{aligned}$$

so that, from the term in α^2, we have

$$\frac{A}{f^2+\theta}\left\{-s - 2q - 2 + \frac{g^2}{g^2+\theta} + \ldots + \frac{h^2}{h^2+\theta}\right\} - \tfrac{1}{2}\kappa A - \frac{Bf^2}{f^2+\theta\,.\,g^2+\theta} - \ldots - \frac{Cf^2}{f^2+\theta\,.\,h^2+\theta} = 0;$$

or, what is the same thing,

$$A\left\{-2q - 3 - \frac{\theta}{g^2+\theta} + \ldots + \frac{\theta}{h^2+\theta} - \tfrac{1}{2}\kappa(f^2+\theta)\right\} - B\frac{f^2}{g^2+\theta} - \ldots - \frac{Cf^2}{h^2+\theta} = 0,$$

with the like equations from $\beta^2, .., \gamma^2$; and from the constant term we have

$$A\frac{1}{f^2+\theta} + B\frac{1}{g^2+\theta} + \ldots + C\frac{1}{h^2+\theta} - \kappa D = 0.$$

118. Multiplying this last by f^2, and adding it to the first, we obtain

$$A\left\{-2q - 2 - \frac{\theta}{f^2+\theta} - \frac{\theta}{g^2+\theta} - \ldots - \frac{\theta}{h^2+\theta} - \tfrac{1}{2}\kappa(f^2+\theta)\right\} - \kappa f^2 D = 0;$$

viz. putting for shortness $\Omega = \theta\left(\frac{1}{f^2+\theta} + \frac{1}{g^2+\theta} + \ldots + \frac{1}{h^2+\theta}\right)$, this is

$$A\{2q + 2 + \Omega + \tfrac{1}{2}\kappa(f^2+\theta)\} + \kappa f^2 D = 0;$$

and similarly

$$\begin{aligned}
&B\{2q + 2 + \Omega + \tfrac{1}{2}\kappa(g^2+\theta)\} + \kappa g^2 D = 0,\\
&\vdots\\
&C\{2q + 2 + \Omega + \tfrac{1}{2}\kappa(h^2+\theta)\} + \kappa h^2 D = 0.
\end{aligned}$$

To these we join the foregoing equation

$$\frac{A}{f^2+\theta}+\frac{B}{g^2+\theta}+\ldots+\frac{C}{h^2+\theta}-\kappa D=0.$$

Eliminating $A, B, .., C, D$, we have an equation which determines κ as a function of θ; and the equations then determine the ratios of $A, B, .., C, D$, so that these quantities will be given as determinate multiples of an arbitrary quantity M. The equation for κ is in fact

$$\frac{f^2}{(f^2+\theta)\{2q+2+\Omega+\tfrac{1}{2}\kappa(f^2+\theta)\}}+\frac{g^2}{(g^2+\theta)\{2q+2+\Omega+\tfrac{1}{2}\kappa(g^2+\theta)\}}+\ldots$$
$$+\frac{h^2}{(h^2+\theta)\{2q+2+\Omega+\tfrac{1}{2}\kappa(h^2+\theta)\}}+1=0;$$

and the values of $A, B, .., C, D$ are then

$$\frac{Mf^2}{2q+2+\Omega+\tfrac{1}{2}\kappa(f^2+\theta)},\ \frac{Mg^2}{2q+2+\Omega+\tfrac{1}{2}\kappa(g^2+\theta)},\ldots,\ \frac{Mh^2}{2q+2+\Omega+\tfrac{1}{2}\kappa(h^2+\theta)},\ -\frac{M}{\kappa},$$

values which seem to be dependent on θ: if they were so, it would be fatal to the success of the process; but they are really independent of θ.

119. That they are independent of θ depends on the theorems; that we have

$$\kappa=\frac{(2q+2+\Omega)\kappa_0}{2q+2-\tfrac{1}{2}\kappa_0\theta},$$

where κ_0 is a quantity independent of θ determined by the equation

$$\frac{1}{2q+2+\tfrac{1}{2}\kappa_0 f^2}+\frac{1}{2q+2+\tfrac{1}{2}\kappa_0 g^2}+\ldots+\frac{1}{2q+2+\tfrac{1}{2}\kappa_0 h^2}+1=0,$$

(κ_0 is in fact the value of κ on writing $\theta=0$): and that, omitting the arbitrary multiplier, the values of $A, B, .., C, D$ then are

$$\frac{f^2}{2q+2+\tfrac{1}{2}\kappa_0 f^2},\ \frac{g^2}{2q+2+\tfrac{1}{2}\kappa_0 g^2},\ldots,\ \frac{h^2}{2q+2+\tfrac{1}{2}\kappa_0 h^2},\ -\frac{1}{\kappa_0};$$

or, what is the same thing, the value of ϕ is

$$=\frac{\tfrac{1}{2}f^2\alpha^2}{2q+2+\tfrac{1}{2}\kappa_0 f^2}+\frac{\tfrac{1}{2}g^2\beta^2}{2q+2+\tfrac{1}{2}\kappa_0 g^2}+\ldots+\frac{\tfrac{1}{2}h^2\gamma^2}{2q+2+\tfrac{1}{2}\kappa_0 h^2}-\frac{1}{\kappa_0}.$$

120. To explain the ground of the assumption

$$\kappa=\frac{(2q+2+\Omega)\kappa_0}{2q+2-\tfrac{1}{2}\kappa_0\theta},$$

observe that, assuming

$$\frac{2q+2+\Omega+\tfrac{1}{2}\kappa(f^2+\theta)}{2q+2+\tfrac{1}{2}\kappa_0 f^2}=\frac{2q+2+\Omega+\tfrac{1}{2}\kappa(g^2+\theta)}{2q+2+\tfrac{1}{2}\kappa_0 g^2},$$

then multiplying out and reducing, we obtain

$$\tfrac{1}{2}\kappa_0(2q+2+\Omega)(g^2-f^2)+(2q+2)\tfrac{1}{2}\kappa(f^2-g^2)+\tfrac{1}{4}\kappa_0\kappa(g^2-f^2)\theta=0;$$

viz. the equation divides out by the factor g^2-f^2, thereby becoming

$$\kappa_0(2q+2+\Omega)-(2q+2)\alpha+\tfrac{1}{2}\kappa\kappa_0\theta=0,$$

that is, it gives for κ the foregoing value: hence clearly, κ having this value, we obtain by symmetry

$$2q+2+\Omega+\tfrac{1}{2}\kappa(f^2+\theta),\ 2q+2+\Omega+\tfrac{1}{2}\kappa(g^2+\theta),\ldots,\ 2q+2+\Omega+\tfrac{1}{2}\kappa(h^2+\theta),$$

proportional to

$$2q+2+\tfrac{1}{2}\kappa_0 f^2,\qquad 2q+2+\tfrac{1}{2}\kappa_0 g^2,\ldots,\qquad 2q+2+\tfrac{1}{2}\kappa_0 h^2;$$

viz. the ratios, not only of $A:B$, but of $A:B:\ldots:C$ will be independent of θ.

121. To complete the transformation, starting with the foregoing value of κ, we have

$$2q+2+\Omega+\tfrac{1}{2}\kappa(f^2+\theta)=(2q+2+\Omega)\frac{2q+2+\frac{1}{2}\kappa_0 f^2}{2q+2-\frac{1}{2}\kappa_0\theta},\ \&\text{c.};$$

so that we have

$$\begin{aligned}
&A\{2q+2+\tfrac{1}{2}\kappa_0 f^2\}+\kappa_0 f^2 D=0,\\
&B\{2q+2+\tfrac{1}{2}\kappa_0 g^2\}+\kappa_0 g^2 D=0,\\
&\quad\vdots\\
&C\{2q+2+\tfrac{1}{2}\kappa_0 h^2\}+\kappa_0 h^2 D=0,
\end{aligned}$$

and

$$\frac{A}{f^2+\theta}+\frac{B}{g^2+\theta}+\ldots+\frac{C}{h^2+\theta}-\frac{(2q+2+\Omega)\kappa_0 D}{2q+2-\frac{1}{2}\kappa_0\theta}=0.$$

Substituting for $A, B,\ldots, C$ their values, this last becomes

$$-\frac{\kappa_0 D}{2q+2+\frac{1}{2}\kappa_0\theta}\left\{\frac{2q+2}{2q+2+\frac{1}{2}\kappa_0 f^2}-\frac{\theta}{f^2+\theta}\right\}-\ldots-\frac{\kappa_0 D}{2q+2-\frac{1}{2}\kappa_0\theta}\left\{\frac{2q+2}{2q+2+\frac{1}{2}\kappa_0 h^2}-\frac{\theta}{h^2+\theta}\right\}$$

$$-\frac{\kappa_0 D}{2q+2-\frac{1}{2}\kappa_0\theta}\{2q+2+\Omega\}=0;$$

viz. this is

$$\left\{\frac{2q+2}{2q+2+\frac{1}{2}\kappa_0 f^2}-\frac{\theta}{f^2+\theta}\right\}+\ldots+\left\{\frac{2q+2}{2q+2+\frac{1}{2}\kappa_0 h^2}-\frac{\theta}{h^2+\theta}\right\}+2q+2+\Omega=0;$$

or, substituting for Ω its value, and dividing out by $2q+2$, we have

$$\frac{1}{2q+2+\frac{1}{2}\kappa_0 f^2}+\frac{1}{2q+2+\frac{1}{2}\kappa_0 g^2}+\ldots+\frac{1}{2q+2+\frac{1}{2}\kappa_0 h^2}+1=0,$$

the equation for the determination of κ_0.

122. The equation for κ_0 is of the order s; there are consequently s functions of the form in question, and each of the terms $\alpha^2, \beta^2, .., \gamma^2$ can be expressed as a linear function of these. It thus appears that any quadric function of $\alpha, \beta, .., \gamma$ can be expressed as a sum of Greenian functions; viz. the form is

$$\begin{aligned}
&A\\
&+ B\alpha + \&c.\\
&+ C\alpha\beta + \&c.\\
&+ D'\left(\frac{\tfrac{1}{2}f^2\alpha^2}{2q+2+\tfrac{1}{2}\kappa_0'f^2} + \frac{\tfrac{1}{2}g^2\beta^2}{2q+2+\tfrac{1}{2}\kappa_0'g^2} + \ldots + \frac{\tfrac{1}{2}h^2\gamma^2}{2q+2+\tfrac{1}{2}\kappa_0'h^2} - \frac{1}{\kappa_0'}\right)\\
&+ D''(\quad ,, \quad ,, \quad ,, \quad)\\
&(s \text{ lines}),
\end{aligned}$$

viz. the terms multiplied by D', D'', &c. respectively are those answering to the roots κ_0', κ_0'',.. of the equation in κ_0.

The general conclusion is that any rational and integral function of $\alpha, \beta, .., \gamma$ can be expressed as a sum of Greenian functions.

123. We have next to integrate the equation

$$4\theta\frac{d^2\Theta}{d\theta^2} + 2\frac{d\Theta}{d\theta}\left(2q+2+\frac{\theta}{f^2+\theta}+\frac{\theta}{g^2+\theta}+\ldots+\frac{\theta}{h^2+\theta}\right) - \kappa\Theta = 0.$$

Suppose $\kappa = 0$, a particular solution is $\Theta = 1$. Next, suppose

$\kappa = \dfrac{1}{f^2+\theta}\left(-2q-2-\dfrac{\theta}{g^2+\theta}-\ldots-\dfrac{\theta}{h^2+\theta}\right)$; a particular solution is $\dfrac{\sqrt{f^2+\theta}}{\sqrt{f^2+g^2+\ldots+h^2}}$:

in fact, omitting the constant denominator, or writing $\Theta = \sqrt{f^2+\theta}$, and therefore

$$\frac{d\Theta}{d\theta} = \frac{1}{2\sqrt{f^2+\theta}},\quad \frac{d^2\Theta}{d\theta^2} = -\frac{1}{4(f^2+\theta)^{\frac{3}{2}}},$$

the equation to be verified is

$$-\frac{\theta}{(f^2+\theta)^{\frac{3}{2}}} + \frac{1}{\sqrt{f^2+\theta}}\left\{\ 2q+2+\frac{\theta}{f^2+\theta}+\frac{\theta}{g^2+\theta}+\ldots+\frac{\theta}{h^2+\theta}\right\}$$

$$+\frac{1}{\sqrt{f^2+\theta}}\left\{-2q-2 \qquad -\frac{\theta}{g^2+\theta}-\ldots-\frac{\theta}{h^2+\theta}\right\} = 0, \text{ which is right.}$$

Again, suppose $\kappa = \dfrac{-2\theta}{f^2+\theta\,.\,g^2+\theta}$ + &c. (value belonging to $\phi = \alpha\beta$, see No. 116); a particular solution is $\dfrac{\sqrt{f^2+\theta}\sqrt{g^2+\theta}}{f^2+g^2+\ldots+h^2}$: in fact, omitting the constant factor, or writing

$$\Theta = \sqrt{f^2+\theta}\sqrt{g^2+\theta},$$

and therefore

$$\frac{d\Theta}{d\theta} = \tfrac{1}{2}\left\{\frac{\sqrt{g^2+\theta}}{\sqrt{f^2+\theta}} + \frac{\sqrt{f^2+\theta}}{\sqrt{g^2+\theta}}\right\},$$

$$\frac{d^2\Theta}{d\theta^2} = \tfrac{1}{4}\left\{-\frac{\sqrt{g^2+\theta}}{(f^2+\theta)^{\frac{3}{2}}} + \frac{2}{\sqrt{f^2+\theta}\sqrt{g^2+\theta}} - \frac{\sqrt{f^2+\theta}}{(g^2+\theta)^{\frac{3}{2}}}\right\},$$

the equation to be verified is

$$\theta\left\{-\frac{\sqrt{g^2+\theta}}{(f^2+\theta)^{\frac{3}{2}}} + \frac{2}{\sqrt{f^2+\theta}\sqrt{g^2+\theta}} - \frac{\sqrt{f^2+\theta}}{(g^2+\theta)^{\frac{3}{2}}}\right\}$$

$$+\left(\frac{\sqrt{g^2+\theta}}{\sqrt{f^2+\theta}} + \frac{\sqrt{f^2+\theta}}{\sqrt{g^2+\theta}}\right)\left\{2q+2+\frac{\theta}{f^2+\theta}+\frac{\theta}{g^2+\theta}+\ldots+\frac{\theta}{h^2+\theta}\right\}$$

$$+\sqrt{f^2+\theta}\sqrt{g^2+\theta}\quad\left\{\frac{-2\theta}{f^2+\theta\,.\,g^2+\theta}+\frac{1}{f^2+\theta}\left(-2q-2-\frac{\theta}{g^2+\theta}-\ldots-\frac{\theta}{h^2+\theta}\right)\right.$$

$$\left.+\frac{1}{g^2+\theta}\left(-2q-2-\frac{\theta}{f^2+\theta}-\ldots-\frac{\theta}{h^2+\theta}\right)\right\}=0;$$

or putting for shortness $\Omega = \frac{\theta}{f^2+\theta}+\frac{\theta}{g^2+\theta}+\ldots+\frac{\theta}{h^2+\theta}$, this is

$$-\frac{\theta\sqrt{g^2+\theta}}{(f^2+\theta)^{\frac{3}{2}}}+\frac{2\theta}{\sqrt{f^2+\theta}\sqrt{g^2+\theta}}-\frac{\theta\sqrt{f^2+\theta}}{(g^2+\theta)^{\frac{3}{2}}}\qquad+\left(\frac{\sqrt{g^2+\theta}}{\sqrt{f^2+\theta}}+\frac{\sqrt{f^2+\theta}}{\sqrt{g^2+\theta}}\right)(2q+2+\Omega)$$

$$-\frac{2\theta}{\sqrt{f^2+\theta}\sqrt{g^2+\theta}}+\frac{\sqrt{g^2+\theta}}{\sqrt{f^2+\theta}}\left(-2q-2+\frac{\theta}{f^2+\theta}-\Omega\right)+\frac{\sqrt{f^2+\theta}}{\sqrt{g^2+\theta}}\left(-2q-2+\frac{\theta}{g^2+\theta}-\Omega\right)=0,$$

which is true. And, generally, the particular solution is deduced from the value of ϕ by writing therein

$$\frac{\sqrt{f^2+\theta}}{\sqrt{f^2+g^2+\ldots+h^2}},\ \frac{\sqrt{g^2+\theta}}{\sqrt{f^2+g^2+\ldots+h^2}},\ldots,\ \frac{\sqrt{h^2+\theta}}{\sqrt{f^2+g^2+\ldots+h^2}}$$

in place of $\alpha, \beta, \ldots, \gamma$ respectively: say the value thus obtained is $\Theta = H$, where H is what ϕ becomes by the above substitution.

124. Represent for a moment the equation in Θ by

$$4\theta\frac{d^2\Theta}{d\theta^2}+2\frac{d\Theta}{d\theta}P+\kappa\Theta=0,$$

and assume that this is satisfied by $\Theta = H\int z\,d\theta$. Then we have

$$4\theta\left(\frac{d^2H}{d\theta^2}\int z\,d\theta+2\frac{dH}{d\theta}z+H\frac{dz}{d\theta}\right)$$

$$+2P\left(\frac{dH}{d\theta}\int z\,d\theta+\qquad Hz\right)$$

$$+\qquad\kappa H\int z\,d\theta\qquad\qquad=0,$$

and therefore

$$\left(8\theta\frac{dH}{d\theta}+2PH\right)z+4\theta H\frac{dz}{d\theta}=0;$$

viz. multiplying by $\frac{H}{4\theta}$, this is

$$\frac{d}{d\theta}(H^2z)+\frac{1}{2\theta}PH^2z=0,$$

or

$$\frac{1}{H^2z}\frac{d}{d\theta}(H^2z)+\frac{1}{2\theta}P=0;$$

viz. substituting for P its value, this is

$$\frac{1}{H^2z}\frac{d}{d\theta}(H^2z)+\frac{1}{2\theta}\left(2q+2+\frac{\theta}{f^2+\theta}+\frac{\theta}{g^2+\theta}+\ldots+\frac{\theta}{h^2+\theta}\right)=0.$$

Hence, integrating,

$$H^2z=\frac{C\theta^{-q-1}}{\sqrt{f^2+\theta\,.\,g^2+\theta\ldots h^2+\theta}},\qquad C \text{ an arbitrary constant,}$$

and

$$\Theta=CH\int_\chi\frac{\theta^{-q-1}\,d\theta}{H^2\sqrt{f^2+\theta\,.\,g^2+\theta\ldots h^2+\theta}},\qquad \chi \text{ arbitrary,}$$

where the constants of integration are C, λ; or, what is the same thing, taking T the same function of t that H is of θ (viz. T is what ϕ becomes on writing therein

$$\frac{\sqrt{f^2+t}}{\sqrt{f^2+g^2+\ldots+h^2}},\ \frac{\sqrt{g^2+t}}{\sqrt{f^2+g^2+\ldots+h^2}},\ldots\ \frac{\sqrt{h^2+t}}{\sqrt{f^2+g^2+\ldots+h^2}},$$

in place of α, β, .., γ respectively), then

$$\Theta=-CH\int_\theta^\chi\frac{t^{-q-1}\,dt}{T^2\sqrt{f^2+t\,.\,g^2+t\ldots h^2+t}},$$

where χ may be taken $=\infty$: we thus have

$$V=\Theta\phi=-CH\phi\int_\theta^\infty\frac{t^{-q-1}\,dt}{T^2\sqrt{f^2+t\,.\,g^2+t\ldots h^2+t}}.$$

Recollecting that

$$1=\frac{a^2}{f^2+\theta}+\frac{b^2}{g^2+\theta}+\ldots+\frac{c^2}{h^2+\theta}+\frac{e^2}{\theta},$$

so that for $\theta=\infty$ we have $a^2+b^2+\ldots+c^2+e^2=\theta$, the assumption $\chi=\infty$ comes to making V vanish for infinite values of $(a, b,.., c, e)$.

125. We have to find the value of ρ corresponding to the foregoing value of V; viz. W being the value of V, on writing therein $(x, y,.., z)$ in place of $(a, b,.., c)$, then (theorem A)

$$\rho=-\frac{\Gamma(\frac{1}{2}s+q)}{2(\Gamma\frac{1}{2})^s\,\Gamma(q+1)}\left(e^{2q+1}\frac{dW}{de}\right)_0.$$

Take λ the same function of $(x, y, .., z, e)$ that θ is of $(a, b, .., c, e)$: viz. take λ the positive root of

$$\frac{x^2}{f^2+\lambda}+\frac{y^2}{g^2+\lambda}+\ldots+\frac{z^2}{h^2+\lambda}-\frac{e^2}{\lambda}=1\,;$$

and let $(\xi, \eta, .., \zeta, \tau)$ correspond to $(\alpha, \beta, .., \gamma, \epsilon)$, viz.

$$\xi=\frac{x}{\sqrt{f^2+\lambda}},\ \eta=\frac{y}{\sqrt{g^2+\lambda}},\ldots,\zeta=\frac{z}{\sqrt{h^2+\lambda}},\ \tau=\sqrt{1-\frac{x^2}{f^2+\lambda}-\frac{y^2}{g^2+\lambda}-\ldots-\frac{z^2}{h^2+\lambda}},$$

so that W is the same function of $(\xi, \eta, .., \lambda)$ that V is of $(\alpha, \beta, .., \theta)$: say this is

$$W=-C\Lambda\psi\int_\lambda^\infty\frac{t^{-q-1}\,dt}{T^2\sqrt{f^2+t\,.\,g^2+t\ldots h^2+t}};$$

then we have for ρ the value

$$\rho=\frac{\Gamma(\tfrac{1}{2}s+q)}{2\,(\Gamma\tfrac{1}{2})^s\,\Gamma(q+1)}\lambda^{q+1}\tau^{2q+2}\left(1-\frac{f^2\xi^2}{f^2+\lambda}-\ldots-\frac{h^2\zeta^2}{h^2+\lambda}\right)^{-1}$$
$$\times\left(\frac{1}{f^2+\lambda}\xi\frac{dW}{d\xi}+\ldots+\frac{1}{h^2+\lambda}\zeta\frac{dW}{d\zeta}-2\frac{dW}{d\lambda}\right),$$

where e is to be put $=0$.

126. Suppose e is $=0$; then, if $\dfrac{x^2}{f^2}+\dfrac{y^2}{g^2}+\ldots+\dfrac{z^2}{h^2}>1$, λ is not $=0$ but is the positive root of $\dfrac{x^2}{f^2+\lambda}+\dfrac{y^2}{g^2+\lambda}+\ldots+\dfrac{z^2}{h^2+\lambda}=1$: $\tau,=\sqrt{1-\dfrac{x^2}{f^2+\lambda}-\dfrac{y^2}{g^2+\lambda}-\ldots-\dfrac{z^2}{h^2+\lambda}}$, is $=0$: and we have $\rho=0$, viz. ρ is $=0$ for all points outside the ellipsoid $\dfrac{x^2}{f^2}+\dfrac{y^2}{g^2}+\ldots+\dfrac{z^2}{h^2}=1$.

But if $\dfrac{x^2}{f^2}+\dfrac{y^2}{g^2}+\ldots+\dfrac{z^2}{h^2}<1$, then, on writing $e=0$, we have $\lambda=0$, $\tau^2=\dfrac{e^2}{\lambda}$,

$$\rho=\frac{\Gamma(\tfrac{1}{2}s+q)}{2\pi^{\frac{1}{2}s}\,\Gamma(q+1)}\cdot\lambda^{q+1}\frac{e^{2q+2}}{\lambda^{q+1}}\cdot\frac{\lambda}{e^2}\left(\frac{1}{f^2}\xi\frac{dW}{d\xi}+\frac{1}{g^2}\eta\frac{dW}{d\eta}+\ldots+\frac{1}{h^2}\zeta\frac{dW}{d\zeta}-2\frac{dW}{d\lambda}\right)_{\lambda=0}$$
$$=\frac{\Gamma(\tfrac{1}{2}s+q)}{2\pi^{\frac{1}{2}s}\,\Gamma(q+1)}\cdot\quad e^{2q}\lambda\ .\ \left(\frac{1}{f^2}\xi\frac{dW}{d\xi}+\frac{1}{g^2}\eta\frac{dW}{d\eta}+\ldots+\frac{1}{h^2}\zeta\frac{dW}{d\zeta}-2\frac{dW}{d\lambda}\right)_{\lambda=0},$$

where the term in ( ) is

$$=-C\Lambda_0\psi_0\left(+\frac{2\lambda^{-q-1}}{\Lambda_0{}^2fg\ldots h}\right)$$
$$=-2C\frac{\psi_0}{\Lambda_0 fg\ldots h}\cdot\frac{1}{\lambda^{q+1}}.$$

Hence

$$\rho=\frac{\Gamma(\tfrac{1}{2}s+q)}{2\pi^{\frac{1}{2}s}\,\Gamma(q+1)}\cdot\frac{2C\psi_0}{\Lambda_0 fg\ldots h}\left(\frac{e^2}{\lambda}\right)^q$$
$$=\frac{-\Gamma(\tfrac{1}{2}s+q)}{2\pi^{\frac{1}{2}s}\,\Gamma(q+1)}\cdot\frac{2C\psi_0}{\Lambda_0 fg\ldots h}\left(1-\frac{x^2}{f^2}-\frac{y^2}{g^2}-\ldots-\frac{z^2}{h^2}\right)^q,$$

where ψ_0, Λ_0 are what ψ, Λ become on writing therein $\lambda=0$. It will be remembered that Λ is what H becomes on changing therein θ into λ; hence Λ_0 is what H becomes on writing therein $\theta=0$.

Moreover ψ is what ϕ becomes on changing therein $\alpha, \beta, .., \gamma$ into $\xi, \eta, .., \zeta$: writing $\lambda=0$, we have $\xi=\frac{x}{f}$, $\eta=\frac{y}{g}, .., \zeta=\frac{z}{h}$; hence ψ_0 is what ϕ becomes on changing therein $\alpha, \beta, .., \gamma$ into $\frac{x}{f}, \frac{y}{g}, .., \frac{z}{h}$. And it is proper in ϕ to restore the original variables by writing $\frac{a}{\sqrt{f^2+\theta}}, \frac{b}{\sqrt{g^2+\theta}}, .., \frac{c}{\sqrt{h^2+\theta}}$ in place of $\alpha, \beta, .., \gamma$.

127. Recapitulating,

$$V=\int\frac{\rho\, dx \ldots dz}{[(a-x)^2+\ldots+(c-z)^2+e^2]^{\frac{1}{2}s+q}},$$

where, since for the value of V about to be mentioned ρ vanishes for points outside the ellipsoid, the integral is to be taken over the ellipsoid

$$\frac{x^2}{f^2}+\ldots+\frac{z^2}{h^2}=1:$$

and then, transferring a constant factor, if

$$V=\frac{(\Gamma\frac{1}{2})^s\,\Gamma(q+1)}{\Gamma(\frac{1}{2}s+q)}\Lambda_0(f\ldots h).H\phi\int_\theta^\infty\frac{t^{-q-1}\,dt}{T^2\sqrt{(t+f^2)\ldots(t+h^2)}},$$

the corresponding value of ρ is

$$\rho=\psi_0\left(1-\frac{x^2}{f^2}-\ldots-\frac{z^2}{h^2}\right)^q,$$

where Λ_0 is what H becomes on writing therein $\theta=0$, and ψ_0 is what ψ becomes on writing

$$\frac{x}{f}, .., \frac{z}{h} \text{ in place of } \alpha, .., \gamma.$$

128. Thus, putting for shortness $\Omega=t^{-q-1}\{(t+f^2)\ldots(t+h^2)\}^{-\frac{1}{2}}$, we have in the three several cases $\phi=1$, $\phi=\frac{a}{\sqrt{f^2+\theta}}$, $\phi=\frac{ab}{\sqrt{f^2+\theta}.\sqrt{g^2+\theta}}$ respectively,

$$H=1, \quad \rho=\left(1-\frac{x^2}{f^2}-\ldots-\frac{z^2}{h^2}\right)^q, \quad V=\frac{(\Gamma\frac{1}{2})^s\Gamma(1+q)}{\Gamma(\frac{1}{2}s+q)}(f\ldots h)\int_\theta^\infty \Omega dt,$$

$$H=\frac{\sqrt{f^2+\theta}}{\sqrt{f^2+\ldots+h^2}}, \quad \rho=x\,(\quad ,, \quad ,, \quad)^q, \quad V= \quad ,, \quad ,, \quad a\int_\theta^\infty\frac{f^2}{f^2+t}\,\Omega dt,$$

$$H=\frac{\sqrt{f^2+\theta}.\sqrt{g^2+\theta}}{f^2+\ldots+h^2}, \quad \rho=xy\,(\quad ,, \quad ,, \quad)^q, \quad V= \quad ,, \quad ,, \quad ab\int_\theta^\infty\frac{f^2g^2}{f^2+t\,.\,g^2+t}\,\Omega dt.$$

For the case last considered

$$\phi = \frac{\frac{1}{2}\frac{f^2a^2}{f^2+\theta}}{2q+2+\frac{1}{2}\kappa_0 f^2} + \ldots + \frac{\frac{1}{2}\frac{h^2c^2}{h^2+\theta}}{2q+2+\frac{1}{2}\kappa_0 h^2} - \frac{1}{\kappa_0},$$

$$H = \frac{\frac{1}{2}f^2(f^2+\theta)}{2q+2+\frac{1}{2}\kappa_0 f^2} + \ldots + \frac{\frac{1}{2}h^2(h^2+\theta)}{2q+2+\frac{1}{2}\kappa_0 h^2} - \frac{1}{\kappa_0}, \quad T \text{ same function with } t \text{ for } \theta,$$

$$\psi_0 = \frac{\frac{1}{2}x^2}{2q+2+\frac{1}{2}\kappa_0 f^2} + \ldots + \frac{\frac{1}{2}z^2}{2q+2+\frac{1}{2}\kappa_0 h^2} - \frac{1}{\kappa_0},$$

$$\Lambda_0 = \frac{\frac{1}{2}f^4}{2q+2+\frac{1}{2}\kappa_0 f^2} + \ldots + \frac{\frac{1}{2}h^4}{2q+2+\frac{1}{2}\kappa_0 h^2} - \frac{1}{\kappa_0},$$

where κ_0 is the root of the equation

$$\frac{1}{2q+2+\frac{1}{2}\kappa_0 f^2} + \ldots + \frac{1}{2q+2+\frac{1}{2}\kappa_0 h^2} + 1 = 0,$$

$$\rho = \left(1 - \frac{x^2}{f^2} - \ldots - \frac{z^2}{h^2}\right)^q \psi_0, \quad V = \frac{(\Gamma\frac{1}{2})^s\,\Gamma(1+q)}{\Gamma(\frac{1}{2}s+q)}(f\ldots h)\,\Lambda_0 H\phi \int_\theta^\infty T^{-2}\,t^{-q-1}\{(t+f^2)\ldots(t+h^2)\}^{-\frac{1}{2}}dt.$$

ANNEX VI. *Examples of Theorem* C. Art. Nos. 129 to 132.

129. First example: relating to the $(s+1)$-coordinal sphere $x^2 + \ldots + z^2 + w^2 = f^2$. Assume

$$V' = \frac{M}{(a^2+\ldots+c^2+e^2)^{\frac{1}{2}(s-1)}}, \quad V'' = \frac{M}{f^{s-1}}, \text{ (a constant)};$$

these values each satisfy the potential equation.

V' is not infinite for any point outside the surface, and for indefinitely large distances it is of the proper form.

V'' is not infinite for any point inside the surface; and at the surface $V' = V''$.

The conditions of the theorem are therefore satisfied. Writing

$$V = \int \frac{\rho\, dS}{\{(a-x)^2+\ldots+(c-z)^2+(e-w)^2\}^{\frac{1}{2}s-\frac{1}{2}}},$$

we have

$$\rho = -\frac{\Gamma(\frac{1}{2}s-\frac{1}{2})}{4(\Gamma\frac{1}{2})^{s+1}}\left(\frac{dW'}{d\mathfrak{s}'} + \frac{dW''}{d\mathfrak{s}''}\right),$$

where

$$W' = \frac{M}{(x^2+\ldots+z^2+w^2)^{\frac{1}{2}s-\frac{1}{2}}}, \quad W'' = \frac{M}{f^{s-1}}; \text{ hence } \frac{dW''}{d\mathfrak{s}''} = 0,$$

$$\frac{dW'}{d\mathfrak{s}'} = \left(\frac{x}{f}\frac{d}{dx} + \ldots + \frac{z}{f}\frac{d}{dz} + \frac{w}{f}\frac{d}{dw}\right)\frac{M}{(x^2+\ldots+z^2+w^2)^{\frac{1}{2}s-\frac{1}{2}}}$$

$$= -\frac{(s-1)\frac{1}{f}(x^2+\ldots+z^2+w^2)\,M}{(x^2+\ldots+z^2+w^2)^{\frac{1}{2}s+\frac{1}{2}}},$$

which at the surface is

$$=\frac{-(s-1)M}{f^s}.$$

Hence

$$\rho=\frac{(s-1)\Gamma(\tfrac{1}{2}s-\tfrac{1}{2})M}{4(\Gamma\tfrac{1}{2})^{s+1}f^s},\quad =\frac{\Gamma(\tfrac{1}{2}s+\tfrac{1}{2})M}{2(\Gamma\tfrac{1}{2})^{s+1}f^s}:\ \text{(viz. }\rho\text{ is constant).}$$

130. Writing for convenience $M=\dfrac{2(\Gamma\frac{1}{2})^{s+1}f^s}{\Gamma(\frac{1}{2}s+1)}\delta f$ (δf a constant which may be put $=1$), also $a^2+\ldots+c^2+e^2=\kappa^2$, we have $\rho=\delta f$, and consequently

$$\int\frac{\delta f\,dS}{\{(a-x)^2+\ldots+(c-z)^2+(e-w)^2\}^{\frac{1}{2}s-\frac{1}{2}}}$$

$$=\frac{2(\Gamma\tfrac{1}{2})^{s+1}f^s\,\delta f}{\Gamma(\tfrac{1}{2}s+\tfrac{1}{2})}\,\frac{1}{\kappa^{s-1}}\ \text{for exterior point }\kappa>f,$$

$$=\frac{2(\Gamma\tfrac{1}{2})^{s+1}f^s\,\delta f}{\Gamma(\tfrac{1}{2}s+\tfrac{1}{2})}\,\frac{1}{f^{s-1}}\ \text{for interior point }\kappa<f.$$

By making $a,\ldots,c,\ e$ all indefinitely large, we find

$$\int\delta f\,dS=\frac{2(\Gamma\tfrac{1}{2})^{s+1}f^s\,\delta f}{\Gamma(\tfrac{1}{2}s+\tfrac{1}{2})},$$

viz. the expression on the right-hand side is here the mass of the shell thickness δf.

Taking $s=3$, we have the ordinary formulæ for the Potential of a uniform spherical shell.

131. Suppose $s=3$, but let the surface be the infinite cylinder $x^2+y^2=f^2$. Take here

$$V'=M\log\sqrt{a^2+b^2},\quad V''=M\log f,$$

each satisfying the potential equation $\dfrac{d^2V}{da^2}+\dfrac{d^2V}{db^2}=0$; but V', instead of vanishing, is infinite at infinity, and the conditions of the theorem are not satisfied; the Potential of the cylinder is in fact infinite. But the failure is a mere consequence of the special value of s, viz. this is such that $s-2$, instead of being positive, is $=0$. Reverting to the general case of $(s+1)$-dimensional space, let the surface be the infinite cylinder $x^2+\ldots+z^2=f^2$; and assume

$$V'=\frac{M}{(a^2+\ldots+c^2)^{\frac{1}{2}(s-2)}},\quad V''=\frac{M}{f^{s-2}}\ \text{(a constant).}$$

These satisfy the potential equation; viz. as regards V', we have

$$\left(\frac{d^2}{da^2}+\ldots+\frac{d^2}{dc^2}+\frac{d^2}{de^2}\right)V'=0,\ \text{that is, }\left(\frac{d^2}{da^2}+\ldots+\frac{d^2}{dc^2}\right)V'=0.$$

V' is not infinite at any point outside the cylinder; and it vanishes at infinity, except indeed when only the coordinate e is infinite, and its form at infinity is not

$$= M \div (a^2 + \dots + c^2 + e^2)^{\frac{1}{2}(s-1)}.$$

V'' is not infinite for any point within the cylinder; and at the surface we have $V' = V''$.

We have

$$\rho = -\frac{\Gamma(\frac{1}{2}s - \frac{1}{2})}{4(\Gamma\frac{1}{2})^{s+1}} \left(\frac{dW'}{d\mathfrak{s}'} + \frac{dW''}{d\mathfrak{s}''}\right),$$

where

$$\frac{dW'}{d\mathfrak{s}'} = \frac{-(s-2)\frac{1}{f}(x^2 + \dots + z^2) M}{(x^2 + \dots + z^2)^{\frac{1}{2}s}}, \quad = \frac{-(s-2) M}{f^{s-1}} \text{ at the surface}; \quad \frac{dW''}{d\mathfrak{s}''} = 0,$$

and therefore

$$\rho = \frac{(s-2)\Gamma(\frac{1}{2}s - \frac{1}{2}) M}{4(\Gamma\frac{1}{2})^{s+1} f^{s-1}} : \text{ (viz. } \rho \text{ is constant)};$$

or, what is the same thing, writing $M = \frac{4(\Gamma\frac{1}{2})^{s+1} f^{s-1}\delta f}{(s-2)\Gamma\frac{1}{2}(s-1)}$, whence $\rho = \delta f$, and writing also $a^2 + \dots + c^2 = \kappa^2$, we have

$$\int \frac{\delta f\, dS}{\{(a-x)^2 + \dots + (c-z)^2 + (e-w)^2\}^{\frac{1}{2}s - \frac{1}{2}}}$$

$$= \frac{4(\Gamma\frac{1}{2})^{s+1} f^{s-1}\delta f}{(s-2)\Gamma(\frac{1}{2}s - \frac{1}{2})} \frac{1}{\kappa^{s-2}} \text{ for an exterior point } \kappa > f,$$

$$= \frac{4(\Gamma\frac{1}{2})^{s+1} f^{s-1}\delta f}{(s-2)\Gamma(\frac{1}{2}s - \frac{1}{2})} \frac{1}{f^{s-2}} \text{ for an interior point } \kappa < f.$$

132. This is right; but we can without difficulty bring it to coincide with the result obtained for the $(s+1)$-dimensional sphere with only $s-1$ in place of s; we may in fact, by a single integration, pass from the cylinder $x^2 + \dots + z^2 = f^2$ to the s-dimensional sphere or circle $x^2 + \dots + z^2 = f^2$, which is the base of this cylinder. Writing first $dS = d\Sigma\, dw$, where $d\Sigma$ refers to the s variables $(x, .., z)$ and the sphere $x^2 + \dots + z^2 = f^2$; or using now dS in this sense, then in place of the original dS we have $dS\, dw$: and the limits of w being $\infty, -\infty$, then in place of $e - w$ we may write simply w. This being so, and putting for shortness $(a-x)^2 + \dots + (c-z)^2 = A^2$, the integral is

$$\int_{-\infty}^{\infty} dw \int \frac{\delta f\, dS\, dw}{(A^2 + w^2)^{\frac{1}{2}(s-1)}};$$

and we have without difficulty

$$\int_{-\infty}^{\infty} \frac{dw}{(A^2 + w^2)^{\frac{1}{2}(s-1)}} = \frac{1}{A^{s-2}} \frac{\Gamma\frac{1}{2}\,\Gamma\frac{1}{2}(s-2)}{\Gamma\frac{1}{2}(s-1)}.$$

To prove it, write $w = A\tan\theta$, then the integral is in the first place converted into $\frac{2}{A^{s-2}}\int_0^{\frac{\pi}{2}}\cos^{s-3}\theta\, d\theta$, which, putting $\cos\theta = \sqrt{x}$ and therefore $\sin\theta = \sqrt{1-x}$, becomes

$$= \frac{1}{A^{s-2}}\int_0^1 x^{\frac{1}{2}-1}(1-x)^{\frac{1}{2}(s-2)-1}\, dx,$$

which has the value in question.

Hence, replacing A by its value, we have

$$\frac{\Gamma\frac{1}{2}\,\Gamma\frac{1}{2}(s-2)}{\Gamma\frac{1}{2}(s-1)}\int\frac{\delta f\, dS}{\{(a-x)^2+\ldots+(c-z)^2\}^{\frac{1}{2}(s-2)}} = \frac{4\pi^{\frac{1}{2}s}\,\Gamma(\frac{1}{2})f^{s-1}\,\delta f}{(s-2)\,\Gamma\frac{1}{2}(s-1)}\left\{\frac{1}{(a^2+\ldots+c^2)^{\frac{1}{2}(s-2)}} \text{ or } \frac{1}{f^{s-2}}\right\};$$

that is,

$$\int\frac{\delta f\, dS}{\{(a-x)^2+\ldots+(c-z)^2\}^{\frac{1}{2}(s-2)}} = \frac{4\pi^{\frac{1}{2}s} f^{s-1}\,\delta f}{(s-2)\,\Gamma\frac{1}{2}(s-2)}\left\{\frac{1}{(a^2+\ldots+c^2)^{\frac{1}{2}(s-2)}} \text{ or } \frac{1}{f^{s-2}}\right\}$$

$$= \frac{2\pi^{\frac{1}{2}s} f^{s-1}\,\delta f}{\Gamma\frac{1}{2}s}\left\{\frac{1}{(a^2+\ldots+c^2)^{\frac{1}{2}(s-2)}} \text{ or } \frac{1}{f^{s-2}}\right\};$$

viz. this is the formula for the sphere with $s-1$ instead of s.

ANNEX VII. *Example of Theorem* D. Art. Nos. 133 and 134.

133. The example relates to the $(s+1)$-dimensional sphere $x^2+\ldots+z^2+w^2=f^2$. Instead of at once assuming for V a form satisfying the proper conditions as to continuity, we assume a form with indeterminate coefficients, and make it satisfy the conditions in question. Write

$$V = \frac{M}{(a^2+\ldots+c^2+e^2)^{\frac{1}{2}s-\frac{1}{2}}} \quad \text{for } a^2+\ldots+c^2+e^2 > f^2;$$

$$= A(a^2+\ldots+c^2+e^2)+B \quad \text{for } a^2+\ldots+c^2+e^2 < f^2.$$

In order that the two values may be equal at the surface, we must have

$$\frac{M}{f^{s-1}} = Af^2+B:$$

in order that the derived functions $\frac{dV}{da}$, &c. may be equal, we must have

$$\frac{-(s-1)\,aM}{f^{s+1}} = 2Aa, \text{ \&c.},$$

viz. these are all satisfied if only $\frac{-(s-1)\,M}{f^{s+1}} = 2A.$

We have thus the values of A and B; or the exterior potential being as above

$$= \frac{M}{(a^2+\ldots+c^2+e^2)^{\frac{1}{2}s-\frac{1}{2}}},$$

the value of the interior potential must be

$$=\frac{M}{f^{s-1}}\left\{(\tfrac{1}{2}s+\tfrac{1}{2})-(\tfrac{1}{2}s-\tfrac{1}{2})\frac{a^2+\ldots+c^2+e^2}{f^2}\right\}.$$

The corresponding values of W are of course

$$\frac{M}{(x^2+\ldots+z^2+w^2)^{\frac{1}{2}s-\frac{1}{2}}} \text{ and } \frac{M}{f^{s-1}}\left\{(\tfrac{1}{2}s+\tfrac{1}{2})-(\tfrac{1}{2}s-\tfrac{1}{2})\frac{x^2+\ldots+z^2+w^2}{f^2}\right\};$$

and we thence find

$$\rho=0, \qquad \text{if } x^2+\ldots+z^2+w^2>f^2,$$

$$\rho=-\frac{\Gamma(\frac{1}{2}s-\frac{1}{2})}{4(\Gamma\frac{1}{2})^{s+1}}\{-4(\tfrac{1}{2}s-\tfrac{1}{2})(\tfrac{1}{2}s+\tfrac{1}{2})\}\frac{M}{f^{s+1}}, \quad =\frac{\Gamma(\frac{1}{2}s+\frac{3}{2})}{(\Gamma\frac{1}{2})^{s+1}}\frac{M}{f^{s+1}},$$

$$\text{if } x^2+\ldots+z^2+w^2<f^2.$$

Assuming for M the value $\dfrac{(\Gamma\frac{1}{2})^{s+1}}{\Gamma(\frac{1}{2}s+\frac{3}{2})}f^{s+1}$, the last value becomes $\rho=1$; writing for shortness $a^2+\ldots+c^2+e^2=\kappa^2$, we have

$$V=\int\frac{dx\ldots dz\,dw}{\{(a-x)^2+\ldots+(c-z)^2+(e-w)^2\}^{\frac{1}{2}s+\frac{1}{2}}} \text{ over } (s+1)\text{-dimensional sphere } x^2+\ldots+z^2+w^2=f^2,$$

$$=\frac{(\Gamma\frac{1}{2})^{s+1}}{\Gamma(\frac{1}{2}s+\frac{3}{2})}\frac{f^{s+1}}{\kappa^{s-1}}, \quad \text{for an exterior point } \kappa>f,$$

$$=\frac{(\Gamma\frac{1}{2})^{s+1}}{\Gamma(\frac{1}{2}s+\frac{3}{2})}\{(\tfrac{1}{2}s+\tfrac{1}{2})f^2-(\tfrac{1}{2}s-\tfrac{1}{2})\kappa^2\}, \text{ for an interior point } \kappa<f.$$

134. The case of the ellipsoid $\dfrac{x^2}{f^2}+\ldots+\dfrac{z^2}{h^2}=1$ for $s+1$-dimensional space may be worked out by the theorem; this is, in fact, what is done in tridimensional space by Lejeune-Dirichlet in his Memoir of 1846 above referred to (p. 321).

ANNEX VIII. *Prepotentials of the Homaloids.* Art. Nos. 135 to 137.

135. We have in tridimensional space the series of figures—the plane, the line, the point; and there is in like manner in $(s+1)$-dimensional space a corresponding series of $(s+1)$ terms; the $(s+1)$-coordinal plane—the line, the point: say these are the homaloids or homaloidal figures. And, taking the density as uniform, or, what is the same thing, $=1$, we may consider the prepotentials of these several figures in regard to an attracted point, which, for greater simplicity, is taken not to be on the figure.

136. The integral may be written

$$V=\int\frac{dw\ldots dt}{\{(a-x)^2+\ldots+(c-z)^2+(d-w)^2+\ldots+(e-t)^2+u^2\}^{\frac{1}{2}s+q}},$$

which still relates to a $(s+1)$-dimensional space: the $(s+1)$ coordinates of the attracted point are $(a,\ldots,c,\,d,\ldots,e,\,u)$, instead of being $(a,\ldots,c,\,e)$; viz. we have the

s' coordinates $(a,\ldots,c)$, the $s-s'$ coordinates $(d,\ldots,e)$, and the $(s+1)$th coordinate u: and the integration is extended over the $(s-s')$-dimensional figure $w=-\infty$ to $+\infty,\ldots,$ $t=-\infty$ to $+\infty$. And it is also assumed that q is positive.

It is at once clear that we may reduce the integral to

$$V=\int\frac{dw\ldots dt}{\{(a-x)^2+\ldots+(c-z)^2+u^2+w^2+\ldots+t^2\}^{\frac{1}{2}s+q}},$$

say for shortness

$$=\int\frac{dw\ldots dt}{(A^2+w^2+\ldots+t^2)^{\frac{1}{2}s+q}},$$

where A^2, $=(a-x)^2+\ldots+(c-z)^2+u^2$, is a constant as regards the integration, and where the limits in regard to each of the $s-s'$ variables are $-\infty,+\infty$.

We may for these variables write $r\xi,\ldots,r\zeta$, where $\xi^2+\ldots+\zeta^2=1$; and we then have $w^2+\ldots+t^2=r^2$, $dw\ldots dt=r^{s-s'-1}\,dr\,dS$, where dS is the element of surface of the $(s-s')$-coordinal unit-sphere $\xi^2+\ldots+\zeta^2=1$. We thus obtain

$$V=\int\frac{r^{s-s'-1}\,dr}{\{A^2+r^2\}^{\frac{1}{2}s+q}}\int dS,$$

where the integral in regard to r is taken from 0 to ∞, and the integral $\int dS$ over the surface of the unit-sphere; hence by Annex I. the value of this last factor is $=\dfrac{2(\Gamma\frac{1}{2})^{s-s'}}{\Gamma\frac{1}{2}(s-s')}$. The integral represented by the first factor will be finite, provided only $\frac{1}{2}s'+q$ be positive; which is the case for any value whatever of s', if only q be positive.

The first factor is an integral such as is considered in Annex II.; to find its value we have only to write $r=A\sqrt{x}$, and we thus find it to be

$$=\frac{1}{(A^2)^{\frac{1}{2}s'+q}}\,\frac{1}{2}\int_0^\infty\frac{x^{\frac{1}{2}s-\frac{1}{2}s'-1}\,dx}{(1+x)^{\frac{1}{2}s+q}},\ \text{viz.}=\frac{1}{A^{s'+2q}}\,\frac{\frac{1}{2}\Gamma\frac{1}{2}(s-s')\,\Gamma(\frac{1}{2}s'+q)}{\Gamma(\frac{1}{2}s+q)},$$

and we thus have

$$V=\frac{1}{A^{s'+2q}}\,\frac{(\Gamma\frac{1}{2})^{s-s'}\,\Gamma(\frac{1}{2}s'+q)}{\Gamma(\frac{1}{2}s+q)},$$

$$=\frac{(\Gamma\frac{1}{2})^{s-s'}\,\Gamma(\frac{1}{2}s'+q)}{\Gamma(\frac{1}{2}s+q)}\,\frac{1}{\{(a-x)^2+\ldots+(c-z)^2+u^2\}^{\frac{1}{2}s'+q}}.$$

137. As a verification, observe that the prepotential equation $\square V=0$, that is,

$$\left(\frac{d^2}{da^2}+\ldots+\frac{d^2}{dc^2}+\frac{d^2}{dd^2}+\ldots+\frac{d^2}{de^2}+\frac{d^2}{du^2}+\frac{2q+1}{u}\frac{d}{du}\right)V=0,$$

for a function V, which contains only the $s'+1$ variables $(a,\ldots,c,u)$, becomes

$$\left(\frac{d^2}{da^2}+\ldots+\frac{d^2}{dc^2}+\frac{d^2}{du^2}+\frac{2q+1}{u}\frac{d}{du}\right)V=0,$$

which is satisfied by V, a constant multiple of $\{(a-x)^2+\ldots+(c-z)^2+u^2\}^{\frac{1}{2}-s'-q}$.

ANNEX IX. *The* GAUSS-JACOBI *Theory of Epispheric Integrals.* Art. No. 138.

138. The formula obtained (Annex IV. No. 110) is proved only for positive values of m; but writing therein $q=0$, $m=-\tfrac{1}{2}$, it becomes

$$\int \frac{dx \ldots dz}{\sqrt{1-\frac{x^2}{f^2}-\ldots-\frac{z^2}{h^2}}\,\{(a-x)^2+\ldots+(c-z)^2+e^2\}^{\frac{1}{2}s}}$$

$$=\frac{(\Gamma\tfrac{1}{2})^s}{\Gamma\tfrac{1}{2}s} f \ldots h \int^{\infty} dt \,.\, t^{-1}\left(1-\frac{a^2}{t+f^2}-\ldots-\frac{c^2}{t+h^2}-\frac{e^2}{t}\right)^{-\frac{1}{2}}\{(t+f^2)\ldots(t+h^2)\}^{-\frac{1}{2}},$$

a formula which is obtainable as a particular case of the more general formula

$$\int \frac{dS}{\{(*\between x, \ldots, z,\ w)^2\}^{\frac{1}{2}s}}=\frac{2\,(\Gamma\tfrac{1}{2})^s}{\Gamma(\tfrac{1}{2}s)}\int_{-\text{A}}^{\infty} dt \frac{1}{\sqrt{-\text{Disct.}\,\{(*\between X, \ldots, Z, W, T)^2+t\,(X^2+\ldots+Z^2+W^2+T^2)\}}},$$

(notation to be presently explained), being a result obtained by Jacobi by a process which is in fact the extension to any number of variables of that used by Gauss* in his Memoir "Determinatio attractionis quam ..... exerceret planeta, &c." (1818). I proceed to develop this theory.

139. Jacobi's process has reference to a class of s-tuple integrals (including some of those here previously considered) which may be termed "epispheric": viz. considering the $(s+1)$ variables $(x, \ldots, z,\ w)$ connected by the equation $x^2+\ldots+z^2+w^2=1$, or say they are the coordinates of a point on a $(s+1)$-tuple unit-sphere, then the form is $\int UdS$, where dS is the element of the surface of the unit-sphere, and U is any function of the $s+1$ coordinates; the integral is taken to be of the form $\int \frac{dS}{\{(*\between x, \ldots, z,\ w,\ 1)^2\}^{\frac{1}{2}s}}$, and we then obtain the general result above referred to.

Before going further it is convenient to remark that, taking as independent variables the s coordinates $x, \ldots, z$, we have $dS=\frac{dx \ldots dz}{dw}$, where w stands for $\pm\sqrt{1-x^2-\ldots-z^2}$; we must in obtaining the integral take account of the two values of w, and finally extend the integral to the values of $x, \ldots, z$ which satisfy $x^2+\ldots+z^2<1$.

If, as is ultimately done, in place of $x, \ldots, z$ we write $\frac{x}{f}, \ldots, \frac{z}{h}$ respectively, then the value of dS is $=\frac{1}{f \ldots h}\frac{dx \ldots dz}{w}$, where w now stands for $\pm\sqrt{1-\frac{x^2}{f^2}-\ldots-\frac{z^2}{h^2}}$; we must, in finding the value of the integral, take account of the two values of w, and finally extend the integral to the values of $x, \ldots, z$ which satisfy $\frac{x^2}{f^2}+\ldots+\frac{z^2}{h^2}<1$.

* [*Ges. Werke*, t. iii, pp. 331—355.]

140. The determination of the integral depends upon formulæ for the transformation of the spherical element dS, and of the quadric function $(x, y,.., z, w, 1)^2$.

First, as regards the spherical element dS; let the $s+1$ variables $x, y,.., z, w$ which satisfy $x^2+y^2+\ldots+z^2+w^2=1$ be regarded as functions of the s independent variables $\theta, \phi,.., \psi$; then we have

$$dS = \begin{vmatrix} x, & y,.., & z, & w \\ \frac{dx}{d\theta}, & \frac{dy}{d\theta},.., & \frac{dz}{d\theta}, & \frac{dw}{d\theta} \\ \frac{dx}{d\phi}, & \frac{dy}{d\phi},.., & \frac{dz}{d\phi}, & \frac{dw}{d\phi} \\ \vdots & & & \\ \frac{dx}{d\psi}, & \frac{dy}{d\psi},.., & \frac{dz}{d\psi}, & \frac{dw}{d\psi} \end{vmatrix} d\theta\, d\phi \ldots d\psi, = \frac{\partial(x, y,.., z, w)}{\partial(\theta, \phi,.., \psi, *)} d\theta\, d\phi \ldots d\psi, \text{ for shortness.}$$

Suppose we effect on the $s+1$ variables $(x, y,.., z, w)$ a transformation

$$x, y,.., z, w = \frac{X}{T}, \frac{Y}{T},.., \frac{Z}{T}, \frac{W}{T},$$

thus introducing for the moment $s+2$ variables $X, Y,.., Z, W, T$, which satisfy identically $X^2+Y^2+\ldots+Z^2+W^2-T^2=0$; then, considering these as functions of the foregoing s independent variables $\theta, \phi,.., \psi$, we have

$$dS = \frac{1}{T^{s+1}} \begin{vmatrix} X, & Y,.., & Z, & W \\ \frac{dX}{d\theta}, & \frac{dY}{d\theta},.., & \frac{dZ}{d\theta}, & \frac{dW}{d\theta} \\ \frac{dX}{d\phi}, & \frac{dY}{d\phi},.., & \frac{dZ}{d\phi}, & \frac{dW}{d\phi} \\ \vdots & & & \\ \frac{dX}{d\psi}, & \frac{dY}{d\psi},.., & \frac{dZ}{d\psi}, & \frac{dW}{d\psi} \end{vmatrix} d\theta\, d\phi \ldots d\psi = \frac{1}{T^{s+1}} \frac{\partial(X, Y,.., Z, W)}{\partial(\theta, \phi,.., \psi, *)} d\theta\, d\phi \ldots d\psi.$$

141. Considering next the $s+2$ variables $X, Y,.., Z, W, T$ as linear functions (with constant terms) of the $s+1$ new variables $\xi, \eta,.., \zeta, \omega$, or say as linear functions of the $s+2$ quantities $\xi, \eta,.., \zeta, \omega, 1$: which implies between them a linear relation

$$aX+bY+\ldots+cZ+dW+eT=1:$$

and assuming that we have *identically*

$$X^2+Y^2+\ldots+Z^2+W^2-T^2=\xi^2+\eta^2+\ldots+\zeta^2+\omega^2-1,$$

so that, in consequence of the left-hand side being $=0$, the right-hand side is also $=0$; viz. $\xi, \eta,.., \zeta, \omega$ are connected by

$$\xi^2+\eta^2+\ldots+\zeta^2+\omega^2=1:$$

let $d\Sigma$ represent the spherical element belonging to the coordinates $\xi, \eta, .., \zeta, \omega$. Considering these as functions of the foregoing s independent variables $\theta, \phi, .., \psi$, we have

$$d\Sigma = \begin{vmatrix} \xi, & \eta, \ldots, & \zeta, & \omega \\ \frac{d\xi}{d\theta}, & \frac{d\eta}{d\theta}, \ldots, & \frac{d\zeta}{d\theta}, & \frac{d\omega}{d\theta} \\ \frac{d\xi}{d\phi}, & \frac{d\eta}{d\phi}, \ldots, & \frac{d\zeta}{d\phi}, & \frac{d\omega}{d\phi} \\ \vdots & & & \\ \frac{d\xi}{d\psi}, & \frac{d\eta}{d\psi}, \ldots, & \frac{d\zeta}{d\psi}, & \frac{d\omega}{d\psi} \end{vmatrix} d\theta\, d\phi \ldots d\psi = \frac{\partial(\xi, \eta, .., \zeta, \omega)}{\partial(\theta, \phi, .., \psi, *)} d\theta\, d\phi \ldots d\psi.$$

142. In this expression we have $\xi, \eta, .., \zeta, \omega$, each of them a linear function of the $s+2$ quantities $X, Y, .., Z, W, T$; the determinant is consequently a linear function of $s+2$ like determinants obtained by substituting for the variables any $s+1$ out of the $s+2$ variables $X, Y, .., Z, W, T$; but in virtue of the equation

$$X^2 + Y^2 + \ldots + Z^2 + W^2 - T^2 = 0,$$

these $s+2$ determinants are proportional to the quantities $X, Y, .., Z, W, T$ respectively, and the determinant thus assumes the form

$$\frac{aX + bY + \ldots + cZ + dW + eT}{T} \Delta,$$

where Δ is the like determinant with $(X, Y, .., Z, W)$, and where the coefficients $a, b, .., c, d, e$ are precisely those of the linear relation $aX + bY + \ldots + cZ + dW + eT = 1$; the last-mentioned expression is thus $= \frac{1}{T}\Delta$, or, substituting for Δ its value, we have

$$d\Sigma = \frac{1}{T} \frac{\partial(X, Y, .., Z, W)}{\partial(\theta, \phi, .., \psi, *)} d\theta\, d\phi \ldots d\psi;$$

viz. comparing with the foregoing expression for dS we have

$$dS = \frac{1}{T^s} d\Sigma,$$

which is the requisite formula for the transformation of dS.

143. Consider the integral

$$\int \frac{dS}{\{(*\between x, y, .., z, w, 1)^2\}^{\frac{1}{2}s}},$$

which, from its containing a single quadric function, may be called "one-quadric." Then effecting the foregoing transformation,

$$x, y, .., z, w = \frac{X}{T}, \frac{Y}{T}, \ldots, \frac{Z}{T}, \frac{W}{T},$$

and observing that

$$(*\between x,\ y,..,\ z,\ w,\ 1)^2 = \frac{1}{T^2}(*\between X,\ Y,..,\ Z,\ W,\ T)^2,$$

the integral becomes

$$= \int \frac{d\Sigma}{\{(*\between X,\ Y,..,\ Z,\ W,\ T)^2\}^{\frac{1}{2}s}},$$

where $X, Y,.., Z, W, T$ denote given linear functions (with constant coefficients) of the $s+1$ variables $\xi, \eta,.., \zeta, \omega$, or, what is the same thing, given linear functions of the $s+2$ quantities $\xi, \eta,.., \zeta, \omega, 1$, such that identically

$$X^2 + Y^2 + \ldots + Z^2 + W^2 - T^2 = \xi^2 + \eta^2 + \ldots + \zeta^2 + \omega^2 - 1.$$

We have then $\xi^2 + \eta^2 + \ldots + \zeta^2 + \omega^2 - 1 = 0$, and $d\Sigma$ as the corresponding spherical element.

144. We may have $X, Y,.., Z, W, T$ such linear functions of $\xi, \eta,.., \zeta, \omega, 1$ that not only

$$X^2 + Y^2 + \ldots + Z^2 + W^2 - T^2 = \xi^2 + \eta^2 + \ldots + \zeta^2 + \omega^2 - 1$$

as above, but also

$$(*\between X,\ Y,..,\ Z,\ W,\ T)^2 = A\xi^2 + B\eta^2 + \ldots + C\zeta^2 + E\omega^2 - L;$$

this being so, the integral becomes

$$\int \frac{d\Sigma}{\{A\xi^2 + B\eta^2 + \ldots + C\zeta^2 + E\omega^2 - L\}^{\frac{1}{2}s}},$$

where the $s+2$ coefficients $A, B,.., C, E, L$ are given by means of the identity

$$\begin{aligned} &-(\theta + A)(\theta + B)\ldots(\theta + C)(\theta + E)(\theta + L) \\ &\quad = \text{Disct. } \{(*\between X,\ Y,..,\ Z,\ W,\ T)^2 + \theta(X^2 + Y^2 + \ldots + Z^2 + W^2 - T^2)\}; \end{aligned}$$

viz. equating the discriminant to zero, we have an equation in θ, the roots whereof are $-A, -B,.., -C, -E, -L$.

The integral is

$$\int \frac{d\Sigma}{\{(A-L)\xi^2 + (B-L)\eta^2 + \ldots + (C-L)\zeta^2 + (E-L)\omega^2\}^{\frac{1}{2}s}},$$

which is of the form

$$\int \frac{d\Sigma}{\{a\xi^2 + b\eta^2 + \ldots + c\zeta^2 + e\omega^2\}^{\frac{1}{2}s}},$$

where I provisionally assume that $a, b,.., c, e$ are all positive.

145. To transform this, in place of the $s+1$ variables $\xi, \eta,.., \zeta, \omega$ connected by $\xi^2 + \eta^2 + \ldots + \zeta^2 + \omega^2 = 1$, we introduce the $s+1$ variables $x, y,.., z, w$, such that

$$x = \frac{\xi\sqrt{a}}{\rho},\quad y = \frac{\eta\sqrt{b}}{\rho},\ \ldots,\quad z = \frac{\zeta\sqrt{c}}{\rho},\quad w = \frac{\omega\sqrt{d}}{\rho},$$

where

$$\rho^2 = a\xi^2 + b\eta^2 + \ldots + c\zeta^2 + e\omega^2,$$

and consequently

$$x^2 + y^2 + \ldots + z^2 + w^2 = 1.$$

Hence, writing dS to denote the spherical element corresponding to the point $(x, y, \ldots, z, w)$, we have, by a former formula,

$$dS = \frac{1}{\rho^{s+1}} \frac{\partial(\xi\sqrt{a}, \eta\sqrt{b}, \ldots, \zeta\sqrt{c}, \omega\sqrt{e})}{\partial(\theta, \phi, \ldots, \psi, *)} d\theta\, d\phi \ldots d\psi$$

$$= \frac{(ab \ldots ce)^{\frac{1}{2}}}{\rho^{s+1}} d\Sigma;$$

or, what is the same thing,

$$\frac{d\Sigma}{\{a\xi^2 + b\eta^2 + \ldots + c\zeta^2 + e\omega^2\}^{\frac{1}{2}(s+1)}} = \frac{1}{(ab \ldots ce)^{\frac{1}{2}}} dS.$$

Hence, integrating each side, and observing that $\int dS$, taken over the whole spherical surface $x^2 + y^2 + \ldots + z^2 + w^2 = 1$, is $= 2(\Gamma\tfrac{1}{2})^{s+1} \div \Gamma(\tfrac{1}{2}s + \tfrac{1}{2})$, we have

$$\int \frac{d\Sigma}{\{a\xi^2 + b\eta^2 + \ldots + c\zeta^2 + e\omega^2\}^{\frac{1}{2}(s+1)}} = \frac{2(\Gamma\frac{1}{2})^{s+1}}{\Gamma(\frac{1}{2}s + \frac{1}{2})} \frac{1}{(ab \ldots ce)^{\frac{1}{2}}}.$$

146. For $a, b, \ldots, c, e$ write herein $a + \theta, b + \theta, \ldots, c + \theta, e + \theta$ respectively, and multiply each side by θ^{q-1}, where q is any positive integer or fractional number less than $\tfrac{1}{2}s$: integrate from $\theta = 0$ to $\theta = \infty$. On the left-hand side, attending to the relation $\xi^2 + \eta^2 + \ldots + \zeta^2 + \omega^2 = 1$, the integral in regard to θ is

$$\int_0^\infty \frac{\theta^{q-1}\, d\theta}{\{\rho^2 + \theta\}^{\frac{1}{2}(s+1)}},$$

where $\rho^2, = a\xi^2 + b\eta^2 + \ldots + c\zeta^2 + e\omega^2$, is independent of θ as before; the value of the definite integral is

$$= \frac{\Gamma\{\frac{1}{2}(s+1) - q\}\, \Gamma(q)}{\Gamma\frac{1}{2}(s+1)} \frac{1}{\rho^{s+1-2q}},$$

which, replacing ρ by its value and multiplying by $d\Sigma$, and prefixing the integral sign, gives the left-hand side; hence, forming the equation and dividing by a numerical factor, we have

$$\int \frac{d\Sigma}{(a\xi^2 + \ldots + c\zeta^2 + e\omega^2)^{\frac{1}{2}(s+1)-q}} = \frac{2(\Gamma\frac{1}{2})^{s+1}}{\Gamma q\, \Gamma\{\frac{1}{2}(s+1) - q\}} \int_0^\infty dt \,.\, t^{-q-1} \{(t+a) \ldots (t+c)(t+e)\}^{-\frac{1}{2}}.$$

In particular, if $q = -\tfrac{1}{2}$, then

$$\int \frac{d\Sigma}{(a\xi^2 + \ldots + c\zeta^2 + e\omega^2)^{\frac{1}{2}s}} = \frac{2(\Gamma\frac{1}{2})^s}{\Gamma\frac{1}{2}s} \int_0^\infty dt \,.\, t^{-\frac{1}{2}} \{(t+a) \ldots (t+c)(t+e)\}^{-\frac{1}{2}};$$

or, if for $a,.., c, e$ we restore the values $A-L,.., C-L, E-L$, then

$$\int \frac{d\Sigma}{(A\xi^2+\ldots+C\zeta^2+E\omega^2-L)^{\frac{1}{2}s}} = \frac{2(\Gamma\tfrac{1}{2})^s}{\Gamma\tfrac{1}{2}s}\int_0^\infty dt\,.\,t^{-\frac{1}{2}}\{(t+A-L)\ldots(t+C-L)(t+E-L)\}^{-\frac{1}{2}},$$

$$= \frac{2(\Gamma\tfrac{1}{2})^s}{\Gamma\tfrac{1}{2}s}\int_{-L}^\infty dt\,\{(t+A)\ldots(t+C)(t+E)(t+L)\}^{-\frac{1}{2}};$$

viz. we thus have

$$\int \frac{dS}{\{(*\between x,.., z, w, 1)^2\}^{\frac{1}{2}s}} = \frac{2(\Gamma\tfrac{1}{2})^s}{\Gamma\tfrac{1}{2}s}\int_{-L}^\infty dt\,\{(t+A)\ldots(t+C)(t+E)(t+L)\}^{-\frac{1}{2}},$$

where $(t+A)\ldots(t+C)(t+E)(t+L)$ is in fact a given rational and integral function of t; viz. it is

$$= -\text{ Disct. }\{(*\between X,.., Z, W, T)^2+t(X^2+\ldots+Z^2+W^2-T^2)\}.$$

147. Consider, in particular, the integral

$$\int \frac{dS}{\{(a-fx)^2+\ldots+(c-hz)^2+(e-kw)^2+l^2\}^{\frac{1}{2}s}};$$

here

$$\begin{aligned}(*\between X,.., Z, W, T)^2+t(X^2+\ldots+Z^2+W^2-T^2)\\ &= (aT-fX)^2+\ldots+(cT-hZ)^2+(eT-kW)^2+l^2T^2+t(X^2+\ldots+Z^2+W^2-T^2)\\ &= (f^2+t)X^2+\ldots+(h^2+t)Z^2+(k^2+t)W^2+(a^2+\ldots+c^2+e^2+l^2-t)T^2\\ &\qquad -2afXT-\ldots-2chZT-2ekWT;\end{aligned}$$

viz. the discriminant taken negatively is

$$\begin{vmatrix} t+f^2,\ldots\ , & -af \\ \vdots & \\ \ldots, t+h^2, & -ch \\ -af,\ldots -ch, & -(a^2+\ldots+c^2+e^2+l^2)+t \end{vmatrix},$$

which is

$$\begin{aligned}&= (t+f^2)\ldots(t+h^2)(t+k^2)\left(t-a^2-\ldots-c^2-e^2-l^2+\frac{a^2f^2}{t+f^2}+\ldots+\frac{c^2h^2}{t+h^2}+\frac{e^2k^2}{t+k^2}\right),\\ &= \{t(t+f^2)\ldots(t+h^2)(t+k^2)\}\left(1-\frac{a^2}{t+f^2}-\ldots-\frac{c^2}{t+h^2}-\frac{e^2}{t+k^2}-\frac{l^2}{t}\right)\\ &= (t+A)\ldots(t+C)(t+E)(t+L);\end{aligned}$$

and consequently $-A,.., -C, -E, -L$ are the roots of the equation

$$1-\frac{a^2}{t+f^2}-\ldots-\frac{c^2}{t+h^2}-\frac{e^2}{t+k^2}-\frac{l^2}{t}=0.$$

148. The roots are all real; moreover there is one and only one positive root. Hence, taking $-L$ to be the positive root, we have $A,.., C, E, -L$ all positive, and therefore *à fortiori* $A-L,.., C-L, E-L$ all positive: which agrees with a foregoing

provisional assumption. Or, writing for greater convenience θ to denote the positive quantity $-L$, that is, taking θ to be the positive root of the equation

$$1-\frac{a^2}{\theta+f^2}-\ldots-\frac{c^2}{\theta+h^2}-\frac{e^2}{\theta+k^2}-\frac{l^2}{\theta}=0,$$

we have

$$\int\frac{dS}{\{(a-fx)^2+\ldots+(c-hz)^2+(e-kw)^2+l^2\}^{\frac{1}{2}s}}$$

$$=\frac{2(\Gamma\tfrac{1}{2})^s}{\Gamma\tfrac{1}{2}s}\int_\theta^\infty dt\,\frac{1}{\sqrt{t(t+f^2)\ldots(t+h^2)(t+k^2)\left(1-\dfrac{a^2}{t+f^2}-\ldots-\dfrac{c^2}{t+h^2}-\dfrac{e^2}{t+k^2}-\dfrac{l^2}{t}\right)}};$$

or, what is the same thing, we have

$$\frac{1}{f\ldots h}\int\frac{dx\ldots dz}{\pm w\{(a-x)^2+\ldots+(c-z)^2+(e\mp kw)^2+l^2\}^{\frac{1}{2}s}}$$

$$=\frac{\Gamma\tfrac{1}{2}s}{2(\Gamma\tfrac{1}{2})^s}\int_\theta^\infty dt\left(1-\frac{a^2}{t+f^2}-\ldots-\frac{c^2}{t+h^2}-\frac{e^2}{t+k^2}-\frac{l^2}{t}\right)^{-\frac{1}{2}}\{t(t+f^2)\ldots(t+h^2)(t+k^2)\}^{-\frac{1}{2}},$$

where on the left-hand side w now denotes $\sqrt{1-\dfrac{x^2}{f^2}-\ldots-\dfrac{z^2}{h^2}}$, and the limiting equation is $\dfrac{x^2}{f^2}+\ldots+\dfrac{z^2}{h^2}=1$.

149. Suppose $l=0$: then, if

$$\frac{a^2}{f^2}+\ldots+\frac{c^2}{h^2}+\frac{e^2}{k^2}>1,$$

the equation

$$1-\frac{a^2}{\theta+f^2}-\ldots-\frac{c^2}{\theta+h^2}-\frac{e^2}{\theta+k^2}=0$$

has a positive root differing from zero, which may be represented by the same letter θ; but if

$$\frac{a^2}{f^2}+\ldots+\frac{c^2}{h^2}+\frac{e^2}{k^2}<1,$$

then the positive root of the original equation becomes $=0$; viz. as l gradually diminishes to zero, the positive root θ also diminishes and becomes ultimately zero.

Hence, writing $l=0$, we have

$$\int\frac{dS}{\{(a-fx)^2+\ldots+(c-hz)^2+(e-kw)^2\}^{\frac{1}{2}s}},$$

or, what is the same thing,

$$\frac{1}{f\ldots h}\int\frac{dx\ldots dz}{\pm w\{(a-x)^2+\ldots+(c-z)^2+(e\mp kw)^2\}^{\frac{1}{2}s}},$$

$$=\frac{2(\Gamma\tfrac{1}{2})^s}{\Gamma\tfrac{1}{2}s}\int_\theta^\infty dt\left(1-\frac{a^2}{t+f^2}-\ldots-\frac{c^2}{t+h^2}-\frac{e^2}{t+k^2}\right)^{-\frac{1}{2}}\{t(t+f^2)\ldots(t+h^2)(t+k^2)\}^{-\frac{1}{2}},$$

θ now denoting either the positive root of the equation

$$1-\frac{a^2}{\theta+f^2}-\ldots-\frac{c^2}{\theta+h^2}-\frac{e^2}{\theta+k^2}=0,$$

or else 0, according as

$$\frac{a^2}{f^2}+\ldots+\frac{c^2}{h^2}+\frac{e^2}{k^2}>1 \text{ or } <1.$$

In the case $\frac{a^2}{f^2}+\ldots+\frac{e^2}{k^2}<1$, the inferior limit being then 0, this is, in fact, Jacobi's theorem (*Crelle*, t. XII. p. 69, 1834); but Jacobi does not consider the general case where l is not $=0$, nor does he give explicitly the formula in the other case

$$l=0,\ \frac{a^2}{f^2}+\ldots+\frac{c^2}{h^2}+\frac{e^2}{k^2}>1.$$

150. Suppose $k=0$, e being in the first instance not $=0$: then the former alternative holds good; and observing, in regard to the form which contains $\pm w$ in the denominator, that we can now take account of the two values by simply multiplying by 2, we have

$$\int\frac{dS}{\{(a-fx)^2+\ldots+(c-hz)^2+e^2\}^{\frac{1}{2}s}},\ =\frac{2}{f\ldots h}\int\frac{dx\ldots dz}{w\,\{(a-x)^2+\ldots+(c-z)^2+e^2\}^{\frac{1}{2}s}},$$

(w on the right-hand side denoting $\sqrt{1-\frac{x^2}{f^2}-\ldots-\frac{z^2}{h^2}}$, and the limiting equation being $\frac{x^2}{f^2}+\ldots+\frac{z^2}{h^2}=1$), each

$$=\frac{2\,(\Gamma\tfrac{1}{2})^s}{\Gamma\tfrac{1}{2}s}\int_\theta^\infty dt\left(1-\frac{a^2}{t+f^2}-\ldots-\frac{c^2}{t+h^2}-\frac{e^2}{t}\right)^{-\frac{1}{2}}t^{-1}\{(t+f^2)\ldots(t+h^2)\}^{-\frac{1}{2}},$$

where θ is here the positive root of the equation $1-\frac{a^2}{\theta+f^2}-\ldots-\frac{c^2}{\theta+h^2}-\frac{e^2}{\theta}=0$, which is the formula referred to at the beginning of the present Annex. We may in the formula write $e=0$, thus obtaining the theorem under two different forms for the cases $\frac{a^2}{f^2}+\ldots+\frac{c^2}{h^2}>1$ and <1 respectively.

ANNEX X. *Methods of* LEJEUNE-DIRICHLET *and* BOOLE. Art. Nos. 151 to 162.

151. The notion, that the density ρ is a discontinuous function vanishing for points outside the attracting mass, has been made use of in a different manner by Lejeune-Dirichlet (1839) and Boole (1857): viz. supposing that ρ has a given value $f(x,.., z)$ within a given closed surface S and is $=0$ outside the surface, these geometers in the expression of a potential or prepotential integral replace ρ by a definite integral which possesses the discontinuity in question, viz. it is $=f(x,.., z)$ for points inside

the surface and $=0$ for points outside the surface; and then in the potential or prepotential integral they extend the integration over the whole of infinite space, thus getting rid of the equation of the surface as a limiting equation for the multiple integral.

152. Lejeune-Dirichlet's paper "Sur une nouvelle méthode pour la détermination des intégrales multiples" is published in *Comptes Rendus*, t. VIII. pp. 155—160 (1839), and *Liouville*, t. IV. pp. 164—168 (same year). The process is applied to the form

$$-\frac{1}{p-1}\frac{d}{da}\int\frac{dx\,dy\,dz}{\{(a-x)^2+(b-y)^2+(c-z)^2\}^{\frac{1}{2}(p-1)}}$$

taken over the ellipsoid $\dfrac{x^2}{\alpha^2}+\dfrac{y^2}{\beta^2}+\dfrac{z^2}{\gamma^2}=1$; but it would be equally applicable to the triple integral itself, or say to the s-tuple integral

$$\int\frac{dx\ldots dz}{\{(a-x)^2+\ldots+(c-z)^2\}^{\frac{1}{2}s+q}},$$

or, indeed, to

$$\int\frac{dx\ldots dz}{\{(a-x)^2+\ldots+(c-z)^2+e^2\}^{\frac{1}{2}s+q}}$$

taken over the ellipsoid $\dfrac{x^2}{f^2}+\ldots+\dfrac{z^2}{h^2}=1$; but it may be as well to attend to the first form, as more resembling that considered by the author.

153. Since $\dfrac{2}{\pi}\displaystyle\int_0^\infty\frac{\sin\phi}{\phi}\cos\lambda\phi\,d\phi$ is $=1$ or 0, according as λ is <1 or >1, it follows that the integral is equal to the real part of the following expression,

$$\frac{2}{\pi}\int_0^\infty d\phi\,\frac{\sin\phi}{\phi}\int e^{i\left(\frac{x^2}{f^2}+\ldots+\frac{z^2}{h^2}\right)}\frac{dx\ldots dz}{\{(a-x)^2+\ldots+(c-z)^2\}^{\frac{1}{2}s+q}},$$

where the integrations in regard to $x,\ldots, z$ are now to be extended from $-\infty$ to $+\infty$ for each variable. A further transformation is necessary: since

$$\frac{1}{\sigma^r}=\frac{1}{\Gamma r}e^{-r\pi i}\int_0^\infty d\psi\,.\,\psi^{r-1}e^{i\sigma\psi},\qquad \sigma \text{ positive, and } r \text{ positive and } <1,$$

writing herein $(a-x)^2+\ldots+(c-z)^2$ for σ, and $\frac{1}{2}s+q$ for r, we have

$$\frac{1}{\{(a-x)^2+\ldots+(c-z)^2\}^{\frac{1}{2}s+q}}=\frac{1}{\Gamma(\frac{1}{2}s+q)}e^{-(\frac{1}{2}s+q)\pi i}\int_0^\infty d\psi\,.\,\psi^{\frac{1}{2}s+q-1}e^{i\psi\{(a-x)^2+\ldots+(c-z)^2\}},$$

and the value is thus

$$=\frac{2}{\pi\Gamma(\frac{1}{2}s+q)}e^{-(\frac{1}{2}s+q)\frac{\pi i}{2}}\int_0^\infty d\phi\,\frac{\sin\phi}{\phi}\int_0^\infty d\psi\,.\,\psi^{\frac{1}{2}s+q-1}\int e^{i\left(\frac{x^2}{f^2}+\ldots+\frac{z^2}{h^2}\right)\phi}e^{-i\psi\{(a-x)^2+\ldots+(c-z)^2\}}dx\ldots dz,$$

where the integral in regard to the variables $(x, .., z)$ is

$$= e^{i\psi(a^2+\ldots+c^2)} \int dx\, e^{i\left\{\left(\psi+\frac{\phi}{f^2}\right)x^2+2a\psi x\right\}} \ldots \int dz\, e^{i\left\{\left(\psi+\frac{\phi}{h^2}\right)z^2-2c\psi z\right\}};$$

and here the x-integral is

$$= e^{\frac{1}{4}i\pi} \sqrt{\frac{f^2\pi}{f^2\psi+\phi}}\, e^{-\frac{a^2f^2\psi^2 i}{f^2\psi+\phi}},$$

and the like for the other integrals up to the z-integral. The resulting value is thus

$$= \frac{2}{\pi\Gamma(\frac{1}{2}s+q)} e^{-\frac{1}{2}q\pi i} \int_0^\infty \frac{\sin\phi}{\phi} d\phi \int_0^\infty d\psi\,.\,\psi^{\frac{1}{2}s+q-1}\, e^{\psi\phi i\left(\frac{a^2}{\phi+f^2\psi}+\ldots+\frac{c^2}{\phi+h^2\psi}\right)} \frac{\pi^{\frac{1}{2}s} f\ldots h}{\sqrt{(\phi+f^2\psi)\ldots(\phi+h^2\psi)}},$$

which, putting therein $\psi = \frac{\phi}{t}$, $d\psi = -\frac{\phi}{t^2}dt$, is

$$= \frac{2\pi^{\frac{1}{2}s-1}}{\Gamma(\frac{1}{2}s+q)} (f\ldots h)\, e^{-\frac{1}{2}q\pi i} \int_0^\infty dt \frac{t^{-q-1}}{\sqrt{(f^2+t)\ldots(h^2+t)}} \int_0^\infty e^{i\phi\left(\frac{a^2}{f^2+t}+\ldots+\frac{c^2}{h^2+t}\right)} \sin\phi\,.\,\phi^{q-1}\, d\phi.$$

154. But we have to consider only the real part of this expression; viz. writing for shortness $\sigma = \frac{a^2}{f^2+t} + \ldots + \frac{c^2}{h^2+t}$, we require the real part of

$$e^{-\frac{1}{2}q\pi i} \int_0^\infty e^{i\sigma\phi}\, \phi^{q-1} \sin\phi\, d\phi.$$

Writing here for $\sin\phi$ its exponential value $\frac{1}{2i}(e^{i\phi}-e^{-i\phi})$, and using the formula

$$\frac{1}{\sigma^q} = \frac{1}{\Gamma q} e^{-q\pi i} \int_0^\infty d\phi\,.\,\phi^{q-1} e^{i\sigma\phi} \qquad (\sigma \text{ positive}),$$

and the like one

$$\frac{1}{(-\sigma)^q} = \frac{1}{\Gamma q} e^{q\pi i} \int_0^\infty d\phi\,.\,\phi^{q-1} e^{i\sigma\phi} \qquad (\sigma \text{ negative}),$$

(in which formulæ q must be positive and less than 1), we see that the real part in question is $= 0$, or is

$$-\frac{\Gamma q \sin(q+1)\pi}{2(1-\sigma)^q}, \quad = \frac{\pi}{2\Gamma(1-q)} \frac{1}{(1-\sigma)^q},$$

according as $\sigma > 1$ or $\sigma < 1$.

155. If the point is interior, $\frac{a^2}{f^2} + \ldots + \frac{c^2}{h^2} < 1$, and consequently also $\sigma < 1$, and the value, writing $(\Gamma\frac{1}{2})^2$ instead of π, is

$$= \frac{(\Gamma\frac{1}{2})^s}{\Gamma(\frac{1}{2}s+q)\Gamma(1-q)} (f\ldots h) \int_0^\infty dt\,.\,t^{-q-1} \{(t+f^2)\ldots(t+h^2)\}^{-\frac{1}{2}} \left(1 - \frac{a^2}{f^2+t} - \ldots - \frac{c^2}{h^2+t}\right)^{-q}.$$

But if the point be exterior, $\frac{a^2}{f^2}+\ldots+\frac{c^2}{h^2}>1$, and hence, writing θ for the positive root of the equation, $\sigma=1$; viz. θ is the positive root of the equation $\frac{a^2}{f^2+\theta}+\ldots+\frac{c^2}{h^2+\theta}=1$; then $t=0$, σ is greater than 1, and continues so as t increases, until, for $t=\theta$, σ becomes $=1$, and for larger values of t we have $\sigma<1$; and the expression thus is

$$=\frac{(\Gamma\tfrac{1}{2})^s}{\Gamma(\tfrac{1}{2}s+q)\,\Gamma(1-q)}(f\ldots h)\int_\theta^\infty dt\,.\,t^{-q-1}\,\{(t+f^2)\ldots(t+h^2)\}^{-\frac{1}{2}}\left(1-\frac{a^2}{f^2+t}-\ldots-\frac{c^2}{h^2+t}\right)^{-q};$$

viz. the two expressions, in the cases of an interior point and an exterior point respectively, give the value of the integral

$$\int\frac{dx\ldots dz}{\{(a-x)^2+\ldots+(c-z)^2\}^{\frac{1}{2}s+q}}.$$

This is, in fact, the formula of Annex IV. No. 110, writing therein $e=0$ and $m=-q$.

156. Boole's researches are contained in two memoirs dated 1846, "On the Analysis of Discontinuous Functions," *Trans. Royal Irish Academy*, vol. XXI. (1848), pp. 124—139, and "On a certain Multiple Definite Integral," do. pp. 140—150 (the particular theorem about to be referred to is stated in the postscript of this memoir), and in the memoir "On the Comparison of Transcendents, with certain applications to the theory of Definite Integrals," *Phil. Trans.* vol. CXLVII. (1857), pp. 745—803, the theorem being the third example, p. 794. The method is similar to, and was in fact suggested by, that of Lejeune-Dirichlet; the auxiliary theorem made use of in the memoir of 1857 for the representation of the discontinuity being

$$\frac{f(x)}{t^i}=\frac{1}{\pi\Gamma i}\int_{-\infty}^\infty\int_0^\infty\int_0^\infty da\,dv\,ds\cos\{(a-x-ts)\,v+\tfrac{1}{2}i\pi\}\,v^is^{i-1}f(a),$$

which is a deduction from Fourier's theorem.

Changing the notation (and in particular writing s and $\frac{1}{2}s+q$ for his n and i), the method is here applied to the determination of the s-tuple integral

$$V=\int dx\ldots dz\,\frac{\phi\left(\frac{x^2}{f^2}+\ldots+\frac{z^2}{h^2}\right)}{\{(a-x)^2+\ldots+(c-z)^2+e^2\}^{\frac{1}{2}s+q}},$$

where ϕ is an arbitrary function, taken over the ellipsoid $\frac{x^2}{f^2}+\ldots+\frac{z^2}{h^2}=1$.

157. The process is as follows: we have

$$\frac{\phi\left(\frac{x^2}{f^2}+\ldots+\frac{z^2}{h^2}\right)}{\{(a-x)^2+\ldots+(c-z)^2+e^2\}^{\frac{1}{2}s+q}}=\frac{1}{\pi\,\Gamma(\frac{1}{2}s+q)}\int_0^1\int_0^\infty\int_0^\infty du\,dv\,d\tau\,v^{\frac{1}{2}s+q}\,t^{\frac{1}{2}s+q-1}$$

$$\cos\left\{\left(u-\frac{x^2}{f^2}-\ldots-\frac{z^2}{h^2}-\tau\,\{(a-x)^2+\ldots+(c-z)^2+e^2\}\right)v+\tfrac{1}{2}(\tfrac{1}{2}s+q)\,\pi\right\}\phi u;$$

viz. the right-hand side is here equal to the left-hand side or is $=0$, according as $\frac{x^2}{f^2}+\ldots+\frac{z^2}{h^2}<1$ or >1. V is consequently obtained by multiplying the right-hand side by $dx \ldots dz$ and integrating from $-\infty$ to $+\infty$ for each variable.

Hence, changing the order of the integration,

$$V=\frac{1}{\pi\,\Gamma(\tfrac{1}{2}s+q)}\int_0^1\int_0^\infty\int_0^\infty du\,dv\,d\tau\,v^{\frac{1}{2}s+q}\,\tau^{\frac{1}{2}s+q-1}\,\phi u\,.\,\Omega,$$

where

$$\Omega=\int dx\ldots dz\cos\left\{\left(u-e^2\tau-\frac{x^2}{f^2}-\ldots-\frac{z^2}{h^2}+\tau\{(a-x)^2+\ldots+(c-z)^2\}\right)v+\tfrac{1}{2}(\tfrac{1}{2}s+q)\pi\right\}.$$

Now

$$\frac{x^2}{f^2}+\tau(a-x)^2=\frac{1+f^2\tau}{f^2}\xi^2+\frac{\tau a^2}{1+f^2\tau},\ \ldots,\frac{z^2}{h^2}+\tau(c-z)^2=\frac{1+h^2\tau}{h^2}\zeta^2+\frac{\tau c^2}{1+h^2\tau},$$

if

$$\xi=x-\frac{f^2\tau a}{1+f^2\tau},\qquad \ldots,\ \zeta=z-\frac{h^2\tau c}{1+h^2\tau}.$$

158. Substituting, and integrating with respect to $\xi,\ldots,\zeta$ between the limits $-\infty$, $+\infty$, we have

$$\Omega=\frac{(f\ldots h)\,\pi^{\frac{1}{2}s}}{\{(1+f^2\tau)\ldots(1+h^2\tau)\}^{\frac{1}{2}}\,v^{\frac{1}{2}s}}\cos\left\{\left(u-e^2\tau-\frac{a^2\tau}{1+f^2\tau}-\ldots-\frac{c^2\tau}{1+h^2\tau}\right)v+\tfrac{1}{2}q\pi\right\};$$

or, what is the same thing, writing $\frac{1}{t}$ in place of τ, this is

$$\Omega=\frac{(f\ldots h)\,\pi^{\frac{1}{2}s}\,t^{\frac{1}{2}}}{\{(f^2+t)\ldots(h^2+t)\}^{\frac{1}{2}}\,v^{\frac{1}{2}s}}\cos\left\{\left(u-\frac{a^2}{f^2+t}-\ldots-\frac{c^2}{h^2+t}-\frac{e^2}{t}\right)v+\tfrac{1}{2}q\pi\right\};$$

that is, writing

$$\sigma=\frac{a^2}{f^2+t}+\ldots+\frac{c^2}{h^2+t}+\frac{e^2}{t},$$

we have

$$V=\frac{\pi^{\frac{1}{2}s-1}(f\ldots h)}{\Gamma(\tfrac{1}{2}s+q)}\int_0^1\int_0^\infty\int_0^\infty du\,dv\,dt\,\frac{t^{-q-1}\,v^q\cos\{(u-\sigma)v+\tfrac{1}{2}q\pi\}\,\phi u}{\{(t+f^2)\ldots(t+h^2)\}^{\frac{1}{2}}};$$

or, writing $\pi^{\frac{1}{2}s-1}=\frac{1}{\pi}(\Gamma\tfrac{1}{2})^s$, this is

$$=\frac{(\Gamma\tfrac{1}{2})^s\,(f\ldots h)}{\Gamma(\tfrac{1}{2}s+q)}\int_0^\infty dt\,.\,t^{-q-1}\{(t+f^2)\ldots(t+h^2)\}^{-\frac{1}{2}}\frac{1}{\pi}\int_0^1\int_0^\infty du\,dv\,.\,v^q\cos\{(u-\sigma)v+\tfrac{1}{2}q\pi\}\,\phi u.$$

159. Boole writes

$$\frac{1}{\pi}\int_0^1\int_0^\infty du\,dv\,v^q\cos\{(u-\sigma)v+\tfrac{1}{2}q\pi\}\,\phi u=\left(-\frac{d}{d\sigma}\right)^q\phi(\sigma);$$

viz. starting from Fourier's theorem,

$$\frac{1}{\pi}\int_0^1\int_0^\infty du\,dv\cos(u-\sigma)v\,.\,\phi u=\phi(\sigma),$$

where $\phi(\sigma)$ is regarded as vanishing except when σ is between the limits 0, 1, and the limits of u are taken to be 1, 0 accordingly, then, according to an admissible theory of general differentiation, we have the result in question. He has in the formula $\frac{1}{s}$ instead of my t; and he proceeds, "Here σ increases continually with s. As s varies from 0 to ∞, σ also varies from 0 to ∞. To any positive limits of σ will correspond positive limits of s; and these, as will hereafter appear—this refers to his note B—, will in certain cases replace the limits 0 and ∞ in the expression for V."

160. It seems better to deal with the result in the following manner, as in part shown p. 803 of Boole's memoir. Writing the integral in the form

$$V=\frac{(\Gamma\tfrac{1}{2})^s (f\ldots h)}{\pi\,\Gamma(\tfrac{1}{2}s+q)}\int_0^1\int_0^\infty du\,dt\,.\,t^{-q-1}\{(t+f^2)\ldots(t+h^2)\}^{-\frac{1}{2}}\phi(u)\int_0^\infty dv\,.\,v^q\cos\{(u-\sigma)v+\tfrac{1}{2}q\pi\},$$

effect the integration in regard to v; viz. according as u is greater or less than σ, then

$$\int_0^\infty dv\,.\,v^q\cos\{(u-\sigma)v+\tfrac{1}{2}q\pi\}=\frac{\Gamma(q+1)\sin(q+1)\pi}{(u-\sigma)^{q+1}},\text{ or }0,$$

$$=\frac{\pi}{\Gamma(-q)(u-\sigma)^{q+1}}\quad,\text{ or }0;$$

and consequently, writing for σ its value,

$$V=\frac{(\Gamma\tfrac{1}{2})^s(f\ldots h)}{\Gamma(-q)\,\Gamma(\tfrac{1}{2}s+q)}\int_0^1\int_0^\infty du\,dt\left\{t^{-q-1}\{(t+f^2)\ldots(t+h^2)\}^{-\frac{1}{2}}\right.$$

$$\left.\left(u-\frac{a^2}{f^2+t}-\ldots-\frac{c^2}{h^2+t}-\frac{e^2}{t}\right)^{-q-1}\phi u\right\},\text{ or }0,\text{ as above.}$$

161. To further explain this, consider t as an x-coordinate and u as a y-coordinate; then, tracing the curve

$$y=\frac{a^2}{f^2+x}+\ldots+\frac{c^2}{h^2+x}+\frac{e^2}{x},$$

for positive values of x this is a mere hyperbolic branch, as shown in the figure, viz. $x=0$, $y=\infty$; and as x continually increases to ∞, y continually decreases to zero.

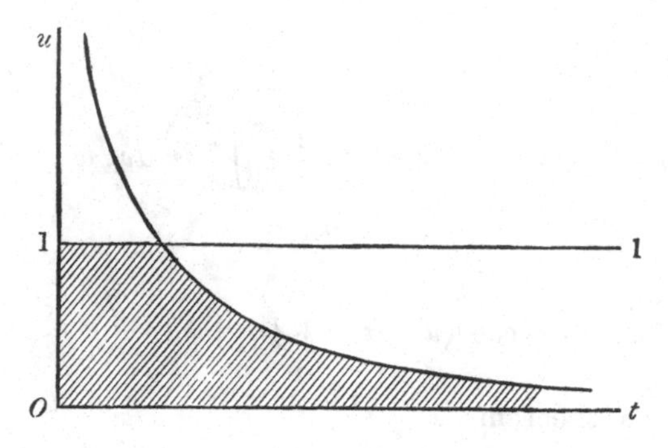

The limits are originally taken to be from $u=0$ to $u=1$ and $t=0$ to $t=\infty$, viz. over the infinite strip bounded by the lines tO, $O1$, 11; but within these limits the function under the integral sign is to be replaced by zero whenever the values u, t are such that u is less than $\frac{a^2}{f^2+t}+\ldots+\frac{c^2}{h^2+t}+\frac{e^2}{t}$, viz. when the values belong to a point in the shaded portion of the strip; the integral is therefore to be extended only over the unshaded portion of the strip; viz. the value is

$$V=\frac{(\Gamma\frac{1}{2})^s(f\ldots h)}{\Gamma(-q)\,\Gamma(\frac{1}{2}s+q)}\iint du\,dt\,.\,t^{-q-1}\{(t+f^2)\ldots(t+h^2)\}^{-\frac{1}{2}}\left(u-\frac{a^2}{f^2+t}-\ldots-\frac{c^2}{h^2+t}-\frac{e^2}{t}\right)^{-q-1}\phi u,$$

the double integral being taken over the unshaded portion of the strip; or, what is the same thing, the integral in regard to u is to be taken from $u=\frac{a^2}{f^2+t}+\ldots+\frac{c^2}{h^2+t}+\frac{e^2}{t}$ (say from $u=\sigma$) to $u=1$, and then the integral in regard to t is to be taken from $t=\theta$ to $t=\infty$, where, as before, θ is the positive root of the equation $\sigma=1$, that is, of $\frac{a^2}{f^2+\theta}+\ldots+\frac{c^2}{h^2+\theta}+\frac{e^2}{\theta}=1$.

162. Write $u=\sigma+(1-\sigma)\,x$, and therefore $u-\sigma=(1-\sigma)\,x$, $1-u=(1-\sigma)\,(1-x)$ and $du=(1-\sigma)\,dx$; then the limits (1, 0) of x correspond to the limits $(1, \sigma)$ of u, and the formula becomes

$$V=\frac{(\Gamma\frac{1}{2})^s(f\ldots h)}{\Gamma(-q)\,\Gamma(\frac{1}{2}s+q)}\int_\theta^\infty dt.t^{-q-1}\{(t+f^2)\ldots(t+h^2)\}^{-\frac{1}{2}}(1-\sigma)^{-q-1}\int_0^1 dx\,.\,x^{-q-1}\,\phi\,\{\sigma+(1-\sigma)\,x\},$$

where σ is retained in place of its value $\frac{a^2}{f^2+t}+\ldots+\frac{c^2}{h^2+t}+\frac{e^2}{t}$. This is, in fact, a form (deduced from Boole's result in the memoir of 1846) given by me, *Cambridge and Dublin Mathematical Journal*, vol. II. (1847), p. 219, [44].

If in particular $\phi u=(1-u)^{q+m}$, then $\phi\,\{\sigma+(1-\sigma)\,x\}=(1-\sigma)^{q+m}\,(1-x)^{q+m}$, and thence

$$\int_0^1 x^{-q-1}\,\{\phi\sigma+(1-\sigma)\,x\}\,dx=(1-\sigma)^m\int_0^1 x^{-q-1}\,(1-x)^{q+m}\,dx,$$

$$=\frac{\Gamma(-q)\Gamma(1+q+m)}{\Gamma(1+m)}\,(1-\sigma)^m;$$

and then, restoring for σ its value, we have

$$V=\frac{(\Gamma\frac{1}{2})^s\,\Gamma(1+q+m)}{\Gamma(\frac{1}{2}s+q)\,\Gamma(1+m)}(f\ldots h)\int_\theta^\infty dt.t^{-q-1}\{(t+f^2)\ldots(t+h^2)\}^{-\frac{1}{2}}\left(1-\frac{a^2}{f^2+t}-\ldots-\frac{c^2}{h^2+t}-\frac{e^2}{t}\right)^m$$

as the value of the integral

$$\int\frac{\left(1-\frac{x^2}{f^2}-\ldots-\frac{z^2}{h^2}\right)^{q+m}dx\ldots dz}{\{(a-x)^2+\ldots+(c-z)^2+e^2\}^{\frac{1}{2}s+q}}$$

taken over the ellipsoid $\frac{x^2}{f^2}+\ldots+\frac{z^2}{h^2}=1$. This is, in fact, the theorem of Annex IV. No. 110 in its general form; but the proof assumes that q is positive.

608.

[EXTRACT FROM A] REPORT ON MATHEMATICAL TABLES.

[From the *Report of the British Association for the Advancement of Science*, (1873), pp. 3, 4.]

It was necessary as a preliminary to form a classification of mathematical (numerical) tables; and the following classification was drawn up by Prof. Cayley and adopted by the Committee.

A. Auxiliary for non-logarithmic computations.

1. Multiplication.

2. Quarter-squares.

3. Squares, cubes, and higher powers, and reciprocals.

B. Logarithmic and circular.

4. Logarithms (Briggian) and antilogarithms (do.); addition and subtraction logarithms, &c.

5. Circular functions (sines, cosines, &c.), natural, and lengths of circular arcs.

6. Circular functions (sines, cosines, &c.), logarithmic.

C. Exponential.

7. Hyperbolic logarithms.

8. Do. antilogarithms (e^x) and h.l tan $(45° + \frac{1}{2}\phi)$, and hyperbolic sines, cosines, &c., natural and logarithmic.

D. Algebraic constants.

9. Accurate integer or fractional values. Bernoulli's Numbers, $\Delta^n 0^m$, &c. Binomial coefficients.

10. Decimal values auxiliary to the calculation of series.

E. 11. Transcendental constants, e, π, γ, &c., and their powers and functions.

F. Arithmological.

12. Divisors and prime numbers. Prime roots. The Canon arithmeticus, &c.

13. The Pellian equation.

14. Partitions.

15. Quadratic forms $a^2 + b^2$, &c., and partition of numbers into squares, cubes, and biquadrates.

16. Binary, ternary, &c. quadratic, and higher forms.

17. Complex theories.

G. Transcendental functions.

18. Elliptic.

19. Gamma.

20. Sine-integral, cosine-integral, and exponential-integral.

21. Bessel's and allied functions.

22. Planetary coefficients for given $\dfrac{a}{a'}$.

23. Logarithmic transcendental.

24. Miscellaneous.

Several of these classes need some little explanation. Thus D 9 and 10 are intended to include the same class of constants, the only difference being that in 9 accurate values are given, while in 10 they are only approximate; thus, for example, the accurate Bernoulli's numbers as vulgar fractions, and the decimal values of the same to (say) ten places are placed in different classes, as the former are of theoretical interest, while the latter are only of use in calculation. It is not necessary to enter into further detail with respect to the classification, as in point of fact it is only very partially followed in the Report.

609.

ON THE ANALYTICAL FORMS CALLED FACTIONS.

[From the *Report of the British Association for the Advancement of Science*, (1875), p. 10.]

A FACTION is a product of differences such that each letter occurs the same number of times; thus we have a quadrifaction where each letter occurs twice, a cubifaction where each letter occurs three times, and so on. A broken faction is one which is a product of factions having no common letter; thus

$$(a-b)^2(c-d)(d-e)(e-c)$$

is a broken quadrifaction, the product of the quadrifactions

$$(a-b)^2 \text{ and } (c-d)(d-e)(e-c).$$

We have, in regard to quadrifactions, the theorem that every quadrifaction is a sum of broken quadrifactions such that each component quadrifaction contains two or else three letters. Thus we have the identity

$$2(a-b)(b-c)(c-d)(d-a)=(b-c)^2 . (a-d)^2-(c-a)^2 . (b-d)^2+(a-b)^2 . (c-d)^2,$$

which verifies the theorem in the case of a quadrifaction of four letters; but the verification even in the next following case of a quadrifaction of five letters is a matter of some difficulty.

The theory is connected with that of the invariants of a system of binary quantics.

610.

ON THE ANALYTICAL FORMS CALLED TREES, WITH APPLICATION TO THE THEORY OF CHEMICAL COMBINATIONS.

[From the *Report of the British Association for the Advancement of Science*, (1875), pp. 257—305.]

I HAVE in two papers "On the Analytical forms called Trees," *Phil. Mag.* vol. XIII. (1857), pp. 172—176, [203], and ditto, vol. XX. (1859), pp. 374—378, [247], considered this theory; and in a paper "On the Mathematical Theory of Isomers," ditto, vol. XLVII. (1874), p. 444, [586], pointed out its connexion with modern chemical theory. In particular, as regards the paraffins C_nH_{2n+2}, we have n atoms of carbon connected by $n-1$ bands, under the restriction that from each carbon-atom there proceed at most 4 bands (or, in the language of the papers first referred to, we have n knots connected by $n-1$ branches), in the form of a tree; for instance, $n=5$, such forms (and the only such forms) are

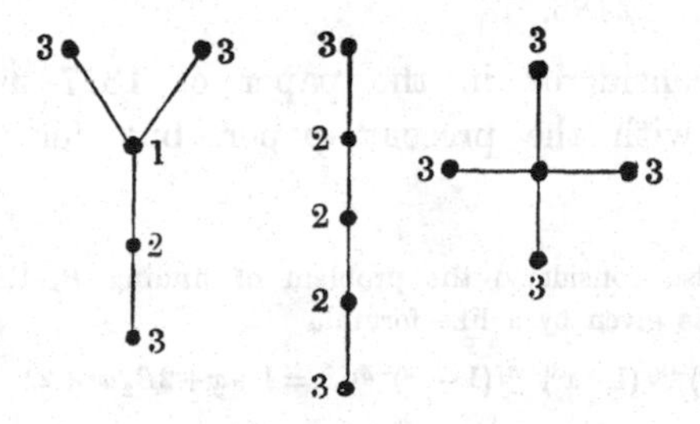

And if, under the foregoing restriction of only 4 bands from a carbon-atom, we connect with each carbon-atom the greatest possible number of hydrogen-atoms, as shown in the diagrams by the affixed numerals, we see that the number of hydrogen-atoms is 12 $(=2.5+2)$; and we have thus the representations of three different paraffins, C_5H_{12}. It should be observed that the tree-symbol of the paraffin is completely determined by means of the tree formed with the carbon-atoms, or say of the carbon-tree, and that the question of the determination of the theoretic number of the paraffins C_nH_{2n+2} is consequently that of the determination of the number of the carbon-trees of n knots, viz. the number of trees with n knots, subject to the condition that the number of branches from each knot is at most $=4$.

In the paper of 1857, which contains no application to chemical theory, the number of branches from a knot was unlimited; and, moreover, the trees were considered as issuing each from one knot taken as a root, so that, $n=5$, the trees regarded as distinct (instead of being as above only 3) were in all 9, viz. these were

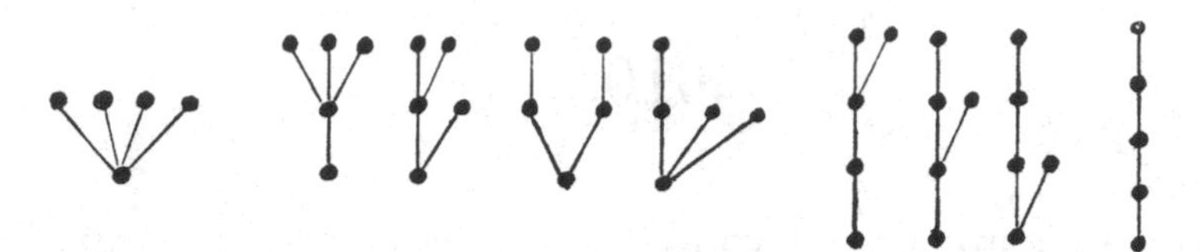

which, regarded as issuing from the bottom knots, are in fact distinct; while, taking them as issuing each from a properly selected knot, they resolve themselves into the above-mentioned 3 forms. The problem considered was in fact that of the "general root-trees with n knots"—*general*, inasmuch as the number of branches from a knot was without limit; *root-trees*, inasmuch as the enumeration was made on the principle last referred to. It was found that for

knots.................	1,	2,	3,	4,	5,	6,	7,	8,......
No. of trees was ...	1,	1,	2,	4.	9,	20,	48,	115,......
=	1,	A_1,	A_2,	A_3,	A_4,	A_5,	A_6,	A_7,...... ;

the law being given by the equation

$$(1-x)^{-1}(1-x^2)^{-A_1}(1-x^3)^{-A_2}(1-x^4)^{-A_3}\ldots = 1 + A_1 x + A_2 x^2 + A_3 x^3 + A_4 x^4 + \ldots;$$

but the next following numbers A_8, A_9, A_{10}, the correct values of which are 286, 719, 1842, were given erroneously as 306, 775, 2009. I have since calculated two more terms, A_{11}, $A_{12} = 4766$, 12486.

The other questions considered in the paper of 1857 and in that of 1859 have less immediate connexion with the present paper, but for completeness I reproduce the results in a Note*.

* In the paper of 1857 I also considered the problem of finding B_r the number with r free branches, with bifurcations at least: this was given by a like formula

$$(1-x)^{-1}(1-x^2)^{-B_2}(1-x^3)^{-B_3}(1-x^4)^{-B_4}\ldots = 1+x+2B_2x^2+2B_3x^3+2B_4x^4\ldots,$$

leading to

$B_r =$	1,	2,	5,	12,	33,	90, ...

for

$r =$	2,	3,	4,	5,	6,	7, ...

In the paper of 1859, the question is to find the number of trees with a given number m of terminal knots: we have here

$$\phi m = 1.2.3\ldots(m-1) \text{ coefficient of } x^{m-1} \text{ in } \frac{1}{2-e^x},$$

To count the trees on the principle first referred to, we require the notions of "centre" and "bicentre," due, I believe, to Sylvester; and to establish these we require the notions of "main branch" and "altitude": viz. in a tree, selecting any knot at pleasure as a root, the branches which issue from the root, each with all the branches that belong to it, are the main branches, and the distance of the furthest knot, measured by the number of intermediate branches, is the altitude of the main

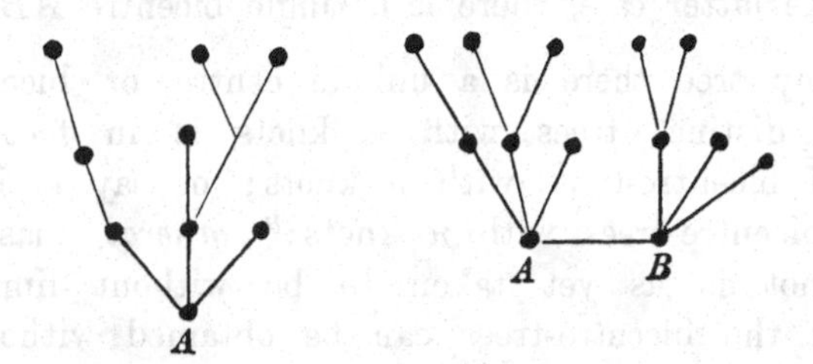

branch. Thus in the left-hand figure, taking A as the root, there are 3 main branches of the altitudes 3, 3, 1 respectively: in the right-hand figure, taking A as the root, there are 4 main branches of the altitudes 2, 2, 1, 3 respectively; and we have then the theorem that in every tree there is either one and only one centre, or else one and only one bicentre; viz. we have (as in the left-hand figure) a centre A which is such that there issue from it two or more main branches of altitudes equal to each other and superior to those of the other main branches (if any); or else (as in the right-hand figure) a bicentre AB, viz. two contiguous knots, such that issuing from A (but not counting AB), and issuing from B (but not counting BA), we have two or more main branches, one at least from A and one at least from B, of altitudes equal to each other and superior to those of the other main branches in question (if any). The theorem, once understood, is proved without difficulty: we consider two terminal knots, the distance of which, measured by the number of intermediate branches, is greater than or equal to that of any other two terminal knots; if, as in the left-hand figure, the distance is even, then the central knot A is the centre of the tree; if, as in the right-hand figure, the distance is odd, then the two central knots AB form the bicentre of the tree.

In the former case, observe that if G, H are the two terminal knots, the distance of which is $=2\lambda$, then the distance of each from A is $=\lambda$, and there cannot be

giving the values

for

$\phi m=$	1,	1,	3,	13,	75,	541,	4683,	47293, ...
$m=$	1,	2,	3,	4,	5,	6,	7,	8, ...

But if from each non-terminal knot there ascend two and only two branches, then in this case $\phi m=$coefficient of x^{m-1} in $\dfrac{1-\sqrt{1-4x}}{2x}$, viz. we have the very simple form

$$\phi m=\frac{1.3.5\ldots 2m-3}{1.2.3\ldots m}2^{m-1},$$

giving

for

$\phi m=$	1,	1,	2,	5,	14,	42, ...
$m=$	1,	2,	3,	4,	5,	7, ...

any other terminal knot I, the distance of which from A is greater than λ (for, if there were, then the distance of I from G or else from H would be greater than 2λ); there cannot be any two terminal knots I, J, the distance of which is greater than 2λ; and if there are any two knots I, J, the distance of which is $=2\lambda$, then these belong to different main branches, the distance of each of them from A being $=\lambda$; whence, starting with I, J (instead of G, H), we obtain the same point A as centre. Similarly, in the latter case, there is a single bicentre AB.

Hence, since in any tree there is a unique centre or bicentre, the question of finding the number of distinct trees with n knots is in fact that of finding the number of centre- and bicentre-trees with n knots; or say it is the problem of the "general centre- and bicentre-trees with n knots:" *general*, inasmuch as the number of branches from a knot is as yet taken to be without limit; or since (as will appear) the number of the bicentre-trees can be obtained without difficulty when the problem of the root-trees is solved, the problem is that of the "general centre-trees with n knots." It will appear that the solution depends upon and is very readily derived from that of the foregoing problem of general root-trees, so that this last has to be considered, not only for its own sake, but with a view to that of the centre-trees. And in each of the two problems we doubly divide the whole system of trees according to the number of the main branches, issuing from the root or centre as the case may be, and according to the altitude of the longest main branch or branches, or say the altitude of the tree; so that the problem really is, for a given number of knots, a given number of main branches, and a given altitude, to find the number of root-trees, or (as the case may be) centre-trees.

We next introduce the restriction that the number of branches from any knot is equal to a given number at most; viz. according as this number is $=2$, 3 or 4, we have, say oxygen-trees, boron-trees*, and carbon-trees respectively; and these are, as before, root-trees or centre- or bicentre-trees, as the case may be. The case where the number is 2 presents no difficulty: in fact, if the number of knots be $=n$, then the number of root-trees is either $\frac{1}{2}(n+1)$ or $\frac{1}{2}n$; viz. $n=3$ and $n=4$, the root-trees are

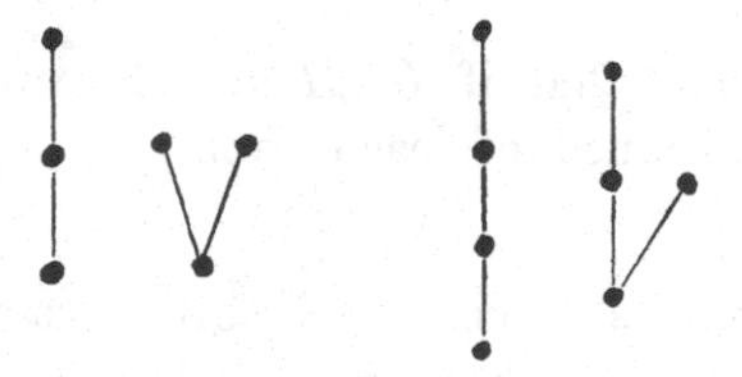

and the number of centre- or bicentre-trees is always $=1$: viz. n odd, there is one centre-tree; and n even, one bicentre-tree; it is only considered as a particular case of the general theorem. The case where the number is $=3$ is analytically interesting: although there may not exist, for any 3-valent element, a series of hydrogen compounds

* I should have said nitrogen-trees; but it appears to me that nitrogen is of necessity 5-valent, as shown by the compound, Ammonium-Chloride, $=NH_4Cl$. Of course, the word boron is used simply to stand for a 3-valent element.

B_nH_{n+2} corresponding to the paraffins. The case, where the number is $=4$ or say the carbon-trees, is that which presents the chief chemical interest, as giving the paraffins C_nH_{2n+2}; and I call to mind here that the theory of the carbon-root trees is established as an analytical result for its own sake and as the foundation for the other case, but that it is the number of the carbon centre- and bicentre-trees which is the number of the paraffins.

The theory extends to the case where the number of branches from a knot is at most $=5$, or $=$ any larger number; but I have not developed the formula.

I pass now to the analytical theory: considering first the case of general root-trees, we endeavour to find for a given altitude N the number of trees of a given number of knots n and main branches α, or say the generating function

$$\Sigma\Omega t^\alpha x^n,$$

where the coefficient Ω gives the number of the trees in question. And we assume that the problem is solved for the cases of the several inferior altitudes $0, 1, 2, 3, .., N-1$.

This being so, observe that a tree of altitude N can be built up as shown in the figure, which I call the edification diagram, by combining one or more trees of altitude $N-1$ with a single tree of altitude not exceeding $N-1$; viz. in the figure, $N=3$, we have the two trees a, b, each of altitude 2, combined, as shown by the

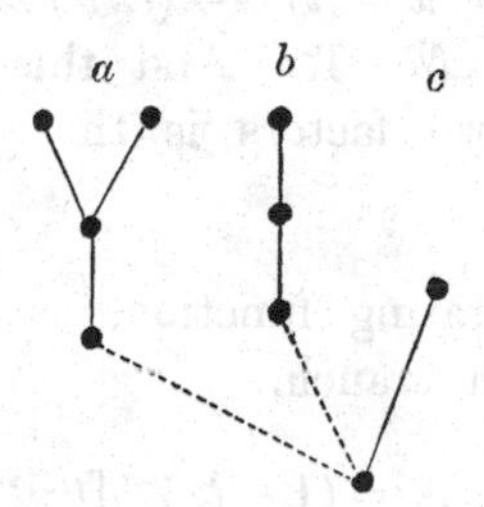

dotted lines, with the tree c of altitude 1: the whole number of knots in the resulting tree is the sum of the number of knots on the three trees a, b, c: the number of main branches is equal to the number of the trees a, b, plus the number of main branches of the tree c. It is to be observed that the tree c may reduce itself to the tree $(\cdot)$ of one knot and of altitude zero; but each of the trees a, b, as being of the altitude $N-1$, must contain at least N knots.

Taking $N=2$ or any larger number, it is hence easy to see that the required generating function $\Sigma\Omega t^\alpha x^n$ is

$$=(1-tx^N)^{-1}(1-tx^{N+1})^{-l_1}(1-tx^{N+2})^{-l_2}\dots[t^{1\dots\infty}] \qquad \text{(first factor)},$$

$$x+(t)\,x^2+(t,\ t^2)\,x^3+(t,\ t^2,\ t^3)\,x^4+\dots \qquad \text{(second factor)}.$$

As regards the first factor, the exponents taken with reversed sign, that is, as positive, are $1=$ no. of trees, altitude $N-1$, of N knots; $l_1=$ ditto, same altitude, of $(N+1)$ knots; $l_2=$ ditto, same altitude, of $N+2$ knots, and so on; and where the symbol $[t^{1 \cdots \infty}]$ denotes that, in the function or product of factors which precedes it, the terms to be taken account of are those in $t^1, t^2, t^3, \ldots$; viz. it denotes that the term in t^0, or constant term ($=1$ in fact), is to be rejected.

In the second factor, the expressions x, $(t)\,x^2$, $(t,\ t^2)\,x^3, \ldots$ represent, for given exponents of t, x, denoting the number of main branches and the number of knots respectively, the number of trees of altitude not exceeding $N-1$: thus x, $=1\,t^0x^1$ represents the number of such trees, 1 knot, 0 main branch, $=1$; and so, if the value of $(t,\ t^2,\ t^3,\ t^4)\,x^5$ be $(\alpha t+\beta t^2+\gamma t^3+\delta t^4)\,x^5$, then for trees of an altitude not exceeding $N-1$, and of 5 knots, α represents the number of trees of 1 main branch, β that of trees of 2 main branches, γ that of trees of 3 main branches, δ that of trees of 4 main branches. It is clear that the number of trees satisfying the given conditions and of an altitude not exceeding $N-1$ is at once obtained by addition of the numbers of the trees satisfying the given conditions, and of the altitudes $0, 1, 2, \ldots, N-1$; all which numbers are taken to be known.

It is to be remarked that the first factor,

$$(1-tx^N)^{-1}\,(1-tx^{N+1})^{-l_1}\,(1-tx^{N+2})^{-l_2}\ldots[t^{1 \cdots \infty}],$$

shows by its development the number of combinations of trees $a, b, \ldots$ of the altitude $N-1$; one such tree at least *must* be taken, and the symbol $[t^{1 \cdots \infty}]$ gives effect to this condition: the second factor $x+(t)\,x^2+(t,\ t^2)\,x^3+\ldots$ shows the number of the trees c of altitude not exceeding $N-1$. And this being so, there is no difficulty in seeing how the product of the two factors is the generating function for the trees of altitude N.

In the case $N=0$, the generating function, or GF, is $=x$; viz. altitude 0, there is only the tree $(\cdot)$, 1 knot, 0 main branch.

When $N=1$, the GF is $=(1-tx)^{-1}\,[t^{1 \cdots \infty}]\,x,\ =tx^2+t^2x^3+t^3x^4\ldots,$

viz. altitude 1, there is 1 tree tx^2, 2 knots, 1 main branch; 1 tree t^2x^3, 3 knots, 2 main branches; and so on.

Hence $N=2$, we obtain

$$GF=(1-tx^2)^{-1}\,(1-tx^3)^{-1}\,(1-tx^4)^{-1}\ldots[t^{1 \cdots \infty}]\,.\,(x+tx^2+t^2x^3+t^3x^4+\ldots);$$

viz. as regards the second factor, altitude not exceeding 1, that is, $=0$ or 1, there is altitude 0, 1 tree x, and altitude 1, 1 tree tx^2, 1 tree t^2x^3, and so on. And we hence derive the GF's for the higher values $N=3$, 4, &c.: the details of the process will be afterwards more fully explained.

So far, we have considered root-trees; but referring to the last diagram, it is at once seen that the assumed root will be a centre, provided only that (instead of, it may be, only a single tree a of the altitude $N-1$), we take always two or more trees of the altitude $N-1$ to form the new tree of the altitude N. And we give effect to this condition by simply writing in place of $[t^{1\cdots\infty}]$ the new symbol $[t^{2\cdots\infty}]$, which denotes that only the terms t^2, t^3, $t^4, \ldots$ are to be taken account of; viz. that the terms in t^0 and t^1 are to be rejected. The component trees of the altitude $N-1$ are, it is to be observed, as before, root-trees; hence the second factor of the generating function is unaltered: the theorem is that for the centre-trees of altitude N we have the same generating function as for the root-trees, writing only $[t^{2\cdots\infty}]$ in place of $[t^{1\cdots\infty}]$. Or, what is the same thing, supposing that the first factor, unaffected by either symbol, is

$$= 1 + x^N(\alpha t + \beta t^2 + \ldots) + x^{N+1}(\alpha' t + \beta' t^2 + \ldots) + \ldots,$$

then, affecting it with $[t^{1\cdots\infty}]$, the value for the root-trees is

$$= x^N(\alpha t + \beta t^2 + \ldots) + x^{N+1}(\alpha' t + \beta' t^2 + \ldots) + \ldots,$$

and, affecting it with $[t^{2\cdots\infty}]$, the value for the centre-trees is

$$= x^N(\beta t^2 + \ldots) + x^{N+1}(\beta' t^2 + \ldots) + \ldots$$

It thus appears how the fundamental problem is that of the root-trees, its solution giving at once that of the centre-trees; whereas we cannot conversely solve the problem of the root-trees by means of that of the centre-trees.

As regards the bicentre-trees, it is to be remarked that, starting from a centre-tree of altitude $N+1$ with two main branches, then by simply striking out the centre, so as to convert into a single branch the two branches which issue from it, we obtain a bicentre-tree of altitude N. Observe that the altitude of a bicentre-tree is measured by that of the longest main branch from A or B, not reckoning AB or BA as a main branch. Hence the number of bicentre-trees, altitude N, is = number of centre-trees of two main branches, altitude $N+1$.

This is, in fact, the convenient formula, provided only the number of centre-trees of two main branches has been calculated up to the altitude $N+1$. But we can find independently the number of bicentre-trees of a given altitude N: the bicentre-tree is, in fact, formed by taking the two connected points A, B each as the root of a root-tree altitude N (the number of knots of the bicentre-tree being thus, it is clear, equal to the sum of the numbers of knots of the two root-trees respectively); and it is thus an easy problem of combinations to find the number of bicentre-trees of a given altitude N. Write

$$x^{N+1}(1 + \beta x + \gamma x^2 + \delta x^3 + \ldots)$$

as the generating function of the root-trees of altitude N; viz. for such trees, $1 =$ no. of trees with $N+1$ knots, $\beta =$ no. with $N+2$ knots, and so on; then the generating function of the bicentre-trees of the same altitude N is

$$= x^{2N+2}(1 + \beta_{,} x + \gamma_{,} x^2 + \delta_{,} x^3 + \ldots),$$

where

$$\beta_{,} = \beta,$$
$$\gamma_{,} = \gamma + \tfrac{1}{2}\beta(\beta+1),$$
$$\delta_{,} = \delta + \beta\gamma,$$
$$\epsilon_{,} = \epsilon + \beta\delta + \tfrac{1}{2}\gamma(\gamma+1),$$
$$\zeta_{,} = \zeta + \beta\epsilon + \gamma\delta,$$

and so on; or, what is the same thing, calling the first generating function ϕx, then the second generating function is $= \tfrac{1}{2}\{(\phi x)^2 + \phi(x^2)\}$.

It will be noticed that the bicentre-trees are not, as were the centre-trees, divided according to the number of their main branches; they might be thus divided according to the sum of the number of the main branches issuing from the two points of the bicentre respectively; a more complete division would be according to the number of main branches issuing from the two points respectively; thus we might consider the bicentre-trees (2, 3), with 2 main branches from one point, and 3 main branches from the other point of the bicentre; but the whole theory of the bicentre-trees is comparatively easy, and I do not go into it further.

We have yet to consider the case of the limited trees where the number of branches from a knot is equal to a given number at most: to fix the ideas, say the carbon-trees, where this number is $=4$. The distinction as to root-trees and centre- and bicentre-trees is as before; and the like theory applies to the two cases respectively. Considering first the case of the root-trees, and referring to the former figure for obtaining the trees of altitude N from those of inferior altitudes, then the trees $a, b, \ldots$ of altitude $N-1$ must be each of them a carbon-tree of not more than $(4-1=)\,3$ main branches: this restriction is necessary, inasmuch as, if for any such tree the number of main branches was $=4$, then there would be from the root of such tree 4 branches *plus* the new branch shown by the dotted line, in all 5 branches; and similarly, inasmuch as there is at least one component tree a contributing one main branch, the number of main branches of the tree c must be $(4-1=)\,3$ at most: the mode of introducing these conditions will appear in the explanation of the actual formation of the generating functions (see explanation preceding Tables III., IV., &c.). The number of main branches is $=4$ at most, and the generating functions have only to be taken up to the terms in t^4; the first factor is consequently in each case affected with a symbol $[t^{1\cdots4}]$, denoting that the only terms to be taken account of are those in t, t^2, t^3, t^4; hence as there is a factor t at least, and the whole is required only up to t^4, the second factor is in each case required only up to t^3.

As regards the centre-trees, the generating functions have here the same expressions as for the root-trees, except that, instead of the symbol $[t^{1\cdots4}]$, we have the symbol $[t^{2\cdots4}]$, denoting that in the first factor the only terms to be taken account of are those in t^2, t^3, t^4; hence as there is a factor t^2 at least, and the whole is required only up to t^4, the second factor is in each case required up to t^2; and we then complete the theory by obtaining the bicentre-trees. The like remarks apply of course to the boron-trees, number of branches $= 3$ at most, and to the oxygen-trees, number $= 2$ at most; but, as already remarked, this last case is so simple, that the general method is applied to it only for the sake of seeing what the general method becomes in such an extreme case.

We thus form the Tables, which I proceed to explain.

Table I. of general root-trees is in fact a Table of triple entry, viz. it gives for any given number of knots from 1 to 13 the number of root-trees corresponding to any given number of main branches and to any given altitude. In each compartment, that is, for any given number of knots, the totals of the columns give the number of the trees for each given altitude, and the totals of the lines give the number of the trees for each given number of main branches: the corner grand totals of these totals respectively show for each given number of knots the whole number of root-trees:—

viz. knots ...	1,	2,	3,	4,	5,	6,	7,	8,	9,	10,	11,	12,	13,
numbers are ...	1,	1,	2,	4,	9,	20,	48,	115,	286,	719,	1842,	4766,	12486,

as already mentioned: these numbers were calculated by an independent method.

Table II. of general centre- and bicentre-trees consists of a centre part and a bicentre part: the centre part is arranged precisely in the same manner as the root-table. As to the bicentre part, where it will be observed there is no division for number of main branches, the calculation of the several columns is effected by the before-mentioned formula,

$$\phi_{,} x = \tfrac{1}{2} \{(\phi x)^2 + \phi (x^2)\};$$

thus column 2, we have by Table I. (totals of column 2)

$$\phi x = x^3 + 2x^4 + 4x^5 + 6x^6 + 10x^7 + 14x^8 + 21x^9 + 29x^{10} + \ldots,$$

and thence

$$\phi_{,} x = x^6 + 2x^7 + 7x^8 + 14x^9 + 32x^{10} + 58x^{11} + 110x^{12} + 187x^{13} + \ldots$$

As already mentioned, each column of Table I. is calculated by means of a generating function given as a product of two factors, each of which is obtained from the columns which precede the column in question; and Table II., the centre part of it, is calculated by means of the same generating functions slightly modified: these generating functions serving for the calculation of the two Tables are given in the table entitled "Subsidiary Table for the calculation of the *GF*'s of Tables I. and II.," which immediately follows these two Tables, and will be further explained.

TABLE I.—General Root-trees.

Index x, or number of knots.	Index t, or number of main branches.	Altitude or number of column.													
		0	1	2	3	4	5	6	7	8	9	10	11	12	13
1	0	1	1												
	Total	1	**1**												
2	1		1	1											
	Total		1	**1**											
3	1			1	1										
	2		1		1										
	Total		1	1	**2**										
4	1			1	1	2									
	2			1		1									
	3		1			1									
	Total		1	2	1	**4**									
5	1			1	2	1	4								
	2			2	1		3								
	3			1			1								
	4		1				1								
	Total		1	4	3	1	**9**								
6	1			1	4	3	1	9							
	2			2	3	1		6							
	3			2	1			3							
	4			1				1							
	5		1					1							
	Total		1	6	8	4	1	**20**							
7	1			1	6	8	4	1	20						
	2			3	8	4	1		16						
	3			3	3	1			7						
	4			2	1				3						
	5			1					1						
	6		1						1						
	Total		1	10	18	13	5	1	**48**						
8	1			1	10	18	13	5	1	48					
	2			3	15	13	5	1		37					
	3			4	9	4	1			18					
	4			3	3	1				7					
	5			2	1					3					
	6			1						1					
	7		1							1					
	Total		1	14	38	36	19	6	1	**115**					
9	1			1	14	38	36	19	6	1	115				
	2			4	30	36	19	6	1		96				
	3			5	19	14	5	1			44				
	4			5	9	4	1				19				
	5			3	3	1					7				
	6			2	1						3				
	7			1							1				
	8		1								1				
	Total		1	21	76	93	61	26	7	1	**286**				

TABLE I. (*continued*).

Index x, or number of knots.	Index t, or number of main branches.	Altitude or number of column.													
		0	1	2	3	4	5	6	7	8	9	10	11	12	13
10	1			1	21	76	93	61	26	7	1	286			
	2			4	51	89	61	26	7	1		239			
	3			7	42	41	20	6	1			117			
	4			6	20	14	5	1				46			
	5			5	9	4	1					19			
	6			3	3	1						7			
	7			2	1							3			
	8			1								1			
	9		1									1			
	Total		1	29	147	225	180	94	34	8	1	**719**			
11	1			1	29	147	225	180	94	34	8	1	719		
	2			5	90	210	180	94	34	8	1		622		
	3			8	79	110	67	27	7	1			299		
	4			9	46	42	20	6	1				124		
	5			7	20	14	5	1					47		
	6			5	9	4	1						19		
	7			3	3	1							7		
	8			2	1								3		
	9			1									1		
	10		1										1		
	Total		1	41	277	528	498	308	136	43	9	1	**1842**		
12	1			1	41	277	528	498	308	136	43	9	1	1842	
	2			5	145	467	493	308	136	43	9	1		1607	
	3			10	152	278	208	101	35	8	1			793	
	4			11	91	115	68	27	7	1				320	
	5			10	47	42	20	6	1					126	
	6			7	20	14	5	1						47	
	7			5	9	4	1							19	
	8			3	3	1								7	
	9			2	1									3	
	10			1										1	
	11		1											1	
	Total		1	55	509	1198	1323	941	487	188	53	10	1	**4766**	
13	1			1	55	509	1198	1323	941	487	188	53	10	1	4766
	2			6	238	1012	1524	941	487	188	53	10	1		4460
	3			12	272	669	376	344	144	44	9	1			1871
	4			15	184	299	213	102	35	8	1				857
	5			13	95	116	68	27	7	1					327
	6			11	47	42	20	6	1						127
	7			7	20	14	5	1							47
	8			5	9	4	1								19
	9			3	3	1									7
	10			2	1										3
	11			1											1
	12		1												1
	Total		1	76	924	2666	3405	2744	1615	728	251	64	11	1	**12486**

TABLE II.—General Centre- and Bicentre-Trees.

Index x, or number of knots.	Index t, or number of main branches.	Centre-Trees. Altitude or number of column.								Centre.	Grand Total.	Bicentre.	Bicentre-Trees. Altitude.					
		0	1	2	3	4	5	6					0	1	2	3	4	5
1	0	1	1							1	1	0						
	Total	1	1															
2										0	1	1	1					
3	2		1	1														
	Total		1	1						1	1	0						
4	2																	
	3		1	1														
	Total		1	1						1	2	1		1				
5	2			1	1													
	3																	
	4		1		1													
	Total		1	1	2					2	3	1		1				
6	2			1	1													
	3			1	1													
	4																	
	5		1		1													
	Total		1	2	3					3	6	3		2	1			
7	2			2	1	3												
	3			2		2												
	4			1		1												
	5																	
	6		1			1												
	Total		1	5	1	7				7	11	4		2	2			
8	2			2	2	4												
	3			3	1	4												
	4			2		2												
	5			1		1												
	6																	
	7		1			1												
	Total		1	8	3	12				12	23	11		3	7	1		
9	2			3	7	1	11											
	3			4	3		7											
	4			4	1		5											
	5			2			2											
	6			1			1											
	7																	
	8		1				1											
	Total		1	14	11	1	27			27	47	20		3	14	3		

TABLE II. (*continued*).

Index x, or number of knots.	Index t, or number of main branches.	Centre-Trees. Altitude or number of column.								Centre.	Grand Total.	Bicentre.	Bicentre-Trees. Altitude.					
		0	1	2	3	4	5	6					0	1	2	3	4	5
10	2			3	14	3	20											
	3			6	11	1	18											
	4			5	3		8											
	5			4	1		5											
	6			2			2											
	7			1			1											
	8																	
	9		1				1											
	Total		1	21	29	4	**55**			55	106	51		4	32	14	1	
11	2			4	32	14	1	51										
	3			7	26	4		37										
	4			8	12	1		21										
	5			6	3			9										
	6			4	1			5										
	7			2				2										
	8			1				1										
	9																	
	10		1					1										
	Total		1	32	74	19	1	**127**		127	235	108		4	58	42	4	
12	2			4	58	42	4	108										
	3			9	63	19	1	92										
	4			10	30	4		44										
	5			9	12	1		22										
	6			6	3			9										
	7			4	1			5										
	8			2				2										
	9			1				1										
	10																	
	11		1					1										
	Total		1	45	167	66	5	**284**		284	551	267		5	110	128	23	1
13	2			5	110	128	23	1	267									
	3			11	132	66	5		214									
	4			14	78	20	1		113									
	5			12	31	4			47									
	6			10	12	1			23									
	7			6	3				9									
	8			4	1				5									
	9			2					2									
	10			1					1									
	11																	
	12		1						1									
	Total		1	65	367	219	29	1	**682**	682	1301	619		5	187	334	88	5

Subsidiary Table for *GF*'s of Tables I. and II.

Index t.	Index of x.														
	0	1	2	3	4	5	6	7	8	9	10	11	12	13	
0		1													*GF*, column 0.
*	.	−1													*GF*, column 1.
0	(1)														First factor.
1		1													
2			1												
3				1											
4					1										
5						1									
6							1								
7								1							
8									1						
9										1					
10											1				
11												1			
12													1		
0		1													Second factor.
*			−1	−1	−1	−1	−1	−1	−1	−1	−1	−1	−1		*GF*, column 2.
0	(1)														First factor.
1			1	1	1	1	1	1	1	1	1	1	1		
2					1	1	2	2	3	3	4	4	5		
3							1	1	2	3	4	5	7		
4									1	1	2	3	5		
5											1	1	2		
6													1		
0		1													Second factor.
1			1												
2				1											
3					1										
4						1									
5							1								
6								1							
7									1						
8										1					
9											1				
10												1			
11													1		
12														1	
*				−1	−2	−4	−6	−10	−14	−21	−29	−41	−55	−76	*GF*, column 3.
0	(1)														First factor.
1		.	.	1	2	4	6	10	14	21	29	41	55		
2							1	2	7	14	32	58	110		
3										1	2	7	18		
4													1		
0		1													Second factor.
1			1	1	1	1	1	1	1	1	1	1	1	1	
2				1	1	2	2	3	3	4	4	5	5	6	
3					1	1	2	3	4	5	7	8	10	12	
4						1	1	2	3	5	6	9	11	15	
5							1	1	2	3	5	7	10	13	
6								1	1	2	3	5	7	11	
7									1	1	2	3	5	7	
8										1	1	2	3	5	
9											1	1	2	3	
10												1	1	2	
11													1	1	
12														1	

Subsidiary Table for *GF*'s of Tables I. and II. (*continued*).

Index t.	Index of x.														
	0	1	2	3	4	5	6	7	8	9	10	11	12	13	
*					-1	-3	-8	-18	-38	-76	-147	-277	-509	-924	*GF*, column 4.
0	(1)														First factor.
1				.	1	3	8	18	38	76	147	277	509		
2									1	3	14	42	128		
3													1		
0		1													Second factor.
1			1	1	2	3	5	7	11	15	22	30	42	56	
2				1	1	3	5	11	18	34	55	95	150	244	
3					1	1	3	6	13	24	49	87	162	284	
4						1	1	3	6	14	26	55	102	199	
5							1	1	3	6	14	27	57	108	
6								1	1	3	6	14	27	58	
7									1	1	3	6	14	27	
8										1	1	3	6	14	
9											1	1	3	6	
10												1	1	3	
11													1	1	
12														1	
*	.				.	-1	-4	-13	-36	-93	-225	-528	-1198	-2666	*GF*, column 5.
0	(1)														First factor.
1						1	4	13	36	93	225	528	1198		
2											1	4	23		
0		1													Second factor.
1			1	1	2	4	8	15	29	53	98	177	319	565	
2				1	1	3	6	15	31	70	144	305	617	1256	
3					1	1	3	7	17	38	90	197	440	953	
4						1	1	3	7	18	40	97	217	498	
5							1	1	3	7	18	41	99	224	
6								1	1	3	7	18	41	100	
7									1	1	3	7	18	41	
8										1	1	3	7	18	
9											1	1	3	7	
10												1	1	3	
11													1	1	
12														1	
*	.	.	.	.	.	.	-1	-5	-19	-61	-180	-498	-1323	-3405	*GF*, column 6.
0	(1)														First factor.
1							1	5	19	61	180	498	1323		
2													1		
0		1													Second factor.
1			1	1	2	4	9	19	42	89	191	402	847	1763	
2				1	1	3	6	16	36	89	205	485	1110	2780	
3					1	1	3	7	18	43	110	264	648	1329	
4						1	1	3	7	19	45	117	285	711	
5							1	1	3	7	19	46	119	292	
6								1	1	3	7	19	46	120	
7									1	1	3	7	19	46	
8										1	1	3	7	19	
9											1	1	3	7	
10												1	1	3	
11													1	1	
12														1	

Subsidiary Table for *GF*'s of Tables I. and II. (*continued*).

Index *t*.	Index of *x*.														
	0	1	2	3	4	5	6	7	8	9	10	11	12	13	
*								-1	-6	-26	-94	-308	-941	-2744	*GF*, column 7.
0	(1)														First factor.
1								1	6	26	94	308	941		
0		1													Second factor.
1			1	1	2	4	9	20	47	108	252	582	1345	3086	
2				1	1	3	6	16	37	95	231	579	1418	3721	
3					1	1	3	7	18	44	116	291	749	1673	
4						1	1	3	7	19	46	123	312	813	
5							1	1	3	7	19	47	125	319	
6								1	1	3	7	19	47	126	
7									1	1	3	7	19	47	
8										1	1	3	7	19	
9											1	1	3	7	
10												1	1	3	
11													1	1	
12														1	
*									-1	-7	-34	-136	-487	-1615	*GF*, column 8.
0	(1)														First factor.
1									1	7	34	136	487	1615	
0		1													Second factor.
1			1	1	2	4	9	20	48	114	278	676	1653	4027	
2				1	1	3	6	16	37	96	238	613	1554	4208	
3					1	1	3	7	18	44	117	298	784	1817	
4						1	1	3	7	19	46	124	319	848	
5							1	1	3	7	19	47	126	326	
6								1	1	3	7	19	47	127	
7									1	1	3	7	19	47	
8										1	1	3	7	19	
9											1	1	3	7	
10												1	1	3	
11													1	1	
12														1	
*										-1	-8	-43	-188	-728	*GF*, column 9.
0	(1)														First factor.
1										1	8	43	188	728	
0		1													Second factor.
1			1	1	2	4	9	20	48	115	285	710	1789	4514	
2				1	1	3	6	16	37	96	239	621	1597	4396	
3					1	1	3	7	18	44	117	299	792	1861	
4						1	1	3	7	19	46	124	320	856	
5							1	1	3	7	19	47	126	327	
6								1	1	3	7	19	47	127	
7									1	1	3	7	19	47	
8										1	1	3	7	19	
9											1	1	3	7	
10												1	1	3	
11													1	1	
12														1	

Subsidiary Table for *GF*'s of Tables I. and II. (*continued*).

Index t.	Index of x.														
	0	1	2	3	4	5	6	7	8	9	10	11	12	13	
*											−1	−9	−53	−251	*GF*, column 10.
0	(1)														First factor.
1											1	9	53	251	
0		1													Second factor.
1			1	1	2	4	9	20	48	115	286	718	1832	4702	
2				1	1	3	6	16	37	96	239	622	1606	4449	
3					1	1	3	7	18	44	117	299	793	1870	
4						1	1	3	7	19	46	124	320	857	
5							1	1	3	7	19	47	126	327	
6								1	1	3	7	19	47	127	
7									1	1	3	7	19	47	
8										1	1	3	7	19	
9											1	1	3	7	
10												1	1	3	
11													1	1	
12														1	
*												−1	−10	−64	*GF*, column 11.
0	(1)														First factor.
1												1	10	64	
0		1													Second factor.
1			1	1	2	4	9	20	48	115	286	719	1841	4755	
2				1	1	3	6	16	37	96	239	622	1607	4459	
3					1	1	3	7	18	44	117	299	793	1871	
4						1	1	3	7	19	46	124	320	857	
5							1	1	3	7	19	47	126	327	
6								1	1	3	7	19	47	127	
7									1	1	3	7	19	47	
8										1	1	3	7	19	
9											1	1	3	7	
10												1	1	3	
11													1	1	
12														1	
*													−1	−11	*GF*, column 12.
0	(1)														First factor.
1													1	11	
0		1													Second factor.
1			1	1	2	4	9	20	48	115	286	719	1842	4765	
2				1	1	3	6	16	37	96	239	622	1607	4460	
3					1	1	3	7	18	44	117	299	793	1871	
4						1	1	3	7	19	46	124	320	857	
5							1	1	3	7	19	47	126	327	
6								1	1	3	7	19	47	127	
7									1	1	3	7	19	47	
8										1	1	3	7	19	
9											1	1	3	7	
10												1	1	3	
11													1	1	
12														1	
*														−1	*GF*, column 13.
0	(1)														First factor.
1														1	
0		1													Second factor.
1			1	1	2	4	9	20	48	115	286	719	1842	4766	
2				1	1	3	6	16	37	96	239	622	1607	4460	
3					1	1	3	7	18	44	117	299	793	1871	
4						1	1	3	7	19	46	124	320	857	
5							1	1	3	7	19	47	126	327	
6								1	1	3	7	19	47	127	
7									1	1	3	7	19	47	
8										1	1	3	7	19	
9											1	1	3	7	
10												1	1	3	
11													1	1	
12														1	

I proceed to explain the Subsidiary Table, first in its application to Table I.

The Subsidiary Table is divided into sections, giving the *GF*'s of the successive columns of Table I., each section being given by means of the preceding columns of Table I.; for instance, that for column 3 by means of columns 0, 1, 2 of Table I.

As regards column 0, the Table shows that the *GF* is $=x$.

As regards column 1, it shows that the *GF* has a first factor,

$$(1-tx)^{-1}, \;=(1)+tx+t^2x^2+t^3x^3+\ldots,$$

which is operated on by the symbol $[t^{1\ldots\infty}]$, viz. the constant term (1) is to be rejected; and that it has a second factor, $=x$: the product of these, viz. $(tx+t^2x^2+t^3x^3+\ldots)\times x$, is the required *GF*, the coefficients of which are accordingly given in column 1 of Table I.

As regards column 2, it shows that the *GF* has a first factor,

$$(1-tx^2)^{-1}(1-tx^3)^{-1}(1-tx^4)^{-1}\ldots,$$

where the indices -1, -1, -1,.. are the sums of the numbers in column 1, Table I., (with their signs changed): which first factor is

$$1+tx^2+tx^3+\begin{pmatrix} t \\ +t^2 \end{pmatrix}x^4+\ldots,$$

and it is as before to be operated on with $[t^{1\ldots\infty}]$, viz. the constant term is to be rejected; and further, that there is a second factor $=x+tx^2+t^2x^3+\ldots$, the coefficients of which are obtained by summation of the numbers in the several lines of columns 0, 1 of Table I. We have thence column 2 of Table I.

As regards column 3, it shows that the *GF* has a first factor,

$$(1-tx^3)^{-1}(1-tx^4)^{-2}(1-tx^5)^{-4}\ldots,$$

where the indices -1, -2, -4,.. are the sums of the numbers in column 2 of Table I., (with their signs changed): which first factor is

$$=1+tx^3+2tx^4+4tx^5+\begin{pmatrix} 6t \\ +t^2 \end{pmatrix}x^6+\ldots,$$

and it is as before to be operated on with $[t^{1\ldots\infty}]$, viz. the constant term is to be rejected; and that there is a second factor

$$=x+tx^2+\begin{pmatrix} t \\ +t^2 \end{pmatrix}x^3+\begin{pmatrix} t \\ +t^2 \\ +t^3 \end{pmatrix}x^4+\ldots,$$

the coefficients of which are obtained by summation of the numbers in the several lines of columns 0, 1, 2 of Table I.: we have thence column 3 of Table I.

And similarly, by means of columns 0, 1, 2, 3 of Table I., we form the *GF* of column 4; that is, we obtain column 4 of Table I., and so on indefinitely.

To apply the Subsidiary Table to the calculation of the *GF*'s of Table II., the only difference is that the first factors are to be taken without the terms in t^1: thus for Table II. column 3, the first factor of the *GF*

$$= t^2x^6 + 2t^2x^7 + 7t^2x^8 + \begin{pmatrix}14t^2\\+t^3\end{pmatrix} x^9 + \&c.,$$

the second factor being as for Table I.

$$= x + tx^2 + \begin{pmatrix}t\\+t^2\end{pmatrix} x^3 + \&c.$$

The remaining Tables are Tables III. and IV., oxygen root-trees and centre- and bicentre-trees, followed by a Subsidiary Table for the calculation of the *GF*'s: Tables V. and VI., boron root-trees and centre- and bicentre-trees, followed by a Subsidiary Table; and Tables VII. and VIII., carbon root-trees and centre- and bicentre-trees, followed by a Subsidiary Table. The explanations given as to Tables I., II. and the Subsidiary Table apply *mutatis mutandis* to these; and but little further explanation is required: that given in regard to the Subsidiary Table of Tables III. and IV. shows how this limiting case comes under the general method. As to the Subsidiary of Tables V. and VI., it is to be observed that each * line of the Table is calculated from a column of Table V., rejecting the numbers which belong to t^3; thus Table V., column 4, the numbers are

t^1	1	3	5	7	8	9...
t^2		1	4	10	21	36...
t^3			1	4	11	26...;

and taking the sums for the first and second lines only, these are

1, 4, 9, 17, 29, 45,..,

which, taken with a negative sign, are the numbers of the line **GF*, column 5.

And so as to the Subsidiary of Tables VII. and VIII., each * line of the Table is calculated from a column of Table VII., rejecting the numbers which belong to t^4; thus Table VII., column 4, the numbers are

t^1	1	3	8	15	27	43...
t^2		1	4	13	33	74...
t^3			1	4	14	38...
t^4				1	4	14...;

and taking the sums for the first, second, and third lines only, these are

1, 4, 13, 32, 74, 155,..,

which, taken with a negative sign, are the numbers of the line **GF*, column 5.

Referring to the foregoing "Edification Diagram," the effect is that we thus introduce the conditions that in a boron-tree the number of component trees $a, b, \ldots$ is at most $(3-1=)\,2$ and that in a carbon-tree the number of component trees $a, b, \ldots$ is at most $(4-1=)\,3$.

TABLE III.—Oxygen Root-Trees.

Index x, or number of knots.	Index t, or number of main branches.	Altitude or number of column.												
		0	1	2	3	4	5	6	7	8	9	10	11	12
1	0	1												
2	1		1											
3	1			1										
	2		1											
4	1				1									
	2			1										
5	1					1								
	2			1	1									
6	1						1							
	2				1	1								
7	1							1						
	2				1	1	1							
8	1								1					
	2					1	1	1						
9	1									1				
	2					1	1	1	1					
10	1										1			
	2						1	1	1	1				
11	1											1		
	2						1	1	1	1	1			
12	1												1	
	2							1	1	1	1	1		
13	1													1
	2							1	1	1	1	1	1	

TABLE IV.—Oxygen Centre- and Bicentre-Trees.

Index x, or number of knots.	Index t, or number of main branches.	Centre-Trees. Altitude or number of column.							Centre.	Total.	Bicentre.	Bicentre-Trees. Altitude.					
		0	1	2	3	4	5	6				0	1	2	3	4	5
1	0	1							1	1	0						
2									0	1	1	1					
3	2		1						1	1	0						
4									0	1	1		1				
5	2			1					1	1	0						
6									0	1	1			1			
7	2				1				1	1	0						
8									0	1	1				1		
9	2					1			1	1	0						
10									0	1	1					1	
11	2						1		1	1	0						
12									0	1	1						1
13	2							1	1	1	0						

Subsidiary Table for *GF*'s of Tables III. and IV.

Index t.	Index of x.														
	0	1	2	3	4	5	6	7	8	9	10	11	12	13	
0															*GF*, column 0.
*		−1													*GF*, column 1.
0	1														First factor.
1		1													
2			1												
0		1													Second factor.
*			−1												*GF*, column 2.
0	1														First factor.
1			1												
2					1										
0		1													Second factor.
1			1												
*				−1											*GF*, column 3.
0	1														First factor.
1				1	.	.	.								
2							1								
0		1													Second factor.
1			1	1											
*					−1										*GF*, column 4.
0	1														First factor.
1					1										
2						.	.	.	1						
0		1													Second factor.
1			1	1	1										
*						−1									*GF*, column 5.
0	1														First factor.
1						1									
2											1				
0		1													Second factor.
1			1	1	1	1									

and so on indefinitely; viz. observing that the first factors, as shown by the Table, are $(1-tx)^{-1}[t^{1\cdot 2}]$, $(1-tx^2)^{-1}[t^{1\cdot 2}]$, &c., the Table in fact shows that as regards Table III. the GF's are for

column 0 : x,
" 1 : $tx + t^2x^2 \,.\, x$,
" 2 : $tx^2 + t^2x^4 \,.\, x + tx^2$,
" 3 : $tx^3 + t^2x^6 \,.\, x + t(x^2 + x^3)$,
" 4 : $tx^4 + t^2x^8 \,.\, x + t(x^2 + x^3 + x^4)$,
" 5 : $tx^5 + t^2x^{10} \,.\, x + t(x^2 + x^3 + x^4 + x^5)$;

viz. developing as far as t^2, that the successive GF's are

column 0 : x,
" 1 : $tx^2 + t^2x^3$,
" 2 : $tx^3 + t^2(x^4 + x^5)$,
" 3 : $tx^4 + t^2(x^5 + x^6 + x^7)$,
" 4 : $tx^5 + t^2(x^6 + x^7 + x^8 + x^9)$,
" 5 : $tx^6 + t^2(x^7 + x^8 + x^9 + x^{10} + x^{11})$;
&c., agreeing with Table III.

And so also it shows that, as regards Table IV. (centre part), the GF's of the successive columns are for

column 0 : x,
" 1 : $t^2x^2 \,.\, x$,
" 2 : $t^2x^4 \,.\, x$,
" 3 : $t^2x^6 \,.\, x$,
" 4 : $t^2x^8 \,.\, x$,
" 5 : $t^2x^{10} \,.\, x$;
⋮

viz. that the successive GF's are x, t^2x^3, t^2x^5, t^2x^7, t^2x^9, $t^2x^{11}, \ldots$, agreeing in fact with Table IV.

Table V.—Boron Root-trees.

Index x, or number of knots.	Index t, or number of main branches.	Altitude or number of column.													
		0	1	2	3	4	5	6	7	8	9	10	11	12	
1	0	1	1												
	Total	1	**1**												
2	1		1	1											
	Total		1	**1**											
3	1			1	1										
	2		1		1										
	Total		1	1	**2**										
4	1			1	1	2									
	2			1		1									
	3		1			1									
	Total		1	2	1	**4**									
5	1				2	1	3								
	2			2	1		3								
	3			1			1								
	Total			3	3	1	**7**								
6	1				2	3	1	6							
	2			1	3	1		5							
	3			2	1			3							
	Total			3	6	4	1	**14**							
7	1				1	5	4	1	11						
	2			1	6	4	1		12						
	3			2	3	1			6						
	Total			3	10	10	5	1	**29**						
8	1				1	7	9	5	1	23					
	2				7	10	5	1		23					
	3			2	7	4	1			14					
	Total			2	15	21	15	6	1	**60**					
9	1					8	17	14	6	1	46				
	2				9	21	15	6	1		52				
	3			1	11	11	5	1			29				
	Total			1	20	40	37	21	7	1	**127**				
10	1					9	29	32	20	7	1	98			
	2				7	36	37	21	7	1		109			
	3			1	18	26	16	6	1			68			
	Total			1	25	71	82	59	28	8	1	**275**			
11	1					7	45	66	53	27	8	1	207		
	2				7	59	82	59	28	8	1		244		
	3				21	53	43	22	7	1			147		
	Total				28	119	170	147	88	36	9	1	**598**		
12	1					7	66	127	125	81	35	9	1	451	
	2				4	82	165	147	88	36	9	1		532	
	3				26	102	105	66	29	8	1			337	
	Total				30	191	336	340	242	125	45	10	1	**1320**	
13	1					4	89	231	274	213	117	44	10	1	983
	2				3	114	316	340	242	125	45	10	1		1196
	3				26	175	236	177	96	37	9	1			757
	Total				29	293	641	748	612	375	171	55	11	1	**2936**

TABLE VI.—Boron Centre- and Bicentre-Trees.

Index x, or number of knots.	Index t, or number of main branches.	Centre-Trees. Altitude or number of column.								Centre.	Grand Total.	Bicentre.	Bicentre-Trees. Altitude.					
		0	1	2	3	4	5	6					0	1	2	3	4	5
1	0	1	1															
	Total	1	1							1	1	0						
2 .																		
										0	1	1	1					
3	2		1	1														
	Total		1	1						1	1	0						
4	2		1	1														
	Total		1	1						1	2	1		1				
5	2			1	1													
	Total			1	1					1	2	1		1				
6	2 3			1 1	1 1													
	Total			2	2					2	4	2		1	1			
7	2 3			1 2	1	2 2												
	Total			3	1	4				4	6	2			2			
8	2 3			 2	2 1	2 3												
	Total			2	3	5				5	11	6			5	1		
9	2 3			 1	5 3	1	6 4											
	Total			1	8	1	10			10	18	8			5	3		
10	2 3			 1	5 9	3 1	8 11											
	Total			1	14	4	19			19	37	18			6	11	1	
11	2 3				6 14	11 4	1	18 18										
	Total				20	15	1	36		36	66	30			4	22	4	
12	2 3				4 21	22 16	4 1	30 38										
	Total				25	38	5	68		68	135	67			3	44	19	1
13	2 3				3 24	44 42	19 5	1	67 71									
	Total				27	86	24	1	138	138	265	127			1	68	53	5

Subsidiary Table for *GF*'s of Tables V. and VI.

Index t.	Index of x.														
	0	1	2	3	4	5	6	7	8	9	10	11	12	13	
0		1													*GF*, column 0.
*		−1													*GF*, column 1.
0	1														First factor.
1		1													
2			1												
3				1											
0		1													Second factor.
*			−1	−1											*GF*, column 2.
0	1														First factor.
1			1	1											
2					1	1	1								
3							1	1	1	1					
0		1													Second factor.
1			1												
2				1											
*				−1	−2	−2	−1	−1							*GF*, column 3.
0	1														First factor.
1				1	2	2	1	1							
2							1	2	5	5	6	4	3		
3										1	2	5	9		
0		1													Second factor.
1			1	1	1										
2				1	1	2	1	1							
*					−1	−3	−5	−7	−8	−9	−7	−7	−4	−3	*GF*, column 4.
0	1														First factor.
1					1	3	5	7	8	9	7	7	4		
2									1	3	11	22	44		
3													1		
0		1													Second factor.
1			1	1	2	2	2	1	1						
2				1	1	3	4	7	7	7	7	7	4	3	
*						−1	−4	−9	−17	−29	−45	−66	−89	−118	*GF*, column 5.
0	1														First factor.
1						1	4	9	17	29	45	66	89		
2											1	4	19		
0		1													Second factor.
1			1	1	2	3	5	6	8	8	9	7	7	4	
2				1	1	3	5	11	17	30	43	66	86	117	
*							−1	−5	−14	−32	−66	−127	−231	−405	*GF*, column 6.
0	1														First factor.
1							1	5	14	32	66	127	231		
2													1		
0		1													Second factor.
1			1	1	2	3	6	10	17	25	38	52	73	93	
2				1	1	3	5	12	22	45	80	148	251	433	

Subsidiary Table for *GF*'s of Tables V. and VI. (*continued*).

Index t.	Index of x.														
	0	1	2	3	4	5	6	7	8	9	10	11	12	13	
*								−1	−6	−20	−53	−125	−274	−571	*GF*, column 7.
0	1														First factor.
1								1	6	20	53	125	274		
0		1													Second factor.
1			1	1	2	3	6	11	22	39	70	118	200	324	
2				1	1	3	5	12	23	51	101	207	398	773	
*									−1	−7	−27	−81	−213	−516	*GF*, column 8.
0	1														First factor.
1									1	7	27	81	213		
0		1													Second factor.
1			1	1	2	3	6	11	23	45	90	171	325	598	
2				1	1	3	5	12	23	52	108	235	486	1015	
*										−1	−8	−35	−117	−338	*GF*, column 9.
0	1														First factor.
1										1	8	35	117		
0		1													Second factor.
1			1	1	2	3	6	11	23	46	97	198	406	811	
2				1	1	3	5	12	23	52	109	243	522	1140	
*											−1	−9	−44	−162	*GF*, column 10.
0	1														First factor.
1											1	9	44		
0		1													Second factor.
1			1	1	2	3	6	11	23	46	98	206	441	928	
2				1	1	3	5	12	23	52	109	244	531	1185	
*												−1	−10	−54	*GF*, column 11.
0	1														First factor.
1												1	10		
0		1													Second factor.
1			1	1	2	3	6	11	23	46	98	207	450	972	
2				1	1	3	5	12	23	52	109	244	532	1195	
*													−1	−11	*GF*, column 12.
0	1														First factor.
1													1		
0		1													Second factor.
1			1	1	2	3	6	11	23	46	98	207	451	982	
2				1	1	3	5	12	23	52	109	244	532	1196	
*														−12	*GF*, column 13.
0	1														First factor.
0		1													Second factor.
1			1	1	2	3	6	11	23	46	98	207	451	983	
2				1	1	3	5	12	23	52	109	244	532	1196	

TABLE VII.—Carbon Root-trees.

Index x, or number of knots.	Index t, or number of main branches.	Altitude or number of column.													
		0	1	2	3	4	5	6	7	8	9	10	11	12	
1	0	1	1												
	Total	1	**1**												
2	1		1	1											
	Total		1	**1**											
3	1			1	1										
	2		1		1										
	Total		1	1	**2**										
4	1			1	1	2									
	2			1		1									
	3		1			1									
	Total		1	2	1	**4**									
5	1			1	2	1	4								
	2			2	1		3								
	3			1			1								
	4		1				1								
	Total		1	4	3	1	**9**								
6	1				4	3	1	8							
	2			2	3	1		6							
	3			2	1			3							
	4			1				1							
	Total			5	8	4	1	**18**							
7	1				4	8	4	1	17						
	2			2	8	4	1		15						
	3			3	3	1			7						
	4			2	1				3						
	Total			7	16	13	5	1	**42**						

TABLE VII. (*continued*).

Index x, or number of knots.	Index t, or number of main branches.	Altitude or number of column.													
		0	1	2	3	4	5	6	7	8	9	10	11	12	
8	1				5	15	13	5	1	39					
	2			1	13	13	5	1		33					
	3			3	9	4	1			17					
	4			3	3	1				7					
	Total			7	30	33	19	6	1	**96**					
9	1				4	27	32	19	6	1	89				
	2			1	22	33	19	6	1		82				
	3			3	17	14	5	1			40				
	4			4	9	4	1				18				
	Total			8	52	78	57	26	7	1	**229**				
10	1				4	43	74	56	26	7	1	211			
	2				29	74	57	26	7	1		194			
	3			3	34	38	20	6	1			102			
	4			4	18	14	5	1				42			
	Total			7	85	169	156	89	34	8	1	**549**			
11	1				3	67	155	151	88	34	8	1	507		
	2				40	154	156	89	34	8	1		482		
	3			2	54	95	63	27	7	1			249		
	4			5	38	39	20	5	1				108		
	Total			7	135	355	394	272	130	43	9	1	**1346**		
12	1				2	97	316	374	267	129	43	9	1	1238	
	2				46	297	389	273	130	43	9	1		1188	
	3			1	88	218	184	96	35	8	1			631	
	4			4	66	100	64	27	7	1				269	
	Total			5	202	712	953	770	439	181	53	10	1	**3326**	
13	1				1	136	612	889	743	432	180	52	10	1	3056
	2				55	550	929	770	439	181	53	10	1		2988
	3			1	127	474	491	309	138	44	9	1			1594
	4			4	117	239	190	97	35	8	1				691
	Total			5	300	1399	2222	2065	1355	665	243	63	11	1	**8329**

TABLE VIII.—Carbon Centre- and Bicentre-Trees.

Index x, or number of knots.	Index t, or number of main branches.	Centre-Trees. Altitude or number of column.								Centre.	Grand Total.	Bicentre.	Bicentre-Trees. Altitude.					
		0	1	2	3	4	5	6					0	1	2	3	4	5
1	0	1	1															
	Total	1	1							1	1	0						
2																		
										0	1	1	1					
3	2		1	1														
	Total		1	1						1	1	0						
4	2		1	1														
	Total		1	1						1	2	1		1				
5	2		1	1	2													
	Total		1	1	2					2	3	1		1				
6	2			1	1													
	3			1	1													
	Total			2	2					2	5	3		2	1			
7	2			2	1	3												
	3			2		2												
	4			1		1												
	Total			5	1	6				6	9	3		1	2			
8	2			1	2	3												
	3			3	1	4												
	4			2		2												
	Total			6	3	9				9	18	9		1	7	1		
9	2			1	7	1	9											
	3			3	3		6											
	4			4	1		5											
	Total			8	11	1	20			20	35	15			12	3		
10	2				12	3	15											
	3			3	11	1	15											
	4			4	3		7											
	Total			7	26	4	37			37	75	38			23	14	1	
11	2				23	14	1	38										
	3			2	24	4		30										
	4			5	12	1		18										
	Total			7	59	19	1	86		86	159	73			30	39	4	
12	2				30	39	4	73										
	3			1	54	19	1	75										
	4			4	27	4		35										
	Total			5	111	62	5	183		183	357	174			42	108	23	1
13	2				42	108	23	1	174									
	3			1	88	63	5		157									
	4			4	63	20	1		88									
	Total			5	193	191	29	1	419	419	799	380			47	244	84	5

Subsidiary Table for GF's of Tables VII. and VIII.

Index t.	Index of x.														
	0	1	2	3	4	5	6	7	8	9	10	11	12	13	
0		1													GF, column 0.
*		−1													GF, column 1.
0	1														First factor.
1		1													
2			1												
3				1											
4					1										
0		1													Second factor.
*			−1	−1	−1										GF, column 2.
0	(1)														First factor.
1			1	1	1										
2					1	1	2	1	1						
3							1	1	2	2	2	1	1		
4									1	1	2	2	3		
0		1													Second factor.
1			1												
2				1											
3					1										
*				−1	−2	−4	−4	−5	−4	−4	−3	−2	−1	−1	GF, column 3.
0	(1)														First factor.
1				1	2	4	4	5	4	4	3	2	1		
2							1	2	4	12	23	30	42		
3										1	2	7	16		
4													1		
0		1													Second factor.
1			1	1	1	1									
2				1	1	2	2	2	1	1					
3					1	1	2	3	3	3	3	2	1	1	
*					−1	−3	−8	−15	−27	−43	−67	−97	−136	−183	GF, column 4.
0	(1)														First factor.
1					1	3	8	15	27	43	67	97	136		
2									1	3	14	39	108		
3													1		
0		1													Second factor.
1			1	1	2	3	4	4	5	4	4	3	2	1	
2				1	1	3	5	10	14	23	29	40	46	55	
3					1	1	3	6	12	20	37	56	89	128	
*						−1	−4	−13	−32	−74	−155	−316	−612	−1160	GF, column 5.
0	(1)														First factor.
1						1	4	13	32	74	155	316	612		
2											1	4	23		
0		1													Second factor.
1			1	1	2	4	7	12	20	31	47	70	99	137	
2				1	1	3	6	14	27	56	103	194	343	605	
3					1	1	3	7	16	34	75	151	307	602	
*							−1	−5	−19	−56	−151	−374	−889	−2032	GF, column 6.
0	1														First factor.
1							1	5	19	56	151	374	889		
2													1		
0		1													Second factor.
1			1	1	2	4	8	16	33	63	121	225	415	749	
2				1	1	3	6	15	32	75	160	350	732	1534	
3					1	1	3	7	17	39	95	214	491	1093	

Subsidiary Table for *GF*'s of Tables VII. and VIII. (*continued*).

Index t.	Index of x.														
	0	1	2	3	4	5	6	7	8	9	10	11	12	13	
*								−1	−6	−26	−88	−267	−743	−1968	*GF*, column 7.
0	(1)														First factor.
1								1	6	26	88	267	743		
0		1													Second factor.
1			1	1	2	4	8	17	38	82	177	376	789	1638	
2				1	1	3	6	15	33	81	186	439	1005	2304	
3					1	1	3	7	17	40	101	241	587	1402	
*									−1	−7	−34	−129	−432	−1320	*GF*, column 8.
0	(1)														First factor.
1									1	7	34	129	432		
0		1													Second factor.
1			1	1	2	4	8	17	39	88	203	464	1056	2381	
2				1	1	3	6	15	33	82	193	473	1135	2743	
3					1	1	3	7	17	40	102	248	622	1540	
*										−1	−8	−43	−180	−657	*GF*, column 9.
0	(1)														First factor.
1										1	8	43	180		
0		1													Second factor.
1			1	1	2	4	8	17	39	89	210	498	1185	2813	
2				1	1	3	6	15	33	82	194	481	1178	2924	
3					1	1	3	7	17	40	102	249	630	1584	
*											−1	−9	−53	−242	*GF*, column 10.
0	(1)														First factor.
1											1	9	53		
0		1													Second factor.
1			1	1	2	4	8	17	39	89	211	506	1228	2993	
2				1	1	3	6	15	33	82	194	482	1187	2977	
3					1	1	3	7	17	40	102	249	631	1593	
*												−1	−10	−63	*GF*, column 11.
0	(1)														First factor.
1												1	10	63	
0		1													Second factor.
1			1	1	2	4	8	17	39	89	211	507	1237	3048	
2				1	1	3	6	15	33	82	194	482	1188	2987	
3					1	1	3	7	17	40	102	249	631	1594	
*													−1	−11	*GF*, column 12.
0	(1)														First factor.
1													1	11	
0		1													Second factor.
1			1	1	2	4	8	17	39	89	211	507	1238	3055	
2				1	1	3	6	15	33	82	194	482	1188	2988	
3					1	1	3	7	17	40	102	249	631	1594	
*														−1	*GF*, column 13.
0	(1)														First factor.
1														1	
0		1													Second factor.
1			1	1	2	4	8	17	39	89	211	507	1238	3056	
2				1	1	3	6	15	33	82	194	482	1188	2988	
3					1	1	3	7	17	40	102	249	634	1594	

I annex the following two Tables of (centre- and bicentre-) trees as far as I have completed them.

TABLE A.

Knots.	Valency not greater than									Gen.
	0	1	2 Oxygen.	3 Boron.	4 Carbon.	5	6	7	8	
1	1	1	1	1	1	1	1	1	1	1
2		1	1	1	1	1	1	1	1	1
3			1	1	1	1	1	1	1	1
4			1	2	2	2	2	2	2	2
5			1	2	3	3	3	3	3	3
6			1	4	5	6	6	6	6	6
7			1	6	9	10	11	11	11	11
8			1	11	18	21	22	23	23	23
9			1	18	35	42	45	46	47	47
10			1	37	75					106
11			1	66	159					235
12			1	135	357					551
13			1	265	799					1301

TABLE B.

Knots.	Actual Valency.								
	0	1	2	3	4	5	6	7	8
1	1								
2		1							
3			1						
4			1	1					
5			1	1	1				
6			1	3	1	1			
7			1	5	3	1	1		
8			1	10	7	3	1	1	
9			1	17	17	7	3	1	1
10			1	36	38				
11			1	65	93				
12			1	134	222				
13			1	264	534				

In A, the columns 2, 3, 4, and the last column are the totals given by the Tables IV., VI., VIII., and II., and the remaining numbers of columns 5, 6, 7, 8 have been found by trial; and, in B, the several columns are the differences of the columns of A. The signification is obvious; for instance, if the number of knots is $=9$, then Table A, if the valency, or the maximum number of branches from a knot,

is =	2,	3,	4,	5,	6,	7,	8 or any greater number,
No. of trees =	1,	18,	35,	42,	45,	46,	47:

viz. with 9 knots the tree can have at most 8 branches from a knot, so that the number of trees having at most 8 branches from a knot is $=47$, the whole number of trees with 9 knots; and so the number of knots being as before $=9$, Table B shows that the number of 47 is made up of the numbers

$$1,\ 17,\ 17,\ 7,\ 3,\ 1,\ 1;$$

viz. 1	is the No. of trees, at most	2	branches from a knot,		
17	„ „	3	„ „	at least one	3-branch knot.
17	„ „	4	„ „	„	4 „
7	„ „	5	„ „	„	5 „
3	„ „	6	„ „	„	6 „
1	„ „	7	„ „	„	7 „
1	„ „	8	„ „	„	8 „

I annex also a plate showing the figures of the $1+1+2+3+6+11+23+47$ trees of 1, 2, 3,.., 9 knots, classified according to their altitudes and number of main branches; and as to the bicentre-trees, according to the number of main branches from *each* point of the bicentre. The affixed numbers show in each case the greatest number of branches from a knot; so that when this is (2), the knots may be oxygen-, boron-, carbon-, &c., atoms; when (3), boron-, carbon-, &c., atoms; when (4), carbon-, &c., atoms; and so on.

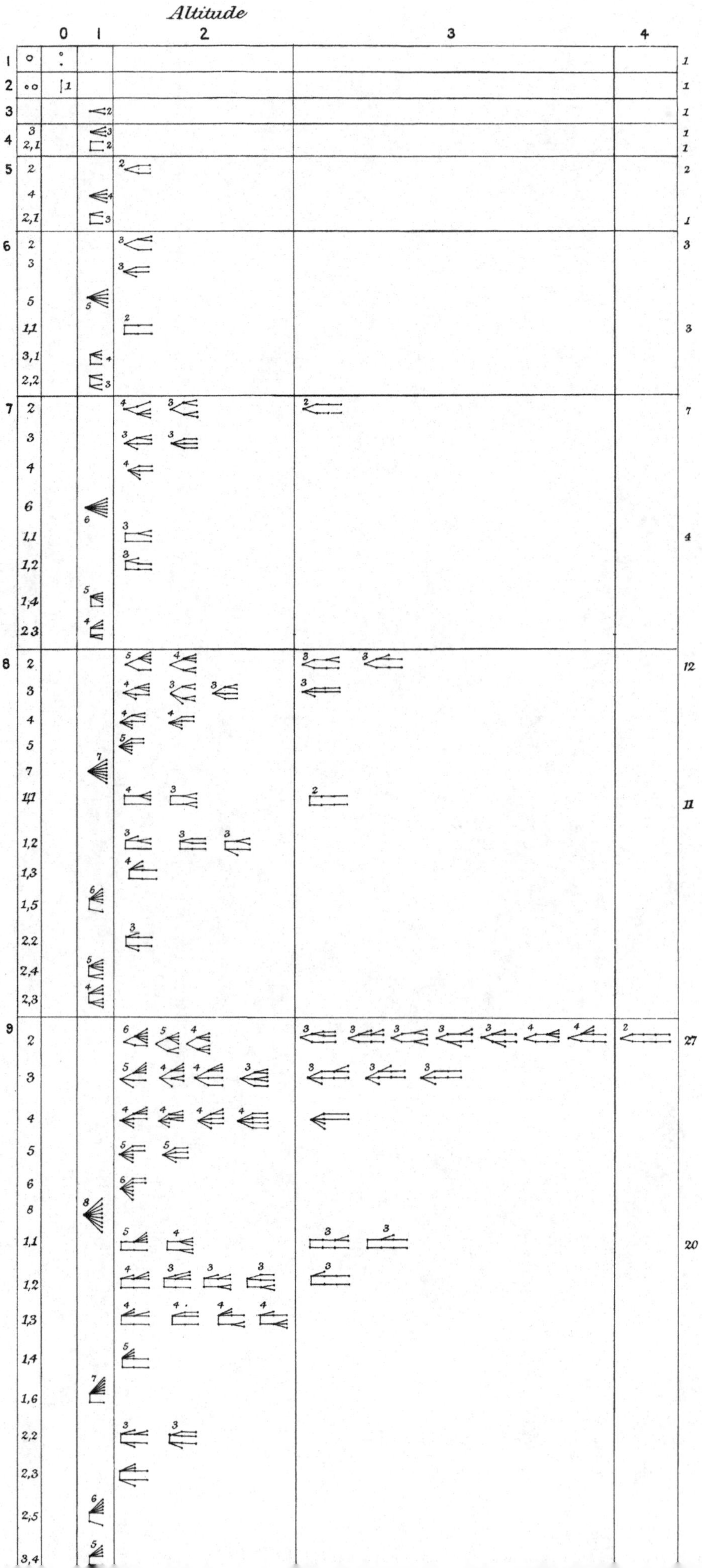

Altitude

		0	1	2	3	4	
1	○						1
2	○○						1
3							1
4	3						1
	2,1						1
5	2						2
	4						
	2,1						1
6	2						3
	3						
	5						
	1,1						3
	3,1						
	2,2						
7	2						7
	3						
	4						
	6						
	1,1						4
	1,2						
	1,4						
	2 3						
8	2						12
	3						
	4						
	5						
	7						
	1,1						11
	1,2						
	1,3						
	1,5						
	2,2						
	2,4						
	2,3						
9	2						27
	3						
	4						
	5						
	6						
	8						
	1,1						20
	1,2						
	1,3						
	1,4						
	1,6						
	2,2						
	2,3						
	2,5						
	3,4						

To face page 460.

611.

REPORT OF THE COMMITTEE ON MATHEMATICAL TABLES: CONSISTING OF PROFESSOR CAYLEY, F.R.S., PROFESSOR STOKES, F.R.S., PROFESSOR SIR W. THOMSON, F.R.S., PROFESSOR H. J. S. SMITH, F.R.S., AND J. W. L. GLAISHER, F.R.S.

[From the *Report of the British Association for the Advancement of Science* (1875), pp. 305—336.]

THE present Report (say Report 1875) is in continuation of that by Mr Glaisher, published in the volume for 1873, and here cited as Report 1873.

Report 1873 extends to all those tables which are at p. 3 (*l. c.*) included under the headings:—

A, auxiliary for non-logarithmic calculation, 1, 2, 3;

B, logarithmic and circular, 4, 5, 6;

C, exponential, 7, 8 (but only partially to C. 8), other than those tables of C referred to as "h. l tan $(45° + \frac{1}{2}\phi)$"; and also partially (see Art. 24, pp. 81—83) to the tables included under the heading "E. 11, transcendental constants ϵ, π, γ, &c., and their powers and functions."

A future Report will comprise the tables, or further tables, included under the headings:—

C. 8. Hyperbolic antilogarithms (e^x) and h. l tan $(45° + \frac{1}{2}\phi)$, and hyperbolic sines, cosines, &c.

D. Algebraic constants.

9. Accurate integer or fractional values. Bernoulli's Numbers, $\Delta^n 0^m$, &c. Binomial coefficients.

10. Decimal values auxiliary to the calculation of series.

E. 11. Transcendental constants ϵ, π, γ, &c., and their powers and functions.

The present Report (1875) comprises the tables included under the headings:—

F. Arithmological.

12. Divisors and prime numbers. Prime roots. The Canon arithmeticus, &c.

13. The Pellian equation.

14. Partitions.

15. Quadratic forms $a^2 + b^2$, &c., and partitions of numbers into squares, cubes, and biquadrates.

16. Binary, ternary, &c., quadratic and higher forms.

17. Complex theories:

which divisions are herein referred to, for instance, as [F. 12. Divisors, &c.].

Report 1873 consists of six sections (§) divided into articles, which are separately numbered (see contents, p. 174); the present Report 1875 forms a single section (§ 7), divided in like manner into articles, which are separately numbered; but besides this the paragraphs are numbered, and that continuously, through the present Report 1875, so that any paragraph may be cited as Report 1875, No. —, as the case may be.

[F. 12. *Divisors, &c.*] *Divisors and Prime Numbers.* Art. I.

1. As to divisors and prime numbers see Report 1873, Art. 8 (Tables of Divisors—factor tables—and Tables of Primes), pp. 34—40. The tables there referred to, such as Chernac, Burckhardt, Dase, Dase and Rosenberg, are chiefly tables running up to very high numbers (the last of them the ninth million): wherein, to save space, multiples of 2, 3, 5 are frequently omitted, and in some of them only the least divisor is given. It would be for many purposes convenient to have a small table, going up say to 10,000, showing in every case all the prime factors of the number. Such a table might be arranged, 500 numbers in a page, in some such form as the following:—

Factor Table | 1 to 500

	0	1	2	3	4	5	6	7	8	9
39	2 . 3 . 5 . 13	17 . 23	$2^3 . 7^2$	3 . 131	2 . 197	5 . 79	$2^2 . 3^2 . 11$	397*	2 . 199	3 . 7 . 19

where the top line shows the units, and the left-hand column the remaining figures, viz. the specimen exhibits the composition of the several numbers from 390 to 399: a prime number, e.g. 397, would be sufficiently indicated by the absence of any decomposition, or it may be further indicated by an asterisk.

It may be noticed that, in the theory of numbers, the decomposition is specially required when the next following number is a prime, viz. that of $p-1$, p being a prime: also, that this is given incidentally, for prime numbers p up to 1000, in Jacobi's *Canon Arithmeticus*, *post*, No. 20, and up to 15,000 in Reuschle's Tables, V. (a, b, c) *post*, No. 22.

2. It may be proper to remark here that any table of a binary form is really a factor-table in the complex theory connected with such binary form. Thus in a table of the form a^2+b^2, a number of this form has a factor $a+bi$ ($i=\sqrt{-1}$ as usual); and the table, in fact, shows the complex factor $a+bi$ of the number in question: a well arranged table would give all the prime complex factors $a+bi$ of the number. But as to this more hereafter; at present, we are concerned with the real theory only, not with any complex theory.

3. Connected with a factor-table, we have (i) a Table of the number of less relative primes; viz. such a table would show, for every number, the number of inferior integers having no common factor with the number itself. The formula is a well-known one: for a number $N=a^\alpha b^\beta c^\gamma \ldots$, ($a$, b, ... the distinct prime factors of N), the number of less relative primes is

$$\varpi(N), = a^{\alpha-1} b^{\beta-1} \ldots (a-1)(b-1)\ldots,$$

or, what is the same thing, $=N\left(1-\frac{1}{a}\right)\left(1-\frac{1}{b}\right)\ldots$ A small table ($N=1$ to 100), occupying half a page, is given by

Euler, *Op. Arith. Coll.* t. II. p. 128; viz. this is $\pi 1=0$, $\pi 2=1, \ldots$, $\pi 100=40$.

4. But it would be interesting to have such a table of the same extent with the proposed factor-table. The table might be of like form; for instance,

Number of less relative Primes Table | 1 to 500

	0	1	2	3	4	5	6	7	8	9
29	112	192	144	292	84	232	144	198	148	264

It would be of still greater interest to have an inverse table showing the values of N which belong to a given value of $\varpi(N)$; for instance,

$\varpi=$	$N=$
40	41, 55, 75, 82, 88, 100, 110,
42	43, 49, 86, 98,
44	69, 92,
46	47, 94,
48	65, 104, 105, 112,
⋮	

where, observe, that ϖ is of necessity even.

5. Again, connected with a factor-table, we have (ii) a Table of the Sum of the divisors of a Number. The formula is also a well-known one; for a number $N = a^{\alpha} b^{\beta} \ldots$, ($a, b, \ldots$ the distinct prime factors of N), the required sum

$$\int N \text{ is } = (1 + a + \ldots + a^{\alpha})(1 + b + \ldots + b^{\beta}) \ldots,$$

or, what is the same thing,

$$= \frac{a^{\alpha+1} - 1}{a - 1} \cdot \frac{b^{\beta+1} - 1}{b - 1} \ldots,$$

where, observe, that the number itself is reckoned as a divisor.

6. Such a table was required by Euler in his researches on Amicable Numbers (see *post*, No. 10), and he accordingly gives one of a considerable extent, viz.

Euler, *Op. Arith. Coll.* t. I. pp. 104—109.

It is to be remarked that, inasmuch as $\int N$ is obviously $= \int a^{\alpha} \int b^{\beta} \ldots$, the function need only be tabulated for the different integer powers a^{α} of each prime number a. The range of Euler's table is as follows:—

$a =$	$\alpha =$
2	1 to 36,
3	1 „ 15,
5	1 „ 9,
7	1 „ 10,
11	1 „ 9,
13	1 „ 7,
17	1 „ 5,
19	1 „ 5,
23	1 „ 4,
29 to 997	1 „ 3,

viz. for the several prime numbers from 29 to 997 the table gives $\int a$, $\int a^2$, and $\int a^3$ And it is to be noticed that the values of the sum are exhibited, decomposed into their prime factors: thus a specimen of the table is

Num.	Summa Divisorum.
139	$2^2 . 5 . 7$
139^2	$3 . 13 . 499$
139^3	$2^3 . 5 . 7 . 9661$

7. The form of the above table is adapted to its particular purpose (the theory of amicable numbers); but Euler gives also,

Euler, *Op. Arith. Coll.* t. I. p. 147—in the paper "Observatio de Summis Divisorum," (1752), pp. 146—154,—a short table of about half a page, $N = 1$ to 100, of the form $\int 1 = 1$, $\int 2 = 3, \ldots, \int 100 = 217$. The paper contains interesting analytical researches on the subject of $\int N$ which connect themselves with the theory of the Partition of Numbers.

8. It would be interesting to carry the last-mentioned table to the same extent as the proposed factor-table; and to add to it an inverse table, as suggested in regard to the number of less relative primes table.

9. *Perfect Numbers.*—A perfect number is a number which is equal to the sum of its divisors, the number itself not being reckoned as a divisor; e.g.

$$6 = 1 + 2 + 3, \text{ and } 28 = 1 + 2 + 4 + 7 + 14.$$

Such numbers are indicated by a table of the sums of divisors $\int 6 = 12$, $\int 28 = 56$, these two being, as appears by the table, Art. 7, the only perfect numbers less than 100.

10. *Amicable Numbers.*—These are pairs of numbers such that each is equal to the sum of the divisors of the other, not reckoning the other number as a divisor; that is, each has the same sum of divisors, the number being here reckoned as a divisor; say $\int' A = B$, $\int' B = A$; or, what is the same thing, $\int A = \int B \,(= A + B)$. Thus for the numbers 220, 284,

$$\int' 220 = (1 + 2 + 4)(1 + 5)(1 + 11) - 220, \; = 284,$$

$$\int' 284 = (1 + 2 + 4)(1 + 71) - 284, \qquad = 220;$$

or, what is the same thing,

$$\int 220 = (1 + 2 + 4)(1 + 5)(1 + 11) = 504 = (1 + 2 + 4)(1 + 71) = \int 284.$$

11. A catalogue of 61 pairs of numbers is given by

Euler, *Op. Arith. Coll.* t. I. pp. 144—145; it occupies about one page. The paper, "De Numeris Amicabilibus," pp. 102—145, contains an elaborate investigation of the theory, by means whereof all but two of the pairs of numbers are obtained. The first pair is the above-mentioned one, $2^2 . 5 . 11$ and $2^2 . 71$ ($= 220$ and 284); and the fifty-ninth pair is the high numbers

$$3^5 . 7^2 . 13 . 19 . 53 . 6959 \text{ and } 3^5 . 7^2 . 13 . 19 . 179 . 2087.$$

The last two pairs are referred to as "formæ diversæ a precedentibus;" viz. these are

$$\begin{cases} 2^3 . 19 . 41 \\ 2^5 . 199 \end{cases} \text{and} \begin{cases} 2^3 . 41 . 467 \\ 2^5 . 19 . 233. \end{cases}$$

12. A Table of the Frequency of Primes is given by

Gauss, Tafel der Frequenz der Primzahlen, *Werke*, t. II. pp. 436—443; viz. this extends to 3,000,000.

The first part, extending to 1,000,000, = 1000 thousand, shows how many primes there are in each thousand: a specimen is

1, 168:
2, 135:
3, 127:
4, 120:
5, 119:
&c.;

viz. in the first thousand there are 168 primes, in the second thousand 135 primes, and so on.

For the second and third millions the frequency is given for each ten thousand: a specimen is

1,000,000 to 1,100,000.

	0	1	2	3	4	5	6	7	8	9	
1	1										1
2		1				1		1	1		4
3		4	2	2	3	1	2	3	3	1	21
4	2	8	5	4	3	6	9	4	5	8	54
5	11	10	8	18	12	10	10	12	15	8	114
6	14	14	18	21	16	22	19	15	17	15	171
7	26	17	23	23	24	24	17	22	20	21	217
8	19	19	21	7	14	15	20	17	15	17	164
9	11	13	9	13	14	14	12	13	11	16	126
10	8	6	8	5	9	5	5	9	7	9	71
11	6	6	4	6	3	1	3	1	4	5	39
12	1	1	2	1	1	1	2	2	1		12
13	1	1			1		1	1	1		6
14											
15											
16											
	752	719	732	700	731	698	713	722	706	737	7210,

$$\int \frac{dx}{\log x} = 7212{\cdot}99;$$

viz. in the interval 1,000,000 to 1,010,000, 100 hundreds, there is 1 hundred containing 1 prime, there are 2 hundreds each containing 4 primes, 11 hundreds each containing 5 primes,.., 1 hundred containing 13 primes, so that, as

$$\begin{array}{rcr} 1\times & 1 = & 1, \\ 4\times & 2 = & 8, \\ 5\times & 11 = & 55, \\ \vdots & & \\ 13\times & 1 = & 13, \\ \hline 100 & & 752, \end{array}$$

the whole 10,000 contains 752 primes; the next 10,000 contains 719 primes, and so on; the whole 100,000 thus containing $752+719+\&c.\ldots=7210$ primes, which number is at the foot compared with the theoretic approximate value

$$\int\frac{dx}{\log x}\ \text{(limits 1,000,000 to 1,010,000)} = 7212{\cdot}99.$$

The integral in question is represented by the notation Li. x or li. x.

p. 443. We have the like tables 1,000,000 to 2,000,000 and 2,000,000 to 3,000,000, showing for each 100,000 how many hundreds there are containing 0 prime, 1 prime, 2 primes, up to (the largest number) 17 primes.

13. It is noticed that

the 26,379th hundred contains no prime,

the 27,050th hundred contains 17 primes.

It may be observed that, if $N=2.3.5\ldots p$, the product of all the primes up to p, then each of the numbers $N+1$ and $N+q$ (if q be the prime next succeeding p) is or is not a prime; but the intermediate numbers $N+2$, $N+3$,.., $N+q-1$ are certainly composite; viz. we thus have at least $q-2$ consecutive composites. To obtain in this manner 99 consecutive composites, the value of N would be $=2.3.5\ldots97$, viz. this is a number far exceeding 2,637,900; but, in fact, the hundred numbers 2,637,901 to 2,638,000 are all composite.

Legendre, in his *Essai sur la Théorie des Nombres* (1st edit., 1798; 2nd edit., 1808, supplement, 1816: references to this edition), gives for the number of primes inferior to a given limit x the approximate formula

$$\frac{x}{\log x - 1{\cdot}08366},$$

and p. 394, and supplement, p. 62, he compares for each 10,000 up to 100,000, and for each 100,000 up to 1,000,000, the values as computed by this formula with the actual numbers of primes exhibited by the tables of Wega and Chernac. Thus for $x=1,000,000$, the computed value is 78,543, the actual value 78,493.

He shows, p. 414, that the number of integers, which are less than n and are not divisible by any of the numbers θ, λ, μ,..., is approximately

$$=n\left(1-\frac{1}{\theta}\right)\left(1-\frac{1}{\lambda}\right)\left(1-\frac{1}{\mu}\right)\ldots;$$

and taking θ, λ, μ,... the successive primes 3, 5, 7,... he gives the values of the function in question, or, say, the function

$$\frac{2}{3}\cdot\frac{4}{5}\cdot\frac{6}{7}\cdot\frac{10}{11}\cdots\frac{\omega-1}{\omega},$$

ω a prime, for the several prime values $\omega=3$ to 1229 in the Table IX. (one page) at the end of the work.

14. A table of frequency is given by

Glaisher, J. W. L., *British Association Report* for 1872, p. 20. This gives for the second and the ninth millions, respectively divided into intervals of 50,000, the actual number of primes in each interval, as compared with the theoretic value $\text{li}\,x'-\text{li}\,x$; and also deduced therefrom, by the formula $\log\frac{1}{2}(x'+x)$, a table of the average interval between two consecutive primes; this average interval increases very slowly: at the beginning and the end of the second million the values are 13·76 and 14·58 (theoretic values 13·84 and 14·50); at the beginning and the end of the ninth million 16·02 and 15·95 (theoretic values 15·90 and 16·01).

15. Coming under the head of Divisor Tables, some tables by Reuschle and Gauss may be here referred to. These are:—

Reuschle, *Mathematische Abhandlung, zahlentheoretische Tabellen sammt einer dieselben treffenden Correspondenz mit der verewigten C. G. J. Jacobi*, 4°, pp. 1—61* (1856). The tables belonging to the present subject are

A. Tafeln zur Zerlegung von a^n-1 (pp. 18—22).

I. Table of the prime factors of 10^n-1, viz.

(a. pp. 18—19.) Complete decomposition of 10^n-1, $n=1$ to 42: and 10^n+1, $n=1$ to 21. Some values of n are omitted.

A specimen is

$$10^{13}-1=3^2.53.79.265371653,$$
$$10^{13}+1=11.189.1058313049.$$

(b. p. 19.) List of the specific prime factors f of 10^n-1, or the prime factors of the residue after separation of the analytical factors, for those values of n for which the complete decomposition is unknown, and omitting those values for which no factor is known, $n=25$ to 243.

* Titlepage missing in my copy; but I find from Prof. Kummer's notice of the work, *Crelle*, t. LIII. (1857), p. 379, that it appeared as a Programm of the Stuttgart Gymnasium, Michaelmas, 1856, and was separately printed by Liesching and Co., Stuttgart.

A specimen is

$$\begin{array}{cc} n & f \\ 25 & 21401. \end{array}$$

The meaning seems to be, residue of $10^{25}-1$ is $1+10^5+10^{10}+10^{15}+10^{20}$, and this contains the prime factor 21401; but it is not clear why this is the "specific prime factor."

II. Prime factors of a^n-1 for different values of a and n.

(a. p. 20) gives for 41 values of a (2, 3, &c. at intervals to 100) and for the following values of n the decompositions of the residues or specific factors of a^n-1; viz. these are

$$\begin{array}{rl} n= 1, & a-1: \\ ,, \ 2, & a+1: \\ ,, \ 3, & a^2+a+1: \\ ,, \ 6, & a^2-a+1: \\ ,, \ 4, & a^2+1: \\ ,, \ 5, & a^4+a^3+a^2+a+1: \\ ,, \ 10, & a^4-a^3+a^2-a+1: \\ ,, \ 8, & a^4+1: \\ ,, \ 12, & a^4-a^2+1. \end{array}$$

A specimen is

a	$a-1$	a^2-1	a^3-1	a^6-1	a^4-1	a^5-1	$a^{10}-1$	a^8-1	$a^{12}-1$
10	3^2	11	$3^3 . 37$	7 . 13	101	41 . 271	9091	73 . 137	9901

(b. p. 21.) Specific prime factors for the numbers 2, 3, 5, 6, 7, 10, (the powers 4, 8, 9 being omitted as coming under 2 and 3), for the exponents 1 to 42.

A specimen is

n	2^n-1	3^n-1	5^n-1	6^n-1	7^n-1	10^n-1
19	524287	1597 . 363889	191 . x	191 . x	419 . x	

where the x denotes that the other factor is not known to be prime. And so, where no number is given, as in $10^{19}-1$, it is not known whether the number $(=1+10^1+10^2+\ldots+10^{18})$ is or is not prime.

Addition, p. 22. For $a=2$, the complete decomposition of the prime factor of 2^n-1 is given for values of n, $=44$, 45,... at intervals to 156.

A specimen is

$$\begin{array}{cc} n & f \\ 44 & 397 . 2113, \end{array}$$

viz.

$$2^{20}-2^{18}+2^{16}-\ldots-2^2+1, =838861=397 . 2113.$$

$n=31$, Fermat's prime. $n=37$, the first case for which the decomposition is not given completely. $n=41$, the first case for which no factor is known.

16. **Gauss,** Tafel zur Cyclotechnie, *Werke*, t. II. pp. 478—495, shows, for 2452 numbers of the several forms a^2+1, a^2+4, $a^2+9, \ldots$, a^2+81, the values of a such that the number in question is a product of prime factors no one of which exceeds 200, and exhibits all the odd prime factors of each such number. The table is in nine parts, zerlegbare a^2+1, zerlegbare a^2+4, &c., with to each part a subsidiary table, as presently mentioned. Thus a specimen is

zerlegbare a^2+9.

1	5
2	13
4	5 . 5
5	17
7	29
8	73
⋮	⋮
1411168679	5 . 5 . 13 . 17 . 17 . 89 . 113 . 157 . 173 . 197 . 197 :

viz.

1^2+9, odd prime factor is 5,

2^2+9, „ „ 13,

4^2+9, „ factors are 5, 5,

and so on.

And the subsidiary table is

5	1, 4, 79
13	2, 11, 41
17	5, 29, 46, 379, 1042
⋮	⋮

showing that the numbers a for which the largest factor is 5 are 1, 4, 79; those for which it is 13 are 2, 11, 41; and so on.

The object of the table is explained in the *Bemerkungen*, (*l. c.*, p. 523), by Schering, the editor of the volume, viz. it is to facilitate the calculation of the circular arcs the cotangents of which are rational numbers. To take a simple example, it appears to be by means of it that Gauss obtained, among other formulæ, the following:

$$\frac{\pi}{4} = 12 \text{ arctan } \tfrac{1}{18} + 8 \text{ arctan } \tfrac{1}{57} - 5 \text{ arctan } \tfrac{1}{239},$$

and

$$= 12 \text{ arctan } \tfrac{1}{38} + 20 \text{ arctan } \tfrac{1}{57} + 7 \text{ arctan } \tfrac{1}{239} + 24 \text{ arctan } \tfrac{1}{268}.$$

[F. 12. *Divisors, &c.*] *continued.* *Prime Roots.* *The Canon Arithmeticus, Quadratic residues.* Art. II.

17. *Prime Roots.*—Let p be a prime number; then there exist $\varpi(p-1)$ inferior integers g, such that all the numbers $1, 2, \ldots, p-1$ are, to the modulus p,

$$\equiv 1,\ g,\ g^2, \ldots,\ g^{p-2}\ (g^{p-1} \text{ is of course } \equiv 1).$$

This being so, g is said to be a prime root of p; and moreover the several numbers g^α, where α is any number whatever less than and prime to $p-1$, constitute the series of the $\varpi(p-1)$ prime roots of p. It may be added that, if β be an integer number less than $p-1$, and having with it a greatest common measure $=k$, so that

$$(g^\beta)^{\frac{p-1}{k}} \equiv g^{\frac{\beta}{k}(p-1)},\ \equiv 1,\ \left(\text{since } \frac{\beta}{k} \text{ is an integer, and } g^{p-1} \equiv 1\right),$$

then g^β has the indicatrix $\dfrac{p-1}{k}$: the prime roots are those numbers which have the indicatrix $p-1$.

The like theory exists as to any number N of the form p^m or $2p^m$. There are here $\varpi(N)$, $= N\left(1-\dfrac{1}{p}\right)$ or $\tfrac{1}{2}N\left(1-\dfrac{1}{p}\right)$, in the two cases respectively, numbers less than N and prime to it; and we have then $\varpi(\varpi(N))$ numbers g such that, to the modulus N, all these numbers are $\equiv 1, g, g^2 \ldots g^{\varpi(N)-1}$ ($g^{\varpi(N)}$ is of course $\equiv 1$). This being so, g may be regarded as a prime root of N ($=p^m$ or $2p^m$, as the case may be); and moreover the several numbers g^α, where α is any number whatever less than and prime to $\varpi(N)$, constitute the series of the $\varpi(\varpi(N))$ prime roots of N. Thus $N = 3^2 = 9$, $\varpi(N) = 6$; we have

$$\begin{array}{ccccccc} & 1, & 2^1, & 2^2, & 2^3, & 2^4, & 2^5, \\ \equiv & 1, & 2, & 4, & 8, & 7, & 5, \text{ mod. } 9; \end{array}$$

or the prime roots of 9 are 2^1 and 2^5, $=2$ and 5.

So also $N = 2 \,.\, 3^2 = 18$, $\varpi(N) = 6$; we have

$$\begin{array}{ccccccc} & 1, & 5^1, & 5^2, & 5^3, & 5^4, & 5^5, \\ \equiv & 1, & 5, & 7, & 17, & 13, & 11, \text{ mod. } 18; \end{array}$$

and 5^1 and 5^5, $=5$ and 11 are the prime roots of 18.

18. A small table of prime roots, $p = 3$ to 37, is given by

Euler, *Op. Arith. Coll.* t. I. pp. 525—526. The Memoir is entitled "Demonstrationes circa residua e divisione potestatum per numeros primos resultantia," pp. 516—537 (1772).

19. A table, p and p^m, 3 to 97, is given by

Gauss, "Disquisitiones Arithmeticæ," 1801, (*Werke*, t. I. p. 468). This gives in each case a prime root, and it shows the exponents in regard thereto of the several prime numbers less than p or p^m. Thus a specimen is

		2	3	5	7	11	13	17	19	23	29	&c.
27	2	1	*	5	16	13	8	15	12	11		
29	10	11	27	18	20	23	2	7	15	24		

viz. for 27 we have 2 a prime root, and $2 \equiv 2^1$, $5 \equiv 2^5$, $7 \equiv 2^{16}$, $11 \equiv 2^{13}$, &c.; and so also for 29 we have 10 a prime root, and $2 \equiv 10^{11}$, $3 \equiv 10^{27}$, $5 \equiv 10^{18}$, &c.

20. Small tables are probably to be found in many other places; but the most extensive and convenient table is **Jacobi's** *Canon Arithmeticus*, the complete title of which is

Canon Arithmeticus sive tabula quibus exhibentur pro singulis numeris primis vel primorum potestatibus infra 1000 *numeri ad datos indices et indices ad datos numeros pertinentes.* Edidit C. G. J. Jacobi. Berolini, 1839. 4°.

The contents are as follows:—

The following is a specimen of the principal tables:—

$$p = 19,\ p - 1 = 2 \,.\, 3^2.$$

Numeri.

I	0	1	2	3	4	5	6	7	8	9
		10	5	12	6	3	11	15	17	18
	9	14	7	13	16	8	4	2	1	

Indices.

N	0	1	2	3	4	5	6	7	8	9
		18	17	5	16	2	4	12	15	10
1	1	6	3	13	11	7	14	8	9	

where the first table gives the values of the powers of the prime root 10 (that 10 is the root appears by its index being given as $=1$) to the modulus 19, viz. $10^1 \equiv 10$, $10^2 \equiv 5$, $10^3 \equiv 12$, &c.; and the second table gives the index of the power to which the same prime root must be raised in order that it may be, to the modulus 19, congruent with a given number: thus $10^{18} \equiv 1$, $10^{17} \equiv 2$, $10^5 \equiv 3$, &c. The units of the index or number, as the case may be, are contained in the top line of the table, and the tens or hundreds and tens in the left-hand column.

21. There is given by

Jacobi, *Crelle,* t. XXX. (1846), pp. 181, 182, a table of m' for the argument m, such that

$$1 + g^m \equiv g^{m'} \text{ (mod. } p), \quad p = 7 \text{ to } 103, \text{ and } m = 0 \text{ to } 102.$$

A specimen is

p	7	11	13	17	19	23	29	31	37 ... to 103
g	3	2	6	10	10	10	10	17	5
m									
11	.	.	6	4	7	*	27	21	34

for instance, $p = 19$, $1 + 10^{11} \equiv 10^7$ (mod. 19).

Jacobi remarks that this table was calculated for him by his class during the winter course of 1836—37; and that, by means of the *Canon Arithmeticus* since published (in 1839), the same might easily be extended to all primes under 1000. In fact, for any such number p, putting any number of the table "Indices" $= m$, the next following number of the table gives the value of m'.

22. We have next, in **Reuschle's** Memoir (*ante,* No. 15), the following relating to prime roots:—

C. Tafeln für primitive Wurzeln und Hauptexponenten, oder V. erweiterte und bereicherte Burkhardtsche Tafel, pp. 41—61, being divided into three parts; viz. these are

a. Table of the Hauptexponenten of the six roots 10, 5, 2, 6, 3, 7 for all prime numbers of the first 1000, together with the least primitive root of each of these numbers (pp. 42—46).

A specimen is as follows:—

		10		5		2		6		3		7		w
p	$p-1$	e	n	e	n	e	n	e	n	e	n	e	n	
53	$2^2.13$	13	4	52	1	52	1	26	2	52	1	26	2	2

where e is the Hauptexponent or indicatrix of the root (10, 5, 2, 6, 3, 7, as the case may be), $n = \dfrac{p-1}{e}$, w the least primitive root; thus

$$p = 53, \quad 10^{13} \equiv 1, \quad 5^{52} \equiv 1, \quad 2^{52} \equiv 1,$$

(2 being accordingly the least prime root),

$$6^{26} \equiv 1, \quad 3^{52} \equiv 1, \quad 7^{26} \equiv 1.$$

The number w of the last column is the least primitive root. It is, of course, not always (as in the present case) one of the numbers 10, 5, 2, 6, 3, 7 to which the table relates: the first exception is $p = 191$, $w = 19$: the highest value of w is $w = 21$, corresponding to $p = 409$.

b. The like table for the roots 10 and 2 for all prime numbers from 1000 to 5000, together with as convenient as possible a prime root (and in some cases two prime roots) for each such number (pp. 47—53).

A specimen is:—

		10		2		
p	$p-1$	e	n	e	n	w
1289	$2^3.7.23$	92	14	161	8	6, 11 ;

viz. here, mod. 1289, $10^{92} \equiv 1$, $2^{161} \equiv 1$; and two prime roots are 6, 11. We have thus by the present tables a prime root for every prime number not exceeding 5000.

c. The like table for the root 10 for all prime numbers between 5000 and 15000, (no column for w, nor any prime root given), pp. 53—61.

A specimen is

p	$p-1$	e	n
9859	2.3.31.53	3286	3:

viz., mod. 9859, we have $10^{3286} \equiv 1$. But in a large number of cases we have $n = 1$, and therefore 10 a prime root. For example,

9887	2.4983	9886	1.

23. For a composite number n, if $N = \varpi(n)$ be the number of integers less than n and prime to it, then if x be any number less than n and prime to it, we have $x^N \equiv 1$ (mod. n). But we have in this case no analogue of a prime root—there is no number x, such that its several powers $x^1, x^2, \ldots, x^{N-1}$ (mod. n) are all different from unity; or, what is the same thing, there is for each value of x some submultiple of N, say N', such that $x^{N'} \equiv 1$ (mod. n). And these several numbers N' have a least common multiple I, which is not $= N$, but is a submultiple of N; and this being so, then for all the several values of x, I is said to be the maximum indicator. For instance, $n = 12$, $N = \varpi(n)$; the numbers less than 12 and prime to it are 1, 5, 7, 11. We have, (mod. 12), $1^1 \equiv 1$, $5^2 \equiv 1$, $7^2 \equiv 1$, $11^2 = 1$, or the values of N' are 1, 2, 2, 2; their least common multiple is 2, and we have accordingly $I = 2$: viz. $x^2 \equiv 1$ (mod. 12) has the $\varpi(12)$ roots 1, 5, 7, 11. So $n = 24$, $\varpi(n) = 8$; the maximum indicator I is in this case also $= 2$.

A table of the maximum indicator $n = 1$ to 1000 is given by

Cauchy, *Exer. d'Analyse &c.*, t. II. (1841), pp. 36—40, contained in the "Mémoire sur la résolution des équations indéterminées du premier degré en nombres entiers," pp. 1—40.

24. It thus appears that for a composite number n, the $\varpi(n)$ numbers less than n and prime to it cannot be expressed as $\equiv$ (mod. n) to the power of a single root; but for the expression of them it is necessary to employ two or more roots. A small table, $n = 1$ to 50, is given by

Cayley, Specimen Table $M \equiv a^{\alpha} b^{\beta}$ (mod. N) for any prime or composite modulus; *Quart. Math. Journ.* vol. IX. (1868), pp. 95, 96, and folding sheet, [397].

A specimen is

Nos.	12
roots	5, 7
Ind.	2, 2
M.I.	2
ϕ	4
1	0, 0
2	
3	
4	
5	1, 0
6	
7	0, 1
8	
9	
10	
11	1, 1

viz. for the modulus 12 the roots are 5, 7, having respectively the indicators 2, 2, viz. $5^2 \equiv 1$ (mod. 12), $7^2 \equiv 1$ (mod. 12). Hence also the maximum indicator is $= 2$. $\phi\ (= \varpi(n)) = 4$ is the number of integers less than 12 and prime to it, viz. these are 1, 5, 7, 11, which in terms of the roots 5, 7 and to mod. 12 are respectively $\equiv 5^0 . 7^0$, $5^1 . 7^0$, $5^0 . 7^1$, and $5^1 . 7^1$.

25. *Quadratic Residues.*—In regard to a given prime number p, a number N is or is not a quadratic residue according as the index of N is even or odd, viz. g being a prime root and $N \equiv g^{\alpha}$, then N is or is not a quadratic residue according as α is even or odd. But the quadratic residues can, of course, be obtained directly without the consideration of prime roots.

A small table, $p = 3$ to 97 and $N = -1$ and (prime values) 3 to 97, is given by

Gauss, "Disquisitiones Arithmeticæ," 1801; Table II. (*Werke*, t. I. p. 469): I notice here a misprint in the top line of the original; it should be $-1, +2, +3$, &c., instead of

1, + 2, + 3, &c.; the − 1 is printed correctly on p. 499 of the French translation *Recherches Arithmétiques*, Paris, 1807 and on p. 469 of vol. I. of *Werke*, (Göttingen, 1870).

A specimen is

	− 1	+ 2	+ 3	+ 5	+ 7	+ 11	+ 13	+ 17	+ 19	+ 23	&c.
19				–	–	–		–	–	–	

viz. − 1, 2, 3, 13 are not, 5, 7, 11, 17 &c. are, quadratic residues of 19. The residues taken positively and less than 19 are, in fact, 1, 4, 5, 6, 7, 11, 16, 17.

The same table carried from $p = 3$ to 503, and prime values $N = 3$ to 997, is given by

Gauss, *Werke*, t. II. pp. 400—409. A specimen is

	2	3	5	7	11	13	17	19	23	&c.;
19			–	–	–		–	–	–	

viz. the arrangement is the same, except only that the − 1 column is omitted.

26. We have also by **Gauss**

"Disquisitiones Arithmeticæ" Table III. (*Werke*, t. I. p. 470), for the conversion into decimals of a vulgar fraction, denominator p or p^μ, not exceeding 100. The explanation is given in Art. 314 *et seq.* of the same work.

But this table, carried to a greater extent, is given by **Gauss,** *Werke*, t. II. pp. 412—434, "Tafel zur Verwandlung gemeiner Brüche mit Nennern aus dem ersten Tausend in Decimalbrüche;" viz. the denominators are here primes or powers of primes, p^μ up to 997.

To explain the table, consider a modulus p^μ (where μ may be $=1$); if 10 is not a prime root of p^μ, consider a prime root r, which is such that $r^e \equiv 10 \pmod{p^\mu}$, e being a submultiple of $p^{\mu-1}(p-1)$; say we have $ef = p^{\mu-1}(p-1)$: then $10^f \equiv 1 \pmod{p^\mu}$. Consider any fraction $\frac{N}{p^\mu}$; then we may write $N \equiv r^{kl+l} \pmod{p^\mu}$, k from 0 to $f-1$ and l from 0 to $e-1$, $\equiv 10^k r^l$, and consequently $\frac{N}{p^\mu}$ and $\frac{10^k r^l}{p^\mu}$ have the same mantissa (decimal part regarded as an integer); hence, in order to know the mantissa of every fraction whatever of $\frac{N}{p^\mu}$, it is sufficient to know the mantissa of $\frac{rl}{p^\mu}$, that is, the mantissæ of $\frac{1}{p^\mu}, \frac{r}{p^\mu}, \frac{r^2}{p^\mu}, \ldots, \frac{r^{e-1}}{p^\mu}$, or, what is the same thing, the mantissæ of $\frac{10}{p^\mu}, \frac{10r}{p^\mu}, \ldots, \frac{10r^{e-1}}{p^\mu}$.

For instance, $p^\mu = 11$, $10^2 \equiv 1 \pmod{11}$, whence $f = 2$, $e = 5$; and taking $r = 2$, we have $10 \equiv r^5 \pmod{11}$.

The required mantissæ, denoted in the table by

$$(0),\quad (1),\quad (2),\quad (3),\quad (4),$$

are those of

$$\frac{10}{11},\ \frac{10\,.\,2}{11},\ \frac{10\,.\,2^2}{11},\ \frac{10\,.\,2^3}{11},\ \frac{10\,.\,2^4}{11},$$

viz. these fractions are respectively =

$$\begin{array}{ccccc}(0), & (1), & (2), & (3), & (4),\\ \cdot 9090\ldots, & 1\cdot 8181\ldots, & 3\cdot 6363\ldots, & 7\cdot 2727\ldots, & 14\cdot 5454\ldots;\end{array}$$

or their mantissæ are 90, 81, 63, 27, 54.

And we accordingly have as a specimen

$$11 \mid (1)\ldots 81,\ (2)\ldots 63,\ (3)\ldots 27,\ (4)\ldots 54,\ (0)\ldots 90.$$

Or again, as another specimen, $r=2$:—

$$27 \mid (1)\ldots 740,\ (2)\ldots 481,\ (3)\ldots 962,\ (4)\ldots 925,\ (5)\ldots 851,\ (0)\ldots 370.$$

The table in this form extends to $p^\mu = 463$; the values of r (not given in the body of the table) are annexed, p. 420.

In the latter part of the table $p^\mu = 467$ to 997, we have only the mantissæ of $\frac{100}{p^\mu}$. A specimen is

	1828153564	8994515539	3053016453	3820840950
547	6398537477	1480804387	5685557586	8372943327
	2394881170	0182815356,		

viz. the fraction $\frac{100}{547} = \cdot 182815\ldots$ has a period of 91, $=\frac{1}{6}546$, figures.

[F. 13. *The Pellian Equation.*] Art. III.

27. The Pellian equation is $y^2 = ax^2 + 1$, a being a given integer number, which is not a square (or rather, if it be, the only solution is $y=1$, $x=0$), and x, y being numbers to be determined: what is required is the least values of x, y, since these, being known, all other values can be found. A small table $a=2$ to 68 is given by

Euler, *Op. Arith. Coll.* t. I. p. 8. The Memoir is "Solutio problematum Diophanteorum per numeros integros," pp. 4—10, 1732—33. The form of the table is

a	$x\,(=p)$	$y\,(=q)$
2	2	3
3	1	2
5	4	9
⋮	⋮	⋮
68	4	33.

Even here, for some of the values of a, the values of x, y are extremely large; thus $a = 61$, $x = 226{,}153{,}980$, $y = 1{,}766{,}399{,}049$.

And probably tables of a like extent may be found elsewhere; in particular, a table of the solution of $y^2 = ax^2 \pm 1$ ($-$ when the value of a is such that there is a solution of $y^2 = ax^2 - 1$, and $+$ for other values of a), $a = 2$ to 135, is given by **Legendre,** *Théorie des Nombres*, 2nd ed. 1808, in the Table X. (one page) at the end of the work. For the before-mentioned number 61, the equation is $y^2 = 61x^2 - 1$, and the values are $x = 3805$, $y = 29718$; much smaller than Euler's values for the equation $y^2 = 61x^2 + 1$.

28. The most extensive table, however, is given by

Degen, *Canon Pellianus, sive Tabula simplicissimam equationis celebratissimæ: $y^2 = ax^2 + 1$, solutionem, pro singulis numeri dati valoribus ab* 1 *usque ad* 1000 *in numeris rationalibus, iisdemque integris exhibens.* Auctore Carolo Ferdinando Degen. Hafn (Copenhagen) apud Gerhardum Bonnarum, 1817. 8vo. pp. iv to xxiv and 1 to 112.

The first table (pp. 3—106) is entitled as "Tabula I. Solutionem Equationis $y^2 - ax^2 - 1 = 0$ exhibens." It, in fact, also gives the expression of $\sqrt{a}$ as a continued fraction; thus a specimen is

209	14	2	5	3	(2)
	1	13	5	8	11
	3220				
	46551				

Here the first line gives the continued fraction, viz.

$$\sqrt{209} = 14 + \frac{1}{2+}\frac{1}{5+}\frac{1}{3+}\frac{1}{2+}\frac{1}{3+}\frac{1}{5+}\frac{1}{2+}\frac{1}{28+}\frac{1}{2+}\text{\&c.},$$

the period being (2, 5, 3, 2, 3, 5, 2) indicated by 2, 5, 3 (2). [The number of terms in the period is here odd, but it may be even; for instance, the period (1, 1, 5, 5, 1, 1) is indicated by 1, 1 (5, 5).]

The second line contains auxiliary numbers presenting themselves in the process; thus, if $R^2 = 239$, we have $R = 14 + \frac{1}{\alpha}$,

$$\alpha = \frac{1}{R-14} = \frac{1\,(R+14)}{209-14^2} = \frac{R+14}{13} = 2 + \frac{1}{\beta},$$

$$\beta = \frac{13}{R-12} = \frac{13\,(R+12)}{209-12^2} = \frac{R+12}{5} = 5 + \frac{1}{\gamma},$$

$$\gamma = \frac{5}{R-13} = \frac{5\,(R+13)}{209-13^2} = \frac{R+13}{8} = 3 + \frac{1}{\delta},$$

&c.,

where the second line 1, 13, 5,... shows the numerical factors of the third column. The value of this second line as a result is not very obvious.

The third line gives x, and the fourth line y.

29. The second table, pp. 109—112, is entitled "Tabula II. Solutionem æquationis $y^2 - ax^2 + 1 = 0$, quotiescunque valor ipsius a talem admiserat, exhibens"; viz. it is remarked that this is only possible (but see *infrà*) for those values of a which in Table I. correspond to a period of an even number of terms, as shown by two equal numbers in brackets; thus $a = 13$, the period of $\sqrt{13}$ given in Table I. is (1, 1, 1, 1) as shown by the top line 3, 1 (1, 1), and accordingly 13 is one of the numbers in Table II.; and we have there 13 | 5 / 18.

Or take another specimen, 241 | 4574225 / 71011068; viz. the first line gives the value of x, and the second line the value of y (least values), for which $y^2 - ax^2 = -1$.

It is to be noticed that $a = 2$ and $a = 5$, for which we have obviously the solutions ($x = 1$, $y = 1$) and ($x = 1$, $y = 2$) respectively, are exceptional numbers not satisfying the test above referred to; and (apparently for this reason) the values in question, 2 and 5, are omitted from the table.

30. **Cayley,** "Table des plus petites solutions impaires de l'équation $x^2 - Dy^2 = \pm 4$, $D \equiv 5$ (mod. 8)." *Crelle,* t. LIII. (1857), page 371 (one page), [231].

As regards the theory of quadratic forms, it is important to know whether for a given value of D ($\equiv 5$, mod. 8) there does or does not exist a solution, in odd numbers, of the equation $x^2 - Dy^2 = 4$. As remarked in the paper, "Note sur l'équation $x^2 - Dy^2 = \pm 4$, $D \equiv 5$ (mod. 8)," pp. 369—371, [231], this can be determined for values of D of the form in question up to $D = 997$ by means of Degen's Table; and the solutions, when they exist, of the equation $x^2 - Dy^2 = 4$, as also of the equation $x^2 - Dy^2 = -4$, can be obtained up to the same value of D. Observe that when the equation $x^2 - Dy^2 = -4$ is possible, the equation $x^2 - Dy^2 = 4$ is also possible, and that its least solution is obtained very readily from that of the other equation; it is therefore sufficient to tabulate the solution of $x^2 - Dy^2 = \pm 4$, the sign being − when the corresponding equation is possible, and being in other cases +. Hence the form of the Table: viz. as a specimen we have

D	$\pm$	x	y
757		imposs.	
765	+	83	3
773	−	139	5
781		imposs.	
⋮			

that is, if $D=757$ or 781, there is no solution of either $x^2-Dy^2=+4$ or $=-4$; if $D=765$, there is a solution $x=83$, $y=3$ of $x^2-Dy^2=+4$, but none of $x^2-Dy^2=-4$; if $D=773$, there is a solution $x=139$, $y=5$ of $x^2-Dy^2=-4$, and therefore also a solution of $x^2-Dy^2=+4$; and so in other cases.

[F. 14. *Partitions.*] Art. IV.

31. The problem of Partitions is closely connected with that of Derivations. Thus if it be asked in how many ways can the number n be expressed as a sum of three parts, the parts being 0, 1, 2, 3, and each part being repeatable an indefinite number of times, it is clear that n is at most $=9$, and that for the values of m, $=0, 1,.., 9$ shown by the top line of the annexed table, the number of partitions has the values shown by the bottom line thereof:—

0	1	2	3	4	5	6	7	8	9
a^3	a^2b	a^2c	a^2d	abd	acd	ad^2	bd^2	cd^2	d^3
		ab^2	abc	ac^2	b^2d	bcd	c^2d		
			b^3	b^2c	bc^2	c^3			
1	1	2	3	3	3	3	2	1	1

But taking a, b, c, d to stand for 0, 1, 2, 3 respectively, the actual partitions of the required form are exhibited by the literal terms of the table (these being obtained, each column from the preceding one, by the method of derivations, or say by the rule of the last and last but one), and the numbers of the bottom line are simply the number of terms in the several columns respectively.

32. A set of such literal tables, say of tables $\begin{pmatrix} a, & b, & c, .., & k \\ =0, & 1, & 2, .., & m \end{pmatrix}^n$, for different values of n and m (where the number of letters is $=m+1$), would be extremely interesting and valuable. The tables for a given value of m and for different values of n are, it is clear, the proper foundation of the theory of the binary quantic $(a, b, c, .., k \between x, 1)^m$, which corresponds to such value of m. Prof. Cayley regrets that he has not in his covariant tables given in every case the complete series of literal terms; viz. the literal terms which have zero coefficients are, for the most part, though not always, omitted in the expressions of the several covariants.

33. But the question at present is as to the *number* of terms in a column, that is, as to the number of the partitions of a given form: the analytical theory has been investigated by Euler and others. The expression for the number of partitions is usually obtained as equal to the coefficient of x^n in the development, in ascending powers of x, of a given rational function of x: for instance, if there is no limitation as to the number of the parts, but if the parts are 1, 2, 3, m (viz. a part may have any value not exceeding m), each part being repeatable an indefinite number of times, then

Number of partitions of $n =$ coefficient of x^n in $\dfrac{1}{(1-x)(1-x^2)(1-x^3)\ldots(1-x^m)}$,

and we can, by actual development, obtain for any given values of m, n the number of partitions.

These have been tabulated $m = 1, 2, \ldots, 20$, and $m = \infty$ (viz. there is in this case no limit as to the largest part), and $n = 1$ to 59, by

Euler, *Op. Arith. Coll.* t. I. pp. 97—101, given in the paper "De Partitione Numerorum," pp. 73—101, (1750); the heading is "Tabula indicans quot variis modis numerus n e numeris 1, 2, 3, 4,.., m, per additionem exhibi potest, seu exhibens valores formulæ $n^{(m)}$." The successive lines are, in fact, the coefficients of the several powers $x^0, x^1, \ldots, x^9$ in the expansions of the functions

$$\frac{1}{1-x}, \frac{1}{1-x\,.\,1-x^2}, \ldots, \frac{1}{1-x\,.\,1-x^2\ldots 1-x^{20}}.$$

34. The generating function for any given value of m is, it is clear, $=\dfrac{1}{1-x^m}$ multiplied by that for the next preceding value of m, and it thus appears how each line of the table is calculated from that which precedes it. The auxiliary numbers are *printed*; thus a specimen is

Valores numeri n.

m	0	1	2	3	4	5	6	7	8	9	10	
					1	1	2	3	5	6	9	
4	1	1	2	3	5	6	9	11	15	18	23	
						1	1	2	3	5	7	
5	1	1	2	3	5	7	10	13	18	23	30	

viz. suppose the numbers in the second 4-line known: then simply moving these each five steps onward we have the (auxiliary) numbers of the first 5-line; and thence by a mere addition the required series of numbers shown by the second 5-line. And similarly from this is obtained the second 6-line, and so on.

35. More extensive tables are contained in the memoir by

Marsano, *Sulle leggi delle derivate generali delle funzioni di funzioni et sulla teoria delle forme di partizione dei numeri intieri,* (4°. Genova, 1870), pp. 1—281; and three tables paged separately, described merely as "Tavole dei numeri $C_{q,r}$, $S_{q,e}$, $S'_{q,e}$ citate nel testo colle indicazioni di Tavole I., II., III., ai n[i] 77, 79, 81"; viz. the reader is referred to these articles for the explanations of what the tabulated functions are; and there is not even then any explicit statement, but the investigation itself has to be studied to make out what the tables are. It is, in fact, easier to make this out from the tables themselves; the explanation is as follows:—

Table I. (16 pages) is, in fact, Euler's table, showing in how many ways the number n can be made up with the parts 1, 2, 3,.., m; but the extent is greater, viz. n is from 1 to 103, and m from 1 to 102. The auxiliary numbers given in Euler's table are omitted, as also certain numbers which occur in each successive line; thus a specimen is

$n=$	1	2	3	4	5	6	7	8	9	10	&c.
$C_{0,n}$	1	0	0	0	0	0	0	0	0	0	
$C_{1,n}$		1	1	1	1	1	1	1	1	1	
$C_{2,n}$			2	2	3	3	4	4	5	5	
$C_{3,n}$				3	4	5	7	8	10	12	
$C_{4,n}$					5	6	9	11	15	18	
&c.											

where the line $C_{4,n}$ (ways of making up $n-1$ with the parts 1, 2, 3, 4) is 1, 1, 2, 3, 5, 6, 9, 11, 15, 18, &c., viz. we read from the corner diagonally downwards as far as the 5, and then horizontally along the line: this saves a large number of figures. The table is printed in ordinary quarto pages, which are taken to come in in tiers of seven, five, and three pages one under the other, as shown by a prefixed diagram; and the necessity of a large folding plate is thus avoided.

The successive lines give, in fact, the coefficients in the expansions of

$$\frac{1}{1-x},\ \frac{1}{1-x\,.\,1-x^2},\ \frac{1}{1-x\,.\,1-x^2\,.\,1-x^3},\ldots,\ \frac{1}{1-x\,.\,1-x^2\ldots1-x^{102}},$$

each expanded as far as x^{103}.

Table II. (6 pages). The successive lines give the coefficients in the expansions of

$$S,\ \frac{S}{1-x},\ \frac{S}{1-x\,.\,1-x^2},\ldots,\ \frac{S}{1-x\,.\,1-x^2\ldots1-x^{35}},$$

where

$$S=\frac{1}{(1-x)(1-x^2)(1-x^3)}\ldots\text{ ad inf.},$$

each expanded as far as x^{53}, and further continued as regards the first ten lines, that is, the expansions of

$$S,\ \frac{S}{1-x},\ \frac{S}{1-x\,.\,1-x^2},\ldots,\ \frac{S}{1-x\,.\,1-x^2\ldots1-x^{9}},$$

each as far as x^{107}.

Table III. (2 pages). The successive lines give the coefficients in the expansions of

$$S^2,\ \frac{S^2}{1-x},\ \frac{S^2}{1-x\,.\,1-x^2},\ldots,\ \frac{S^2}{1-x\,.\,1-x^2\ldots1-x^{5}},$$

each expanded as far as x^{55}.

36. As regards Tables II. and III., the analytical explanations have been given in the first instance; but it is easy to see that the tables give numbers of partitions. Thus, in Table II., the second line gives the coefficients in the development of

$$\frac{1}{(1-x)^2(1-x^2)(1-x^3)\ldots};$$

viz. these are 1, 2, 4, 7, 12, 19, 30, ..., being the number of ways in which the numbers 0, 1, 2, 3, 4, &c. respectively can be made up with the parts 1, 1′, 2, 3, 4, &c.; thus

	Partitions.	No. =
1	1	2
	1′	
2	2	4
	1, 1	
	1, 1′	
	1′, 1′	
3	3	7
	2, 1	
	2, 1′	
	1, 1, 1	
	1, 1, 1′	
	1, 1′, 1′	
	1′, 1′, 1′	
&c.		&c.

Similarly, the third line shows the number of ways in which these numbers respectively can be made up with the parts 1, 1′, 2, 2′, 3, 4, 5, &c.; the fourth line with the parts 1, 1′, 2, 2′, 3, 3′, 4, 5, &c.; and so on.

And in like manner in Table III., the first line shows the number of ways when the parts are 1, 1′, 2, 2′, 3, 3′, ...; the second line when they are 1, 1′, 1″, 2, 2′, 3, 3′, &c.; the third when they are 1, 1′, 1″, 2, 2′, 2″, 3, 3′, &c.; and so on.

It is clear that the series of tables might be continued indefinitely, viz. there might be a Table IV. giving the developments of

$$S^3, \frac{S^3}{1-x}, \frac{S^3}{1-x\,.\,1-x^2}; \text{ and so on.}$$

An interesting table would be one composed of the first lines of the above series, viz. a table giving in its successive lines the developments of S, S^2, S^3, S^4, &c.

There are throughout the work a large number of numerical results given in a quasi-tabular form; but the collection of these, with independent explanations of the significations of the tabulated numbers, would be a task of considerable labour.

[F. 15. *Quadratic forms* a^2+b^2, *&c., and Partitions of Numbers into squares, cubes, and biquadrates.*] Art. V.

37. The forms here referred to present themselves in the various complex theories. Thus $N=a^2+b^2$, $=(a+bi)(a-bi)$; this means that, in the theory of the complex numbers $a+bi$ (a and b integers), N is not a prime but a composite number. It is well known that an ordinary prime number $\equiv 3$, mod. 4, is not expressible as a sum a^2+b^2, being, in fact, a prime in the complex theory as well as in the ordinary one: but that an ordinary prime number $\equiv 1$, mod. 4, is (in one way only) $=a^2+b^2$; so that it is in the complex theory a composite number. A number whose prime factors are each of them $\equiv 1$, mod. 4, or which contains, if at all, an even number of times any prime factor $\equiv 3$, mod. 4, can be expressed in a variety of ways in the form a^2+b^2; but these are all easily deducible from the expressions in the form in question of its several factors $\equiv 1$, mod. 4, so that the required table is a table of the form $p=a^2+b^2$, p an ordinary prime number $\equiv 1$, mod. 4: a and b are one of them odd, the other even; and to render the decomposition definite a is taken to be odd.

$p=a^2+b^2$; viz. decomposition of the primes of the form $4n+1$ into the sum of two squares: a table extending from $p=5$ to 11981 (calculated by *Zornow*) is given by

Jacobi, *Crelle*, t. XXX. (1846), pp. 174—176.

This is carried by Reuschle, as presently mentioned, up to $p=24917$. Reuschle notices that $2713=3^2+52^2$ is omitted, also $6997=39^2+74^2$, and that 8609 should be $=47^2+80^2$.

38. Similarly, primes of the form $6n+1$ are expressible in the form $p=a^2+3b^2$. Observe that, ω being an imaginary cube root of unity, this is connected with $p'=(a+b\omega)(a+b\omega^2)$, $=a^2-ab+b^2$, viz. we have $4p'=(2a-b)^2+3b^2$; or the form a^2+3b^2 is connected with the theory of the complex numbers composed of the cube roots of unity.

$p=a^2+3b^2$; viz. decomposition of the primes of the form $6n+1$ into the form a^2+3b^2: a table extending from $p=7$ to 12007 (calculated also by *Zornow*) is given by

Jacobi, *Crelle*, t. XXX. (1846), *ut suprà*, pp. 177—179.

This is carried by Reuschle up to $p=13369$, and for certain higher numbers up to 49999, as presently mentioned. Reuschle observes that $6427=80^2+3.3^2$ is by accident omitted, and that 6481 should be $=41^2+3.40^2$.

39. Again, primes of the form $8n+1$ are expressible in the form $p=a^2+2b^2$ (or say $=c^2+2d^2$), the theory being connected with that of the complex numbers composed with the 8th roots of unity $\left(\text{fourth root of } -1, =\dfrac{1+i}{\sqrt{2}}\right)$.

$p=c^2+2d^2$; viz. decomposition of primes of the form $8n+1$ into the form c^2+2d^2:

a table extending from $p = 16$ to 5943 (extracted from a MS. table calculated by *Struve*) is given by

Jacobi, *Crelle,* t. XXX. (1846), *ut suprà,* p. 180.

This is carried by Reuschle up to $p = 12377$, and for certain higher numbers up to 24889, as presently mentioned.

40. Reuschle's tables of the forms in question are contained in the work:—

Reuschle, *Mathematische Abhandlung,* &c. (see *ante* No. 15), under the heading "B. Tafeln zur Zerlegung der Primzahlen in Quadrate" (pp. 22—41). They are as follows:—

Table III. for the primes $6n + 1$.

The first part gives $p = A^2 + 3B^2$ and $4p = L^2 + 27M^2$, from $p = 7$ to 5743. The table gives A, B, L, M; those numbers which have 10 for a cubic residue are distinguished by an asterisk. A specimen is

p	A	B	L	M
37*	5	2	11	1

viz. $37 = 5^2 + 3 \cdot 2^2$, $148 = 11^2 + 27 \cdot 1^2$; the asterisk shows that $x^3 \equiv +10$ (mod. 37) is possible: in fact $34^3 \equiv 10$ (mod. 37).

The second part gives $p = A^2 + 3B^2$ only, from $p = 5749$ to 13669. The table gives A, B; and the asterisk implies the same property as before.

The third part gives $p = A^2 + 3B^2$, but only for those values of p which have 10 for a cubic residue, viz. for which $x^3 \equiv 10$ (mod. p) is possible, from $p = 13689$ to 49999. The table gives A, B; the asterisk, as being unnecessary, is not inserted.

Table IV. for the primes $4n + 1$ in the form $A^2 + B^2$, and for those which are also $8n + 1$ in the form $C^2 + 2D^2$.

The first part gives $p = A^2 + B^2, = C^2 + 2D^2$, from $p = 5$ to 12377. The table gives A, B, C, D; those numbers which have 10 for a biquadratic residue, viz. for which $x^4 \equiv 10$ (mod. p) is possible, are distinguished by an asterisk; those which have also 10 for an octic residue, viz. for which $x^8 \equiv 10$ (mod. p) is possible, by a double asterisk. A specimen is

p	A	B	C	D
229	15	2	—	—
233	13	8	15	2
241**	15	4	13	6

The second part gives $p = A^2 + B^2$, from $p = 12401$ to 24917 for all those values of p which have 10 for a biquadratic residue ($x^4 \equiv 10$ (mod. p) possible). The table gives A, B; those values of p which have 10 for an octic residue, viz. for which $x^8 \equiv 10$ (mod. p) is possible, are distinguished by an asterisk.

The third part gives $p = C^2 + 2D^2$, from $p = 12641$ to 24889 for all those values of p which have 10 for a biquadratic residue. The table gives C, D; those values of p which have 10 as an octic residue are distinguished by an asterisk.

41. A table by **Zornow,** *Crelle*, t. XIV. (1835), pp. 279, 280 (belonging to the Memoir "De Compositione numerorum e Cubis integris positivis," pp. 276—280), shows for the numbers 1 to 3000 the least number of cubes into which each of these numbers can be decomposed. Waring gave, without demonstration, the theorem that every number can be expressed as the sum of at most 9 cubes. The present table seems to show that 23 is the only number for which the number of cubes is $= 9\,(= 2\,.\,2^3 + 7\,.\,1^3)$; that there are only fourteen numbers for which the number of cubes is $= 8$, the largest of these being 454; and hence that every number greater than 454 can be expressed as a sum of at most 7 cubes; and further, that every number greater than 2183 can be expressed as a sum of at most 6 cubes. A small subsidiary table (p. 276) shows that the number of numbers requiring 6 cubes gradually diminishes—e.g. between 12^3 and 13^3 there are seventy-five such numbers, but between 13^3 and 14^3 only sixty-four such numbers; and the author conjectures "that for numbers beyond a certain limit every number can be expressed as a sum of at most 5 cubes."

42. For the decomposition of a number into biquadrates we have

Bretschneider, "Tafeln für die Zerlegung der Zahlen bis 4100 in Biquadrate," *Crelle*, t. XLVI. (1853), pp. 3—23.

Table I. gives the decompositions, thus:—

N	1^4,	2^4,	3^4,	4^4,	5^4,
696	6	1	2	2	
	3	2	5	1	
	0	3	8		

viz. $696 = 6\,.\,1^4 + 1\,.\,2^4 + 2\,.\,3^4 + 2\,.\,4^4$, &c.

And Table II. enumerates the numbers which are sums of at least 2, 3, 4,.., 19 biquadrates. There is at the end a summary showing for the first 4100 numbers how many numbers there are of these several forms respectively: 28 numbers are each of them a sum of 2 biquadrates, 75 a sum of 3,..., 7 a sum of 19 biquadrates. The seven numbers, each of them a sum of 19 biquadrates, are 79, 159, 239, 319, 399, 479, 559.

[F. 16. *Binary, Ternary, &c. quadratic and higher forms.*] Art. VI.

43. Euler worked with the quadratic forms $ax^2 \pm cy^2$ (p and q integers), particularly in regard to the forms of the divisors of such numbers. It will be sufficient to refer to his memoir:—

Euler, "Theoremata circa divisores numerorum in hac forma $pa^2 \pm qb^2$ contentorum," (*Op. Arith. Coll.* pp. 35—61, 1744), containing fifty-nine theorems, exhibiting in a quasi-tabular form the linear forms of the divisors of such numbers. As a specimen:—

"Theorema 13. Numerorum in hac forma $a^2 + 76b^2$ contentorum divisores primi omnes sunt vel 2, vel 7, vel in una sex formularum

$$28m+1, \quad 28m+11,$$
$$28m+9, \quad 28m+15,$$
$$28m+25, \quad 28m+23,$$

seu in una harum trium

$$14m+1,$$
$$14m+9,$$
$$14m+11,$$

sunt contenti"; viz. the forms are the three $14m+1$, $14m+9$, $14m+11$.

But Euler did not consider, or if at all very slightly, the trinomial forms $ax^2 + bxy + cy^2$, nor attempt the theory of the reduction of such forms. This was first done by Lagrange in the memoir

Lagrange, *Mém. de Berlin*, 1773. And the theory is reproduced by

Legendre, *Théorie des Nombres*, Paris, 1st ed. 1798; 2nd ed. 1808, § 8, "Réduction de la formule $Ly^2 + Myz + Nz^2$ à l'expression la plus simple," (2nd ed. pp. 61—67).

44. But the classification of quadratic forms, as established by Legendre, is defective as not taking account of the distinction between proper and improper equivalence; and the ulterior theory as to orders and genera, and the composition of forms (although in the meantime established by Gauss), are not therein taken into account; for this reason the **Legendre's** Tables I. to VIII. relating to quadratic forms, given after p. 480 (thirty-two pages not numbered), are of comparatively little value, and it is not necessary to refer to them in detail.

The complete theory was established by

Gauss, *Disquisitiones Arithmeticæ*, 1801.

It is convenient to refer also to the following memoir:

Lejeune Dirichlet, "Recherches sur diverses applications de l'Analyse à la théorie des Nombres," *Crelle*, t. XIX. (1839), p. 338, [*Ges. Werke*, t. I. p. 427], as giving a succinct statement of the principle of classification, and in particular a table of the characters of the genera of the properly primitive order, according to the four forms $D=PS^2$, $P\equiv 1$ or 3 (mod. 4), and $D\equiv 2PS^2$, $P\equiv 1$ or 3 (mod. 4), of the determinant.

45. Tables of quadratic forms arranged on the Gaussian principle are given by

Cayley, *Crelle*, t. LX. (1862), pp. 357—372, [335]; viz. the tables are—

Table I. des formes quadratiques binaires ayant pour déterminants les nombres négatifs depuis $D=-1$ jusqu'à $D=-100$. (Pp. 360—363: [*Coll. Math. Papers*, t. V. pp. 144—147].)

A specimen is

D	Classes	α	β	δ	ϵ	$\delta\epsilon$	Cp
= 26	1, 0, 26	+				+	1
	3, − 1, 9	+				+	g^2
	3, 1, 9	+				+	g^4
	5, 2, 6	−				−	g
	2, 0, 13	−				−	g^3
	5, − 2, 6	−				−	g^5

where α, β denote, as there explained, the characters in regard to the odd prime factors of D; δ, ϵ, $\delta\epsilon$ those in regard to the numbers 4 and 8. The last column shows that the forms in the two genera respectively are 1, g^2, g^4 and g, g^3, g^5, where $g^6 = 1$, viz. the form g, six times compounded, gives the principal form (1, 0, 26).

Table II. des formes quadratiques binaires ayant pour déterminants les nombres positifs non-carrés depuis $D = 2$ jusqu'à $D = 99$. (Pp. 364—369: [*l.c.*, pp. 148—153].)

The arrangement is the same, except that there is a column "Périodes" showing, in an easily understood abbreviated form, the period of each form. Thus $D = 7$, the period of the principal form (1, 0, − 7), is given as 1, $_2$, − 3, $_1$, 2, $_1$, − 3, $_2$, 1, which represents the series of forms (1, 2, − 3), (− 3, 1, 2) (2, 1, − 3), (− 3, 2, 1).

Table III. des formes quadratiques binaires pour les treize déterminants négatifs irréguliers du premier millier. (Pp. 370—372: [*l.c.*, pp. 154—156].)

The arrangement is the same as in Table I. It may be mentioned that the thirteen numbers, and the forms for the principal genus for these numbers, respectively are:—

$-D=$	Principal genus
576, 580, 820, 900	$(1,\ e^2)\,(1,\ e_1^2)$
884	$(1,\ e^2)\,(1,\ i^2,\ i^4,\ i^6)$
243, 307, 339, 459, 675, 891	$(1,\ d,\ d^2)\,(1,\ d_1,\ d_1^2)$
755, 974	$(1,\ d,\ d^2)\,(1,\ d_1,\ d_1^2)\,(1,\ e^2)$,

where $d^3 = d_1^3 = 1$, $e^4 = e_1^4 = 1$, $i^8 = 1$, viz. $(1,\ e^2)(1,\ e_1^2)$ denotes four forms, 1, e^2, e_1^2, $e^2e_1^2$; and so in the other cases.

46. **Gauss** must have computed quadratic forms to an enormous extent; but, for the reasons (rather amusing ones) mentioned in a letter of May 17, 1841, to Schumacher (quoted in Prof. Smith's Report on "The Theory of Numbers," *Brit. Assoc. Report* for 1862, p. 526, [and Smith's *Coll. Math. Papers*, t. I. p. 261]), he did not preserve his results in detail, but only in the form appearing in the

"Tafel der Anzahl der Classen binärer quadratischer Formen," *Werke*, t. II. pp. 449—476; see editor's remarks, pp. 521—523.

This relates almost entirely to negative determinants, only three quarters of p. 475 and p. 476 to positive ones; for negative determinants, it gives the number of genera and classes, as also the index of irregularity for the determinants of the hundreds 1 to 30, 43, 51, 61, 62, 63, 91 to 100, 117 to 120; then, in a different arrangement, for the thousands 1, 3 and 10, for the first 800 numbers of the forms $-(15n+7)$ and $-(15n+13)$; also for some very large numbers, and for positive determinants of the hundreds 1, 2, 3, 9, 10, and for some others.

A specimen is

Centas I.

G II.		(58) ...		(280)
	1.	5,	6,	8,
		9,	10,	12,
		13,	15,	16,
		18,	22,	25,
		28,	37,	58,
	2.	14,	17,	20,
		....	....	....
Summa		233		477
Irreg.		0	Impr.	74;

viz. this shows, as regards the negative determinants 1 to 100, that the determinants belonging to G II. 1, viz. those which have two genera each of one class, are 5, 6, 8, 9, &c., in all fifteen determinants; those belonging to G II. 2, viz. those which have two genera each of two classes, are 14, 17, 20, &c.; and so on. The head numbers (58) ... (280) show the number of determinants, each having two genera, and the number of classes; thus,

G II.	$1 \times 15 =$	15
	$2 \times 17 =$	34
	$3 \times 17 =$	51
	$4 \times 6 =$	24
	$5 \times 2 =$	10
	$6 \times 1 =$	6
	58	140
		$\times 2$
		$= 280$;

and the bottom numbers show the total number of genera and of classes, thus

G I.	$17 \times 1 =$	17	61
II.	$58 \times 2 =$	116	280
IV.	$25 \times 4 =$	100	136
	100	233	477;

viz. seventeen determinants, each of one genus, and together of sixty-one classes; fifty-eight determinants, each of two genera, and together of 280 classes: and twenty-five determinants, each of four genera, and together 136 classes, give in all 233 genera and 477 classes. These are exclusive of 74 classes belonging to the improperly primitive order; and the number of irregular determinants (in the first hundred) is $= 0$.

The irregular determinants are indicated thus:

243(*3*),
307(*3*), 339(*3*),
459(*),
576(*2*), 580(*2*),
675(*3*),
755(*3*),
891(*3*), 820(*2*), 900(*2*), 884(*2*), 974(*3*),
3 243, 307, 339, 459 ?, 675, 755, 891,
2 576, 589, 820, 884, 900, 974,

which is a notation not easily understood.

As regards the positive determinants, a specimen is

Centas I.
Excedunt determinantis
quadrati 10.
G I. ... (12),
1. 2, 5, 13,
17, 29, 41,
53, 61, 73,
89, 97,
3. 37;

viz. in the first hundred, the positive determinants having one genus of one class are 2, 5, 13, &c.... (eleven in number); that having one genus of three classes is 37, (one in number); $11 + 1 = 12$. The irregular determinants, if any, are not distinguished.

47. *Binary cubic forms.*—The earliest table is given by

Arndt, "Tabelle der reducirten binären kubischen Formen und Klassen für alle negativen Determinanten $-D$ von $D = 3$ bis $D = 2000$," *Grunert's Archiv,* t. XXXI. 1858, pp. 369—448.

The memoir is a sequel to one in t. XVII. (1851). The binary cubic form (a, b, c, d), of determinant $-D\left(= (bc - ad)^2 - 4(b^2 - ac)(c^2 - bd)\right)$, is said to be reduced when its characteristic ϕ, $= (A, B, C)$, $= \left(2(b^2 - ac),\ bc - ad,\ 2(c^2 - bd)\right)$, is a reduced quadratic form, that is, when in regard to absolute values B is not $> \frac{1}{2}A$, C not $< A$.

A specimen is

D	Reduced forms, with characters		Classes	
44	(0, 1, 0, − 11) (2, 0, 22	(1, − 1, − 2, 0) 6, 2, 8)	(0, − 1, 0, 11)	(0, − 2, ± 1, 1)

Two subsidiary tables are given, pp. 351, 352, and 353—368.

48. It appeared suitable to remodel a part of this table in the manner made use of for quadratic forms in my tables above referred to; and it is accordingly divided into the three tables given by

Cayley, *Quart. Math. Journ.* t. XI. (1871), where the notation &c. is explained, pp. 251—261, [496]; viz. these are:—

Table I. of the binary cubic forms, the determinants of which are the negative numbers ≡ 0 (mod. 4) from − 4 to − 400 (pp. 251—258; [*Coll. Math. Papers*, t. VIII., pp. 55—61]).

A specimen is

Det. 4 ×	Classes.	Order.	Charact.	Comp.
11.	0, − 1, 0, 11	on	1, 0, 11	1
	0, − 2, − 1, 1	*pp pp*	3, 1, 4	d
	0, − 2, 1, 1		3, − 1, 4	d^2.

Table II. of the binary cubic forms the determinants of which (taken positively) are ≡ 1 (mod. 4) from − 3 to − 99, the original heading is here corrected, [*l.c.*, pp. 61, 62]; and

Table III. of the binary cubic forms the determinants of which are the negative numbers − 972, − 1228, − 1336, − 1836, and − 2700, [*l.c.*, pp. 63, 64]; viz. − 972 = 4 × − 243,.., − 2700 = 4 × − 675, where − 243,.., − 675 are the first six irregular numbers for quadric forms.

4 × − 675, = − 2700 is beyond the limits of Arndt's tables, and for this number the calculation had to be made anew; the table gives nine classes $(1, d, d^2)$ $(1, d_1, d_1^2)$ of the order *ip* on *pp*, but it is remarked that there may possibly be other cubic classes based on a non-primitive characteristic; the point was left unascertained.

49. The theory of ternary quadratic forms was discussed and partially established by Gauss in the *Disquisitiones Arithmeticæ.* It is proper to recall that a ternary quadratic form is either determinate, viz. always positive, such as $x^2+y^2+z^2$, or always negative, such as $-x^2-y^2-z^2$; or else it is indeterminate, such as $x^2+y^2-z^2$. But as regards determinate forms, the negative ones are derived from the positive ones by simply reversing the signs of all the coefficients, so that it is sufficient to attend to the positive forms; and practically the two cases are positive forms (meaning thereby positive determinate forms) and indeterminate forms; but the theory for positive forms was first established completely, and so as to enable the formation of tables, in the work

Seeber, *Ueber die Eigenschaften der positiven ternären quadratischen Formen,* (4to. Freiburg, 1831), which is reviewed by Gauss in the *Gött. Gelehrte Anzeigen*, 1831, July 9 (see Gauss, *Werke*, t. II. pp. 188—193). The author gives (pp. 220—243) tables "of the classes of positive ternary forms represented by means of the corresponding reduced forms" for the determinants 1 to 100. A specimen is

$$\text{Det. } 6 \qquad \begin{pmatrix} 1, & 1, & 2 \\ 0, & 0, & -1 \end{pmatrix}, \quad \begin{pmatrix} 1, & 1, & 2 \\ -1, & -1, & 0 \end{pmatrix},$$

$$\begin{matrix}\text{Zugeordnete} \\ \text{Formen}\end{matrix} \begin{pmatrix} 8, & 8, & 3 \\ 0, & 0, & 8 \end{pmatrix}, \quad \begin{pmatrix} 7, & 7, & 4 \\ 4, & 4, & 2 \end{pmatrix},$$

where it is to be observed that Seeber admits odd coefficients for the terms in yz, zx, xy; viz. his symbol $\begin{pmatrix} a, & b, & c \\ f, & g, & h \end{pmatrix}$ denotes

$$ax^2 + by^2 + cz^2 + fyz + gzx + hxy,$$

and his determinant is

$$4abc - af^2 - bg^2 - ch^2 + fgh.$$

Also his adjoint form is

$$\begin{pmatrix} 4bc - f^2, & 4ca - g^2, & 4ab - h^2 \\ 2gh - 4af, & 2hf - 4bg, & 2fg - 4ch \end{pmatrix}, = (4bc - f^2)\, x^2 + \ldots + (2gh - 4af)\, yz + \ldots$$

In the notation of the *Disquisitiones Arithmeticæ*, followed by Eisenstein and others, the symbol $\begin{pmatrix} a, & b, & c \\ f, & g, & h \end{pmatrix}$ denotes

$$ax^2 + by^2 + cz^2 + 2fyz + 2gzx + 2hxy;$$

the determinant is

$$= -(abc - af^2 - bg^2 - ch^2 + 2fgh),$$

a positive form having thus always a negative determinant. And the adjoint form is

$$-\begin{pmatrix} bc - f^2, & ca - g^2, & ab - h^2 \\ gh - af, & hf - bg, & fg - ch \end{pmatrix}, = -(bc - f^2)\, x^2 - \ldots - 2\,(gh - af)\, yz - \ldots$$

Hence Seeber's determinant is $= -4$ multiplied by that of Gauss, and his tables really extend between the values -1 and -25 of the Gaussian determinant.

50. Tables of greater extent, and in the better form just referred to, are given by

Eisenstein, *Crelle*, t. XLI. (1851), pp. 169—190; viz. these are

I. "Tabelle der eigentlich primitiven positiven ternären Formen für alle negativen Determinanten von -1 bis -100," (pp. 169—185).

A specimen is

D	Anzahl	Reducirte Formen für $-D$
10	3	$\begin{pmatrix} 1, & 1, & 10 \\ 0, & 0, & 0 \end{pmatrix}$, $\begin{pmatrix} 1, & 2, & 5 \\ 0, & 0, & 0 \end{pmatrix}$, $\begin{pmatrix} 2, & 2, & 3 \\ 0, & -1, & 0 \end{pmatrix}$.
		$\delta = 8$ $\quad \delta = 4 \quad$ $\delta = 4$

II. "Tabelle der *uneigentlich* primitiven positiven ternären Formen für alle negativen Determinanten von -2 bis -100," (pp. 186—189).

A specimen is

D	Anzahl	Reducirte Formen für $-D$
10	1	$\begin{pmatrix} 2, & 2, & 4 \\ 1, & 1, & 1 \end{pmatrix}$. $\delta = 6$.

And there is given (p. 190) a table of the reduced forms for the determinant $-385 \,(= -5 \,.\, 7 \,.\, 11)$, selected merely as a largish number with three factors; viz. there are in all fifty-nine forms, corresponding to values 1, 2, 4, 6, 8 of δ.

It may be remarked that δ denotes, for any given form, the number of ways in which this is linearly transformable into itself, this number being always 1, 2, 4, 6, 8, 12, or 24. The theory as to this and other points is explained in the memoir (pp. 141—168), and various subsidiary tables are contained therein and in the *Anhang* (pp. 227—242); and there is given a small table relating to *indeterminate forms*, viz. this is

"C. Versuch einer Tabelle der nicht äquivalenten unbestimmten (indifferenten) ternären quadratischen Formen für die Determinanten ohne quadratischen Theiler unter 20," (pp. 239, 240).

A specimen is

D	Indifferente ternäre quadratische Formen
10	$\begin{pmatrix} 0, & 1, & 10 \\ 0, & 0, & 1 \end{pmatrix}$, $\begin{pmatrix} 1, & 2, & -5 \\ 0, & 0, & 0 \end{pmatrix}$, $\begin{pmatrix} 0, & 0, & 10 \\ 0, & 0, & 1 \end{pmatrix}$,

where, when the determinant is even, the forms in the second line are always improperly primitive forms.

[F. 17. *Complex Theories.*] Art. VII.

51. The theory of binary quadratic forms (a, b, c), with complex coefficients of the form $\alpha + \beta i$, $(i = \sqrt{-1}$ as usual, α and β integers), has been studied by Lejeune Dirichlet, Prof. H. J. S. Smith, and possibly others; but no tables have, it is believed, been calculated. The calculations would be laborious; but tables of a small extent only would be a sufficient illustration of the theory, and would, it is thought, be of great interest.

The theory of complex numbers of the last-mentioned form $\alpha+\beta i$, or say of the numbers formed with the fourth root of unity, had previously been studied by Gauss; and the theory of the numbers formed with the cube roots of unity ($\alpha+\beta\omega$, $\omega^2+\omega+1=0$, α and β integers) was studied by Eisenstein; but the general theory of the numbers involving the nth roots of unity (n an odd prime) was first studied by Kummer. It will be sufficient to refer to his memoir,

Kummer, "Zur Theorie der complexen Zahlen," *Berl. Monatsb.*, March, 1845; and *Crelle*, t. XXXV. (1847), pp. 319—326; also "Ueber die Zerlegung der aus Wurzeln der Einheit gebildeten complexen Zahlen in ihre Primfactoren," same volume, pp. 327—367, where the astonishing theory of "Ideal Complex Numbers" is established.

52. It may be recalled that, p being an odd prime, and ρ denoting a root of the equation $\rho^{p-1}+\rho^{p-2}+\ldots+\rho+1=0$, then the numbers in question are those of the form $a+b\rho+\ldots+k\rho^{n-2}$, where $(a, b, .., k)$ are integers; or (what is in one point of view more, and in another less, general) if $\eta, \eta_1, .., \eta_{e-1}$ are "periods" composed with the powers of ρ (e any factor of $p-1$), then the form considered is $a\eta+b\eta_1+\ldots+h\eta_{e-1}$. For any value of p or e there is a corresponding complex theory. A number (real or complex) is in the complex theory prime or composite, according as it does not, or does, break up into factors of the form under consideration. For p a prime number under 23, if in the complex theory N is a prime, then any power of N (to fix the ideas say N^3) has no other factors than N or N^2; but if $p=23$ (and similarly for higher values of p), then N may be such that, for instance, N^3 has complex factors other than N or N^2 (for $p=23$, $N=47$ is the first value of N, viz. 47^3 has factors other than 47 and 47^2); say N^3 has a complex prime factor A, or we have $\sqrt[3]{A}$ as an ideal complex factor of N. Observe that by hypothesis N is not a perfect cube, viz. there is no complex number whose cube is $=A$. In the foregoing general statement, made by way of illustration only, all reference to the complex factors of unity is purposely omitted, and the statement must be understood as being subject to correction on this account.

What precedes is by way of introduction to the account of **Reuschle's** Tables (*Berliner Monatsberichte*, 1859—60), which give in the different complex theories $p=5$, 7, 11, 13, 17, 19, 23, 29 the complex factors of the decomposable real primes up to in some cases 1000.

It should be remarked that the form of a prime factor is to a certain extent indeterminate, as the factor can without injury be modified by affecting it with a complex factor of unity; but in the tables the choice of the representative form is made according to definite rules, which are fully explained, and which need not be here referred to.

53. The following synopsis is convenient:—

	Theory of the pth roots	Form of real prime to mod. $p \equiv$	No. of factors in complex theory	Extent of table; all primes under	Equation of periods
1859. pp. 488–491.	5	1 4 2, 3	4 2 (not tabulated) prime.	2500	$\alpha^4+\ldots+\alpha+1=0.$ $y^2+y-1=0.$
694–697.	7	1 6 2, 4 3, 5	6 3 2 prime.	1000	$\alpha^6+\ldots+\alpha+1=0.$ $y^3+y^2-2y-1=0.$ $y^2+y+2=0.$
1860. pp. 190–194.	11	1 10 3, 4, 5, 9 2, 6, 7, 8	10 5 2 1	1000	$\alpha^{10}+\ldots+\alpha+1=0.$ $y^5+y^4-4y^3-3y^2+3y+1=0.$ $y^2+y+3=0.$
194–199.	13	1 12 3, 9 5, 8 4, 10 2, 6, 7, 11	12 6 4 3 2 prime.	1000	$\alpha^{12}+\ldots+\alpha+1=0.$ $y^6+y^5-5y^4-4y^3+6y^2+3y-1=0.$ $y^4+y^3+2y^2-4y+3=0.$ $y^3+y^2-4y+1=0.$ $y^2+y-3=0.$
714–719.	17	1 16 4, 13 2, 8, 9, 15 3, 5, 6, 7, 11, 12, 14	16 8 4 2 prime.	1000	$\alpha^{16}+\ldots+\alpha+1=0.$ $y^8+y^7-7y^6-6y^5+15y^4+10y^3-10y^2-4y+1=0.$ $y^4+y^3-6y^2+1=0.$ $y^2+y-4=0.$
719–725.	19	1 18 7, 11 8, 12 4, 5, 6, 9, 13, 17 2, 3, 10, 13, 14, 15	18 9 6 3 2 prime.	1000	$\alpha^{18}+\ldots+\alpha+1=0.$ $y^9+y^8-8y^7-7y^6+21y^5+15y^4-20y^3-10y^2+5y+1=0.$ $y^6+y^5+2y^4-8y^3-y^2+5y+7=0.$ $y^3+y^2-6y-7=0.$ $y^2+y+5=0.$
725–729.	23	1 22 2, 3, 4, 6, 8, 9, 12, 13, 16, 18 ?5, 7, 10, 11, 14, 15, 17, 19, 20, 21	22 11 2 prime.	1000	$\alpha^{22}+\ldots+\alpha+1=0.$ $y^{11}+y^{10}-10y^9-9y^8+36y^7+28y^6-56y^5-35y^4+35y^3+15y^2-6y-1=0.$ $y^2+y+6=0.$
729–734.	29	1 28 12, 17 7, 16, 20, 23, 24, 25 4, 5, 6, 9, 13, 22 2, 3, 8, 10, 11, 14, 15, 18, 19, 21, 26, 27	28 14 7 4 2 prime.	1000	$\alpha^{28}+\ldots+\alpha+1=0.$ $y^{14}+y^{13}-13y^{12}-12y^{11}+66y^{10}+55y^9-165y^8-120y^7+210y^6+126y^5-126y^4-56y^3+28y^2+7y-1=0.$ $y^7+y^6-12y^5-7y^4+28y^3+14y^2-9y+1=0.$ $y^4+y^3+4y^2+20y+23=0.$ $y^2+y-7=0.$

The foregoing synopsis of Reuschle's tables in the *Berliner Monatsberichte* was written previous to the publication of Reuschle's far more extensive work. It is allowed to remain, but some explanations which were given have been struck out, and were instead given in reference to the larger work, which is

Reuschle, *Tafeln complexer Primzahlen, welche aus Wurzeln der Einheit gebildet sind.* Berlin, 4° (1875), pp. iii—vi and 1—671.

This work (the mass of calculation is perfectly wonderful) relates to the roots of unity, the degree being any prime or composite number, as presently mentioned, having all the values up to and a few exceeding 100; viz. the work is in five divisions, relating to the cases:

I. (pp. 1—171), degree any odd prime of the first 100, viz. 3, 5, 7, 11, 13, 17, 19, 23, 29, 31, 37, 41, 43, 47, 53, 59, 61, 67, 71, 73, 79, 83, 89, 97;

II. (pp. 173—192), degree the power of an odd prime 9, 25, 27, 49, 81;

III. (pp. 193—440), degree a product of two or more odd primes or their powers, viz. 15, 21, 33, 35, 39, 45, 51, 55, 57, 63. 65, 69, 75, 77, 85, 87, 91, 93, 95, 99, 105;

IV. (pp. 441—466), degree an even power of 2, viz. 4, 8, 16, 32, 64, 128;

V. (pp. 467—671), degree divisible by 4, viz. 12, 20, 24, 28, 36, 40, 44, 48, 52. 56, 60, 68, 72, 76, 80, 84, 88, 92, 96, 100, 120;

the only excluded degrees being those which are the double of an odd prime, these, in fact, coming under the case where the degree is the odd prime itself.

It would be somewhat long to explain the specialities which belong to degrees of the forms II., III., IV., V.; and what follows refers only to Division I., degree an odd prime.

For instance, if $\lambda = 7$, $\lambda - 1 = 2.3$; the factors of 6 being 6, 3, 2, 1, there are accordingly four divisions, viz.

I. α a prime seventh root, that is, a root of $\alpha^6 + \alpha^5 + \alpha^3 + \alpha^2 + \alpha + 1 = 0$;

II. $\eta_0 = \alpha + \alpha^{-1}$, $\eta_1 = \alpha^2 + \alpha^{-2}$, $\eta_2 = \alpha^3 + \alpha^{-3}$, or η a root of $\eta^3 + \eta^2 - 2\eta - 1 = 0$, $\left\{\begin{array}{l}\eta_0^2 = 2 + \eta_1,\ \eta_1^2 = 2 + \eta_2, \text{ \&c.} \\ \eta_0\eta_1 = \eta_0 + \eta_2, \text{ \&c.} \\ \eta_0\eta_2 = \eta_1 + \eta_2, \text{ \&c.;}\end{array}\right.$

III. $\eta_0 = \alpha + \alpha^2 + \alpha^4$, $\eta_1 = \alpha^3 + \alpha^5 + \alpha^6$, or η a root of $\eta^2 + \eta + 2 = 0$;

IV. Real numbers.

I. $p = 7m + 1$. First, it gives for the several prime numbers of this form 29, 43,.., 967 the congruence roots, mod. p; for instance,

p	α	α^2	α^3	α^4	α^5	α^6
29	− 5	− 4	− 9	− 13	+ 7	− 6
43	+ 11	− 8	− 2	+ 21	+ 16	+ 4.
⋮						

This means that, if $\alpha \equiv -5$ (mod. 29), then $\alpha^2 \equiv 25, \equiv -4$, $\alpha^3 \equiv 20, \equiv -9$, &c., values which satisfy the congruence $\alpha^6 + \alpha^5 + \alpha^4 + \alpha^3 + \alpha^2 + \alpha + 1 \equiv 0$ (mod. 29).

Secondly, it gives, under the simple and the primary forms, the prime factors $f(\alpha)$ of these same numbers 29, 43,.., 967; for instance,

p	$f(\alpha)$ simple.	$f(\alpha)$ primary.
29	$\alpha + \alpha^2 - \alpha^3$	$2 + 3\alpha - \alpha^2 + 5\alpha^3 - 2\alpha^4 + 4\alpha^5$
43	$\alpha^2 + 2\alpha^6$	$2\alpha - 2\alpha^2 + 4\alpha^4 - \alpha^5 - 5\alpha^6$.

The definition of a primary form is a form for which $f(\alpha) f(\alpha^{-1}) \equiv f(1)^2$ mod. λ, and $f(\alpha) \equiv f(1)$ mod. $(1-\alpha)^2$. The simple forms are also chosen so as to satisfy this *last* condition; thus $f(\alpha) = \alpha + \alpha^2 - \alpha^3$, then $f(1) - f(\alpha) = 1 - \alpha - \alpha^2 + \alpha^3 = (1-\alpha)^2 (1+\alpha)$, $\equiv 0$ mod. $(1-\alpha)^2$.

II. $p = 7m - 1$. First, it gives for the several prime numbers of this form 13, 41,.., 937 the congruence roots, mod. p; for instance,

p	η_0	η_1	η_2
13	-3	-6	-5
41	-4	$+14$	-11;
⋮			

and secondly, it gives, under the simple and the primary forms, the prime factors $f(\eta)$ of these same numbers 13, 41,.., 937; for instance,

p	$f(\eta)$ simple.	$f(\eta)$ primary.
13	$\eta_0 + 2\eta_2$	$3 + 7\eta_1$
41	$4 + \eta_0$	$-11 + 7\eta_1 - 7\eta_2$.
⋮		

Thus $13 = (\eta_0 + 2\eta_2)(\eta_1 + 2\eta_0)(\eta_2 + 2\eta_1)$, as is easily verified; the product of first and second factors is $= 4 + 3\eta_0 + 8\eta_1 + 5\eta_2$, and then multiplying by the third factor, the result is $42 + 29(\eta_0 + \eta_2)$, $= 13$.

III. $p = 7m + 2$ or $7m + 4$. First, it gives for the several prime numbers of this form 2, 11,.., 991 the congruence roots, mod. p; for instance,

p	η_0	η_1
2	0	-1
11	4	-5;
⋮		

and secondly, it gives the primary prime factors $f(\eta)$ of these same numbers; for instance,

p	$f(\eta)$
2	η_0
11	$1 - 2\eta_1$.

IV. $p = 7m + 3$ or $7m + 5$. The prime numbers of these forms, viz. 3, 5, 17, 19,.., 997, are primes in the complex theory, and are therefore simply enumerated.

The arrangement is the same for the higher prime numbers $\lambda = 23$, &c., for which ideal factors make their appearance; but it presents itself under a more complicated form. Thus $\lambda = 23$, $\lambda - 1 = 2.11$, and the factors of 22 are 22, 11, 2, 1. There are thus four sections.

I. α a prime root, or $\alpha^{22} + \alpha^{21} + \ldots + \alpha^2 + \alpha + 1 = 0$:

II. $\eta_0 = \alpha + \alpha^{-1}, \ldots, \eta_{10} = \alpha^{11} + \alpha^{-11}$, or η a root of $\eta^{11} + \eta^{10} - 10\eta^9 + \ldots + 15\eta^2 - 6\eta - 1 = 0$;

III. $\eta_0 = \alpha + \alpha^2$, $\eta_1 = \alpha^{-1} + \alpha^{-2}$, or η a root of $\eta^2 + \eta + 6 = 0$;

IV. Real numbers.

I. $p = 23m + 1$. First, it gives for the prime numbers of this form 47, 139,.., 967 congruence roots, mod. p, and *also* congruence roots, mod. p^3*; these last in the form $a + bp + cp^2$, where a is given in the former table; thus first table:—

p	α	α^2	$\alpha^3 \ldots$	α^{22}
47	6	-11	$-19\ldots$	$+8$;

and second table—

p	α	α^2	α^3	$\ldots$	α^{22}
47	$+p-2p^2$	$+13p-23p^2$	$+19p-8p^2\ldots$		$+22p+22p^2$.

The meaning is that, $p = 47$, the roots of the congruence

$$\alpha^{22} + \alpha^{21} + \ldots + \alpha^2 + \alpha + 1 \equiv 0 \text{ (mod. } 47^3\text{)}$$

are

$$\alpha = 6 + p - 2p^2,\ \alpha^2 = -11 + 13p - 23p^2,\ \text{\&c.}$$

Secondly, it then gives $f(\alpha)$, the actual ideal prime factor of these same primes 47, 139,.., 967; viz. the whole of this portion of the table $\lambda = 23$, I. (2) is,

having actual prime factors,

p	$f(\alpha)$
599	$\alpha + \alpha^{16} - \alpha^{17}$
691	$\alpha^3 + \alpha^{21} + \alpha^{22}$
829	$\alpha^2 + \alpha^5 + \alpha^{46}$;

having ideal factors, their third powers actual,

p	$f^3(\alpha)$
47	$\alpha^4 + \alpha^5 + \alpha^9 + \alpha^{10} + \alpha^{16} - \alpha^{20} + \alpha^{22}$
139	$1 - \alpha^3 - \alpha^7 + \alpha^9 + \alpha^{11} + \alpha^{14} + \alpha^{15} + \alpha^{17} + \alpha^{18} + \alpha^{20} + \alpha^{21}$
277	$\alpha^2 - \alpha^4 - \alpha^6 + \alpha^7 - \alpha^{10} - \alpha^{15} - \alpha^{17} + \alpha^{21} + \alpha^{22}$
461	$\alpha - \alpha^2 + \alpha^3 - \alpha^9 + \alpha^{14} - 2\alpha^{15}$
967	$\alpha^2 - \alpha^3 - \alpha^5 + \alpha^{10} + \alpha^{15} - 2\alpha^{16} + \alpha^{17} + \alpha^{19}$.

I repeat the explanation that, for the number 47, this means $f(\alpha) f(\alpha^2) \ldots f(\alpha^{22}) = 47^3$.

* Where, as presently appearing, 3 is the index of ideality or power to which the ideal factors have to be raised in order to become actual.

And the like further complication presents itself in the part III. of the same table, $\lambda = 23$ (not, as it happens, in part II., nor of course in the concluding part IV., which is a mere enumeration of real primes). Thus III. (1), we have congruences, (mod. p^3),

$$p=2, \quad \eta \equiv -2, \quad p=3, \quad \eta_0 = +12, \text{ \&c.};$$

and having actual prime factors,

p	$f(\eta)$
59	$5-2\eta_1$
101	$1-4\eta_1$;
⋮	

and having ideal prime factors, their third powers actual,

p	$f^3(\eta)$
2	$1-\eta_1$
3	$1-2\eta_1$;
⋮	

as regards these last the signification being

$$2^3 = (1-\eta_0)(1-\eta_1), \quad \eta_0+\eta_1 = -1, \quad \eta_0\eta_1 = 6 \text{ (as is at once verified)},$$
$$3^3 = (1-2\eta_0)(1-2\eta_1);$$

but the simple numbers 2, 3 are neither of them of the form $(a+b\eta_0)(a+b\eta_1)$.

Contents of Report 1875 on Mathematical Tables.

§ 7. *Tables* F. *Arithmological.*

612.

NOTE SUR UNE FORMULE D'INTÉGRATION INDÉFINIE.

[From the *Comptes Rendus de l'Académie des Sciences de Paris*, tom. LXXVIII. (*Janvier—Juin*, 1874), pp. 1624—1629.]

En étudiant les Mémoires de M. Serret (*Journal de Liouville*, t. x., 1845) par rapport à la représentation géométrique des fonctions elliptiques, avec les remarques de M. Liouville sur ce sujet, je suis parvenu à une formule d'intégration indéfinie qui me paraît assez remarquable, savoir: en prenant θ entier positif quelconque, je dis que l'intégrale

$$\int \frac{(x+p)^{m+n-\theta}(x+q)^\theta\,dx}{x^{m+1}(x+p+q)^{n+1}}$$

a une valeur algébrique

$$(x+p)^{m+n-\theta+1}(x+p+q)^{-n}x^{-m}(A+Bx+Cx^2+\ldots+Kx^{\theta-1}),$$

pourvu qu'une seule condition soit satisfaite par les quantités m, n, p, q. Cette condition s'écrit sous la forme symbolique

$$([m]\,p^2+[n]\,q^2)^\theta=0,$$

en dénotant ainsi l'équation

$$[m]^\theta p^{2\theta}+\frac{\theta}{1}[m]^{\theta-1}[n]^1p^{2\theta-2}q^2+\ldots+[n]^\theta q^{2\theta}=0,$$

où, comme à l'ordinaire, $[m]^\theta$ signifie $m(m-1)\ldots(m-\theta+1)$.

Je rappelle que les formules de M. Serret ne contiennent que des exposants entiers, et celles de M. Liouville qu'un seul exposant quelconque: la nouvelle formule contient deux exposants quelconques, m, n. Je remarque aussi l'analogie de la condition $([m]\,p^2+[n]\,q^2)^\theta=0$ avec celle-ci

$$\frac{1}{\zeta^{n-m}}\left(\frac{d}{d\zeta}\right)^m\zeta^n(\zeta-1)^m=0$$

(m étant un entier positif), qui figure dans les Mémoires cités.

Pour démontrer la formule, j'écris

$$u = x^{-m}(A + Bx + Cx^2 + \ldots. + Kx^{\theta-1}),$$

et aussi pour abréger

$$X = (x+p)^{m+n-\theta+1}(x+p+q)^{-n},$$

ce qui donne

$$\frac{X}{(x+p)(x+p+q)} = (x+p)^{m+n-\theta}(x+p+q)^{-n-1}.$$

L'équation à vérifier est donc

$$Xu = \int \frac{X(x+q)^\theta\,dx}{x^{m+1}(x+p)(x+p+q)},$$

ou, en différentiant et divisant par X,

$$\frac{X'}{X}u + u' = \frac{(x+q)^\theta}{x^{m+1}(x+p)(x+p+q)},$$

ou enfin

$$[(m+n-\theta+1)(x+p+q) - n(x+p)]\,u + (x+p)(x+p+q)\,u' = \frac{(x+q)^\theta}{x^{m+1}},$$

où u' dénote $\dfrac{du}{dx}$. Il ne s'agit donc que d'exprimer que cette équation ait une intégrale

$$u = x^{-m}(A + Bx + Cx^2 + \ldots + Kx^{\theta-1}).$$

En supposant que cela soit ainsi, et en effectuant la substitution, les termes en $x^{-m+\theta}$ se détruisent, et l'on obtient une équation qui contient des termes en x^{-m-1}, $x^{-m}, \ldots, x^{-m+\theta-1}$, savoir $(\theta+1)$ termes. On a ainsi, entre les θ coefficients $A, B, C, \ldots, K$ un système de $(\theta+1)$ équations linéaires, ce qui implique une condition entre les constantes m, n, p, q; mais, cette condition satisfaite, les équations se réduisent à θ équations indépendantes, et les coefficients seront ainsi déterminés.

Par exemple, soit $\theta = 2$; l'équation différentielle est

$$[\overline{m-1}\,p + \overline{m+n-1}\,q + \overline{m-1}\,x]\,u + [p^2 + pq + x(2p+q) + x^2]\,u' = x^{-m-1}(q+x)^2,$$

laquelle doit être satisfaite par $u = Ax^{-m} + Bx^{-m+1}$. Cela donne

$$\begin{array}{cccc} \overbrace{\hspace{6em}}^{x^{-m-1}} & \overbrace{\hspace{8em}}^{x^{-m}} & \overbrace{\hspace{8em}}^{x^{-m+1}} & \overbrace{\hspace{5em}}^{x^{-m+2}} \end{array}$$

$$\left|\begin{array}{cccc} & (\overline{m-1}\,p + \overline{m+n-1}\,q)\,A, & (\overline{m-1}\,p + \overline{m+n-1}\,q)\,B, & \\ \text{,,}\quad, & \text{,,}\quad, & (m-1)\,A, & (m-1)\,B \\ -m(p^2+pq)\,A, & -(m-1)(p^2+pq)\,B, & \text{,,}\quad, & \text{,,} \\ \text{,,}\quad, & -m(2p+q)\,A, & -(m-1)(2p+q)\,B, & \text{,,} \\ \text{,,}\quad, & \text{,,}\quad, & -mA, & -(m-1)\,B \\ -q^2\quad, & -2q\quad, & -1\quad, & \text{,,} \end{array}\right| = 0,$$

c'est-à-dire

$$\begin{aligned}
&1 + [(m-1)p - nq]\,B && + 1A = 0,\\
&2q + (m-1)(p^2+pq)\,B + [(m+1)p - (n-1)\,q]\,A = 0,\\
&q^2 && + m\,(p^2+pq)\,A = 0,
\end{aligned}$$

ce qui donne une condition, déterminant $=0$, et l'on a une condition de cette même forme pour une valeur quelconque de θ.

En formant, puis en réduisant les expressions de ces déterminants, on obtient pour toutes les valeurs $\theta = 1, 2, 3, 4, \ldots$ respectivement la suite d'équations que voici:

$$0 = \begin{vmatrix} 1\,, & mp - nq \\ q\,, & m(p^2+pq) \end{vmatrix} = \begin{aligned}&[m]^1\,p\,(p+q) \\ -1.\; & \qquad q\,([m]\,p - [n]\,q)^2,\end{aligned} = ([m]\,p^2 + [n]\,q^2)^1,$$

$$0 = \begin{vmatrix} 1\,, & \overline{m-1}\,p - nq\,, & 1 \\ 2q\,, & \overline{m-1}\,(p^2+pq), & \overline{m+1}\,p - \overline{n-1}\,q \\ q^2, & . & , m\,(p^2+pq) \end{vmatrix} = \begin{aligned}&[m]^2 p^2\,(p+q)^2 \\ -2.\;&[m]^1\,p\,(p+q)\,q\,([m-1]\,p - [n]\,q)^1 \\ +1.\;& \qquad q^2\,([m] \quad p - [n]\,q)^2\end{aligned} = ([m]\,p^2 + [n]\,q^2)^2,$$

$$0 = \begin{vmatrix} 1\,, & \overline{m-2}\,p - nq\,, & 1 & , . \\ 3q\,, & \overline{m-2}\,(p^2+pq), & mp - \overline{n-1}\,q & , 2 \\ 3q^2, & . & , \overline{m-1}\,(p^2+pq) & , \overline{m+2}\,p - \overline{n-2}\,q \\ q^3, & . & , . & , m\,(p^2+pq) \end{vmatrix} = \begin{aligned}&[m]^3 p^3\,(p+q)^3 \\ -3.\;&[m]^2 p^2\,(p+q)^2\,q\,([m-2]\,p - [n]\,q)^1 \\ +3.\;&[m]^1\,p\,(p+q)\,q^2\,([m-1]\,p - [n]\,q)^2 \\ -1.\;& \qquad q^3\,([m] \quad p - [n]\,q)^3,\end{aligned} = ([m]\,p^2 + [n]\,q^2)^3,$$

$$0 = \begin{vmatrix} 1, & \overline{m-3}\,p - nq & , 1 & , . & , . \\ 4q, & \overline{m-3}\,(p^2+pq), & \overline{m-1}\,p - \overline{n-1}\,q & , 2 & , . \\ 6q^2, & . & , \overline{m-2}\,(p^2+pq) & , \overline{m+1}\,p - \overline{n-2}\,q & , 3 \\ 4q^3, & . & , . & , \overline{m-1}\,(p^2+pq) & , \overline{m+3}\,p - \overline{n-3}\,q \\ q^4, & . & , . & , . & , m\,(p^2+pq) \end{vmatrix} = \begin{aligned}&[m]^4 p^4\,(p+q)^4 \\ -4.\;&[m]^3 p^3\,(p+q)^3\,q\,([m-3]\,p - [n]\,q)^1 \\ +6.\;&[m]^2 p^2\,(p+q)^2\,q^2\,([m-2]\,p - [n]\,q)^2 \\ -4.\;&[m]^1\,p\,(p+q)\,q^3\,([m-1]\,p - [n]\,q)^3 \\ +1.\;& \qquad q^4\,([m] \quad p - [n]\,q)^4,\end{aligned} = ([m]\,p^2 + [n]\,q^2)^4;$$

et ainsi de suite. Les notations $([m]p-[n]q)^1$, $([m]p-[n]q)^2, \ldots$ ont des significations semblables à celles de $([m]p^2+[n]q^2)^1$, $([m]p^2+[n]q^2)^2, \ldots$, auparavant expliquées. On a, par exemple,

$$([m]p-[n]q)^2=[m]^2p^2-2[m]^1[n]^1pq+[n]^2q^2.$$

Considérons, par exemple, le deuxième déterminant: ceci contient trois termes en 1, $2q$, q^2 respectivement; le premier terme est

$$1\,.\,(m-1)(p^2+pq)\,.\,m(p^2+pq),$$

c'est-à-dire

$$[m]^2p^2(p+q)^2;$$

le deuxième terme est

$$2q\,.-m(p^2+pq)[(m-1)p-nq],$$

c'est-à-dire

$$-2[m]^1p(p+q)q([m-1]p-[n]q)^1;$$

le troisième terme est

$$q^2[(\overline{m-1}\,p-nq)(\overline{m+1}\,p-\overline{n-1}\,q)-(m-1)(p^2+pq)],$$

c'est-à-dire

$$q^2[(m^2-m)p^2-2mnpq+(n^2-n)q^2]=q^2([m]p-[n]q)^2.$$

Et de même le troisième déterminant est composé de quatre termes en 1, $3q$, $3q^2$, q^3 respectivement, lesquels sont les quatre termes de la première expression transformée; et ainsi pour le quatrième déterminant, etc. Au moyen de ces premières transformées, on obtient sans peine les expressions finales $([m]p^2+[n]q^2)^1$, $([m]p^2+[n]q^2)^2, \ldots$.

En écrivant $z-\frac{1}{2}(p+q)$ au lieu de x, et puis $\frac{1}{2}(p+q)=\alpha$, $\frac{1}{2}(p-q)=a$, la formule devient

$$\int\frac{(z-a)^\theta(z+a)^{m+n-\theta}\,d\theta}{(z-\alpha)^{m+1}(z+\alpha)^{n+1}},$$

et la valeur algébrique

$$=(z+a)^{m+n-\theta-1}(z-\alpha)^{-m}(z+\alpha)^{-n}(A'+B'z+\ldots+K'z^{\theta-1}),$$

pourvu qu'on ait entre les quantités m, n, a, α la relation

$$\{[m](\alpha+a)^2+[n](\alpha-a)^2\}^\theta=0.$$

En écrivant $\theta=m$, on a la formule de MM. Serret et Liouville, laquelle, en y écrivant $\dfrac{(\alpha+a)^2}{4a\alpha}=\zeta$ et $\dfrac{(\alpha-a)^2}{4a\alpha}=\zeta-1$, peut s'écrire sous la forme $\{[m]\zeta+[n](\zeta-1)\}^\theta=0$. Je remarque que l'équation en ζ ne donne pas *toujours* pour ζ des valeurs réelles, positives et plus grandes que l'unité: par exemple, pour $\theta=1$, on a $\zeta=\dfrac{n}{m+n}$, valeur qui ne peut pas satisfaire à ces conditions. Je n'ai pas cherché dans quel cas ces conditions (qui ont rapport à l'application des formules à la représentation des fonctions elliptiques) subsistent.

613.

ON THE GROUP OF POINTS G_4^1 ON A SEXTIC CURVE WITH FIVE DOUBLE POINTS.

[From the *Mathematische Annalen*, vol. VIII. (1875), pp. 359—362.]

THE present note relates to a special group of points considered incidentally by MM. Brill and Nöther in their paper "Ueber die algebraischen Functionen und ihre Anwendung in der Geometrie," *Math. Annalen*, t. VII. pp. 268—310 (1874).

I recall some of the fundamental notions. We have a basis-curve which to fix the ideas may be taken to be of the order n, $=p+1$, with $\frac{1}{2}p(p-3)$ dps, and therefore of the "Geschlecht" or deficiency p; any curve of the order $n-3$, $=p-2$ passing through the $\frac{1}{2}p(p-3)$ dps is said to be an adjoint curve. We may have, on the basis-curve, a special group G_Q^q of Q points ($Q \not> 2p-2$); viz. this is the case when the Q points are such that every adjoint curve through $Q-q$ of them—that is, every curve of the order $p-2$ through $\frac{1}{2}p(p-3)$ dps and the $Q-q$ points—passes through the remaining q points of the group: the number q may be termed the "speciality" of the group: if $q=0$, the group is an ordinary one.

It may be observed that a special group G_Q^q is chiefly noteworthy in the case where $Q-q$ is so small that the adjoint curve is not completely determined: thus if $p=5$, viz. if the basis-curve be a sextic with 5 dps, then we may have a special group G_6^2, but there is nothing remarkable in this; the 6 points are intersections with the sextic of an arbitrary cubic through the 5 dps—the cubic of course intersects the sextic in the 5 dps counting as 10 points, and in 8 other points—and such cubic is completely determined by means of the 5 dps and any 4 of the 6 points. But contrariwise, there is something remarkable in the group G_4^1 about to be considered: viz. we have here on the sextic 4 points, such that every cubic through the 5 dps and through 3 of the 4 points (through 8 points in all) passes through the remaining one of the 4 points.

The whole number of intersections of the basis-curve with an adjoint, exclusive of the dps counting as $p(p-3)$ points, is of course $=2p-2$: hence an adjoint through the Q points of a group G_Q^q meets the basis-curve besides in R, $=2p-2-Q$, points; we have then the "Riemann-Roch" theorem that these R points form a special group G_R^r, where

$$Q + R = 2p - 2,$$

as just mentioned, and

$$Q - R = 2q - 2r;$$

viz. dividing in any manner the $2p-2$ intersections of the basis-curve by an adjoint into groups of Q and R points respectively, these will be special groups, or at least one of them will be a special group, G_Q^q, G_R^r, such that their specialities q, r are connected by the foregoing relation $Q - R = 2q - 2r$.

The Authors give (*l.c.*, p. 293) a Table showing for a given basis-curve, or given value of p, and for a given value of r, the least value of R and the corresponding values of q, Q: this table is conveniently expressed in the following form.

The least value of

$$R = p - \frac{p}{r+1} + r;$$

and then

$$q = \frac{p}{r+1} - 1,$$

$$Q = p + \frac{p}{r+1} - r - 2,$$

where $\frac{p}{r+1}$ denotes the integer equal to or next less than the fraction.

It is, I think, worth while to present the table in the more developed form:

n	p	Dps	$r=$					
			1	2	3	4	5	6
4	3	0	G_3^1 G_1^0	G_4^2 G_0^0	. .	. .	. .	. .
5	4	2	G_3^1 G_3^1	G_5^2 G_1^0	G_6^3 G_0^0	. .	. .	. .
6	5	5	G_4^1 G_4^1	G_6^2 G_2^0	G_7^3 G_1^0	G_8^4 G_0^0	. .	. .
7	6	9	G_4^1 G_6^2	G_6^2 G_4^1	G_8^3 G_2^0	G_9^4 G_1^0	G_{10}^5 G_0^0	. .
8	7	14	G_5^1 G_7^2	G_7^2 G_5^1	G_9^3 G_3^0	G_{10}^4 G_2^0	G_{11}^5 G_1^0	G_{12}^6 G_0^0
⋮								

where the table shows the values of $\begin{matrix} G_R^r \\ G_Q^q \end{matrix}$ for any given values of p, r.

I recur to the case $p = 5$ and the group G_4^1, which is the subject of the present note: viz. we have here a sextic curve with 5 dps, and on it a group of 4 points G_4^1, such that every cubic through the 5 dps and through 3 points of the group, 8 points in all, passes through the remaining 1 point.

MM. Brill and Nöther show (by consideration of a rational transformation of the whole figure) that, given 2 points of the group, it is possible, and possible in 5 different ways, to determine the remaining 2 points of the group.

I remark that the 5 dps and the 4 points of the group form "an ennead" or system of the nine intersections of two cubic curves: and that the question is, given the 5 dps and 2 points on the sextic, to show how to determine on the sextic a pair of points forming with the 7 points an ennead: and to show that the number of solutions is $= 5$.

We have the following "Geiser-Cotterill" theorem:

If seven of the points of an ennead are fixed, and the eighth point describes a curve of the order n passing $a_1, a_2, .., a_7$ times through the seven points respectively, then will the ninth point describe a curve of the order ν passing $\alpha_1, \alpha_2, .., \alpha_7$ times through the seven points respectively: where

$$\begin{aligned}\nu &= 8n - 3\Sigma a,\\ \alpha_1 &= 3n - a_1 - \Sigma a,\\ &\vdots\\ \alpha_7 &= 3n - a_7 - \Sigma a,\end{aligned}$$

and conversely

$$\begin{aligned}n &= 8\nu - 3\Sigma\alpha,\\ a_1 &= 3\nu - \alpha_1 - \Sigma\alpha,\\ &\vdots\\ a_7 &= 3\nu - \alpha_7 - \Sigma\alpha.\end{aligned}$$

(Geiser, *Crelle-Borchardt*, t. LXVII. (1867), pp. 78—90; the complete form, as just stated, and which was obtained by Mr Cotterill, has not I believe been published): and also Geiser's theorem "the locus of the coincident eighth and ninth points is a sextic passing twice through each of the seven points."

The sextic and the curve n intersect in $6n$ points, among which are included the seven points counting as $2\Sigma a$ points: the number of the remaining points is $= 6n - 2\Sigma a$. Similarly, the sextic and the curve ν intersect in 6ν points, among which are included the seven points counting as $2\Sigma\alpha$ points: the number of the remaining points is $6\nu - 2\Sigma\alpha$ $(= 6n - 2\Sigma a)$. The points in question are, it is clear, common intersections of the sextic, and the curves n, ν: viz. of the intersections of the curves n, ν, a number $6n - 2\Sigma a$, $= 6\nu - 2\Sigma\alpha$, $= 3n + 3\nu - \Sigma a - \Sigma\alpha$ lie on the sextic.

The curves n, ν intersect in $n\nu$ points, among which are included the seven points counting $\Sigma a\alpha$ times: the number of the remaining intersections is therefore

$n\nu - \Sigma a\alpha$, but among these are included the $3n + 3\nu - \Sigma a - \Sigma \alpha$ points on the sextic; omitting these, there remain $n\nu - 3(n+\nu) - \Sigma a\alpha + \Sigma a + \Sigma \alpha$ points, or, what is the same thing, $(n-3)(\nu-3) - \Sigma(a-1)(\alpha-1) - 2$ points: it is clear that these must form pairs such that, the eighth point being either point of a pair, the ninth point will be the remaining point of the pair: the number of pairs is of course

$$\tfrac{1}{2}[(n-3)(\nu-3) - \Sigma(a-1)(\alpha-1) - 2],$$

and we have thus the solution of the question, given the seven points to determine the number of pairs of points on the curve n (or on the curve ν) such that each pair may form with the seven points an ennead.

In particular, if $n = 6$; $a_1, a_2, a_3, a_4, a_5, a_6, a_7 = 2, 2, 2, 2, 2, 1, 1$ respectively, viz. if the curve be a sextic having 5 of the points for dps, and the remaining two for simple points, then we find $\nu = 12$; $\alpha_1, \alpha_2, \alpha_3, \alpha_4, \alpha_5, \alpha_6, \alpha_7 = 4, 4, 4, 4, 4, 5, 5$ respectively, and the number of pairs is

$$= \tfrac{1}{2}[3.9 - 5(2-1)(4-1) - 2], = \tfrac{1}{2}(27 - 15 - 2), = 5,$$

viz. starting with the 5 dps and any 2 points of the group G_4^1 we can, in 5 different ways, determine the remaining 2 points of the group.

In reference to the number $3p - 3$ of parameters in the curves belonging to a given value of p, it may be remarked as follows. Such a curve is rationally transformable into a curve of the order $p + 1$ with $\tfrac{1}{2}p(p-3)$ dps, and therefore containing $\tfrac{1}{2}(p+1)(p+4)$, $-\tfrac{1}{2}p(p-3)$, $= 4p + 2$ parameters. Employing an arbitrary homographic transformation to establish any assumed relations between the parameters, the number is diminished to $4p + 2 - 8$, $= 4p - 6$; and again employing a rational transformation by means of adjoint curves of the order $p - 2$ drawn through the dps and $p - 3$ points of the curve—thereby transforming the curve into one of the same order $p + 1$ and deficiency p—then, assuming that the $p - 3$ parameters (or constants on which depend the positions of the $p - 3$ points) can be disposed of so as to establish $p - 3$ relations between the parameters and so further diminish the number by $p - 3$, the required number of parameters will finally be $4p - 6 - (p - 3) = 3p - 3$.

Cambridge, 26th October, 1874.

614.

ON A PROBLEM OF PROJECTION.

[From the *Quarterly Journal of Pure and Applied Mathematics*, vol. XIII. (1875), pp. 19—29.]

I MEASURE off on three rectangular axes the distances $\Omega X = \Omega Y = \Omega Z, = \theta$; and then, in a plane through Ω drawing in arbitrary directions the three lines ΩA, ΩB, ΩC, $= a, b, c$ respectively, I assume that A, B, C (fig. 1) are the parallel projections of X, Y, Z respectively; viz. taking ΩO as the direction of the projecting lines, then ΩA, ΩB, ΩC being given in position and magnitude, we have to find θ, and the position of the line ΩO.

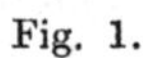

Fig. 1.

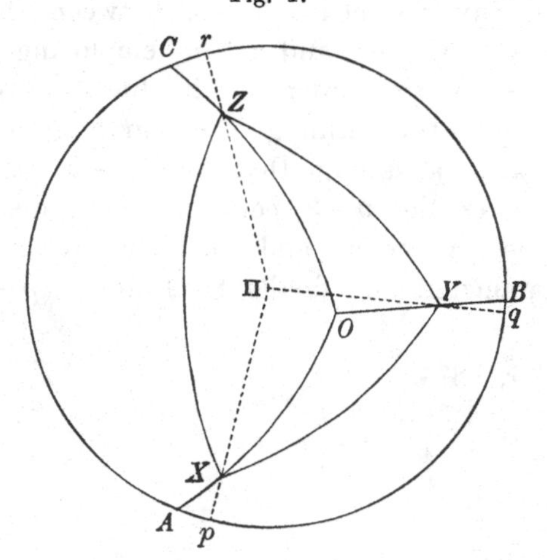

This is in fact a case of a more general problem solved by Prof. Pohlke in 1853, (see the paper by Schwarz, "Elementarer Beweis des Pohlke'schen Fundamentalsatzes der Axonometrie," *Crelle*, t. LXIII. (1864), pp. 309—314), viz. the three lines ΩX, ΩY, ΩZ may be any three axes given in magnitude and direction, and their parallel projection is to be similar to the three lines ΩA, ΩB, ΩC. Schwarz obtains a very elegant construction, which I will first reproduce. We may imagine through Ω a plane cutting at right angles the projecting lines, say in the points X', Y', Z'; we have then in plano a triad of lines $\Omega X'$, $\Omega Y'$, $\Omega Z'$ which are an orthogonal projection of ΩX, ΩY, ΩZ; and are also an orthogonal projection of a plane triad similar to ΩA, ΩB, ΩC; quà such last-mentioned projection, the triangles $\Omega Y'Z'$, $\Omega Z'X'$, $\Omega X'Y'$, must be proportional to the triangles ΩBC, ΩCA, ΩAB; that is, we have to find an orthogonal projection of ΩX, ΩY, ΩZ, such that the triangles $\Omega Y'Z'$, $\Omega Z'X'$, $\Omega X'Y'$, which are the projections of ΩYZ, ΩZX, ΩXY respectively, shall be in given ratios. There is no difficulty in the solution of this problem; referring everything to a sphere centre Ω, let the normals to the planes ΩYZ, ΩZX, ΩXY, meet the sphere in the points X'', Y'', Z'' respectively, and the projecting line through Ω meet the sphere in the point O, then the projection of ΩYZ is to ΩYZ as $\cos OX'' : 1$; and the like as to the projections of ΩZX and ΩXY; that is, in the given spherical triangle $X''Y''Z''$, we have to find a point O, such that the cosines of the distances OX'', OY'', OZ'' are in given ratios: we have at once, through X'', Y'', Z'' respectively, three arcs meeting in the required point O.

The projecting lines being thus obtained, say these are the three parallel lines X', Y', Z', we have next to draw through Ω a plane meeting these in the points A', B', C' such that the triangle $A'B'C'$ is similar to the given triangle ABC; for this being so, the triangles $\Omega B'C'$, $\Omega C'A'$, $\Omega A'B'$ being the projections of, and therefore proportional to $\Omega Y'Z'$, $\Omega Z'X'$, $\Omega X'Y'$, that is, proportional to ΩBC, ΩCA, ΩAB, will, it is clear, be similar to these triangles respectively; that is, we have the triad $\Omega A'$, $\Omega B'$, $\Omega C'$, a projection of ΩX, ΩY, ΩZ, and similar to the triad ΩA, ΩB, ΩC, which is what was required.

It remains only to show how the given three parallel lines X', Y', Z', not in the same plane, can be cut by a plane in a triangle similar to a given triangle ABC.

Fig. 2.

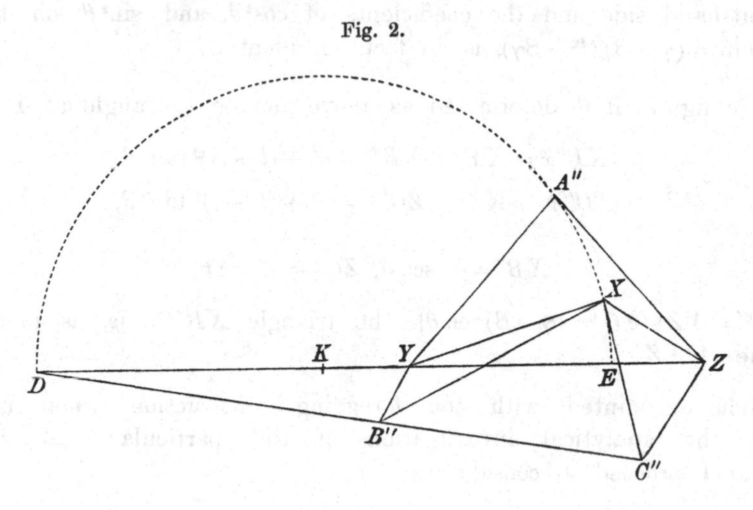

Imagine the three lines at right angles to the plane of the paper, meeting the plane of the paper in the given points X, Y, Z (fig. 2) respectively. On the base

YZ describe a triangle $A''YZ$ similar to the given triangle ABC; and through A'', X with centre on the line YZ, describe a circle meeting this line in the points D and E. Then in the plane, through YZ at right angles to the plane of the paper, we may draw a line meeting the lines Y, Z in the points B'', C'' respectively, such that joining XB'', XC'' we obtain a triangle $XB''C''$ similar to $A''YZ$, that is, to the given triangle ABC.

Taking K the centre of the circle, suppose that its radius is $=l$, and that we have $KY=\beta$, $KZ=\gamma$; also $YX=\sigma$, $ZX=\tau$; $YA''=\sigma''$, $ZA''=\tau''$. If for a moment x, y denote the coordinates of X, then

$$\sigma^2=(x-\beta)^2+y^2,\ =l^2+\beta^2-2\beta x,$$

$$\tau^2=(x-\gamma)^2+y^2,\ =l^2+\gamma^2-2\gamma x,$$

and thence

$$\gamma\sigma^2-\beta\tau^2=\gamma(l^2+\beta^2)-\beta(l^2+\gamma^2),$$

that is,

$$\gamma\sigma^2-\beta\tau^2=(\gamma-\beta)(l^2-\beta\gamma);$$

viz. this is the equation of the circle in terms of the vectors σ, τ; we have therefore in like manner

$$\gamma\sigma''^2-\beta\tau''^2=(\gamma-\beta)(l^2-\beta\gamma).$$

We may determine θ so as to satisfy the two equations

$$\sigma''^2=\sigma^2\cos^2\theta+(l+\beta)^2\sin^2\theta,$$

$$\tau''^2=\tau^2\cos^2\theta+(l+\gamma)^2\sin^2\theta;$$

in fact, these equations give

$$\gamma\sigma''^2-\beta\tau''^2=(\gamma\sigma^2-\beta\tau^2)\cos^2\theta+\{\gamma(l^2+\beta^2)-\beta(l^2+\gamma^2)\}\sin^2\theta,$$

which, the left-hand side and the coefficients of $\cos^2\theta$, and $\sin^2\theta$ on the right-hand side being each $=(\gamma-\beta)(l^2-\beta\gamma)$, is, in fact, an identity.

But in the figure, if θ, determined as above, denote the angle at D, then

$$(XB'')^2=XY^2+YB''^2=\sigma^2+(l+\beta)^2\tan^2\theta,$$

$$(ZC'')^2=XZ^2+ZC''^2=\tau^2+(l+\gamma)^2\tan^2\theta,$$

that is,

$$XB''=\sigma''\sec\theta,\ ZC''=\tau''\sec\theta,$$

or, since $B''C''=YZ\sec\theta\ \{=(\gamma-\beta)\sec\theta\}$, the triangle $XB''C''$ is, as mentioned, similar to the triangle $A''YZ$.

I was not acquainted with the foregoing construction when my paper was written; but the analytical investigation of the particular case is nevertheless interesting, and I proceed to consider it.

Taking (fig. 1) Ω as the centre of a sphere and projecting on this sphere, we have A, B, C given points on a great circle; and we have to find the point O, such that there may be a trirectangular triangle XYZ, the vertices of which lie in OA, OB, OC respectively, and for which

$$\frac{\sin OX}{\sin OA}=\frac{a}{\theta},\ \frac{\sin OY}{\sin OB}=\frac{b}{\theta},\ \frac{\sin OZ}{\sin OC}=\frac{c}{\theta}.$$

I take the arcs BC, CA, $AB=\alpha$, β, γ respectively, $\alpha+\beta+\gamma=2\pi$; and the required arcs OA, OB, OC are taken to be ξ, η, ζ respectively; these are connected by the relation

$$\sin\alpha\cos\xi+\sin\beta\cos\eta+\sin\gamma\cos\zeta=0,$$

to obtain which, observe that from the triangles OAB, OAC, we have

$$\cos A=\frac{\cos\eta-\cos\xi\cos\gamma}{\sin\xi\sin\gamma}=-\frac{\cos\zeta-\cos\xi\cos\beta}{\sin\xi\sin\beta},$$

that is,

$$\sin\beta(\cos\eta-\cos\xi\cos\gamma)+\sin\gamma(\cos\zeta-\cos\xi\cos\beta)=0,$$

which, with $\sin\alpha=-\sin(\beta+\gamma)$, gives the required relation. We have

$$\sin OX=\frac{a}{\theta}\sin\xi,\ \sin OY=\frac{b}{\theta}\sin\eta,\ \sin OZ=\frac{c}{\theta}\sin\zeta;$$

and then from the triangles OBC, OCA, OAB, and the quadrantal triangles OYZ, OZX, OXY, we have

$$\cos BOC=\frac{\cos\alpha-\cos\eta\cos\zeta}{\sin\eta\sin\zeta}=-\frac{\sqrt{\left(1-\frac{b^2}{\theta^2}\sin^2\eta\right)}\sqrt{\left(1-\frac{c^2}{\theta^2}\sin^2\zeta\right)}}{\frac{bc}{\theta^2}\sin\eta\sin\zeta},\ \text{\&c.};$$

that is,

$$\begin{aligned}
bc(\cos\alpha-\cos\eta\cos\zeta)&=-\sqrt{(\theta^2-b^2\sin^2\eta)}\sqrt{(\theta^2-c^2\sin^2\zeta)},\\
ca(\cos\beta-\cos\zeta\cos\xi)&=-\sqrt{(\theta^2-c^2\sin^2\zeta)}\sqrt{(\theta^2-a^2\sin^2\xi)},\\
ab(\cos\gamma-\cos\xi\cos\eta)&=-\sqrt{(\theta^2-a^2\sin^2\xi)}\sqrt{(\theta^2-b^2\sin^2\eta)},
\end{aligned}$$

which, when rationalized, are quadric equations in $\cos\xi$, $\cos\eta$, $\cos\zeta$. The first equation, in fact, gives

$$b^2c^2(\cos\alpha-\cos\eta\cos\zeta)^2=(\theta^2-b^2+b^2\cos^2\eta)(\theta^2-c^2+c^2\cos^2\zeta),$$

that is,

$$(\theta^2-b^2)(\theta^2-c^2)-b^2c^2\cos^2\alpha+(\theta^2-b^2)c^2\cos^2\zeta+(\theta^2-c^2)b^2\cos^2\eta+2b^2c^2\cos\alpha\cos\eta\cos\zeta=0,$$

or, what is the same thing,

$$-\left(1-\frac{b^2c^2\cos^2\alpha}{b^2-\theta^2\,.\,c^2-\theta^2}\right)+\frac{c^2}{c^2-\theta^2}\cos^2\zeta+\frac{b^2}{b^2-\theta^2}\cos^2\eta-\frac{2b^2c^2}{(b^2-\theta^2)(c^2-\theta^2)}\cos\alpha\cos\eta\cos\zeta=0.$$

Completing the system, we have

$$-\left(1-\frac{c^2a^2\cos^2\beta}{c^2-\theta^2\,.\,a^2-\theta^2}\right)+\frac{a^2}{a^2-\theta^2}\cos^2\xi+\frac{c^2}{c^2-\theta^2}\cos^2\zeta-\frac{2c^2a^2}{(c^2-\theta^2)(a^2-\theta^2)}\cos\beta\cos\zeta\cos\xi=0,$$

$$-\left(1-\frac{a^2b^2\cos^2\gamma}{a^2-\theta^2\,.\,b^2-\theta^2}\right)+\frac{b^2}{b^2-\theta^2}\cos^2\eta+\frac{a^2}{a^2-\theta^2}\cos^2\xi-\frac{2a^2b^2}{(a^2-\theta^2)(b^2-\theta^2)}\cos\gamma\cos\xi\cos\eta=0,$$

and, as above,

$$\sin\alpha\cos\xi+\sin\beta\cos\eta+\sin\gamma\cos\zeta=0.$$

It seems difficult from these equations to eliminate ξ, η, ζ, so as to obtain an equation in θ; but I employ some geometrical considerations.

Taking Π as the pole of the circle ABC, and drawing ΠX, ΠY, ΠZ to meet the circle in p, q, r respectively, then, if α'', β'', γ'' are the cosine-inclinations of O to X, Y, Z respectively, we have

$$\sin Xp,\quad \sin Yq,\quad \sin Zr=\alpha'',\ \beta'',\ \gamma''.$$

From the right parallel triangles BYq and CZr, we have

$$\sin Yq=\sin BY\sin B,$$
$$\sin Zr=\sin CZ\ \sin C,$$

and, thence,

$$\frac{\sin Yq}{\sin Zr}=\frac{\sin BY}{\sin CZ}\cdot\frac{\sin OC}{\sin OB};$$

or, since

$$BY=OB-OY,\quad CZ=OC-OZ,$$

and thence

$$\sin BY=\frac{\sin\eta}{\theta}\{\sqrt{(\theta^2-b^2\sin^2\eta)}-b\cos\eta\},$$

$$\sin CZ=\frac{\sin\zeta}{\theta}\{\sqrt{(\theta^2-c^2\sin^2\zeta)}-c\cos\zeta\},$$

we obtain

$$\frac{\beta''}{\gamma''}=\frac{\sqrt{(\theta^2-b^2\sin^2\eta)}-b\cos\eta}{\sqrt{(\theta^2-c^2\sin^2\zeta)}-c\cos\zeta}.$$

We have thence

$$\beta''\sqrt{(\theta^2-c^2\sin^2\zeta)}-\gamma''\sqrt{(\theta^2-b^2\sin^2\eta)}=\beta''c\cos\zeta-\gamma''b\cos\eta,$$

or, squaring and reducing

$$\beta''^2(\theta^2-c^2)+\gamma''^2(\theta^2-b^2)+2\beta''\gamma''\{-\sqrt{(\theta^2-c^2\sin^2\zeta)}\sqrt{(\theta^2-b^2\sin^2\eta)}+bc\cos\eta\cos\zeta\}=0.$$

that is,

$$\beta''^2(\theta^2-c^2)+\gamma''^2\ (\theta^2-b^2)+2\beta''\gamma''\,.\,bc\cos\alpha\ =0\,;$$

and, similarly,

$$\gamma''^2(\theta^2-a^2)+\alpha''^2\ (\theta^2-c^2)+2\gamma''\alpha''\ .\ ca\cos\beta=0,$$
$$\alpha''^2(\theta^2-b^2)+\beta''^2(\theta^2-a^2)+2\alpha''\beta''\ .\ ab\cos\gamma=0,$$

or, what is the same thing,

$$\frac{\beta''^2}{b^2-\theta^2}+\frac{\gamma''^2}{c^2-\theta^2}-\frac{2bc\cos\alpha}{b^2-\theta^2\,.\,c^2-\theta^2}\beta''\gamma''=0,$$

$$\frac{\gamma''^2}{c^2-\theta^2}+\frac{\alpha''^2}{a^2-\theta^2}-\frac{2ca\cos\beta}{c^2-\theta^2\,.\,a^2-\theta^2}\gamma''\alpha''=0,$$

$$\frac{\alpha''^2}{a^2-\theta^2}+\frac{\beta''^2}{b^2-\theta^2}-\frac{2ab\cos\gamma}{a^2-\theta^2\,.\,b^2-\theta^2}\alpha''\beta''=0;$$

writing

$$\alpha'',\ \beta'',\ \gamma''=X\sqrt{(a^2-\theta^2)},\quad Y\sqrt{(b^2-\theta^2)},\quad Z\sqrt{(c^2-\theta^2)},$$

and

$$\frac{bc\cos\alpha}{\sqrt{(b^2-\theta^2\,.\,c^2-\theta^2)}},\ \frac{ca\cos\beta}{\sqrt{(c^2-\theta^2\,.\,a^2-\theta^2)}},\ \frac{ab\cos\gamma}{\sqrt{(a^2-\theta^2\,.\,b^2-\theta^2)}}=f,\ g,\ h,$$

the equations are

$$\begin{aligned}Y'^2+Z'^2-2fY'Z'&=0,\\ Z'^2+X'^2-2gZ'X'&=0,\\ X'^2+Y'^2-2hX'Y'&=0.\end{aligned}$$

Writing the last two under the form

$$\begin{aligned}X'^2-2gZ'X'+Z'^2&=0,\\ X'^2-2hY'X'+Y'^2&=0,\end{aligned}$$

and eliminating X', we have

$$-4(1-g^2)(1-h^2)Y'^2Z'^2+(Y'^2+Z'^2-2ghY'Z')^2=0,$$

which, in virtue of the first equation, is

$$-4(1-g^2)(1-h^2)Y'^2Z'^2+4(gh-f)^2Y'^2Z'^2=0,$$

that is,

$$(1-g^2)(1-h^2)-(gh-f)^2=0;$$

or, what is the same thing,

$$1-f^2-g^2-h^2+2fgh=0.$$

I remark that we may write

$$\begin{aligned}gh-f&=\sqrt{(1-g^2)}\sqrt{(1-h^2)},\\ hf-g&=\sqrt{(1-h^2)}\sqrt{(1-f^2)},\\ fg-h&=\sqrt{(1-f^2)}\sqrt{(1-g^2)},\end{aligned}$$

the signs on the right-hand side being either all +, or else one + and two −, so that the product is +. In fact, multiplying the assumed equations, we have

$$f^2g^2h^2-fgh(f^2+g^2+h^2)+g^2h^2+h^2f^2+f^2g^2-fgh=1-f^2-g^2-h^2+g^2h^2+h^2f^2+f^2g^2-f^2g^2h^2,$$

that is,

$$1-f^2-g^2-h^2+fgh(1+f^2+g^2+h^2)-2f^2g^2h^2=0,$$

or,

$$(1-f^2-g^2-h^2+2fgh)(1-fgh)=0,$$

which is right; but with a different combination of signs the result would not have been obtained.

Substituting for f, g, h their values, we have

$$(a^2-\theta^2)(b^2-\theta^2)(c^2-\theta^2)-b^2c^2(a^2-\theta^2)\cos^2\alpha-c^2a^2(b^2-\theta^2)\cos^2\beta$$
$$-a^2b^2(c^2-\theta^2)\cos^2\gamma+2a^2b^2c^2\cos\alpha\cos\beta\cos\gamma=0,$$

where the term independent of θ is

$$a^2b^2c^2(1-\cos^2\alpha-\cos^2\beta-\cos^2\gamma+2\cos\alpha\cos\beta\cos\gamma),$$

which is $=0$ in virtue of $\alpha+\beta+\gamma=2\pi$. We have, therefore, for θ^2 the quadric equation

$$b^2c^2\sin^2\alpha+c^2a^2\sin^2\beta+a^2b^2\sin^2\gamma-(a^2+b^2+c^2)\theta^2+\theta^4=0,$$

giving for θ^2 the two real positive values

$$\theta^2=\tfrac{1}{2}\{a^2+b^2+c^2\pm\sqrt{(\Omega)}\},$$

where

$$\begin{aligned}\Omega^2&=(a^2+b^2+c^2)^2-4(b^2c^2\sin^2\alpha+c^2a^2\sin^2\beta+a^2b^2\sin^2\gamma)\\&=a^4+b^4+c^4+2b^2c^2\cos 2\alpha+2c^2a^2\cos 2\beta+2a^2b^2\cos 2\gamma\\&=(a^2+b^2\cos 2\gamma+c^2\cos 2\beta)^2+(b^2\sin 2\gamma-c^2\sin 2\beta)^2.\end{aligned}$$

I write now

$$\frac{a\cos\xi}{\sqrt{(a^2-\theta^2)}},\ \frac{b\cos\eta}{\sqrt{(b^2-\theta^2)}},\ \frac{c\cos\zeta}{\sqrt{(c^2-\theta^2)}}=X,\ Y,\ Z,$$

and also

$$\frac{\sqrt{a^2-\theta^2}}{a}\sin\alpha,\ \frac{\sqrt{b^2-\theta^2}}{b}\sin\beta,\ \frac{\sqrt{c^2-\theta^2}}{c}\sin\gamma=A,\ B,\ C.$$

The equations for $\cos\xi$, $\cos\eta$, $\cos\zeta$ become

$$\begin{aligned}Y^2+Z^2-2fYZ-(1-f^2)&=0,\\Z^2+X^2-2gZX-(1-g^2)&=0,\\X^2+Y^2-2hXY-(1-h^2)&=0,\end{aligned}$$

and

$$AX+BY+CZ=0,$$

in virtue of the relation between f, g, h. The first three equations are satisfied by a two-fold relation between X, Y, Z; viz. treating these as coordinates, the equations represent three quadric cylinders having a common conic.

To prove this, I write

$$1-f^2,\ 1-g^2,\ 1-h^2,\ gh-f,\ hf-g,\ fg-h=\mathrm{a},\ \mathrm{b},\ \mathrm{c},\ \mathrm{f},\ \mathrm{g},\ \mathrm{h}.$$

We have, as usual,

$$\mathrm{bc}-\mathrm{f}^2,\ \mathrm{ca}-\mathrm{g}^2,\ \mathrm{ab}-\mathrm{h}^2,\ \mathrm{gh}-\mathrm{af},\ \mathrm{hf}-\mathrm{bg},\ \mathrm{fg}-\mathrm{ch},\ \text{each }=0:$$

the equations

$$\mathrm{a}X+\mathrm{h}Y+\mathrm{g}Z=0,\quad \mathrm{h}X+\mathrm{b}Y+\mathrm{f}Z=0,\quad \mathrm{g}X+\mathrm{f}Y+\mathrm{c}Z=0,$$

represent each of them one and the same plane, which I say is that of the conic in question.

The three given equations are

$$Y^2 + Z^2 - 2fYZ \quad - \mathrm{a} = 0,$$
$$Z^2 + X^2 - 2gZX \quad - \mathrm{b} = 0,$$
$$X^2 + Y^2 - 2hXY - \mathrm{c} = 0,$$

say these are $U=0$, $V=0$, $W=0$; it is to be shown that $\mathrm{c}V - \mathrm{b}W$, $\mathrm{a}W - \mathrm{c}U$, $\mathrm{b}U - \mathrm{a}V$, each contain the linear factor in question. We have

$$\mathrm{c}V - \mathrm{b}W = (\mathrm{c}-\mathrm{b})\,X^2 - \mathrm{b}Y^2 + \mathrm{c}Z^2 - 2\mathrm{c}gZX + 2\mathrm{b}hXY;$$

or, what is the same thing,

$$\mathrm{a}(\mathrm{c}V - \mathrm{b}W) = \mathrm{a}(\mathrm{c}-\mathrm{b})\,X^2 - \mathrm{h}^2Y^2 + \mathrm{g}^2Z^2 - 2g\mathrm{g}^2ZX + 2h\mathrm{h}^2XY.$$

Assuming this

$$= (\mathrm{a}X + \mathrm{h}Y + \mathrm{g}Z)(\lambda X - \mathrm{h}Y + \mathrm{g}Z),$$

we have

$$\mathrm{a}\lambda = \quad \mathrm{a}(\mathrm{c}-\mathrm{b}),$$
$$\mathrm{g}(\quad \mathrm{a} + \lambda) = -2g\mathrm{g}^2,$$
$$\mathrm{h}(-\mathrm{a} + \lambda) = \quad 2h\mathrm{h}^2,$$

that is,

$$\lambda = \mathrm{c} - \mathrm{b}, \quad \mathrm{a} + \lambda = -2g\mathrm{g}, \quad -\mathrm{a} + \lambda = 2h\mathrm{h}:$$

but $\lambda = \mathrm{c} - \mathrm{b}, = -\mathrm{h}^2 + \mathrm{g}^2$, and the other two equations are $\mathrm{a} + \mathrm{c} - \mathrm{b} + 2g\mathrm{g} = 0$, $\mathrm{a} + \mathrm{b} - \mathrm{c} + 2h\mathrm{h} = 0$, which are identically true.

The values of X, Y, Z are thus determined as the coordinates of the intersection of the conic with the plane $AX + BY + CZ = 0$; or, what is the same thing, of the line

$$AX + BY + CZ = 0,$$
$$\mathrm{a}X + \mathrm{h}Y + \mathrm{g}Z = 0,$$

with any one of the three cylinders.

We may, however, complete the analytical solution in a different manner as follows:

Assuming as above $\sqrt{(\mathrm{bc})} = \mathrm{f}$, $\sqrt{(\mathrm{ca})} = \mathrm{g}$, $\sqrt{(\mathrm{ab})} = \mathrm{h}$, and thence $\mathrm{h}\sqrt{(\mathrm{c})} - \mathrm{g}\sqrt{(\mathrm{b})} = 0$, we obtain from the second and the third equations

$$Y = hX + \sqrt{(\mathrm{c})}\,\sqrt{(1 - X^2)}, \quad Z = gX - \sqrt{(\mathrm{b})}\,\sqrt{(1 - X^2)},$$

(the signs are one + the other −, in order that this may consist with the equation $\mathrm{a}X + \mathrm{h}Y + \mathrm{g}Z = 0$). Substituting in $AX + BY + CZ = 0$, we have

$$(A + Bh + Cg)\,X + \{B\sqrt{(\mathrm{c})} - C\sqrt{(\mathrm{b})}\}\,\sqrt{(1 - X^2)} = 0,$$

that is,

$$(A + Bh + Cg)^2\,X^2 - (B^2\mathrm{c} + C^2\mathrm{b} - 2BC\mathrm{f})\,(1 - X^2) = 0,$$

or say

$$(A + Bh + Cg)^2 X^2 + \{B^2(1-h^2) + C^2(1-g^2) - 2BC(gh-f)\}(X^2-1) = 0,$$

that is,

$$(A^2 + B^2 + C^2 + 2BCf + 2CAg + 2ABh) X^2 = \{B^2 + C^2 + 2BCf - (Bh + Cg)^2\},$$

or writing

$$A^2 + B^2 + C^2 + 2BCf + 2CAg + 2ABh = \Delta,$$

say we have

$$\Delta X^2 = B^2 + C^2 + 2BCf - (Bh + Cg)^2,$$

$$\Delta Y^2 = C^2 + A^2 + 2CAg - (Cf + Ah)^2,$$

$$\Delta Z^2 = A^2 + B^2 + 2ABh - (Ag + Bf)^2.$$

Now attending to the values of A, B, C, f, g, h, we have

$$BCf,\ CAg,\ ABh = \sin\beta\sin\gamma\cos\alpha,\quad \sin\gamma\sin\alpha\cos\beta,\quad \sin\alpha\sin\beta\cos\gamma,$$

and thence

$$\Delta = \sin^2\alpha\left(1-\frac{\theta^2}{a^2}\right) + \sin^2\beta\left(1-\frac{\theta^2}{b^2}\right) + \sin^2\gamma\left(1-\frac{\theta^2}{c^2}\right)$$

$$+ 2(\sin\beta\sin\gamma\cos\alpha + \sin\gamma\sin\alpha\cos\beta + \sin\alpha\sin\beta\cos\gamma);$$

in virtue of $\alpha+\beta+\gamma=2\pi$, the last term is

$$= 2(\cos\alpha\cos\beta\cos\gamma - 1),$$

whence

$$\Delta = -\theta^2\left(\frac{\sin^2\alpha}{a^2} + \frac{\sin^2\beta}{b^2} + \frac{\sin^2\gamma}{c^2}\right),\ \text{say this is} = -\theta^2\Lambda.$$

Moreover $Bh + Cg = \dfrac{a\sin\alpha}{\sqrt{(a^2-\theta^2)}}$, whence the value of ΔX^2 is

$$= \sin^2\beta\left(1-\frac{\theta^2}{b^2}\right) + \sin^2\gamma\left(1-\frac{\theta^2}{c^2}\right) + 2\sin\beta\sin\gamma\cos\alpha - \left(1-\frac{\theta^2}{a^2}\right)\sin^2\alpha.$$

Here the constant term is

$$= \sin^2\beta + \sin^2\gamma + 2\sin\beta\sin\gamma\cos\alpha,$$

that is,

$$= 1 - (1-\sin^2\beta)(1-\sin^2\gamma) + \sin^2\beta\sin^2\gamma + 2\sin\beta\sin\gamma\cos\alpha$$

$$= 1 - \cos^2\beta\cos^2\gamma - \cos^2\alpha + (\cos\alpha + \sin\beta\sin\gamma)^2$$

$$= 1 - \cos^2\alpha,\ = \sin^2\alpha,$$

or the whole is

$$\sin^2\alpha\left(1-\frac{a^2}{a^2-\theta^2}\right) - \theta^2\left(\frac{\sin^2\beta}{b^2} + \frac{\sin^2\gamma}{c^2}\right),$$

which is

$$= -\theta^2\left(\frac{\sin^2\alpha}{a^2-\theta^2}+\frac{\sin^2\beta}{b^2}+\frac{\sin^2\gamma}{c^2}\right),$$

so that we have

$$\Lambda X^2 = \left(\frac{\sin^2\alpha}{a^2-\theta^2}+\frac{\sin^2\beta}{b^2}+\frac{\sin^2\gamma}{c^2}\right).$$

Similarly,

$$\Lambda Y^2 = \left(\frac{\sin^2\alpha}{a^2}+\frac{\sin^2\beta}{b^2-\theta^2}+\frac{\sin^2\gamma}{c^2}\right),$$

$$\Lambda Z^2 = \left(\frac{\sin^2\alpha}{a^2}+\frac{\sin^2\beta}{b^2}+\frac{\sin^2\gamma}{c^2-\theta^2}\right);$$

and hence also

$$\Lambda(1-X^2) = \frac{-\theta^2\sin^2\alpha}{a^2(a^2-\theta^2)},\quad \Lambda(1-Y^2) = \frac{-\theta^2\sin^2\beta}{b^2(b^2-\theta^2)},\quad \Lambda(1-Z^2) = \frac{-\theta^2\sin^2\gamma}{c^2(c^2-\theta^2)},$$

where

$$\Lambda = \frac{\sin^2\alpha}{a^2}+\frac{\sin^2\beta}{b^2}+\frac{\sin^2\gamma}{c^2}.$$

The equation in X is

$$\Lambda\left(1-\frac{a^2\cos^2\xi}{a^2-\theta^2}\right) = \frac{-\theta^2\sin^2\alpha}{a^2(a^2-\theta^2)},$$

that is,

$$\Lambda(a^2\sin^2\xi-\theta^2) = -\theta^2\frac{\sin^2\alpha}{a^2},$$

or

$$a^2\sin^2\xi = \theta^2\left(1-\frac{\sin^2\alpha}{a^2\Lambda}\right),$$

and the like for η, ζ. Writing for greater convenience $\dfrac{\sin^2\alpha}{a^2}$, $\dfrac{\sin^2\beta}{b^2}$, $\dfrac{\sin^2\gamma}{c^2}=p, q, r$, then $\Lambda = p+q+r$, and we have

$$\sin^2\xi = \frac{\theta^2}{a^2}\frac{q+r}{p+q+r},\quad \sin^2\eta = \frac{\theta^2}{b^2}\frac{r+p}{p+q+r},\quad \sin^2\zeta = \frac{\theta^2}{c^2}\frac{p+q}{p+q+r},$$

(whence also $a^2\sin^2\xi + b^2\sin^2\eta + c^2\sin^2\zeta = 2\theta^2$: as a simple verification, observe that, if the projection is rectangular, the axes being all equally inclined to the plane of projection, then $\xi=\eta=\zeta=90°$, $a=b=c=\theta\sin s$, and the equation is $3\sin^2 s = 2$; s, s are here the sides of an isosceles quadrantal triangle, the included angle being 120°, that is, we have $\cos 120°\ (=-\tfrac{1}{2}) = -\cot^2 s$, that is, $\cot^2 s = \tfrac{1}{2}$, or $\sin^2 s = \tfrac{2}{3}$, which is right).

I remark, that a geometrical solution may be obtained upon very different principles. We have on a sphere the trirectangular triangle XYZ, which by parallel lines is projected into ABC. Every great circle of the sphere is projected into an ellipse having double contact at the extremities of a diameter with the ellipse which is the apparent contour of the sphere. Moreover, if the arc of great circle XY is a quadrant, then the radius through X and the tangent at Y are parallel to each other, whence, if Ω be the projection of the centre, and AB the projection of the arc XY, then in the projection the line ΩA and the tangent at B are parallel to each other. It is now easy to derive a construction: with centre Ω, and conjugate semi-axes $(\Omega B, \Omega C)$, $(\Omega C, \Omega A)$, $(\Omega A, \Omega B)$ respectively, describe three ellipses; and find a concentric ellipse having double contact with each of these (there are in fact two such ellipses, one touching the three ellipses internally, and giving an imaginary solution; the other touching them externally, which is the ellipse intended). Drawing then through the ellipse a right cylinder (there are two such cylinders, but only one of them is real), and inscribing in it a sphere, and projecting on to the surface of the sphere by lines parallel to the axis of the cylinder, the three ellipses are projected into three great circles cutting at right angles, or, say, the elliptic arcs BC, CA, AB are projected into the trirectangular triangle XYZ.

615.

ON THE CONIC TORUS.

[From the *Quarterly Journal of Pure and Applied Mathematics*, vol. XIII. (1875), pp. 127—129.]

THE equation $p+\sqrt{(qr)}+\sqrt{(st)}=0$, where p, q, r, s, t are linear functions of the coordinates (x, y, z, w), and as such are connected by a linear relation, belongs to a quartic surface having a nodal conic ($p=0$, $qr-st=0$); and four nodes (conical points), viz. these are the intersections of the line $q=0$, $r=0$ with the quadric surface $p^2-qr-st=0$, and of the line $r=0$, $s=0$ with the same surface. The quartic surface has also four tropes (planes which touch the surface along a conic); viz. these are the planes $q=0$, $r=0$, $s=0$, $t=0$, the conic of contact or tropal conic in each plane being the intersection of the plane with the before-mentioned quadric surface $p^2-qr-st=0$. The planes $q=0$, $r=0$, and also the conics in these planes pass through two of the nodes, say A, C; and the planes $s=0$, $t=0$, and also the conics in these planes pass through the remaining two nodes, say B, D; so that the relations of the surface are as is shown in fig. 1. It is to be added that AB, BC, CD, DA (but not AC or BD) are lines on the surface.

The planes $q=0$, $r=0$, which contain the tropal conics through A, C, are in general distinct from the planes ABC, ADC which contain the line-pairs BA, BC and DA, DC respectively: and so also the planes $s=0$, $t=0$, which contain the tropal conics through B, D, are in general distinct from the planes ABD, CBD which contain the line-pairs AB, AD and CB, CD respectively.

If, however, the identical linear relation contain only p, s, t, then the planes $q=0$, $r=0$ will be the planes ABC, ADC respectively: and the tropal conics in these planes will consequently be the line-pairs BA, BC, and DA, DC respectively. But the planes $s=0$, $t=0$ will continue to be distinct from the planes ABD, CBD: and the tropal conics in the planes $s=0$, $t=0$ will remain proper conics.

A surface of the last-mentioned form is

$$mz + \sqrt{(xy)} + \sqrt{(w^2 - z^2)} = 0,$$

viz. this has the nodal conic $z = 0$, $xy - w^2 = 0$, the nodes $\{x = 0, y = 0, (m^2+1)z^2 - w^2 = 0\}$, and $(z = 0, w = 0, x = 0)$, $(z = 0, w = 0, y = 0)$, and the tropes $x = 0$, $y = 0$, $z + w = 0$, $z - w = 0$; but the planes $z + w = 0$ and $z - w = 0$ are ordinary tropal planes each touching the surface in a proper conic; the planes $x = 0$, $y = 0$ special planes each touching along a line-pair.

Fig. 1.

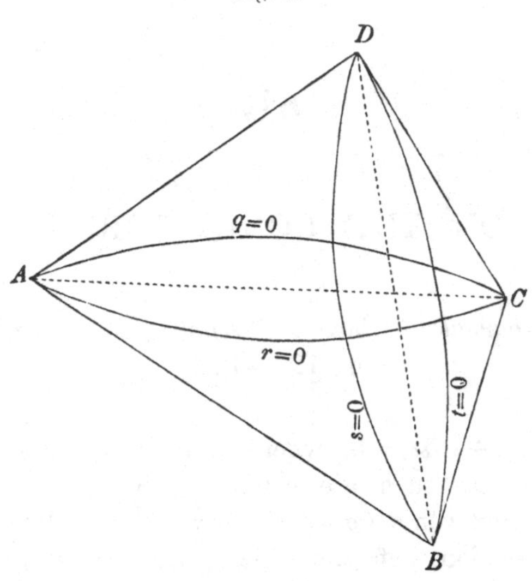

The equation in question, writing therein $w = 1$ and $x + iy$, $x - iy$ in place of (x, y) respectively, is

$$\{\sqrt{(x^2 + y^2)} + mz\}^2 = 1 - z^2,$$

which is derived from

$$(x + mz)^2 = 1 - z^2,$$

by the change of x into $\sqrt{(x^2 + y^2)}$; and the surface is consequently the torus generated by the rotation of the conic $(x + mz)^2 = 1 - z^2$ about its diameter. Or, what is the same thing, the surface

$$mz + \sqrt{(xy)} + \sqrt{(w^2 - z^2)} = 0,$$

regarding therein (x, y) as circular coordinates and w as being $= 1$, is a torus. The rational equation is $U = 0$, where we have

$$\begin{aligned} U &= \{(m^2+1) z^2 - w^2 + xy\}^2 - 4m^2z^2xy \\ &= \{xy + (1 - m^2) z^2 - w^2\}^2 + 4m^2z^2 (z^2 - w^2) \\ &= x^2y^2 + (m^2+1)^2 z^4 + w^4 + (2 - 2m^2) z^2xy - (2 + 2m^2) z^2w^2 - 2xyw^2. \end{aligned}$$

I find that the Hessian H of this function U contains the factor $xy + (1 - m^2) z^2 - w^2$, viz. that we have

$$H = \{xy + (1 - m^2) z^2 - w^2\} H',$$

where

$$\begin{aligned} H' = &\quad x^3y^3\ (1-m^2) \\ &+ x^2y^2\{(3+8m^2+m^4)\,z^2+(-3+m^2)\,w^2\} \\ &+ xy\ \ \{(3+11m^2+9m^4+m^6)\,z^4+(-6-12m^2+6m^4)\,z^2w^2+(3+m^2)\,w^4\} \\ &+ \quad (1+m^2)\,\{(1-m^2)\,z^2-w^2\}\,\{(1+m^2)\,z^2-w^2\}^2, \end{aligned}$$

giving without much difficulty

$$\begin{aligned} H' = &\quad z^6\ (1+m^2)^3\,(1-m^2) \\ &+ 2z^4\,[(1+4m^2+m^4)\,xy-(1-m^4)\,w^2]\,(1+m^2) \\ &+\ z^2\ (xy-w^2)\,[(1+12m^2-m^4)\,xy-(1-m^4)\,w^2] \\ &+ \quad [(1-m^2)\,xy-(1+m^2)\,w^2]\ U; \end{aligned}$$

say this is

$$= z^2H''+[(1-m^2)\,xy-(1+m^2)\,w^2]\ U,$$

where

$$\begin{aligned} H'' = &\quad z^4\,(1+m^2)^3\ \ (1-m^2) \\ &+ 2z^2\,(1+m^2)\ [(1+\ 4m^2+m^4)\,xy-(1-m^4)\,w^2] \\ &+ \quad (xy-w^2)\,[(1+12m^2-m^4)\,xy-(1-m^4)\,w^2], \end{aligned}$$

or, what is the same thing,

$$\begin{aligned} H'' = &\quad x^2y^2\,(1+12m^2-m^4) \\ &+ 2xy\,[(1+4m^2+m^4)\,(1+m^2)\,z^2+(-1+6m^2)\,w^2] \\ &+ \quad (1-m^4)\,\{(1+m^2)\,z^2-w^2\}^2. \end{aligned}$$

It consequently appears that the complete spinode curve or intersection of the quartic surface and its Hessian, being of order 4×8, $=32$, breaks up into

$$U=0,\ xy+(1-m^2)\,z^2-w^2=0,$$

that is,

conic $z=0,\ xy-w^2=0$ twice, order		4
conic $z+w=0,\ xy-m^2w^2=0,$	„	2
conic $z-w=0,\ xy-m^2w^2=0,$	„	2

and

$$U=0,\ z^2H''=0,$$

that is,

$U=0,\ z^2=0$, conic $z=0,\ xy-w^2=0$ four times,	„	8
proper spinode curve $U=0,\ H''=0,$	„	16
		32;

viz. the intersection is made up of the conic $z=0,\ xy-w^2=0$ six times, the conics $z\pm w=0,\ xy-m^2w^2=0$ each twice, and the proper spinode curve of the order 16.

616.

A GEOMETRICAL ILLUSTRATION OF THE CUBIC TRANSFORMATION IN ELLIPTIC FUNCTIONS.

[From the *Quarterly Journal of Pure and Applied Mathematics*, vol. XIII. (1875), pp. 211—216.]

CONSIDER the cubic curve

$$x^3 + y^3 + z^3 + 6lxyz = 0.$$

If through one of the inflexions $z = 0$, $x + y = 0$, we draw an arbitrary line $z = u(x + y)$, we have at the other intersections of this line with the curve

$$u\{u^2(x+y)^2 + 6lxy\} + x^2 - xy + y^2 = 0;$$

that is,

$$(u^3 + 1)(x^2 + y^2) + 2xy(u^3 + 3lu - \tfrac{1}{2}) = 0;$$

and from this equation it appears that the ratio $x : y$ is given as a function involving the square root of

$$(u^3 + 3lu - \tfrac{1}{2})^2 - (u^3 + 1)^2,$$

which, rejecting a factor 3, is

$$= (2u^3 + 3lu + \tfrac{1}{2})(lu - \tfrac{1}{2}).$$

It may be noticed that $lu - \frac{1}{2} = 0$ gives the value of u, which in the equation $z = u(x + y)$ belongs to the tangent at the inflexion; and $2u^3 + 3lu + \frac{1}{2} = 0$ gives the values which belong to the three tangents from the inflexion.

It thus appears that the coordinates x, y, z of any point of the curve can be expressed as proportional to functions of u involving the radical

$$\sqrt{\{(lu - \tfrac{1}{2})(2u^3 + 3lu + \tfrac{1}{2})\}},$$

and the theory of the curve is connected with that of a quasi-elliptic integral depending on this radical.

Taking ω an imaginary cube root of unity, write

$$\omega x + \omega^2 y - 2lz = x',$$
$$\omega^2 x + \omega y - 2lz = y',$$
$$x + y - 2lz = z':$$

then we have

$$x'y'z' = x^3 + y^3 - 8l^3z^3 + 6lxyz = x^3 + y^3 + z^3 + 6lxyz - (1 + 8l^3) z^3.$$

Also

$$-6lz = x' + y' + z', \quad z^3 = \frac{-1}{216l^3}(x' + y' + z')^3,$$

whence

$$(x' + y' + z')^3 - \frac{216l^3}{1 + 8l^3} x'y'z' = \frac{216l^3}{1 + 8l^3}(x^3 + y^3 + z^3 + 6lxyz);$$

so that, putting

$$m^3 = \frac{-l^3}{1 + 8l^3},$$

or, what is the same thing,

$$8l^3m^3 + l^3 + m^3 = 0,$$

the equation of the curve is

$$(x' + y' + z')^3 + 216m^3x'y'z' = 0;$$

and if we write

$$x' : y' : z' = X^3 : Y^3 : Z^3,$$

then the original curve is transformed into

$$(X^3 + Y^3 + Z^3)^3 + 216m^3X^3Y^3Z^3 = 0,$$

a curve of the ninth order breaking up into three cubic curves, one of which is

$$X^3 + Y^3 + Z^3 + 6mXYZ = 0,$$

and for the other two we write herein $m\omega$ and $m\omega^2$ respectively in place of m. Attending only to the first curve, we have

$$x^3 + y^3 + z^3 + 6lxyz = 0,$$
$$X^3 + Y^3 + Z^3 + 6mXYZ = 0,$$

as corresponding curves, the corresponding points being connected by the relation

$$\omega x + \omega^2 y - 2lz : \omega^2 x + \omega y - 2lz : x + y - 2lz = X^3 : Y^3 : Z^3,$$

or, for convenience, we may write

$$\begin{aligned} \omega x + \omega^2 y - 2lz &= X^3, \text{ giving} & 3x &= \omega^2 X^3 + \omega Y^3 + Z^3, \\ \omega^2 x + \omega y - 2lz &= Y^3, & 3y &= \omega X^3 + \omega^2 Y^3 + Z^3, \\ x + y - 2lz &= Z^3, & -6lz &= X^3 + Y^3 + Z^3. \end{aligned}$$

This is a (1, 3) correspondence; viz. to a given point on the curve (m), there corresponds one point on (l); but to a given point on (l), three points on (m). As to the first case, this is obvious. As to the second case, if the point (x, y, z) is given, then the corresponding point (X, Y, Z) on the other curve will lie on one of the three lines

$$Y^3(\omega x+\omega^2 y-2lz)-X^3(\omega^2 x+\omega y-2lz)=0\,;$$

each of these intersects the curve (m) in three points: but of the points in the same line it is only one which is a corresponding point of (x, y, z), and the number of the corresponding points is consequently the same as the number of lines, viz. it is $=3$.

We infer that the above equations lead to a cubic transformation of the quasi-elliptic integral

$$\int du\div\sqrt{\{(lu-\tfrac{1}{2})(2u^3+3lu+\tfrac{1}{2})\}},$$

into one of the like form

$$\int dv\div\sqrt{\{(mv-\tfrac{1}{2})(2v^3+3mv+\tfrac{1}{2})\}}\,;$$

and this is now to be verified.

We have, as before, the line $z=u(x+y)$ meeting the curve (l) in the points

$$(u^3+1)(x^2+y^2)+2xy(u^3+3lu-\tfrac{1}{2})=0\,;$$

and if similarly through an inflexion of the curve (m) we take the line $Z=v(X+Y)$, this meets the curve in the points

$$(v^3+1)(X^2+Y^2)+2XY(v^3+3mv-\tfrac{1}{2})=0.$$

Then if (x, y, z), (X, Y, Z) are taken to be the corresponding points as above, we can obtain v as a function of u. We, in fact, have

$$-2lu=\frac{-2lz}{x+y}=\frac{X^3+Y^3+Z^3}{-X^3-Y^3+2Z^3}=\frac{X^3+Y^3+v^3(X+Y)^3}{-(X^3+Y^3)+2v^3(X+Y)^3}$$

$$=\frac{X^2-XY+Y^2+v^3(X+Y)^2}{-X^2+XY-Y^2+2v^3(X+Y)^2},$$

$$=\frac{(v^3+1)(X^2+Y^2)+(2v^3-1)XY}{(2v^3-1)(X^2+Y^2)+(4v^3+1)XY};$$

or, since we have

$$(v^3+1)(X^2+Y^2)+2XY(v^3+3mu-\tfrac{1}{2})=0,$$

that is,

$$X^2+Y^2 : XY=-2v^3-6mv+1 : v^3+1,$$

the equation becomes

$$-2lu=\frac{-6mv(v^3+1)}{(2v^3-1)(-2v^3-6mv+1)+(4v^3+1)(v^3+1)}$$

$$=\frac{-6mv(v^3+1)}{-3v(4mv^3-3v^2-2m)};$$

or say,

$$-lu = m(v^3+1)(\div), \text{ where the denominator } = 4mv^3 - 3v^2 - 2m.$$

This may also be written

$$-(lu - \tfrac{1}{2}) = 3v^2(mv - \tfrac{1}{2}) \div .$$

Proceeding to calculate $2u^3 + 3lu + \frac{1}{2}$, omitting the denominator $(4mv^3 - 3v^2 - 2m)^3$, this is

$$-\frac{2m^3}{l^3}(v^3+1)^3 - 3m(v^3+1)(4mv^3 - 3v^2 - 2m)^2 + \tfrac{1}{2}(4mv^3 - 3v^2 - 2m)^3;$$

or, observing that

$$m^3 = \frac{-l^3}{1+8l^3},$$

that is,

$$l^3 = \frac{-m^3}{1+8m^3} \text{ or } -\frac{m^3}{l^3} = 1 + 8m^3,$$

the numerator is

$$= 2(1+8m^3)(v^3+1)^3 - 3m(v^3+1)(4mv^3 - 3v^2 - 2m)^2 + \tfrac{1}{2}(4mv^3 - 3v^2 - 2m)^3,$$

which is found to be identically

$$= (2v^3 + 3mv + \tfrac{1}{2})(v^3 + 6mv - 2)^2;$$

viz. we have

$$2u^3 + 3lu + \tfrac{1}{2} = (2v^3 + 3mv + \tfrac{1}{2})(v^3 + 6mv - 2)^2 \div (4mv^3 - 3v^2 - 2m)^3,$$

and hence

$$(lu - \tfrac{1}{2})(2u^3 + 3lu + \tfrac{1}{2}) = -3(mv - \tfrac{1}{2})(2v^3 + 3mv + \tfrac{1}{2})(v^3 + 6mv - 2)^2 v^2 \div (4mv^3 - 3v^2 - 2m)^4.$$

Moreover, we find

$$ldu = 3mdv \,.\, v(v^3 + 6mv - 2) \div (4mv^3 - 3v^2 - 2m)^2,$$

and we thence have

$$\frac{ldu}{\sqrt{\{(lu - \frac{1}{2})(2u^3 + 3lu + \frac{1}{2})\}}} = \sqrt{(-3)}\,\frac{mdv}{\sqrt{\{(mv - \frac{1}{2})(2v^3 + 3mv + \frac{1}{2})\}}};$$

viz. this differential equation corresponds to the integral equation

$$-lu = m(v^3+1) \div (4mv^3 - 3v^2 - 2m),$$

where $8l^3m^3 + l^3 + m^3 = 0$, which corresponds to the modular equation.

It may be remarked that, if v is the same function of u', l, m that u is of v, m, l; viz. if

$$-mv = l(u'^3 + 1) \div (4lu'^3 - 3u'^2 - 2m'),$$

then

$$\frac{mdv}{\sqrt{\{(mv - \frac{1}{2})(2v^3 + 3mv + \frac{1}{2})\}}} = \sqrt{(-3)}\,\frac{-ldu'}{\sqrt{\{(lu' - \frac{1}{2})(2u'^3 + 3lu' + \frac{1}{2})\}}},$$

and consequently

$$\frac{du}{\sqrt{\{(lu - \frac{1}{2})(2u^3 + 3lu + \frac{1}{2})\}}} = \frac{-3du'}{\sqrt{\{(lu' - \frac{1}{2})(2u'^3 + 3lu' + \frac{1}{2})\}}},$$

which accords with the general theory of the cubic transformation.

We may inquire into the relation between the absolute invariants of the two curves. Taking the absolute invariant to be

$$\Omega = \frac{64S^3 - T^2}{64S^3},$$

where S and T bear the usual significations, we have for the one curve

$$\Omega = \frac{(1+8l^3)^3}{64l^3(1-l^3)^3},$$

and for the other curve

$$\Omega' = \frac{(1+8m^3)^3}{64m^3(1-m^3)^3},$$

and, as above, $8l^3m^3 + l^3 + m^3 = 0$: writing herein

$$l^3 = -\frac{1}{8\alpha'},\quad m^3 = -\frac{1}{8\beta'},$$

the relation between α', β' is simply $\alpha' + \beta' = 1$; and the values of Ω, Ω' are found to be

$$\Omega = \frac{64\alpha'(1-\alpha')^3}{(1+8\alpha')^3},\quad \Omega' = \frac{64\beta'(1-\beta')^3}{(1+8\beta')^3};$$

viz. the required relation is given by the elimination of α', β' from these three equations. Or, what is the same thing, writing $\alpha' = \frac{1}{2} + \theta$, and therefore $\beta' = \frac{1}{2} - \theta$, we have

$$(5+8\theta)^3\,\Omega = 4(1+2\theta)(1-2\theta)^3,$$

$$(5-8\theta)^3\,\Omega' = 4(1+2\theta)^3(1-2\theta),$$

and the elimination of θ from these equations gives the required relation between Ω and Ω'.

It of course follows that, if we have a *cubic* transformation

$$\frac{dx}{\sqrt{\{(a, b, c, d, e \between x, 1)\}^4}} = \frac{Cdx'}{\sqrt{\{(a', b', c', d', e' \between x', 1)\}^4}},$$

then the absolute invariants Ω, Ω' of the two quartic functions are connected by the above relation. I have obtained this result, by reducing the radicals to the standard forms

$$\sqrt{(1-x^2 . 1-k^2x^2)},\quad \sqrt{(1-x'^2 . 1-\lambda^2x'^2)},$$

from the known modular equation as represented by the equations

$$\lambda^2 = \frac{\alpha^3(2+\alpha)}{1+2\alpha},\quad k^2 = \frac{\alpha(2+\alpha)^3}{(1+2\alpha)^3};$$

viz. the values of the absolute invariants

$$\left(=1-\frac{27J^2}{I^3},\ 1-\frac{27J'^2}{I'^3}\right),$$

are

$$\Omega = \frac{108k^2(1-k^2)^4}{(k^4+14k^2+1)^3},\quad \Omega' = \frac{108\lambda^2(1-\lambda^2)^4}{(\lambda^4+14\lambda^2+1)^3},$$

but the method of effecting this is by no means obvious.

617.

ON THE SCALENE TRANSFORMATION OF A PLANE CURVE.

[From the *Quarterly Journal of Pure and Applied Mathematics*, vol. XIII. (1875), pp. 321—328.]

THE transformation by reciprocal radius vectors can be effected mechanically by Sylvester's Peaucellier-cell. But, employing a more general cell (considered incidentally by him) which may be called the scalene-cell, we have the scalene transformation in question*; viz. if, in two curves, r, r' are radius vectors belonging to the same angle (or say opposite angles) θ, then the relation between r, r' is

$$rr'(r+r')+(m^2-l^2)r+(m^2-n^2)r'=0;$$

or, as this may also be written,

$$r^2+\left(r'-\frac{l^2-m^2}{r'}\right)r+m^2-n^2=0.$$

The transformation is, it will be seen, an interesting one for its own sake, independently of the remarkably simple mechanical construction, viz. the scalene cell is simply a system of 3 pairs of equal rods PA, QA; PB, QB; PC, QC (fig. 1, p. 528), jointed together at and capable of rotating about the points P, Q, A, B, C; the three lengths PA, PB, PC (say these are $=l$, m, n) are all of them unequal: in the case of any two of them equal, we have Peaucellier or isosceles cell. The effect of the arrangement is that the points A, B, C are retained in a right line, the distances BA, $=r'$, and BC, $=r$, being connected by the above-mentioned equation; so that taking B as a fixed point, if the point A describe any given curve, the point C will describe the corresponding or transformed curve.

In the case where the given curve is a right line or a circle, we may through B draw at right angles to the curve the axis $x'Bx$: viz. in the case of the circle,

* The transformation itself, and doubtless many of the results obtained by means of it, are familiar to Prof. Sylvester; and I abandon all claim to priority.

the axis $x'Bx$ passes through its centre; and we measure the angle θ from this line, viz. we write $\angle xBC = \angle x'BA = \theta$.

Fig. 1.

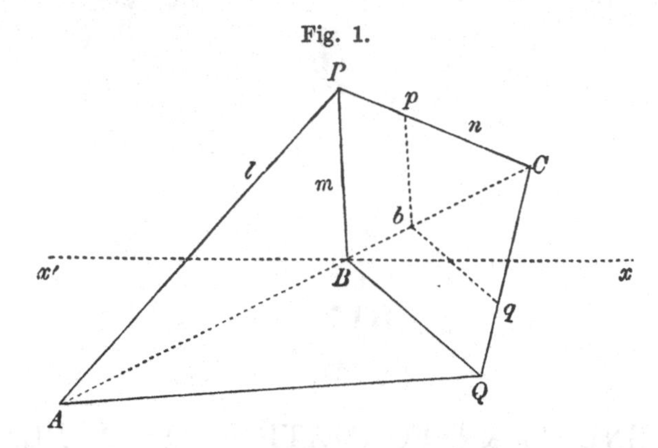

Suppose, first, that the locus of A is a right line, or a circle passing through B. Its equation is $r' = \frac{c}{\cos\theta}$ or $= c\cos\theta$; and we accordingly have for the transformed curve

$$r^2 + \left(\frac{c}{\cos\theta} - \frac{l^2 - m^2}{c\cos\theta}\right) r + m^2 - n^2 = 0,$$

or else

$$r^2 + \left(c\cos\theta - \frac{l^2 - m^2}{c\cos\theta}\right) r + m^2 - n^2 = 0;$$

viz. multiplying in each case by $r\cos\theta$, and then writing $r\cos\theta = x$, $r^2 = x^2 + y^2$, the equations become

$$x(x^2 + y^2) + c(x^2 + y^2) - \frac{l^2 - m^2}{c} a^2 \qquad + (m^2 - n^2)x = 0,$$

and

$$x(x^2 + y^2) + cx^2 \qquad - \frac{l^2 - m^2}{c}(x^2 + y^2) + (m^2 - n^2)x = 0;$$

viz. in each case the curve is a circular cubic passing through the origin B and having an asymptote parallel to the axis of y. The curve is nodal, if $m = n$, viz. in this case the origin is a node: or if $c = \sqrt{(l^2 - n^2)} + \sqrt{(m^2 - n^2)}$.

Suppose next that the locus of A is a circle, centre at a distance $= \gamma$ along Bx' and radius $= h$: we have

$$r'^2 - 2\gamma r'\cos\theta + \gamma^2 - h^2 = 0,$$

viz. if

$$\gamma^2 - h^2 = -(l^2 - m^2),$$

or, what is the same thing,

$$h^2 + m^2 = \gamma^2 + l^2,$$

then we have

$$r' - \frac{l^2 - m^2}{r'} = 2\gamma\cos\theta,$$

and the transformed curve is

$$r^2 + 2\gamma r \cos\theta + m^2 - n^2 = 0,$$

or, as this may be written,

$$r^2 + 2\gamma r \cos\theta + \gamma^2 - f^2 = 0,$$

where $\gamma^2 - f^2 = m^2 - n^2$, that is, $f^2 + m^2 = \gamma^2 + n^2$; viz. this is a concentric circle radius f.

The theorem may be presented as follows. Consider two concentric circles, centre O and radii h, f respectively; take an arbitrary point B, distance $OB = \gamma$; and taking m arbitrary, determine l, n by the equations

$$l^2 = m^2 + h^2 - \gamma^2,\ n^2 = m^2 + f^2 - \gamma^2;$$

then drawing through B an arbitrary line to meet the circles in A, C respectively; also describing a circle, centre B and radius $= m$; and through O drawing a line perpendicular to ABC to meet the last-mentioned circle in two points P, Q: for these points, the distances from the points A, B, C are $= l, m, n$ respectively.

To verify this, take O as the origin, OB for the axis of x, θ the inclination of ABC to this axis, $BA = r'$, $BC = r$; the coordinates of C, B, A are

$$\begin{array}{ll} \gamma + r\cos\theta, & r\sin\theta, \\ \gamma \quad , & 0 \quad , \\ \gamma + r'\cos\theta, & -r'\sin\theta, \end{array}$$

whence, taking (x, y) for the coordinates of P (or Q), the equations to be verified are

$$(x - \gamma - r\cos\theta)^2 + (y - r\sin\theta)^2 = n^2,$$
$$(x - \gamma)^2 + y^2 = m^2,$$
$$(x - \gamma + r'\cos\theta)^2 + (y + r'\sin\theta)^2 = l^2.$$

By means of the second equation, the other two become

$$-2(x - \gamma) r\cos\theta - 2yr\sin\theta + r^2 = n^2 - m^2,$$
$$2(x - \gamma) r'\cos\theta + 2yr'\sin\theta + r'^2 = l^2 - m^2;$$

or, substituting for $n^2 - m^2$, $l^2 - m^2$ the values $f^2 - \gamma^2$ and $h^2 - \gamma^2$, the equations are

$$-2xr\cos\theta - 2yr\sin\theta + r^2 + 2\gamma r\cos\theta + \gamma^2 - f^2 = 0,$$
$$2xr'\cos\theta + 2yr'\sin\theta + r'^2 - 2\gamma r'\cos\theta + \gamma^2 - h^2 = 0,$$

viz. in virtue of the equations of the two circles, these reduce themselves each of them to

$$x\cos\theta + y\sin\theta = 0,$$

which equation, together with the second equation

$$(x - \gamma)^2 + y^2 = m^2,$$

determine (x, y) as above.

Reverting to the case where the locus of A is the circle

$$r'^2 - 2\gamma r' \cos\theta + \gamma^2 - h^2 = 0,$$

this gives

$$r' = \gamma \cos\theta + \sqrt{(h^2 - \gamma^2 \sin^2\theta)},$$

$$\frac{1}{r'} = \frac{\gamma\cos\theta - \sqrt{(h^2 - \gamma^2\sin^2\theta)}}{\gamma^2 - h^2};$$

so that for the transformed curve we have

$$r^2 + r\left(1 - \frac{l^2 - m^2}{\gamma^2 - h^2}\right)\gamma\cos\theta + r\left(1 + \frac{l^2 - m^2}{\gamma^2 - h^2}\right)\sqrt{(h^2 - \gamma^2\sin^2\theta)} + m^2 - n^2 = 0.$$

Putting for shortness $\dfrac{l^2 - m^2}{\gamma^2 - h^2} = \lambda$, and for r, $r\cos\theta$, $r\sin\theta$, writing $\sqrt{(x^2 + y^2)}$, x, y respectively, this is

$$x^2 + y^2 + (1 - \lambda)\gamma x + (1 + \lambda)\sqrt{\{h^2(x^2 + y^2) - \gamma^2 y^2\}} + m^2 - n^2 = 0,$$

or, what is the same thing,

$$\{x^2 + y^2 + (1 - \lambda)\gamma x + m^2 - n^2\}^2 = (1 + \lambda)^2\{h^2(x^2 + y^2) - \gamma^2 y^2\},$$

a bicircular quartic. In the case $\lambda = -1$, it reduces itself to the circle

$$x^2 + y^2 + 2\gamma x + m^2 - n^2 = 0$$

twice, which is the case considered above; and in the case $\lambda = 1$, or $l^2 + h^2 = m^2 + \gamma^2$, the equation is

$$(x^2 + y^2 + m^2 - n^2)^2 = 4\{h^2(x^2 + y^2) - \gamma^2 y^2\},$$

so that the curve is symmetrical in regard to each axis. In the case $\gamma = 0$, the locus is a pair of concentric circles, centre B.

The equation

$$\{x^2 + y^2 + (1 - \lambda)\gamma x + m^2 - n^2\}^2 = (1 + \lambda)^2\{h^2(x^2 + y^2) - \gamma^2 y^2\},$$

which contains the four constants λ, γ, h and $m^2 - n^2$, may be written in the form

$$(x^2 + y^2 + Ax + B)^2 = ax^2 + ey^2,$$

(where the constants A, B, a, e are also arbitrary). This is, in fact, the equation of the general symmetrical bicircular quartic, referred to a properly-selected point on the axis as origin, viz. the origin is the centre of any one of the three involutions formed by the vertices (or points on the axis); say it is any one of the three involution-centres of the curve.

To show this, assume

$$(x - \alpha)(x - \beta)(x - \gamma)(x - \delta) = x^4 - px^3 + qx^2 - rx + s:$$

then, taking B arbitrary, the equation of the symmetrical bicircular quartic having for vertices the points $x=\alpha$, $x=\beta$, $x=\gamma$, $x=\delta$, is

$$(x^2+y^2-\tfrac{1}{2}px+B)^2=(2B+\tfrac{1}{4}p^2-q)\,x^2+(r-pB)\,x+(-s+B^2);$$

in fact, this is the form of the general equation, and writing therein $y=0$, it becomes $x^4-px^3+qx^2-rx+s=0$, that is, $(x-\alpha)(x-\beta)(x-\gamma)(x-\delta)=0$. Hence, writing for convenience

$$\begin{aligned} A &= -\tfrac{1}{2}p, \\ a &= 2B+\tfrac{1}{4}p^2-q, \\ b &= r-pB, \\ c &= -s+B^2, \end{aligned}$$

the equation is

$$(x^2+y^2+Ax+B)^2=ax^2+bx+c.$$

This may be written

$$(x^2+y^2+Ax+B+\theta)^2=(a+2\theta)\,x^2+2\theta y^2+(b+2\theta A)\,x+c+2B\theta+\theta^2,$$

viz. assuming $\theta=-\dfrac{b}{2A}$ in order on the right-hand side to destroy the term in x, the equation is

$$\left(x^2+y^2+Ax+B-\frac{b}{2A}\right)^2=\left(a-\frac{b}{A}\right)x^2-\frac{b}{A}y^2+\frac{1}{4A^2}(b^2-4ABb+4A^2c),$$

which is of the form

$$(x^2+y^2+Ax+B)^2=ax^2+ey^2+f;$$

and if $f=0$, that is, if $b^2-4ABb+4A^2c=0$, then it is of the required form

$$(x^2+y^2+Ax+B)^2=ax^2+ey^2.$$

We have

$$\begin{aligned} b^2-4ABb+4A^2c &= (r-pB)^2+2pB(r-pB)+p^2(-s+B^2) \\ &= r^2-p^2s, \end{aligned}$$

or the required condition is $r^2-p^2s=0$. But we have

$$p^2s-r^2=(\alpha\delta-\beta\gamma)(\beta\delta-\gamma\alpha)(\gamma\delta-\alpha\beta),$$

as is easily verified by writing

$$p=\delta+p_0,\quad q=\delta p_0+q_0,\quad r=\delta q_0+r_0,\quad s=\delta r_0,$$

where p_0, q_0, r_0 stand for

$$\alpha+\beta+\gamma,\quad \beta\gamma+\gamma\alpha+\alpha\beta,\quad \alpha\beta\gamma,$$

respectively. The required condition thus is

$$(\alpha\delta-\beta\gamma)(\beta\delta-\gamma\alpha)(\gamma\delta-\alpha\beta)=0,$$

viz. the origin (that is, the fixed point B of the cell) must be at one of the three involution-centres.

Comparing the equation

$$\{x^2+y^2+(1-\lambda)\gamma x+m^2-n^2\}^2=(1+\lambda)^2\{h^2(x^2+y^2)-\gamma^2y^2\}$$

with the equation

$$\{x^2+y^2+Ax+B\}^2=ax^2+ey^2,$$

we have

$$\begin{aligned} A &= (1-\lambda)\gamma,\\ B &= m^2-n^2,\\ a &= (1+\lambda)^2h^2,\\ e &= (1+\lambda)^2(h^2-\gamma^2),\end{aligned}$$

and thence $a-e=(1+\lambda)^2\gamma^2$. Consequently $\dfrac{a-e}{A^2}=\left(\dfrac{1+\lambda}{1-\lambda}\right)^2$, which gives λ: and then $h^2=\dfrac{a}{(1+\lambda)^2}$, $\gamma^2=\dfrac{a-e}{(1+\lambda)^2}$, $m^2-n^2=B$; viz. we thus have the values of λ, h, γ and m^2-n^2 for the description of a given curve $(x^2+y^2+Ax+B)^2=ax^2+ey^2$. In order that the description may be possible, a and $a-e$ must be each of them positive.

For the Cartesian a is $=e$, whence $1+\lambda=0$, and the equation becomes

$$(x^2+y^2+2\gamma x+m^2-n^2)^2=0,$$

which is a twice repeated circle; hence the Cartesian cannot be constructed by means of a cell as above.

To obtain a construction of the Cartesian, it may be remarked that, if a symmetrical bicircular quartic be inverted in regard to an axial focus, viz. if the focus be taken as the centre of inversion, we obtain a Cartesian. The axial foci of the curve

$$(x^2+y^2+Ax+B)^2=ax^2+ey^2$$

are points on the axis, the abscissa $x=\theta$ being determined by the equation

$$e(\theta^2+A\theta+B)^2-a(\theta^2-B)^2-ae\theta^2=0.$$

The equation referred to a focus as origin is therefore

$$\{x^2+y^2+(A+2\theta)x+B+\theta^2\}^2=ax^2+ey^2+2a\theta x+\theta^2;$$

then inverting, viz. for x, y writing $\dfrac{k^2x}{r^2}$, $\dfrac{k^2y}{r^2}$ (k arbitrary), we have, as may be verified the equation of a Cartesian.

The inversion can be performed mechanically by an ordinary Peaucellier-cell; the complete apparatus for the construction of a Cartesian is therefore as in fig. 2, viz. we have a cell BAC as before, B a fixed point, locus of A a circle (for convenience of drawing, the arrangement has been made BAC instead of ABC), and we connect with C a Peaucellier-cell $CA'B'$, arms n', n', m', the fixed point B' being on the axis, which is the line joining B with the centre of the circle described by A. This being so, then A describing a circle, C will describe a symmetrical bicircular quartic, and A' will describe the inverse of this, being in general a like curve; but if the position of B' be properly determined, viz. if B' be at a focus of the first-mentioned quartic,

Fig. 2.

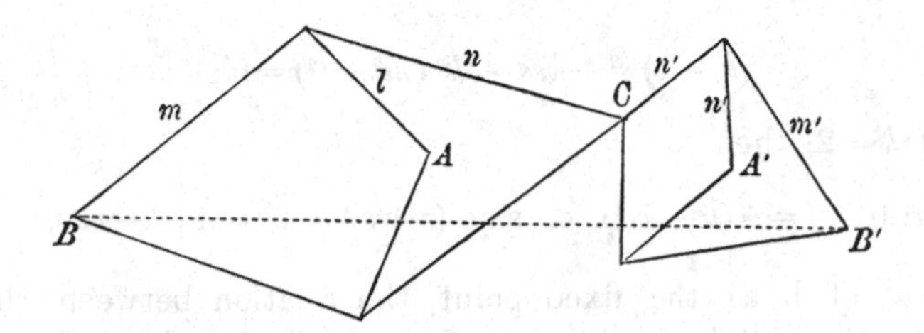

then A' will describe a Cartesian. A further investigation would be necessary in order to determine how to adapt the apparatus to the description of a given Cartesian.

A more convenient mechanical description of a Cartesian is, however, that given in the paper which follows the present one [618].

The equation

$$\{x^2+y^2+(1-\lambda)\gamma x+m^2-n^2\}^2=(1+\lambda)^2\{h^2(x^2+y^2)-\gamma^2y^2\}$$

may also be written

$$\{x^2+y^2+(1-\lambda)\gamma x-\tfrac{1}{2}(1+\lambda)^2(h^2-\gamma^2)+m^2-n^2\}^2$$
$$=(1+\lambda)^2\{\gamma^2x^2-(1-\lambda)(h^2-\gamma^2)\gamma x+\tfrac{1}{4}(1+\lambda)^2(h^2-\gamma^2)^2-(m^2-n^2)(h^2-\gamma^2)\},$$

viz. the equation is now brought into the form

$$(x^2+y^2+Ax+B)^2=ax^2+bx+c.$$

Expressing the coefficients A, B, a, b, c in terms of λ, γ, h, m^2-n^2, it appears by what precedes, that we should have identically $b^2-4ABb+4A^2c=0$, viz. this is the equation which expresses that the origin is an involution-centre.

If, instead of the original cell, we consider a new cell obtained by substituting for the arms PB, BQ, the arms pb, bq, jointed on to the points p, q on the arms CP, CQ respectively, and instead of B, making b the fixed point; then writing $C\text{p}=kn$, $\text{pb}=km$, so that the parameters of the cell are l, m, n, k, and taking $C\text{b}=s$, $\text{b}A=s'$, we have $s = kr$, $s + s' = r + r'$, that is, $r = \frac{s}{k}$, $r' = \frac{k-1}{k}s + s'$. Substituting in the equation between r, r', written for greater convenience in the form

$$(r + r')(rr' + m^2 - l^2) + (l^2 - n^2)\, r' = 0,$$

the relation between s, s' is found to be

$$(s + s')\left(\frac{k-1}{k^2}s^2 + \frac{ss'}{k} + m^2 - l^2\right) + (l^2 - n^2)\left(\frac{k-1}{k}s + s'\right) = 0.$$

On account of the term in s^3, this equation in its general form does not, it would appear, give rise to transformations of much elegance. If, however, $l = n$, then the relation becomes

$$(k - 1)\, s^2 + kss' + k^2 (m^2 - l^2) = 0\,;$$

and in particular, if $k = 2$, then

$$s^2 + 2ss' = 4\,(l^2 - m^2), \text{ or say } (s + s')^2 - s'^2 = 4\,(l^2 - m^2),$$

viz. taking A instead of b as the fixed point, the relation between the radii AC, Ab is $\rho^2 - \rho'^2 = 4\,(l^2 - m^2)$; the cell is in this case Sylvester's "quadratic-binomial extractor."

618.

ON THE MECHANICAL DESCRIPTION OF A CARTESIAN.

[From the *Quarterly Journal of Pure and Applied Mathematics*, vol. XIII. (1875), pp. 328—330.]

SUPPOSE that in two different curves the radius vectors r, r', which belong to the same angle θ, are connected by the equation

$$r^2 + \left(Mr' + N + \frac{P}{r'}\right) r + B = 0\,;$$

then, taking one of the curves to be the circle

$$Mr' + \frac{P}{r'} = A \cos\theta,$$

the other curve is

$$r^2 + (A\cos\theta + N)\, r + B = 0,$$

viz. this is a Cartesian. It perhaps would not be difficult to contrive a mechanical arrangement to connect the radius vectors in accordance with the foregoing equation; but the required result may be obtained equally well by means of a particular case of the relation in question; viz. taking this to be

$$r^2 + (-r' + N)\, r + B = 0,$$

then, taking the one curve to be the circle $r' = -A\cos\theta$, the other curve is the Cartesian,

$$r^2 + (2l\cos\theta + B)\, r + D = 0,\ \text{that is,}\ r^2 + (A\cos\theta + N)\, r + B = 0.$$

The relation between the radius vectors may in this case be written

$$r' = N + r + \frac{B}{r},$$

which can be constructed mechanically by a simple addition to the Peaucellier-cell, viz. if we joint on to C (fig. 1, p. 536) a rod CDA, having a slot, working on a pin at A, so that the rod is thereby kept always in the line BAC, then, making B the fixed point and taking $BA = r$, we have $AC = \frac{m^2 - l^2}{r}$, whence $BC = r + \frac{m^2 - l^2}{r}$, or D being a point at the distance $CD, = \alpha$, from the point C, and denoting BD by r', we have $r' = r + \frac{m^2 - l^2}{r} + \alpha$, which is an equation of the required form; whence, if the point D describe a circle passing through B, then the point A will describe a Cartesian.

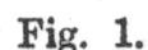

Fig. 1.

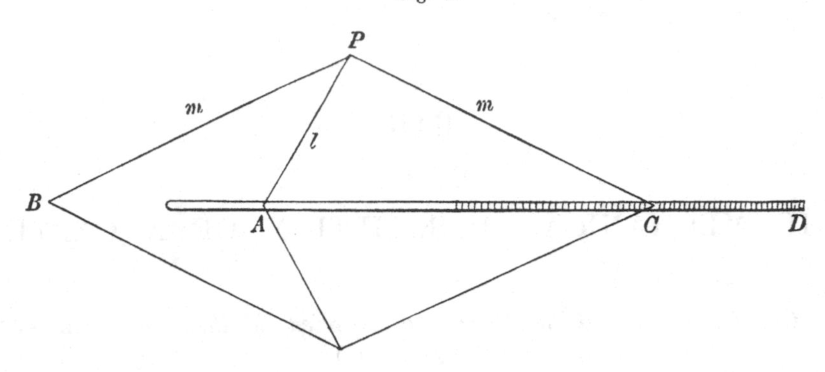

The equation of the Cartesian is $r^2 + (A \cos\theta + N) r + B = 0$, viz. this is $x^2 + y^2 + Ax + B = -N\sqrt{(x^2 + y^2)}$, or writing $N^2 = a$, it is

$$(x^2 + y^2 + Ax + B)^2 = ax^2 + ay^2,$$

which is the form considered in the preceding paper. It may be further observed in regard to it that, starting from the focal equation $r = ls + m$, where r, s are the distances of a point (x, y) of the Cartesian from any two of its three foci, this equation gives $r^2 - l^2s^2 + m^2 = 2mr$, or writing $r^2 = x^2 + y^2$, $s^2 = (x - \alpha)^2 + y^2$, the function on the left-hand is of the form $(1 - l^2)(x^2 + y^2 + Ax + B)$, whence, assuming $\frac{2m}{1 - l^2} = \sqrt{(a)}$, the equation becomes as above

$$(x^2 + y^2 + Ax + B)^2 = a(x^2 + y^2).$$

Taking the distance $r = \sqrt{(x^2 + y^2)}$ to be measured from a given focus, it is easy to see that, no matter which of the other two foci we associate with it, we obtain the *same* equation $(x^2 + y^2 + Ax + B)^2 = a(x^2 + y^2)$; viz. starting with any one focus, we connect with it a determinate circle $x^2 + y^2 + Ax + B = 0$, and a determinate coefficient a, such that taking this focus as the origin, the equation of the curve is

$$(x^2 + y^2 + Ax + B)^2 = a(x^2 + y^2);$$

but there are for the given curve three such forms of equation, according as the origin is taken at one or other of the three foci.

(Addition, Feb. 1875.) It is obviously the same thing, but I find that it is mechanically more convenient to derive the Cartesian from the Limaçon $r' = -N - A\cos\theta$, by the transformation $r' = r + \frac{B}{r}$: I have on this principle constructed an apparatus whereby the Cartesian is described on a rotating board by a pencil moving in a fixed line.

619.

ON AN ALGEBRAICAL OPERATION.

[From the *Quarterly Journal of Pure and Applied Mathematics*, vol. XIII. (1875), pp. 369—375.]

I CONSIDER

$$\Omega F(a, x),$$

an operation Ω performed upon $F(a, x)$ a rational function of (a, x); viz. F being first expanded or regarded as expanded in ascending powers of a, the coefficients of the several powers are then to be expanded or regarded as expanded in ascending powers of x, and the operation consists in the rejection of all negative powers of x.

In the cases intended to be considered, F contains only positive powers of a: but this restriction is not necessary to the theory.

The investigation has reference to the functions $A(x)$ of my "Ninth Memoir on Quantics," *Phil. Trans.*, t. CLXI. (1871), pp. 17—50, [462]; for instance, we there have as regards the covariants of a quadric

$$A(x) - \frac{1}{x^2} A\left(\frac{1}{x}\right) = \frac{1 - x^{-2}}{1 - ax^2 \,.\, 1 - a \,.\, 1 - ax^{-2}},$$

and consequently, in the present notation,

$$A(x) \qquad = \Omega \frac{1 - x^{-2}}{1 - ax^2 \,.\, 1 - a \,.\, 1 - ax^{-2}};$$

by a process of development and summation, the value of this expression was found to be

$$= \frac{1}{1 - ax^2 \,.\, 1 - a^2},$$

and in the other more complicated cases the value of $A(x)$ was found only by trial and verification. What I purpose now to show is that the operation Ω can be performed without any development in an infinite series; or say that it depends on finite algebraical operations only.

It is clear that if $F(a, x)$, considered to be developed as above contains only positive powers of x, then

$$\Omega F(a, x) = F(a, x);$$

and if it contains only negative powers of x, then $\Omega F(a, x) = 0$.

Consider now $\Omega \dfrac{\phi(x)}{x-a}$, where $\phi(x)$ is a function containing only positive powers of x; we have

$$\frac{\phi(x)}{x-a} = \frac{\phi(x)-\phi(a)}{x-a} + \frac{\phi(a)}{x-a},$$

and thence

$$\Omega \frac{\phi(x)}{x-a} = \Omega \frac{\phi(x)-\phi(a)}{x-a} + \Omega \frac{\phi(a)}{x-a}$$

$$= \frac{\phi(x)-\phi(a)}{x-a},$$

since $\dfrac{\phi(x)-\phi(a)}{x-a}$ is a rational and integral function of a, which when developed contains only positive powers of x, and $\dfrac{\phi(a)}{x-a}$ when developed contains only negative powers of x.

Consider next $\Omega \dfrac{\phi(x)}{x^2-a}$, where $\phi(x)$ is a rational and integral function of x; writing this $= f(x^2) + xg(x^2)$, we have

$$\Omega \frac{\phi(x)}{x^2-a} = \Omega \frac{f(x^2)}{x^2-a} + \Omega \frac{xg(x^2)}{x^2-a}$$

$$= \frac{f(x^2)-f(a)}{x^2-a} + \frac{x\{g(x^2)-g(a)\}}{x^2-a}.$$

As regards the last term, notice that

$$\Omega \frac{xg(x^2)}{x^2-a} = \Omega \frac{x\{g(x^2)-g(a)\}}{x^2-a} + \Omega \frac{xg(a)}{x^2-a},$$

in which $\dfrac{x\{g(x^2)-g(a)\}}{x^2-a}$ is a rational and integral function of (a, x), and therefore when developed contains only positive powers of x, while $\dfrac{xg(a)}{x^2-a}$ when developed contains only negative powers of x.

We thus have

$$\Omega \frac{\phi(x)}{x^2-a} = \frac{f(x^2)+xg(x^2)-f(a)-xg(a)}{x^2-a}$$

$$= \frac{\phi(x)-f(a)-xg(a)}{x^2-a}.$$

Similarly, if $\phi(x) = f(x^3) + xg(x^3) + x^2h(x^3)$, then

$$\Omega \frac{\phi(x)}{x^3 - a} = \frac{\phi(x) - f(a) - xg(a) - x^2h(a)}{x^3 - a};$$

and so on.

Consider now the above-mentioned function

$$A(x), = \Omega \frac{1 - x^{-2}}{1 - ax^2 . 1 - a . 1 - ax^{-2}}.$$

Writing

$$\frac{1 - x^{-2}}{1 - ax^2 . 1 - a . 1 - ax^{-2}} = \frac{P}{1 - ax^2} + \frac{Q}{1 - a} + \frac{R}{1 - ax^{-2}},$$

we have

$$P = \left(\frac{1 - x^{-2}}{1 - a . 1 - ax^{-2}}\right)_{a=x^{-2}}, = \frac{1}{1 - x^{-4}}, \quad = \frac{-x^4}{1 - x^4},$$

$$Q = \left(\frac{1 - x^{-2}}{1 - ax^2 . 1 - ax^{-2}}\right)_{a=1}, = \frac{1}{1 - x^2},$$

$$R = \left(\frac{1 - x^{-2}}{1 - ax^2 . 1 - a}\right)_{a=x^2}, = \frac{1 - x^{-2}}{1 - x^4 . 1 - x^2}, \quad = \frac{-1}{x^2(1 - x^4)};$$

that is,

$$\frac{1 - x^{-2}}{1 - ax^2 . 1 - a . 1 - ax^{-2}} = \frac{-x^4}{1 - x^4} \frac{1}{1 - ax^2} + \frac{1}{1 - x^2} \frac{1}{1 - a} - \frac{1}{1 - x^4} \frac{1}{x^2 - a},$$

and thence

$$\Omega \frac{1 - x^{-2}}{1 - ax^2 . 1 - a . 1 - ax^{-2}} = \frac{-x^4}{1 - x^4} \frac{1}{1 - ax^2} + \frac{1}{1 - x^2 . 1 - a} + \Omega \frac{-1}{1 - x^4} \frac{1}{x^2 - a}.$$

Here, as regards the last term,

$$\phi(x^2) = \frac{-1}{1 - x^4}, = f(x^2), \ g(x^2) = 0,$$

$$f(a) = -\frac{1}{1 - a^2},$$

$$\frac{\phi(x^2) - f(a)}{x^2 - a} = \frac{1}{x^2 - a}\left(\frac{-1}{1 - x^4} + \frac{1}{1 - a^2}\right) = -\frac{x^4 - a^2}{x^2 - a . 1 - x^4 . 1 - a^2} = -\frac{x^2 + a}{1 - x^4 . 1 - a^2},$$

and we have

$$\Omega \frac{1 - x^{-2}}{1 - ax^2 . 1 - a . 1 - ax^{-2}} = \frac{-x^4}{1 - x^4 . 1 - ax^2} + \frac{1}{1 - x^2 . 1 - a} - \frac{x^2 + a}{1 - x^4 . 1 - a^2}.$$

The second term is $= \frac{1 + x^2 . 1 + a}{1 - x^4 . 1 - a^2}$: combining this with the third term, the two together are $= \frac{1 + ax^2}{1 - x^4 . 1 - a^2}$.

Hence the value is

$$=\frac{1}{1-x^4}\left(\frac{-x^4}{1-ax^2}+\frac{1+ax^2}{1-a^2}\right),$$

which is

$$=\frac{1}{1-ax^2\,.\,1-a^2},$$

being, in fact, the expression for this function when decomposed into partial fraction of the denominators $1-ax^2$ and $1-a^2$ respectively. Hence finally

$$A(x)=\Omega\,\frac{1-x^{-2}}{1-ax^2\,.\,1-a\,.\,1-ax^{-2}}=\frac{1}{1-ax^2\,.\,1-a^2},$$

as it should be.

For the cubic function, we have

$$A(x)=\Omega\,\frac{1-x^{-2}}{1-ax^3\,.\,1-ax\,.\,1-ax^{-1}\,.\,1-ax^{-3}};$$

the function operated upon, when decomposed into partial fractions, is

$$=\frac{x^{10}}{1-x^4\,.\,1-x^6}\,\frac{1}{1-ax^3}-\frac{x^4}{1-x^2\,.\,1-x^4}\,\frac{1}{1-ax}$$

$$+\frac{x}{1-x^2\,.\,1-x^4}\,\frac{1}{x-a}+\frac{-x}{1-x^4\,.\,1-x^6}\,\frac{1}{x^3-a}.$$

Hence we require

$$\Omega\,\frac{x}{1-x^2\,.\,1-x^4}\,\frac{1}{x-a}+\Omega\,\frac{-x}{1-x^4\,.\,1-x^6}\,\frac{1}{x^3-a}.$$

The first of these is

$$=\frac{1}{x-a}\left\{\frac{x}{1-x^2\,.\,1-x^4}-\frac{a}{1-a^2\,.\,1-a^4}\right\},$$

which is

$$=\frac{1}{1-x^2\,.\,1-x^4\,.\,1-a^2\,.\,1-a^4}\left\{\begin{array}{l}1\\+x\,(a+a^3-a^5)\\+x^2\,(a^2-a^4)\\+x^3\,(a-a^3)\\-x^4a^2\\-x^5a\end{array}\right\}.$$

As regards the second, the function operated on may be expressed in the form

$$\frac{-x^9-x-x^5}{1-x^{12}\,.\,1-x^6}\,\frac{1}{x^3-a},$$

whence $f(a)$, $g(a)$, $h(a)$, and therefore $f(a)+xg(a)+x^2h(a)$, respectively, are $=-a^3$, -1, $-a$, $-a^3-x-ax^2$, each divided by $1-a^2.1-a^4$; or the term is

$$=\frac{1}{x^3-a}\left\{\frac{-x}{1-x^4.1-x^6}+\frac{a^3+x+ax^2}{1-a^2.1-a^4}\right\},$$

which is

$$=\frac{1}{1-x^4.1-x^6.1-a^2.1-a^4}\left\{\begin{array}{l}-a^2\\+x(-a-a^3+a^5)\\+x^2.-1\\+x^3.-a\\+x^4.(a^4-1)\\+x^5.0\\+x^6.a^2\\+x^7.a^3\\+x^8.1\\+x^9.a\end{array}\right\}.$$

To combine the two terms, we multiply the numerator and denominator of the first by $1+x^2+x^4$, thereby reducing its denominator to $1-x^4.1-x^6.1-a^2.1-a^4$, the denominator of the second term; then the sum of the numerators is found to be

$$\begin{array}{l}=1-a^2\\+x\ (a^2-a^4)\\+x^3(a\ -a^3)\\+x^5(a\ -a^3)\\+x^6(a^2-a^4)\\+x^8(1\ -a^2),\end{array}\qquad \text{viz. this is } =(1-a^2)\left\{\begin{array}{l}\ 1\\+a^2x^2\\+(a+a^3)\,x^3\\+(a+a^3)\,x^5\\+a^2x^6\\+x^8.\end{array}\right.$$

Hence we have

$$A(x)=\Omega\frac{1-x^{-2}}{1-ax^3.1-ax.1-ax^{-1}.1-ax^{-3}}$$

$$=\frac{x^{10}}{1-x^4.1-x^6}\frac{1}{1-ax^3}$$

$$+\frac{-x^4}{1-x^2.1-x^4}\frac{1}{1-ax}$$

$$+\frac{1+x^8+(x^3+x^5)\,a+(x^2+x^6)\,a^2+(x^3+x^5)\,a^3}{1-x^4.1-x^6}\frac{1}{1-a^4},$$

which is, in fact, the expression for $\dfrac{1-ax+a^2x^2}{1-ax^3.1-ax.1-a^4}$ decomposed into partial fractions with the denominators $1-ax^3$, $1-ax$, $1-a^4$ respectively. This is most easily seen by completing the decomposition, viz. we have

$$4\{1+x^8+(x^3+x^5)a+(x^2+x^6)a^2+(x^3+x^5)a^3\}$$

$$=(1+x^3)^2(1+x^2)(1+a)(1+a^2)+(1-x^3)^2(1+x^2)(1-a)(1+a^2)+2(1-x^2)(1-x^6)(1-a^2),$$

and thence the expression is

$$\begin{aligned}
&= \frac{x^{10}}{1-x^4\,.\,1-x^6}\,\frac{1}{1-ax^3}\\
&\quad+\frac{-x^4}{1-x^2\,.\,1-x^4}\,\frac{1}{1-ax}\\
&\quad+\tfrac{1}{4}\,\frac{1+x^3}{1-x^2\,.\,1-x^3}\,\frac{1}{1-a}+\tfrac{1}{4}\,\frac{1-x^3}{1-x^2\,.\,1+x^3}\,\frac{1}{1+a}+\tfrac{1}{2}\,\frac{1}{1+x^2}\,\frac{1}{1+a^2}\\
&= \frac{1-ax+a^2x^2}{1-ax^3\,.\,1-ax\,.\,1-a^4},
\end{aligned}$$

as above. Hence finally

$$\begin{aligned}
A(x) &= \Omega\,\frac{1-x^{-2}}{1-ax^3\,.\,1-ax\,.\,1-ax^{-1}\,.\,1-ax^{-3}}\\
&= \frac{1-ax+a^2x^2}{1-ax^3\,.\,1-ax\,.\,1-a^4}\\
&= \frac{1+a^3x^3}{1-ax^3\,.\,1-a^2x^2\,.\,1-a^4}\\
&= \frac{1-a^6x^6}{1-ax^3\,.\,1-a^2x^2\,.\,1-a^3x^3\,.\,1-a^4}.
\end{aligned}$$

620.

CORRECTION OF TWO NUMERICAL ERRORS IN SOHNKE'S PAPER RESPECTING MODULAR EQUATIONS.

[From the *Journal für die reine und angewandte Mathematik* (Crelle), t. LXXXI. (1876), p. 229.]

IN Sohnke's paper "Aequationes modulares pro transformatione functionum ellipticarum," *Crelle*, t. XVI. (1837), there is, on p. 113, an obvious error in the expression of u^6, viz. the term q^{18} is given with the same numerical coefficient as it had in u^5: this remark was made to me by Mr W. Barrett Davis, who finds that the term of u^6 should be

$$+ 13569463\, q^{18}.$$

In the expression of u^{18} (*l. c.*, p. 115), I had remarked that, in the coefficient of q^{16}, a figure must have dropped out. Mr Davis has verified this, and finds that the figure omitted is a 1 in the unit's place, and thus that the correct value* is

$$+ 80177033781\, q^{16}.$$

Cambridge, 26 October, 1875.

* [The former of these numbers should replace the number 15063859 in the Table, p. 128 of this volume; the latter has been introduced in the Table, p. 129.]

621.

ON THE NUMBER OF THE UNIVALENT RADICALS C_nH_{2n+1}.

[From the *Philosophical Magazine,* series 5, vol. III. (1877), pp. 34, 35.]

I HAVE just remarked that the determination is contained in my paper "On the Analytical Forms called Trees, &c.," *British Association Report*, 1875, [610]; in fact, in the form C_nH_{2n+1}, there is one carbon atom distinguished from the others by its being combined with (instead of 4, only) 3 other atoms; viz. these are 3 carbon atoms, 2 carbon atoms and 1 hydrogen atom, or else 1 carbon atom and 2 hydrogen atoms (CH_3, methyl, is an exception; but here the number is $=1$). The number of carbon atoms thus combined with the first-mentioned atom is the number of main branches, which is thus $=3$, 2, or 1; hence we have, number of radicals C_nH_{2n+1} is $=$

No. of carbon root-trees C_n with one main branch,
+ No. of " " with two main branches,
+ No. of " " with three main branches;

and the three terms for the values $n=1$ to 13 are given in Table VII. (pp. 454, 455 of this volume) of the paper referred to.

Thus, if $n=5$, an extract from the Table (p. 454 of this volume), is

Index x, or number of knots	Index t, or number of main branches	Altitude					
		0	1	2	3	4	
5	1			1	2	1	4
	2			2	1		3
	3			1			1
	4		1				1
	Total ...		1	4	3	1	9

and the number of the radicals C_5H_{11} (isomeric amyls) is $4+3+1=8$: or, what is the same thing, it is $9-1$, the corner-total less the number immediately above it. The tree-forms corresponding to the numbers 1, 2, 1; 2, 1; 1 in the body of the Table are the trees 2 to 9 in the figure, p. 428 of this volume.

The numbers of the radicals C_nH_{2n+1}, as obtained from the Table in the manner just explained, are :—

$n=$	Number of radicals C_nH_{2n+1}.				
1	1		=	1	Methyl.
2	1			1	Ethyl.
3	1			1	Propyl.
4	4			4	Butyls.
5	9	− 1		8	Amyls.
6	18	− 1		17	Hexyls.
7	42	− 3		39	Heptyls.
8	96	− 7		89	Octyls.
9	229	− 18		211	Nonyls.
10	549	− 42		507	Decyls.
11	1346	− 108		1238	Undecyls.
12	3326	− 269		3057	Dodecyls.
13	8329	− 691		7638	Tridecyls.

The question next in order, that of the determination of the number of the bivalent radicals C_nH_{2n}, might be solved without much difficulty.

Cambridge, November 20, 1876.

622.

ON A SYSTEM OF EQUATIONS CONNECTED WITH MALFATTI'S PROBLEM.

[From the *Proceedings of the London Mathematical Society*, vol. VII. (1875—1876), pp. 38—42. Read December 9, 1875.]

I CONSIDER the equations

$$\begin{aligned} X, &= by^2 + cz^2 - 2fyz - a(bc - f^2), = 0, \\ Y, &= cz^2 + ax^2 - 2gzx - b(ca - g^2), = 0, \\ Z, &= ax^2 + by^2 - 2hxy - c(ab - h^2), = 0, \end{aligned}$$

where the constants (a, b, c, f, g, h) are such that

$$K, = abc - af^2 - bg^2 - ch^2 + 2fgh, = 0.$$

Hence, writing as usual (A, B, C, F, G, H) to denote the inverse coefficients

$$(bc - f^2,\ ca - g^2,\ ab - h^2,\ gh - af,\ hf - bg,\ fg - ch),$$

we have $(A, B, C, F, G, H \between x, y, z)^2 =$ the square of a linear function, $= (\alpha x + \beta y + \gamma z)^2$ suppose; that is,

$$(A, B, C, F, G, H) = (\alpha^2, \beta^2, \gamma^2, \beta\gamma, \gamma\alpha, \alpha\beta).$$

It is to be shown that the three quadric surfaces $X = 0$, $Y = 0$, $Z = 0$ intersect in a conic Θ lying in the plane $a\alpha x + b\beta y + c\gamma z = 0$, and in two points I, J; or more completely, that

the surfaces Y, Z meet in the conic Θ and a conic P,
" Z, X " " " Q,
" X, Y " " " R,

where the conics P, Q, R each pass through the two points I, J, and meet the conic Θ in two points, viz.,

the conics P, Θ meet in two points P_1, P_2,
,, Q, Θ ,, ,, Q_1, Q_2,
,, R, Θ ,, ,, R_1, R_2.

For this purpose, writing

$$\nabla = Aa - Ff, \ = Bb - Gg, \ = Cc - Hh, \ = abc - fgh,$$
$$\Omega = \tfrac{1}{2}(X + Y + Z), \ = ax^2 + by^2 + cz^2 - fyz - gzx - hxy - \nabla,$$
$$\theta = a\alpha x + b\beta y + c\gamma z,$$
$$\xi = \frac{Aa}{\alpha}x + \frac{Ff}{\beta}y + \frac{Ff}{\gamma}z,$$
$$\eta = \frac{Gg}{\alpha}x + \frac{Bb}{\beta}y + \frac{Gg}{\gamma}z,$$
$$\zeta = \frac{Hh}{\alpha}x + \frac{Hh}{\beta}y + \frac{Cc}{\gamma}z,$$

then we have identically

$$aA\Omega - \nabla X = \theta\xi,$$
$$bB\Omega - \nabla Y = \theta\eta,$$
$$cC\Omega - \nabla Z = \theta\zeta.$$

In fact, the first of these equations, written at full length, is

$$aA(ax^2 + by^2 + cz^2 - fyz - gzx - hxy - \nabla) - \nabla(by^2 + cz^2 - 2fyz - aA)$$
$$= (a\alpha x + b\beta y + c\gamma z)\left(\frac{Aa}{\alpha}x + \frac{Ff}{\beta}y + \frac{Ff}{\gamma}z\right),$$

where on the left-hand side the constant term is $= 0$. Comparing, first, the coefficients of x^2, y^2, z^2, on the two sides respectively, these are Aa^2, $(Aa - \nabla)b$, $(Aa - \nabla)c$, and Aa^2, Ffb, Ffc, which are equal. Comparing the coefficients of yz, zx, xy, the equations which remain to be verified are

$$-(aA - 2\nabla)f = Ff\left(c\frac{\gamma}{\beta} + b\frac{\beta}{\gamma}\right),$$
$$-aAg = Faf\frac{\alpha}{\gamma} + Aac\frac{\gamma}{\alpha},$$
$$-aAh = Faf\frac{\alpha}{\beta} + Aab\frac{\beta}{\alpha};$$

or, as these may be written,

$$-(aA - 2\nabla)\beta\gamma = F(c\gamma^2 + b\beta^2),$$
$$-Ag\gamma\alpha = Ff\alpha^2 + Ac\gamma^2,$$
$$-Ah\alpha\beta = Ff\alpha^2 + Ab\beta^2;$$

and, substituting for α^2, β^2, γ^2, $\beta\gamma$, $\gamma\alpha$, $\alpha\beta$ their values, these may be verified without difficulty.

It thus appears that the equations of the three quadric surfaces may be written in the form

$$aA\Omega - \theta\xi = 0,\ bB\Omega - \theta\eta = 0,\ cC\Omega - \theta\zeta = 0;$$

and we thus obtain the conics Θ, P, Q, R as the intersections of the surface $\Omega = 0$ by the four planes

$$\theta = 0,\quad \frac{\eta}{Bb} - \frac{\zeta}{Cc} = 0,\quad \frac{\zeta}{Cc} - \frac{\xi}{Aa} = 0,\quad \frac{\xi}{Aa} - \frac{\eta}{Bb} = 0,$$

respectively. There is no difficulty in verifying that the conics intersect as mentioned above, and that the coordinates of their points of intersection are

$$P,\ P_1 : \left(\sqrt{bc},\ \frac{ch}{\sqrt{bc}},\ \frac{bg}{\sqrt{bc}}\right),\ \left(-\sqrt{bc},\ -\frac{ch}{\sqrt{bc}},\ -\frac{bg}{\sqrt{bc}}\right);$$

$$Q,\ Q_1 : \left(\frac{ch}{\sqrt{ca}},\ \sqrt{ca},\ \frac{af}{\sqrt{ca}}\right),\ \left(-\frac{ch}{\sqrt{ca}},\ -\sqrt{ca},\ -\frac{af}{\sqrt{ca}}\right);$$

$$R,\ R_1 : \left(\frac{bg}{\sqrt{ab}},\ \frac{af}{\sqrt{ab}},\ \sqrt{ab}\right),\ \left(-\frac{bg}{\sqrt{ab}},\ -\frac{af}{\sqrt{ab}},\ -\sqrt{ab}\right);$$

$$I,\ J\ : (f,\ g,\ h),\qquad (-f,\ -g,\ -h).$$

In a paper "On a system of Equations connected with Malfatti's Equation and on another Algebraical System," *Camb. and Dublin Math. Journal*, vol. IV. (1849), pp. 270—275, [79], I considered a system of equations which, writing therein $\theta = 1$, and changing the signs of (f, g, h), are the equations here considered, $X = 0$, $Y = 0$, $Z = 0$: only the constants (a, b, c, f, g, h) are not connected by the equation $K = 0$, but are perfectly arbitrary. The three quadric surfaces intersect therefore in 8 points, the coordinates of which are obtained in the paper just referred to, viz. making the above changes of notation, the values are

$$x^2 = \frac{1}{2a}(abc + fgh - f\sqrt{BC} + g\sqrt{CA} + h\sqrt{AB}),$$

$$y^2 = \frac{1}{2b}(abc + fgh + f\sqrt{BC} - g\sqrt{CA} + h\sqrt{AB}),$$

$$z^2 = \frac{1}{2c}(abc + fgh + f\sqrt{BC} + g\sqrt{CA} - h\sqrt{AB}),$$

$$yz = \tfrac{1}{2}(gh + af + \sqrt{BC}),$$

$$zx = \tfrac{1}{2}(hf + bg + \sqrt{CA}),$$

$$xy = \tfrac{1}{2}(fg + ch + \sqrt{AB});$$

where the radicals are such that $\sqrt{BC}\,.\,\sqrt{CA}\,.\,\sqrt{AB} = ABC$, so that the system $(x^2, y^2, z^2, yz, zx, xy)$ has four values only, and consequently (x, y, z) has eight values.

It is very remarkable that, introducing the foregoing relation $K=0$, there is not in the solution any indication that the intersection has become a conic and two points, but the solution gives eight determinate points, viz. the before-mentioned points P, P_1, Q, Q_1, R, R_1, and I, J.

To develope the solution, remark that, in virtue of the relation in question, we have

$$\sqrt{BC} = \pm F, \quad \sqrt{CA} = \pm G, \quad \sqrt{AB} = \pm H,$$

where the signs must be such that the product is $= FGH$ (viz. they must be all positive, or else one positive and the other two negative); for, taking the product to be $+FGH$, the equations give

$$0 = ABC - FGH,$$

that is,

$$0 = A(BC - F^2) - F(GH - AF), \ = K(Aa - Ff), \ = K\nabla,$$

which is true in virtue of the relation $K=0$. Taking the signs all positive, we have for $x^2, y^2, z^2, yz, zx, xy$, the values $f^2, g^2, h^2, gh, hf, fg$, viz. we have thus the points

$$(f, g, h), \quad (-f, -g, -h),$$

which are the points I, J. Taking the signs one positive and the other two negative, say $\sqrt{BC} = F$, $\sqrt{CA} = -G$, $\sqrt{AB} = -H$, we find for $x^2, y^2, z^2, yz, zx, xy$ the values $bc, \frac{ch^2}{b}, \frac{bg^2}{c}, gh, bg, ch$, viz. we have thus the points

$$\left(\sqrt{bc}, \frac{ch}{\sqrt{bc}}, \frac{bg}{\sqrt{bc}}\right), \quad \left(-\sqrt{bc}, -\frac{ch}{\sqrt{bc}}, -\frac{bg}{\sqrt{bc}}\right),$$

which are the points P, P_1; and the other two combinations of sign give of course the points Q, Q_1 and R, R_1 respectively.

If the coefficients (a, b, c, f, g, h), instead of the foregoing relation $K=0$, satisfy the relation

$$abc - af^2 - bg^2 - ch^2 - 2fgh = 0, \text{ say } K' = 0,$$

the quadric surfaces intersect in 8 points, the coordinates of which are given by the general formulæ: but the expressions assume a very simple form. Writing for shortness

$$F' = gh + af, \quad G' = hf + bg, \quad H' = fg + ch,$$

then, in virtue of the assumed relation,

$$\sqrt{BC} = \pm F', \quad \sqrt{CA} = \pm G', \quad \sqrt{AB} = \pm H',$$

where the signs are such that the product of the three terms is positive, viz. they must be all positive, or else one positive and the other two negative. For, assuming it to be so, we have

$$0 = ABC - F'G'H',$$

that is,

$$0 = A(BC - F'^2) - F'(G'H' - AF'),$$
$$= K'(Aa + F'f), \ = K'(abc + fgh);$$

which is right, in virtue of the relation $K' = 0$. Taking the signs all positive, we find for $(x^2, y^2, z^2, yz, zx, xy)$ the values (A, B, C, F', G', H'), giving two points of intersection

$$\left(\sqrt{A}, \frac{H'}{\sqrt{A}}, \frac{G'}{\sqrt{A}}\right) \text{ and } \left(-\sqrt{A}, -\frac{H'}{\sqrt{A}}, -\frac{G'}{\sqrt{A}}\right).$$

Taking the signs one positive and the other two negative, say

$$\sqrt{BC} = F', \quad \sqrt{CA} = -G', \quad \sqrt{AB} = -H',$$

we find for $(x^2, y^2, z^2, yz, zx, xy)$ the values

$$\left(0, \frac{Cc}{b}, \frac{Bb}{c}, F', 0, 0\right),$$

viz. we have thus two intersections

$$\left(0, \sqrt{\frac{Cc}{b}}, F'\sqrt{\frac{b}{Cc}}\right), \ \left(0, -G'\sqrt{\frac{Cc}{b}}, -F'\sqrt{\frac{b}{Cc}}\right);$$

and the other combinations of signs give the remaining two pairs of intersections

$$\left(G'\sqrt{\frac{c}{Aa}}, 0, \sqrt{\frac{Aa}{c}}\right), \ \left(-G'\sqrt{\frac{c}{Aa}}, 0, -\sqrt{\frac{Aa}{c}}\right),$$

and

$$\left(\sqrt{\frac{Bb}{a}}, H'\sqrt{\frac{a}{Bb}}, 0\right), \ \left(-\sqrt{\frac{Bb}{a}}, -H'\sqrt{\frac{a}{Bb}}, 0\right).$$

But the most convenient statement of the result is that the values of $(ax^2, by^2, cz^2, yz, zx, xy)$, for the four pairs of points respectively, are

$$\begin{array}{llllll} (aA, & bB, & cC, & F', & G', & H'), \\ (\ 0\ , & cC, & bB, & F', & 0, & 0\ ), \\ (cC, & 0\ , & aA, & 0, & G', & 0\ ), \\ (bB, & aA, & 0\ , & 0, & 0, & H'); \end{array}$$

there is no difficulty in substituting these values in the original equations, and in verifying that the equations are in each case satisfied.

623.

ON THREE-BAR MOTION.

[From the *Proceedings of the London Mathematical Society*, vol. VII. (1875—1876), pp. 136—166. Read March 10, 1876.]

THE discovery by Mr Roberts of the triple generation of a Three-Bar Curve, throws a new light on the whole theory, and is a copious source of further developments *. The present paper gives in its most simple form the theorem of the triple generation; it also establishes the relation between the nodes and foci; and it contains other researches. I have made on the subject a further investigation, which I give in a separate paper, "On the Bicursal Sextic," [624]; but the two papers are intimately related and should be read in connection.

The Three-Bar Curve is derived from the motion of a system of three bars of given lengths pivoted to each other, and to two fixed points, so as to form the three sides of a quadrilateral, the fourth side of which is the line joining the two fixed points; the curve is described by a point rigidly connected with the middle bar; or, what is more convenient, we take the middle bar to be a triangle pivoted at the extremities of the base to the other two bars (say, the radial bars), and having its vertex for the describing point.

Including the constants of position and magnitude, the Three-Bar Curve thus depends on nine parameters; viz. these are the coordinates of the two fixed points, the lengths of the connecting bars, and the three sides of the triangle. It is known that the curve is a tricircular trinodal sextic, and the equation of such a curve contains $27 - 6 - 6 - 3, = 12$ constants. Imposing on the curve the condition that the three nodes lie upon a given curve, the number of constants is reduced to $12 - 3, = 9$: and it is in this way that the Three-Bar Curve is distinguished from the general tricircular

* See his paper "On Three-Bar Motion in Plane Space," *l.c.*, vol. VII., pp. 15—23, which contains more than I had supposed of the results here arrived at. There is no question as to Mr Roberts' priority in all his results.

trinodal sextic; viz. in the Three-Bar Curve the two fixed points are foci, and they determine a third focus*; and the condition is that the nodes are situate on the circle through the 3 foci.

The nodes are two of them arbitrary points on the circle; and the third of them is a point such that, measuring the distances along the circle from any fixed point of the circumference, the sum of the distances of the nodes is equal to the sum of the distances of the foci. Considering the two fixed points as given, the curve depends upon five parameters, viz. the lengths of the connecting bars and the sides of the triangle. Taking the *form* of the triangle as given, there are then only three parameters, say the lengths of the connecting bars and the base of the triangle; in this case the third focus is determined, and therefore the circle through the three foci; we may then take two of the nodes as given points on this circle, and thereby establish two relations between the three parameters, in fact, we thereby determine the differences of the squares of the lengths in question: but the third node is then an absolutely determined point on the circle, and we cannot make use of it for completing the determination of the parameters; viz. one parameter remains arbitrary. Or, what is the same thing, given the three foci and also the three nodes, consistently with the foregoing conditions, viz. the nodes lie in the centre through the three foci, the sum of the distances of the nodes being equal to the sum of the distances of the foci: we have a singly infinite series of three-bar curves.

In reference to the notation proper for the theorem of the triple generation, I shall, when only a single node of generation is attended to, take the curve to be generated as shown in the annexed Figure 1; viz. O is the generating point, OC_1B_1 the triangle, C, B the fixed points, CC_1 and BB_1 the radial bars. The sides of the

Fig. 1.

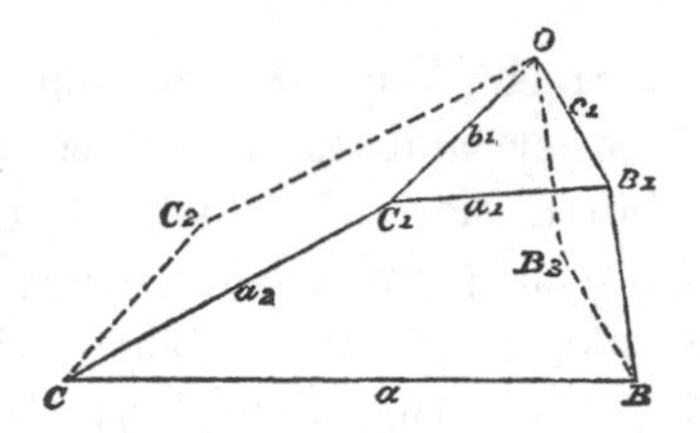

triangle are a_1, b_1, c_1; its angles are $O, = A$, $B_1, = B$, $C_1, = C$: the bars CC_1 and BB_1 are $= a_2$ and a_3 respectively, and the distance CB is $= a$. The sides a_1, b_1, c_1 may be put $= k_1(\sin A, \sin B, \sin C)$, and the lines a_1, a_2, $a_3 = (k_1, k_2, k_3)\sin A$, viz. the original data a_1, b_1, c_1, a_1, a_2, a_3, may be replaced by the angles A, B, C $(A+B+C=\pi)$ and the lines k_1, k_2, k_3. And it is convenient to mention at once that the third focus A is then a point such that ABC is a triangle similar and congruent to OB_1C_1.

* A focus is a point, given as the intersection of a tangent to the curve from one circular point at infinity with a tangent from the other circular point at infinity; if the circular points are simple or multiple points on the curve, then the tangent or tangents *at* a circular point should be excluded from the tangents *from* the point; and the intersection of two such tangents at the two circular points respectively is not an ordinary focus; but, as the points in question are the only kind of foci occurring in the present paper, I have in the text called them foci.

It may be remarked that, producing CC_1 and BB_1 to meet in a point α, this is the centre of instantaneous rotation of the triangle, and therefore αO is the normal to the curve at O.

I proceed to show that the three nodes F, G, H are in the circle circumscribed about ABC, and that their positions are such that (the distances being measured along the circle as before) we have the property, Sum of the distances of F, G, H is equal to the Sum of the distances of A, B, C.

Supposing O to be at a node F, we have then the two equal triangles FB_1C_1, $FB_1'C_1'$, such that C_1, C_1' are equidistant from C, and B_1, B_1' equidistant from B. Hence the angles B_1FB_1', C_1FC_1' are equal; consequently the halves of these angles CFC_1' and

Fig. 2.

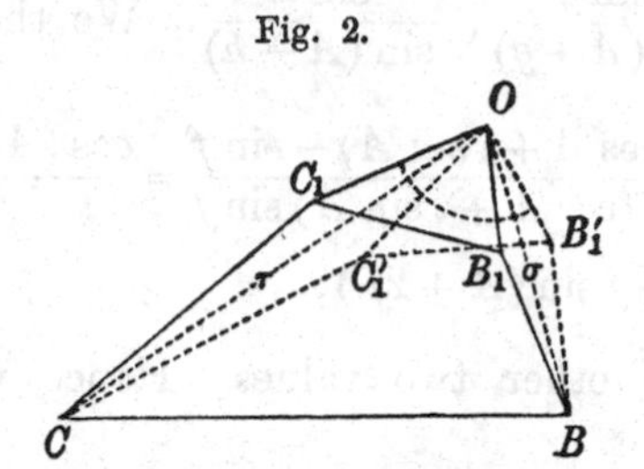

BFB_1' are equal; whence the angle CFB is equal to the angle $C_1'FB_1'$, that is, to the angle A; or F lies on a circle through B, C such that the segment upon BC contains the angle A, that is, upon the circle through A, B, C. To complete the investigation of the nodes, suppose $CF = \tau$, $BF = \sigma$: then the condition $\angle CFC_1' = \angle BFB_1'$ gives

$$\frac{b_1^2 + \tau^2 - a_2^2}{2b_1\tau} = \frac{c_1^2 + \sigma^2 - a_3^2}{2c_1\sigma},$$

that is,

$$c_1\sigma (b_1^2 + \tau^2 - a_2^2) - b_1\tau (c_1^2 + \sigma^2 - a_3^2) = 0;$$

and the condition that F is on the circle gives

$$\sigma^2 + \tau^2 - 2\sigma\tau \cos A = a^2.$$

These equations give six values of (σ, τ) corresponding in pairs to each other; viz. if (σ_1, τ_1) is a solution, then $(-\sigma_1, -\tau_1)$ is also a solution; and to each pair of solutions corresponds a single point on the circle, viz. we have thus the three nodes F, G, H.

Writing the foregoing equation in the form

$$\{c_1 (b_1^2 - a_2^2)\, \sigma - b_1 (c_1^2 - a_3^2)\, \tau\} (\sigma^2 + \tau^2 - 2\sigma\tau \cos A) + a^2 (c_1\sigma\tau^2 - b_1\sigma^2\tau) = 0,$$

and putting the left-hand side $= M(\sigma - p_1\tau)(\sigma - p_2\tau)(\sigma - p_3\tau)$; then, if α, β, γ denote $\cos A + i \sin A$, $\cos B + i \sin B$, $\cos C + i \sin C$ respectively, putting first $\sigma = \alpha\tau$ and next $\sigma = \frac{\tau}{\alpha}$, and dividing one of these results by the other, we find

$$\frac{c_1 - b_1\alpha}{c_1\alpha - b_1} = \frac{\alpha - p_1 \,.\, \alpha - p_2 \,.\, \alpha - p_3}{1 - p_1\alpha \,.\, 1 - p_2\alpha \,.\, 1 - p_3\alpha}.$$

The left-hand side is here

$$=\frac{\sin C-\alpha\sin B}{\alpha\sin C-\sin B}=\frac{\sin C-\sin B(\cos A+i\sin A)}{\sin C(\cos A+i\sin A)-\sin B}$$

$$=\frac{\sin A(\cos B-i\sin B)}{-\sin A(\cos C-i\sin C)}=-\frac{\gamma}{\beta},$$

or the equation is

$$\frac{\alpha-p_1\,.\,\alpha-p_2\,.\,\alpha-p_3}{1-p_1\alpha\,.\,1-p_2\alpha\,.\,1-p_3\alpha}=-\frac{\gamma}{\beta}.$$

Also, writing f for the angle FCB, we have $\sigma=\frac{\sin f}{\sin(A+f)}\tau$, viz. the values of p_1, p_2, p_3 are $\frac{\sin f}{\sin(A+f)}$, $\frac{\sin g}{\sin(A+g)}$, $\frac{\sin h}{\sin(A+h)}$. We thence find

$$\frac{\alpha-p_1}{1-\alpha p_1}=\frac{\sin(A+f)(\cos A+i\sin A)-\sin f}{\sin(A+f)-(\cos A+i\sin A)\sin f}=\frac{\cos(A+f)+i\sin(A+f)}{\cos f-i\sin f}$$

$$=\cos(A+2f)+i\sin(A+2f);$$

with the like values for the other two values. Hence, writing also

$$-\frac{\gamma}{\beta}=-\cos(C-B)-i\sin(C-B)=\cos(\pi+C-B)+i\sin(\pi+C-B),$$

the equation becomes

$$\cos(3A+2f+2g+2h)+i\sin(3A+2f+2g+2h)=\cos(\pi+C-B)+i\sin(\pi+C-B),$$

that is,

$$3A+2f+2g+2h=\pi+C-B,$$

or, what is the same thing,

$$2f+2g+2h=\pi+C-B-3A.$$

Fig. 3.

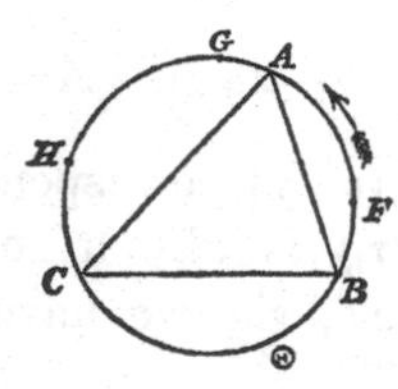

Reckoning the angles round the centre from a point Θ on the circumference, if A', B', C', F', G', H' are the angles belonging to the points A, B, C, F, G, H respectively, then

$$A'=\lambda+2C,\qquad F'=\lambda+2f,$$
$$B'=\lambda,\qquad G'=\lambda+2g,$$
$$C'=\lambda+2C+2B,\quad H'=\lambda+2h;$$

and therefore

$$A'+B'+C'=3\lambda+4C+2B,\quad F'+G'+H'=3\lambda+2f+2g+2h,\quad =3\lambda+\pi+C-B-3A;$$

that is, $A'+B'+C'-F'-G'-H'=-\pi+3(A+B+C)$; or, omitting an angle 2π, this is $A'+B'+C'=F'+G'+H'$, the equation which determines the relation between the three nodes on the circle ABC.

Reverting to the equation $c_1\sigma(b_1^2+\tau^2-a_2^2)-b_1\tau(c_1^2+\sigma^2-a_3^2)=0$, which belongs to a node: if we consider the form of the triangle as given, and write b_1, $c_1=k_1\sin B$, $k_1\sin C$, this becomes

$$\sigma\sin C(b_1^2-a_2^2)-\tau\sin B(c_1^2-a_3^2)+\sigma\tau(\tau\sin C-\sigma\sin B)=0;$$

viz. considering the node as given, then the values of σ, τ are given, and the equation establishes a relation between the values of $b_1^2-a_2^2$ and $c_1^2-a_3^2$*. If a second node be given, we have a second relation between these same quantities, and the two equations give the values of the two quantities, viz. the values of $k_1^2\sin^2 B-k_2^2\sin^2 A$, $k_1^2\sin^2 C-k_3^2\sin^2 A$, or, what is the same thing, the value of $\dfrac{k_1^2}{\sin^2 A}-\dfrac{k_2^2}{\sin^2 B}$, $\dfrac{k_1^2}{\sin^2 A}-\dfrac{k_3^2}{\sin^2 B}$. It thus appears that, if l_1, l_2, l_3 are any values of k_1, k_2, k_3 belonging to a given system of three nodes, the general values of k_1, k_2, k_3 belonging to the same system of three nodes are

$$k_1^2=l_1^2+u\sin^2 A,\quad k_2^2=l_2^2+u\sin^2 B,\quad k_3^2=l_3^2+u\sin^2 C,$$

where u is an arbitrary constant.

It may be added that there will be a node at B, if the equation is satisfied by $\tau=0$, $\sigma=a$, for the condition is $b_1^2-a_2^2=0$: that is, if $\dfrac{k_1}{\sin A}=\dfrac{k_2}{\sin B}$; similarly, there will be a node at C, if $c_1^2-a_3^2=0$, that is, if $\dfrac{k_1}{\sin A}=\dfrac{k_3}{\sin B}$; and a node at A, if $\dfrac{k_2}{\sin C}=\dfrac{k_3}{\sin B}$. If two of these equations are satisfied, the third equation is also satisfied, viz. we then have $\dfrac{k_1}{\sin A}=\dfrac{k_2}{\sin B}=\dfrac{k_3}{\sin C}$; and the three nodes coincide with the three foci respectively.

If, in Figure 2 (p. 553), the points C_1, C_1' coincide on the line CF, and therefore also the points B_1, B_1' coincide on the line BF, then, instead of a node at F, we have a cusp. We have in this case a triangle the sides of which are a_2+b_1, a_0+c_1, a, and the included angle between the first two sides is $=A$: we have, therefore, the relation

$$a^2=(a_2+b_1)^2+(a_3+c_1)^2-2(a_2+b_1)(a_3+c_1)\cos A.$$

Substituting herein for a, a_2, b_1, &c., the values $k\sin A$, $k_2\sin A$, $k_1\sin B$, &c., the equation is

$$k^2\sin^2 A=(k_1\sin B+k_2\sin A)^2+(k_1\sin C+k_3\sin A)^2$$
$$-2(k_1\sin B+k_2\sin A)(k_1\sin C+k_3\sin A)\cos A.$$

* Considering, in the equation, a_2 and a_3 as the distances of a variable point P from the points C and B respectively, the equation represents a circle having its centre on the line CB. Similarly, when a second node is given, the corresponding equation represents another circle, having its centre on the line CB, and the intersections of the two circles determine a_2 and a_3, the lengths of the radial bars, in order that the curve may have the given nodes.

Expanding the right-hand side and reducing by means of $A+B+C=\pi$, the whole becomes divisible by $\sin^2 A$, and we have

$$k^2 = k_1^2 + k_2^2 + k_3^2 - 2k_2k_3 \cos A + 2k_3k_1 \cos B + 2k_1k_2 \cos C;$$

viz. considering A, B, C, k_1, k_2, k_3 as given, this equation determines k so that the curve may have a cusp. The equation is one of the system of four equations

$$k^2 = k_1^2 + k_2^2 + k_3^2 - 2k_2k_3 \cos A + 2k_3k_1 \cos B + 2k_1k_2 \cos C,$$
$$k^2 = k_1^2 + k_2^2 + k_3^2 + 2k_2k_3 \cos A - 2k_3k_1 \cos B + 2k_1k_2 \cos C,$$
$$k^2 = k_1^2 + k_2^2 + k_3^2 + 2k_2k_3 \cos A + 2k_3k_1 \cos B - 2k_1k_2 \cos C,$$
$$k^2 = k_1^2 + k_2^2 + k_3^2 - 2k_2k_3 \cos A - 2k_3k_1 \cos B - 2k_1k_2 \cos C,$$

which belong to the different arrangements CC_1F or CFC_1, BB_1F or BFB_1, of the three points on the lines BF and CF; if k has any of these four values, the curve will have a cusp. If two of the equations subsist together, we have a curve with two cusps. Taking k_1, k_2, k_3, and also $\cos A$, $\cos B$, $\cos C$, as positive, viz. assuming that the triangle is acute-angled, the fourth equation cannot subsist with any one of the others: but two of the others may subsist together, for instance, the first and second will do so, if $k_2k_3 \cos A = k_3k_1 \cos B$, that is, if $\dfrac{k_1}{\cos A} = \dfrac{k_2}{\cos B}$, and then $k^2 = k_1^2 + k_2^2 + k_3^2 + 2k_1k_2 \cos C$: the curve has then two cusps. Similarly, the three equations may subsist together, viz. we must then have

$$\frac{k_1}{\cos A} = \frac{k_2}{\cos B} = \frac{k_3}{\cos C}, \quad k^2 = k_1^2 + k_2^2 + k_3^2 + 2k_2k_3 \cos A:$$

writing herein $k_1, k_2, k_3 = \lambda \cos A, \lambda \cos B, \lambda \cos C$, we find

$$k^2 = \lambda^2 (\cos^2 A + \cos^2 B + \cos^2 C + 2 \cos A \cos B \cos C) = \lambda^2;$$

viz. if k_1, k_2, k_3 are respectively $= k \cos A$, $k \cos B$, $k \cos C$, the curve has then three cusps. It will be recollected that, if

$$k_1 : k_2 : k_3 = \sin A : \sin B : \sin C,$$

the nodes coincide with the foci; the two sets of conditions subsist together, if $A = B = C = 60°$; $k_1 = k_2 = k_3 = \tfrac{1}{2}k$, viz. we have then a curve with three cusps coinciding with the three foci respectively.

Before going further, I will establish the theorem for the triple generation of the curve.

The theorem which gives the triple generation may be stated as follows. See Figures 4, 5, 6*.

Imagine a triangle ABC and a point O, through which point are drawn lines parallel to the sides dividing the triangle into three triangles OB_1C_1, OC_2A_2, OA_3B_3,

* Figure 6 (substantially the same as Fig. 5) belongs to the same curve as Figures 1 and 2, and it exhibits the triple generation of this curve: the generating point O being taken at a node (the same node as in Figure 2), and the two positions OB_1C_1 and $OB_1'C_1'$ of one of the triangles being shown in the figure.

similar *inter se* and to the original triangle, and into three parallelograms OA_2AA_3, OB_3BB_1, OC_1CC_2. Then, considering the three triangles as pivoted together at the point O, and replacing the exterior sides of the parallelograms by pairs of bars A_2AA_3, B_3BB_1, C_1CC_2 pivoted together at A, B, C, and to the triangles at A_2, A_3, B_3, B_1, C_1, C_2, the figure thus consisting of the three triangles and the six bars; let the

Fig. 4. Fig. 5. Fig. 6.

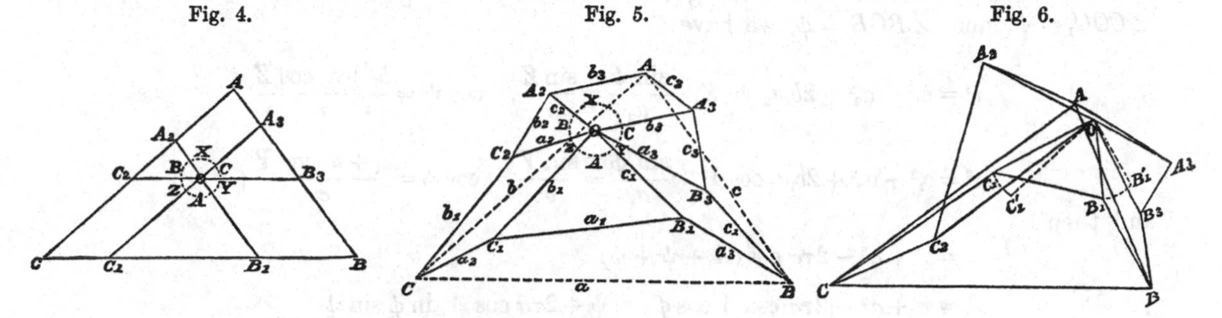

three triangles be turned at pleasure about the point O, so as to displace in any manner the points A, B, C: we have the theorem that the triangle ABC will remain always similar to the original triangle ABC, that is, to each of the three triangles OB_1C_1, OC_2A_2, OA_3B_3: and further, that, starting from any given positions of the three triangles, we may so move them as not to alter the triangle ABC in magnitude: whence, conversely, fixing the three points A, B, C, the point O will be moveable in a curve.

Assuming this, it is clear that the locus of the point O is simultaneously the locus given by

The triangle	OB_1C_1,	connected by bars	B_1B and	C_1C	to fixed points	B, C,
"	OC_2A_2,	"	C_2C "	A_2A	"	C, A,
"	OA_3B_3,	"	A_3A "	B_3B	"	A, B;

or, that we have a triple generation of the same three-bar curve. It may be remarked that the intersection of the lines BB_1 and CC_1 is the axis of the instantaneous rotation of the triangle OB_1C_1, so that, joining this intersection with the point O, we have the normal at O to the locus; and similarly for the other two triangles. It of course follows that the intersections of BB_1 and CC_1, of CC_2 and AA_2, and of AA_3 and BB_3, lie on a line through O, viz. this line is the normal at O.

The result depends on the following theorem: viz. starting with the similar triangles OB_1C_1, A_2OC_2, A_3B_3O, say, the angles of these are A, B, C, so that the sides are

$$k_1(\sin A,\ \sin B,\ \sin C),\ k_2(\sin A,\ \sin B,\ \sin C),\ k_3(\sin A,\ \sin B,\ \sin C);$$

then it follows that the sides of the triangle ABC are

$$k(\sin A,\ \sin B,\ \sin C),$$

the value of k being given by the equation

$$k^2 = k_1^2 + k_2^2 + k_3^2 + 2k_2k_3\cos(X - A) + 2k_3k_1\cos(Y - B) + 2k_1k_2\cos(Z - C),$$

where X, Y, Z denote the angles A_2OA_3, B_3OB_1, C_1OC_2 respectively: whence, since $A+B+C=\pi$, we have also $X+Y+Z=\pi$. If therefore the angles X, Y, Z vary in any manner subject to this last relation and to the equation $k^2=$ const., the triangle ABC will be constant in magnitude.

There is no difficulty in proving the theorem. Writing $OC=\tau$, and $OB=\sigma$, also $\angle COC_1=\psi$, and $\angle BOB_1=\phi$, we have

$$\tau^2 = b_1^2 + a_2^2 + 2b_1a_2 \cos Z, \quad \frac{\sin\psi}{a_2} = \frac{\sin Z}{\tau}, \quad \cos\psi = \frac{b_1 + a_2\cos Z}{\tau},$$

$$\sigma^2 = c_1^2 + a_3^2 + 2c_1a_3 \cos Y, \quad \frac{\sin\phi}{a_3} = \frac{\sin Y}{\sigma}, \quad \cos\phi = \frac{c_1 + a_3\cos Y}{\sigma};$$

and then

$$\begin{aligned}
a^2 &= \tau^2 + \sigma^2 - 2\tau\sigma\cos(A+\psi+\phi)\\
&= \tau^2 + \sigma^2 - 2\tau\sigma\cos A\cos\phi\cos\psi + 2\tau\sigma\cos A\sin\phi\sin\psi\\
&\qquad + 2\tau\sigma\sin A\sin\psi\cos\phi + 2\tau\sigma\sin A\cos\psi\sin\phi\\
&= b_1^2 + c_1^2 + a_2^2 + a_3^2 + 2b_1a_2\cos Z + 2c_1a_3\cos Y\\
&\quad - 2\cos A\,(b_1 + a_2\cos Z)(c_1 + a_3\cos Y)\\
&\quad + 2\cos A\,.\,a_2a_3\sin Y\sin Z\\
&\quad + 2\sin A\sin Z\,.\,a_2(c_1 + a_3\cos Y)\\
&\quad + 2\sin A\sin Y\,.\,a_3(b_1 + a_2\cos Z)\\
&= b_1^2 + c_1^2 - 2b_1c_1\cos A + a_2^2 + a_3^2\\
&\quad + 2a_2a_3\,[-\cos A\,(-\sin Y\sin Z + \cos Y\cos Z)\\
&\qquad\qquad + \sin A\,(\sin Y\cos Z + \cos Y\sin Z)]\\
&\quad + 2a_3\,[(c_1 - b_1\cos A)\cos Y + b_1\sin A\sin Y]\\
&\quad + 2a_2\,[(b_1 - c_1\cos A)\cos Z + c_1\sin A\sin Z].
\end{aligned}$$

We have here $b_1^2+c_1^2-2b_1c_1\cos A = a_1^2$: the second line is $=-2a_2a_3\cos(A+Y+Z)$ which, by virtue of $Y+Z=\pi-X$, is $=2a_2a_3\cos(X-A)$: and in the third and fourth lines

$$c_1 - b_1\cos A = a_1\cos B, \quad b_1\sin A = a_1\sin B,$$

$$b_1 - c_1\cos A = a_1\cos C, \quad c_1\sin A = a_1\sin C;$$

whence these lines are $2a_3a_1\cos(Y-B)$, $2a_1a_2\cos(Z-C)$: the equation therefore is

$$a^2 = a_1^2 + a_2^2 + a_3^2 + 2a_2a_3\cos(X-A) + 2a_3a_1\cos(Y-B) + 2a_1a_2\cos(Z-C),$$

which, putting therein for a_1, a_2, a_3 the values $k_1\sin A$, $k_2\sin A$, $k_3\sin A$, and assuming as above

$$k^2 = k_1^2 + k_2^2 + k_3^2 + 2k_2k_3\cos(X-A) + 2k_3k_1\cos(Y-B) + 2k_1k_2\cos(Z-C),$$

becomes $a^2 = k^2\sin^2 A$, or say $a = k\sin A$; and similarly $b = k\sin B$, $c = k\sin C$, that is, $(a, b, c) = k(\sin A, \sin B, \sin C)$, the required theorem.

Before proceeding to find the equation of the curve, I insert, by way of lemma, the following investigation:—

Three triads (A, B, C), (F, G, H), (I, J, K) of points in a line, or of lines through a point, may be in cubic involution; viz. representing A, B, &c. by the equations $x - ay = 0$, $x - by = 0$, &c., then this is the case when the cubic functions

$$(x-ay)(x-by)(x-cy),\ (x-fy)(x-gy)(x-hy),\ (x-iy)(x-jy)(x-ky),$$

are connected by a linear equation. Regarding I, J, K as given, the condition establishes between (A, B, C) and (F, G, H) two relations: viz. these are

$$\begin{aligned}&(i-a)(i-b)(i-c) : (j-a)(j-b)(j-c) : (k-a)(k-b)(k-c)\\ &=(i-f)(i-g)(i-h) : (j-f)(j-g)(j-h) : (k-f)(k-g)(k-h).\end{aligned}$$

But, if K be regarded as indeterminate, then the condition establishes only the single relation

$$\begin{aligned}&(i-a)(i-b)(i-c) : (j-a)(j-b)(j-c)\\ &=(i-f)(i-g)(i-h) : (j-f)(j-g)(j-h),\end{aligned}$$

which relation, if $i = 0$, $j = \infty$, takes the form $abc = fgh$. When K is thus indeterminate, we may say that the triads (A, B, C), (F, G, H) are in cubic involution with the duad I, J.

If A, B, &c. are points on a conic, then, considering the pencils obtained by joining these points with a point Θ on the conic, if the cubic involution exists for any particular position of Θ, it will exist for every position whatever of Θ; hence, considering triads of points on a conic, we may have a cubic involution between three triads, or between two triads and a duad, as above.

Taking $x = 0$, $y = 0$ for the equations of the tangents at the points I, J respectively, and $z = 0$ for the equation of the line joining these two points, the equation of the conic may be taken to be $xy - z^2 = 0$, and consequently the coordinates of any point A on the conic may be taken to be $x : y : z = \alpha : \frac{1}{\alpha} : 1$. It is then readily shown that α, β, γ, f, g, h referring to the points A, B, C, F, G, H respectively, the condition for the cubic involution of (A, B, C), (F, G, H) with the duad (I, J) is $\alpha\beta\gamma = fgh$.

And we thence at once prove the theorem, that there exists a cubic curve $J_A I_B I_C FGH$, viz. a cubic curve passing through J, and having there the tangent JA, having at I a node with the tangents IB, IC to the two branches respectively, and passing through the points F, G, H; viz. that the triads (A, B, C), (F, G, H) being in cubic involution with (I, J) as above, there exists a cubic curve satisfying these $2 + 5 + 3, = 10$ conditions. In fact, the equation of the cubic curve is

$$\begin{aligned}J_A I_B I_C FGH\,;\ &\left(y-\frac{z}{\alpha}\right)(x-\beta z)(x-\gamma z)\\ &+\frac{z}{\alpha x}\{(x-\alpha z)(x-\beta z)(x-\gamma z)-(x-fz)(x-gz)(x-hz)\}=0,\end{aligned}$$

where observe that second term is an integral function $\frac{1}{\alpha}z^2(-Mx+Nz)$, if, for shortness,

$$M=\alpha+\beta+\gamma-f-g-h,$$
$$N=\beta\gamma+\gamma\alpha+\alpha\beta-gh-hf-fg.$$

In fact, the equations of the lines JA, IB, IC are $y-\frac{z}{\alpha}=0$, $x-\beta z=0$, $x-\gamma z=0$, respectively, and we at once see that these lines are tangents at the points I, J respectively; moreover, at the point F, we have $x, y, z=f, \frac{1}{f}, 1$. Substituting these values, the equation becomes

$$\left(\frac{1}{f}-\frac{1}{\alpha}\right)(f-\beta)(f-\gamma)+\frac{1}{\alpha f}(f-\alpha)(f-\beta)(f-\gamma)=0,$$

viz. the equation is satisfied identically, or the curve passes through F; and similarly the curve passes through G and H.

In precisely the same manner there exists a cubic curve $I_AJ_BJ_CFGH$; viz. this is

$$I_AJ_BJ_CFGH;\ (x-\alpha z)\left(y-\frac{z}{\beta}\right)\left(y-\frac{z}{\gamma}\right)$$
$$+\frac{\alpha z}{y}\left\{\left(y-\frac{z}{\alpha}\right)\left(y-\frac{z}{\beta}\right)\left(y-\frac{z}{\gamma}\right)-\left(y-\frac{z}{f}\right)\left(y-\frac{z}{g}\right)\left(y-\frac{z}{h}\right)\right\}=0,$$

where the second term is an integral function, $\alpha z^2(-M'y+N'z)$; if, for shortness,

$$M'=\frac{1}{\alpha}+\frac{1}{\beta}+\frac{1}{\gamma}-\frac{1}{f}-\frac{1}{g}-\frac{1}{h}=\frac{1}{\alpha\beta\gamma}N,$$
$$N'=\frac{1}{\beta\gamma}+\frac{1}{\gamma\alpha}+\frac{1}{\alpha\beta}-\frac{1}{gh}-\frac{1}{hf}-\frac{1}{fg}=\frac{1}{\alpha\beta\gamma}M,$$

in virtue of the relation $\alpha\beta\gamma=fgh$; so that the second term is in fact $=\frac{z^2}{\beta\gamma}(-Cx+Bz)$.

Writing for shortness J_A, I_A to denote these two cubics respectively, we have four other like cubics, $J_B(=J_BI_CI_AFGH)$, $I_B(=I_BJ_CJ_AFGH)$, $J_C(=J_CI_AI_BFGH)$, and $I_C(=I_CJ_AJ_BFGH)$; the equations being

$$J_A;\ \left(y-\frac{z}{\alpha}\right)(x-\beta z)(x-\gamma z)+\frac{z^2}{\alpha}(-Mx+Nz)=0,$$

$$J_B;\ \left(y-\frac{z}{\beta}\right)(x-\gamma z)(x-\alpha z)+\frac{z^2}{\beta}(-Mx+Nz)=0,$$

$$J_C;\ \left(y-\frac{z}{\gamma}\right)(x-\alpha z)(x-\beta z)+\frac{z^2}{\gamma}(-Mx+Nz)=0,$$

$$I_A;\ (x-\alpha z)\left(y-\frac{z}{\beta}\right)\left(y-\frac{z}{\gamma}\right)+\frac{z^2}{\beta\gamma}(-Ny+Mz)=0,$$

$$I_B;\ (x-\beta z)\left(y-\frac{z}{\gamma}\right)\left(y-\frac{z}{\alpha}\right)+\frac{z^2}{\gamma\alpha}(-Ny+Mz)=0,$$

$$I_C;\ (x-\gamma z)\left(y-\frac{z}{\alpha}\right)\left(y-\frac{z}{\beta}\right)+\frac{z^2}{\alpha\beta}(-Ny+Mz)=0.$$

We require the differences of the products $I_A J_A$, $I_B J_B$, $I_C J_C$. We find

$$I_B J_B = (x-\alpha z)(x-\beta z)(x-\gamma z)\left(y-\frac{z}{\alpha}\right)\left(y-\frac{z}{\beta}\right)\left(y-\frac{z}{\gamma}\right)+\frac{z^4}{\alpha\beta\gamma}(-Mx+Nz)(-Ny+Mz)$$

$$+\frac{z^2}{\gamma\alpha}(-Ny+Mz)\left(y-\frac{z}{\beta}\right)(x-\gamma z)(x-\alpha z)$$

$$+\frac{z^2}{\beta}(-Mx+Nz)(x-\beta z)\left(y-\frac{z}{\gamma}\right)\left(y-\frac{z}{\alpha}\right);$$

let Ω denote the sum of the two expressions in the first line. Similarly, we have

$$I_C J_C = \Omega + \frac{z^2}{\alpha\beta}(-My+Nz)\left(y-\frac{z}{\gamma}\right)(x-\alpha z)(x-\beta z)$$

$$+\frac{z^2}{\gamma}(-Mx+Nz)(x-\gamma z)\left(y-\frac{z}{\gamma}\right)\left(y-\frac{z}{\beta}\right).$$

We have thence

$$I_B J_B - I_C J_C = z^2\left\{\frac{1}{\alpha}(x-\alpha z)(-Ny+Mz)-\left(y-\frac{z}{\alpha}\right)(-Mx+Nz)\right\}$$

$$\times\left\{\frac{1}{\gamma}\left(y-\frac{z}{\beta}\right)(x-\gamma z)-\frac{1}{\beta}\left(y-\frac{z}{\gamma}\right)(x-\beta z)\right\}:$$

the factors in { } are respectively

$$=\left(\frac{1}{\gamma}-\frac{1}{\beta}\right)(xy-z^2)\text{ and }\left(M-\frac{N}{\alpha}\right)(xy-z^2),$$

so that we have

$$I_B J_B - I_C J_C = \left(\frac{1}{\beta}-\frac{1}{\gamma}\right)\left(\frac{N}{\alpha}-M\right)z^2(xy-z^2)^2.$$

The constant factor

$$\left(\frac{1}{\beta}-\frac{1}{\gamma}\right)\left(\frac{N}{\alpha}-M\right)$$

is

$$=\left(\frac{N}{\alpha\beta}+\frac{M}{\gamma}\right)-\left(\frac{N}{\gamma\alpha}+\frac{M}{\beta}\right),\ =P_3-P_2,$$

if P_1, P_2, P_3 denote respectively the functions

$$\frac{N}{\beta\gamma}+\frac{M}{\alpha},\quad \frac{N}{\gamma\alpha}+\frac{M}{\beta},\quad \frac{N}{\alpha\beta}+\frac{M}{\gamma}.$$

Attending to the equation $\alpha\beta\gamma = fgh$, it appears that we have

$$P_1 = \frac{1}{\alpha}(\alpha+\beta+\gamma-f-g-h)+\alpha\left(\frac{1}{\alpha}+\frac{1}{\beta}+\frac{1}{\gamma}-\frac{1}{f}-\frac{1}{g}-\frac{1}{h}\right),$$

with like values for P_2 and P_3.

We have thus

$$I_B J_B - I_C J_C = -(P_2 - P_3)\, z^2 (xy - z^2)^2,$$

and similarly

$$I_C J_C - I_A J_A = -(P_3 - P_1)\, z^2 (xy - z^2)^2,$$

$$I_A J_A - I_B J_B = -(P_1 - P_2)\, z^2 (xy - z^2)^2.$$

Any function $I_A J_A + \lambda z^2 (xy - z^2)^2$, where λ is arbitrary, can of course be expressed in the form $I_A J_A + (\theta + P_1)\, z^2 (xy - z^2)^2$, where θ is arbitrary, and therefore in the three equivalent forms

$$I_A J_A + (\theta + P_1)\, z^2 (xy - z^2)^2,$$

$$I_B J_B + (\theta + P_2)\, z^2 (xy - z^2)^2,$$

$$I_C J_C + (\theta + P_3)\, z^2 (xy - z^2)^2.$$

We have $z = 0$, the line IJ: and $xy - z^2 = 0$, the conic $IJABCFGH$. The equation $I_A J_A + \lambda z^2 (xy - z^2)^2 = 0$ may thus be written in the more complete form

$$I_A J_B J_C FGH \,.\, J_A I_B I_C FGH + \lambda\, (IJ)^2 (IJABCFGH)^2 = 0,$$

and we hence see that it is the equation of a sextic curve, having a triple point at I, the tangents there being IA, IB, IC; having a triple point at J, the tangents there being JA, JB, JC; and having a node (double point) at each of the points F, G, H. There are thus in all $(6 + 3) + (6 + 3) + 3 + 3 + 3$, $= 27$ conditions, and these would in general be sufficient to determine the sextic. The data are, however, related in a special manner; viz. regarding the points I, J, F, G, H as arbitrary, the lines IA, IB, IC, JA, JB, JC are not arbitrary, but satisfy the conditions that A, B are arbitrary points, and C a determinate point, on the conic $IJABC$. And the foregoing result shows that, this being so, there exists a sextic satisfying the foregoing conditions, but containing in its equation an arbitrary constant λ or θ, and that the equation may be presented under the three forms

$$I_A J_B J_C FGH \,.\, J_A I_B I_C FGH + (\theta + P_1)\, (IJ)^2 (IJABCFGH)^2 = 0, \text{ \&c.},$$

corresponding to the partitions A, BC; B, CA; C, AB of the three points A, B, C.

In the case where I, J are the circular points at infinity, the conic $IJABCFGH$ is a circle passing through the six points A, B, C, F, G, H; and the condition of the cubic involution of the triads (A, B, C) and (F, G, H) with the points (I, J) is easily seen to be equivalent to the following relation, viz. the sum of the distances (measured along the circle from any fixed point of the circumference) of the three points A, B, C is equal to the sum of the distances of the three points F, G, H.

The sextic is a tricircular sextic having the three points A, B, C for foci, and having three nodes F, G, H, on the circle ABC, two of them being arbitrary points, and the third of them a determinate point on this circle. And it appears that there exists a sextic satisfying the foregoing conditions, and containing in its equation an arbitrary parameter.

I proceed to find the equation of the curve.

Consider the curve (see Fig. 1, p. 552) as generated by the point O, the vertex of the triangle OC_1B_1, connected by the bars C_1C and B_1B with the fixed points C and B respectively; and suppose, as before, $CB = a$, $C_1C = a_2$, $B_1B = a_3$, $B_1C_1 = a_1$, $OC_1 = b_1$, $OB_1 = c_1$; and draw as in the figure the parallelograms C_1CC_2O and B_1BB_3O; then O may be considered as the intersection of a circle, centre C_2 and radius C_2O, with a circle, centre B_3 and radius B_3O. Take $\angle C_2CB = \theta$, $\angle B_3BC = \phi$: the lines CC_2, BB_3 are parallel to OC_1, OB_1 respectively, and consequently $\theta + \phi = \pi - A$, a relation between the two variable angles θ, ϕ.

Taking the origin at C and the axis of x along the line CB, that of y being at right angles to it: the coordinates of C_2 are $(b_1 \cos\theta,\ b_1 \sin\theta)$, and those of B_3 are $(a - c_1 \cos\phi,\ c_1 \sin\phi)$; the equations of the circles thus are

$$(x - b_1 \cos\theta)^2 + (y - b_1 \sin\theta)^2 = a_2^2,$$
$$(x - a + c_1 \cos\phi)^2 + (y - c_1 \sin\phi)^2 = a_3^2;$$

whence

$$+ 2b_1 x \cos\theta + 2b_1 y \sin\theta = x^2 + y^2 + b_1^2 - a_2^2,$$
$$- 2c_1 (x - a) \cos\phi + 2c_1 y \sin\phi = (x - a)^2 + y^2 + c_1^2 - a_3^2,$$

which equations, writing therein for θ its value $= \pi - A - \phi$ and eliminating the single parameter ϕ, give the equation of the curve.

We in fact have

$$- 2b_1 x \cos(A + \phi) + 2b_1 y \sin(A + \phi) = x^2 + y^2 + b_1^2 - a_2^2,$$
$$- 2c_1 (x - a) \cos\phi + 2c_1 y \sin\phi = (x - a)^2 + y^2 + c_1^2 - a_3^2;$$

or say these are

$$P \cos\phi + Q \sin\phi = R,$$
$$P' \cos\phi + Q' \sin\phi = R',$$

where

$$P = - 2b_1 x \cos A + 2b_1 y \sin A,\quad P' = - 2c_1 (x - a),$$
$$Q = 2b_1 x \sin A + 2b_1 y \cos A,\quad Q' = 2c_1 y,$$
$$R = x^2 + y^2 + b_1^2 - a_2^2,\quad R' = (x - a)^2 + y^2 + c_1^2 - a_3^2.$$

The equations give therefore

$$\cos\phi : \sin\phi : -1 = QR' - Q'R : RP' - R'P : PQ' - P'Q,$$

whence

$$(QR' - Q'R)^2 + (RP' - R'P)^2 = (PQ' - P'Q)^2;$$

and it hence follows that the nodes are the common intersections of the three curves

$$QR' - Q'R = 0,\ RP' - R'P = 0,\ PQ' - P'Q = 0.$$

We have, retaining R and R' to denote their values,

$$QR' - Q'R = -2\,[(Rc_1 - R'b_1 \cos A)\,y - R'b_1 \sin A \,.\, x],$$

$$RP' - R'P = -2\,[(Rc_1 - R'b_1 \cos A)\,(x-a) + R'b_1 \sin A\,(y - a \cot A)],$$

$$PQ' - P'Q = -4\,b_1c_1\,[x(x-a) + y\,(y - a \cot A)].$$

Observing that $R = 0$, $R' = 0$ are circles; the equation $QR' - Q'R = 0$ is a circular cubic through the point $x = 0$, $y = 0$; the equation $RP' - R'P = 0$, a circular cubic through the point $x = a$, $y = a \cot A$; and the equation $PQ' - P'Q = 0$, a circle through these two points (and also the points $x = 0$, $y = a \cot A$; $x = a$, $y = 0$). Hence the first and third curves intersect in the point ($x = 0$, $y = 0$), in the circular points at infinity, and in three other points which are the nodes; viz. the curve has three nodes, say these are F, G, H. The second and third curves intersect in the point ($x = 0$, $y = a \cot A$), in the circular points at infinity, and in the three nodes. As regards the first and second curves, it is readily shown that these touch at the circular points at infinity; viz. they intersect in these points each twice, in the two finite intersections of the circles $R = 0$, $R' = 0$, and in the three nodes.

The three nodes F, G, H thus lie in the circle

$$x(x-a) + y\,(y - a \cot A) = 0,$$

which passes through the points ($x = 0$, $y = 0$) and ($x = a$, $y = 0$), that is, the points C and B. Assuming $b = \dfrac{a \sin A}{\sin B}$, the circle also passes through the point $x = b \cos C$, $y = b \sin C$, that is, the point A of the figure. Thus the three nodes F, G, H lie in the circle circumscribed about the triangle ABC.

Writing, for greater convenience,

$$R = x^2 + y^2 - e^2,\ R' = x^2 + y^2 - 2ax - f^2,$$

the nodes F, G, H lie on the two curves

$$c_1y\,(x^2 + y^2 - e^2) - b_1 \sin A\,(x + y \cot A)\,(x^2 + y^2 - 2ax - f^2) = 0,$$

$$x^2 + y^2 = a\,(x + y \cot A).$$

The first of these is

$$\begin{aligned}&[c_1y - b_1 \sin A\,(x + y \cot A)]\,(x^2 + y^2)\\ &+ [b_1 \sin A\,(x + y \cot A)\,f^2 - c_1e^2y]\\ &+ 2ab_1 \sin A\,(x + y \cot A)\,x = 0.\end{aligned}$$

We may combine these equations so as to obtain the equation of the triad of lines CF, CG, CH; viz. multiplying the second and the third terms of the first equation by $\dfrac{(x^2+y^2)^2}{a^2\,(x + y \cot A)^2}$ and $\dfrac{x^2+y^2}{a\,(x + y \cot A)}$ (each $= 1$ in virtue of the second equation), the equation becomes divisible by $x^2 + y^2$: and, throwing this out, the equation is

$$\begin{aligned}&c_1y - b_1 \sin A\,(x + y \cot A)\\ &+ [b_1 \sin A\,(x + y \cot A)\,f^2 - c_1e^2y]\,\frac{x^2+y^2}{a^2\,(x + y \cot A)^2}\\ &+ 2b_1x \sin A = 0,\end{aligned}$$

where the first and the third terms together are $=(c_1 - b_1 \cos A)\, y + b_1 x \sin A$, viz. this is $= a_1 \sin B\,(x + y \cot B)$. Hence, writing also in the second term $a_1 \sin B$ for $b_1 \sin A$, the equation is

$$(x + y\cot A)^2 (x + y \cot B) + \frac{1}{a^2}\left\{(x + y \cot A) f^2 - \frac{c_1 e^2}{a_1 \sin B} y\right\}(x^2 + y^2) = 0\,;$$

or say this is

$$(x \sin A + y \cos A)^2 (x \sin B + y \cos B)$$
$$+ \frac{\sin A \sin B}{a^2}\left\{(x \sin A + y \cos A) f^2 - \frac{c_1 e^2}{b_1} y\right\}(x^2 + y^2) = 0\,;$$

viz. there is a term in $x^2 + y^2$, and another term

$$(x \sin A + y \cos A)^2 (x \sin B + y \cos B).$$

Suppose for a moment that the angles FBC, GBC, HBC are called F, G, H; then the function on the left hand must be

$$= M(x \sin F - y \cos F)(x \sin G - y \cos G)(x \sin H - y \cos H).$$

Writing in the identity $x = iy$, we have

$$(\cos A + i \sin A)^2 (\cos B + i \sin B) = - M(\cos F - i \sin F)(\cos G - i \sin G)(\cos H - i \sin H)\,;$$

and similarly, writing $x = -iy$, we have the like equation with $-i$ instead of $+i$; whence, dividing the two equations and taking the logarithms,

$$4A + 2B = 2m\pi - F - G - H,$$

which leads as before to the relation $A' + B' + C' = F' + G' + H'$.

In completion of the investigation, observe that M is determinately $+1$ or -1: and that

$$\frac{\sin A \sin B}{a^2}\left\{(x \sin A + y \cos A) f^2 - \frac{c_1}{b_1} e^2 y\right\}$$

is the linear factor of

$$M(x \sin F - y \cos F)(x \sin G - y \cos G)(x \sin H - y \cos H)$$
$$- (x \sin A + y \cos A)^2 (x \sin B + y \cos B),$$

which remains after throwing out the factor $x^2 + y^2$. Calling this linear factor $px + qy$, we have

$$\frac{a^2 p}{\sin A \sin B} = f^2 \sin A, \quad \frac{a^2 q}{\sin A \sin B} = f^2 \cos A - \frac{c_1}{b_1} e^2,$$

or, as this last equation may be written,

$$\frac{a^2 q}{\sin A \sin B} = f^2 \cos A - \frac{\sin C}{\sin B} e^2.$$

Hence, writing $a = k \sin A$, we have

$$f^2 = \frac{k^2 p}{\sin B}, \quad e^2 = \frac{k^2}{\sin C}(p \cos A - q \sin A)\,;$$

substituting for f^2 and e^2 their values, we have

$$-k_1^2 \sin^2 C + k_3^2 \sin^2 A = \frac{k^2 p}{\sin B} + k^2 \sin^2 A,$$

$$-k_1^2 \sin^2 B + k_2^2 \sin^2 A = \frac{k^2}{\sin C}(p \cos A - q \sin A),$$

or, what is the same thing,

$$-\frac{k_1^2}{\sin^2 A} + \frac{k_3^2}{\sin^2 C} = \frac{k^2}{\sin^2 A \sin B \sin^2 C}(p + \sin^2 A \sin B)$$

$$= \frac{k^2 \sin F \sin G \sin H}{\sin A \sin C \,.\, \sin A \sin B \sin C},$$

$$-\frac{k_1^2}{\sin^2 A} + \frac{k_2^2}{\sin^2 B} = \frac{k^2}{\sin^2 A \sin^2 B \sin C}(p \cos A - q \sin A)$$

$$= \frac{k^2 \sin (A - F) \sin (A - G) \sin (A - H)}{\sin A \sin B \,.\, \sin A \sin B \sin C},$$

which are the relations connecting k_1, k_2, k_3, when the foci and nodes are given.

It is to be remarked that if, for instance, $F = 0$ and $G = A$, then $k_1 : k_2 : k_3 = \sin A : \sin B : \sin C$; the nodes in this case coincide with the foci. A simple example is when $A = B = C$; the three triangles are here equal equilateral triangles. The general equations show that, if l_1, l_2, l_3 are values of k_1, k_2, k_3 belonging to a given set of nodes and foci, then the values $k_1^2 = l_1^2 + u \sin^2 A$, $k_2^2 = l_2^2 + u \sin^2 B$, $k_3^2 = l_3^2 + u \sin^2 C$ (where u is arbitrary) will belong to the same set of nodes and foci.

I write the equation of the curve in the form

$$\{(QR' - Q'R) + i(RP' - R'P)\}\{QR' - Q'R - i(RP' - R'P)\} - (PQ' - P'Q)^2 = 0,$$

where

$$(QR' - Q'R) + i(RP' - R'P) = (Rc_1 - R'b_1 \cos A)\, i\, (x - a - iy) - R'b_1 \sin A \{x - i(y - a \cot A)\}.$$

Calling I, J the circular points (∞, $x + iy = 0$) and (∞, $x - iy = 0$), this is a nodal circular cubic having I for an ordinary point, but J for a node. Moreover, one of the tangents at J is the line $x - iy = 0$, that is, the line JB; in fact, writing as before

$$R = x^2 + y^2 - e^2, \quad R' = x^2 + y^2 - 2ax - f^2,$$

then, when $x - iy = 0$, we have $R = -e^2$, $R' = -2ax - f^2$, and the equation becomes

$$\{-ce^2 + b_1 \cos A\,(2ax + f^2)\}(-ia) + b_1 \sin A\,(2ax + f^2)(ia \cot A) = 0;$$

viz. the term in x here disappears, or the three intersections are at infinity. The other tangent at J is the line $x - a - iy = 0$, that is, the line JC; in fact, when $x - a - iy = 0$, that is, $y = -i(x - a)$, we have $R = 2ax - a^2 - e^2$, $R' = -a^2 - f^2$, and the equation becomes

$$\{c_1(2ax - a^2 - e^2) + b_1 \cos A\,(a^2 + f^2)\} \,.\, 0 + b_1 \sin A\,(a^2 + f^2) \,.\, a(1 + i \cot A) = 0,$$

viz. the three intersections are here at infinity. The tangent at I is the line $x - b\cos C + i(y - b\sin C) = 0$, that is, the line IA; in fact, writing this in the form

$$y = ix - ib(\cos C + i\sin C) = ix - ib\gamma,$$

(if for a moment $\cos C + i\sin C = \gamma$, and similarly $\cos A + i\sin A = \alpha$, $\cos B + i\sin B = \beta$); then, y having this value, we find

$$R = 2bxy - b^2\gamma^2 - e^2,\ R' = 2(b\gamma - a)x - b^2\gamma^2 - f^2,$$

$$= -\frac{2c}{\beta}x - b^2\gamma^2 - f^2;$$

and the equation becomes

$$\left\{\begin{array}{l} C_1(2b\gamma x - b^2\gamma^2 - e^2) \\ -b_1\cos A\left(-\dfrac{2c}{\beta}x - b^2\gamma^2 - e^2\right)\end{array}\right\} i(2x - a - b\gamma)$$

$$-b_1\sin A\left(-\frac{2c}{\beta}x - b^2\gamma^2 - e^2\right)(2x + ia\cot A - b\gamma) = 0.$$

The coefficient of x^2 is here

$$2i\left(bc_1\gamma + \frac{b_1c}{\beta}\cdot\frac{1}{\alpha}\right),$$

or, since $b_1c = bc_1$, this is

$$= 2ibc_1\left(\gamma + \frac{1}{\alpha\beta}\right),\ = 0,$$

in virtue of the relation $A + B + C = \pi$, giving $\alpha\beta\gamma = -1$: hence there is only one finite intersection, or the line IA is a tangent.

The cubic in question

$$QR' - Q'R + i(RP' - R'P) = 0$$

is thus a nodal circular cubic which it is convenient to represent in the form

$$(I_AJ_BJ_CFGH) = 0;$$

viz. this is a cubic, through I with the tangent IA, having J as a node with the tangents JB, JC, and through the points F, G, H. Observe that, if F, G, H were arbitrary, this would be $2 + 5 + 3$, $= 10$ conditions. The before-mentioned relation is, in fact, the condition in order to the existence of the cubic.

Similarly the cubic

$$QR' - Q'R - i(RP' - R'P) = 0$$

is the cubic

$$(J_AI_BI_CFGH) = 0.$$

The circle $PQ' - P'Q = 0$ is the conic through I, J, A, B, C, F, G, H; or it may in like manner be written $(IJABCFGH) = 0$; and we may write $(IJ) = 0$, as the equation of the line infinity. The functions denoted as above contain implicitly constant multipliers which give, in the equation of the three-bar curve, one arbitrary parameter—and the equation thus is

$$(I_A J_B J_C FGH)(J_A I_B I_C FGH) - \theta (IJ)^2 (IJABCFGH)^2 = 0,$$

a form which puts in evidence that I, J are triple points having the tangents IA, IB, IC, and JA, JB, JC respectively (whence also A, B, C are foci), and that F, G, H are nodes; viz. the result is as follows:—

Taking A, B, C, F, G, H points in a circle, such that, Sum of the distances (being the angular distances from a fixed point in the circumference) of A, B, C is equal to the sum of the distances of F, G, H: then there exist the cubics $(I_A J_B J_C FGH) = 0$, $(J_A I_B I_C FGH) = 0$, and the sextic is as above.

Writing for shortness

$$(I_A J_B J_C FGH) = I_A, \quad (J_A I_B I_C FGH) = J_A,$$

then the above form is clearly one of three equivalent forms

$$\begin{aligned} U &= I_A J_A - \theta_1 \Omega^2, \\ &= I_B J_B - \theta_2 \Omega^2, \\ &= I_C J_C - \theta_3 \Omega^2. \end{aligned}$$

This implies an identical linear relation between the functions $I_A J_A$, $I_B J_B$, $I_C J_C$; whence also U and Ω^2 are each of them a linear function of any two of these quantities.

I originally obtained the equation of the curve in a form which, though far less valuable than the preceding one, is nevertheless worth preserving; viz. the equation

$$(QR' - Q'R)^2 + (RP' - R'P)^2 = (PQ' - P'Q)^2$$

may be written

$$(R^2 - P^2 - Q^2)(R'^2 - P'^2 - Q'^2) - (RR' - PP' - QQ')^2 = 0,$$

which equation, substituting therein for P, Q, R, P', Q', R' their values, gives the form in question.

Proceeding to the reduction, we have

$$\begin{aligned} R^2 - P^2 - Q^2 &= (x^2 + y^2 + b_1^2 - a_2^2)^2 - 4b_1^2(x^2 + y^2) \\ &= (x^2 + y^2)^2 - 2(b_1^2 + a_2^2)(x^2 + y^2) + (b_1^2 - a_2^2)^2 \\ &= (x^2 + y^2 - \overline{b_1 + a_2}^2)(x^2 + y^2 - \overline{b_1 - a_2}^2); \end{aligned}$$

$$\begin{aligned} R'^2 - P'^2 - Q'^2 &= (\overline{x - a}^2 + y^2 + c_1^2 - a_3^2)^2 - 4c_1^2(\overline{x - a}^2 + y^2) \\ &= (\overline{x - a}^2 + y^2)^2 - 2(c_1^2 + a_3^2)(\overline{x - a}^2 + y^2) + (c_1^2 - a_3^2)^2 \\ &= (\overline{x - a}^2 + y^2 - \overline{c_1 + a_3}^2)(\overline{x - a}^2 + y^2 - \overline{c_1 - a_3}^2). \end{aligned}$$

But the reduction of $RR' - PP' - QQ'$ is somewhat longer. We have

$$RR' - PP' - QQ' = (x^2 + y^2 + b_1^2 - a_2^2)(\overline{x-a}^2 + y^2 + c_1^2 - a_3^2)$$
$$- 4b_1c_1(x\,\overline{x-a} + y^2)\cos A - 4b_1c_1\,ay\sin A:$$

and here

$$2b_1c_1\cos A = b_1^2 + c_1^2 - a_1^2,\quad 2x(x-a) + 2y^2 = x^2 + y^2 + (x-a)^2 + y^2 - a^2,$$

also $b_1c_1\sin A = a_1p_1$, if p_1 be the perpendicular distance of O from the base B_1C_1. Hence the second line is

$$-(b_1^2 + c_1^2 - a_1^2)(x^2 + y^2 + \overline{x-a}^2 + y^2 - a^2) - 4aa_1p_1y,$$

and the whole is

$$\begin{aligned} = \;& (x^2+y^2)(\overline{x-a}^2+y^2) \\ & + (x^2+y^2)(a_1^2 - b_1^2 - a_3^2) \\ & + (\overline{x-a}^2 + y^2)(a_1^2 - c_1^2 - a_2^2) \\ & + (b_1^2 - a_2^2)(c_1^2 - a_3^2) + a^2(b_1^2 + c_1^2 - a_1^2) - 4aa_1p_1y; \end{aligned}$$

whence, finally, we have

$$RR' - PP' - QQ' = (x^2 + y^2 + a_1^2 - a_2^2 - c_1^2)(\overline{x-a}^2 + y^2 + a_1^2 - a_3^2 - b_1^2)$$
$$+ (a^2 + a_1^2 - a_2^2 - a_3^2)(b_1^2 + c_1^2 - a_1^2) - 4aa_1p_1y.$$

Hence the equation of the curve is

$$(x^2 + y^2 - \overline{b_1 + a_2}^2)(x^2 + y^2 - \overline{b_1 - a_2}^2)(\overline{x-a}^2 + y^2 - \overline{c_1 + a_3}^2)(\overline{x-a}^2 + y^2 - \overline{c_1 - a_3}^2)$$
$$- \{(x^2 + y^2 + a_1^2 - a_2^2 - c_1^2)(\overline{x-a}^2 + y^2 + a_1^2 - a_3^2 - b_1^2)$$
$$+ (a^2 + a_1^2 - a_2^2 - a_3^2)(b_1^2 + c_1^2 - a_1^2) - 4aa_1p_1y\}^2 = 0,$$

where p_1 is given in terms of the constants a_1, b_1, c_1 by the equation

$$2a_1p_1 = \sqrt{2b_1^2c_1^2 + 2c_1^2a_1^2 + 2a_1^2b_1^2 - a_1^4 - b_1^4 - c_1^4}.$$

There are in the equation two terms, $(x^2 + y^2)^2$, $(\overline{x-a}^2 + y^2)^2$, which destroy each other, and the remaining terms are of the order 6 at most. Hence the curve is a sextic; and it is, moreover, readily seen that the curve is tricircular. Assuming this, it appears at once that the lines $x + iy = 0$, $x - iy = 0$ are tangents to the curve at the two circular points at infinity. In fact, assuming either of these equations, we have $x^2 + y^2 = 0$, and the equation becomes

$$(b_1^2 - a_2^2)(-2ax + a^2 - \overline{c_1 + a_3}^2)(-2ax + a^2 - \overline{c_1 - a_3}^2)$$
$$- \{(a_1^2 - a_2^2 - c_1^2)(-2ax + a^2 + a_1^2 - a_3^2 - b_1^2)$$
$$+ (a^2 + a_1^2 - a_2^2 - a_3^2)(b_1^2 + c_1^2 - a_1^2) - 4aa_1p_1y\}^2 = 0,$$

a quadric equation. Hence there are on each of the two lines only two finite intersections, or the number of intersections at infinity is $= 4$; viz. the line is a tangent to one of the branches at the triple point. Similarly, the lines $x-a+iy=0$, $x-a-iy=0$ are tangents. Thus the points C and B are foci. It might with somewhat more difficulty be shown from the equation that the point $x=b\cos C$, $y=b\sin C$ $\left(\text{where, as before, } b=\dfrac{a\sin A}{\sin B}\right)$, viz. the point A of the figure, is a focus; but I have not verified this directly. It clearly follows, if we generate the curve by means of the triangle OA_2C_2 and the fixed points C, A. Hence A, B, C are a triad of foci, and the theorem as to the nodes is that these lie on the circle drawn through the three foci A, B, C.

I prove in a somewhat different manner, for the sake of the further theory which arises, the theorem of the triple generation; for this purpose, constructing the foregoing Figure 2 (p. 553) by means of the three triangles OB_1C_1, OC_2A_2, OA_3B_3, but without assuming anything as to the form or position of the triangle ABC, I draw through O a line Ox, the position of which is in the first instance arbitrary, say its inclination to OC_2 is $=v$; and drawing Oy at right angles to Ox, I proceed, in regard to these axes, to find the coordinates of the points C, B. We have, for C,

$$x=a_2\cos v+b_1\cos(v+Z),\quad y=a_2\sin v+b_1\sin(v+Z);$$

for B,

$$x=c_1\cos(v+A+Z)+a_3\cos(v+A+Z+Y),$$
$$y=c_1\sin(v+A+Z)+a_3\sin(v+A+Z+Y);$$

or, writing for $Y+Z$ the value $\pi-X$, so that

$$v+A+Z+Y=\pi+v+A-X,$$

the coordinates of B are

$$x=c_1\cos(v+A+Z)-a_3\cos(v+A-X),$$
$$y=c_1\sin(v+A+Z)-a_3\sin(v+A-X).$$

Taking the two values of y equal to each other, the equation to determine v is

$$a_2\sin v+b_1\sin(v+Z)-c_1\sin(v+A+Z)+a_3\sin(v+A-X)=0.$$

We make the line Ox parallel to BC, so that, writing

$$x=a_2\cos v\qquad\qquad +b_1\cos(v+Z),$$
$$x-a=c_1\cos(v+A+Z)-a_3\cos(v+A-X),$$

we have

$$a=a_2\cos v+b_1\cos(v+Z)-c_1\cos(v+A+Z)+a_3\cos(v+A-X),$$

which determines the distance BC, $=a$. And moreover, writing

$$y=a_2\sin v\qquad\qquad +b_1\sin(v+Z),$$
$$=c_1\sin(v+A+Z)-a_3\sin(v+A-X),$$

we have y as the perpendicular distance of O from BC, and x and $(a-x)$ as the two parts into which BC is divided by the foot of this perpendicular. In the reduction of the formulæ we assume that the three triangles are similar; viz. we write

$$(a_1,\ b_1,\ c_1),\ (a_2,\ b_2,\ c_2),\ (a_3,\ b_3,\ c_3)$$
$$= k_1(\sin A,\ \sin B,\ \sin C),\ k_2(\sin A,\ \sin B,\ \sin C),\ k_3(\sin A,\ \sin B,\ \sin C);$$

and we use when required the relation $A+B+C=\pi$.

The equation for v becomes

$$k_1 \sin(v - C + Z) + k_2 \sin v + k_3 \sin(v + A - X) = 0,$$

which may be written

$$L \sin v - M \cos v = 0,$$

where

$$L = k_2 + k_1 \cos(Z - C) + k_3 \cos(X - A),$$
$$M = \quad - k_1 \sin(Z - C) + k_3 \sin(X - A);$$

hence, putting

$$k^2 = k_1^2 + k_2^2 + k_3^2 + 2k_2k_3 \cos(X - A) + 2k_3k_1 \cos(Y - B) + 2k_1k_2 \cos(Z - C),$$

we have $L^2 + M^2 = k^2$, that is, $\sqrt{L^2 + M^2} = k$, and therefore

$$k \sin v = M, \quad k \cos v = L,$$

which gives the value of v; and then, after all reductions,

$$\begin{aligned} kx = {} & k_1^2 \sin B \cos C + k_2^2 \sin A + k_3^2 . 0 + k_2k_3 \sin A \cos(X - A) \\ & + k_3k_1[-\sin B \cos(Y + A)] \\ & + k_1k_2[\sin(B - A)\cos(Z + A) + 2 \sin A \sin B \sin(Z + A)], \end{aligned}$$

$$\begin{aligned} k(a - x) = {} & k_1^2 \sin C \cos B + k_2^2 . 0 + k_3^2 \sin A + k_2k_3 \sin A \cos(X - A) \\ & + k_3k_1[\sin(C - A)\cos(Y + A) + 2 \sin A \sin C \sin(Y + A)] \\ & + k_1k_2[-\sin C \cos(Z + A)], \end{aligned}$$

and

$$ky = k_1^2 \sin B \sin C + k_2k_3 \sin A \sin(X - A) + k_3k_1 \sin B \sin(Y + A) + k_1k_2 \sin C \sin(Z + A).$$

The first and second equations give $ka = k^2 \sin A$, that is, $a = k \sin A$; and, similarly, $b = k \sin B$, $c = k \sin C$; viz. we have

$$(a,\ b,\ c) = k(\sin A,\ \sin B,\ \sin C),$$

or the triangle ABC is similar to the other three triangles, its magnitude being given by the foregoing equation for k^2. These are the properties which give the triple generation.

Changing the notation of the coordinates, and writing (x, y, z) for the perpendicular distances from O on the sides of the triangle ABC, we have, as above,

$$kx = k_1^2 \sin B \sin C + k_2k_3 \sin A \sin (X - A) + k_3k_1 \sin B \sin (Y + A) + k_1k_2 \sin C \sin (Z + A),$$

and therefore

$$ky = k_2^2 \sin B \sin C + k_2k_3 \sin A \sin (X + B) + k_3k_1 \sin B \sin (Y - B) + k_1k_2 \sin C \sin (Z + B),$$

$$kz = k_3^2 \sin C \sin A + k_2k_3 \sin A \sin (X + C) + k_3k_1 \sin B \sin (Y + C) + k_1k_2 \sin C \sin (Z - C),$$

values which give, as they should do,

$$x \sin A + y \sin B + z \sin C = k^2 \sin A \sin B \sin C.$$

Taking (x, y, z) as simply proportional (instead of equal) to the perpendicular distances, then (x, y, z) will be a system of trilinear coordinates in which the equation of the line infinity is

$$x \sin A + y \sin B + z \sin C = 0;$$

and considering (x, y, z) as proportional to the foregoing values, and in these X, Y, Z as connected by the equation $X + Y + Z = \pi$ and by the equation which determines k^2, the coordinates (x, y, z) are given as proportional to functions of a single parameter, so that the equations in effect determine the curve which is the locus of O.

But to determine the order, &c., the trigonometrical functions must be expressed algebraically; and this is done most readily by introducing instead of X, Y, Z the functions

$$\cos X + i \sin X, \quad \cos Y + i \sin Y, \quad \cos Z + i \sin Z, \quad = \xi, \eta, \zeta;$$

and we may at the same time, in place of A, B, C, introduce the functions

$$\cos A + i \sin A, \quad \cos B + i \sin B, \quad \cos C + i \sin C, = \alpha, \beta, \gamma.$$

The relation $X + Y + Z = \pi$ gives $\xi\eta\zeta = -1$; and similarly $A + B + C = \pi$ gives $\alpha\beta\gamma = -1$.

We have

$$\cos (X - A) = \tfrac{1}{2}\left(\frac{\xi}{\alpha} + \frac{\alpha}{\xi}\right), \quad i \sin (X - A) = \tfrac{1}{2}\left(\frac{\xi}{\alpha} - \frac{\alpha}{\xi}\right), \text{ \&c.};$$

the equation $k^2 = k_1^2 + \text{\&c.}$ becomes

$$k^2 = k_1^2 + k_2^2 + k_3^2 + k_2k_3\left(\frac{\xi}{\alpha} + \frac{\alpha}{\xi}\right) + k_3k_1\left(\frac{\eta}{\beta} + \frac{\beta}{\eta}\right) + k_1k_2\left(\frac{\zeta}{\gamma} + \frac{\gamma}{\zeta}\right),$$

or, as this may be written,

$$(-k^2 + k_1^2 + k_2^2 + k_3^2) + k_2k_3\left(\frac{\xi}{\alpha} - \alpha\eta\zeta\right) + k_3k_1\left(\frac{\eta}{\beta} - \beta\zeta\xi\right) + k_1k_2\left(\frac{\zeta}{\gamma} - \gamma\xi\eta\right) = 0.$$

Also the value of x is proportional to

$$k_1^2\left(\beta - \frac{1}{\beta}\right)\left(\gamma - \frac{1}{\gamma}\right) + k_2k_3\left(\alpha - \frac{1}{\alpha}\right)\left(\frac{\xi}{\alpha} - \frac{\alpha}{\xi}\right) + k_3k_1\left(\beta - \frac{1}{\beta}\right)\left(\alpha\eta - \frac{1}{\alpha\eta}\right) + k_1k_2\left(\gamma - \frac{1}{\gamma}\right)\left(\alpha\zeta - \frac{1}{\alpha\zeta}\right),$$

or, what is the same thing, to

$$k_1^2\left(\beta-\frac{1}{\beta}\right)\left(\gamma-\frac{1}{\gamma}\right)+k_2k_3\left(\alpha-\frac{1}{\alpha}\right)\left(\frac{\xi}{\alpha}+\alpha\eta\zeta\right)+k_3k_1\left(\beta-\frac{1}{\beta}\right)\left(\alpha\eta+\frac{\zeta\xi}{\alpha}\right)+k_1k_2\left(\gamma-\frac{1}{\gamma}\right)\left(\alpha\zeta+\frac{\xi\eta}{\alpha}\right);$$

with the like expressions as to the values of y and z. Introducing for homogeneity a quantity ω, viz. writing $\frac{\xi}{\omega}$, $\frac{\eta}{\omega}$, $\frac{\zeta}{\omega}$ in place of ξ, η, ζ, we have the parameters $(\xi, \eta, \zeta, \omega)$ connected by the homogeneous equations

$$\xi\eta\zeta+\omega^3=0,$$

$$(-k^2+k_1^2+k_2^2+k_3^2)\,\omega^2+k_2k_3\left(\frac{\xi\omega}{\alpha}-\alpha\eta\zeta\right)+k_3k_1\left(\frac{\eta\omega}{\beta}-\beta\zeta\xi\right)+k_1k_2\left(\frac{\zeta\omega}{\gamma}-\gamma\xi\eta\right)=0,$$

and the ratios of the coordinates are

$$\begin{aligned}x : y : z = {} & k_1^2\left(\beta-\frac{1}{\beta}\right)\left(\gamma-\frac{1}{\gamma}\right)\omega^2+k_2k_3\left(\alpha-\frac{1}{\alpha}\right)\left(\frac{\xi\omega}{\alpha}+\alpha\eta\zeta\right)\\ & \qquad +k_3k_1\left(\beta-\frac{1}{\beta}\right)\left(\alpha\eta\omega+\frac{\xi\zeta}{\alpha}\right)+k_1k_2\left(\gamma-\frac{1}{\gamma}\right)\left(\alpha\zeta\omega+\frac{\xi\eta}{\alpha}\right)\\ & : k_2^2\left(\gamma-\frac{1}{\gamma}\right)\left(\alpha-\frac{1}{\alpha}\right)\omega^2+k_2k_3\left(\alpha-\frac{1}{\alpha}\right)\left(\beta\xi\omega+\frac{\eta\zeta}{\beta}\right)\\ & \qquad +k_3k_1\left(\beta-\frac{1}{\beta}\right)\left(\frac{\eta\omega}{\beta}+\beta\zeta\xi\right)+k_1k_2\left(\gamma-\frac{1}{\gamma}\right)\left(\beta\zeta\omega+\frac{\xi\eta}{\beta}\right)\\ & : k_3^2\left(\alpha-\frac{1}{\alpha}\right)\left(\beta-\frac{1}{\beta}\right)\omega^2+k_2k_3\left(\alpha-\frac{1}{\alpha}\right)\left(\gamma\xi\omega+\frac{\eta\zeta}{\gamma}\right)\\ & \qquad +k_3k_1\left(\beta-\frac{1}{\beta}\right)\left(\gamma\eta\omega+\frac{\zeta\xi}{\gamma}\right)+k_1k_2\left(\gamma-\frac{1}{\gamma}\right)\left(\frac{\zeta\omega}{\gamma}+\gamma\xi\eta\right).\end{aligned}$$

Suppose, for shortness, these are $x : y : z = P : Q : R$. Observe that the form of the equations is $\xi\eta\zeta+\omega^3=0$, $\Omega=0$, and $x : y : z = P : Q : R$, where Ω and P, Q, R are each of them a quadric function of the form $(\omega^2, \omega\xi, \omega\eta, \omega\zeta, \eta\zeta, \zeta\xi, \xi\eta)$, the terms in ξ^2, η^2, ζ^2 being wanting.

Treating $(\xi, \eta, \zeta, \omega)$ as the coordinates of a point in space, the equation $\xi\eta\zeta+\omega^3=0$ is a cubic surface having a binode at each of the points $(\xi=0, \omega=0)$, $(\eta=0, \omega=0)$, $(\zeta=0, \omega=0)$, and the second equation is that of a quadric surface passing through these three points; hence the two equations together represent a sextic in space, or say a skew sextic, having a node at each of these three points. The equations $x : y : z = P : Q : R$ establish a (1, 1) correspondence between the locus of O and this skew sextic. To find the degree of the locus we intersect it by the arbitrary line $ax+by+cz=0$; viz. we intersect the skew sextic by the quadric surface $aP+bQ+cR=0$. This is a surface passing through the three nodes of the skew sextic, and it therefore besides intersects the skew sextic in $12-2.3$, $=6$ points. Hence the locus is (as it should be) a sextic.

I consider the point $\eta=0$, $\zeta=0$, $\omega=0$, or say the point (1, 0, 0, 0), of the skew sextic. This is a node, and for the consecutive point on one branch we have $\eta' : \zeta : \omega = m\epsilon : l\epsilon^2 : n\epsilon$, where ϵ is infinitesimal. The equation of the cubic surface gives $lm+n^3=0$, and the equation of the quadric surface gives $k_2k_3 . \frac{\omega}{\alpha} - k_1k_2\gamma\eta = 0$, that is, $k_3\omega = \alpha\gamma k_1\eta$, which, in fact, determines the ratio $l : m$; but it will be convenient to retain the equation in this form. For the corresponding values of (x, y, z) we have

$$x : y : z = k_3\left(\alpha-\frac{1}{\alpha}\right)\frac{\omega}{\alpha} + k_1\left(\gamma-\frac{1}{\gamma}\right)\frac{\eta}{\alpha}$$
$$: k_3\left(\alpha-\frac{1}{\alpha}\right)\beta\omega + k_1\left(\gamma-\frac{1}{\gamma}\right)\frac{\eta}{\beta}$$
$$: k_3\left(\alpha-\frac{1}{\alpha}\right)\gamma\omega + k_1\left(\gamma-\frac{1}{\gamma}\right)\gamma\eta,$$

which, writing for $k_3\omega$ its value $=\alpha\gamma k_1\eta$, become

$$x : y : z = \left(\alpha-\frac{1}{\alpha}\right)\gamma + \left(\gamma-\frac{1}{\gamma}\right)\frac{1}{\alpha} = \alpha\gamma - \frac{\gamma}{\alpha} + \frac{\gamma}{\alpha} - \frac{1}{\gamma\alpha}$$
$$: \left(\alpha-\frac{1}{\alpha}\right)\alpha\beta\gamma + \left(\gamma-\frac{1}{\gamma}\right)\frac{1}{\beta} \quad : -\alpha + \frac{1}{\alpha} - \alpha\gamma^2 + \alpha$$
$$: \left(\alpha-\frac{1}{\alpha}\right)\alpha\gamma^2 + \left(\gamma-\frac{1}{\gamma}\right)\gamma \quad : \alpha^2\gamma^2 - \gamma^2 + \gamma^2 - 1,$$

the last set of values being obtained by aid of the relation $\alpha\beta\gamma=-1$; viz. we thus have

$$x : y : z = \frac{1}{\alpha\gamma}(-1+\alpha^2\gamma^2) : -\frac{1}{\alpha}(-1+\alpha^2\gamma^2) : (-1+\alpha^2\gamma^2),$$

that is,

$$x : y : z = \frac{1}{\gamma} : -1 : \alpha,$$

which are, in fact, the values belonging to one of the circular points at infinity. For the consecutive point on the other branch we should obtain in like manner $x : y : z = \gamma : -1 : \frac{1}{\alpha}$, which are the values belonging to the other circular point at infinity; viz. the node (1, 0, 0, 0) of the skew sextic corresponds to the circular points at infinity. But, in like manner, the other two nodes (0, 1, 0, 0) and (0, 0, 1, 0) each correspond to the circular points at infinity, or say we have in the skew sextic the three nodes each corresponding to one circular point at infinity, and the same three nodes each corresponding to the other circular point at infinity; viz. we thus prove that each of the circular points at infinity is a triple point on the locus of O.

In order not to interrupt the demonstration, I have assumed the formulæ which, in the system of coordinates defined by taking x, y, z proportional to the perpendiculars on the sides of a triangle ABC, or where the equation of the line infinity is

$$x \sin A + y \sin B + z \sin C = 0,$$

give the circular points at infinity; viz. writing

$$\cos A + i\sin A,\ \cos B + i\sin B,\ \cos C + i\sin C = \alpha,\ \beta,\ \gamma,$$

the coordinates for the two points respectively are

$$\begin{aligned} x : y : z &= -1 : \gamma : \frac{1}{\beta} &\quad\text{and}\quad x : y : z &= -1 : \frac{1}{\gamma} : \beta \\ &= \frac{1}{\gamma} : -1 : \alpha & &= \gamma : -1 : \frac{1}{\alpha} \\ &= \beta : \frac{1}{\alpha} : -1, & &= \frac{1}{\beta} : \alpha : -1, \end{aligned}$$

the three values for each point being equivalent in virtue of the relation $\alpha\beta\gamma = -1$. This is, in fact, under a different form, the theorem given in my Smith's Prize paper for 1875; viz. the theorem was: If λ, μ, ν are the inclinations to a fixed line of the perpendiculars let fall from an interior point on the sides of the fundamental triangle ABC, then, in the system of trilinear coordinates in which the coordinates of a point P are proportional to the triangles PBC, PCA, PAB (or where the equation of the line infinity is $x+y+z=0$), the coordinates of the circular points at infinity are proportional, those of the one point to $e^{i\lambda}\sin(\mu-\nu)$, $e^{i\mu}\sin(\nu-\lambda)$, $e^{i\nu}\sin(\lambda-\mu)$, and those of the other point to $e^{-i\lambda}\sin(\mu-\nu)$, $e^{-i\mu}\sin(\nu-\lambda)$, $e^{-i\nu}\sin(\lambda-\mu)$.

In the plane curve, the lines drawn from A, B, C to the circular points at infinity are:

	To the one point.	To the other point.
From A,	$\alpha y + z = 0$,	$y + \alpha z = 0$;
„ B,	$\beta z + x = 0$,	$z + \beta x = 0$;
„ C,	$\gamma x + y = 0$,	$x + \gamma y = 0$.

Each of these lines, *quà* tangent at a triple point, meets the curve in the circular point at infinity counted four times, and in two other points. The corresponding points on the skew sextic should be a node counted twice, the two other nodes counted each once, and two other points. The proof that this is so would show that the points A, B, C are a triad of foci. There is also the question of the determination of the values of $(\xi, \eta, \zeta, \omega)$ which correspond to the nodes of the plane curve. But I have not further pursued the theory.

ADDITION.—Since writing the foregoing paper, I have found that the relation between the nodes and foci (sum of angular distances of the foci = sum of angular distances of the nodes) may be expressed in a different form; viz. the triangle of the foci and the triangle of the nodes are circumscribed to a parabola (having its focus on the circle); and I have made in relation to the question the following further investigations:—

Considering a circle: and a parabola having its focus at K, a point of the circle; then if, as usual, I, J are the circular points at infinity, we have IJK a triangle inscribed in the circle and circumscribed to the parabola; hence there exists a singly-infinite series of in- and circumscribed triangles, so that, drawing from a point A of the circle tangents to the parabola again meeting the circle in the points B and C respectively, BC will be a tangent to the parabola; or, what is the same thing, starting with the triangle ABC inscribed in the circle, we can, with the arbitrary point K on the circle as focus, describe a parabola touching the three sides of the triangle ABC; viz. the parabola described to touch two of the sides of the triangle will touch the third side.

Taking, then, a circle radius $\frac{1}{2}k$, and upon it the three points A, B, C determined by the angles 2α, 2β, 2γ respectively (viz. the coordinates of A are $x, y = \frac{1}{2}k\cos 2\alpha$, $\frac{1}{2}k\sin 2\alpha$, &c.), and a point K determined by the angle 2κ (suppose for a moment the origin is at K), the equation of a parabola having K for its focus will be

$$x^2 + y^2 = (x\cos 2\theta + y\sin 2\theta - p)^2,$$

or, what is the same thing,

$$(x\sin 2\theta - y\cos 2\theta)^2 + 2p\,(x\cos 2\theta + y\sin 2\theta) - p^2 = 0,$$

where θ, p are in the first instance arbitrary; and the condition in order that $\xi x + \eta y + \zeta = 0$ may be a tangent is easily found to be

$$p\,(\xi^2 + \eta^2) + 2\xi\cos 2\theta + 2\eta\sin 2\theta = 0.$$

It is to be shown that p, θ can be determined so that the parabola shall touch each of the lines BC, CA, AB.

Taking the origin at the centre, the equation of BC is

$$x\cos(\beta+\gamma) + y\sin(\beta+\gamma) - \tfrac{1}{2}k\cos(\beta-\gamma) = 0,$$

as is at once verified by showing that this equation is satisfied by the values

$$x,\ y = \tfrac{1}{2}k\cos 2\beta,\ \tfrac{1}{2}k\sin 2\beta,\ \text{and } = \tfrac{1}{2}k\cos 2\gamma,\ \tfrac{1}{2}k\sin 2\gamma.$$

Hence, transforming to the point K as origin, the equation is

$$[x + \tfrac{1}{2}k\cos 2\kappa]\cos(\beta+\gamma) + [y + \tfrac{1}{2}k\sin 2\kappa]\sin(\beta+\gamma) - \tfrac{1}{2}k\cos(\beta-\gamma) = 0;$$

viz. this is

$$x\cos(\beta+\gamma) + y\sin(\beta+\gamma) - \tfrac{1}{2}k[\cos(\beta-\gamma) - \cos(\beta+\gamma-2\kappa)] = 0;$$

or, finally, it is

$$x\cos(\beta+\gamma) + y\sin(\beta+\gamma) - k\sin(\kappa-\beta)\sin(\kappa-\gamma) = 0.$$

Hence the condition of contact with the line BC is

$$p = 2k\sin(\kappa-\beta)\sin(\kappa-\gamma)\cos(2\theta-\beta-\gamma);$$

and, similarly, the condition of contact with the line CA is

$$p = 2k\sin(\kappa-\gamma)\sin(\kappa-\alpha)\cos(2\theta-\gamma-\alpha);$$

viz. these conditions determine the unknown quantities p, θ. It is at once seen that we have

$$2\theta - \beta - \gamma = \tfrac{1}{2}\pi - (\kappa - \alpha), \text{ that is, } 2\theta = \tfrac{1}{2}\pi - \kappa + \alpha + \beta + \gamma;$$

and then

$$p = 2k \sin(\kappa - \alpha) \sin(\kappa - \beta) \sin(\kappa - \gamma);$$

from symmetry, we see that the parabola touches also the side AB.

Suppose, next, F, G are points on the circle determined by the angles $2f$, $2g$; retaining p and θ to denote their values,

$$p = 2k \sin(\kappa - \alpha) \sin(\kappa - \beta) \sin(\kappa - \gamma), \text{ and } 2\theta = \tfrac{1}{2}\pi - \kappa + \alpha + \beta + \gamma,$$

the condition, in order that FG may be a tangent, is

$$p = 2k \sin(\kappa - f) \sin(\kappa - g) \cos(2\theta - f - g);$$

viz. determining h by the equation

$$\alpha + \beta + \gamma = f + g + h,$$

this is

$$p = 2k \sin(\kappa - f) \sin(\kappa - g) \sin(\kappa - h),$$

or, what is the same thing,

$$\sin(\kappa - \alpha) \sin(\kappa - \beta) \sin(\kappa - \gamma) = \sin(\kappa - f) \sin(\kappa - g) \sin(\kappa - h);$$

viz. this equation, considering therein h as standing for $\alpha + \beta + \gamma - f - g$, is the relation which must subsist between f and g, in order that the line FG may be a tangent to the parabola. And then, h being determined as above, and the point H on the circle being determined by the angle $2h$, it is clear that the lines GH, HF will also be tangents to the parabola; viz. FGH will be an in- and circumscribed triangle, provided only f, g, h satisfy the above-mentioned two equations. The latter of these, if f, g, h satisfy only the relation $\alpha + \beta + \gamma = f + g + h$, serves to determine κ; and then, θ and κ denoting as above, the equation of the parabola is

$$x^2 + y^2 = (x \cos 2\theta + y \sin 2\theta - p)^2;$$

and it thus appears that the condition in question, $\alpha + \beta + \gamma = f + g + h$, is equivalent to the condition that the triangles ABC, FGH shall be circumscribed to the same parabola.

It is to be remarked that the distances KA, KB, &c. are equal to $k \sin(\kappa - \alpha)$, $k \sin(\kappa - \beta)$, &c.; hence the condition

$$\sin(\kappa - \alpha) \sin(\kappa - \beta) \sin(\kappa - \gamma) = \sin(\kappa - f) \sin(\kappa - g) \sin(\kappa - h)$$

becomes

$$KA \,.\, KB \,.\, KC = KF \,.\, KG \,.\, KH;$$

viz. the focus K is a point on the circle such that the product of its (linear) distances from the foci A, B, C is equal to the product of its (linear) distances from the nodes F, G, H.

It is to be remarked that the foregoing equation in κ determines a single position of the point K; viz. it determines $\tan\kappa$, and therefore $\sin 2\kappa$ and $\cos 2\kappa$, linearly. The equation is, in fact, a cubic equation in $\tan\kappa$, satisfied identically by $\tan\kappa = i$ and $\tan\kappa = -i$, and therefore reducible to a linear equation.

Write for a moment $\tan\kappa = \omega$, and

$$(\tan\kappa - \tan\alpha)(\tan\kappa - \tan\beta)(\tan\kappa - \tan\gamma) = \omega^3 - p\omega^2 + q\omega - r,$$

$$(\tan\kappa - \tan f)(\tan\kappa - \tan g)(\tan\kappa - \tan h) = \omega^3 - p'\omega^2 + q'\omega - r';$$

also

$$M = \cos f \cos g \cos h \div \cos\alpha \cos\beta \cos\gamma.$$

Then we have

$$\omega^3 - p\omega^2 + q\omega - r = M(\omega^3 - p'\omega^2 + q'\omega - r'),$$

where

$$p = r + M(r' - p), \quad q = 1 + M(q' - 1).$$

Substituting these values, the equation becomes

$$\omega^3 - r\omega^2 + \omega - r = M(\omega^3 - r'\omega^2 + \omega - r'),$$

viz. dividing by $\omega^2 + 1$, this is $\omega - r = M(\omega - r')$; or substituting for r, r', M their values,

$$(\cos\alpha\cos\beta\cos\gamma - \cos f\cos g\cos h)\tan\kappa = (\sin\alpha\sin\beta\sin\gamma - \sin f\sin g\sin h),$$

which is the value of $\tan\kappa$, and then

$$\sin 2\kappa = \frac{2\tan\kappa}{1 + \tan^2\kappa}, \quad \cos 2\kappa = \frac{1 - \tan^2\kappa}{1 + \tan^2\kappa}.$$

It may be further noticed that, if the parabola intersect the circle in a point L, and the tangent at L to the parabola again meet the circle in M, then, if $2l$, $2m$ are the angles for the points L, M, we have l, m, m for values of f, g, h, whence l, m are determined by the equations

$$l + 2m = \alpha + \beta + \gamma, \quad \sin(\kappa - l)\sin^2(\kappa - m) = \sin(\kappa - \alpha)\sin(\kappa - \beta)\sin(\kappa - \gamma);$$

but as the circle intersects the parabola not only in two real points, but in two other imaginary points, there is no simple formula for the determination of l and m.

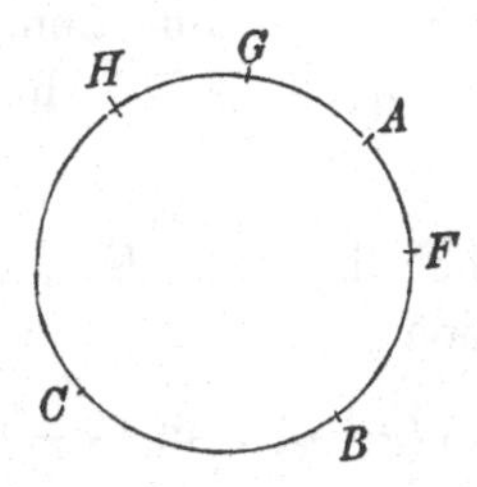

To determine the linkage when the nodes are given, suppose that, in the generation by O, the vertex of the triangle OB_1C_1, we have O at the node F: then, if τ, σ are the distances of C, B from the node in question, we have, as in the memoir,

$$(b_1^2 + \tau^2 - a_2^2)\, c_1\sigma = (c_1^2 + \sigma^2 - a_3^2)\, b_1\tau,$$

that is,

$$(b_1^2 - a_2^2) c_1\sigma - (c_1^2 - a_3^2) b_1\tau + \sigma\tau (c_1\tau - b_1\sigma) = 0,$$

or, what is the same thing,

$$(b_1^2 - a_2^2) c\sigma - (c_1^2 - a_3^2) b\tau + \sigma\tau (c\tau - b\sigma) = 0.$$

Suppose, as in the figure, that F is between B and A; then, if $AF = \rho$, we have $c\tau = b\sigma + a\rho$, and the equation becomes

$$(b_1^2 - a_2^2) c\sigma - (c_1^2 - a_3^2) b\tau + a\rho\sigma\tau = 0.$$

Similarly, if, as in the figure, G is on the other side of A, that is, between A and C, and if ρ', σ', τ' be the distances AG, BG, CG, then $b\sigma' = c\tau' + a\rho'$, that is, $c\tau' - b\sigma' = - a\rho'$, and the corresponding equation is

$$(b_1^2 - a_2^2) c\sigma' - (c_1^2 - a_3^2) b\tau' - a\rho'\sigma'\tau' = 0.$$

We hence find

$$(b_1^2 - a_2^2) c (\sigma\tau' - \sigma'\tau) + a\tau\tau' (\rho\sigma + \rho'\sigma') = 0,$$
$$(c_1^2 - a_3^2) b (\sigma\tau' - \sigma'\tau) + a\sigma\sigma' (\rho\tau + \rho'\tau') = 0.$$

But we have

$$BG \,.\, CF = BF \,.\, CG + BC \,.\, FG,$$

that is,

$$\sigma'\tau = \sigma\tau' + a \,.\, FG, \text{ or } \sigma\tau' - \sigma'\tau = - a \,.\, FG;$$

also,

$$\rho\sigma + \rho'\sigma' = AF \,.\, BF + AG \,.\, BG, \; = FG \,.\, CH, \; = FG \,.\, \tau'',$$

and

$$\rho\tau + \rho'\tau' = AF \,.\, CF + AG \,.\, CG, \; = FG \,.\, BH, \; = FG \,.\, \sigma'',$$

as may be shown without difficulty, ρ'', σ'', τ'' being the distances AH, BH, CH. Hence the equations become

$$c (b_1^2 - a_2^2) - \tau\tau'\tau'' = 0,$$
$$b (c_1^2 - a_3^2) - \sigma\sigma'\sigma'' = 0,$$

showing that, the foci being as in the figure, $b_1^2 - a_2^2$ and $c_1^2 - a_3^2$ are each of them positive; viz. that, in the generation by the triangle OC_1B_1, the radial bars a_2, a_3 are shorter than the sides b_1, c_1 respectively. Substituting for b_1, &c. the values $k_1 \sin B$, &c.; also, instead of k_1, k_2, k_3, introducing the quantities λ_1, λ_2, λ_3, where

$$k_1, \; k_2, \; k_3 = \lambda_1 \sin A, \; \lambda_2 \sin B, \; \lambda_3 \sin C,$$

these equations become

$$c (\lambda_1^2 - \lambda_2^2) \sin^2 A \sin^2 B = \tau\tau'\tau'',$$
$$b (\lambda_1^2 - \lambda_3^2) \sin^2 A \sin^2 C = \sigma\sigma'\sigma'';$$

or, as these may be written, putting for shortness $M = \sin A \sin B \sin C$,

$$M^2k^2 (\lambda_1^2 - \lambda_2^2) = c\, \tau\tau'\tau'',$$
$$M^2k^2 (\lambda_1^2 - \lambda_3^2) = b\, \sigma\sigma'\sigma''.$$

All the quantities have so far been regarded as positive, and the formulæ are applicable to the particular figure; but, to present them in a form applicable to any order of the nodes and foci, we have only to write the equations in the forms

$$M^2(\lambda_1^2 - \lambda_2^2) = k^2 \sin(\alpha - \beta) \sin(f - \gamma) \sin(g - \gamma) \sin(h - \gamma),$$
$$M^2(\lambda_1^2 - \lambda_3^2) = k^2 \sin(\alpha - \gamma) \sin(f - \beta) \sin(g - \beta) \sin(h - \beta);$$

and these may be replaced by the system

$$M^2(\lambda_2^2 - \lambda_3^2) = k^2 \sin(\beta - \gamma) \sin(f - \alpha) \sin(g - \alpha) \sin(h - \alpha),$$
$$M^2(\lambda_3^2 - \lambda_1^2) = k^2 \sin(\gamma - \alpha) \sin(f - \beta) \sin(g - \beta) \sin(h - \beta),$$
$$M^2(\lambda_1^2 - \lambda_2^2) = k^2 \sin(\alpha - \beta) \sin(f - \gamma) \sin(g - \gamma) \sin(h - \gamma),$$

since the first of these equations is implied in the other two; and then, reverting to the original form, we may write

$$M^2k^2(\lambda_2^2 - \lambda_3^2) = BC \,.\, FA \,.\, GA \,.\, HA,$$
$$M^2k^2(\lambda_3^2 - \lambda_1^2) = CA \,.\, FB \,.\, GB \,.\, HB,$$
$$M^2k^2(\lambda_1^2 - \lambda_2^2) = AB \,.\, FC \,.\, GC \,.\, HC,$$

it being understood that the distances BC, FA, &c., which enter into these equations, are not all positive, but that they stand for $k\sin(\beta - \gamma)$, $k\sin(f - \alpha)$, &c., and that their signs are to be taken accordingly. Or, again, these may be written

$$BC(c_2^2 - b_3^2) = FA \,.\, GA \,.\, HA,$$
$$CA(a_3^2 - c_1^2) = FB \,.\, GB \,.\, HB,$$
$$AB(b_1^2 - a_2^2) = FC \,.\, GC \,.\, HC,$$

where the signs are as just mentioned. We may say that $\pm(c_2^2 - b_3^2)$ is the modulus for the focus A; and the formula then shows that this modulus, taken positively, is equal to the product of the distances FA, GA, HA of A from the three nodes respectively, divided by BC, the distance of the other two foci from each other.

624.

ON THE BICURSAL SEXTIC.

[From the *Proceedings of the London Mathematical Society*, vol. VII. (1875—1876), pp. 166—172. Read March 10, 1876.]

IN the paper "On the mechanical description of certain sextic curves," *Proceedings of the London Mathematical Society*, vol. IV. (1872), pp. 105—111, [504], I obtained the bicursal sextic as a rational transformation of a binodal quartic. The theory was in effect as follows: taking Ω, P, Q, R, each of them a function of λ, μ of the form $(*\between 1, \lambda)^2(1, \mu)^2$, and considering (λ, μ) as connected by the equation $\Omega = 0$, (viz. λ, μ being coordinates, this represents a binodal quartic), then, if we assume $x : y : z = P : Q : R$, the locus of the point (x, y, z) is a curve rationally connected with the binodal quartic, viz. the points of the two curves have with each other a $(1, 1)$ correspondence; whence the locus in question, say the curve $U = 0$, is bicursal. The degree is obtained as the number of the intersections of the curve by an arbitrary line, or, what is the same thing, the number of the variable intersections of the corresponding $\lambda\mu$-curves

$$\Omega = 0, \quad \alpha P + \beta Q + \gamma R = 0,$$

viz. each of these being a quartic curve having the same two nodes, the nodes each count as 4 intersections, and the number of the remaining intersections is $4 . 4 - 2 . 4, = 8$, and thus the curve $U = 0$ is in general of the order 8. But if the curves $\Omega = 0$, $P = 0$, $Q = 0$, $R = 0$ have (besides the nodes) k common intersections, then these are also fixed intersections of the two curves $\Omega = 0$, $\alpha P + \beta Q + \gamma R = 0$, and the number of variable intersections is reduced to $8 - k$; we have thus $8 - k$ as the order of the curve $U = 0$. In particular, if $k = 2$, then the curve is a bicursal sextic.

The theory assumes a different and more simple form if, in the several functions Ω, P, Q, R, we suppose that the terms in λ^2, μ^2 are wanting. The curves $\Omega = 0$, $P = 0$, $Q = 0$, $R = 0$ are here cubics having two common points; the curve $U = 0$, *quà* rational transformation of the cubic $\Omega=0$, is still a bicursal curve; but its order is given as the number of the variable intersections of the cubics

$$\Omega=0,\quad \alpha P+\beta Q+\gamma R=0,$$

viz. this is $=3.3-2$, $=7$. But if the curves $\Omega=0$, $P=0$, $Q=0$, $R=0$ have (besides the before-mentioned two common points) k other common points, then the number of the variable intersections is $=7-k$: and this is therefore the order of the curve $U=0$. In particular, if $k=1$, then the curve is a bicursal sextic. And, in the present paper, I consider the binodal sextic as thus obtained, viz. as given by the equations $\Omega=0$, $x:y:z=P:Q:R$, where $\Omega=0$, $P=0$, $Q=0$, $R=0$ are cubics, having (in all) three common points.

The bicursal sextic has in general 9 nodes; but 3 of these may unite together into a triple point: this will be the case if, in the series of curves $\alpha P+\beta Q+\gamma R=0$, there are any two curves which have 3 common intersections with the curve $\Omega=0$. (Observe that we throughout disregard the 3 common points of the curves $\Omega=0$, $P=0$, $Q=0$, $R=0$, and attend only to the 6 variable points of intersection of the curves $\Omega=0$ and $\alpha P+\beta Q+\gamma R=0$,—the meaning is, that there are two curves of the series such that, attending only to the 6 variable intersections of each of them with the curve $\Omega=0$, there are three common intersections.) For, supposing the two curves to be $\alpha P+\beta Q+\gamma R=0$ and $\alpha' P+\beta' Q+\gamma' R=0$, then any curve whatever

$$\alpha P+\beta Q+\gamma R+\theta(\alpha' P+\beta' Q+\gamma' R)=0$$

has the same three intersections with the curve $\Omega=0$, say these are the points A_1, A_2, A_3, the coordinates of which are independent of θ. Hence the line

$$(\alpha x+\beta y+\gamma z)+\theta(\alpha' x+\beta' y+\gamma' z)=0$$

intersects the curve $U=0$ in six points, three of which, as corresponding to the points A_1, A_2, A_3, are independent of θ, viz. they are the same three points for any line whatever of the series; and this means that the curve $U=0$ has at the point

$$(\alpha x+\beta y+\gamma z=0,\quad \alpha' x+\beta' y+\gamma' z=0)$$

a triple point; and that to this triple point correspond the three points A_1, A_2, A_3.

We may, in the series of lines $\alpha x+\beta y+\gamma z+\theta(\alpha' x+\beta' y+\gamma' z)=0$, rationally determine θ so that one of the three variable points of intersection shall correspond to A_1, A_2, or A_3; viz. θ must be such that the curve $\alpha P+\beta Q+\gamma R+\theta(\alpha' P+\beta' Q+\gamma' R)=0$ shall touch the curve $\Omega=0$ at one of the points A_1, A_2, A_3. The three lines thus determined are the three tangents to the curve at the triple point: and the three branches may be considered as corresponding to the three points A_1, A_2, A_3, respectively.

There is no loss of generality in assuming that the triple point is the point $(x=0,\ z=0)$; the condition then simply is that the curves $P=0$, $R=0$ shall have three common intersections with the curve $\Omega=0$; and the tangents at the triple point are $x+\theta z=0$, θ being so determined that one of the three variable points of intersection shall correspond to one of the three points A_1, A_2, A_3: in particular, if this is the case for the line $x=0$, then this line will be one of the tangents at the triple point.

The bicursal sextic may have a second triple point, viz. three other nodes may unite together into a triple point. The theory is precisely the same: we must have two other curves $\alpha P + \beta Q + \gamma R = 0$, $\alpha' P + \beta' Q + \gamma' R = 0$, having with the curve $\Omega = 0$ three common intersections B_1, B_2, B_3: there is then a second triple point

$$(\alpha x + \beta y + \gamma z = 0, \quad \alpha' x + \beta' y + \gamma' z = 0);$$

and, to find the tangents at this point, we must determine θ so that one of the variable points of intersection of the line

$$\alpha x + \beta y + \gamma z + \theta (\alpha' x + \beta' y + \gamma' z) = 0$$

with the sextic shall correspond with B_1, B_2, or B_3; viz. θ must be such that the curve $\alpha P + \beta Q + \gamma R + \theta (\alpha' P + \beta' Q + \gamma' R) = 0$ shall touch the curve $\Omega = 0$ at one of the points B_1, B_2, B_3. In particular, if, as before, the curves $P = 0$, $R = 0$ have three common intersections with the curve $\Omega = 0$, and if, moreover, the curves $Q = 0$, $R = 0$ have three common intersections with the curve $\Omega = 0$, then the bicursal sextic will have the two triple points $(x = 0, z = 0)$ and $(y = 0, z = 0)$; and it may further happen that the line $x = 0$ is a tangent at the first triple point, and the line $y = 0$ a tangent at the second triple point. The sextic may in like manner have a third triple point, but this is a special case which I do not at present consider.

I write for greater convenience $\frac{\lambda}{\nu}$, $\frac{\mu}{\nu}$ in place of λ, μ, so as to make Ω, P, Q, R each of them a homogeneous cubic function of (λ, μ, ν); and I give to these functions, not the most general values belonging to a bicursal sextic with two triple points, but the values in the form obtained for them, as appearing further on, in the problem of three-bar motion; viz. the equations $\Omega = 0$, $x : y : z = P : Q : R$ are respectively taken to be

$$\nu (h\nu\lambda + f\lambda^2) + \mu (g\nu^2 + e\nu\lambda + g\lambda^2) + \mu^2 (f\nu + h\lambda) = 0,$$

$$x : y : z = \lambda\mu (a\lambda + b\mu) : \nu^2 (c\lambda + d\mu) :: \lambda\mu\nu.$$

The four curves $\Omega = 0$, $P = 0$, $Q = 0$, $R = 0$ have thus the three common intersections

$$(\mu = 0, \nu = 0), \ (\nu = 0, \lambda = 0), \ (\lambda = 0, \mu = 0),$$

represented in the figure by the points A, B, C; the curve drawn in the figure is the curve $\Omega = 0$, and the points F, G, H are the third points of intersection of the cubic with the lines BC, CA, AB respectively.

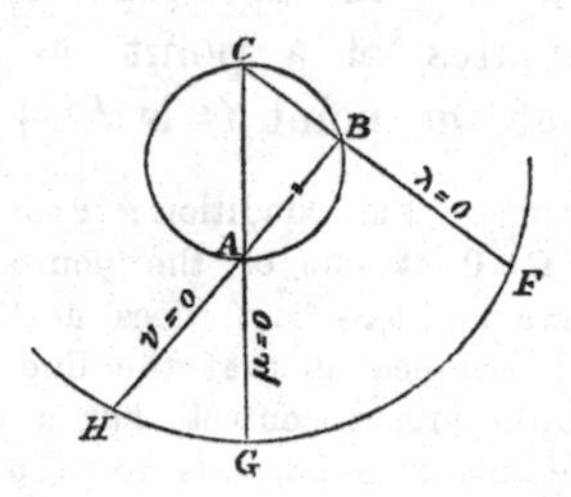

The equation $P + \theta R = 0$ is here $\lambda\mu (a\lambda + b\mu + \theta\nu) = 0$, which intersects $\Omega = 0$ in the points C, A, G, C, B, F, and the three intersections by the line $a\lambda + b\mu + \theta\nu = 0$;

viz. excluding the fixed points A, B, C, the six intersections are C, F, G, and the three intersections by the line. Hence, of the six intersections, we have C, F, G independent of θ, or we have $(x=0,\ z=0)$ a triple point, say I, corresponding to the three points C, F, G, viz. these are the points

$$(\lambda,\ \mu,\ \nu)=(0,\ 0,\ 1),\quad (0,\ g,\ -f),\quad (h,\ 0,\ -f),\qquad (C,\ F,\ G).$$

The equation $Q+\theta R=0$ is $\nu(c\lambda\nu+d\mu\nu+\theta\lambda\mu)=0$; viz. the line $\nu=0$ meets $\Omega=0$ in the three points A, B, H, and the conic $c\lambda\nu+d\mu\nu+\theta\lambda\mu=0$ meets $\Omega=0$ in the points A, B, C, and three other points: hence, rejecting the points A, B, C, the six points of intersection are the points A, B, H, and the three variable points of intersection by the conic; or we have $(y=0,\ z=0)$ a triple point, say J, corresponding to the three points A, B, H, viz. these are the points

$$(\lambda,\ \mu,\ \nu)=(1,\ 0,\ 0),\quad (0,\ 1,\ 0),\quad (h,\ -g,\ 0),\qquad (A,\ B,\ H).$$

To find the tangents at the triple point I, these are $x+\theta z=0$, where θ is to be successively determined by the conditions that the line $a\lambda+b\mu+\theta\nu=0$ shall pass through the points C, F, G*; viz. we thus have

$$\begin{array}{lll} \theta=0, & x=0, \text{ the tangent corresponding to the point } & C,\ (0,\ 0,\ 1),\\ \theta=+\dfrac{bg}{f}, & fx+bgz=0, \quad \text{,,} \qquad \text{,,} \qquad \text{,,} \qquad \text{,,} & F,\ (0,\ g,\ -f),\\ \theta=+\dfrac{ah}{f}, & fx+ahz=0, \quad \text{,,} \qquad \text{,,} \qquad \text{,,} \qquad \text{,,} & G,\ (h,\ 0,\ -f). \end{array}$$

And similarly, at the triple point J, the tangents are $y+\theta z=0$, where θ is to be successively determined by the conditions that the conic $c\nu\lambda+d\nu\mu+\theta\lambda\mu=0$ shall pass through the point H, and shall touch the cubic at the points A, B; viz. we thus have

$$\begin{array}{lll} \theta=0, & y=0, \text{ the tangent corresponding to the point } & H,\ (h,\ -g,\ 0),\\ \theta=\dfrac{cg}{f}, & fy+cgz=0, \quad \text{,,} \qquad \text{,,} \qquad \text{,,} \qquad \text{,,} & A,\ (1,\ 0,\ 0),\\ \theta=\dfrac{dh}{f}, & fy+dhz=0, \quad \text{,,} \qquad \text{,,} \qquad \text{,,} \qquad \text{,,} & B,\ (0,\ 1,\ 0). \end{array}$$

The two last values of θ are obtained by the consideration that the equations of the tangents to $\Omega=0$ at the points A, B respectively, are $g\mu+f\nu=0$, $h\lambda+f\nu=0$, where λ, μ, ν are current coordinates of a point on the tangent: it may be added that the equation of the tangent at the point C is $h\lambda+g\mu=0$.

* Observe the somewhat altered form of the condition: θ is to be determined so that the cubic $\lambda\mu(a\lambda+b\mu+\theta\nu)=0$ shall touch the cubic $\Omega=0$ at one of the points C, F, G: but, as the first-mentioned cubic breaks up, and the component curve $a\lambda+b\mu+\theta\nu=0$ does not pass through any one of these points, this can only mean that θ shall be so determined as that the line shall pass through one of these points, viz. that there shall be at the point, not a proper contact, but a double intersection, arising from a node of the cubic $\lambda\mu(a\lambda+b\mu+\theta\nu)=0$. And the like case happens for the other triple point; viz. there the cubic $\nu(c\nu\lambda+d\nu\mu+\theta\lambda\mu)=0$ is to touch the cubic $\Omega=0$ at one of the points A, B, H; the component conic $c\nu\lambda+d\nu\mu+\theta\lambda\mu=0$ passes through the points A and B but not through H; hence the conditions for θ are, that the conic shall touch the cubic at A or B, or that it shall pass through H.

The three-bar curve may be represented by means of a system of equations of the last-mentioned form, viz. $x : y : z = \lambda\mu(a\lambda + b\mu) : \nu^2(c\lambda + d\mu) : \lambda\mu\nu$, where λ, μ, ν are connected as above; or, taking X, Y as ordinary rectangular coordinates, x, y, and z are here the circular coordinates $\frac{x}{z} = X + iY$, $\frac{y}{z} = X - iY$, and $z = 1$; and the parameters $\frac{\lambda}{\nu}$, $\frac{\mu}{\nu}$ denote like functions $\cos\theta + i\sin\theta$, $\cos\phi + i\sin\phi$ of angles which are the inclinations of two bars to a fixed line. Using, for convenience, Figure 2 of my paper on Three-bar Motion, (p. 553 of this volume), the curve is considered as the locus of the vertex O of the triangle OC_1B_1, connected by the bars C_1C and B_1B with the fixed points B and C respectively; and we have $CC_1 = a_2$, $OC_1 = b_1$, $OB_1 = c_1$, $C_1B_1 = a_1$, $B_1B = a_3$. Also, to avoid confusion with the foregoing notation of the present paper, instead of calling it a, I take $BC = a_0$: the angle OC_1B_1 is $= C_1$, and $\cos C_1 + i\sin C_1$ is taken $= \gamma$.

Hence, taking the origin at C, the axis of X coinciding with CB and that of Y being at right angles to it: taking also θ, ϕ, ψ for the inclinations of CC_1, C_1B_1, and B_1B to CB, we have

$$a_2\cos\theta + a_1\cos\phi - a = -a_3\cos\psi,$$

$$a_2\sin\theta + a_1\sin\phi \quad = \quad a_3\sin\psi;$$

viz. writing $\cos\theta + i\sin\theta = \lambda$, $\cos\phi + i\sin\phi = \mu$, these give

$$a_2\lambda + a_1\mu - a_0 = -a_3(\cos\psi - i\sin\psi),$$

$$a_2\frac{1}{\lambda} + a_1\frac{1}{\mu} - a_0 = -a_3(\cos\psi + i\sin\psi),$$

that is,

$$(a_2\lambda + a_1\mu - a)\left(a_2\frac{1}{\lambda} + a_1\frac{1}{\mu} - a_0\right) - a_3^2 = 0;$$

viz.

$$(a_0^2 + a_1^2 + a_2^2 - a_3^2) + a_1a_2\left(\frac{\mu}{\lambda} + \frac{\lambda}{\mu}\right) - a_0a_2\left(\lambda + \frac{1}{\lambda}\right) - a_0a_1\left(\mu + \frac{1}{\mu}\right) = 0,$$

for the relation between the parameters λ, μ. And then

$$X = a_2\cos\theta + b_1\cos(\phi + C_1),$$

$$Y = a_2\sin\theta + b_1\sin(\phi + C_1);$$

viz. if x, $y = X + iY$, $X - iY$, then

$$\begin{cases} x = a_2\lambda + b_2\gamma_1\mu, \\ y = a_2\dfrac{1}{\lambda} + b_2\dfrac{1}{\gamma_1\mu}, \end{cases}$$

which equations determine the coordinates (x, y) in terms of the parameters λ, μ connected by the foregoing relation.

Writing for homogeneity $\frac{\lambda}{\nu}$, $\frac{\mu}{\nu}$ in place of λ, μ, and $\frac{x}{z}$, $\frac{y}{z}$ in place of x, y, the equations become

$$(a_0^2+a_1^2+a_2^2-a_3^2)\lambda\mu\nu+a_1a_2(\lambda^2+\mu^2)\nu-a_0a_2\mu(\nu^2+\lambda^2)-a_0a_1\lambda(\mu^2+\nu^2)=0,$$

and

$$x:y:z=(a_2\lambda+b_2\gamma_1\mu)\lambda\mu : \left(\frac{b_2}{\gamma_1}\lambda+a_2\mu\right)\nu^2 : \lambda\mu\nu.$$

Comparing with the foregoing equations

$$e\lambda\mu\nu+f(\lambda^2+\mu^2)\nu+g(\nu^2+\lambda^2)\mu+h(\mu^2+\nu^2)\lambda=0,$$

and

$$x:y:z=(a\lambda+b\mu)\lambda\mu : (c\lambda+d\mu)\nu^2 : \lambda\mu\nu,$$

the equations agree together, and we have

$$\begin{cases} e=a_0^2+a_1^2+a_2^2-a_3^2, \\ f=+a_1a_2, \\ g=-a_0a_2, \\ h=-a_0a_1, \\ a=\quad a_2, \\ b=\quad b_1\gamma_1, \\ c=\quad \dfrac{b_1}{\gamma_1}, \\ d=\quad a_2. \end{cases}$$

The tangents at the triple points thus are

$$\begin{array}{ll} x=0, & y=0, \\ a_1x-a_0b_1\gamma_1z=0, & a_1y-\dfrac{a_0b_1}{\gamma_1}z=0, \\ x-a_0z=0, & y-a_0z=0; \end{array}$$

viz. restoring the rectangular coordinates, and for γ substituting the value $\cos C+i\sin C$, for a_0 writing a, and taking $b=\frac{ab_1}{a_1}$, we have

$$\begin{array}{ll} X+iY=0, & X-iY=0, \\ X+iY=b(\cos C+i\sin C), & X-iY=b(\cos C-i\sin C), \\ X+iY=a_0, & X-iY=a_0; \end{array}$$

viz. the first two intersect in the point $(0, 0)$, the second two in the point $(b\cos C, b\sin C)$, the third two in the point $(a, 0)$: the first and third of these are the points B and C, the second of them is the point A of the figure; viz. the formulæ give the point A, forming, with B and C, a triad of foci.

625.

ON THE CONDITION FOR THE EXISTENCE OF A SURFACE CUTTING AT RIGHT ANGLES A GIVEN SET OF LINES.

[From the *Proceedings of the London Mathematical Society*, vol. VIII. (1876—1877), pp. 53—57. Read December 14, 1876.]

IN a congruency or doubly infinite system of right lines, the direction-cosines α, β, γ of the line through any given point (x, y, z), are expressible as functions of x, y, z; and it was shown by Sir W. R. Hamilton in a very elegant manner that, in order to the existence of a surface (or, what is the same thing, a set of parallel surfaces) cutting the lines at right angles, $\alpha dx + \beta dy + \gamma dz$ must be an exact differential: when this is so, writing $V = \int (\alpha dx + \beta dy + \gamma dz)$, we have $V = c$, the equation of the system of parallel surfaces each cutting the given lines at right angles.

The proof is as follows:—If the surface exists, its differential equation is $\alpha dx + \beta dy + \gamma dz = 0$, and this equation must therefore be integrable by a factor. Now the functions α, β, γ are such that $\alpha^2 + \beta^2 + \gamma^2 = 1$, and they besides satisfy a system of partial differential equations which Hamilton deduces from the geometrical notion of a congruency; viz. passing from the point (x, y, z) to the consecutive point on the line, that is, to the point whose coordinates are $x + \rho\alpha$, $y + \rho\beta$, $z + \rho\gamma$ (ρ infinitesimal), the line belonging to this point is the original line; and consequently α, β, γ, considered as functions of x, y, z, must remain unaltered when these variables are changed into $x + \rho\alpha$, $y + \rho\beta$, $z + \rho\gamma$, respectively. We thus obtain the equations

$$\alpha \frac{d\alpha}{dx} + \beta \frac{d\alpha}{dy} + \gamma \frac{d\alpha}{dz} = 0,$$

$$\alpha \frac{d\beta}{dx} + \beta \frac{d\beta}{dy} + \gamma \frac{d\beta}{dz} = 0,$$

$$\alpha \frac{d\gamma}{dx} + \beta \frac{d\gamma}{dy} + \gamma \frac{d\gamma}{dz} = 0.$$

Combining herewith the equations obtained by differentiation of $\alpha^2+\beta^2+\gamma^2=1$, viz.

$$\alpha\frac{d\alpha}{dx}+\beta\frac{d\beta}{dx}+\gamma\frac{d\gamma}{dx}=0,$$

$$\alpha\frac{d\alpha}{dy}+\beta\frac{d\beta}{dy}+\gamma\frac{d\gamma}{dy}=0,$$

$$\alpha\frac{d\alpha}{dz}+\beta\frac{d\beta}{dz}+\gamma\frac{d\gamma}{dz}=0,$$

and subtracting the corresponding equations, we obtain three equations which may be written

$$\alpha:\beta:\gamma=\frac{d\beta}{dz}-\frac{d\gamma}{dy}:\frac{d\gamma}{dx}-\frac{d\alpha}{dz}:\frac{d\alpha}{dy}-\frac{d\beta}{dx},$$

or, what is the same thing,

$$\frac{d\beta}{dz}-\frac{d\gamma}{dy},\ \frac{d\gamma}{dx}-\frac{d\alpha}{dz},\ \frac{d\alpha}{dy}-\frac{d\beta}{dx}=k\alpha,\ k\beta,\ k\gamma,$$

and, multiplying by α, β, γ, and adding,

$$k=\alpha\left(\frac{d\beta}{dz}-\frac{d\gamma}{dy}\right)+\beta\left(\frac{d\gamma}{dx}-\frac{d\alpha}{dz}\right)+\gamma\left(\frac{d\alpha}{dy}-\frac{d\beta}{dx}\right).$$

We thus see that, if the function on the right-hand vanish, then $k=0$, and consequently also

$$\frac{d\beta}{dz}-\frac{d\gamma}{dy},\ \frac{d\gamma}{dx}-\frac{d\alpha}{dz},\ \frac{d\alpha}{dy}-\frac{d\beta}{dx}\ \text{each}\ =0;$$

viz. if the equation $\alpha dx+\beta dy+\gamma dz=0$ be integrable, then $\alpha dx+\beta dy+\gamma dz$ is an exact differential; which is the theorem in question.

But it is interesting to obtain the first mentioned set of differential equations from the analytical equations of a congruency, viz. these are $x=mz+p$, $y=nz+q$, where m, n, p, q are functions of two arbitrary parameters, or, what is the same thing, p, q are given functions of m, n; and therefore, from the three equations, m, n are given functions of x, y, z. And it is also interesting to express in terms of these quantities m, n, considered as functions of x, y, z, the condition for the existence of the set of surfaces.

We have

$$\alpha,\ \beta,\ \gamma=\frac{m}{R},\ \frac{n}{R},\ \frac{1}{R},\ \text{where}\ R=\sqrt{1+m^2+n^2};$$

and thence without difficulty

$$\left(\alpha\frac{d}{dx}+\beta\frac{d}{dy}+\gamma\frac{d}{dz}\right)\alpha=\frac{1}{R^4}\left[(1+n^2)\left(m\frac{dm}{dx}+n\frac{dm}{dy}+\frac{dm}{dz}\right)\quad -mn\quad\left(m\frac{dn}{dx}+n\frac{dn}{dy}+\frac{dn}{dz}\right)\right],$$

$$\left(\alpha\frac{d}{dx}+\beta\frac{d}{dy}+\gamma\frac{d}{dz}\right)\beta=\frac{1}{R^4}\left[\quad -mn\quad\left(\quad ,, \quad\right)+(1+m^2)\left(\quad ,, \quad\right)\right],$$

$$\left(\alpha\frac{d}{dx}+\beta\frac{d}{dy}+\gamma\frac{d}{dz}\right)\gamma=\frac{1}{R^4}\left[\quad -\quad\left(\quad ,, \quad\right)\quad -\quad\left(\quad ,, \quad\right)\right],$$

so that the required equations in α, β, γ will be satisfied if only

$$m\frac{dm}{dx}+n\frac{dm}{dy}+\frac{dm}{dz}=0,$$

$$m\frac{dn}{dx}+n\frac{dn}{dy}+\frac{dn}{dz}=0,$$

and it is to be shown that these equations hold good.

Writing for shortness $dp=Adm+Bdn$, $dq=Cdm+Ddn$, the equations of the line give

$$\begin{array}{l|l} 1=z\dfrac{dm}{dx}+A\dfrac{dm}{dx}+B\dfrac{dn}{dx}, & 0=z\dfrac{dn}{dx}+C\dfrac{dm}{dx}+D\dfrac{dn}{dx}, \\ 0=z\dfrac{dm}{dy}+A\dfrac{dm}{dy}+B\dfrac{dn}{dy}, & 1=z\dfrac{dn}{dy}+C\dfrac{dm}{dy}+D\dfrac{dn}{dy}, \\ -m=z\dfrac{dm}{dz}+A\dfrac{dm}{dz}+B\dfrac{dn}{dz}, & -n=z\dfrac{dn}{dz}+C\dfrac{dm}{dz}+D\dfrac{dn}{dz}; \end{array}$$

or, writing

$$\lambda,\ \mu,\ \nu=\frac{dm}{dy}\frac{dn}{dz}-\frac{dm}{dz}\frac{dn}{dy},\quad \frac{dm}{dz}\frac{dn}{dx}-\frac{dm}{dx}\frac{dn}{dz},\quad \frac{dm}{dx}\frac{dn}{dy}-\frac{dm}{dy}\frac{dn}{dx},$$

so that identically

$$\lambda\frac{dm}{dx}+\mu\frac{dm}{dy}+\nu\frac{dm}{dz}=0,$$

$$\lambda\frac{dn}{dx}+\mu\frac{dn}{dy}+\nu\frac{dn}{dz}=0,$$

then in each set, multiplying by λ, μ, ν and adding, so as to eliminate A, B, C, D, we find

$$\lambda-m\nu=0,\quad \mu-n\nu=0.$$

Substituting these values of λ, μ in the last preceding equations, ν divides out, and we have the two equations in question.

The foregoing equations give further

$$A,\ B,\ C,\ D=-z+\frac{1}{\nu}\frac{dn}{dy},\quad -\frac{1}{\nu}\frac{dm}{dy},\quad -\frac{1}{\nu}\frac{dn}{dx},\quad -z+\frac{1}{\nu}\frac{dm}{dx}.$$

Taking for α, β, γ the before-mentioned values, we find

$$\frac{d\alpha}{dy}-\frac{d\beta}{dx}=\frac{1}{R}\left(\frac{dm}{dy}-\frac{dn}{dx}\right)-\frac{m}{R^3}\left(m\frac{dm}{dy}+n\frac{dn}{dy}\right)-\frac{n}{R^3}\left(m\frac{dm}{dx}+n\frac{dn}{dx}\right)$$

$$=\frac{1}{R^3}\left\{(1+n^2)\frac{dm}{dy}-(1+m^2)\frac{dn}{dx}+mn\left(\frac{dm}{dx}-\frac{dn}{dy}\right)\right\};$$

and similarly, but using the equations

$$m\frac{dm}{dx}+n\frac{dm}{dy}+\frac{dm}{dz}=0,\quad m\frac{dn}{dx}+n\frac{dn}{dy}+\frac{dn}{dz}=0,$$

to eliminate the coefficients $\frac{dm}{dz}, \frac{dn}{dz}$ which in the first instance present themselves, we find

$$\frac{d\beta}{dz}-\frac{d\gamma}{dy}=\frac{m}{R^3}\left\{(1+n^2)\frac{dm}{dy}-(1+m^2)\frac{dn}{dx}+mn\left(\frac{dm}{dx}-\frac{dn}{dy}\right)\right\},$$

$$\frac{d\gamma}{dx}-\frac{d\alpha}{dz}=\frac{n}{R^3}\left\{\quad ,, \quad ,, \quad ,, \quad\right\}:$$

whence, multiplying by γ, α, β, and adding,

$$\alpha\left(\frac{d\beta}{dz}-\frac{d\gamma}{dy}\right)+\beta\left(\frac{d\gamma}{dx}-\frac{d\alpha}{dz}\right)+\gamma\left(\frac{d\alpha}{dy}-\frac{d\beta}{dx}\right)$$

$$=\frac{1}{1+m^2+n^2}\left\{(1+n^2)\frac{dm}{dy}-(1+m^2)\frac{dn}{dx}+mn\left(\frac{dm}{dx}-\frac{dn}{dy}\right)\right\};$$

or we have

$$(1+n^2)\frac{dm}{dy}-(1+m^2)\frac{dn}{dx}+mn\left(\frac{dm}{dx}-\frac{dn}{dy}\right)=0$$

as the condition for the existence of the set of surfaces.

It is clear that the condition is satisfied when the lines are the normals of a given surface: seeking the surfaces which cut the lines at right angles, we obtain the parallel surfaces; and we are led to the theorem that any parallel surface is the locus of the extremity of a line of constant length measured off from each point of the surface along the normal—or, what is equivalent thereto, the parallel surface is the envelope of a sphere of constant radius having its centre on the surface. I will verify the theorem for the case of the ellipsoid. Taking X, Y, Z as the coordinates of a point on the ellipsoid $\frac{X^2}{a^2}+\frac{Y^2}{b^2}+\frac{Z^2}{c^2}=1$, and x, y, z as current coordinates, the equations of the normal are

$$\frac{a^2}{X}(x-X)=\frac{b^2}{Y}(y-Y)=\frac{c^2}{Z}(z-Z),\ (=\lambda \text{ suppose}).$$

We have therefore

$$X,\ Y,\ Z=\frac{a^2x}{a^2+\lambda},\ \frac{b^2y}{b^2+\lambda},\ \frac{c^2z}{c^2+\lambda},$$

and thence

$$\frac{a^2x^2}{(a^2+\lambda)^2}+\frac{b^2y^2}{(b^2+\lambda)^2}+\frac{c^2z^2}{(c^2+\lambda)^2}=1,$$

an equation which determines λ as a function of x, y, z.

The direction-cosines α, β, γ of the normal are proportional to $\frac{X}{a^2}, \frac{Y}{b^2}, \frac{Z}{c^2}$, that is, to $\dfrac{x}{a^2+\lambda}$, $\dfrac{y}{b^2+\lambda}$, $\dfrac{z}{c^2+\lambda}$, and the equation $\alpha^2+\beta^2+\gamma^2=1$ then determines their absolute magnitudes: the equation $\alpha dx+\beta dy+\gamma dz=dV$ thus is

$$\frac{\dfrac{x\,dx}{a^2+\lambda}+\dfrac{y\,dy}{b^2+\lambda}+\dfrac{z\,dz}{c^2+\lambda}}{\sqrt{\dfrac{x^2}{(a^2+\lambda)^2}+\dfrac{y^2}{(b^2+\lambda)^2}+\dfrac{z^2}{(c^2+\lambda)^2}}}=dV,$$

viz. the left-hand side, considering therein λ as a given function of V, is an exact differential. We verify this by finding the value of V, viz. writing down the two equations

$$\frac{x^2}{(a^2+\lambda)^2}+\frac{y^2}{(b^2+\lambda)^2}+\frac{z^2}{(c^2+\lambda)^2}-\frac{V^2}{\lambda^2}=0,$$

$$\frac{x^2}{a^2+\lambda}+\frac{y^2}{b^2+\lambda}+\frac{z^2}{c^2+\lambda}-\frac{V^2}{\lambda}=1,$$

these are equivalent in virtue of the equation that determines λ; and it is to be shown that, regarding V as given by either of them, say by the second equation, we have for dV its foregoing value. In fact, differentiating the second equation, the term in $d\lambda$ disappears by virtue of the first equation, and the result is

$$\frac{x\,dx}{a^2+\lambda}+\frac{y\,dy}{b^2+\lambda}+\frac{z\,dz}{c^2+\lambda}-\frac{V\,dV}{\lambda}=0,$$

in which substituting for $\dfrac{V}{\lambda}$ its value from the first equation, we have for dV the value in question. Regarding V as a given constant, the two equations give, by elimination of λ, an equation $\phi(x, y, z, V)=0$, which is, in fact, the surface parallel to the ellipsoid and at a constant normal distance $=V$ from it.

626.

ON THE GENERAL DIFFERENTIAL EQUATION $\frac{dx}{\sqrt{X}}+\frac{dy}{\sqrt{Y}}=0$, WHERE X, Y ARE THE SAME QUARTIC FUNCTIONS OF x, y RESPECTIVELY.

[From the *Proceedings of the London Mathematical Society*, vol. VIII. (1876—1877), pp. 184—199. Read February 8, 1877.]

WRITE $\Theta = a + b\theta + c\theta^2 + d\theta^3 + e\theta^4$, the general quartic function of θ; and let it be required to integrate by Abel's theorem the differential equation

$$\frac{dx}{\sqrt{X}}+\frac{dy}{\sqrt{Y}}=0.$$

We have

$$\begin{vmatrix} x^2, & x, & 1, & \sqrt{X} \\ y^2, & y, & 1, & \sqrt{Y} \\ z^2, & z, & 1, & \sqrt{Z} \\ w^2, & w, & 1, & \sqrt{W} \end{vmatrix} = 0,$$

a particular integral of

$$\frac{dx}{\sqrt{X}}+\frac{dy}{\sqrt{Y}}+\frac{dz}{\sqrt{Z}}+\frac{dw}{\sqrt{W}}=0;$$

and consequently the above equation, taking therein z, w as constants, is the general integral of

$$\frac{dx}{\sqrt{X}}+\frac{dy}{\sqrt{Y}}=0,$$

viz. the two constants z, w must enter in such wise that the equation contains only a single constant; whence also, attributing to w any special value, we have the general integral with z as the arbitrary constant.

Take $w=\infty$; the equation becomes

$$\begin{vmatrix} x^2, & x, & 1, & \sqrt{X} \\ y^2, & y, & 1, & \sqrt{Y} \\ z^2, & z, & 1, & \sqrt{Z} \\ 1, & 0, & 0, & \sqrt{e} \end{vmatrix}=0,$$

a relation between x, y, z which may be otherwise expressed by means of the identity

$$e(\theta^2+\beta\theta+\gamma)^2-(e\theta^4+d\theta^3+c\theta^2+b\theta+a)=(2\beta e-d)(\theta-x)(\theta-y)(\theta-z),$$

or, what is the same thing,

$$\begin{aligned} e\,(2\gamma+\beta^2)-c &= -(2\beta e-d)(x+y+z),\\ e\ 2\beta\gamma \qquad -b &= \quad (2\beta e-d)(yz+zx+xy),\\ e\ \gamma^2 \qquad -a &= -(2\beta e-d)\,xyz, \end{aligned}$$

where β, γ are indeterminate coefficients which are to be eliminated.

Write

$$x^2-\frac{\sqrt{X}}{\sqrt{e}}=P,\quad y^2-\frac{\sqrt{Y}}{\sqrt{e}}=Q;$$

then we have

$$\beta x+\gamma+P=0,\quad \beta y+\gamma+Q=0;$$

giving

$$\beta:\gamma:1=Q-P:Py-Qx:x-y.$$

Substituting these values in the first of the preceding three equations, we have

$$e\,\frac{2(Py-Qx)(x-y)+(Q-P)^2}{(x-y)^2}-c=-\left\{\frac{2(Q-P)e}{x-y}-d\right\}(x+y+z),$$

that is,

$$e\left\{\frac{2(Qy-Px)}{x-y}+\frac{(Q-P)^2}{(x-y)^2}+\frac{2(Q-P)}{x-y}z\right\}=c+d(x+y+z);$$

or, reducing by

$$Qy-Px=y^3-x^3+\frac{x\sqrt{X}-y\sqrt{Y}}{\sqrt{e}},$$

$$Q-P=y^2-x^2+\frac{\sqrt{X}-\sqrt{Y}}{\sqrt{e}},\quad =y^2-x^2+(y-x)\frac{M}{\sqrt{e}},\ \text{if}\ M=\frac{\sqrt{X}-\sqrt{Y}}{x-y},$$

this is

$$e\left\{\frac{2(x\sqrt{X}-y\sqrt{Y})}{\sqrt{e}(x-y)}+2xy+\frac{M^2}{e}-2(x+y)\frac{M}{\sqrt{e}}-2(x+y)z+2z\frac{M}{\sqrt{e}}\right\}$$

$$=c+d(x+y+z)+e(x+y)^2.$$

We have Euler's solution in the far more simple form

$$M^2=C+d(x+y)+e(x+y)^2,$$

where C is the arbitrary constant. It is to be observed that, in the particular case where $e=0$, the first equation becomes

$$M^2=c+d(x+y+z);$$

and the two results for this case agree on putting $C=c+dz$.

But it is required to identify the two solutions in the general case where e is not $=0$. I remark that I have, in my *Treatise on Elliptic Functions*, Chap. XIV., further developed the theory of Euler's solution, and have shown that, regarding C as variable, and writing

$$\mathfrak{C}=ad^2+b^2e-2bcd+C[-4ae+bd+(C-c)^2],$$

then the given equation between the variables x, y, C corresponds to the differential equation

$$\frac{dx}{\sqrt{X}}+\frac{dy}{\sqrt{Y}}+\frac{dC}{\sqrt{\mathfrak{C}}}=0,$$

a result which will be useful for effecting the identification. The Abelian solution may be written

$$e\left\{\frac{2(x\sqrt{X}-y\sqrt{Y})}{\sqrt{e}(x-y)}-x^2-y^2+\frac{M^2}{e}-2(x+y)\frac{M}{\sqrt{e}}\right\}-c-d(x+y)=z\{d+2e(x+y)-2M\sqrt{e}\};$$

and substituting for M its value, and multiplying by $(x-y)^2$, the equation becomes

$$\begin{aligned}2\sqrt{e}(x-y)(x\sqrt{X}-y\sqrt{Y})-e(x^2+y^2)(x-y)^2+(\sqrt{X}-\sqrt{Y})^2&\\ -2(x^2-y^2)(\sqrt{X}-\sqrt{Y})\sqrt{e}-c(x-y)^2-d(x+y)(x-y)^2&\\ =z(x-y)\{d(x-y)+2e(x^2-y^2)-2(\sqrt{X}-\sqrt{Y})\sqrt{e}\}.&\end{aligned}$$

On the left-hand side, the rational part is

$$X+Y+c(-x^2+2xy-y^2)+d(-x^3+x^2y+xy^2-y^3)+e(-x^4+2x^3y-2x^2y^2+2xy^3-y^4),$$

which, substituting therein for X, Y their values, becomes

$$=2a+b(x+y)+c\,.\,2xy+d\,xy(x+y)+e\,.\,2xy(x^2-xy+y^2);$$

and the irrational part is at once found to be

$$=2\sqrt{e}(x-y)(x\sqrt{Y}-y\sqrt{X})-2\sqrt{XY}.$$

The equation thus is

$$z=\frac{\left\{\begin{array}{c}2a+b(x+y)+c\,.\,2xy+d\,xy(x+y)+e\,.\,2xy(x^2-xy+y^2)\\ +2\sqrt{e}(x-y)(x\sqrt{Y}-y\sqrt{X})-2\sqrt{XY}\end{array}\right\}}{(x-y)\{d(x-y)+2e(x^2-y^2)-2(\sqrt{X}-\sqrt{Y})\sqrt{e}\}},$$

which equation is thus a form of the general integral of $\frac{dx}{\sqrt{X}}+\frac{dy}{\sqrt{Y}}=0$, and also a particular integral of $\frac{dx}{\sqrt{X}}+\frac{dy}{\sqrt{Y}}+\frac{dz}{\sqrt{Z}}=0$.

Multiplying the numerator and the denominator by

$$d(x-y)+2e(x^2-y^2)+2(\sqrt{X}-\sqrt{Y})\sqrt{e},$$

the denominator becomes

$$=(x-y)^3\left[\{d+2e(x+y)\}^2-4e\left(\frac{\sqrt{X}-\sqrt{Y}}{x-y}\right)^2\right],$$

which, introducing herein the C of Euler's equation, is

$$=(x-y)^3(d^2-4eC).$$

We have therefore

$$z(x-y)^3(d^2-4eC)=\{2a+b(x+y)+c\,.\,2xy+d\,xy(x+y)+e\,.\,2xy(x^2-xy+y^2)$$
$$+2\sqrt{e}(x-y)(x\sqrt{Y}-y\sqrt{X})-2\sqrt{XY}\}\times\{d(x-y)+2e(x^2-y^2)+2\sqrt{e}(\sqrt{X}-\sqrt{Y})\}.$$

Using $\mathfrak{C}$ to denote the same value as before, the function on the right-hand is, in fact,

$$=(x-y)^3\{2be-cd+dC+2\sqrt{e}\sqrt{\mathfrak{C}}\};$$

and, this being so, the required relation between z, C is

$$z(d^2-4eC)=\{2be-cd+dC+2\sqrt{e}\sqrt{\mathfrak{C}}\}.$$

To prove this, we have first, from the equation

$$\left(\frac{\sqrt{X}-\sqrt{Y}}{x-y}\right)^2=C+d(x+y)+e(x+y)^2,$$

to express $\mathfrak{C}$ as a function of x, y. This equation, regarding therein C as a variable, gives

$$\frac{dx}{\sqrt{X}}+\frac{dy}{\sqrt{Y}}+\frac{dC}{\sqrt{\mathfrak{C}}}=0;$$

and we have therefore

$$-\sqrt{\mathfrak{C}}=\sqrt{X}\frac{dC}{dx}=\sqrt{Y}\frac{dC}{dy},$$

viz. $\sqrt{X}\frac{dC}{dx}$ will be a symmetrical function of x, y. Putting, as before

$$M=\frac{\sqrt{X}-\sqrt{Y}}{x-y},$$

we have

$$C=M^2-d(x+y)-e(x+y)^2,$$

and thence

$$\frac{dC}{dx}=2M\frac{dM}{dx}-d-2e(x+y).$$

We have

$$\frac{dM}{dx}=\frac{1}{x-y}\,\frac{X'}{2\sqrt{X}}-\frac{\sqrt{X}-\sqrt{Y}}{(x-y)^2},$$

and hence

$$\sqrt{\mathfrak{C}}\,(x-y)^3 = -\sqrt{X}\,(x-y)^3\left\{2M\frac{dM}{dx} - d - 2e\,(x+y)\right\}$$

$$= -(x-y)\,X'(\sqrt{X}-\sqrt{Y}) + 2\,(X+Y-2\sqrt{XY})\sqrt{X}$$
$$+ (d+2e\,\overline{x+y})\,(x-y)^3\sqrt{X}$$

$$= \;\;[(x-y)\,X' + 2X + 2Y + (d+2e\,\overline{x+y})\,(x-y)^3]\sqrt{X}$$
$$+ [(x-y)\,X' - 4X]\sqrt{Y}.$$

We obtain at once the coefficient of $\sqrt{Y}$, and with little more difficulty that of $\sqrt{X}$; and the result is

$$\sqrt{\mathfrak{C}}\,(x-y)^3 = -[4a+3bx+2cx^2+dx^3+y\,(b+2cx+3dx^2+4ex^3)]\sqrt{Y}$$
$$+[4a+3by+2cy^2+dy^3+x\,(b+2cy+3dy^2+4ey^3)]\sqrt{X}.$$

We have also

$$C\,(x-y)^2 = (\sqrt{X}-\sqrt{Y})^2 - d\,(x+y)\,(x-y)^2 - e\,(x+y)^2\,(x-y)^2$$
$$= X+Y-d\,(x^3-x^2y-xy^2+y^3) - e\,(x^4-2x^2y^2+y^4) - 2\sqrt{XY}$$
$$= 2a+b\,(x+y)+c\,(x^2+y^2)+d\,xy\,(x+y)+2e\,x^2y^2-2\sqrt{XY},$$

or, say

$$C\,(x-y)^3 = 2a\,(x-y)+b\,(x^2-y^2)+c\,(x^3-x^2y+xy^2-y^3)+d\,xy\,(x^2-y^2)$$
$$+2e\,x^2y^2\,(x-y)-2\,(x-y)\sqrt{XY}.$$

We can hence form the expression of

$$(x-y)^3\,\{2be-cd+dC+2\sqrt{e}\sqrt{\mathfrak{C}}\},$$

viz. this is

$$=(2be-cd)\,(x-y)^3+2ad\,(x-y)+bd\,(x^2-y^2)+cd\,(x^3-x^2y+xy^2-y^3)+d^2\,xy\,(x^2-y^2)$$
$$+2de\,x^2y^2\,(x-y)-2d\,(x-y)\sqrt{XY}$$
$$+2\sqrt{e}\,\{[-(4a+3bx+2cx^2+dx^3)-y\,(b+2cx+3dx^2+4ex^3)]\sqrt{Y}$$
$$+[(4a+3by+2cy^2+dy^3)+x\,(b+2cy+3dy^2+4ey^3)]\sqrt{X}\},$$

and this should be

$$=\{2a+b\,(x+y)+c\,.\,2xy+d\,xy\,(x+y)+e\,.\,2xy\,(x^2-xy+y^2)$$
$$+2\sqrt{e}\,(x-y)\,(x\sqrt{Y}-y\sqrt{X})-2\sqrt{XY}\}\times\{d\,(x-y)+2e\,(x^2-y^2)+2\sqrt{e}\,(\sqrt{X}-\sqrt{Y})\}.$$

The function on the right-hand is, in fact,

$$=\{2a+b\,(x+y)+c\,.\,2xy+d\,xy\,(x+y)+e\,.\,2xy\,(x^2-xy+y^2)-2\sqrt{XY}\}$$
$$\times\{d\,(x-y)+2e\,(x^2-y^2)\}+4e\,(x-y)\,(\sqrt{X}-\sqrt{Y})\,(x\sqrt{Y}-y\sqrt{X})$$
$$+2\sqrt{e}\,(\sqrt{X}-\sqrt{Y})\,\{2a+b\,(x+y)+c\,.\,2xy+d\,xy\,(x+y)+e\,.\,2xy\,(x^2-xy+y^2)-2\sqrt{XY}\}$$
$$+2\sqrt{e}\,(x-y)\,(x\sqrt{Y}-y\sqrt{X})\,\{d\,(x-y)+2e\,(x^2-y^2)\},$$

viz. this is

$$\begin{aligned}
&= \{2a + b(x+y) + c\,.\,2xy + d\,xy(x+y) + e\,.\,2xy(x^2-xy+y^2)\} \\
&\qquad\qquad \times \{d(x-y) + 2e(x^2-y^2)\} + 4e(x-y)(-xY-yX) \\
&\quad - 2\sqrt{XY}\{d(x-y)+2e(x^2-y^2)\} + 4e(x-y)(x+y)\sqrt{XY} \\
&\quad + 2\sqrt{e}\left\{\begin{array}{l} \sqrt{X}\{2a+b(x+y)+c\,.\,2xy+d\,xy(x+y)+e\,.\,2xy(x^2-xy+y^2) \\ \qquad + 2Y - (x-y)\,y\,[d(x-y)+2e(x^2-y^2)]\} \\ -\sqrt{Y}\{2a+b(x+y)+c\,.\,2xy+d\,xy(x+y)+e\,.\,2xy(x^2-xy+y^2) \\ \qquad + 2X - (x-y)\,x\,[d(x-y)+2e(x^2-y^2)]\} \end{array}\right\},
\end{aligned}$$

which is, in fact, equal to the expression on the left-hand side.

To complete the theory, we require to express $\sqrt{Z}$ as a function of x, y. It would be impracticable to effect this by direct substitution of the foregoing value of z; but, observing that the value in question is a solution of $\frac{dx}{\sqrt{X}}+\frac{dy}{\sqrt{Y}}+\frac{dz}{\sqrt{Z}}=0$, or, what is the same thing, that $\frac{1}{\sqrt{X}}+\frac{1}{\sqrt{Z}}\frac{dz}{dx}=0$, $\frac{1}{\sqrt{Y}}+\frac{1}{\sqrt{Z}}\frac{dz}{dy}=0$, we can from either of these equations, considering therein z as a given function of x, y, calculate $\sqrt{Z}$.

Writing for shortness

$$z=\frac{J-2\sqrt{e}\,y(x-y)\sqrt{X}+2\sqrt{e}\,x(x-y)\sqrt{Y}-2\sqrt{XY}}{R-2\sqrt{e}(x-y)\sqrt{X}+2\sqrt{e}(x-y)\sqrt{Y}},$$

where

$$R=(x-y)^2\{d+2e(x+y)\},$$

$$J=2a+b(x+y)+2c\,xy+d\,xy(x+y)+2e\,xy(x^2-xy+y^2);$$

or, if for a moment $z=\frac{N}{D}$, then

$$\frac{dz}{dx}=\frac{1}{D^2}\left(D\frac{dN}{dx}-N\frac{dD}{dx}\right)=-\frac{\sqrt{Z}}{\sqrt{X}},$$

that is,

$$\sqrt{Z}=\frac{\sqrt{X}}{D^2}\left(N\frac{dD}{dx}-D\frac{dN}{dx}\right),\ =\frac{\Omega}{D^2}\text{ suppose};$$

or, writing for shortness X', R', J to denote the derived functions $\frac{dX}{dx}$, $\frac{dR}{dx}$, $\frac{dJ}{dx}$, (Y' is afterwards written to denote $\frac{dY}{dy}$, but as the final formulæ contain only X', $=\frac{dX}{dx}$, and Y', $=\frac{dY}{dy}$, this does not occasion any defect of symmetry), we find

$$\begin{aligned}
\Omega = {}& N\{R'\sqrt{X}-2\sqrt{e}\,X-\sqrt{e}(x-y)X'+2\sqrt{e}\sqrt{XY}\} \\
& -D\{J'\sqrt{X}-2\sqrt{e}\,yX-\sqrt{e}(x-y)\,yX'+2\sqrt{e}(2x-y)\sqrt{XY}-X'\sqrt{Y}\};
\end{aligned}$$

and substituting herein for N, D their values, and arranging the terms, we find

$$\Omega=\sqrt{e}\,\mathfrak{A}+\mathfrak{B}\sqrt{X}+\mathfrak{C}\sqrt{Y}+\sqrt{e}\,\mathfrak{D}\sqrt{XY},$$

where

$$\begin{aligned}\mathfrak{A}=&-J\{2X+(x-y)X'\}\\&-2(x-y)yR'X\\&-4XY\\&+Ry\{2X+(x-y)X'\}\\&+2(x-y)XJ'\\&+2(x-y)X'Y,\end{aligned}\qquad\begin{aligned}\mathfrak{B}=&\ \ JR'\\&+2e(x-y)y\{2X+(x-y)X'\}\\&+4ex(x-y)Y\\&-RJ'\\&-2e(x-y)y\{2X+(x-y)X'\}\\&-4e(x-y)(2x-y)Y,\end{aligned}$$

$$\begin{aligned}\mathfrak{C}=&-4ey(x-y)X\\&-2e(x-y)x\{2X+(x-y)X'\}\\&-2R'X\\&+RX'\\&+2e(x-y)y\{2X+(x-y)X'\}\\&+4e(x-y)(2x-y)X,\end{aligned}\qquad\begin{aligned}\mathfrak{D}=&\ \ 2J\\&+2(x-y)xR'\\&+2\{2X+(x-y)X'\}\\&-2(2x-y)R\\&-2(x-y)X'\\&-2(x-y)J',\end{aligned}$$

where the terms have been written down as they immediately present themselves; but, collecting and arranging, we have

$$\begin{aligned}\mathfrak{A}&=\ \ 2X(-J+Ry-2Y)+(x-y)\{2XJ'+2X'Y-X'J-2yR'X+yRX'\},\\\mathfrak{B}&=\ \ JR'-J'R-4e(x-y)^2Y,\\\mathfrak{C}&=-2XR'+X'R+4e(x-y)^2X-2e(x-y)^3X',\\\mathfrak{D}&=\ \ 2J+4X-2Rx+2(x-y)(xR'-R-J').\end{aligned}$$

To reduce these expressions, writing

$$M=d+2e(x+y),$$
$$\Lambda=c+\ d(x+y)+e(x^2+y^2),$$

we have $R=(x-y)^2M$, and therefore $R'=2(x-y)M+2e(x-y)^2$; also

$$J=X+Y-(x-y)^2\Lambda;$$

also, from the original form,

$$J'=b+2cy+d(2xy+y^2)+e(6x^2y-4xy^2+2y^3).$$

The final values are

$$\begin{aligned}\mathfrak{A}&=\quad -X^2-6XY-Y^2+(x-y)^4\{\Lambda^2+(-b+dxy)M+xyM^2\},\\\mathfrak{B}&=\ \ (x-y)M\{4Y+(x-y)Y'\}+2e(x-y)^3Y',\\\mathfrak{C}&=-(x-y)M\{4X-(x-y)X'\}-2e(x-y)^3X',\\\mathfrak{D}&=\qquad 4(X+Y)+4e(x-y)^4,\end{aligned}$$

which, once obtained, may be verified without difficulty.

Verification of 𝔄.—The equation is

$$-X^2-6XY-Y^2+(x-y)^4\{\Lambda^2+(-b+dxy)M+xyM^2\}$$
$$=2X(-J+Ry-2Y)+(x-y)\{2XJ'+2X'Y-X'J-2yR'X+yRX'\};$$

or, putting for shortness

$$\Lambda^2+(-b+dxy)M+xyM^2=\nabla,$$

this is

$$\begin{aligned}(x-y)^4\nabla= &\ X^2+6XY+Y^2\\ &+2X\{-X-3Y+(x-y)^2\Lambda+(x-y)^2yM\}\\ &+(x-y)\{2XJ'+2X'Y-X'J-2yR'X+yRX'\},\\ =&-X^2+Y^2+2(x-y)^2X\Lambda+2(x-y)^2yXM\\ &+(x-y)\{2XJ'+2X'Y-X'J-2yR'X+yRX'\};\end{aligned}$$

we have $-X^2+Y^2=-(X-Y)(X+Y)$, where $X-Y$ divides by $x-y$, $=(x-y)\Omega$ suppose; hence, throwing out the factor $x-y$, the equation becomes

$$\begin{aligned}(x-y)^3\nabla=&-\Omega(X+Y)+2(x-y)X\Lambda+2(x-y)yXM\\ &+2XJ'+2X'Y-X'\{X+Y-(x-y)^2\Lambda\}\\ &-2yX\{2(x-y)M+2(x-y)^2e\}+(x-y)^2yMX',\\ =&-\Omega(X+Y)+2XJ'-X'(X-Y)\\ &+2(x-y)X\Lambda-2(x-y)yXM\\ &+(x-y)^2X'\Lambda-4(x-y)^2eyX+(x-y)^2yMX'.\end{aligned}$$

We have $2XJ'=J'(X+Y)+J'(X-Y)$, and hence the first line is

$$=(-\Omega+J')(X+Y)+J'(X-Y);$$

$-\Omega+J'$, as will be shown, divides by $x-y$, or say it is $=(x-y)\Phi$, and, as before, $X-Y$ is $=(x-y)\Omega$; hence, throwing out the factor $x-y$, the equation becomes

$$(x-y)^2\nabla=\Phi(X+Y)+\Omega(J'-X')+2X\Lambda-2yXM+(x-y)\{X'\Lambda-4eyX+yMX'\}.$$

We have

$$\Omega=b+c(x+y)+d(x^2+xy+y^2)+e(x^3+x^2y+xy^2+y^3),$$

and thence

$$-\Omega+J'=c(-x+y)+d(-x^2+xy)+e(-x^3+5x^2y-5xy^2+y^3);$$

or, dividing this by $(x-y)$, we find

$$\Phi=-c-dx-e(x^2-4xy+y^2),$$

or, as this may be written,

$$\Phi=-\Lambda+dy+4exy.$$

We find, moreover,

$$J'-X'=2c\,(-x+y)+d\,(-3x^2+2xy+y^2)+e\,(-4x^3+6x^2y-4xy^2+2y^3),$$

which divides by $(x-y)$, the quotient being

$$-2c-d\,(3x+y)-e\,(4x^2-2xy+2y^2),$$

viz. this is

$$=-2\Lambda-(x-y)\,(d+2ex).$$

Hence the equation now is

$$\begin{aligned}(x-y)^2\,\nabla=&(X+Y)\{-\Lambda+dy+4exy\}+2X\Lambda-2yXM\\&+(x-y)\,\Omega\,\{-2\Lambda-(x-y)\,(d+2ex)\}\\&+(x-y)\quad\{\;X'\Lambda-4eyX+yMX'\;\}.\end{aligned}$$

The first line is

$$(X+Y)\{-\Lambda+yM+2\,(x-y)\,ye\}+2X\Lambda-2yXM,$$

which is

$$=(\Lambda-yM)\,(X-Y)+2\,(x-y)\,ey\,(X+Y);$$

hence, throwing out the factor $x-y$, the equation becomes

$$\begin{aligned}(x-y)\,\nabla&=(\Lambda-yM)\,\Omega+2ey\,(X+Y)-2\Lambda\Omega+X'\Lambda-4eyX+yMX'-(x-y)\,\Omega\,(d+2ex)\\&=(\Lambda+yM)\,(-\Omega+X')-2ey\,(X-Y)-(x-y)\,\Omega\,(d+2ex).\end{aligned}$$

We have

$$-\Omega+X'=c\,(x-y)+d\,(2x^2-xy-y^2)+e\,(3x^3-x^2y-xy^2-y^3),$$

which is $=(x-y)\,(\Lambda+xM)$: also $(X-Y)=(x-y)\,\Omega$, as before; whence, throwing out the factor $x-y$, the equation is

$$\nabla=(\Lambda+xM)\,(\Lambda+yM)-2ey\Omega-(d+2ex)\,\Omega,$$

that is,

$$\nabla=(\Lambda+xM)\,(\Lambda+yM)-M\Omega\,;$$

viz. substituting for ∇ its value, reducing, and throwing out the factor M, the equation becomes

$$-b+dxy=(x+y)\,\Lambda-\Omega,$$

which is right.

Verification of $\mathfrak{B}$.—The equation is

$$\begin{aligned}J\,\{2\,(x-y)\,M+2e\,(x-y)^2\}-J'\,(x-y)^2\,M-4e\,(x-y)^2\,Y&\\=4\,(x-y)\,MY+(x-y)^2\,MY'+2e\,(x-y)^3\,Y',&\end{aligned}$$

which, throwing out the factor $x-y$, is

$$0=2M\,(-J+2Y)+(x-y)\,M\,(J'+Y')+2e\,(x-y)\,(-J+2Y)+2e\,(x-y)^2\,Y'.$$

Here $-J+2Y, =-(X-Y)+(x-y)^2\Lambda$, is divisible by $(x-y)$: hence, throwing out the factor $x-y$, the equation is

$$0=M\{-2b-2c(x+y)-2d(x^2+xy+y^2)-2e(x^3+x^2y+xy^2+y^3)\}$$
$$+M(J'+Y')+2M(x-y)\Lambda+2e(-J+2Y)+2e(x-y)Y'.$$

In the first and second terms, the factor which multiplies M is

$$c(-2x+2y)+d(-2x^2+2y^2)+e(-2x^3+4x^2y-6xy^2+4y^3),$$

which is divisible by $x-y$; also $-J+2Y, =-(X-Y)+(x-y)^2\Lambda$, is divisible by $(x-y)$: hence, throwing this factor out, the equation is

$$0=M\{-2c+d(-2x-2y)+e(-2x^2+2xy-4y^2)\}+2M\Lambda$$
$$+2e\{-b-c(x+y)-d(x^2+xy+y^2)-e(x^3+x^2y+xy^2+y^3)\}$$
$$+2e(x-y)\Lambda+2eY'.$$

Here in the first line the coefficient of M is $=e(2xy-2y^2)$: hence, throwing out the constant factor $2e$, the equation is

$$0=-b-c(x+y)-d(x^2+xy+y^2)-e(x^3+x^2y+xy^2+y^3)+Y'+(x-y)yM+(x-y)\Lambda.$$

The first five terms are

$$=c(-x+y)+d(-x^2-xy+2y^2)+e(-x^3-x^2y-xy^2+3y^3),$$

which is divisible by $x-y$; throwing out this factor, the equation is

$$0=-c-d(x+2y)-e(x^2+2xy+3y^2)+\Lambda+yM,$$

which is right.

Verification of $\mathfrak{C}$.—We have

$$-2X\{2(x-y)M+2e(x-y)^2\}+(x-y)^2X'M+4e(x-y)^2X-2e(x-y)^3X'$$
$$=-(x-y)M\{4X-(x-y)X'\}-2e(x-y)^3X',$$

which is, in fact, an identity.

Verification of $\mathfrak{D}$.—The equation may be written

$$4X+4Y+4e(x-y)^4$$
$$=\quad 2X+2Y-2(x-y)^2\Lambda$$
$$+4X-2x(x-y)^2M$$
$$+2(x-y)\{2(x-y)xM+2ex(x-y)^2-M(x-y)^2-J'\},$$

viz. this is

$$0=2X-2Y-4e(x-y)^4-2(x-y)^2\Lambda+2x(x-y)^2M$$
$$+4ex(x-y)^3-2M(x-y)^3-2(x-y)J'.$$

The first term $2(X - Y)$ is divisible by $2(x - y)$; throwing this factor out, the equation becomes

$$0 = b + c(x + y) + d(x^2 + xy + y^2) + e(x^3 + x^2y + xy^2 + y^3) - J'$$
$$- 2e(x - y)^3 - (x - y)\Lambda + x(x - y)M + 2ex(x - y)^2 - M(x - y)^2.$$

Substituting for J' its value, the first line becomes

$$c(x - y) + d(x^2 - xy) + e(x^3 - 5x^2y + 5xy^2 - y^3),$$

which is divisible by $(x - y)$; hence, throwing out this factor, the equation is

$$0 = c + dx + e(x^2 - 4xy + y^2) - \Lambda + xM - 2e(x - y)^2 + 2ex(x - y) - M(x - y),$$

where the sum of all the terms but the last is $= d(x - y) + e(2x^2 - 2xy)$: hence, again throwing out the factor $x - y$, the equation becomes

$$0 = d + 2ex - 2e(x - y) + 2ex - M,$$

which is right.

Recapitulating, we have for the general integral of $\frac{dx}{\sqrt{X}} + \frac{dy}{\sqrt{Y}} = 0$, or for a particular integral of $\frac{dx}{\sqrt{X}} + \frac{dy}{\sqrt{Y}} + \frac{dz}{\sqrt{Z}} = 0$,

$$z = \frac{J - 2\sqrt{e}(x - y)y\sqrt{X} + 2\sqrt{e}(x - y)x\sqrt{Y} - 2\sqrt{XY}}{(x - y)^2 M - 2\sqrt{e}(x - y)\sqrt{X} + 2\sqrt{e}(x - y)\sqrt{Y}},$$

the corresponding value of $\sqrt{Z}$ being

$$\sqrt{Z} = \frac{\begin{array}{l}\sqrt{e}\,[-X^2 - 6XY - Y^2 + (x - y)^4\{\Lambda^2 + (-b + dxy)M + xyM^2\}] \\ + [\{4Y + (x - y)Y'\}M + 2e(x - y)^2 Y'](x - y)\sqrt{X} \\ - [\{4X - (x - y)X'\}M + 2e(x - y)^2 X'](x - y)\sqrt{Y} \\ + [\quad 4(X + Y) + 4e(x - y)^4] \quad \sqrt{XY}\end{array}}{\{(x - y)^2 M - 2\sqrt{e}(x - y)\sqrt{X} + 2\sqrt{e}(x - y)\sqrt{Y}\}^2},$$

where, as before,

$$M = d + 2e(x + y),$$
$$\Lambda = c + d(x + y) + e(x^2 + y^2),$$
$$J = 2a + b(x + y) + 2cxy + dxy(x + y) + exy(x^2 - xy + y^2):$$

also X is the general quartic function $a + bx + cx^2 + dx^3 + ex^4$, and Y, Z are the same functions of y, z respectively.

In connexion with what precedes, I give some investigations relating to the more simple form $\Theta = a + c\theta^2 + e\theta^4$, or, as it will be convenient to write it, $\Theta = 1 - l\theta^2 + \theta^4$.

We have

$$\left.\begin{array}{l}\begin{vmatrix} x, & \sqrt{X} \\ y, & \sqrt{Y} \end{vmatrix} = 0 \text{ a particular integral} \\ \begin{vmatrix} x^3, & x, & \sqrt{X} \\ y^3, & y, & \sqrt{Y} \\ z^3, & z, & \sqrt{Z} \end{vmatrix} = 0 \text{ the general integral} \end{array}\right\} \text{of } \frac{dx}{\sqrt{X}}+\frac{dy}{\sqrt{Y}}=0,$$

$$\left.\begin{array}{l}\begin{vmatrix} x^3, & x, & \sqrt{X} \\ y^3, & y, & \sqrt{Y} \\ z^3, & z, & \sqrt{Z} \end{vmatrix} = 0 \text{ a particular integral} \\ \begin{vmatrix} x^3, & x, & x^2\sqrt{X}, & \sqrt{X} \\ y^3, & y, & y^2\sqrt{Y}, & \sqrt{Y} \\ z^3, & z, & z^2\sqrt{Z}, & \sqrt{Z} \\ w^3, & w, & w^2\sqrt{W}, & \sqrt{W} \end{vmatrix} = 0 \text{ the general integral} \end{array}\right\} \text{of } \frac{dx}{\sqrt{X}}+\frac{dy}{\sqrt{Y}}+\frac{dz}{\sqrt{Z}}=0,$$

$$\left.\begin{array}{l}\begin{vmatrix} x^3, & x, & x^2\sqrt{X}, & \sqrt{X} \\ y^3, & y, & y^2\sqrt{Y}, & \sqrt{Y} \\ z^3, & z, & z^2\sqrt{Z}, & \sqrt{Z} \\ w^3, & w, & w^2\sqrt{W}, & \sqrt{W} \end{vmatrix} = 0 \text{ a particular integral} \\ \quad \dots \quad \dots \end{array}\right\} \text{of } \frac{dx}{\sqrt{X}}+\frac{dy}{\sqrt{Y}}+\frac{dz}{\sqrt{Z}}+\frac{dw}{\sqrt{W}}=0,$$

and so on; viz. in taking

$$\begin{vmatrix} x^3, & x, & \sqrt{X} \\ y^3, & y, & \sqrt{Y} \\ z^3, & z, & \sqrt{Z} \end{vmatrix} = 0 \text{ as the general integral of } \frac{dx}{\sqrt{X}}+\frac{dy}{\sqrt{Y}}=0,$$

we consider z as the constant of integration: and so in other cases.

It is to be remarked that it is an essentially different problem to verify a particular integral and to verify a general integral, and that the former is the more difficult one. In fact, if $U=0$ is a particular integral of the differential equation $Mdx+Ndy=0$, then we must have $N\frac{dU}{dx}-M\frac{dU}{dy}=0$, not identically but in virtue of the relation $U=0$, or we have to consider whether two given relations between x and y are in fact one and the same relation. In the case of a general solution, this is theoretically reducible to the form $c=U$, c being the constant of integration, and we have then the equation $N\frac{dU}{dx}-M\frac{dU}{dy}=0$, satisfied identically, or, what is the same thing, U a solution of this partial differential equation.

Hence it is theoretically easier to verify that

$$\begin{vmatrix} x^3, & x, & \sqrt{X} \\ y^3, & y, & \sqrt{Y} \\ z^3, & z, & \sqrt{Z} \end{vmatrix} = 0$$

is a general solution, than to verify that

$$\begin{vmatrix} x, & \sqrt{X} \\ y, & \sqrt{Y} \end{vmatrix} = 0$$

is a particular solution of the differential equation $\frac{dx}{\sqrt{X}}+\frac{dy}{\sqrt{Y}}=0$. Moreover, taking the first equation in the before mentioned form

$$-z=\frac{x^2-y^2}{x\sqrt{Y}-y\sqrt{X}},$$

and writing therein $z=\infty$, we see that the second equation

$$\begin{vmatrix} x, & \sqrt{X} \\ y, & \sqrt{Y} \end{vmatrix}=0$$

is, in fact, a particular case of the first equation, so that we only require to verify the first equation; or, what is the same thing, to verify that

$$-z=\frac{x^2-y^2}{x\sqrt{Y}-y\sqrt{X}}$$

is the general integral of

$$\frac{dx}{\sqrt{X}}+\frac{dy}{\sqrt{Y}}.$$

To verify this, we have to show that $dz=\Omega\left(\frac{dx}{\sqrt{X}}+\frac{dy}{\sqrt{Y}}\right)$, viz. that $\sqrt{X}\frac{dz}{dx}=\Omega$, a symmetrical function of (x, y); for then $\sqrt{Y}\frac{dz}{dy}=\Omega$, and we have the relation in question.

We have

$$(x\sqrt{Y}-y\sqrt{X})^2\sqrt{X}\frac{dz}{dx}=\sqrt{X}\left\{(x^2-y^2)\left(\sqrt{Y}-\frac{yX'}{2\sqrt{X}}\right)-2x(x\sqrt{Y}-y\sqrt{X})\right\}$$

$$=\sqrt{X}\left\{(x^2-y^2-2x^2)\sqrt{Y}-\frac{(x^2-y^2)yX'}{2\sqrt{X}}+2xy\sqrt{X}\right\}$$

$$=-(x^2+y^2)\sqrt{XY}+2xyX-\tfrac{1}{2}(x^2-y^2)yX'.$$

Writing here $X=1-lx^2+x^4$, then $X'=-2lx+4x^3$, and we have the last two terms

$$=2xy(1-lx^2+x^4)+(x^2-y^2)xy(l-2x^2)$$

$$=xy\{2-2lx^2+2x^4+(x^2-y^2)(l-2x^2)\}$$

$$=xy\{2-l(x^2+y^2)+2x^2y^2\}.$$

Hence the equation is

$$(x\sqrt{Y}-y\sqrt{X})^2\sqrt{X}\frac{dz}{dx}=-(x^2+y^2)\sqrt{XY}+xy\{2-l(x^2+y^2)+2x^2y^2\},$$

or we have

$$\Omega=\frac{1}{(x\sqrt{Y}-y\sqrt{X})^2}\{-(x^2+y^2)\sqrt{XY}+xy(2-l(x^2+y^2)+2x^2y^2)\},$$

which is symmetrical in (x, y), as it should be. And observe, further, that since the equation is a particular solution of $\frac{dx}{\sqrt{X}}+\frac{dy}{\sqrt{Y}}+\frac{dz}{\sqrt{Z}}=0$, we must have $\Omega=-\sqrt{Z}$; viz. we have

$$\sqrt{Z}\,(x\sqrt{Y}-y\sqrt{X})^2=-(x^2+y^2)\sqrt{XY}+xy\,\{2-l\,(x^2+y^2)+2x^2y^2\}.$$

Proceeding to the next case, where we have between x, y, z, w a relation which may be written

$$(x^3,\ x,\ x^2\sqrt{X},\ \sqrt{X})=0,$$

then here a, b, c, d can be determined so that

$$(c\theta^2+d)^2\,(1+\beta\theta^2+\gamma\theta^4)-(a\theta^3+b\theta)^2=c^2\gamma\,(\theta^2-x^2)\,(\theta^2-y^2)\,(\theta^2-z^2)\,(\theta^2-w^2),$$

viz. we have $d^2=c^2\gamma\,x^2y^2z^2w^2$, or say $d=c\sqrt{\gamma}\,xyzw$. And, supposing the ratios of a, b, c, d determined by the three equations which contain (x, y, z) respectively, we have

$$a:b:c:d=(x,\ x^2\sqrt{X},\ \sqrt{X}):-(x^3,\ x^2\sqrt{X},\ \sqrt{X}):(x^3,\ x,\ \sqrt{X}):-(x^3,\ x,\ x^2\sqrt{X}),$$

or in particular

$$\frac{d}{c}=\frac{-(x^3,\ x,\ x^2\sqrt{X})}{(x^3,\ x,\ \sqrt{X})},\quad=\frac{-xyz\,(x^2,\ 1,\ x\sqrt{X})}{(x^3,\ x,\ \sqrt{X})};$$

whence we have

$$w=-\frac{(x^2,\ 1,\ x\sqrt{X})}{(x^3,\ x,\ \sqrt{X})}$$

as a new form of the integral equation; viz. written at full length, this is

$$-w=\begin{vmatrix} x^2, & 1, & x\sqrt{X} \\ y^2, & 1, & y\sqrt{Y} \\ z^2, & 1, & z\sqrt{Z}\end{vmatrix}\div\begin{vmatrix} x^3, & x, & \sqrt{X} \\ y^3, & y, & \sqrt{Y} \\ z^3, & z, & \sqrt{Z}\end{vmatrix};$$

and taking $w=0$ and $=\infty$ respectively, we thus see how

$$\begin{vmatrix} x^2, & 1, & x\sqrt{X} \\ y^2, & 1, & y\sqrt{Y} \\ z^2, & 1, & z\sqrt{Z}\end{vmatrix}=0,\qquad\begin{vmatrix} x^3, & x, & \sqrt{X} \\ y^3, & y, & \sqrt{Y} \\ z^3, & z, & \sqrt{Z}\end{vmatrix}=0,$$

are each of them a particular integral of

$$\frac{dx}{\sqrt{X}}+\frac{dy}{\sqrt{Y}}+\frac{dz}{\sqrt{Z}}=0.$$

Reverting to the general form

$$w=-\frac{(x^2,\ 1,\ x\sqrt{X})}{(x^3,\ x,\ \sqrt{X})},$$

this will be a general integral if only

$$dw=\Omega\left(\frac{dx}{\sqrt{X}}+\frac{dy}{\sqrt{Y}}+\frac{dz}{\sqrt{Z}}\right),$$

viz. if we have

$$-\sqrt{X}\,\frac{d}{dx}\,\frac{(x^2,\ 1,\ x\sqrt{X})}{(x^3,\ x,\ \sqrt{X})}=\Omega,\ \text{a symmetrical function of } (x,\ y,\ z).$$

The expression is

$$\Omega=\frac{1}{(x^3,\ x,\ \sqrt{X})^2}\left\{(x^2,\ 1,\ x\sqrt{X})\sqrt{X}\,\frac{d}{dx}(x^3,\ x,\ \sqrt{X})-(x^3,\ x,\ \sqrt{X})\sqrt{X}\,\frac{d}{dx}(x^2,\ 1,\ x\sqrt{X})\right\},$$

or, writing for shortness

$$\begin{array}{ll} \alpha=x\,(y^2-z^2), & a=yz\,(y^2-z^2), \\ \beta=y\,(z^2-x^2), & b=zx\,(z^2-x^2), \\ \gamma=z\,(x^2-y^2), & c=xy\,(x^2-y^2), \end{array}$$

we have

$$\begin{aligned} (x^2,\ 1,\ x\sqrt{X})&=\alpha\sqrt{X}+\beta\sqrt{Y}+\gamma\sqrt{Z}, \\ (x^3,\ x,\ \ \sqrt{X})&=a\sqrt{X}+b\sqrt{Y}+c\sqrt{Z}; \end{aligned}$$

and the formula is

$$\begin{aligned} &(x^3,\ x,\ \sqrt{X})^2\,\Omega \\ &\quad=(\alpha\sqrt{X}+\beta\sqrt{Y}+\gamma\sqrt{Z})\,\{(y^3z-yz^3)\tfrac{1}{2}X'+(-3x^2z+z^3)\sqrt{XY}+(3x^2y-y^3)\sqrt{XZ}\} \\ &\quad\quad-(a\sqrt{X}+b\sqrt{Y}+c\sqrt{Z})\,\{(y^2-z^2)(X+\tfrac{1}{2}X'x)-2xy\sqrt{XY}-2xz\sqrt{XZ}\} \\ &\quad=(\alpha\sqrt{X}+\beta\sqrt{Y}+\gamma\sqrt{Z})(L+M\sqrt{XY}+N\sqrt{XZ}) \\ &\quad\quad-(a\sqrt{X}+b\sqrt{Y}+c\sqrt{Z})(P+Q\sqrt{XY}+R\sqrt{XZ}),\ \text{suppose,} \end{aligned}$$

$$=\ \begin{array}{llll} \sqrt{X} & +\sqrt{Y} & +\sqrt{Z} & +\sqrt{XYZ} \\ \hline \alpha L & +\alpha MX & +\alpha NX & \\ +\beta MY & +\beta L & & +\beta N \\ +\gamma NZ & & +\gamma L & +\gamma M \\ -aP & -aQX & -aRX & \\ -bQY & -bP & & -bR \\ -cRZ & & -cP & -cQ \end{array}\ ;$$

viz. this is

$$\begin{aligned} =\ &\{\alpha L-aP+Y(\beta M-bQ)+Z(\gamma N-cR)\}\sqrt{X} \\ &+\{X(\alpha M-aQ)+\beta L-bP\}\sqrt{Y} \\ &+\{X(\alpha N-aR)+\gamma L-cP\}\sqrt{Z} \\ &+(\beta N+\gamma M-bR-cQ)\sqrt{XYZ}. \end{aligned}$$

The coefficient of $\sqrt{XYZ}$ is here

$$\begin{array}{ll} = \quad y(z^2-x^2)(3x^2y-y^3) & = \quad y^2(z^2-x^2)(3x^2-y^2) \\ + z(x^2-y^2)(-3x^2z+z^3) & + z^2(x^2-y^2)(-3x^2+z^2) \\ - zx(z^2-x^2)(2xz) & - 2x^2z^2(z^2-x^2) \\ - xy(x^2-y^2)(-2xy) & + 2x^2y^2(x^2-y^2), \end{array}$$

which is

$$= 6x^2y^2z^2 - y^2z^4 - y^4z^2 - z^2x^4 - z^4x^2 - x^4y^2 - x^2y^4.$$

The coefficient of $\surd Y$ is

$$\begin{aligned} &= [x(y^2-z^2)(-3x^2z+z^3)+yz(y^2-z^2)\,2xy]\,X \\ &\quad + y(z^2-x^2)\,\tfrac{1}{2}X'(y^3z-yz^3) - zx(z^2-x^2)(y^2-z^2)(X+\tfrac{1}{2}X'x) \\ &= -2xz(x^2-y^2)(y^2-z^2)X - z(x^2-y^2)(y^2-z^2)(z^2-x^2)\tfrac{1}{2}X' \\ &= -(x^2-y^2)(y^2-z^2)\,z\,\{2xX+\tfrac{1}{2}(z^2-x^2)X'\}, \end{aligned}$$

where the term in $\{\ \}$ is

$$\begin{aligned} &= 2x(1-lx^2+x^4)+(z^2-x^2)(-lx+2x^3), \\ &= x\{2-l(z^2+x^2)+2z^2x^2\}, \end{aligned}$$

or the whole coefficient is

$$= -(x^2-y^2)(y^2-z^2)\,zx\,\{2-l(z^2+x^2)+2z^2x^2\}.$$

We obtain in like manner the coefficient of $\surd Z$, and with a little more trouble that of $\surd X$; and the final result is

$$\begin{aligned} \Omega(x^3, x, \surd X)^2 = &-(z^2-x^2)(x^2-y^2)\,yz\,\{2-l(y^2+z^2)+2y^2z^2\}\surd X \\ &-(x^2-y^2)(y^2-z^2)\,zx\,\{2-l(z^2+x^2)+2z^2x^2\}\surd Y \\ &-(y^2-z^2)(z^2-x^2)\,xy\,\{2-l(x^2+y^2)+2x^2y^2\}\surd Z \\ &+(6x^2y^2z^2-y^2z^4-y^4z^2-z^2x^4-z^4x^2-x^2y^4-x^4y^2)\sqrt{XYZ}. \end{aligned}$$

And inasmuch as the equation is a solution of

$$\frac{dx}{\surd X}+\frac{dy}{\surd Y}+\frac{dz}{\surd Z}+\frac{dw}{\surd W}=0,$$

it follows that $\Omega = -\surd W$, viz. that $\surd W$ is by the foregoing equation expressed as a function of x, y, z.

The equation $(x^3, x, x^2\surd X, \surd X)=0$, that is,

$$\begin{vmatrix} x^3, & x, & x^2\surd X, & \surd X \\ y^3, & y, & y^2\surd Y, & \surd Y \\ z^3, & z, & z^2\surd Z, & \surd Z \\ w^3, & w, & w^2\surd W, & \surd W \end{vmatrix} = 0,$$

gives

$$w=\frac{(x^2,\ 1,\ x\surd X)}{(x^3,\ x,\ \surd X)},$$

where the numerator and the denominator are determinants formed with the variables x, y, z.

Writing $\frac{1}{w}$ for w, it follows that the equation

$$\begin{vmatrix} x^3, & x, & x^2\surd X, & \surd X \\ y^3, & y, & y^2\surd Y, & \surd Y \\ z^3, & z, & z^2\surd Z, & \surd Z \\ w, & w^3, & \surd W, & w^2\surd W \end{vmatrix}=0$$

gives

$$w=\frac{(x^3,\ x,\ \surd X)}{(x^2,\ 1,\ x\surd X)},$$

which last equation is a transformation of

$$\begin{vmatrix} x^4, & x^2, & 1, & x\surd X \\ y^4, & y^2, & 1, & y\surd Y \\ z^4, & z^2, & 1, & z\surd Z \\ w^4, & w^2, & 1, & w\surd W \end{vmatrix}=0.$$

The two equations, involving these determinants of the order 4, are consequently equivalent equations.

627.

GEOMETRICAL ILLUSTRATION OF A THEOREM RELATING TO AN IRRATIONAL FUNCTION OF AN IMAGINARY VARIABLE.

[From the *Proceedings of the London Mathematical Society*, vol. VIII. (1876—1877), pp. 212—214. Read May 11, 1876.]

IF we have v, a function of u, determined by an equation $f(u, v) = 0$, then to any given imaginary value $x + iy$ of u there belong two or more values, in general imaginary, $x' + iy'$ of v: and for the complete understanding of the relation between the two imaginary variables, we require to know the series of values $x' + iy'$ which correspond to a given series of values $x + iy$, of v, u respectively. We must for this purpose take x, y as the coordinates of a point P in a plane Π, and x', y' as the coordinates of a corresponding point P' in another plane Π'. The series of values $x + iy$ of u is then represented by means of a curve in the first plane, and the series of values $x' + iy'$ of v by means of a corresponding curve in the second plane. The correspondence between the two points P and P' is of course established by the two equations into which the given equation $f(x + iy, x' + iy') = 0$ breaks up, on the assumption that x, y, x', y' are all of them real. If we assume that the coefficients in the equation are real, then the two equations are

$$f(x + iy, x' + iy') + f(x - iy, x' - iy') = 0,$$

$$f(x + iy, x' + iy') - f(x - iy, x' - iy') = 0;$$

viz. if in these equations we regard either set of coordinates, say (x, y), as constants, then the other set (x', y') are the coordinates of any real point of intersection of the curves represented by these equations respectively.

I consider the particular case where the equation between u, v is $u^2 + v^2 = a^2$: we have here $(x + iy)^2 + (x' + iy')^2 = a^2$: so that, to a given point P in the first plane, there correspond in general two points P_1', P_2' in the second plane: but to each of the points A and B, coordinates $(a, 0)$ and $(-a, 0)$, there corresponds only a single point in the second plane.

We have here a particular case of a well-known theorem: viz. if from a given point P we pass by a closed curve, not containing within it either of the points A or B, back to the initial point P, we pass in the other plane from P_1' by a closed curve back to P_1'; and similarly from P_2' by a closed curve back to P_2': but if the closed curve described by P contain within it A or B, then, in the other plane, we pass continuously from P_1' to P_2'; and also continuously from P_2' to P_1'.

The relations between (x, y), (x', y') are

$$x'^2 - y'^2 = a^2 - (x^2 - y^2),$$

$$x'y' = -xy,$$

whence also

$$(x'^2 + y'^2)^2 = a^4 - 2a^2(x^2 - y^2) + (x^2 + y^2)^2.$$

And if the point (x, y) describe a curve $x^2 + y^2 = \phi(x^2 - y^2)$, then will the point (x', y') describe a curve $x'^2 + y'^2 = \psi(x'^2 - y'^2)$, obtained by the elimination of $x^2 - y^2$ from the two equations

$$x'^2 - y'^2 = a^2 - (x^2 - y^2),$$

$$(x'^2 + y'^2)^2 = a^4 - 2a^2(x^2 - y^2) + \phi(x^2 - y^2);$$

viz. this is

$$(x'^2 + y'^2)^2 = -a^4 + 2a^2(x'^2 - y'^2) + \phi\{a^2 - (x'^2 - y'^2)\}.$$

In particular, if the one curve be $(x^2 + y^2)^2 = \alpha + \beta(x^2 - y^2)$; then the other curve is

$$(x'^2 + y'^2)^2 = -a^4 + 2a^2(x'^2 - y'^2) + \alpha + \beta\{a^2 - (x'^2 - y'^2)\},$$

that is,

$$(x'^2 + y'^2)^2 = \alpha' + \beta'(x'^2 - y'^2),$$

where

$$\alpha' = -a^4 + \beta a^2 + \alpha, \quad \beta' = 2a^2 - \beta.$$

Writing for greater simplicity $a = 1$, then $\alpha' = -1 + \alpha + \beta$, $\beta' = 2 - \beta$; in particular, if $\alpha = 0$, then $\alpha' = -1 + \beta$, $\beta' = 2 - \beta$.

Supposing successively $\beta < 1$, $\beta = 1$, and $\beta > 1$, then in each case P describes a closed curve or half figure-of-eight, as shown in the annexed P-figure; but in the first case the point A is inside the curve, in the second case on it, and in the third case outside it, as shown by the letters A, A, A of the figure; and, corresponding to the three cases respectively, we have the three P'-figures, the curve in the first of them consisting of two ovals, in the second of them being a figure of eight, and in the third a twice-indented or pinched oval: the small figures 1, 2, 3, 4 in the P-figure, and 1, 2, 3, 4 and 1′, 2′, 3′, 4′ in the P'-figures serve to show the corresponding positions of the points P and P_1', P_2' respectively; and the courses are further indicated by the arrows. And we thus see how the two separate closed curves described

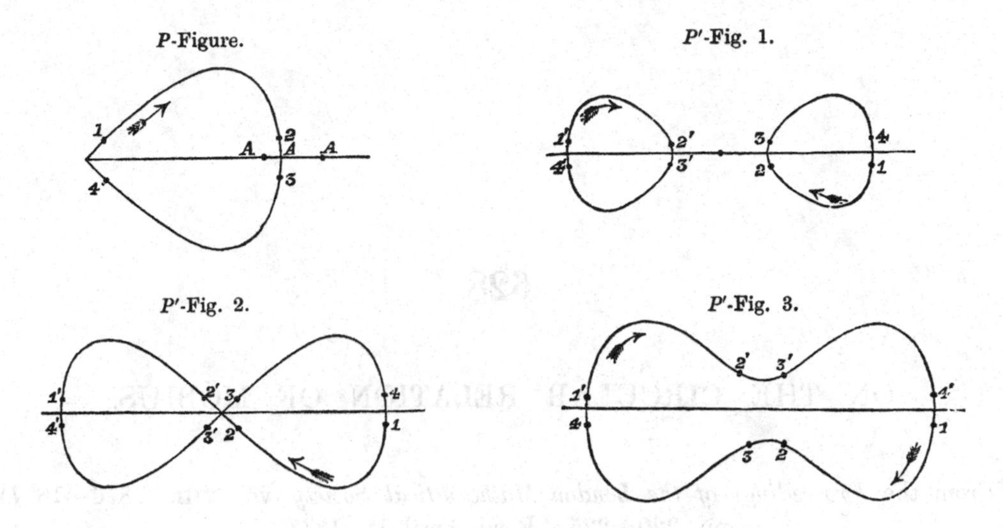

by P_1' and P_2', as in figure 1, change into the single closed curve described one half of it by P_1' and the other half of it by P_2' as in figure 3.

628.

ON THE CIRCULAR RELATION OF MÖBIUS.

[From the *Proceedings of the London Mathematical Society*, vol. VIII. (1876—1877), pp. 220—225. Read April 12, 1877.]

IN representing a given imaginary or complex quantity $u, =x+iy$, by means of the point whose coordinates are x, y, we assume in the first instance that x, y are real,—but in the results this restriction may be abandoned—for instance, if the imaginary quantities u, u', c are connected by the equation $u^2+u'^2=c^2$; then, writing $u=x+iy$, $u'=x'+iy'$, $c=a+bi$, we have $x^2-y^2+x'^2-y'^2=a^2-b^2$, $xy+x'y'=ab$, equations connecting the points U, U', C which serve to represent the quantities u, u', c, and which (regarding C as a fixed point) establish a correspondence between the two variable points U, U': any given value $u, =x+iy$, is represented by the point U, and corresponding hereto we have (in the present case) two points U', viz. these are the real intersections of the curves $x'^2-y'^2=a^2-b^2-(x^2-y^2)$, $x'y'=ab-xy$, and then the coordinates x', y' of either of these give the value $x'+iy'$ of u'.

But, the two curves once arrived at, we may for other purposes be concerned with their intersections as well imaginary as real; or, still more generally, all the quantities entering into the two equations may be regarded as imaginary.

Theoretically we seem to require two imaginary roots of unity, incommensurable and convertible, viz. taking these to be i, I, then $i^2=-1$, $I^2=-1$, $iI=Ii$, but without any relation between i, I: thus, in what precedes, writing I instead of i, viz. $u, u', c=x+Iy, x'+Iy', a+bI$, here each of the quantities x, y, x', y', a, b can be *ab initio* an imaginary quantity of the form $\lambda+\mu i$. But, conforming to the ordinary practice, I use i only, writing for instance $u=x+iy$, without any express statement that x, y are real; on the understanding that any equation containing such quantities, and therefore ultimately of the form $P+iQ=0$, denotes the two equations $P=0$, $Q=0$ (or, what is the same thing, that we have not only the original equation, but, in addition to it, the like equation with each such quantity $x+iy$ replaced by the conjugate quantity $x-iy$): and the further understanding that, in the pair of equations, each of the quantities x, y, &c. entering therein may itself be considered as an imaginary quantity of the form $\lambda+\mu i$.

The foregoing explanation is required, for otherwise it would appear as if the circular relation of Möbius * about to be explained was of necessity a relation between real points: I hold that this is *not* the case. But in all that follows I do, in fact, consider primarily the case of real points; and indeed the occasion does not arise for any explicit consideration of the case of imaginary points.

The circular relation is as follows. If in the first instance we have four points U, A, B, C on a line, and u, a, b, c their distances from any fixed point on that line; and again, U', A', B', C' four other points on a line (the same or a different line), and u', a', b', c' their distances from any fixed point on that line; then the same equation between u, a, b, c, u', a', b', c' which expresses the homographic relation between the two ranges of points U, A, B, C and U', A', B', C', expresses, when differently interpreted, the circular relation between the four points U, A, B, C in a plane, and the four other points U', A', B', C' in the same or a different plane—viz. for the new interpretation, u is used as denoting $x+iy$, the linear function of the coordinates x, y of the point U, and the like as regards the remaining quantities a, b, c, u', a', b', c'.

As in the homographic theory (but of course without the condition of being in a line), we have A, A'; B, B'; C, C' given pairs of corresponding points: the equation now represents two equations; and these, when either of the points U, U' is given, determine the corresponding point U' or U.

The homographic relation may be written in the forms

$$\begin{vmatrix} 1, & u, & u', & uu' \\ 1, & a, & a', & aa' \\ 1, & b, & b', & bb' \\ 1, & c, & c', & cc' \end{vmatrix} = 0,$$

$$u-a\,.\,b-c : u-b\,.\,c-a : u-c\,.\,a-b = u'-a'\,.\,b'-c' : u'-b'\,.\,c'-a' : u'-c'\,.\,a'-b',$$

viz. these are forms of one and the same equation: and it may be added that, if ω in the first system corresponds to ∞ in the second system, and ω' in the second system to ∞ in the first system (of course ω, ω' are not corresponding values in the two systems respectively); so that

$$\begin{vmatrix} & 1, & \omega & \\ 1, & a, & a', & aa' \\ 1, & b, & b', & bb' \\ 1, & c, & c', & cc' \end{vmatrix} = 0, \qquad \begin{vmatrix} & 1, & & \omega' \\ 1, & a, & a', & aa' \\ 1, & b, & b', & bb' \\ 1, & c, & c', & cc' \end{vmatrix} = 0,$$

$$\omega-a\,.\,b-c : \omega-b\,.\,c-a : \omega-c\,.\,a-b = b'-c' : c'-a' : a'-b';$$

$$b-c : c-a : a-b = \omega'-a'\,.\,b'-c' : \omega'-b'\,.\,c'-a' : \omega'-c'\,.\,a'-b';$$

* [Möbius, *Ges. Werke;* t. II., pp. 243—314, and elsewhere.]

whence also, if

$$\Lambda = \begin{vmatrix} 1, & 1, & 1 \\ a, & b, & c \\ a', & b', & c' \end{vmatrix}, \quad = bc' - b'c + ca' - c'a + ab' - a'b,$$

then

$$\begin{aligned} -\Lambda \,.\, \omega - a &= b - c \,.\, c' - a' \,.\, a' - b', & \Lambda \,.\, \omega' - a' &= b' - c' \,.\, c - a \,.\, a - b, \\ -\Lambda \,.\, \omega - b &= c - a \,.\, a' - b' \,.\, b' - c', & \Lambda \,.\, \omega' - b' &= c' - a' \,.\, a - b \,.\, b - c, \\ -\Lambda \,.\, \omega - c &= a - b \,.\, b' - c' \,.\, c' - a', & \Lambda \,.\, \omega' - c' &= a' - b' \,.\, b - c \,.\, c - a. \end{aligned}$$

Then, ω, ω' being thus determined, we have

$$\omega - a \,.\, \omega' - a' = \omega - b \,.\, \omega' - b' = \omega - c \,.\, \omega' - c' = \omega - u \,.\, \omega' - u'$$

$$= -\frac{b - c \,.\, c - a \,.\, a - b \,.\, b' - c' \,.\, c' - a' \,.\, a' - b'}{\Lambda^2} \;(= \Delta \text{ suppose}),$$

viz we have $\omega - u \,.\, \omega' - u' =$ a given value; which is the most simple form of the relation between u, u'.

Interpreting everything in the first instance in regard to the homographic ranges, the equations show that there is in the first range a point O, and in the second range a point O', such that OA, &c. denoting distances, we have

$$OA \,.\, O'A' = OB \,.\, O'B' = OC \,.\, O'C' = OU \,.\, O'U' \;(= \Delta);$$

or, what is the same thing, if in the line of the first range we construct A_1, B_1, C_1, U_1 by the formulæ

$$OA \,.\, OA_1 = OB \,.\, OB_1 = OC \,.\, OC_1 = OU \,.\, OU_1 = \Delta,$$

that is, invert the first range in regard to the centre O and squared radius Δ, then we have a range O, A_1, B_1, C_1, U_1 *equal* to the range O', A', B', C', U'; viz. the distances of corresponding points are equal in the two cases: or say a range O, A_1, B_1, C_1, U_1 imposable upon O', A', B', C', U'.

The like result holds for the circular relation, but the interpretation must be explained more in detail. And, first, it is to be remarked that O in the first figure is the point corresponding to any point whatever at infinity in the second figure; viz. writing $u' = \xi' + i\eta'$, $= \infty$, then, whatever value we give to the ratio of the two infinite quantities ξ', η', we obtain the same complex value of ω, that is, the same coordinates for the point O. And, similarly, O' in the second figure is the point corresponding to any point whatever at infinity in the first figure.

To determine O, we have the equation

$$\frac{\omega - a}{\omega - b} = \frac{b' - c'}{c' - a'} \cdot \frac{c - a}{a - b}.$$

Any such equation gives at once the geometrical construction, viz. $\omega - a = OAe^{iOAx}$, where OA is the distance of the points O, A regarded as positive, and OAx is the inclination of the line OA regarded as drawn from A to O to the line Ax, such angle being measured in the sense Ax to Ay; where Ax, Ay are the lines drawn from A in the senses x positive and y positive respectively: and so in other cases.

The equation is therefore equivalent to the two equations

$$\frac{OA}{OB} = \frac{B'C'}{C'A'} \cdot \frac{CA}{AB},$$

and

$$\angle OAx - \angle OBx = \angle B'C'x - \angle C'A'x + \angle CAx - \angle ABx.$$

The former of these expresses that O is in a certain circle which, having its centre on the line AB, cuts AB and AB produced in the one or the other sense; the latter that it is in the segment described on a determinate side of AB and containing a given angle: hence O, as the intersection of the segment with the first-mentioned circle, is a uniquely determined point. Similarly O' is a uniquely determined point.

It is not obvious how to construct Λ, from its original value as given above (but, ω being known, we can without difficulty construct it from the value

$$-\Lambda \,.\, \omega - a = b - c \,.\, c' - a' \,.\, a' - b'),$$

nor consequently Δ from its expression in terms of Λ: but, ω and ω' being known, we can construct Δ from the expression $\omega - a \,.\, \omega' - a' = \Delta$; supposing it thus constructed, $= ke^{i\theta}$ suppose, then if, with centre O and squared radius k, we invert the first figure, thereby obtaining the points A_1, B_1, C_1, U_1 such that

$$OA \,.\, OA_1 = OB \,.\, OB_1 = OC \,.\, OC_1 = OU \,.\, OU_1 = k,$$

(the points A_1, B_1, C_1, U_1 being on the lines OA, OB, OC, OU respectively,) then the equations

$$\omega - a \,.\, \omega' - a' = \omega - b \,.\, \omega' - b' = \omega - c \,.\, \omega' - c' = \omega - u \,.\, \omega' - u' = ke^{i\theta}$$

give

$$\omega - a \,.\, \omega' - a' = OA \,.\, OA_1 e^{i\theta},$$

that is,

$$OA \,.\, O'A' = OA \,.\, OA_1, \text{ or, simply, } O'A' = OA_1,$$

and

$$\angle AOx + \angle AO'x' = \theta,$$

or, what is the same thing,

$$\angle A_1Ox + \angle A'O'x' = \theta,$$

and so for the other letters, viz. we have

$$O'A',\ O'B',\ O'C',\ O'U' = OA_1,\ OB_1,\ OC_1,\ OU_1,$$

respectively; and further

$$\angle\text{'s } A_1Ox,\ B_1Ox,\ C_1Ox,\ U_1Ox = \theta - A'O'x',\ \theta - B'O'x',\ \theta - C'O'x',\ \theta - U'O'x',$$

respectively; viz. the system of points O, A_1, B_1, C_1, U_1 is *equal* to the system O', A', B', C', U', that is, the distances of corresponding points and magnitudes of corresponding angles are severally equal—but the angles A_1Ox and $A'O'x'$, &c. are in opposite senses, as appears by the just mentioned equations $A_1Ox = \theta - A'O'x'$, &c.; that is, the two figures are symmetrically equal: but the one of them is not, except by a turning over of its plane, imposable upon the other.

The conclusion is, the two figures A, B, C, U and A', B', C', U' are each of them equal by symmetry, but not superimposably, to a figure which is the inverse of the other of them; viz. there exists in the first figure a point O, and in the second figure a point O', such that, inverting say the first figure, with centre O and a squared radius of determinate magnitude, we obtain the points A_1, B_1, C_1, U_1, forming with O a figure equal by symmetry, but not superimposably, to the second figure A', B', C', U', O'. Hence also to any line in the first figure corresponds in the second figure a circle through O', and to any line in the second figure there corresponds in the first figure a circle through O; or, more generally, to any circle in either figure there corresponds a circle in the other figure.

There is a particular case of peculiar interest, viz. writing for greater convenience d, d' as corresponding values in place of u, u', the system a, b, c, d corresponds homographically to itself in three different ways; that is, we may have

$$(a', b', c', d') = (b, a, d, c), (c, d, a, b) \text{ or } (d, c, b, a).$$

To fix the ideas, attending to the first case, we have thus the range of points (A, B, C, D) corresponding homographically to (B, A, D, C), viz. here $\omega' = \omega$, and $\omega - a \,.\, \omega - b = \omega - c \,.\, \omega - d$, that is, the corresponding points U, U' belong to the involution where A and B and also C and D are corresponding points. The like theory applies to the circular transformation: viz. the points (A, B, C, D) may correspond to (B, A, D, C), viz. there exists a point O (or say O_1) and squared radius k_1, such that, inverting the figure and marking the inverse points of A, B, C, D as B_1, A_1, D_1, C_1 respectively, the new figure O_1, A_1, B_1, C_1, D_1 is equal by symmetry, but not superimposably, to the original figure $OABCD$. The equation $\omega_1 - a \,.\, \omega_1 - b = \omega_1 - c \,.\, \omega_1 - d$ gives the geometrical definition of the point O_1, viz. this is a point such that $O_1A \,.\, O_1B = O_1C \,.\, O_1D$ and further that AB and CD subtend at O_1 equal angles: we have $\omega_1 = \dfrac{ab - cd}{a + b - c - d}$, giving for $\omega_1 - a$, $\omega_1 - b$, $\omega_1 - c$, $\omega_1 - d$ convenient expressions the first of which is $\omega_1 - a = \dfrac{c - a \,.\, d - a}{c + d - a - b}$. We hence obtain a convenient construction for O, viz. taking M for the middle point of AB and N for the middle point of CD, and drawing from A in the sense M to N a line AP, $= 2MN$, then this equation may be written $\omega_1 - a = \dfrac{c - a \,.\, d - a}{p - a}$ (p the function $x + iy$ which belongs to the point P); thence $O_1A = \dfrac{CA \,.\, DA}{PA}$ and

$$\angle O_1Ax = \angle CAx + \angle DAx - \angle PAx,$$

conditions which determine uniquely the position of O_1.

We may have (A, B, C, D) corresponding to (C, D, A, B) and (D, C, B, A), the inversions for these depending on the points O_2 and O_3 respectively: I annex a figure showing the three inversions of the same four points A, B, C, D.

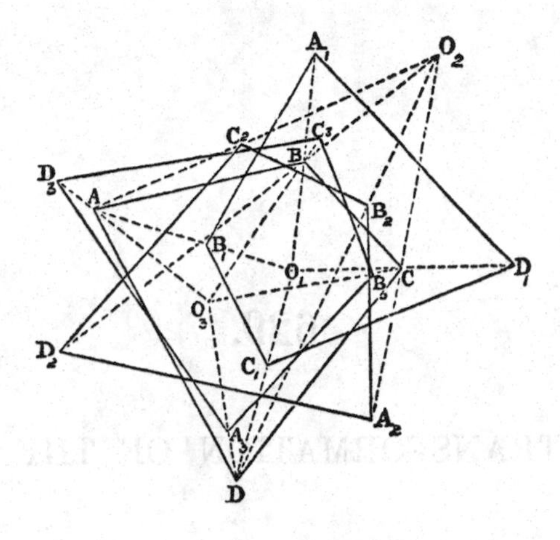

629.

ON THE LINEAR TRANSFORMATION OF THE INTEGRAL $\int \frac{du}{\sqrt{U}}$.

[From the *Proceedings of the London Mathematical Society*, vol. VIII. (1876—1877), pp. 226—229. Read April 12, 1877.]

THE quartic function U is taken to be $= \epsilon \,.\, u-a \,.\, u-b \,.\, u-c \,.\, u-d$, where a, b, c, d are imaginary values represented in the usual manner by means of the points A, B, C, D; viz. if $a = \alpha_0 + \alpha_1 i$, then A is the point whose rectangular coordinates are α_0, α_1; and the like as regards B, C, D. And I consider chiefly the definite integrals such as $\int_a^b \frac{du}{\sqrt{U}}$ where the path is taken to be the right line from A to B. There is here nothing to fix the sign of the radical; but if at any particular point of the path we assign to it at pleasure one of its two values, then (the radical varying continuously) this determines the value at every other point of the path; and the integral defined as above is completely determinate except as to its sign, which might be fixed as above, but which is better left indeterminate. The integral, thus determinate except as to its sign, is denoted by (AB).

I wish to establish the theorem that, if the points A, B, C, D taken in this order form a convex quadrilateral, then

$$(AB) = \pm (CD), \quad (AD) = \pm (BC), \quad \text{but not} \quad (AC) = \pm (BD);$$

whereas, if the four points form a triangle and interior point, then the three equations all hold good. I regard the theorem as the precise statement of Bouquet and Briot's theorem, $A - B + C - D = 0$, or say $(OA) - (OB) + (OC) - (OD) = 0$, where the four terms are the rectilinear integrals taken from a point O to the four points A, B, C, D respectively. The two cases may be called, for shortness, the convex and the reentrant cases respectively.

To prove in the case of a convex quadrilateral that (AC) is not $= \pm (BD)$, it is sufficient to consider the integral $\int \frac{du}{\sqrt{u^4-1}}$, where A, B, C, D are the points (1, 0), (0, 1), (− 1, 0), and (0, − 1) respectively, and where, writing $v = iu$, it at once appears that we have

$$\int_{-1}^{1} \frac{du}{\sqrt{u^4-1}} = \pm i \int_{-i}^{i} \frac{du}{\sqrt{u^4-1}},$$

that is,

$$(AC) = \pm i\,(BD), \text{ not } (AC) = \pm\,(BD).$$

But I consider the general question of the linear transformation. If a', b', c', d' correspond homographically to a, b, c, d, then to represent these values a', b', c', d' we have the points A', B', C', D', connected with A, B, C, D according to the circular relation of Möbius; and then, making u', a', b', c', d' correspond homographically to u, a, b, c, d, and representing in like manner the variables u, u' by the points U, U' respectively, we have the circular relation between the two systems U, A, B, C, D and U', A', B', C', D'.

Before going further I remark that the distinction of the convex and reentrant cases is not an invariable one; the figures are transformable the one into the other. Thus, taking C on the line BD (that is, between B and D, not on the line produced), there is not this relation between B', C', D', and the figure $A'B'C'D'$ is convex or reentrant as the case may be. Giving to C an infinitesimal displacement to the one side or the other of the line BD, we have in the one case a convex figure, in the other case a reentrant figure $ABCD$; but the corresponding displacement of C' being infinitesimal, the figure $A'B'C'D'$ remains for either displacement, convex or reentrant, as it originally was; that is, we have a convex figure $ABCD$ and a reentrant figure $ABCD$, each corresponding to the figure $A'B'C'D'$ (which is convex, or else reentrant, as the case may be).

Writing for convenience

$$\begin{aligned} \mathrm{a, b, c, f, g, h} &= b-c,\ c-a,\ a-b,\ a-d,\ b-d,\ c-d, \\ \mathrm{a', b', c', f', g', h'} &= b'-c',\ c'-a',\ a'-b',\ a'-d',\ b'-d',\ c'-d', \end{aligned}$$

so that identically

$$\mathrm{af + bg + ch} = 0, \quad \mathrm{a'f' + b'g' + c'h'} = 0,$$

then the homographic relation between (a, b, c, d), (a', b', c', d') may be written in the forms

$$\mathrm{af : bg : ch = a'f' : b'g' : c'h'},$$

or, what is the same thing, there exists a quantity N such that

$$\frac{\mathrm{a'f'}}{\mathrm{af}} = \frac{\mathrm{b'g'}}{\mathrm{bg}} = \frac{\mathrm{c'h'}}{\mathrm{ch}} = N^2.$$

The relation between u, u' may be written in the forms

$$\frac{u'-a'}{u'-d'} = P\frac{u-a}{u-d}, \quad \frac{u'-b'}{u'-d'} = Q\frac{u-b}{u-d}, \quad \frac{u'-c'}{u'-d'} = R\frac{u-c}{u-d};$$

and then, writing for u, u' their corresponding values, we find

$$P = \frac{\mathrm{b'h}}{\mathrm{bh'}} = \frac{\mathrm{c'g}}{\mathrm{cg'}}, \quad Q = \frac{\mathrm{c'f}}{\mathrm{cf'}} = \frac{\mathrm{a'h}}{\mathrm{ah'}}, \quad R = \frac{\mathrm{a'g}}{\mathrm{ag'}} = \frac{\mathrm{b'f}}{\mathrm{bf'}},$$

giving

$$\mathrm{f}^2PN^2 = \mathrm{f'}^2QR, \quad \mathrm{g}^2QN^2 = \mathrm{g'}^2RP, \quad \mathrm{h}^2RN^2 = \mathrm{h'}^2PQ, \quad \sqrt{PQR} = \frac{\mathrm{fgh}}{\mathrm{f'g'h'}}N^3.$$

Differentiating any one of the equations in (u, u'), for instance the first of them, we find

$$\frac{\mathrm{f'}du'}{(u'-d')^2} = \frac{\mathrm{f}Pdu}{(u-d)^2};$$

then, forming the equation

$$\frac{\sqrt{\epsilon \,.\, u'-a' \,.\, u'-b' \,.\, u'-c' \,.\, u'-d'}}{(u'-d')^2} = \pm\frac{\sqrt{PQR}\sqrt{\epsilon \,.\, u-a \,.\, u-b \,.\, u-c \,.\, u-d}}{(u-d)^2},$$

and attending to the relation $\mathrm{f}^2PN^2 = \mathrm{f'}^2QR$, we obtain

$$\pm\frac{Ndu'}{\sqrt{U'}} = \frac{du}{\sqrt{U}},$$

which is the differential relation between u, u'.

We have in connection with A, B, C, D the point O, and in connection with A', B', C', D' the point O'. As U describes the right line AB, U' describes the arc not containing O' of the circle $A'B'O'$; for observe that O' corresponds in the second figure to the point at infinity on the line AB, viz. as U passes from A to B, not passing through the point at infinity, U' must pass from A' to B', not passing through the point O', that is, it must describe, not the arc $A'O'B'$, but the remaining arc $2\pi - A'O'B'$, say this is the arc $\overset{\frown}{A'B'}$. The integral in regard to u' is thus not the rectilinear integral $(A'B')$, but the integral along the just-mentioned circular arc, say this is denoted by $(\overset{\frown}{A'B'})$; and we thus have

$$(AB) = \pm N(\overset{\frown}{A'B'}).$$

But we have $(\overset{\frown}{A'B'})$ = or not $= (A'B')$, according as the chord $A'B'$ and the arc $\overset{\frown}{A'B'}$ do not include between them either of the points C', D', or include between them one or both of these points; and in the same cases respectively

$$(AB) = \text{or not} = \pm N(A'B').$$

Of course we may in any way interchange the letters, and write under the like circumstances

$$(AC) = \text{or not} = \pm N(A'C'), \text{ \&c.}$$

Suppose now that $ABCD$ is a convex quadrilateral, and consider first in regard to (AB), and next in regard to (AC), the three transformations $A'B'C'D' = BADC$, $= CDAB$, and $= DCBA$, respectively. We have here a figure as in the paper "On the circular relation of Möbius," [628], p. 617 of this volume, the points O_1, O_2, O_3 belonging to the three cases respectively. It will be observed in the figure, and it is easy to see generally, that the points O_1 and O_3 are interior, the point O_2 exterior. We have $N = 1$, and therefore

$$(AB) = \text{ or not } = \pm(AB), = \text{ or not } = \pm(CD), = \text{ or not } = \pm(CD),$$

according as

(1) the chord AB and the arc $\breve{AB}$ of ABO_1 do not or do inclose C and D or either of them;

(2) the chord CD and the arc $\breve{CD}$ of CDO_2 do not or do inclose A and B or either of them;

(3) the chord CD and the arc $\breve{CD}$ of CDO_3 do not or do inclose A and B or either of them.

The first test gives merely the identity $(AB) = \pm(AB)$; the other two each of them give $(AB) = \pm(CD)$, as is seen from the positions of the points O_1, O_2, O_3.

Next, apply the test to AC; we have

$$(AC) = \text{ or not } = \pm(BD), = \text{ or not } = \pm(AC), = \text{ or not } = \pm(BD),$$

according as

(1) the chord AC and the arc $\breve{AC}$ of ACO_1 do not or do inclose B and D or either of them;

(2) the chord BD and the arc $\breve{BD}$ of BDO_2 do not or do inclose A and C or either of them;

(3) the chord BD and the arc $\breve{BD}$ of BDO_3 do not or do inclose A and C or either of them.

In the second case, neither A nor C is inclosed, but we have merely the identity $(AC) = \pm(AC)$; in the first case, B is inclosed and, in the third case, C is inclosed; and the tests each give (AC) not $= \pm(BD)$.

I have not taken the trouble of drawing the figure for a reentrant quadrilateral $ABCD$; the mere symmetry is here enough to show that, having one, we have all three, of the relations in question

$$(AD) = \pm(BC), \quad (BD) = \pm(CA), \quad (CD) = \pm(AB).$$

END OF VOL. IX.